Springer Series in Materials Science

Volume 245

More information about this series at http://www.springer.com/series/856

The Springer Series in Materials Science covers the complete spectrum of materials physics, including fundamental principles, physical properties, materials theory and design. Recognizing the increasing importance of materials science in future device technologies, the book titles in this series reflect the state-of-the-art in understanding and controlling the structure and properties of all important classes of materials.

Christopher R. Weinberger • Garritt J. Tucker
Editors

Multiscale Materials Modeling for Nanomechanics

Editors
Christopher R. Weinberger
Drexel University
Philadelphia, PA, USA

Garritt J. Tucker
Drexel University
Philadelphia, PA, USA

ISSN 0933-033X ISSN 2196-2812 (electronic)
Springer Series in Materials Science
ISBN 978-3-319-81524-4 ISBN 978-3-319-33480-6 (eBook)
DOI 10.1007/978-3-319-33480-6

Printed on acid-free paper

This Springer imprint is published by Springer Nature
The registered company is Springer International Publishing AG Switzerland

We would like to dedicate this book to our wives, Allison Weinberger and Ashley Tucker, who made all of this possible.

Preface

Multiscale Materials Modeling for Nanomechanics is an interesting title for a book as the terms "nanomechanics" and "multiscale modeling" have different definitions and meaning to various scientific communities. Therefore, we feel it is important to inform the reader up front of our thoughts on this topic and why we chose to include the subjects discussed herein. We think of nanomechanics as the mechanics that occur at the smallest length scales in materials, down to nanometers. Thus, the study of nanomechanics should focus on materials where this regime of mechanics dominates or controls behavior, usually when a governing length scale is below a micron. In addition, it is our intent with a book on nanomechanics to focus equally on the materials as on the mechanics, because they are intrinsically linked at these size scales.

Multiscale modeling is also an intriguing term because of different ideas about what multiscale modeling encompasses. One might immediately assume that a book on multiscale modeling would focus on methods that couple scales or simulation methodologies, such as continuum and atomistic methods. While this is certainly part of multiscale modeling, this does leave out one other important way to conduct multiscale modeling: information passing between scales. The former is called *concurrent multiscale modeling* and the latter *hierarchical multiscale modeling*. In this book, we will cover both types of multiscale modeling approaches providing some breadth on the uses of the different types of multiscale modeling as applied to the mechanics of materials and structures at the nanoscale.

Finally, we would like to point out that our intent in writing this book was to cover a range of topics that span from fundamental methods to applications. The beginning of the book covers fundamental techniques like atomistic simulations, dislocation dynamics, continuum methods, and density functional theory calculations. These chapters are meant to provide the reader with a broad perspective on computational techniques that have been fundamental to materials modeling at various length and time scales. However, we realize that for those who are a completely unfamiliar with these techniques, the chapters in this book will provide a good introduction, but more detailed books will likely be needed to master the techniques. The self-containment of this book was sacrificed in order to allow for the later chapters of the book that provide examples of both concurrent and hierarchical applications of multiscale modeling. We hope that the brevity of the first few chapters is accounted for by the number and depth of the applications provided later.

With this in mind, the book is organized in three parts. The first part, which contains Chaps. 1–4, covers what may be considered as fundamental or basic materials modeling methods. These chapters introduce the methods of molecular simulations, dislocation dynamics, continuum approximations, and electronic density functional theory calculations. These methods provide some of the basic modeling approaches that are used extensively in all of the chapters that follow and span from angstroms to meters in length scale. The second part, Chaps. 5–8, provides several more recently developed methods that are scale bridging including both length and time scales. The specific methods include the quasicontinuum method (Chap. 5), accelerated molecular dynamics (Chap. 6), the concurrent atomistic-continuum method (Chap. 8), and the "atoms to continuum" method. Chapter 6 is the only chapter that deals with time-scale bridging, while the rest of these chapters present length-scale bridging techniques which can be largely classified as concurrent multiscale techniques. It is important to note that there are many other length, and time-scale bridging methods, but our intent is to provide a few here that have been or can be used for modeling nanomechanics in materials. Chapters 9 and 10 provide methods to analyze and visualize the modeling results of particle systems (e.g., molecular simulations). This is of particular importance because the analysis and visualization of data are precisely what can be used to conduct more efficient multiscale modeling, either through concurrent schemes or hierarchical methods.

The third part, covering Chaps. 11–17, provides case studies in the use of multiscale materials modeling as applied to nanomechanics. The topics are diverse in the problems and how they solve them. The chapters include the study of the mechanical properties of nanowires and nanopillars made of metals by studying dislocation nucleation (Chap. 12) or dislocation motion of preexisting dislocations (Chap. 11) or their statistical nature in nano-grained materials using the quantized crystal plasticity method (Chap. 13). The rest of the chapters each deal with the mechanics of different systems including amorphous materials (Chap. 14), ferroelectric thin films (Chap. 15), silicon electrodes (Chap. 16), and thin liquid films (Chap. 17).

Philadelphia, PA, USA — Christopher R. Weinberger
Philadelphia, PA, USA — Garritt J. Tucker
February 2016

Acknowledgments

The compilation of any book requires the effort of a large number of people, and this work would not have been possible without their help. We would first like to acknowledge the hard work and dedication of all of the chapter authors. These individuals worked tirelessly not only to write their contributions but in many cases to help improve the contributions of the other authors. Additionally, we would like to thank Ian Bakst, Matthew Guziewski, Hang Yu, and Ryan Sills for the additional help in reviewing book chapters and providing timely comments.

Contents

1 **Introduction to Atomistic Simulation Methods** 1
Reese E. Jones, Christopher R. Weinberger, Shawn P. Coleman, and Garritt J. Tucker

2 **Fundamentals of Dislocation Dynamics Simulations** 53
Ryan B. Sills, William P. Kuykendall, Amin Aghaei, and Wei Cai

3 **Continuum Approximations** 89
Joseph E. Bishop and Hojun Lim

4 **Density Functional Theory Methods for Computing and Predicting Mechanical Properties** 131
Niranjan V. Ilawe, Marc N. Cercy Groulx, and Bryan M. Wong

5 **The Quasicontinuum Method: Theory and Applications** 159
Dennis M. Kochmann and Jeffrey S. Amelang

6 **A Review of Enhanced Sampling Approaches for Accelerated Molecular Dynamics** 195
Pratyush Tiwary and Axel van de Walle

7 **Principles of Coarse-Graining and Coupling Using the Atom-to-Continuum Method** 223
Reese E. Jones, Jeremy Templeton, and Jonathan Zimmerman

8 **Concurrent Atomistic-Continuum Simulation of Defects in Polyatomic Ionic Materials** 261
Shengfeng Yang and Youping Chen

9 **Continuum Metrics for Atomistic Simulation Analysis** 297
Garritt J. Tucker, Dan Foley, and Jacob Gruber

10 Visualization and Analysis Strategies for Atomistic Simulations 317
Alexander Stukowski

11 Advances in Discrete Dislocation Dynamics Modeling of Size-Affected Plasticity 337
Jaafar A. El-Awady, Haidong Fan, and Ahmed M. Hussein

12 Modeling Dislocation Nucleation in Nanocrystals 373
Matthew Guziewski, Hang Yu, and Christopher R. Weinberger

13 Quantized Crystal Plasticity Modeling of Nanocrystalline Metals 413
Lin Li and Peter M. Anderson

14 Kinetic Monte Carlo Modeling of Nanomechanics in Amorphous Systems 441
Eric R. Homer, Lin Li, and Christopher A. Schuh

15 Nanomechanics of Ferroelectric Thin Films and Heterostructures ... 469
Yulan Li, Shengyang Hu, and Long-Qing Chen

16 Modeling of Lithiation in Silicon Electrodes 489
Feifei Fan and Ting Zhu

17 Multiscale Modeling of Thin Liquid Films 507
Han Hu and Ying Sun

Appendix A Available Software and Codes 537

Index 541

Contributors

Amin Aghaei Stanford University, Stanford, CA, USA

Jeffrey S. Amelang Harvey Mudd College, Claremont, CA, USA

Peter M. Anderson Department of Materials Science and Engineering, The Ohio State University, Columbus, OH, USA

Joseph E. Bishop Sandia National Laboratories, Engineering Sciences Center, Albuquerque, NM, USA

Wei Cai Stanford University, Stanford, CA, USA

Long-Qing Chen Pennsylvania State University, University Park, PA, USA

Youping Chen Department of Mechanical and Aerospace Engineering, University of Florida, Gainesville, FL, USA

Shawn P. Coleman Army Research Laboratory, Aberdeen Proving Ground, Aberdeen, MD, USA

Jaafar A. El-Awady Department of Mechanical Engineering, Johns Hopkins University, Baltimore, MD, USA

Feifei Fan Department of Mechanical Engineering, University of Nevada, Reno, NV, USA

Haidong Fan Department of Mechanics, Sichuan University, Chengdu, Sichuan, China

Dan Foley Department of Materials Science and Engineering, Drexel University, Philadelphia, PA, USA

Marc N. Cercy Groulx University of California-Riverside, Riverside, CA, USA

Jacob Gruber Department of Materials Science and Engineering, Drexel University, Philadelphia, PA, USA

Matthew Guziewski Drexel University, Philadelphia PA, USA

Eric R. Homer Department of Mechanical Engineering, Brigham Young University, Provo, UT, USA

Han Hu Department of Mechanical Engineering and Mechanics, Drexel University, Philadelphia, PA, USA

Shengyang Hu Pacific Northwest National Laboratory, Richland, WA, USA

Ahmed M. Hussein Department of Mechanical Engineering, Johns Hopkins University, Baltimore, MD, USA

Niranjan V. Ilawe University of California-Riverside, Riverside, CA, USA

Reese E. Jones Sandia National Laboratories, Livermore, CA, USA

Dennis M. Kochmann California Institute of Technology, Graduate Aerospace Laboratories, Pasadena, CA, USA

William P. Kuykendall Stanford University, Stanford, CA, USA
Sandia National Laboratories, Livermore, CA, USA

Lin Li Department of Metallurgical and Materials Engineering, The University of Alabama, Tuscaloosa, AL, USA

Yulan Li Pacific Northwest National Laboratory, Richland, WA, USA

Hojun N. Lim Sandia National Laboratories, Engineering Sciences Center, Albuquerque, NM, USA

Christopher A. Schuh Department of Materials Science and Engineering, Massachusetts Institute of Technology, Cambridge, MA, USA

Ryan B. Sills Sandia National Laboratories, Livermore, CA, USA

Alexander Stukowski Technische Universität Darmstadt, Darmstadt, Germany

Ying Sun Department of Mechanical Engineering and Mechanics, Drexel University, Philadelphia, PA, USA

Jeremy Templeton Sandia National Laboratories, Livermore, CA, USA

Pratyush Tiwary Department of Chemistry, Columbia University, New York, NY, USA

Garritt J. Tucker Drexel University, Philadelphia, PA, USA Department of Materials Science and Engineering, Drexel University, Philadelphia, PA, USA

Axel van de Walle School of Engineering, Brown University, Providence, RI, USA

Christopher R. Weinberger Drexel University, Philadelphia, PA, USA

Bryan M. Wong University of California-Riverside, Riverside, CA, USA

Shengfeng Yang Department of Mechanical and Aerospace Engineering, University of Florida, Gainesville, FL, USA

Hang Yu Drexel University, Philadelphia PA, USA

Jonathan Zimmerman Sandia National Laboratories, Livermore, CA, USA

Ting Zhu Woodruff School of Mechanical Engineering, Georgia Institute of Technology, Atlanta, GA, USA

Chapter 1
Introduction to Atomistic Simulation Methods

Reese E. Jones, Christopher R. Weinberger, Shawn P. Coleman, and Garritt J. Tucker

1.1 Introduction

Atomistic and molecular methods play a central role in multiscale modeling for nanomechanics since the applicable length scales of atomistic methods span a few to hundreds of nanometers. In fact, many of the applied research chapters presented later in this book employ atomistic methods, either by directly coupling other methods to atomistics, by creating information that is passed to or from atomistics, or by using atomistics to directly investigate nano-mechanical phenomena.

In this chapter, we will introduce the basic principles of atomistic and molecular modeling as applied to nanomechanics. It is worth pointing out that the terms *atomistic modeling*, *molecular modeling*, and *molecular dynamics* are often used interchangeably. Within this chapter, we consider *atomistic* and *molecular modeling* as essentially the same concept, only distinguished by whether we are modeling atoms or molecules; however, we reserve *dynamics* for a specific type of simulation, notably the direct integration of the equation of motions, in contrast to *molecular statics* which employs energy minimization.

Molecular simulation methods are generally more efficient than ab initio methods mainly due to the fact that they describe atomic interactions empirically using

R.E. Jones (✉)
Sandia National Laboratories, Livermore, CA, USA
e-mail: rjones@sandia.gov

C.R. Weinberger • G.J. Tucker
Drexel University, Philadelphia, PA, USA
e-mail: cweinberger@coe.drexel.edu; gtucker@coe.drexel.edu

S.P. Coleman
Army Research Laboratory, Aberdeen Proving Ground, Aberdeen, MD, USA
e-mail: shawn.p.coleman8.ctr@mail.mil

C.R. Weinberger, G.J. Tucker (eds.), *Multiscale Materials Modeling for Nanomechanics*, Springer Series in Materials Science 245,
DOI 10.1007/978-3-319-33480-6_1

classical mechanics. Thus, this class of methods are well suited for simulating the structure and mechanical properties of materials at the length scales in the nanoscale regime. For example, atomistic methods can be used effectively to study: the mechanics of defects in materials and how they interact with the large number of free surfaces in nanostructures, and the physics of phase transformation at the nanoscale.

There are some important limitations of atomistic methods that should be noted as well. As most atomistic methods use classical empirical potentials, they are unable to directly investigate properties associated with the electronic structure of materials such as band gaps, optical properties, and magnetism, and how these properties interact with nanostructures. Simulations using electronic structure theory, such as density functional theory, are better suited to investigate these properties. In addition, the accuracy of atomistic simulations is limited by the accuracy of the interatomic potentials used. Classical atomistic methods are also limited both in length and timescales due to numerical cost. The length scale associated with atomistic simulations of condensed matter systems is often below 100 nm even for large simulations. Simulations above this length scale, at the time of the writing of this book, are prohibitively expensive on generally accessible computers. In addition, the timescales for classical molecular dynamics simulations are generally less than tens to hundreds of nanoseconds even for long simulations due to time-steps on the order of femtoseconds that are needed to directly integrate the equations of motion. Both length scale and timescale challenges can be addressed with multiscale methods, some of which will be described in this book.

Molecular methods, in general, model a system of idealized particles that follow the laws of classical mechanics. Two physically distinct strategies are used to evolve the atoms in an atomistic simulation. The first method, *molecular statics*, attempts to evolve the state to a local or global energy minimum based on external forces and is discussed in Sect. 1.2. The second method, *molecular dynamics*, simulates the dynamic trajectory of the atoms by directly integrating the equations of motion. This method is discussed in Sect. 1.4. Section 1.5 describes the appropriate statistical thermodynamics required to connect the ensemble of atoms to physical properties in classical thermodynamics described at the continuum scale. All these methods are predicated on empirical interatomic potentials, which characterizes the potential energy of the system and hence the forces on all the atoms. These potentials are discussed in detail in Sect. 1.6. The resulting system potential energy governs the behavior of the system and hence both its equilibrium and non-equilibrium properties. Finally, in Sect. 1.9 we give an overview how atomistic methods can be used in modeling nanomechanics, as a prelude to the other chapters in this book, as well as a discussion of the specific software that can be used to conduct molecular simulations.

1.2 Molecular Statics

Molecular statics simulations seek to find minimum energy conformations, transition states, and/or reaction pathways in atomic systems in the absence of thermal vibrations. In classical atomistic simulations, this process starts with constructing the potential energy surface for the system using an appropriate interatomic potential. More details about interatomic potentials will be given in Sect. 1.6; however, for now, one needs to know that the interatomic potential relates the potential energy of the system to the atomic positions, i.e., $\Phi = \Phi(\mathbf{R})$ where $\mathbf{R} = \{\mathbf{x}_i\}$ represents the positions of all the atoms (indexed by i). A two-dimensional model for a potential energy surface that resembles the contours of a physical surface is shown in Fig. 1.1. While simple, this two-dimensional potential energy surface illustrates common features that are often described by more complex interatomic potentials in higher dimensions including many local energy minima and transition states nearby a unique global minimum energy configuration.

A goal for molecular statics simulations is to gain further understanding of the potential energy surface using knowledge of the interatomic potential and its gradients. Given that the forces are the first derivatives of the system potential with respect to atomic positions, minimum and maximum energy configurations occur when the forces acting on all atoms are zero. The forces acting on each atom i are

$$\mathbf{f}_i = -\nabla_{\mathbf{x}_i}\Phi \tag{1.1}$$

where the subscript $\mathbf{x}_i$ denotes that the gradient is taken with respect to $\mathbf{x}_i$ holding all other atomic positions as fixed. Alternatively, we can define the total force vector $\mathbf{F}$ as the gradient of the potential energy with respect to $\mathbf{R}$:

$$\mathbf{F} = -\nabla_{\mathbf{R}}\Phi. \tag{1.2}$$

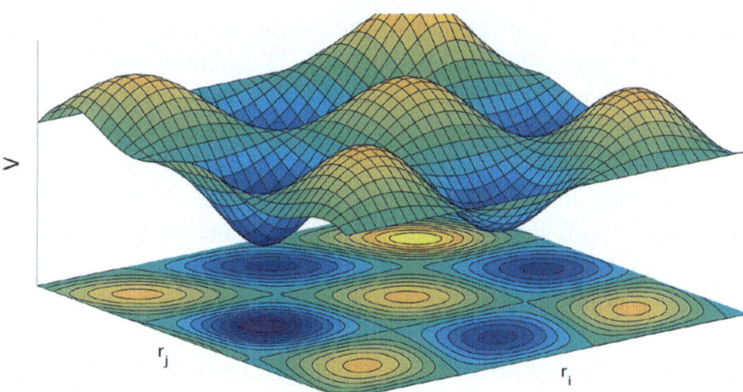

Fig. 1.1 A representative two-dimensional potential energy surface

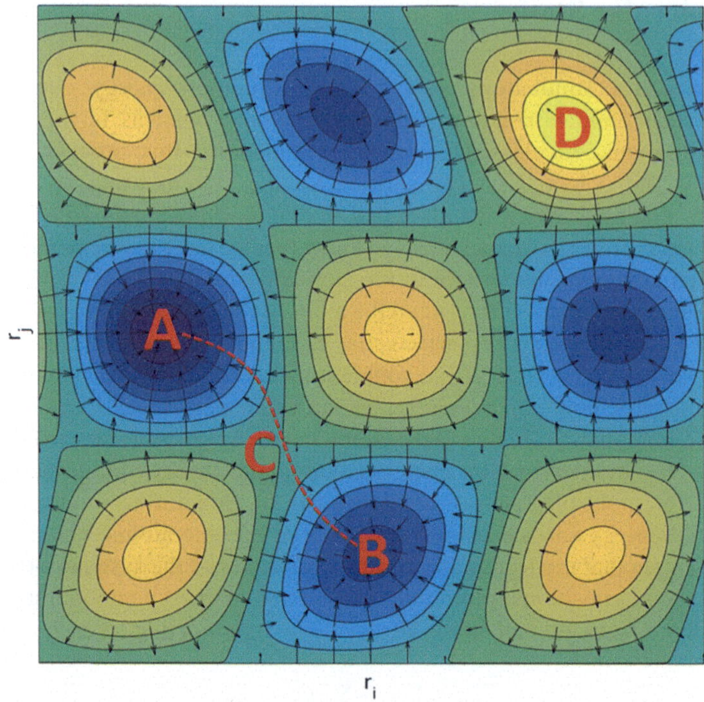

Fig. 1.2 Contour plot of the potential energy surface shown in Fig. 1.1 with an overlay of *arrows* representing the forces. The contour plot shows the location of (*A*) global minima, (*B*) local minima, (*C*) transition state, and (*D*) global maxima

Figure 1.2 shows a contour plot of the previous potential energy surface with forces represented by arrows. When the atomic forces are equal to zero, the minimum and maximum energy configuration occurs when the Hessian, i.e., the matrix of second order derivatives with respect to the atomic positions $\mathbf{H} = \nabla_{\mathbf{R}} \nabla_{\mathbf{R}} \Phi$, is strictly positive or negative definite, respectively. Transition states occur at atomic configurations corresponding to saddle-points along the potential energy surface. These saddle-points are located at unstable equilibrium states where the Hessian is not positive nor negative definite and typically are at points where a few eigenvalues of the Hessian are negative and the rest are positive. Reaction pathways describe the minimum energy path along the potential energy surface between two adjacent energy minima. All the aforementioned features are highlighted in Fig. 1.2. In the following sections, common methods used to determine these features along the potential energy surface are introduced and overall advantages and limitations of molecular statics simulations are discussed. For much more detail on the algorithms discussed herein, we suggest examining the texts by Leach [1] and Nocedal [2].

In the practical implementation of many molecular simulation codes, scaled coordinates are used in place of real space coordinates. The use of scaled coordinates is particularly advantageous when simulation domains are allowed to change shape, such as in stress-controlled simulations discussed later, since the variations in the coordinates of all the atoms are reduced to a few explicit parameters. In addition, scaled coordinates are also important in the practical use of periodic boundary conditions as they provide a convenient way to map atoms back to the simulation domain as they move across box bounds.

Scaled coordinates are a set of fractional coordinates relative to the simulation domain geometry and can be directly related to the real space coordinates through the linear transformation:

$$\mathbf{x}_i = \mathbf{h}\mathbf{s}_i \tag{1.3}$$

where $\mathbf{h}$ is a 3×3 matrix that defines the three simulation direction vectors that span the box and are not necessarily orthogonal. The volume of the simulation domain is the determinant of $\mathbf{h}$. Clearly, from Eq. (1.3), the simulation domain, which is sometimes called the simulation box, can change its shape through $\mathbf{h}$, which changes the real space coordinates of the particles, $\mathbf{x}_i$.

1.2.1 Energy Minimization

Energy minimization, resulting in structure optimization, is one of the most important tools in atomistic simulations, and is used to discover the zero temperature stable structures of a system. During energy minimization, numerical algorithms search for the set of atom coordinates, $\mathbf{R}$, that minimize a defined objective function, E, related to the potential energy of the system. In the case where the volume of the simulation box and the number of atoms, N, are held constant, the objective function is equal to the total potential energy of the system:

$$E(\mathbf{R}) = \Phi(\mathbf{R}). \tag{1.4}$$

where Φ is the interatomic potential. However, if the system box size is allowed to change in order to achieve and maintain a targeted system stress $\bar{\boldsymbol{\sigma}}$, the objective function must also include terms that account for the strain energy introduced by the changes in the system volume and shape. In this case, the objective function for an energy minimization with box relaxation is the enthalpy:

$$E(\mathbf{R}) = \Phi(\mathbf{R}) + \bar{p}(V - V_0) + \underbrace{V_0(\bar{\boldsymbol{\sigma}} - \bar{p}\mathbf{I}) \cdot \boldsymbol{\varepsilon}}_{E_d}, \tag{1.5}$$

derived by Parrinello and Rahman [3]. In the second term, which is due to work done in changing the volume of the system V, $\bar{p} = \operatorname{tr} \bar{\boldsymbol{\sigma}}$ is the target pressure and

V_0 the volume of the stress-free reference configuration. The third term, E_d, is the distortional strain energy that is associated with changing the box shape, but not its volume and the strain $\boldsymbol{\varepsilon}$ is defined as

$$\boldsymbol{\varepsilon} = \frac{1}{2}\left(\left(\mathbf{h}\mathbf{h}_0^{-1}\right)^T \mathbf{h}\mathbf{h}_0^{-1} - \mathbf{I}\right) \tag{1.6}$$

with $\mathbf{h}_0$ being the shape matrix $\mathbf{h}$ for the stress-free reference configuration of the system.

Iterative algorithms, such as line search and damped dynamics-based methods, are commonly used to find the atomic configuration that minimizes the associated objective function. The robustness of these iterative techniques is ideal for the complex, high-order potential energy surfaces common in atomistic simulations; however, they do not guarantee convergence to a global minimum energy structure. If a local energy minimum configuration is nearby the starting configuration, these iterative minimization algorithms typically converge to this state instead (see [4] for an in-depth exploration for the deformation of a nanowire). Techniques to uncover the atomic configurations of global energy minima, such as direct Monte-Carlo sampling and simulated annealing, sample many initial configurations or perturb the simulation during these minimization processes in order to escape local energy minima enroute to lower energy states. A selection of common energy minimization methods are described in the following sections.

1.2.1.1 Line Search Minimization Methods

Line search minimization algorithms, such as steepest descent, conjugate gradient, and Newton–Raphson methods, use one-dimensional searches along specified search directions to optimize an objective function. In the context of molecular statics these algorithms iteratively displace the atoms within a simulation in order to find a lower energy state. During line search minimizations, the configuration of the system $\mathbf{R}_k$ at iteration k is adjusted by $\alpha_k \mathbf{d}_k$ in order to reduce its energy via,

$$\mathbf{R}_{k+1} = \mathbf{R}_k + \alpha_k \mathbf{d}_k \tag{1.7}$$

where α_k is the step size and $\mathbf{d}_k$ is the search direction. These parameters must depend on the gradient of the minimization objective function in order to reduce the system energy; however, the details are unique to each line search minimization algorithm. In practice, line search minimizations will continue to adjust the atomic coordinates until energy minimization criteria (Wolfe conditions) are met. Usually the criterion is the sum of the atomic forces in the system is zero within a defined numerical tolerance or the energy deviation from successive iterations is sufficiently small.

The simplest of the line search algorithms is the method of steepest descent. In this method, the search direction $\mathbf{d}_k$ is equal to the negative gradient of the objective function. In the case where the simulation box does not change, the direction of steepest descent is the force **F**. At each iteration, atoms are moved along the search direction until a minimum in the objective function along this line has been reached. At which point, if the defined energy minimization criteria are not met, a subsequent line search is made along the new total force direction orthogonal to the previous. This stepwise descent continues until the energy minimization criteria are met. In general, the steepest descent method is extremely robust because it relies only on the gradient of the objective function, $E = \Phi$, for the minimization; however, convergence near the minimum can be slow as this gradient approaches zero. Convergence also slows using steepest descent methods as the path oscillates near the energy minima due to the orthogonality of successive search directions.

The method of conjugate gradients is a second type of line search algorithm that can avoid oscillations near the energy minima by choosing better search directions based on knowledge from prior iterations. At each iteration, the search direction is determined by

$$\mathbf{d}_{k+1} = \begin{cases} -\nabla_{\mathbf{R}} E_0 & k = 0 \\ -\nabla_{\mathbf{R}} E_{k+1} + \beta_{k+1}\mathbf{d}_k & k > 0 \end{cases}, \tag{1.8}$$

where β is an update parameter that scales the input from the previous search direction. If $\beta = 0$, no input from the previous search direction is included and the method reverts to steepest descent. In general, however, the method of conjugate gradients chooses β such that each new search direction conjugates to all prior directions. Conjugate directions are orthogonal with respect to the Hessian of the potential, such that

$$\mathbf{d}_k \mathbf{H} \mathbf{d}_{k+1} = 0. \tag{1.9}$$

There are several formulations of β that update the search direction to an approximate conjugate direction (without computing the **H** directly), which exhibits varying degrees of effectiveness. Two common β values used in conjugate gradient energy minimization were developed by Fletcher and Reeves [5]:

$$\beta_{k+1}^{\mathrm{FR}} = \frac{\nabla_{\mathbf{R}} E_{k+1}^T \nabla_{\mathbf{R}} E_{k+1}}{\nabla_{\mathbf{R}} E_k^T \nabla_{\mathbf{R}} E_k} \tag{1.10}$$

and Polak and Ribière [6]:

$$\beta_{k+1}^{\mathrm{PR}} = \frac{-\nabla_{\mathbf{R}} E_{k+1}^T (-\nabla_{\mathbf{R}} E_{k+1} + \nabla_{\mathbf{R}} E_k)}{\nabla_{\mathbf{R}} E_k^T \nabla_{\mathbf{R}} E_k} \tag{1.11}$$

Figure 1.3 shows an example of the steepest descent and conjugate gradient energy minimization methods using a portion of the potential energy surface introduced previously. This example mimics the addition of an adatom onto a

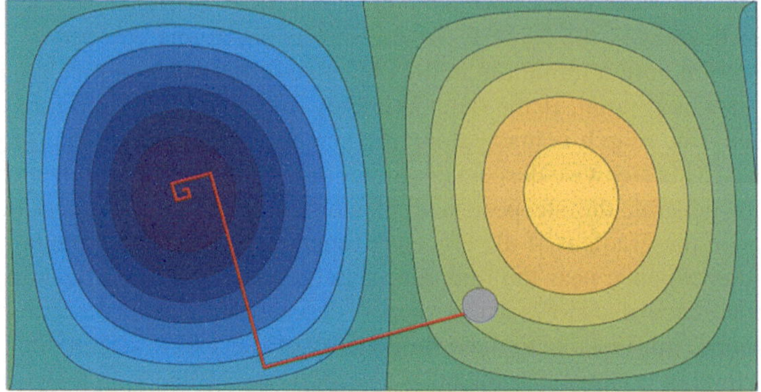
Steepest Descent Minimization

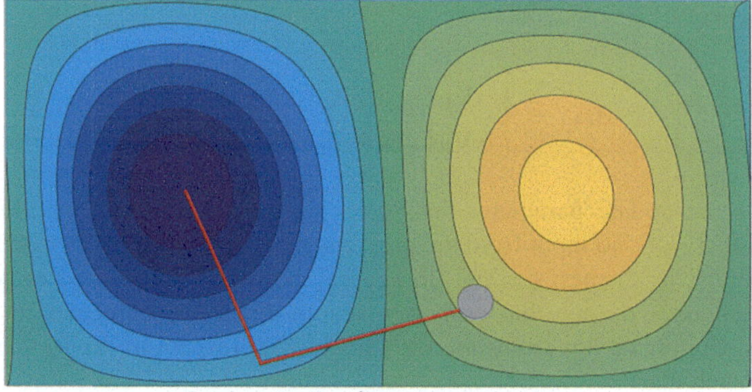
Conjugate Gradient Minimization

Fig. 1.3 Steepest descent and conjugate gradient minimization methods reduce the system energy using a line search steps. The method of steepest descent is robust, but orthogonal line search directions tend to slow convergence. Using conjugate gradient step directions can improve convergence

surface at an unfavorable site, but near the minimum energy configuration. The lines in the figure trace the energy minimization steps used by each line search method. Both minimization methods eventually relax the position of the adatom to the minimum energy configuration; however, the orthogonal directions used in the steepest descent method increase the number of steps required to reach the energy minima as compared to the energy conjugate directions of the conjugate gradient method.

Newton–Raphson methods are a third type of line search energy minimization that adds even more complexity in order to gain even better convergence rates. Newton–Raphson methods also consider the second order gradient, the Hessian $\mathbf{H} = \nabla_{\mathbf{R}}\nabla_{\mathbf{R}}E$, of the energy minimization objective function, E, when choosing the appropriate step size and direction during minimization. The Hessian gives information regarding the curvature of the potential energy surface and implicitly creates a local second order approximation of the system energy. Newton–Raphson methods use the inverse of the Hessian matrix applied to the current forces as search direction. Using the Hessian, Newton–Raphson methods generally converge to the energy minimized configuration quadratically and hence more quickly than steepest descent and conjugate gradient methods, which are linear; however, the increased convergence efficiency comes with the high computational cost for computing the Hessian and its inverse.

1.2.1.2 Damped Dynamics Minimization Methods

Damped dynamics minimization methods such as the quickmin [7] and fast inertial relaxation engine [8] use time integration schemes driven by the same atomic forces governing the line search methods to move the atoms to a lower energy state. These integration schemes, e.g., the velocity Verlet algorithm, are also used in molecular dynamics simulation (discussed in Sect. 1.4) in order to update the atomic velocities and positions by solving the equations of motion. However, unlike molecular dynamics simulations, the motion of the atoms under these methods does not resolve high-frequency thermal vibrations. To enable rapid convergence, these methods add damping parameters to modify the velocity terms in the equations of motion driven by internal forces due to the gradient of the interatomic potential. In addition, both methods are capable of rapidly adjusting the course of the energy minimization in order to avoid higher energy configurations. In some cases, energy convergence of these methods has been shown to be on par or quicker than traditional line search methods, e.g., steepest descent and conjugate gradient; thus, making these valuable alternative methods for energy minimization.

1.2.2 Transition States and Reaction Pathways

Transition state and reaction pathway (minimum energy path) calculations are another important class of molecular statics simulations for understanding the rate of conformational transformations that occur at the atomic level, for example, during diffusion events or chemical reactions. These events can sometimes be explored directly using molecular dynamics; however, often the rate of transformation can be much slower than what can be observed through direct simulation. For slow events, knowledge of the transition state and the reaction pathway can be incorporated into transition state theory in order to determine the rate of the event spanning a long time frame. There are several algorithms to find transition states and reaction pathways; all of which rely upon energy minimization computations as discussed in the previous sections.

Transition state and reaction pathway calculations are most commonly determined by modeling a set of connected conformations or *images*. The images detail a

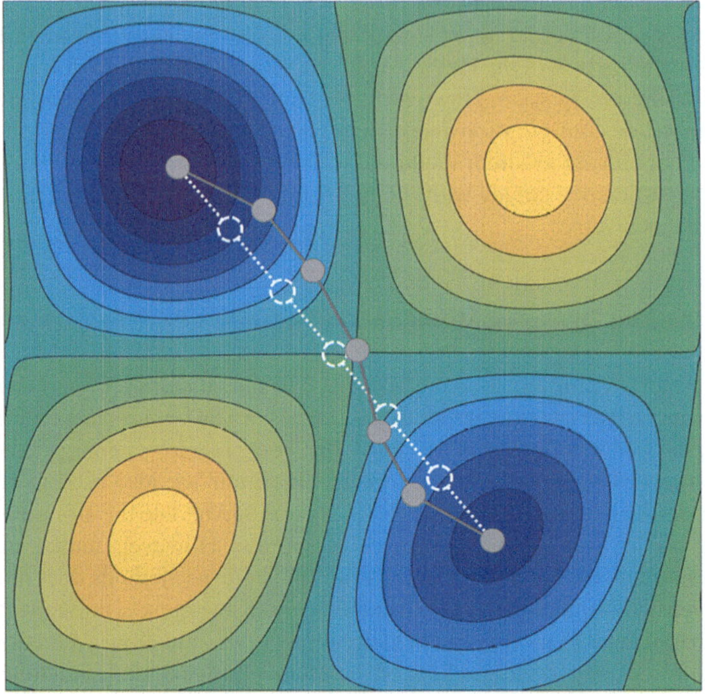

Fig. 1.4 Nudged elastic band method relaxes a series of connected images to find the reaction pathway (minimum energy path) between two minima. The *white outline and gray spheres* indicate the initial and minimized set of images, respectively

series of conformational states between two known energy states in order to model the reaction pathway. Typically, the initial series of images connecting the two states is constructed as a linear interpolation between the points in configurational space; however, in some cases more careful attention is needed. For example, if the atomic configurations within an image created by this linear interpolation become too close or even overlap, one might choose to add a repulsive force to push these initial atomic configurations into a lower energy, spread out arrangement before starting a transition state and reaction pathway calculation.

Nudged elastic band (NEB) [9, 10] is one of the most common methods for determining transition states and reaction pathways. NEB methods connect each image with a set of springs to drive the collection of images toward the minimum energy path. Figure 1.4 illustrates an example NEB calculation between two minima on the potential energy surface introduced previously. During NEB, spring forces act only parallel to the vector connecting adjacent images, while the forces defined by the potential within each image are projected perpendicularly from this vector. Mathematically, the NEB force on each image j can be represented by

$$\mathbf{F}_j^{\mathrm{NEB}} = \mathbf{F}_j^{\parallel} + \mathbf{F}_j^{\perp}. \tag{1.12}$$

If $\boldsymbol{\tau}_j$ is the unit vector tangent to the springs connecting each state and $\mathbf{R}_j$ is the set of atom coordinates within the image, the parallel force due to the spring, $\mathbf{F}_j^{\parallel}$ is

$$\mathbf{F}_j^{\parallel} = k\left(\left\|\mathbf{R}_{j+1} - \mathbf{R}_j\right\| - \left\|\mathbf{R}_j - \mathbf{R}_{j-1}\right\|\right)\boldsymbol{\tau}_j, \tag{1.13}$$

where k is the chosen NEB spring constant. The perpendicular force $\mathbf{F}_j^{\perp}$ is evaluated from the potential Φ as

$$\mathbf{F}_j^{\perp} = \nabla_{\mathbf{R}}\Phi(\mathbf{R}_j) - \left(\nabla_{\mathbf{R}}\Phi(\mathbf{R}_j)\cdot\boldsymbol{\tau}_j\right)\boldsymbol{\tau}_j. \tag{1.14}$$

The orthogonal projection of these forces decouples the minimization of the individual image from the minimization of the path itself and ensures that the spring forces do not effect the relaxation of the images.

Often NEB calculations do not relax a single image directly at the transition state configuration; instead a pair of images may straddle each side saddle-point along the potential energy surface. In these cases, climbing image NEB [11] can be invoked to find a better approximation of the transition state along the reaction pathway. Climbing image NEB determines the transition state configuration by finding the highest energy configuration along the computed reaction pathway. After performing a traditional NEB calculation, climbing image NEB systematically searches for the transition state using the highest energy image. During this search, the spring forces on this one image are turned off so that it is constrained only by forces computed using the potential. To make the image climb to a higher energy state along the previously computed minimum energy path, the potential forces aligned with the minimum energy path are inverted. Climbing image NEB calculations have shown to substantially improve predictions in transition state conformation and energies, especially when using a limited number of images.

1.3 Boundary Conditions

Different boundary conditions, e.g., fixed, free, or periodic, can be imposed on the simulation boxes which encase atomistic simulations. These boundary conditions represent specific interactions with the surrounding environment and allow one to model different phenomena. If necessary, one can specify a unique boundary condition to the different faces of the simulation box.

Fixed boundary conditions are the simplest but most restrictive conditions. The normal faces associated with the fixed boundary directions are exposed to an infinite vacuum, the dimensions along the fixed direction are not allowed to change, and any atoms that pass this dimension are lost from the system. Fixed boundary conditions, effected by the addition of layers of atoms with prescribed motion, are used often when mechanically loading a system. The related free (*shrink-wrapped*) boundary conditions also expose the associated boundary surface to an infinite vacuum. However, the atoms are allowed to move freely in the specified direction, thus emulating a free surface. The dimensions of simulation box are adjusted to always provide a bounding box of the deforming system.

On the other hand, periodic boundary conditions do not create free surfaces. These boundary conditions, introduced by Born and von Kármán in 1912 [12], mimic an infinite set of systems while modeling a finite collection of atoms. Fully periodic boundary conditions can be thought of as being constructed from $3^3 - 1$ replicated simulation cells (in three-dimensions) surrounding a primary simulation box. The atom positions within each replica are identical after making appropriate adjustments for the periodic length defined by the dimensions of the primary simulation box. Atom velocities within each cell are also identical among the replicas and atom motion across replica boundaries results in corresponding motion of atoms across the mirrored boundaries.

1.4 Molecular Dynamics

Molecular dynamics (MD), as mentioned in the introduction, is based on the direct integration of Newton's law with forces $\mathbf{f}_i$ derived from the potential energy Φ as:

$$\mathbf{f}_i = -\nabla_{\mathbf{x}_i} \Phi\left(\mathbf{R}\right) \tag{1.15}$$

and hence the particles in an atomistic system simulated with MD behave classically. So, in contrast to molecular statics where the atomic positions $\mathbf{R} = \{\mathbf{x}_i\}$ are solved via an energy minimization process, the atomic positions in MD evolve according to:

$$m_i \ddot{\mathbf{x}}_i = \mathbf{f}_i \tag{1.16}$$

given the atomic masses m_i as well as initial positions $\{\mathbf{x}_i(0)\}$ and velocities $\{\mathbf{v}_i(0)\}$.

In its simplest application, MD is a simulation of the dynamics of particles in classical force field; however, applications of MD where the intent is to reproduce the statistics of a large number of atoms with connections to continuum quantities through statistical mechanics are where the deep value of MD appears. These connections are often better made if we consider the particles following either Hamilton's equations of motion or Lagrange's equations of motion based on a phase space composed of positions and momenta or positions and velocities, respectively. The set of positions $\mathbf{R} = \{\mathbf{x}_i\}$ and momenta $\mathbf{Q} = \{\mathbf{p}_i = m_i \dot{\mathbf{x}}_i\}$ of the atoms $i = 1 \ldots N$ as they evolve with time t describe the state of the system in this high-dimensional phase space (see Fig. 1.5 for an illustration of an atomic trajectory through phase space). The Hamiltonian $\mathscr{H}$ is the phase function, i.e., a function of $\mathbf{R}$ and $\mathbf{Q}$, equal to the total energy of the system E; and, in the case the system is closed, is simply the sum of the potential and kinetic energy:

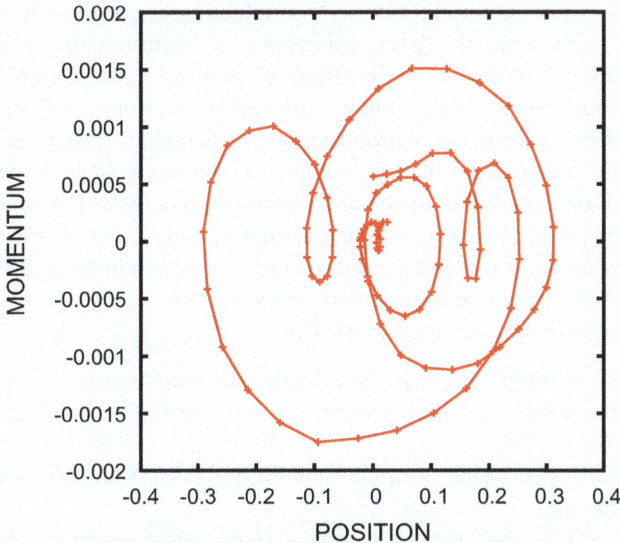

Fig. 1.5 The constant energy trajectory in phase space of a single atom in many atom system

$$\mathscr{H} = \mathscr{H}(\mathbf{R}, \mathbf{Q}) = \Phi(\mathbf{R}) + K(\mathbf{Q}) \tag{1.17}$$

In general, the kinetic energy K has the quadratic form

$$K = \sum_{i=1}^{N} \frac{1}{2m_i} \mathbf{p}_i \cdot \mathbf{p}_i \tag{1.18}$$

With a Hamiltonian, the dynamics are simply:

$$\dot{\mathbf{R}} = \nabla_{\mathbf{Q}} \mathscr{H} \tag{1.19}$$

$$\dot{\mathbf{Q}} = -\nabla_{\mathbf{R}} \mathscr{H} \tag{1.20}$$

If the system evolves according to Hamilton's equations of motion, then the volume occupied by the system in phase space remains constant over time. Furthermore, if the system is isolated, the Hamiltonian and total energy are conserved.

1.4.1 Time Integrators

Newton's law, Eq. (1.16), or alternately Hamilton's equations, Eqs. (1.19) and (1.20), are systems of ordinary differential equations (ODEs) and hence their basic properties can be analyzed with general methods such as using $\mathscr{H}$ as a

Lyapunov function to determine stability. Symplectic integrators [13], in particular, are designed to preserve the basic properties of Hamiltonian system, such as incompressibility of flows in phase space as well as conservation of the total energy in a closed system. These integrators still have errors associated with their trajectories but these errors are considered more acceptable, in that individual atom trajectories may deviate from the true trajectories, but the system properties should be accurately represented due to the conservation properties of these methods. In particular, unlike other integration schemes that may have better truncation/short-time errors, symplectic integrators due to their representation of the underlying Hamiltonian of the system have bounded errors and hence long time stability.

The integrators commonly used in MD are

- Verlet, which is simple, explicit, symplectic, and second order accurate [14];
- Runge–Kutta, which is a higher-order accurate multi-step method that can be made symplectic [15];
- Gear, another higher-order accurate method based on backward differences but is not symplectic;
- SHAKE [16, 17] condenses stiff bonds with high-frequency vibrations into constraints; and
- rRESPA [18], which handles multiple timescales hierarchically.

Generally, these integrators are *explicit*, in that the updates only involve previously calculated $\{\mathbf{x}_i, \mathbf{p}_i\}$ making them relatively cheap but the tradeoff is that they are conditional stable.

This raises the important question, how do we select an appropriate time-step? Loosely speaking, in the sense of the Nyquist frequency criterion, the time-step of the integrator must be much less than the period of highest frequency $\Delta t \ll 1/\omega$. The Einstein frequency, $\omega = \sqrt{\frac{\nabla_{\mathbf{x}_i} \nabla_{\mathbf{x}_i} \Phi}{m_i}}$, for atom i, is the frequency of an atom oscillating in a potential well with the rest of the atoms being fixed, is typically the highest frequency in the spectrum. Alternatively, the phonon density of states (DOS) gives information regarding all of the frequencies available in the system, an example of which is shown in Fig. 1.6. For this case, we can see the largest frequency is roughly 16 THz, which corresponds to a period of 62 femtoseconds (fs). Given that for many materials, the maximum frequency is of this order magnitude, typically time-steps are on the order of 1 fs. An alternative and popular method to evaluate time-steps size is to use a closed system and evaluate the conservation of energy to establish a sufficiently small time-step. However, testing is required and light atoms with stiff bonds require smaller time-steps, such as in hydrocarbons.

1.4.1.1 Verlet

The most commonly used time integration algorithm, Verlet, is just three steps

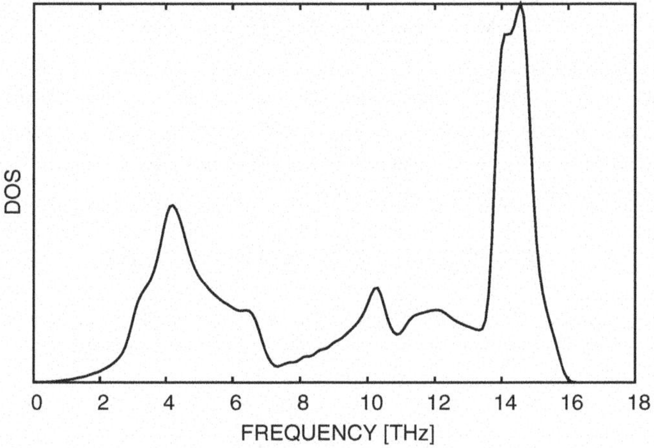

Fig. 1.6 The phonon density of states of Si

$$\begin{aligned} \mathbf{V}_{j+1/2} &= \mathbf{V}_j + \frac{1}{2m}\Delta t\,\mathbf{F}_j \\ \mathbf{R}_{j+1} &= \mathbf{R}_j + \Delta t\,\mathbf{V}_{j+1/2} \\ \mathbf{V}_{j+1} &= \mathbf{V}_{j+1/2} + \frac{1}{2m}\Delta t\,\mathbf{F}_{j+1} \end{aligned} \tag{1.21}$$

where the subscript j denotes the time-step, $\mathbf{V} = \{\mathbf{v}_i\}$ and, here, $m_i = m$ for simplicity. Note: (a) the positions, and hence the forces, are updated with the velocity mid-step $i + 1/2$, and (b) there are two velocity updates, the first is explicit and the second is *implicit* and yet there are no linear systems to solve. To see the (short-time) accuracy of the method, add the Taylor series for the positions:

$$\mathbf{R}_{j+1} = \mathbf{R}_j + \Delta t\,\mathbf{V}_j + \frac{\Delta t^2}{2!m}\mathbf{F}_j + \frac{\Delta t^3}{3!}\mathbf{B}_j + \mathcal{O}(\Delta t^4) \tag{1.22}$$

$$\mathbf{R}_{j-1} = \mathbf{R}_j - \Delta t\,\mathbf{V}_j + \frac{\Delta t^2}{2!m}\mathbf{F}_j - \frac{\Delta t^3}{3!}\mathbf{B}_j + \mathcal{O}(\Delta t^4) \tag{1.23}$$

the third order terms involving $\mathbf{B}(\mathbf{R}, \mathbf{V})$ cancel, to obtain

$$\mathbf{R}_{j+1} = 2\mathbf{R}_j - \mathbf{R}_{j-1} + \frac{\Delta t^2}{2!m}\mathbf{F}_j + \mathcal{O}(\Delta t^4) \tag{1.24}$$

which is why the scheme is also named the central difference scheme. The velocities, being derivatives of the positions, are only accurate up to $\mathcal{O}(\Delta t^2)$.

We can also investigate the ability of the Verlet algorithm to conserve momentum and energy. From Eq. (1.21) we can obtain

$$m\mathbf{V}_{j+1} = m\mathbf{V}_j + \Delta t \frac{1}{2} \left(\mathbf{F}_{j+1} + \mathbf{F}_j\right) \tag{1.25}$$

which when summed over all the atoms shows that the total momentum does not change for an isolated system where the forces sum to zero. To examine energy conservation, first assume linear forces $\mathbf{F} = -\mathbf{K}\mathbf{R}$ with $\mathbf{K} = \mathbf{K}^T$ and $\sum_i K_{ij} = 0$. Now the change in energy over a time-step is

$$\Delta E = E_{j+1} - E_j = \bar{\mathbf{R}} \cdot \mathbf{K}\, \Delta\mathbf{R} + \bar{\mathbf{V}} \cdot m\, \Delta\mathbf{V} \tag{1.26}$$

where $\bar{\mathbf{V}}$ is the average $\bar{\mathbf{V}} = \frac{1}{2}(\mathbf{V}_{j+1} + \mathbf{V}_j)$ and $\Delta\mathbf{V}$ is the difference $\Delta\mathbf{V} = \mathbf{V}_{j+1} - \mathbf{V}_j$. Substitute

$$\Delta\mathbf{R} = \Delta t\, \bar{\mathbf{V}} - \frac{1}{4}\frac{\Delta t^2}{m} \Delta\mathbf{F} \tag{1.27}$$

$$\Delta\mathbf{V} = \Delta t \frac{1}{m} \bar{\mathbf{F}} \tag{1.28}$$

to obtain

$$\begin{aligned} \Delta E &= \Delta t\, \bar{\mathbf{R}} \cdot \mathbf{K}\, \bar{\mathbf{V}} + \Delta t\, \bar{\mathbf{V}} \cdot \bar{\mathbf{F}} + +\bar{\mathbf{R}} \cdot \mathbf{K} \left(-\frac{1}{4}\frac{\Delta t^2}{m} \Delta\mathbf{F} \right) \\ &= \frac{1}{4}\frac{\Delta t^2}{m} \bar{\mathbf{F}} \cdot \Delta\mathbf{F} = -\frac{1}{8}\frac{\Delta t^2}{m} \left(\mathbf{F}_{j+1} \cdot \mathbf{F}_{j+1} - \mathbf{F}_j \cdot \mathbf{F}_j\right) \end{aligned} \tag{1.29}$$

So energy is not preserved exactly step-to-step. Exact energy conserving algorithms can be constructed [13]; however, in practice, the stability limit for the time-step of the Verlet algorithm also gives satisfactory energy conservation in most cases. Moreover, due to Verlet being a symplectic integrator, it preserves a non-canonical Hamiltonian that shadows the traditional one, Eq. (1.17).

1.4.1.2 SHAKE

In order not to suffer from time-step restrictions associated with stiff (typically covalent) bonds, the SHAKE integration algorithm treats them as geometric constraints. There are two standard ways of handling distance constraints: generalized coordinates for the remaining degrees of freedom or Lagrange multipliers. SHAKE uses the latter. For example, for a set $\mathscr{B}$ of pairwise constraints for bond length

$$\|\mathbf{x}_{ij}\|^2 = \ell_{ij} \quad ij \in \mathscr{B} \tag{1.30}$$

where $\mathbf{x}_{ij} = \mathbf{x}_i - \mathbf{x}_j$, SHAKE forms an amended Verlet update for the positions

$$\mathbf{x}_i^+ = \mathbf{x}_i^* + \frac{\Delta t^2}{m_i} \sum_j \lambda_{ij} \mathbf{x}_{ij} \tag{1.31}$$

where $\mathbf{x}_i^*$ is the Verlet update, Eq. (1.21). The second term in Eq. (1.31) is the gradient of the set of constraints with respect to $\mathbf{x}_i$ times the multiplier λ_{ij}. The multipliers $\{\lambda_{ij}\}$ are solved for using the constraints applied to the new positions $\mathbf{x}_i^+$.

$$\left\| \mathbf{x}_{ij}^* + \Delta t^2 \left(\frac{1}{m_i} + \frac{1}{m_j} \right) \sum_j \lambda_{ij} \mathbf{x}_{ij} \right\|^2 = \ell_{ij} \tag{1.32}$$

These equations are quadratic in λ_j but the SHAKE algorithm ignores the quadratic terms and relaxes the linear equations using a Gauss–Siedel procedure.

1.4.2 Ensembles and Thermostats

Now that means of creating stable, accurate trajectories have been established, we can turn to using these trajectories to compute quantities of interest. In traditional MD, *ensembles* represent macroscopic conditions for which numerous atomic states of the system are probable. For example:

- NVE is an ensemble with constant number, volume, and energy. This is the micro-canonical ensemble of an isolated system.
- NVT is an ensemble with constant number, volume, and temperature, i.e., the canonical ensemble of a system in contact with a thermal reservoir.
- NPT is a constant number, pressure, and temperature ensemble.
- NPH is a constant number, pressure, and enthalpy ensemble,
- μPT is the constant chemical potential, pressure, temperature, and grand canonical ensemble (which is difficult to simulate in MD).

In a practical sense, these ensembles represent conserved quantities that act as constraints on the dynamics, e.g., NVE: $\{\mathbf{R}(t), \mathbf{Q}(t) \mid \mathscr{H}(\mathbf{R}, \mathbf{Q}) = E\}$ where E is a constant. Many integration algorithms are designed to preserve energy (and momentum) but typically additional control algorithms are needed to maintain consistency of the dynamics of any given system with a desired macroscopic state such as constant temperature.

In particular, the NVT ensemble is represented by the Maxwell–Boltzmann distribution

$$\text{pdf}(\mathbf{R}, \mathbf{Q}) \propto \exp\left(-\frac{\mathscr{H}(\mathbf{R}, \mathbf{Q})}{k_B T} \right) \tag{1.33}$$

where k_B is Boltzmann's constant. The proportionality constant (normalization factor) for this probability density function (PDF) is called the *partition function*,

$$\mathscr{P} = \frac{1}{N!h^{3N}} \int \exp\left(-\frac{\mathscr{H}(\mathbf{R},\mathbf{Q})}{k_B T}\right) d\mathbf{R} d\mathbf{Q} \tag{1.34}$$

and is related to the volume occupied in phase space. Here, N is the number of atoms in the system and h is Planck's constant.

To estimate, for example, the pressure of a material at a specific temperature, we need to form an average over all the possible states in the relevant ensemble. An ensemble average is a probability weighted average of a phase function A:

$$\langle A \rangle = \int A(\mathbf{R},\mathbf{Q}) \; \mathrm{pdf}(\mathbf{R},\mathbf{Q}) \, \mathrm{d}\mathbf{R}\mathrm{d}\mathbf{Q} \tag{1.35}$$

and represents the expected value of A. The connection between expected values of macroscopic observables and samples from atomic trajectories is, generally speaking, ergodicity. Ergodicity is tantamount to the assumption that long time averages equal ensemble averages

$$\bar{A} \equiv \lim_{\tau \to \infty} \frac{1}{\tau} \int_0^{\tau} A(\mathbf{R}(t), \mathbf{Q}(t)) \, \mathrm{d}t = \langle A \rangle \tag{1.36}$$

For this to be true the trajectory of the system $(\mathbf{R}(t), \mathbf{Q}(t))$ must traverse phase space with the probability corresponding to the PDF. In certain systems ergodicity can be proven but it generally assumed for all but degenerate systems. The concept of ergodicity is the main link between the trajectories that MD can produce and observable, macroscale properties that we want to predict.

In order to produce trajectories consistent with a given temperature (in an NVT ensemble, for instance) and hence predict material properties at finite temperature, a thermostat is used in addition to the basic time integration scheme. Thermostats are time integration schemes that control the dynamics so that the temperature T is conserved (instead of energy E). Popular thermostats include: *Nosé–Hoover*, which is deterministic and has the property that time averages equal ensemble averages; *Langevin*, which uses a stochastic force and the fluctuation–dissipation theorem to generate Brownian dynamics and emulate momentum-exchanging collisions of atoms with fictitious reservoir atoms; *Gaussian isokinetic*, which employs a Lagrange multiplier to enforce the temperature constraint and only generates the NVT ensemble in the thermodynamic (large system size) limit; and *rescale*, which is a very simple method where velocities are rescaled to enforce a constraint on K, but does not generate the NVT ensemble. As illustrated in Fig. 1.7, the thermostat does indeed affect the phase space dynamics; it samples a larger volume of phase space than a corresponding constant energy simulation.

The need for a thermostat to control temperature begs the question: what is temperature at the atomic scale? This is not a trivial question but one that can

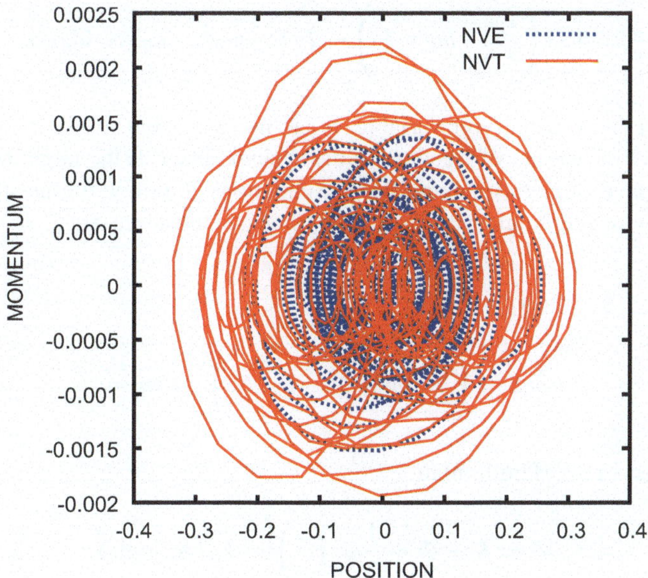

Fig. 1.7 Trajectories of an atom in an NVE and an NVT environment

be answered using statistical thermodynamics. Thermodynamics relates the heat capacity ρc, temperature T, and the average energy of the system:

$$\rho c T = \langle E \rangle = \langle K \rangle + \langle \Phi \rangle \tag{1.37}$$

In equilibrium, equipartition of energy implies that

$$\langle K \rangle = \langle \Phi \rangle \tag{1.38}$$

(assuming a harmonic solid, see [19] for the more general statements). The Dulong–Petit law $\rho c = 3k_B N/V$ relates the heat capacity and the Boltzmann constant, i.e., each atom has a heat capacity of $3k_B$, for a classical system. The result is the kinetic definition for temperature

$$3k_B T V = 2\langle K \rangle \tag{1.39}$$

Note that $K(t)$ fluctuates and the central limit theorem implies that the variance of this fluctuation is proportional to $1/N$.

With a computable definition of temperature in hand, the Nosé–Hoover thermostat uses feedback control and an extended system with Lagrangian (the Legendre transform of the Hamiltonian $\mathscr{H}$):

$$\mathscr{L} = s^2 \left(\frac{1}{2} \sum_i m_i \|\mathbf{v}_i\|^2 \right) - \Phi + \frac{1}{2} m_s \dot{s}^2 - k_B \bar{T} N_f \log(s) \tag{1.40}$$

where s is a control/reservoir variable, m_s is its associated mass, $N_f = 3N + 1$ is the number of degrees of freedom in the system, and $\bar{T}$ is the target temperature [20], [21, Sect. 6.2], [22, Appendix B]. The equations of motion for the atoms can be derived from the Lagrangian, assuming the forces on the atoms are $\mathbf{f}_i^* = -\nabla_{\mathbf{x}_i} \Phi$, as:

$$m_i \dot{\mathbf{v}}_i = \mathbf{f}_i^* - \frac{\dot{s}}{s} m_i \mathbf{v}_i \tag{1.41}$$

$$m_s \ddot{s} = \left(\sum_i m_i \|\mathbf{v}_i\|^2 - k_B \bar{T} N_f \right) s + m_s \frac{\dot{s}^2}{s} \tag{1.42}$$

which conserves the Hamiltonian

$$\mathscr{H} = K + \Phi + \frac{1}{2} m_s \left(\frac{\dot{s}}{s} \right)^2 - k_B \bar{T} N_f \log(s) \tag{1.43}$$

Since s is finite, $\langle \ddot{s} \rangle \to 0$; so

$$\left\langle \frac{d}{dt} k_B \bar{T} N_f \log(s) \right\rangle = \left\langle \frac{1}{s} \sum_i m_i \|\mathbf{v}_i\|^2 \right\rangle \tag{1.44}$$

In most implementations, the thermostat is reformulated so

$$m_i \dot{\mathbf{v}}_i = \mathbf{f}_i^* - \lambda m_i \mathbf{v}_i \tag{1.45}$$

$$\dot{\lambda} = \frac{1}{\tau} \left(\frac{T}{\bar{T}} - 1 \right) \tag{1.46}$$

but in this version time t does not correspond to physical time. Also, the characteristic period of fluctuation $\tau = 2\pi \sqrt{\frac{m_s \langle s \rangle}{2 N_f k_B \bar{T}}}$ is function of the parameter m_s and the particular system dynamics.

In addition to being deterministic, the Nosé–Hoover thermostat has the important property that its partition function (marginalized over s) is the partition function of the canonical ensemble. This means that sampling Nosé–Hoover dynamics uniformly in t should result in the NVT ensemble average.

To effect dynamics consistent with the NPT ensemble, a Nosé–Hoover-like barostat can be formulated where the volume of the system V is also controlled in order to obtain a target pressure P:

$$m_i \dot{\mathbf{v}}_i = \frac{1}{V^{1/3}} \mathbf{f}_i^* - m \left(\frac{\dot{s}}{s} + \frac{2\dot{V}}{3V} \right) \mathbf{v}_i \tag{1.47}$$

$$\ddot{s} = \frac{\dot{s}^2}{s} + \frac{g_s}{m_s} s \tag{1.48}$$

$$\ddot{V} = \frac{\dot{s}\dot{V}}{s} + \frac{g_V}{3 m_V V} s^2 \tag{1.49}$$

where the ensemble constraints are

$$g_s = 2V^{2/3} K - k_B \bar{T} N_f \tag{1.50}$$

$$g_V = 2V^{2/3} K + V^{1/3} \sum_{i<j} \mathbf{f}_{ij} \cdot \mathbf{x}_{ij} - 3PV \tag{1.51}$$

(Recognize the virial in the second constraint.) For the full stress ensemble, the system box needs also to change shape [23], as shown in Sect. 1.2.

1.4.3 Initial Conditions and Replicas

Now that we have introduced ensembles, we can discuss appropriate initial conditions to start the dynamics. Typically, the atomic positions in solid systems are taken to be perfect (zero temperature) lattice sites since determining more appropriate initial conditions for the positions consistent with the ensemble's distribution is exceedingly difficult. Given that the kinetic energy is quadratic in the velocities, the initial velocities can be sampled from the Boltzmann distribution appropriate for a desired temperature (which defines the variance of Gaussian-like Boltzmann distribution). Since the chosen initial positions are extremely improbable, and the initial dynamics describe a relaxation to a set of equilibrium states where both $\{\mathbf{x}_i\}$ and $\{\mathbf{p}_i\}$ are consistent with the Boltzmann distribution based on the potential and kinetic energies. As shown in Fig. 1.8, this results in a net transfer of energy from the kinetic energy to the potential energy. It is interesting to note that kinetic energy relaxes to half its initial value and the potential energy is approximately equal to the kinetic energy, roughly satisfying the equipartition theorem due to the small displacements solid state atoms make in a room temperature environment. On a practical note, even using a thermostat, starting the system with twice the desired temperature can expedite the thermal equilibration process.

With the thermostats and barostats we have introduced, molecular dynamics can now be used to compute the trajectory of a large number of atoms whose time average will approach a continuum value. For example, the temperature can be related to the average of the kinetic energy, which can be sampled directly in molecular dynamics. However, using a single large system or running a single

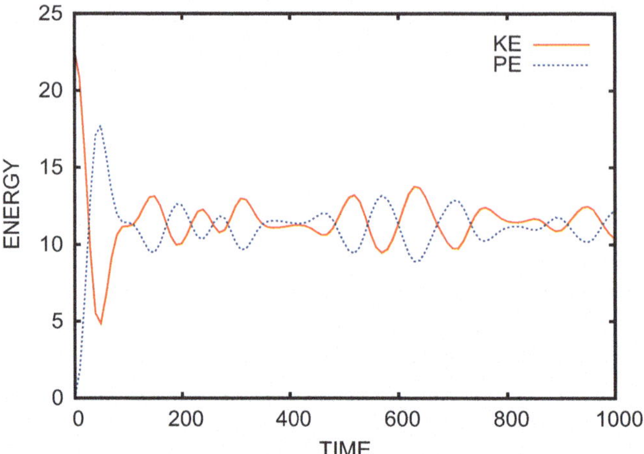

Fig. 1.8 Relaxation of a small Lennard-Jones system that is initially a perfect face centered cubic crystal with a Gaussian distribution of velocities

system for a sufficiently long time to reduce statistical errors is often impractical. Instead it is common practice to combine time averages of independent systems constructed from the probability distribution associated with the ensemble of choice. These replica systems are typically generated by sampling velocities from the appropriate distribution (and then allowing relaxation).

As an illustration of why ensemble averaging through replicas can result in more accurate and efficient averages, take the phase average of a harmonic function $\langle \cos(t) \rangle \equiv 0$ as a simple surrogate for an atomic trajectory or system fluctuation, via: (a) one long run

$$\langle \sin(\omega t) \rangle \approx \frac{1}{n\tau} \int_0^{n\tau} \cos(t)\, dt = \frac{\sin(\tau)}{n\tau}$$

or (b) n short runs with random initial conditions (and hence phases ϕ_α)

$$\langle \sin(\omega t) \rangle \approx \frac{1}{n\tau} \sum_{\alpha=1}^{n} \int_0^{\tau} \cos(t + \phi_\alpha)\, dt = \frac{1}{n\tau} \sum_{\alpha=1}^{n} \sin(\tau + \phi_\alpha) - \sin(\phi_\alpha) \quad (1.52)$$

With one long run the average will suffer from reinforcement of the phase error (aka n perfectly correlated phase errors), whereas with the same dynamics with initial conditions taken from the appropriate distribution the average benefits from cancellation of errors (de-phasing).

In summary, given the interatomic forces, the molecular dynamics method results in a set of non-linear ODEs that can be numerically integrated to reproduce the trajectory of all the atoms in the system. The presence of coupled non-linear ODEs requires the construction of stable and accurate time integrators and preference is given to symplectic integrators that preserve the symmetry and hence properties of the Hamiltonian. In MD, simulating just the Hamilton's equations of motion in an isolated system only samples from the micro-canonical ensemble, which is usually not of interest. To simulate the more interesting ensembles, such as the canonical ensemble, either thermostats or barostats or both must be introduced to control temperature and pressure, respectively. The thermostats and barostats alter the equations of motion and thus the Hamiltonian itself. Once we are assured that the dynamics of the simulation are consistent with the ensemble of interest, material properties can be predicted via time and/or ensemble averages of the appropriate atomic quantities. Many of the atomic level representations of these quantities will be discussed in detail in the next section.

1.5 Observables, Properties, and Continuum Fields

As we have discussed, material properties can be calculated from molecular statics and dynamics simulations using well-designed potentials and atomic structure. These quantities range from static to dynamic, and from equilibrium to non-equilibrium properties. For example, surface energies, defect energies, enthalpies of formation of crystal structures, mass densities, structure functions like the radial distribution function, heat capacities, elastic constants, and viscosity and thermal conductivity are regularly computed with now standard methods. However, since electrons are not treated explicitly, electronic, magnetic, and many properties involving charge transfer cannot be computed directly (see density function theory and other ab initio methods), but computing ionic conductivity of liquids is possible.

1.5.1 Equilibrium Properties

To begin, the mass density, ρ, of a system with volume V is simply

$$\rho = \frac{1}{V} \sum_i m_i \tag{1.53}$$

where the mass of atom i is m_i. For an NVT system $\rho \equiv \frac{N}{V} m_i$; however, for an NPT system $\rho = \rho(P, T)$ is non-trivial since $V = V(P, T)$.

Another density-like quantity, the heat capacity is given by the Dulong–Petit law that states that $\rho c = 3 k_B \frac{N}{V}$ for classical systems. The heat capacity can also be calculated from the variance of fluctuations in the energy E

$$c = \left. \frac{\partial E}{\partial T} \right|_V = \frac{\langle E^2 \rangle - \langle E \rangle^2}{k_B T^2} \tag{1.54}$$

using a derivation based on the partition function. (The idea of material properties being related to fluctuations will come up again in the following discussion of Green–Kubo methods.)

At low temperatures, MD still treats particle motion classically, which results in a failure in its ability to predict quantum effects on lattice vibrations and incorrect low temperature heat capacities. To correct for this, the internal energy of the phonons (lattice vibrations) can be modeled as a collection of harmonic oscillators

$$E = \sum_{\mathbf{k},p} \frac{\hbar\omega_{\mathbf{k},p}}{\exp\left(\hbar\omega_{\mathbf{k},p}/k_B T\right) - 1} \approx \sum_{p} \int k_B T \epsilon(\omega)\, \varphi(\omega)\, D(\omega)\, d\omega \tag{1.55}$$

where the first sum is over wave-vectors $\mathbf{k}$ and polarizations p. The (normalized) energy of oscillator is $\epsilon(\omega) = \frac{\hbar\omega}{k_B T}$, the Bose–Einstein distribution is $\varphi(\epsilon) = (\exp(\epsilon) - 1)^{-1}$, and $\varphi d\epsilon \sim \varphi\, D\, d\omega$. This requires an accurate description of the DOS, $D(\omega)$, which can be obtained from molecular statics or molecular dynamics methods (see Fig. 1.6). The DOS, i.e., the density of phonons in any given interval of frequency as determined by the dispersion relation $\omega(\mathbf{k})$, is typically estimated from the Fourier transform of the velocity auto-correlation $\mathscr{A}_{\mathbf{v}}(t)$

$$D(\omega) = \int_0^\infty \mathscr{A}_{\mathbf{v}}(t) \cos(\omega t)\, dt = \int_0^\infty \langle \mathbf{v}(0) \cdot \mathbf{v}(t) \rangle \cos(\omega t)\, dt \tag{1.56}$$

where

$$\mathscr{A}_{\mathbf{v}}(\alpha\, \Delta t) \equiv \langle \mathbf{v}(0) \cdot \mathbf{v}(\alpha\, \Delta t) \rangle = \frac{1}{N N_t} \sum_{i,\beta} \mathbf{v}_i(\alpha\, \Delta t) \cdot \mathbf{v}_i((\alpha + \beta)\, \Delta t) \tag{1.57}$$

and is usually normalized by $\langle \mathbf{v}(0) \cdot \mathbf{v}(0) \rangle$ so that $\mathscr{A}_{\mathbf{v}}(0) = 1$. Here, N is the number of atoms, N_t is the number of time samples, and we use the time invariance of the product, i.e., stationary statistics (the joint probability of $\mathbf{v}(s)$ and $\mathbf{v}(s + t)$ does not change with time s). Notice that this is essentially the Wiener–Khinchin theorem relating spectral density to the auto-correlation. Finally, using the chain rule $\frac{\partial}{\partial T} = \frac{\partial \epsilon}{\partial T} \frac{\partial}{\partial \epsilon}$ leads to

$$\rho c = \frac{\partial E}{\partial T} = k_B \sum_p \int \frac{\epsilon^2(\omega) \exp(\epsilon(\omega))}{(\exp(\epsilon(\omega)) - 1)^2} D(\omega)\, d\omega \tag{1.58}$$

An example of this calculation for silicon is shown in Fig. 1.9.

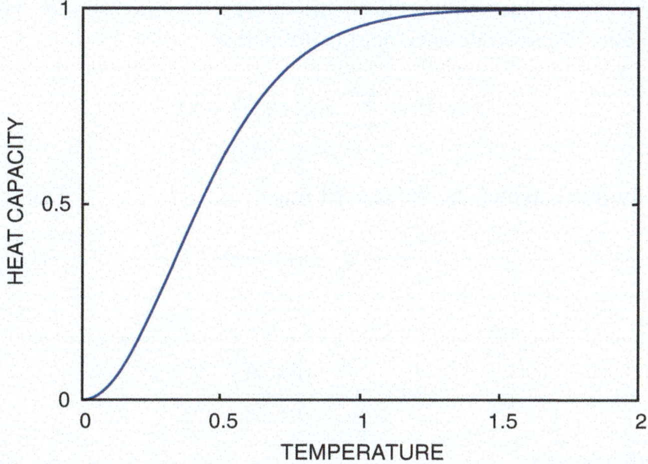

Fig. 1.9 Heat capacity of Si per atom normalized by $3k_B$ as a function of temperature in multiples of the Debye temperature computed using the DOS given in Fig. 1.6

1.5.2 Transport Properties

The properties discussed so far have been properties of equilibrium systems. Equilibrium processes are characterized as having zero average dissipative fluxes and the relevant statistics, such as the energy or temperature and average structural properties, are steady. Many properties of interest, however, are related to transport and, hence, to non-equilibrium but potentially steady processes. In these processes, system level fluxes of mass, charge, energy, etc., are not zero on average. So what is a flux in general? It is a field conjugate to a generalized coordinate through a thermodynamic potential, e.g., consider the Helmholtz free energy Ψ. Its rate

$$\dot{\Psi} = \frac{\partial \Psi}{\partial \mathbf{F}} \dot{\mathbf{F}} = \mathbf{P}\dot{\mathbf{F}} \tag{1.59}$$

can be written in terms of the rate-of-deformation tensor, $\dot{\mathbf{F}}$, the kinematic variable, and $\mathbf{P}$ is the (first Piola–Kirchhoff) stress, the flux of momentum. Similarly, we can define the heat flux $\mathbf{q}$ by its conjugacy with temperature T through the entropy S:

$$\dot{S} = \frac{\partial S}{\partial \mathbf{X}} \dot{\mathbf{X}} = \nabla \left(\frac{1}{T} \right) \mathbf{q} \tag{1.60}$$

Unlike the stress, the heat flux $\mathbf{q}$ is a *dissipative* flux i.e., one that directly contributes to entropy production [24].

To make the proper microscopic definition of the various fluxes, Irving and Kirkwood [25] made a correspondence between Newton's law for the atoms and the Euler balances of mass, momentum, and energy for a continuum. For instance, in this formalism, the mass density field is defined as:

$$\rho(\mathbf{x}, t) = \sum_i m_i \Delta(\mathbf{x}_i(t) - \mathbf{x}) \tag{1.61}$$

since this definition satisfies the balance of mass

$$\frac{\partial}{\partial t}\rho + \nabla \cdot (\rho \mathbf{v}) = 0 \tag{1.62}$$

with velocity

$$\mathbf{v}(\mathbf{x}, t) = \frac{\mathbf{p}(\mathbf{x}, t)}{\rho(\mathbf{x}, t)} \tag{1.63}$$

and momentum density

$$\mathbf{p}(\mathbf{x}, t) = \sum_i m_i \mathbf{v}_i(t) \Delta(\mathbf{x}_i(t) - \mathbf{x}) \tag{1.64}$$

Here, $\Delta(\mathbf{x})$ is a smoothing kernel with some regularity properties that enable coarse-graining point-wise atomic data into continuous fields.

To see the connection, start with the time derivative of the momentum density $\mathbf{p}$

$$\dot{\mathbf{p}} = \sum_i m_i \dot{\mathbf{v}}_i \Delta(\mathbf{x}_i - \mathbf{x}) - m_i \mathbf{v}_i \nabla_{\mathbf{x}} \Delta(\mathbf{x}_i - \mathbf{x}) \cdot \mathbf{v}_i \,. \tag{1.65}$$

After using Newton's law $\mathbf{f}_i = m_i \dot{\mathbf{v}}_i$ this can be written as:

$$\dot{\mathbf{p}} = \sum_i \mathbf{f}_i \Delta(\mathbf{x}_i - \mathbf{x}) + m_i \mathbf{v}_i \otimes \mathbf{v}_i \nabla_{\mathbf{x}} \Delta(\mathbf{x}_i - \mathbf{x}) \,. \tag{1.66}$$

Defining $B(\mathbf{x}, \mathbf{y})$ by $-(\mathbf{x}_i - \mathbf{x}_j) \cdot \nabla B(\mathbf{x}_i - \mathbf{x}, \mathbf{x}_j - \mathbf{x}) = \Delta(\mathbf{x}_i) - \Delta(\mathbf{x}_j)$ the momentum balance becomes

$$\begin{aligned} \dot{\mathbf{p}} = \nabla_{\mathbf{x}} \cdot \Bigg[& \sum_{i \neq j} \mathbf{f}_{ij} \otimes (\mathbf{x}_i - \mathbf{x}_j) B(\mathbf{x}_i - \mathbf{x}, \mathbf{x}_j - \mathbf{x}) \\ & + \sum_i m_i \mathbf{v}_i \otimes \mathbf{v}_i \Delta(\mathbf{x}_i - \mathbf{x}) \Bigg] = \nabla \cdot \boldsymbol{\sigma} \end{aligned} \tag{1.67}$$

where we can identify the stress $\boldsymbol{\sigma}$ after specializing to pair forces $\mathbf{f}_i = \sum_j \mathbf{f}_{ij}$ and $\mathbf{f}_{ij} = -\mathbf{f}_{ji}$. For the whole system occupying a region with volume V, the stress is given by

$$\sigma = \frac{1}{V} \sum_i \left(m_i \mathbf{v}_i \otimes \mathbf{v}_i + \frac{1}{2} \sum_j \mathbf{f}_{ij} \otimes \mathbf{x}_{ij} \right) \tag{1.68}$$

(this is essentially the formula for the virial and corresponds to the Cauchy stress), and heat flux by

$$\mathbf{q} = \frac{1}{V} \sum_i \left(\phi_i \mathbf{I} + \sigma_i^T \right) \mathbf{v}_i \tag{1.69}$$

where the potential energy $\phi_i = \frac{1}{N_j} \sum_j \phi(r_{ij})$ is partitioned per atom (not per bond which would be more natural). In the expression for $\mathbf{q}$, Eq. (1.69), the first, convective, term tends to dominate in fluids, whereas the second, non-convective, term tends to dominate in solids. These formulas for macroscopic observables in terms of atomic data can now be used in estimating transport coefficients such as elastic moduli and thermal conductivities with molecular simulation.

To make the macroscale–microscale connection for transport and dissipative processes, first acknowledge that macroscale laws do apply in aggregate at the atomic scale but there can be strong size effects. One of the foundations of this observation is the Onsager regression hypothesis (1931) which states: the equilibrium fluctuations in a phase variable are governed by the same transport coefficients as is the relaxation of that same phase variable to equilibrium. In other words, the decay of an equilibrium fluctuation is indistinguishable from that of a (small) external perturbation. We will see that Green–Kubo theory makes this discovery computable.

As an example of the macro–micro connection take Fick's law as a model of the flux in concentration as a function of its gradient

$$\mathbf{J} = -D\nabla c \tag{1.70}$$

This is an example of linear response where the flux is linearly related to the (external) force. Fick's law, together with the conservation of species

$$\frac{\partial}{\partial t} c + \nabla \mathbf{J} = 0 \,, \tag{1.71}$$

gives the usual diffusion equation

$$\frac{\partial}{\partial t} c = D\nabla^2 c \tag{1.72}$$

The question is: can we determine the diffusion coefficient D from MD? Take the governing conservation partial differential equation (PDE) Eq. (1.72) and multiply by the initial conditions and ensemble average:

$$\begin{aligned} \left\langle c(\mathbf{0},0)\cdot\frac{\partial}{\partial t}c(\mathbf{x},t)\right\rangle &= \left\langle c(\mathbf{0},0)\cdot D\nabla^2 c(\mathbf{x},t)\right\rangle \\ \frac{\partial}{\partial t}\mathscr{A}(\mathbf{x},t) &= D\nabla^2\mathscr{A}(\mathbf{x},t) \end{aligned} \tag{1.73}$$

So the correlation $\mathscr{A}(\mathbf{x},t)$ satisfies the PDE, which is a general result [26]. Next, recognize that the Green's function

$$G(\mathbf{x},t) = \left(\frac{1}{\sqrt{4\pi Dt}}\right)^3 \exp\left(-\frac{\mathbf{x}\cdot\mathbf{x}}{4Dt}\right) \tag{1.74}$$

for the PDE is such that $G \sim \mathscr{A}$ and the transport coefficient D is embedded in both which hints at the underpinnings of the Green–Kubo method.

There are three main types of methods for determining transport properties:

- Analytical, which usually involves derivatives of potential and idealized structure. These methods can be complex or tedious and have limited applicability due to strong assumptions.
- Direct, which is based on a macroscale analogue. They usually involve non-equilibrium with unphysically large gradients and large systems. Variants include the Müller-Plathe for thermal conductivity [27].
- Green–Kubo, which employs equilibrium fluctuations and small systems. (There are non-equilibrium variants such as Evans' method and SLLOD [28].) The main issue with these methods is that the noise/errors inherent in the estimates must be handled correctly.

1.5.2.1 Analytical Methods

First, let us examine a practical and common method of computing the elastic constants of a crystal. We can define the elastic constants as derivatives of the internal energy with respect to the deformation gradient: $\mathbb{B} = \nabla_\mathbf{F}\mathbf{P} = \nabla_\mathbf{F}\nabla_\mathbf{F}\Phi$, where $\mathbf{P}$ is the first Piola–Kirchhoff stress measure. If we assume that the potential energy is a simple pair potential, then $\Phi = \sum_{i\neq j}\phi(r_{ij})$ where $r_{ij} = |\mathbf{x}_i - \mathbf{x}_j|$. To employ the Cauchy–Born rule we assume a homogeneous (local) deformation from a perfect lattice reference $\mathbf{x}_i = \mathbf{F}\mathbf{X}_i$ so that stress is given by:

$$\mathbf{P} = \nabla_\mathbf{F}\Phi = \frac{1}{V}\sum\frac{\phi'}{r_{ij}}\mathbf{x}_{ij}\otimes\mathbf{x}_{ij} = \frac{1}{V}\sum\mathbf{f}_{ij}\otimes\mathbf{X}_{ij} \tag{1.75}$$

which is basically identical to a referential form of the virial. The elastic constants can also be calculated directly:

$$\begin{aligned}\mathbb{B} &= \nabla_{\mathbf{F}}\nabla_{\mathbf{F}}\Phi = \nabla_{\mathbf{u}}\nabla_{\mathbf{u}}\Phi\, \nabla_{\mathbf{F}}\mathbf{u}\, \nabla_{\mathbf{F}}\mathbf{u} \\ &= \frac{1}{2V}\sum_i \left(\phi'' - \frac{\phi'}{r_i}\right) r_i^2 \frac{\mathbf{x}_i}{r_i} \otimes \frac{\mathbf{X}_i}{r_i} \otimes \frac{\mathbf{x}_i}{r_i} \otimes \frac{\mathbf{X}_i}{r_i} \\ &+ \frac{1}{r_i}\phi'\delta_{ij}\,\mathbf{e}_i \otimes \frac{\mathbf{X}_i}{r_i} \otimes \mathbf{e}_j \otimes \frac{\mathbf{X}_i}{r_i}\end{aligned} \tag{1.76}$$

where we have simplified the notation by assuming all positions $\mathbf{X}_i$ are relative to a central atom at the origin and $r_i = \|\mathbf{X}_I\|$.

The zero temperature Cauchy–Born rule can be extended to finite temperatures using the same harmonic oscillator model discussed in the context of the heat capacity. Since the stress at finite temperature is derived from the Helmholtz free energy, $\Psi = \Phi - TS$, the added difficulty arises in evaluating the entropy S. The necessary entropy S can be constructed via the (linearized) bond stiffness matrix

$$\mathbb{D}_{ij} = \frac{1}{\sqrt{m_i m_j}} \frac{\partial^2 \Phi}{\partial \mathbf{x}_i \partial \mathbf{x}_j} \tag{1.77}$$

embedded in Eq. (1.76), so that

$$\Psi = \Phi + \frac{k_B T N}{V} \log\left(\left(\frac{\hbar}{k_B T}\right)^3 \sqrt{\det \mathbb{D}}\right) . \tag{1.78}$$

where $\det \mathbb{D} = \prod_i \omega_i$. With this expression the stress and elastic constants can be derived via the first and second derivatives of Ψ with respect to deformation gradient $\mathbf{F}$ at constant T.

1.5.2.2 Direct Methods

In contrast to purely analytical methods, direct methods employ simulation of an atomistic analogue of a continuum test, e.g. a tensile test. The basic issue is how to set up and support a gradient and therefore a measurable flux via boundary conditions which mimic the external environment.

One of the simplest direct methods is used to extract elastic constants. By imposing a deformation gradient, $\mathbf{F}$, to the boundary of the cell allowing the cell to relax or thermalize depending on desired temperature and measure the full system stress or energy. If the atoms were to be held fixed this essentially enforces the Cauchy–Born rule; however, the atoms should be allowed to relax to their equilibrium positions, especially for complex lattices which can have internal relaxation modes. If the stress is measured, finite difference perturbations about a desired strain and temperature state are used to extract the relevant elastic constants. Typically, both stretches and shear strains need to be imposed to obtain the full matrix of elastic moduli. Likewise, if the energy is measured enough perturbations of the base configuration need to be constructed to evaluate the quadratic variation of the energy with respect to the deformation gradient, after which the necessary derivatives can be taken analytically. While it is conceivable to use the energy variation at finite temperature, the Helmholtz free energy as opposed to the potential energy must be evaluated, which is not immediately available (see the quasi-harmonic model from the previous section). In this case, the elastic constants must be extracted from the fluctuations of the system at equilibrium [29].

As an illustrative example, consider the direct determination of the bulk modulus K, which is the constant of proportionality between the pressure $p = \frac{1}{3}\operatorname{tr}\boldsymbol{\sigma}$ and the dilational strain $p = K \operatorname{tr}\boldsymbol{\epsilon} \approx K\frac{\Delta V}{V}$ for a material behaving linearly. To obtain this elastic constant we simply stretch the periodic box equally in all directions and take the derivative of the resulting pressure–volume curve at any particular volume strain.

The direct determination of thermal conductivity κ is more involved than estimating elastic constants since an inhomogeneous, non-equilibrium state must be set-up. In this case, a steady-state temperature gradient must be supported by injecting energy in a hot region and extracting the same amount from a cold region via: (a) rescaling velocities to achieve kinetic energy fluxes, (b) a thermostat to attain constant temperature regions, or (c) swapping hot atoms for cold ones. Once a region with a linear temperature profile is established the conductivity can be estimated by the ratio of the energy flux needed to support the gradient and the temperature gradient. Unfortunately there are strong size effects due to the fact that the reservoir regions are comparable in size with the regions used to determine the temperature gradient. The usual remedy is to perform the extrapolation of $\frac{1}{\kappa}$ vs $\frac{1}{L}$ to the bulk limit of a sequence of simulations of increasing size L using Matthiesen's rule. This rule assumes that the system behavior is due to independent scattering mechanisms related to phonon interactions in the test region, e.g., Umklapp processes, and the artificial reservoir-test transition region [30].

1.5.2.3 Green–Kubo Methods

Finally, we examine methods to determine non-equilibrium transport properties from correlations of phase variables from equilibrium atomic systems. To begin, let us consider a well-known example of how the fluctuations of a system are connected to continuum transport properties. Einstein derived a relationship between the diffusion constant for a single particle in a fluid and its mean squared displacement $\langle r^2 \rangle \equiv \frac{1}{3}\langle \mathbf{x}(0) \cdot \mathbf{x}(0) \rangle$:

$$\frac{\partial}{\partial t}\langle r^2 \rangle = D \int r^2 \nabla c \, dV = 2D \int \nabla \cdot \mathbf{x} \, c \, dr = 6D \tag{1.79}$$

via integration by parts and a normalized concentration $\int c\,dV = 1$. Using this Einstein relation and $\mathbf{x} = \int_0^t \mathbf{v}\,dt$, the constant D can be connected to a correlation:

$$\begin{aligned} 6D &= \lim_{t\to\infty} \frac{\partial}{\partial t}\langle r^2\rangle \\ &= \lim_{t\to\infty} \frac{\partial}{\partial t}\int_0^t \int_0^t \mathbf{v}(t_1)\cdot\mathbf{v}(t_2)\,dt_1 dt_2 = \lim_{t\to\infty}\int_0^t \mathbf{v}(0)\cdot\mathbf{v}(s)\,ds \end{aligned} \tag{1.80}$$

using the symmetry and time invariance of the correlation $\mathbf{v}(t_1)\cdot\mathbf{v}(t_2)$. (Note the limit is necessary to eliminate a $1-\frac{t_1}{t}$ factor that is significant at short times.)

The Green–Kubo method provides similar estimates to those derived via Einstein relations but is more generally applicable. As we have seen in Eq. (1.73) the auto-correlation function $\mathscr{A}_{\mathrm{v}}$ satisfies same PDE as phase variable $\mathbf{v}$. Based on this, Evans [26] derives the general Green–Kubo relation as:

$$L(F_e = 0) = \frac{V}{k_B T}\int_0^\infty \langle J(0)J(s)\rangle_{F_e=0}\,ds \tag{1.81}$$

where L is the linear response of the generic flux J to the external force F_e using the fluctuation and central limit theorems. Notice the flux auto-correlation $\langle J(0)J(s)\rangle$ must decay in a reasonable amount of time to obtain an estimate, see Fig. 1.10 for a typical well-behaved flux correlation. For instance, the self-diffusion constant D can be calculated from the auto-correlation of atomic velocities:

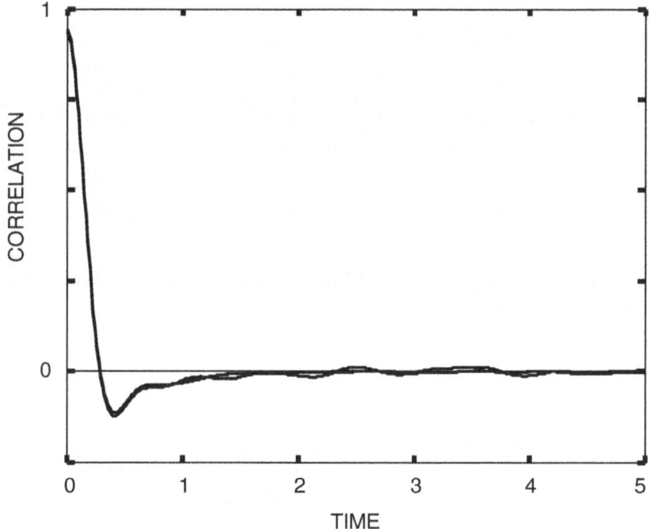

Fig. 1.10 The normalized thermal flux correlation of a liquid system showing the typical decay to statistical noise around zero

$$D = \frac{1}{3}\int_0^\infty \langle \mathbf{v}(0) \otimes \mathbf{v}(t) \rangle \, dt \approx \frac{1}{3N_sN_a} \sum_{k=1}^{N_c} \sum_{j=1}^{N_s} \sum_{i=1}^{N} \mathbf{v}_i(k\Delta s) \otimes \mathbf{v}_i((k+j)\Delta s)\Delta s \tag{1.82}$$

where N_s is the number of time samples used for averaging, N_c is the number of time samples between zero and the maximum time correlations, N_a is the number of atoms sampled and averaged, and Δs is the sample interval. Specific Green–Kubo formulas exist for most transport coefficients, e.g. viscosity,

$$\nu = \frac{V}{k_B T} \int_0^\infty \langle \boldsymbol{\varsigma}(0) \otimes \boldsymbol{\varsigma}(t) \rangle \, dt \,, \tag{1.83}$$

where $\boldsymbol{\varsigma}$ is a vector of off-diagonal components of stress, and thermal conductivity

$$\kappa = \frac{V}{k_B T^2} \int_0^\infty \langle \mathbf{q}(0) \otimes \mathbf{q}(t) \rangle \, dt \,. \tag{1.84}$$

where $\mathbf{q}$ is the heat flux (whose thermodynamic affinity is $\frac{1}{T}$ thus the additional factor of $1/T$).

In conclusion, with a well-verified potential and appropriate atomic structure, we can use molecular dynamics to predict a variety of static and transport properties via a variety of methods with distinct advantages. For further reading see the texts by: Allen and Tildesley [31], Evans and Morriss [26], and Frenkel and Smit [22].

1.6 Interatomic Potentials

Up to this point, we have neglected giving the specifics of the form of the interatomic potential. As mentioned earlier, interatomic potentials are one of the most important components of molecular simulations since they determine the system behavior. In fact, an appropriately constructed simulation can be meaningless if the chosen interatomic potential does not describe the physics of interest. Thus, the selection of the interatomic models must be done with care. In this section, we describe some of the basics of interatomic models including their form, the materials systems which they are believed to be appropriate, as well as some of the limitations of these types of models in simulating certain properties.

The interatomic potential describes the potential energy of the system and is generally assumed to be a function of the interatomic positions only, i.e., $\Phi = \Phi(\{\mathbf{x}_i\}) = \Phi(\mathbf{R})$, where $\mathbf{x}_i$ are the positions of the atoms, in order to be compatible with Hamiltonian dynamics. Ultimately, $\Phi(\mathbf{R})$ is best described (implicitly) by quantum mechanical calculations. For example, the use of density functional theory (DFT) and tight-binding (TB) methods can be used to both evaluate the potential energy and its derivatives. In the sense of classical molecular dynamics, one must assume that the electrons move sufficiently fast (they typically do in common states of matter) such that they reach their ground state much faster than the timescale of the motion of the nucleus, and thus we can evaluate the potential energy and forces corresponding to the ground state of the electrons. This is the Born–Oppenheimer approximation and is useful in establishing what is typically called ab initio molecular dynamics. The drawback of this approach is computational cost. DFT and TB can be very expensive, allowing the simulations of hundreds (DFT) to possibly thousands (TB) of atoms on modern computer architectures, but not the hundreds of thousands to millions of atoms required to model most nanostructures. Thus, there is a need to describe the interatomic interactions in a more approximate and efficient manner. This method, sometimes called empirical potential modeling, differs from the purely ab-initio approach in that interatomic interaction is usually constructed with functional forms based on theory, intuition, and fits to both experimental and ab-initio data. The functional forms are often chosen to be also numerically expedient at the cost of some of the underlying physical motivation.

In this section, we will continue to use the notation where Latin indices will denote the atoms. We will also use Greek letters to denote Cartesian components so that $\alpha \in [x, y, z]$. Also note that we do not specifically call out atom types in the interatomic interactions. This is implied through the interaction between the atom numbers i, j, etc.; however, in practice, the atom types must be specified.

1.6.1 Pair Potentials

Pair potentials were first introduced as a means to model molecular systems, initially with an emphasis on fluids but this approach has been widely used to model solids as well. With a pair potential the potential energy can be written solely as a function of the (pairwise) distance between the atoms:

$$\Phi(\mathbf{R}) = \sum_{i<j}^{N} \phi(r_{ij}) \tag{1.85}$$

where $r_{ij} = |\mathbf{x}_j - \mathbf{x}_i|$.

1.6.1.1 Lennard-Jones

The most common pair potential is the Lennard-Jones potential [32], which in its most general attractive–repulsive form is

$$\phi(r) = \frac{A}{r^m} - \frac{B}{r^n} \tag{1.86}$$

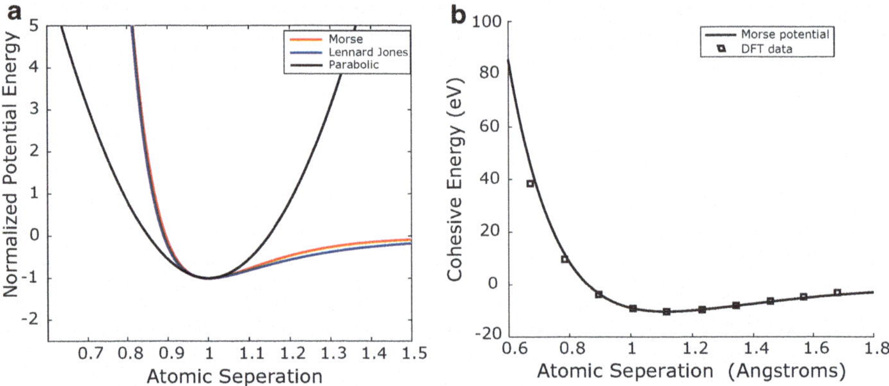

Fig. 1.11 (**a**) A comparison of the two classical pair potentials: the 12-6 Lennard Jones potential and the Morse potential. The potential parameters were selected so that they have the same minimum energy value and separation distance as well as the same curvature. These potentials are then compared with a pure harmonic potential, highlighting the anharmonic parts of the potentials. One thing that is important to note is that the two potentials are very similar over a wide range of separations, making the differences between the Morse and Lennard-Jones often minimal. However, the Morse potential is more flexible as the curvature can be changed while holding the location and value of the minimum fixed, which cannot be done with the 12-6 Lennard Jones. (**b**) The energy-separation curve for an N_2 molecule computed using density functional theory and fit using a Morse potential. The parameters are $D = 10.36\,\text{eV}$, $r_0 = 1.127\,\text{Å}$, and $a = 2.7\,\text{Å}^{-1}$

where typically $A, B > 0$ and $m > n > 4$ was proposed by Lennard-Jones [32]. One particular form of this potential has become well established, which is called the 12-6 form of the potential:

$$\phi(r) = \epsilon\left(\frac{\sigma}{r^{12}} - \frac{\sigma}{r^6}\right) \tag{1.87}$$

where ϵ and σ are fitting parameters with units of energy and length, respectively. The potential has a minimum of $-\epsilon$ that occurs at $2^{1/6}\sigma$, and hence ϵ is referred to as the *well-depth*. The parameter σ determines the equilibrium lattice spacing. A plot of the potential normalized such that $\epsilon = 1$ and $\sigma = 2^{1/6}$ is shown in Fig. 1.11. It is interesting to note that Lennard-Jones introduced the potential as a model for the interaction of molecules in a gas, which is where the potential has its most use. However, this and other pair potentials are often the base for constructing more complicated many-body potentials and thus understanding this potential does have practical value beyond modeling gases.

1.6.1.2 Morse

In the study of the binding between two molecules in a gas, such as the O–O bond in oxygen, Morse introduced an empirical formula for the binding energy between two molecules. Similar to the Lennard-Jones potential, the Morse form is the difference of two exponentials with different decay rates:

$$\phi(r) = D\left[\exp\left(-2a(r - r_0)\right) - 2\exp\left(-a(r - r_0)\right)\right]$$

where D is the binding energy (minimum) at a distance $r = r_0$ and the value a defines the curvature for the potential well.

Figure 1.11a shows a comparison of the Morse and Lennard-Jones pair potentials where the energy minimum and equilibrium separation were chosen to be the same for each potential. The Morse potential was further confined to have the same curvature as the Lennard-Jones potential at the equilibrium separation distance. If comparing the two, it is important to match the curvature because it sets the vibrational frequency of diatomic molecules in the bonded state. We can see that the two potentials match very closely within this range and the discrepancy becomes larger at larger atomic separations, although these large equilibrium separations are of less physical importance. Also plotted in Fig. 1.11b is a harmonic potential that also matches the three chosen parameters: minimum energy, equilibrium separation, and curvature. The harmonic potential differs significantly from both the Lennard-Jones and Morse potentials, most significantly by the fact that it does not have a finite cohesive energy.

Finally, to show that the Morse potential does a good job of replicating the pair potential of diatomic molecules, we computed the energy of an N_2 molecule as a function of separation distance using density functional theory, see Fig. 1.11b. The Morse potential was fit to the data using a simple scheme. The bonding energy and the equilibrium spacing were chosen to match the as-computed data directly and the free parameter a was selected to match the rest of the data as closely as possible. Overall the fit is good; however, improvement can probably be made with full non-linear regression, allowing all the unknowns to be adjustable.

1.6.2 Coulombic Potentials

Another classic pair potential is the interaction between two charged ions which can be represented using the classic Coulombic interaction potential:

$$V(r_{ij}) = \frac{1}{4\pi\epsilon_0}\sum_{i<j}\frac{q_i q_j}{|\mathbf{r}_i - \mathbf{r}_j|} \tag{1.88}$$

where q_i is the charge on ion i and ϵ_0 is the permittivity of free space. Note that there are no free parameters in this interaction unless a permittivity other than ϵ_0 is used to model a background dielectric and the interaction, by itself, only represents the behavior of well-separated atoms since at some distance repulsive forces will dominate. Typically in the modeling of covalent materials like polymers and biomolecules the Coulomb potential added to the Lennard-Jones potential to capture both long-range charge–charge and short-range van der Waals effects (see CHARMM [33], AMBER [34], and related parameterizations).

In classical electrostatics, this interaction energy can be easily evaluated but it is problematic in molecular systems, especially those that are conducted with periodic boundary conditions. First, due to the summation, it is important to realize that the energy will only have a finite value if the system under periodic boundaries is charge neutral. Second, because of the long-ranged nature of the $1/r$ interaction, the energy will be slow to converge and will do so conditionally, i.e., the value it converges to will depend on the way the interaction energies are summed over the periodic images. Thus, a number of ways have been introduced to evaluate the absolutely convergent Coulomb sum and to do it efficiently. The classic method is the Ewald sum [35], which splits the electrostatic interaction into a long-ranged and short-ranged potentials where the short-ranged interactions are summed in real space and converge rapidly. The long-range interactions are converted into reciprocal space and summed, where they also converge rapidly. The sum of the two converge absolutely and efficiently in $\mathcal{O}(N^{3/2})$ operations as compared to the $\mathcal{O}(N^2)$ required for the direct sum. More efficient forms of the Ewald sum have been formulated which utilize a fast-Fourier transform in the particle mesh Ewald (PME) and particle-particle particle-mesh Ewald sum (PPPM) [36], which scale as $\mathcal{O}(N \ln N)$.

1.6.2.1 Advantages and Disadvantages of Pair Potentials

An advantage of pair potentials is that they are relatively inexpensive. The potential energy is summed over just pairs of atoms, as opposed to triplets or local groups, which make the sum particularly fast. Despite this numerical advantage, pair potentials can have a number of drawbacks. One of the classic drawbacks of pair potentials is their inability to correctly represent the three independent elastic constants in a cubic crystals: C_{11}, C_{12}, and C_{44}. Pair potentials predict that the *Cauchy pressure*, $C_p = C_{12} - C_{44}$, is zero and Poisson's ratio is exactly 1/4. It is interesting to note that $C_{12} \approx C_{44}$ in noble gas crystals at low pressures, suggesting that pair potentials may be appropriate for these materials in the solid state. In fact, Gilman has provided some fits of the Lennard-Jones potential to the noble gas crystals [37]. However, the Cauchy pressure of most other cubic materials such as metals and ceramics is not zero. Thus, interactions beyond simple pair potentials are needed to accurately model most materials in the solid state.

1.6.3 The Embedded Atom Method

Since pair potentials can be insufficient to model real materials in the solid state, there has been extensive work in developing accurate and expedient many-body potentials. One of the most widely used is the embedded atom method (EAM) and its related forms. The EAM potential has the form:

$$\Phi = \sum_{i<j} \phi(r_{ij}) + \sum_{i} F(\rho_i) \tag{1.89}$$

where $\phi(r)$ is a pair potential that describes pairwise interaction between two atoms and $F(\rho)$ is the energy required to embed an atom at site with an electron density ρ. The surrounding atoms contribute to the electron density at the site i, ρ_i, in a pairwise fashion:

$$\rho_i = \sum_{j \neq i} f(r_{ij}) \tag{1.90}$$

This form requires the formulation of three different functions: $\phi(r)$, $F(\rho)$, and $f(r)$. The interatomic potential becomes many body in nature due to the complex form of the embedding function $F(\rho)$. However, the EAM, due to its basis in pairwise functions, is relatively computationally efficient. First, the electron density at each atom is computed via the pairwise form. Then, the potential energy can be computed from the pair potential and the embedding function making the cost only moderately more expensive than pure pair potentials; and, the many-body nature of the interactions eliminate the many of the obvious drawbacks of pair potentials including the zero Cauchy pressure.

1.6.4 Extensions of the EAM Formalism

The EAM formulation works well in metals where directional effects of the bonding are minimal. This gives rise to some problems with BCC metals where the crystal is typically stabilized by directional bonding. This, in conjunction with the desire to model mixed covalent and metallic bonding, has lead numerous authors to present extensions to the EAM framework. Two particular formulations which we will discuss here have gained some use in the wider literature.

The first and by far the most popular extension of the EAM method is the modified EAM (MEAM) [38–40]. The purpose of this generalization was to include directional bonding and provide a general framework where both metals and ceramics could be modeled. The MEAM is similar to the EAM in that the energy of an atom can be formulated as a sum of pairwise interactions and an embedding function. However, in the MEAM formalism a multibody screening function, S_{ij} is introduced that allows atoms between the bonded atoms to screen their interaction. Thus, the potential energy of an atom can be written as:

$$\Phi = \sum_{i<j} S_{ij}\phi(r_{ij}) + \sum_{i} F(\rho_i)$$

$$\rho_i = \sum_{j \neq i} f(r_{ij})$$

Another alteration of the EAM method is the angular dependent potential (ADP) [41, 42] to include some aspects of covalent bonding in intermetallics, which has also been extended to include BCC metals. The potential within the ADP formalism can be written as:

$$\Phi = \sum_{i<j} \phi(r_{ij}) + \sum_{i} F(\rho_i) + \frac{1}{2}\sum_{i,\alpha} \left(\mu_i^{\alpha}\right)^2 + \frac{1}{2}\sum_{i,\alpha\beta} \left(\lambda_i^{\alpha,\beta}\right)^2 - \frac{1}{6}\sum_{i} \nu_i^2$$

$$\rho_i = \sum_{j \neq i} f(r_{ij}) \qquad \mu_i^{\alpha} = \sum_{i \neq j} u(r_{ij}) r_{ij}^{\alpha}$$

$$\lambda_i^{\alpha,\beta} = \sum_{i \neq j} w(r_{ij}) r_{ij}^{\alpha} r_{ij}^{\beta} \qquad \nu_i = \sum_{\alpha} \lambda_i^{\alpha\alpha}$$

where the Latin indices refer to the atom indices and the Greek indices refer to Cartesian components. Typically, the functions $u(r)$ and $w(r)$ are simple tabulated functions and behave similarly to that of the standard EAM, making the ADP just slightly slower than the EAM. The μ and λ functions can be thought of as dipole and quadruple functions that penalize the energy for deviations from cubic symmetry and introduce an angular component in the potential.

In practice, the ADP potential behaves similarly to the EAM, its angular term is generally weak and contributes to the energy only when an atom is perturbed from a non-centrosymmetric position. The functions in the ADP can be tabulated just like EAM and the computations are nearly as fast. However, it is unclear how much better the ADP is in modeling metals with angular dependence in their bonding as it has been applied to a limited number of systems.

In comparison, the MEAM formalism is able to represent both covalent and metallic systems. One of the potential advantages of the MEAM framework is that with the development of single component potentials, cross-potentials can be created with a small number of parameters. This has allowed a great number of authors to create binary compounds that include both metallic and covalent materials like gold–silicon [43] and iron–carbon [44, 45] systems.

Despite the general formula given MEAM above, its general implementation follows a methodology that is different from the EAM. The MEAM formulation typically specifies an equation of state, relating the cohesive energy to the equilibrium lattice constant, bulk modulus, and cohesive energy. It also assumes specific functional forms of the embedding function and electron density functions. These three analytical functions can then be used to numerically determine the pair interaction function ϕ. This is in contrast to the way the EAM is usually parameterized with the pair function and electron density functions being specified and the embedding function either being specified or determined by fitting the values to an equation of state. The result is a set of functions that are typically tabulated and read into computer codes with splines being used to interpolate between values. There are certainly exceptions to these rules of thumb. Notably, the reference-less MEAM was introduced as a tabulated form of the MEAM where the pair function and embedding function are tabulated and read into computer codes in similar fashion to EAM. One of the large drawbacks of MEAM is that it can be nearly an order of magnitude more expensive than the EAM method.

1.6.5 Other Many-Body Functions

To model covalent bonded materials, more sophisticated many-body functions are needed. The Stillinger–Weber potential was introduced as a simple 3-body potential to model semiconducting materials, notably silicon [46]. The intent of the three body term was to penalize deviations from the 104° bond angle in tetrahedrally bonded materials. This potential has the form:

$$\Phi = \sum_{i<j} \phi_2(r_{ij}) + \sum_i \sum_{j\neq i} \sum_{k>j} \phi_3(r_{ij}, r_{ik}, \theta_{ijk})$$

$$\phi_2(r_{ij}) = A\epsilon \left[B_{ij} \left(\frac{\sigma_{ij}}{r_{ij}}\right)^{p_{ij}} - \left(\frac{\sigma_{ij}}{r_{ij}}\right)^{q_{ij}} \right] \exp\left(\frac{\sigma_{ij}}{r_{ij} - a_{ij}\sigma_{ij}}\right)$$

$$\phi_3(r_{ij}, r_{ik}, \theta_{ijk}) = \lambda_{ijk}\epsilon_{ijk} \left[\cos\theta_{ijk} - \cos\theta_{ijk,0}\right]^2$$
$$\exp\left(\frac{\gamma_{ij}}{r_{ij} - a_{ij}\sigma_{ij}}\right) \exp\left(\frac{\gamma_{ik}}{r_{ik} - a_{ik}\sigma_{ik}}\right)$$

The two-body term is a generalized Lennard-Jones potential with and exponential cutoff function. The three-body term is simply an energy penally for deviation from a specified bond angle. The simplicity of this many-body potential is appealing and it has been used extensively to model semiconductors. Parameterizations for Si [46–48], Ge [48, 49], group III–V semiconductors [50–55], and MoS_2 [56] exist in the open literature.

One of the most popular potentials is a simple bond-order potential (BOP). There is a general framework for this type of potential and it goes by many names including the analytic BOP [57, 58], the Tersoff potential [59–61], or Abell–Tersoff potential [62]. While a careful study of the literature would suggest these potentials are all perhaps slightly different, we can combine them into one class of BOPs by writing the potential energy as:

$$\Phi = \sum_{i<j} f_c(r_{ij}) \left[A \exp(-\lambda_1 r) - b_{ij} B \exp(-\lambda_2 r)\right]$$

the function f_c is a cutoff function that makes the potential short-ranged and b_{ij} is the bond function which controls how much a third atom k affects the bond between atoms i and j. The two exponential functions look similar to the Morse potential but provide even more flexibility in selecting the parameters. A very general framework for this class of bond-order functions is utilized by large atomistic massively parallel simulator (LAMMPS) [63] which allows for the general use of these functions as:

$$b_{ij} = \left(1 + \beta^n \zeta_{ij}^n\right)^{-\frac{1}{2n}}$$

$$\zeta_{ij} = \sum_{k \neq i,j} f_c(r_{ik}) g(\theta_{ijk}) \exp\left[\lambda_3^m (r_{ij} - r_{ik})^m\right]$$

$$g(\theta) = \gamma_{ijk} \left(1 + \frac{c^2}{d^2} - \frac{c^2}{[d^2 + (\cos\theta - \cos\theta_0)^2]}\right)$$

The Tersoff type potentials have been used extensively to model covalently bonded materials as well as materials with mixed metallic-covalent bonding. Tersoff originally parameterized his potential for silicon [59, 64] and Brenner parameterized a well-known carbon potential [65]. However, a large number of parameterizations have arisen for this type of potential. There are numerous parameterizations for group IVA elements including silicon [66], carbon, germanium [61], gallium-nitride [58, 67], gallium-arsenide [57, 67], silicon-carbon [61, 68–70], silicon-nitride [71], silicon-oxygen [72], tungsten-carbon-hydrogen [73, 74], iron-carbon [75, 76], platinum-carbon [57], zinc-oxygen [77], gold [78], among many others.

Another popular, more expensive and accurate bond-order type potential are the BOPs introduced by Pettifor [79–81]. There can be some confusion as the Tersoff type potentials are sometimes referred to as analytic BOPs, while Pettifor's original papers also use the term analytic BOPs. It is therefore useful to call the former potentials "Tersoff" and the latter BOP and avoid the use of "analytic bond-order potentials" altogether. We refer the interested reader to the ideas behind BOPs to a number of reviews on the subject [82–84]. These BOPs have been used to model hydro-carbons [85], semiconductors [86–88, 88], and BCC transition metals [73, 89–91].

1.6.6 Ionic Many-Body Potentials

In the section on pair potentials, the idea of Coulombic potentials was introduced. These potentials are key for modeling materials where ionic bonding is important; however, they lack the ability to describe the angular bonding present in mixed ionic-covalent (i.e., polar covalent) materials. To address this issue, it is quite possible to mix fix-charged pair potentials to capture the ionic nature of the bonding and use a many-body potential to capture the angular dependent nature of the bonding.

These types of many-body potentials are sufficient for modeling oxides as long as the composition does not change and the oxidation state of the atoms remains the same. However, in many materials the charge state of the ions should be change as the atoms interact. The oxidation of aluminum is a classical example of such a process where nominally charge-free aluminum comes into contact with elemental oxygen, which bonds with the aluminum forming amorphous aluminum oxide.

This has created the need for variable charge potentials where the charge is able to evolve in the simulations. In molecular dynamics simulations, the charge is typically assumed to equilibrate instantaneously at each classical MD time-step to minimize the total Coulombic potential energy. Thus, a charge equilibration step must be taken in between each ionic step in which the positions of the atoms are updated, which can increase the cost of performing the molecular dynamics simulations substantially.

One set of potentials that can be grouped together are the Yasukawa and the charge optimized many body (COMB) type potentials. These potentials use modified form of the Tersoff potential to describe the many-body nature of the intermolecular forces and a variable charge model of the charge on the ions. Yasukawa originally introduced these potentials to model oxides. As an extension of the Yasukawa potentials, the first generation COMB potentials were introduced to improve upon the Yasukawa potentials with parameterizations of the Si–O systems and Al–O systems. The potentials have been further improved to more accurately model the host metal structure by incorporating bond-bending terms to capture stacking fault energies. A newer form of the potentials, called COMB3, have emerged and the parameterizations for the original COMB and COMB3 are being created for a number of oxides including Al–O, Si–O, Ti–O, Mg–O and more exotic compounds like Ti–N.

An alternative variable charge anybody potential that is used extensively is the reactive force field (REAX) potentials. The REAX type potentials are a very general potential designed to be applicable to many systems and is therefore the most complicated potential discussed here. The potential was originally developed by van Duin et al. to describe bonding in hydro-carbons [92], but is has been subsequently extended to model a wide range of solid state materials [93–97].

1.7 Available Software and Potentials

There is a wide variety of codes available to construct atomistic/molecular configurations or structures, perform atomistic simulations, as well as visualize and post-process molecular simulations. Furthermore, as the interatomic potential is crucial for a successful atomistic simulation, there are a number of available potentials in various databases. Here, we provide the reader with an overview of the available software, codes, and potentials that are commonly used in atomistic and molecular simulations. This overview is by no means exhaustive.

To begin an atomistic or molecular simulation, an initial or starting configuration of the material or structure of interest is required. Many researchers opt to create their own starting atomic or molecular configurations using in-house programs that are not publicly distributed. However, many common molecular simulation codes can successfully construct some starting structures. Among those are LAMMPS [98], visual molecular dynamics (VMD) [99], Avogadro [100], and GROMACS [101, 102] (among many others). In addition to the starting configuration, one or more interatomic potentials must also be chosen.

The use of interatomic potentials requires accessing both the parameters for the potential itself and a code that has implemented the specific interatomic form. LAMMPS [98], which is distributed freely by Sandia National Laboratories, is a widely used molecular dynamics code that has a large majority of the potentials discussed here. It has the capability to implement all of the empirical potentials discussed in this chapter with full capability except for the BOP potentials as LAMMPS has not implemented a form that can handle bonding of the transition metals. EAM potentials in LAMMPS are read in as tables as described on the LAMMPS website [63] and generally does not provide the ability to use analytic EAM potentials. The freely available LAMMPS distribution has a large number of potential files included with the simulation package including COMB, REAX, EAM, Tersoff and Stillinger–Weber, amongst others.

One of the difficulties many users faces is getting parametrizations of the potentials in the literature. This is particularly challenging when one notes the frequencies of typographical errors in data in published manuscripts. This difficulty is compounded by the fact that many potential developers frequently need to update their parameterizations and thus published values may not be up to date and it becomes a challenge for users of interatomic models to keep up to date with developers. This has created the need to store and disseminate interatomic models and there are now a few locations where potentials can be found. First, the National Institute for Standards and Technology has a program underway for cataloguing and distributing interatomic models where users can readily download parameters for interatomic models [103, 104] (http://www.ctcms.nist.gov/potentials/). Another resource that is available for obtaining interatomic model parameterizations as well as testing them is through the knowledge base of interatomic models (KIM) [105, 106] (http://openkim.org/). In addition, we direct the reader to the EAM parameterizations at the website http://sites.google.com/site/eampotentials/. In addition, one of the developers of REAX, Adri van Duin, distributes parameterizations of the potentials he develops. These parameterizations and files can be obtained by directly contacting his research group (http://www.engr.psu.edu/adri/).

There is a growing list of available molecular simulation codes or software available. LAMMPS [98] is one of the most widely used codes to conduct atomistic and molecular modeling [63]. This simulation software is widely used by the community especially in the simulation of condensed matter. However, it has the capability to also model polymers and other soft matter. It has a large number of contributors from both government laboratories and academic institutions. It can conduct both molecular dynamics and molecular statics and includes accelerated MD techniques and methods to find minimum energy paths. One advantage of LAMMPS is its ability to be easily modified or extended to encompass capabilities not included in the standard release. That said, the LAMMPS software package includes a large collection of capabilities to model and analyze a variety of systems.

There are a number of other molecular dynamics codes that are open-source and freely available. IMD [107, 108], The ITAP Molecular Dynamics Program [109], is a freely distributed and can handle a number of potentials and thermodynamic ensembles and is intended to simulate condensed matter. Many of the other widely used MD codes target primarily soft matter simulations, especially biological systems. Two worth noting are NAMD [110] (http://www.ks.uiuc.edu/research/namd/) and GROMACS [101, 102] (http://www.gromacs.org/).

After successfully performing an atomistic simulation or modeling approach, the results can be extracted and presented in a number of useful formats. For example, data files containing the thermodynamic output or simulation information, such as stress–strain or energy–volume, can be obtained, but one of the more useful outputs of an atomistic simulation is the atomic data files. Usually, the atomic data files contain information about each atom in the simulation (such as position, energy, bonds, etc.) and the simulation cell at a given time-step of the simulation. Most atomistic simulations software that is publicly available allows a user to specify the time intervals for output of the atomic configuration data. There is a large variety in the types of information that can be included for each atom in these files based on the user's need. Furthermore, many researchers have developed post-processing algorithms for calculating a myriad of additional atomic information not directly output from the simulation software.

One of the most common visualization tools for molecular dynamics simulations is VMD [99], especially for soft materials. Developed within the Theoretical and Computational Biophysics Group at the University of Illinois at Urbana-Champaign, VMD was first applied to biomolecular systems, but has since expanded greatly for use into other atomistic research areas, such as nanomechanics of crystalline systems. VMD is capable of dealing with large atomistic datasets from a variety of standard file formats and is able to output professional quality images in a large array of formats. It also has extra features for data analysis, GPU-accelerated computing, and constructing common atomistic structures, such as carbon nanotubes and graphene. Another publicly available visualization tool is Paraview [111, 112]. Developed through a collaborative effort by Kitware, Advanced Simulation and Computing, Sandia National Laboratories, Los Alamos National Laboratories, and Army Research Laboratories, Paraview provides users with an open-source data analysis and visualization tool that is also designed to run on multiple platforms. Large datasets and the ability to deal with a variety of file formats including finite element data also make Paraview a beneficial tool for the atomistic simulation community. AtomEye [113] is another publicly available software that has been leveraged extensively within the materials nanomechanics community. Developed for quick visualization and analysis, AtomEye has successfully incorporated many useful abilities and features that make it valuable to any atomistic simulation researcher. AtomEye utilizes fast rendering capabilities and can read in some of the more common file formats from atomistic simulations. More recently, researchers within the nanomechanics community have been using Open Visualization Tool (OVITO) [114].

1.8 Atomistic Simulation Analysis and Visualization

For researchers in the nanomechanics community, a variety of defects are commonly studied or encountered in atomistic simulations. In the study of defects in materials, the dimensionality of the defect is used to help separate and identify the various types of defects. For example, there are zero-dimensional defects (e.g., point defects such as vacancies and interstitials), one-dimensional defects (e.g., dislocations), two-dimensional defects (e.g., grain boundaries, twins, and other interfaces), and three-dimensional defects (e.g., embedded particles and voids). As defects are paramount in material behavior, correctly identifying and quantifying the role of defects is important. In the discussion below and shown in Fig. 1.12, we use one of the previously mentioned visualization software (OVITO) [114] to illustrate how a grain boundary and a dislocation can be identified and visualized from atomistic simulations using different scalar methods. The different methods or metrics that will be used to identify the defects are atomic potential energy, centrosymmetry, and the common neighbor analysis (CNA) method. Atomic potential energy is calculated directly from atomic positions and the interatomic potential, centrosymmetry provides insight into the symmetry of the atomic neighborhood around each atom, and the CNA method classifies different crystalline structures with integer values (e.g., FCC is 1, HCP is 2, and non-12 coordinated atoms are 5). Please refer to the chapters on atomistic simulation analysis and visualization for further discussion on these scalar metrics and their usefulness.

As shown in Fig. 1.12, scalar metrics can be useful for identifying various defects and different metrics provide different information about the defect or structure. In Fig. 1.12a–c, a grain boundary is shown where atoms are colored according to (a) potential energy, (b) centrosymmetry, and (c) the CNA method. Notice how each metric provides various levels of structural or energetic information about the atoms, and that the perfect FCC lattice away from the grain boundary is clearly delineated from the atoms in the grain boundary. This is true for a dislocation as well, as shown in Fig. 1.12d–f. In this case, a perfect dislocation emitted from a triple junction in nanocrystalline copper has dissociated into two partial dislocations bounding a short stacking fault. The stacking fault is clearly identified using the CNA method in Fig. 1.12f as the HCP atomic region (light blue), where the two bounding partial dislocations are identified at non-12 coordinated (red). In addition to these commonly used metrics, additional metrics can also be calculated for each atom based on a local neighborhood calculation to provide estimates of deformation, rotation, and strain fields, see chapter on continuum metrics. Other methods to extract more precise information about defects are available, see chapter on analysis strategies.

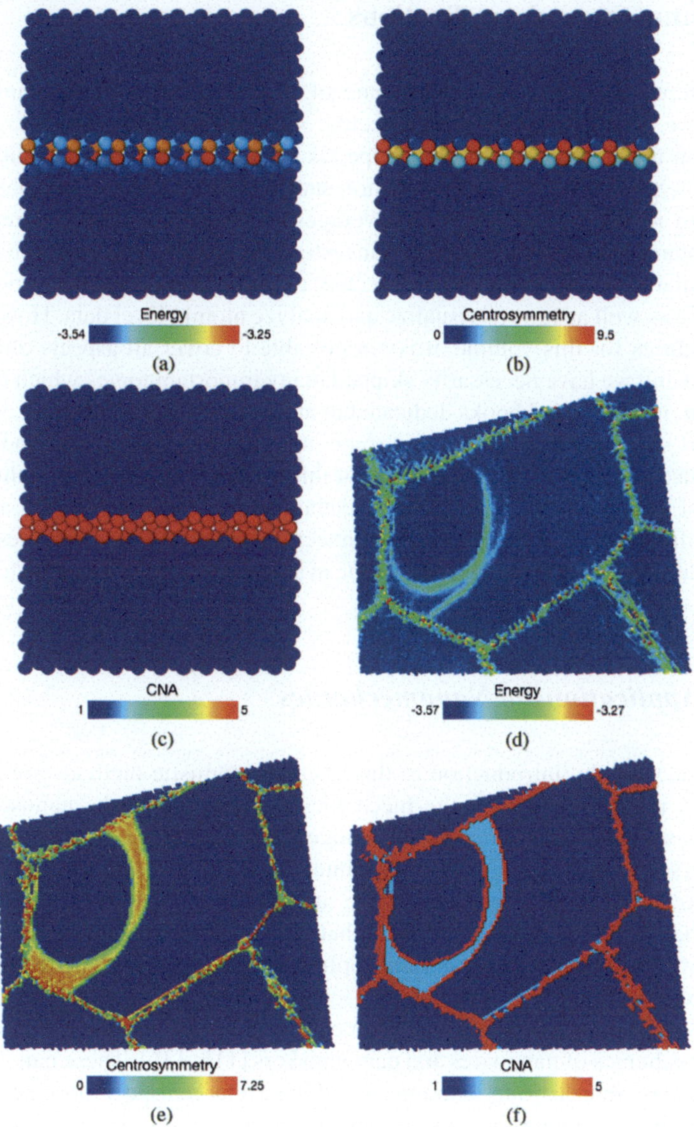

Fig. 1.12 (**a**)–(**c**) A Σ5 (210) grain boundary in copper where atoms are colored according to (**a**) energy, (**b**) centrosymmetry, and (**c**) CNA. (**d**)–(**f**) A dislocation within a nanograined copper structure where atoms are colored according to (**d**) energy, (**e**) centrosymmetry, and (**f**) CNA

1.9 Summary and Applications

In this chapter, we have presented some of the basics of atomistic simulations. Notably, we have discussed both molecular statics, which involves searching for extrema on a potential energy landscape, as well as molecular dynamics, which directly integrates the equations of motion such that the atoms sample phase space in order to reproduce the statistical averages associated with the corresponding thermodynamics ensemble. We have investigated how the information obtained in molecular dynamics can be connected to thermodynamic quantities at the macroscale as well as how to visualize and analyze atomic level data. However, due size limitations for this volume, it is not possible to cover all aspects of atomistic simulation and we have necessarily skipped many important topics which the reader can find in more detailed books dedicated to atomistic modeling [21, 22, 31]. In the rest of this section, we will point out some other atomistic-based methods that we have omitted in the preliminary sections of this chapter. Then we highlight some of the uses of atomistic modeling in nanomechanics and comment on how atomistics fit into multiscale modeling in terms of connections with other length scales. Finally, we provide a short overview of some of the available software and codes to perform atomistic simulations.

1.9.1 Applications to Nanomechanics

As mentioned in the introduction to this chapter, atomistic methods are poised to play a key role in simulating the mechanics of materials at the nanoscale since the length-scales of atomistics, which range from angstroms to about a hundred nanometers, covers the length scales of interest in nanomechanics. Here, we will attempt to point out a few areas where atomistics have been used to simulate nanomechanics. We note from the onset that this list will necessarily be incomplete and probably will miss some important application areas, but such oversights cannot be avoided.

Atomistic simulations have been and continue to be used extensively to investigate the mechanics of nanowires and nanowhiskers [115, 116]. These nanostructures are usually less than 100 nm in diameter and have a large aspect ratio, i.e., the ratio of the length to diameter. Nanowires have been proposed to be used in a number of applications including nano-electromechanical systems. Atomistic simulations have been used to investigate the mechanics of semiconductor, metallic as well as insulating nanowires. Similar to nanowires, the mechanics of nanoparticles have also been investigated extensively using direct atomistic methods. Nanoparticles can be viewed as crystals that are of nanoscale size (less than 100 nm) with similar dimensions in all directions. Atomistics have been useful in describing both the mechanical and structural properties of nanoparticles.

Recently, the mechanical behavior of nanoporous metallic foams has also been studied using atomistics. These foam nanostructures are composed of both solid ligaments and ligament junctions, in addition to the open pores. Often, the pores are not isolated and form a bicontinuous topology. In particular, the role of nanostructure and surfaces on the strength and mechanics has been highlighted with recent atomistic efforts [117]. In related studies, the interaction of nanoporous metal-organic frameworks with fluids has been evaluated with grand canonical (open system) molecular dynamics, see, e.g., [118].

Atomistic simulations have also been used to investigate the nanoscale origins of fracture [119] and plasticity [120] in general. Atomistics can be used to investigate dislocation nucleation, see Chap. 12, or dislocation mobility for use in discrete dislocation dynamics, see Chap. 2. In addition, atomistics can be used to directly investigate dislocation core effects as well as dislocation interactions. Similarly, atomistic methods can be used to investigate the local stress fields around cracks or how they respond to applied stress intensity factors.

Grain boundaries in metallic materials have received significant attention from the atomistic simulation community. Many of these studies have focused on highlighting the nanomechanics of grain boundaries under various loading conditions and as a function of structure and composition in a variety of systems. Atomistic simulations have proven to be a valuable tool to study and potentially predict interfacial mechanics at the nanoscale. Furthermore, the stability and thermodynamic properties of grain boundaries have also been investigated with atomistics.

Finally, we will point out where atomistic methods are used in this book. Chapters 5 (Quasicontinuum), and 8 all present methods in which an atomistic domain is directly coupled with a continuum region which helps extend the size of the atomistic region beyond the nanoscale. Chapter 6 also presents a review of accelerated timescale methods applied to atomistic simulations. Chapter 10 presents methods to visualize data from atomistic methods as well as identify atomistic structures and Chap. 9 describes a method to analyze atomistic data and extract continuum metrics. Chapter 12 uses atomistics to model dislocation nucleation in nanowires and Chap. 13 uses atomistics as the basis for deriving a new quantum crystal plasticity model. Chapter 16 uses atomistics to investigate the fundamentals of the lithiation if nanoscale silicon and Chap. 17 uses atomistics to investigate the properties of thin liquid films at the nanoscale.

Acknowledgements Sandia is a multiprogram laboratory managed and operated by Sandia Corporation, a wholly owned subsidiary of Lockheed Martin Corporation, for the US Department of Energy's National Nuclear Security Administration under contract No. DE-AC04-94AL85000.

References

1. A. Leach, *Molecular Modelling: Principles and Applications*, 2nd edn. (Prentice Hall, New York, 2001)
2. J. Nocedal, S. Wright, *Numerical Optimization*. Springer Series in Operations Research and Financial Engineering (Springer, New York, 2006)
3. M. Parrinello, A. Rahman, Crystal structure and pair potentials: a molecular-dynamics study. Phys. Rev. Lett. **45**(14), 1196–1199 (1980)

4. S. Pattamatta, R.S. Elliott, E.B. Tadmor, Mapping the stochastic response of nanostructures. Proc. Natl. Acad. Sci. **111**(17), E1678–E1686 (2014)
5. R. Fletcher, Function minimization by conjugate gradients. Comput. J. **7**(2), 149–154 (1964)
6. E. Polak, G. Ribiere, Note sur la convergence de méthodes de directions conjuguées. Modélisation Mathématique et Analyse Numérique (ESAIM: Mathematical Modelling and Numerical Analysis) **3**(R1), 35–43 (1969)
7. H. Jónsson, G. Mills, K.W. Jacobsen, Nudged elastic band method for finding minimum energy paths of transitions, in Classical and Quantum Dynamics in Condensed Phase Simulations, vol. 1 (World Scientific, Singapore, 1998), pp. 385–404
8. E. Bitzek, P. Koskinen, F. Gähler, M. Moseler, P. Gumbsch, Structural relaxation made simple. Phys. Rev. Lett. **97**(17), 170201 (2006)
9. G. Mills, H. Jónsson, Quantum and thermal effects in H2 dissociative adsorption: evaluation of free energy barriers in multidimensional quantum systems. Phys. Rev. Lett. **72**(7), 1124–1127 (1994)
10. G. Mills, H. Jónsson, G.K. Schenter, Reversible work transition state theory: application to dissociative adsorption of hydrogen. Surf. Sci. **324**(2–3), 305–337 (1995)
11. G. Henkelman, B.P. Uberuaga, H. Jónsson, A climbing image nudged elastic band method for finding saddle points and minimum energy paths. J. Chem. Phys. **113**(22), 9901 (2000)
12. V.M. Born, Th. von Karman, Uber schwingungen in raumgittern. Phys. Z. **13**(8), 297 (1912)
13. E. Hairer, C. Lubich, G. Wanner, *Geometric Numerical Integration: Structure-Preserving Algorithms for Ordinary Differential Equations*, vol. 31 (Springer Science and Business Media, New York, 2006)
14. R.D. Ruth, A canonical integration technique. IEEE Trans. Nucl. Sci. **30**(CERN-LEP-TH-83-14), 2669–2671 (1983)
15. E. Forest, R.D. Ruth, Fourth order symplectic integration. Physica **43**(LBL-27662), 105–117 (1989)
16. J.-P. Ryckaert, G. Ciccotti, H.J.C. Berendsen, Numerical integration of the Cartesian equations of motion of a system with constraints: molecular dynamics of n-alkanes. J. Comput. Phys. **23**(3), 327–341 (1977)
17. B.J. Leimkuhler, R.D. Skeel, Symplectic numerical integrators in constrained Hamiltonian systems. J. Comput. Phys. **112**(1), 117–125 (1994)
18. M. Tuckerman, B.J Berne, G.J. Martyna, Reversible multiple time scale molecular dynamics. J. Chem. Phys. **97**(3), 1990–2001 (1992)
19. K. Huang, *Introduction to Statistical Mechanics* (World Scientific, Singapore, 2001)
20. S. Nose, Constant-temperature molecular dynamics. J. Phys.: Condens. Matter **2**(S), SA115 (1990)
21. D.C. Rapaport, *The Art of Molecular Dynamics Simulation* (Cambridge University Press, Cambridge, 2004)
22. D. Frenkel, B. Smit, *Understanding Molecular Simulation: From Algorithms to Applications*, vol. 1 (Academic, San Diego, 2001)
23. M. Parrinello, A. Rahman, Polymorphic transitions in single crystals: a new molecular dynamics method. J. Appl. Phys. **52**(12), 7182–7190 (1981)
24. S.R. De Groot, P. Mazur, *Non-Equilibrium Thermodynamics* (Courier Corporation, North Chelmsford, 2013)
25. J.H. Irving, J.G. Kirkwood, The statistical mechanical theory of transport processes. IV. The equations of hydrodynamics. J. Chem. Phys. **18**(6), 817–829 (1950)
26. D.J. Evans, G. Morriss, *Statistical Mechanics of Nonequilibrium Liquids* (Cambridge University Press, Cambridge, 2008)
27. F. Müller-Plathe, A simple nonequilibrium molecular dynamics method for calculating the thermal conductivity. J. Chem. Phys. **106**(14), 6082–6085 (1997)
28. D.J. Evans, Homogeneous NEMD algorithm for thermal conductivity—application of non-canonical linear response theory. Phys. Lett. A **91**(9), 457–460 (1982)
29. L.T. Kong, Phonon dispersion measured directly from molecular dynamics simulations. Comput. Phys. Commun. **182**(10), 2201–2207 (2011)

30. P.K. Schelling, S.R. Phillpot, P. Keblinski, Comparison of atomic-level simulation methods for computing thermal conductivity. Phys. Rev. B **65**(14), 144306 (2002)
31. M.P. Allen, D.J. Tildesley, *Computer Simulation of Liquids* (Oxford University Press, New York, 1989)
32. J.E. Lennard-Jones, On the determination of molecular fields. II. From the equation of state of a gas. Proc. R. Soc. Lond. A: Math. Phys. Eng. Sci. **106**, 463–477 (1924)
33. A.D. MacKerell, D. Bashford, M. Bellott, R.L. Dunbrack, J.D. Evanseck, M.J. Field, S. Fischer, J. Gao, H. Guo, S. Ha, D. Joseph-McCarthy, L. Kuchnir, K. Kuczera, F.T.K. Lau, C. Mattos, S. Michnick, T. Ngo, D.T. Nguyen, B. Prodhom, W.E. Reiher, B. Roux, M. Schlenkrich, J.C. Smith, R. Stote, J. Straub, M. Watanabe, J. Wiorkiewicz-Kuczera, D. Yin, M. Karplus, All-atom empirical potential for molecular modeling and dynamics studies of proteins. J. Phys. Chem. B **102**(18), 3586–3616 (1998)
34. J. Wang, R.M. Wolf, J.W. Caldwell, P.A. Kollman, D.A. Case, Development and testing of a general amber force field. J. Comput. Chem. **25**(9), 1157–1174 (2004)
35. N. Karasawa, W.A. Goddard III, Acceleration of convergence for lattice sums. J. Phys. Chem. **93**(21), 7320–7327 (1989)
36. R.W. Hockney, J.W. Eastwood, *Computer Simulation Using Particles* (CRC Press, Boca Raton, 2010)
37. J.J. Gilman, *Electronic Basis of the Strength of Materials* (Cambridge University Press, Cambridge, 2003)
38. M.I. Baskes, Modified embedded-atom potentials for cubic materials and impurities. Phys. Rev. B **46**(5), 2727 (1992)
39. B.-J. Lee, M.I. Baskes, Second nearest-neighbor modified embedded-atom-method potential. Phys. Rev. B **62**(13), 8564 (2000)
40. B.-J. Lee, M.I. Baskes, H. Kim, Y.K. Cho, Second nearest-neighbor modified embedded atom method potentials for bcc transition metals. Phys. Rev. B **64**(18), 184102 (2001)
41. Y. Mishin, M.J. Mehl, D.A. Papaconstantopoulos, Phase stability in the Fe–Ni system: investigation by first-principles calculations and atomistic simulations. Acta Mater. **53**(15), 4029–4041 (2005)
42. Y. Mishin, A.Y. Lozovoi, Angular-dependent interatomic potential for tantalum. Acta Mater. **54**(19), 5013–5026 (2006)
43. S. Ryu, W. Cai, A gold–silicon potential fitted to the binary phase diagram. J. Phys.: Condens. Matter **22**(5), 055401 (2010)
44. B.-J. Lee, A modified embedded-atom method interatomic potential for the Fe-C system. Acta Mater. **54**(3), 701–711 (2006)
45. L.S.I. Liyanage, S.-G. Kim, J. Houze, S. Kim, M.A. Tschopp, M.I. Baskes, M.F. Horstemeyer, Structural, elastic, and thermal properties of cementite (Fe3C) calculated using a modified embedded atom method. Phys. Rev. B **89**(9), 094102 (2014)
46. F.H. Stillinger, T.A. Weber, Computer simulation of local order in condensed phases of silicon. Phys. Rev. B **31**(8), 5262 (1985)
47. R.L.C. Vink, G.T. Barkema, W.F. Van der Weg, N. Mousseau, Fitting the Stillinger–Weber potential to amorphous silicon. J. Non-Cryst. Solids **282**(2), 248–255 (2001)
48. Z. Jian, Z. Kaiming, X. Xide, Modification of Stillinger-Weber potentials for Si and Ge. Phys. Rev. B **41**(18), 12915 (1990)
49. K. Ding, H.C. Andersen, Molecular-dynamics simulation of amorphous germanium. Phys. Rev. B **34**(10), 6987 (1986)
50. A. Béré, A. Serra, On the atomic structures, mobility and interactions of extended defects in GaN: dislocations, tilt and twin boundaries. Philos. Mag. **86**(15), 2159–2192 (2006)
51. N. Aıchoune, V. Potin, P. Ruterana, A. Hairie, G. Nouet, E. Paumier, An empirical potential for the calculation of the atomic structure of extended defects in wurtzite GaN. Comput. Mater. Sci. **17**(2), 380–383 (2000)
52. H.P. Lei, J. Chen, S. Petit, P. Ruterana, X.Y. Jiang, G. Nouet, Stillinger–Weber parameters for In and N atoms. Superlattice. Microst. **40**(4), 464–469 (2006)
53. M. Ichimura, Stillinger-Weber potentials for III–V compound semiconductors and their application to the critical thickness calculation for InAs/GaAs. Phys. Status Solidi (a) **153**(2), 431–437 (1996)

54. J.E. Angelo, M.J. Mills, Investigations of the misfit dislocation structure at a CdTe(001)/GaAs(001) interface using Stillinger-Weber potentials and high-resolution transmission electron microscopy. Philos. Mag. A **72**(3), 635–649 (1995)
55. X.W. Zhou, D.K. Ward, J.E. Martin, F.B. van Swol, J.L. Cruz-Campa, D. Zubia, Stillinger-Weber potential for the II-VI elements Zn-Cd-Hg-S-Se-Te. Phys. Rev. B **88**(8), 085309 (2013)
56. J.-W. Jiang, H.S. Park, T. Rabczuk, Molecular dynamics simulations of single-layer molybdenum disulphide (mos2): Stillinger-Weber parametrization, mechanical properties, and thermal conductivity. J. Appl. Phys. **114**(6), 064307 (2013)
57. K. Albe, K. Nordlund, J. Nord, A. Kuronen, Modeling of compound semiconductors: analytical bond-order potential for Ga, As, and GaAs. Phys. Rev. B **66**(3), 035205 (2002)
58. J. Nord, K. Albe, P. Erhart, K. Nordlund, Modelling of compound semiconductors: analytical bond-order potential for gallium, nitrogen and gallium nitride. J. Phys.: Condens. Matter **15**(32), 5649 (2003)
59. J. Tersoff, New empirical approach for the structure and energy of covalent systems. Phys. Rev. B **37**(12), 6991 (1988)
60. J. Tersoff, Empirical interatomic potential for carbon, with applications to amorphous carbon. Phys. Rev. Lett. **61**(25), 2879 (1988)
61. J. Tersoff, Modeling solid-state chemistry: interatomic potentials for multicomponent systems. Phys. Rev. B **39**(8), 5566 (1989)
62. G.C. Abell, Empirical chemical pseudopotential theory of molecular and metallic bonding. Phys. Rev. B **31**(10), 6184 (1985)
63. LAMMPS: http://lammps.sandia.gov/ (2016). Retrieved in 2016
64. J. Tersoff, New empirical model for the structural properties of silicon. Phys. Rev. Lett. **56**(6), 632 (1986)
65. D.W. Brenner, Empirical potential for hydrocarbons for use in simulating the chemical vapor deposition of diamond films. Phys. Rev. B **42**(15), 9458 (1990)
66. T. Kumagai, S. Izumi, S. Hara, S. Sakai, Development of bond-order potentials that can reproduce the elastic constants and melting point of silicon for classical molecular dynamics simulation. Comput. Mater. Sci. **39**(2), 457–464 (2007)
67. D. Powell, M.A. Migliorato, A.G. Cullis, Optimized Tersoff potential parameters for tetrahedrally bonded III-V semiconductors. Phys. Rev. B **75**(11), 115202 (2007)
68. J. Tersoff, Carbon defects and defect reactions in silicon. Phys. Rev. Lett. **64**(15), 1757 (1990)
69. J. Tersoff, Chemical order in amorphous silicon carbide. Phys. Rev. B **49**(23), 16349 (1994)
70. P. Erhart, K. Albe, Analytical potential for atomistic simulations of silicon, carbon, and silicon carbide. Phys. Rev. B **71**(3), 035211 (2005)
71. F. de Brito Mota, J.F. Justo, A. Fazzio, Structural properties of amorphous silicon nitride. Phys. Rev. B **58**(13), 8323 (1998)
72. S. Munetoh, T. Motooka, K. Moriguchi, A. Shintani, Interatomic potential for Si–O systems using Tersoff parameterization. Comput. Mater. Sci. **39**(2), 334–339 (2007)
73. M. Mrovec, R. Gröger, A.G. Bailey, D. Nguyen-Manh, C. Elsässer, V. Vitek, Bond-order potential for simulations of extended defects in tungsten. Phys. Rev. B **75**(10), 104119 (2007)
74. N. Juslin, P. Erhart, P. Träskelin, J. Nord, K.O.E. Henriksson, K. Nordlund, E. Salonen, K. Albe, Analytical interatomic potential for modeling nonequilibrium processes in the W–C–H system. J. Appl. Phys. **98**(12), 123520 (2005)
75. M. Müller, P. Erhart, K. Albe, Analytic bond-order potential for bcc and fcc iron—comparison with established embedded-atom method potentials. J. Phys.: Condens. Matter **19**(32), 326220 (2007)
76. K.O.E. Henriksson, K. Nordlund, Simulations of cementite: an analytical potential for the Fe-C system. Phys. Rev. B **79**(14), 144107 (2009)
77. P. Erhart, N. Juslin, O. Goy, K. Nordlund, R. Müller, K. Albe, Analytic bond-order potential for atomistic simulations of zinc oxide. J. Phys.: Condens. Matter **18**(29), 6585 (2006)
78. M. Backman, N. Juslin, K. Nordlund, Bond order potential for gold. Eur. Phys. J. B **85**(9), 1–5 (2012)

79. D.G. Pettifor, I.I. Oleinik, Analytic bond-order potentials beyond Tersoff-Brenner. I. Theory. Phys. Rev. B **59**(13), 8487 (1999)
80. D.G. Pettifor, I.I. Oleinik, Bounded analytic bond-order potentials for σ and π bonds. Phys. Rev. Lett. **84**(18), 4124 (2000)
81. D.G. Pettifor, I.I. Oleinik, Analytic bond-order potential for open and close-packed phases. Phys. Rev. B **65**(17), 172103 (2002)
82. M.W. Finnis, Bond-order potentials through the ages. Prog. Mater. Sci. **52**(2), 133–153 (2007)
83. M. Aoki, D. Nguyen-Manh, D.G. Pettifor, V. Vitek, Atom-based bond-order potentials for modelling mechanical properties of metals. Prog. Mater. Sci. **52**(2), 154–195 (2007)
84. D.G. Pettifor, I.I. Oleynik, Interatomic bond-order potentials and structural prediction. Prog. Mater. Sci. **49**(3), 285–312 (2004)
85. I.I. Oleinik, D.G. Pettifor, Analytic bond-order potentials beyond Tersoff-Brenner. II. Application to the hydrocarbons. Phys. Rev. B **59**(13), 8500 (1999)
86. D.A. Murdick, X.W. Zhou, H.N.G. Wadley, D. Nguyen-Manh, R. Drautz, D.G. Pettifor, Analytic bond-order potential for the gallium arsenide system. Phys. Rev. B **73**(4), 045206 (2006)
87. R. Drautz, X.W. Zhou, D.A. Murdick, B. Gillespie, H.N.G. Wadley, D.G. Pettifor, Analytic bond-order potentials for modelling the growth of semiconductor thin films. Prog. Mater. Sci. **52**(2), 196–229 (2007)
88. D.K. Ward, X.W. Zhou, B.M. Wong, F.P. Doty, J.A. Zimmerman, Analytical bond-order potential for the Cd-Zn-Te ternary system. Phys. Rev. B **86**(24), 245203 (2012)
89. M. Mrovec, D. Nguyen-Manh, D.G. Pettifor, V. Vitek, Bond-order potential for molybdenum: application to dislocation behavior. Phys. Rev. B **69**(9), 094115 (2004)
90. R. Gröger, A.G. Bailey, V. Vitek, Multiscale modeling of plastic deformation of molybdenum and tungsten: I. Atomistic studies of the core structure and glide of 1/2<111> screw dislocations at 0k. Acta Mater. **56**(19), 5401–5411 (2008)
91. M. Mrovec, D. Nguyen-Manh, C. Elsässer, P. Gumbsch, Magnetic bond-order potential for iron. Phys. Rev. Lett. **106**(24), 246402 (2011)
92. A.C.T. Van Duin, S. Dasgupta, F. Lorant, W.A. Goddard, Reaxff: a reactive force field for hydrocarbons. J. Phys. Chem. A **105**(41), 9396–9409 (2001)
93. A.C.T. Van Duin, A. Strachan, S. Stewman, Q. Zhang, X. Xu, W.A. Goddard, Reaxffsio reactive force field for silicon and silicon oxide systems. J. Phys. Chem. A **107**(19), 3803–3811 (2003)
94. D. Fantauzzi, J. Bandlow, L. Sabo, J.E Mueller, A.C.T. van Duin, T. Jacob, Development of a reaxff potential for Pt–O systems describing the energetics and dynamics of Pt-oxide formation. Phys. Chem. Chem. Phys. **16**(42), 23118–23133 (2014)
95. Q. Zhang, T. Çağın, A. van Duin, W.A. Goddard III, Y. Qi, L.G. Hector Jr., Adhesion and nonwetting-wetting transition in the Al/α-Al2O3 interface. Phys. Rev. B **69**(4), 045423 (2004)
96. T. T Järvi, A.C.T. van Duin, K. Nordlund, W.A. Goddard III, Development of interatomic reaxff potentials for Au–S–C–H systems. J. Phys. Chem. A **115**(37), 10315–10322 (2011)
97. A.C.T. van Duin, B.V. Merinov, S.S. Jang, W.A. Goddard, Reaxff reactive force field for solid oxide fuel cell systems with application to oxygen ion transport in yttria-stabilized zirconia. J. Phys. Chem. A **112**(14), 3133–3140 (2008)
98. S. Plimpton, Fast parallel algorithms for short-range molecular dynamics. J. Comput. Phys. **117**(1), 1–19 (1995)
99. W. Humphrey, A. Dalke, K. Schulten, VMD - visual molecular dynamics. J. Mol. Graph. **14**, 33–38 (1996)
100. M.D. Hanwell, D.E. Curtis, D.C. Lonie, T. Vandermeersch, E. Zurek, G.R. Hutchison, Avogadro: an advanced semantic chemical editor, visualization, and analysis platform. J. Chem. **4**(1), 17 (2012)
101. D. van der Spoel, E. Lindahl, B. Hess, A.R. Van Buuren, E. Apol, P.J. Meulenhoff, D.P. Tieleman, A.L.T.M. Sijbers, K.A. Feenstra, R. van Drunen, et al., Gromacs User Manual Version 3.3. (2008). http://www.gromacs.org/Documentation/Manual
102. S. Pronk, S. Páll, R. Schulz, P. Larsson, P. Bjelkmar, R. Apostolov, M.R. Shirts, J.C. Smith, P.M. Kasson, D. van der Spoel, et al., Gromacs 4.5: a high-throughput and highly parallel open source molecular simulation toolkit. Bioinformatics **29**, 845–854 (2013). doi:10.1093/bioinformatics/btt055

103. C.A. Becker, F. Tavazza, Z.T. Trautt, R.A. Buarque de Macedo, Considerations for choosing and using force fields and interatomic potentials in materials science and engineering. Curr. Opin. Solid State Mater. Sci. **17**(6), 277–283 (2013)
104. C.A. Becker, Atomistic simulations for engineering: potentials and challenges, in *Models, Databases and Simulation Tools Needed for Realization of Integrated Computational Mat. Eng. (ICME 2010)* (2014), p. 91
105. E.B. Tadmor, R.S. Elliott, J.P. Sethna, R.E. Miller, C.A. Becker, Knowledgebase of interatomic models (kim) (2011). http://www.gromacs.org/Documentation/Manual
106. E.B. Tadmor, R.S. Elliott, J.P. Sethna, R.E. Miller, C.A. Becker, The potential of atomistic simulations and the knowledgebase of interatomic models. JOM J. Miner. Met. Mater. Soc. **63**(7), 17–17 (2011)
107. J. Roth, Imd: a typical massively parallel molecular dynamics code for classical simulations–structure, applications, latest developments, in *Sustained Simulation Performance 2013* (Springer, New York, 2013), pp. 63–76
108. J. Stadler, R. Mikulla, H.-R. Trebin, Imd: a software package for molecular dynamics studies on parallel computers. Int. J. Mod. Phys. C **8**(05), 1131–1140 (1997)
109. IMD: http://imd.itap.physik.uni-stuttgart.de/ (2016). Retrieved in 2016
110. J.C. Phillips, R. Braun, W. Wang, J. Gumbart, E. Tajkhorshid, E. Villa, C. Chipot, R.D. Skeel, L. Kale, K. Schulten, Scalable molecular dynamics with NAMD. J. Comput. Chem. **26**(16), 1781–1802 (2005)
111. J. Ahrens, B. Geveci, C. Law, in *ParaView: An End-User Tool for Large Data Visualization, Visualization Handbook* (Elsevier, San Diego, 2005)
112. U. Ayachit, *The ParaView Guide: A Parallel Visualization Application* (Kitware, Clifton Park, 2015)
113. J. Li, Atomeye: an efficient atomistic configuration viewer. Model. Simul. Mater. Sci. Eng. **11**, 173–177 (2003)
114. A. Stukowski, Visualization and analysis of atomistic simulation data with OVITO - the open visualization tool. Model. Simul. Mater. Sci. Eng. **18**, 015012 (2010)
115. H.S. Park, W. Cai, H.D. Espinosa, H. Huang, Mechanics of crystalline nanowires. MRS Bull. **34**(03), 178–183 (2009)
116. C.R. Weinberger, W. Cai, Plasticity of metal nanowires. J. Mater. Chem. **22**(8), 3277–3292 (2012)
117. D. Farkas, A. Caro, E. Bringa, D. Crowson, Mechanical response of nanoporous gold. Acta Mater. **61**(9), 3249–3256 (2013)
118. J.A. Greathouse, M.D. Allendorf, The interaction of water with MOF-5 simulated by molecular dynamics. J. Am. Chem. Soc. **128**(33), 10678–10679 (2006)
119. M.J. Buehler, *Atomistic Modeling of Materials Failure* (Springer Science and Business Media, New York, 2008)
120. V. Bulatov, W. Cai, *Computer Simulations of Dislocations*, vol. 3 (Oxford University Press, Oxford, 2006)

Chapter 2
Fundamentals of Dislocation Dynamics Simulations

Ryan B. Sills, William P. Kuykendall, Amin Aghaei, and Wei Cai

2.1 Overview

When crystalline solids undergo plastic deformation, line defects known as dislocations move, multiply, and react with one another. The overall mechanical properties of the crystal in this plastic regime are governed by these dislocation processes. Dislocation dynamics (DD) is a modeling approach that aims to simulate the motion and interaction of these dislocation lines to gain insights concerning the mechanical properties of the material.

Dislocation lines are defects whose core widths are at the scale of the crystal lattice. The length scale over which dislocation structures evolve is, however, many orders of magnitude larger than the interatomic distance. A classical example is the formation of dislocation cells; at moderate to large amounts of plastic deformation, dislocation networks are known to form cellular structures, with an average cell size on the order of $1\,\mu m$ (see Fig. 2.1a). Hence, any model which hopes to inform our understanding of bulk plastic deformation—for example, understanding the temperature dependence of the stress–strain curves shown in Fig. 2.1b—must

R.B. Sills (✉)
Sandia National Laboratories, Livermore, CA 94550, USA

Stanford University, Stanford, CA 94305, USA
e-mail: rbsills@sandia.gov

W.P. Kuykendall
Stanford University, Stanford, CA 94305, USA

Sandia National Laboratories, Livermore, CA 94550, USA
e-mail: wpkuyken@sandia.gov

A. Aghaei • W. Cai
Stanford University, Stanford, CA 94305, USA
e-mail: am.aghaei@gmail.com; caiwei@stanford.edu

C.R. Weinberger, G.J. Tucker (eds.), *Multiscale Materials Modeling for Nanomechanics*, Springer Series in Materials Science 245,
DOI 10.1007/978-3-319-33480-6_2

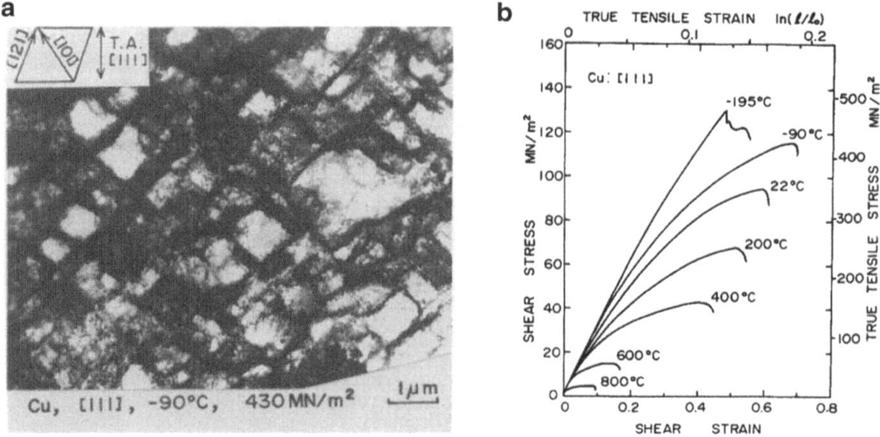

Fig. 2.1 Examples of dislocation plasticity. (**a**) Cellular dislocation structure in single crystal copper after tensile loading in the [1 1 1] direction at −90°C and (**b**) stress–strain curves for single crystal copper at various temperatures. Reproduced from [48] with permission from Elsevier

simulate a material volume above this length scale. Using an atomistic approach would require the simulation of more than 10^{10} atoms, a simulation size which is prohibitively expensive even for the most modern computational tools. This gap in scale necessitates a new model at the so-called mesoscale: dislocation dynamics.

The idea behind the DD approach is that because plastic deformation is dominated by the motion and interaction of dislocation lines, one only needs to consider the dislocation lines, rather than the locations of all of the atoms, to understand the plastic behavior of a material. Taking such an approach enables simulations with length scales of 10 μm and time scales of 1 ms. As with all mesoscale approaches, DD requires the input of multiple physical models to describe the various behaviors of the dislocation lines, meaning that much information must be provided either from experiments or more fundamental models. Unlike other mesoscale models of plasticity which consider the dislocation density in terms of a homogenized field, in DD dislocation lines are treated explicitly so that individual dislocation–dislocation interactions can be properly captured.

Much of the theory that feeds into the models that describe the dislocation lines has been established for many decades, as has the concept of DD itself [34, 53]. However, only recently have large-scale simulations been made possible with the inception of modern computational tools. Despite these many advances, DD remains a challenging tool to use, often requiring hundreds of computer cores for a single simulation of a short duration of physical time relative to experiments.

The remainder of the chapter will be organized as follows. First, in Sect. 2.2 we will discuss the basic features of the DD formulation. In Sect. 2.3, we will then discuss how to run a DD simulation all the way from inputs to outputs, and show a few examples. Section 2.4 will discuss DD's place in the hierarchy of material models. Finally, Sect. 2.5 will present topics of current research and challenges that the DD community need to overcome to enable more widespread use of the tool.

2.2 Fundamentals

In order to simulate the motion and interaction of dislocation lines, a number of algorithms, rules, and procedures have been developed. In this section, we break these features into two groups. First, we discuss the most basic ingredients necessary to conduct a DD simulation: how driving forces are exerted on dislocations (2.2.2.1), how to determine dislocation velocities given these forces (2.2.2.2), discretization and adaptive remeshing of the dislocation lines (2.2.2.3), time integration of the equations of motion (2.2.2.4), and how dislocations can collide and react (2.2.2.5). We will then introduce more advanced aspects of DD simulations: how to handle dislocation junctions and intersections (2.2.3.1), different types of boundary conditions (2.2.3.2), how screw dislocations can change their glide plane through cross-slip (2.2.3.3), and a brief discussion of two-dimensional DD simulations (2.2.3.4). These features are presented in the flowchart shown in Fig. 2.3. We begin, however, with a discussion of the overall problem formulation.

2.2.1 Problem Formulation

The basic idea behind DD is to embed the physics of dislocations into a set of governing equations that can be solved for the positions of a network of dislocation lines, given an initial dislocation configuration, boundary conditions, and loading conditions. The positions of the lines are described by the vector $\mathbf{r}(s, t)$, where s is a scalar parameter dictating the location along the lines, as shown in Fig. 2.2a, and t denotes time. Because we seek a tool that can obtain a solution in arbitrary settings (e.g., many dislocation lines loaded multiaxially), we will need to discretize our system in both space and time, and employ numerical methods to solve the governing equations. Figure 2.2b shows an example of discretization in space.

As we will discuss, many things can exert forces on dislocations. These forces can be broken into drag forces, which resist dislocation motion, and driving forces, which promote it. Additionally, dislocation lines are known to have effective masses, giving rise to inertial forces [51]. In many crystalline materials under a broad range of conditions, however, drag forces intrinsic to the crystal lattice are orders of magnitude larger than the inertial forces, making dislocation motion overdamped [51]. This means that in the overall equations of motion, we can neglect inertial terms altogether, and simply require that the total driving force balance the total drag force, i.e.,

$$\sum \mathbf{F}_{\text{drag}}(\mathbf{v}, s) + \sum \mathbf{F}_{\text{drive}}(s) = 0 \tag{2.1}$$

where $\mathbf{v}$ is the dislocation velocity

$$\mathbf{v} = \frac{\partial \mathbf{r}(s, t)}{\partial t} \tag{2.2}$$

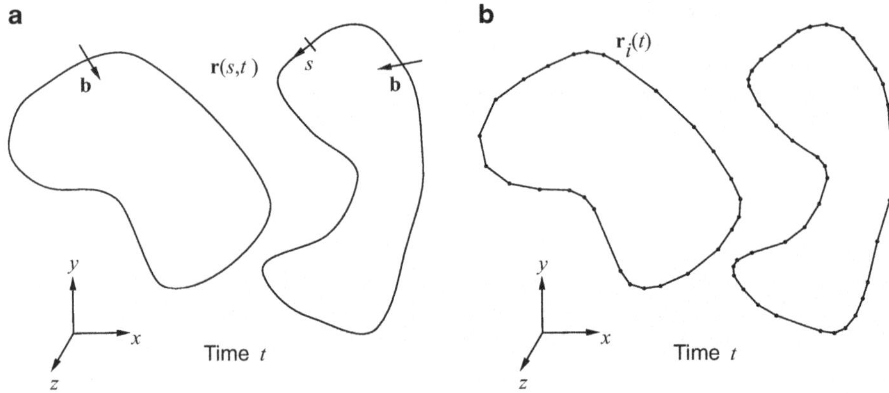

Fig. 2.2 Position of a pair of dislocation loops at time t. (**a**) Continuous representation described by position vector $\mathbf{r}$ as a function of parameter s. (**b**) Discrete representation using node-based discretization (see Sect. 2.2.2.3) described by position vectors of nodes $\mathbf{r}_i$

and the summations are over all drag and driving force contributions. Usually, in dislocation dynamics, Eq. (2.1) can be explicitly solved for $\mathbf{v}$ and restated as

$$\mathbf{v} = \mathbf{M}(\mathbf{F}^{\text{tot}}_{\text{drive}}) \tag{2.3}$$

where $\mathbf{F}^{\text{tot}}_{\text{drive}}[\mathbf{r}(s), \boldsymbol{\sigma}_{\text{ext}}, \ldots] = \sum \mathbf{F}_{\text{drive}}$ is the total driving force as a function of parameter s, dependent upon the dislocation position $\mathbf{r}(s)$, the externally applied stress $\boldsymbol{\sigma}_{\text{ext}}$, and any other features which exert driving forces. The function $\mathbf{M}(\cdot)$, which provides the velocity given a total driving force, is called the *mobility law*. The final governing equation of motion can be written as

$$\frac{\partial \mathbf{r}(s,t)}{\partial t} = \mathbf{g}\left[\mathbf{r}(s), \boldsymbol{\sigma}_{\text{ext}}, \ldots\right], \tag{2.4}$$

where $\mathbf{g} \equiv \mathbf{M}\left(\mathbf{F}^{\text{tot}}_{\text{drive}}\left[\mathbf{r}(s), \boldsymbol{\sigma}_{\text{ext}}, \ldots\right]\right)$ is an operator which computes the velocity $\mathbf{v}(s)$ from a given dislocation structure $\mathbf{r}(s)$ and loading condition.

2.2.2 Basic Features

A flowchart depicting the major steps in a DD simulation is presented in Fig. 2.3. We will now discuss each of these steps in turn.

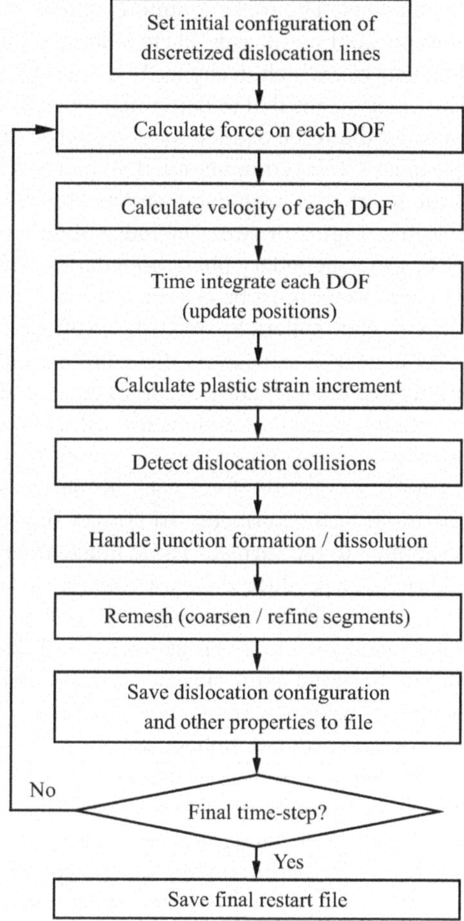

Fig. 2.3 Flowchart showing the basic steps for a dislocation dynamics code. Note that the force and velocity of each degree of freedom (DOF) may be computed multiple times per time step depending on the time integration scheme used (see Sect. 2.2.2.4)

2.2.2.1 Driving Forces

Many features of crystalline solids can apply driving forces to dislocation lines. These forces can be divided into two categories: forces arising from local stress fields (Peach–Koehler forces) and forces due to dislocation self interaction.

To determine the driving force exerted on a dislocation by a stress field $\boldsymbol{\sigma}(s)$ applied at position s, the Peach–Koehler expression is commonly used. It gives that the force per unit length, $\mathbf{F}(s)$, is [42]

$$\mathbf{F}(s) = (\boldsymbol{\sigma}(s) \cdot \mathbf{b}) \times \boldsymbol{\xi}(s), \tag{2.5}$$

where $\mathbf{b}$ is the Burgers vector of the dislocation and $\boldsymbol{\xi}(s)$ is the direction of the dislocation line at s (which varies with position for curved dislocations). Hence, any feature of a crystalline solid that results in a stress field can exert forces on dislocations. The most common sources of stress in dislocation dynamics simulations are applied stresses due to loading of the simulation cell and stresses from other dislocations, which decay as $1/r$, where r is the distance from the dislocation [1]. The latter means that to determine the total force on a dislocation segment, we must consider the force exerted by every other segment in the simulation cell. This gives rise to an $\mathcal{O}(N^2)$ computation if all pair-wise interactions are computed explicitly (N is the number of segments), and makes DD difficult to implement efficiently. Other possible origins of stress include solute atoms, precipitates or inclusions, and free surfaces or secondary phase boundaries. The material system of interest will decide which of these must be considered.

With nanomaterials, free surface effects are especially important, since the small specimen size means every dislocation is near a free surface. For this reason, we will briefly discuss the nature of forces generated by free surfaces. In elasticity theory, for simplicity, stress expressions for dislocations are usually derived in a homogeneous, infinite medium. When these expressions are then used in finite media, they result in nonzero traction forces at the surfaces, violating the traction-free boundary condition at the surfaces. To correct this, a set of so-called image tractions must be applied to the surface. These image tractions render the surface traction-free, but additionally produce their own *image stress field*, which can also exert forces on dislocations. Thus, the problem of a finite solid requires that the image stresses be determined for the given geometry and distribution of dislocations; this generally has to be done numerically, and we defer further discussion of image solvers (which compute the image field) to Sect. 2.2.3.2. We will discuss an example with a cylindrical specimen in Sect. 2.3.5.

The above discussion applies to forces arising externally from the dislocation line. In addition to these effects, the dislocation line can exert a force on itself. This self-force can be thought of as resulting from the energy of the dislocation line, and has two contributions. The first contribution is elastic, and can be computed using a number of approaches, such as the non-singular theory of dislocations [15]. The second contribution is due to nonlinear interatomic interactions at the dislocation core, and we shall refer to it as the *core force*. Core forces can influence the dislocation line in two ways. First, the core force will try to reduce the length of the dislocation line, since the total core energy scales with the line length. Second, because the core energy varies with line character (i.e., edge and screw dislocations have different energies), the core force will exert a torque on the line, trying to rotate it into its orientation of lowest energy. One approach for determining the core force is to derive it from the core energy. The core energy per unit length, E_c, of a dislocation line can be calculated using atomistic or first-principles methods as a function of the character angle θ (the angle between the Burgers vector and line direction). Alternatively, it is common in DD simulations to use an approximate analytical model to describe the core energy. For example, in the deWit and Koehler model [25] the core energy varies as

$$E_c(\theta) = \mathcal{E} b^2 \left(\frac{1}{1-\nu} \sin^2\theta + \cos^2\theta \right) \tag{2.6}$$

where ν is Poisson's ratio, b is the magnitude of the Burgers vector, and $\mathscr{E}$ is a parameter that controls the magnitude of the core energy; this is the same way the line energy varies according to elasticity theory for an isotropic solid. Often $\mathscr{E}$ is approximated as $\mathscr{E} \approx \alpha\mu$, where μ is the shear modulus and α is a material parameter in the range $0.1 - 0.5$ [45]. Given this function $E_c(\theta)$, the core force can be determined using a number of approaches. Our preference is to calculate the core force after the dislocation lines have been discretized, and hence we postpone further discussion of core forces until Sect. 2.2.2.3.

2.2.2.2 Mobility Laws

As we discussed in Sect. 2.2.1, mobility laws serve as constitutive equations in dislocation dynamics simulations, relating the total driving force per unit length acting on a dislocation line to its velocity. Since the movement of a dislocation is strongly material dependent, mobility laws must be constructed with a specific material system in mind [9, 14]. The mobility of a dislocation line is commonly dependent upon the dislocation character, direction of motion, the crystallographic plane on which the dislocation can move conservatively—known as the *glide plane*—and the temperature. The goal of a mobility law is to express these dependencies in terms of an explicit function for the velocity given a total driving force per unit length. Usually, this means determining the drag force exerted by the crystal lattice on a dislocation. In this section we explain how mobility laws can be obtained, and provide an example of a mobility law for face-centered cubic (FCC) crystals.

Linear mobility laws are commonly used. The viscous drag forces experienced by dislocations in crystalline solids, due, for instance, to phonon dispersion, are often proportional to the dislocation velocity [45]. Hence, a linear mobility model can be written as

$$\mathbf{M}(\mathbf{F}^{\text{tot}}_{\text{drive}}) = \mathscr{B}^{-1}(s) \cdot \mathbf{F}^{\text{tot}}_{\text{drive}}(s), \tag{2.7}$$

where $\mathscr{B}(s)$ is a drag coefficient tensor (with dimensions [mass]/([length][time])) and is strongly material dependent. The components of $\mathscr{B}(s)$ account for the various features affecting the dislocation drag coefficient. If more than one mechanism exerts linear drag on a dislocation, the net drag coefficient is the sum of the drag coefficients for each mechanism.

As an example of a linear mobility law, we consider the case of FCC crystals (using the same model as [12]). Excluding the possibility of cross-slip (to be discussed separately in Sect. 2.2.3.3), dislocations in FCC metals are confined to glide on $\{1\,1\,1\}$ planes; climb motion out of the glide plane requires the diffusion of vacancies into or out of the core, and is generally negligible at temperatures less than one-third of the melting point [42]. This glide confinement is a reflection of the dissociated core structure in FCC metals. The glide constraint can be enforced by setting the components of $\mathscr{B}$ coupled to out-of-plane motion to very large values.

This can lead to an ill-conditioned system, however, and it is numerically easier to project out climb motion by simply zeroing the velocity components in the direction of the glide plane normal; we will represent motion within the glide plane with the superscript ^{g}. Additionally, we often find with FCC metals that the drag coefficient is isotropic with respect to dislocation character (screw versus edge). Therefore, we can write the FCC mobility law as

$$\mathbf{v} = \mathbf{v}^g = \frac{\mathbf{F}^g}{B}, \tag{2.8}$$

where B is the isotropic drag coefficient and is typically between 10^{-5} and 10^{-4} Pa s for FCC metals [52]. With other materials, such as body-centered cubic (BCC) crystals, the drag coefficient is not isotropic and the glide constraint is not as strictly obeyed (for screw or near-screw dislocations), so that $\mathscr{B}$ will have to take a more complex form [12].

In many settings, a linear mobility law is inappropriate. For example, at low-to-moderate temperatures with BCC metals, the motion of screw dislocations is a thermally activated process; it occurs by the formation and movement of so-called kink pairs in the dislocation line. In this case, thermal activation theory should be used [52], which generally leads to a nonlinear mobility law. Nonlinear mobility laws have also been proposed to incorporate material effects besides lattice friction. For instance, solute atoms are known to exert drag forces on dislocations. A number of researchers have proposed nonlinear mobility laws that incorporate these effects [62, 95], and DD simulations have been conducted by approximating solute drag as a constant "back stress" which is subtracted from the driving force [69] (i.e., a ramp function mobility law).

2.2.2.3 Line Discretization and Remeshing

To employ numerical methods, we need to discretize the dislocation lines so that the overall dislocation structure is characterized by a set of nodes (or segments) and a data structure defining the connectivity between them. Discretization allows us to focus on a finite number of degrees of freedom (DOF), rather than an infinite number of points along the dislocation lines. Since dislocation lines can change their shape significantly during a simulation, and the total length of dislocation lines often increases, we also need to implement remeshing algorithms to modify the discretization when necessary. Dislocation lines can be discretized in a number of ways. Across the major DD codes, there are two general approaches to line discretization: lattice-based discretization and node-based discretization. Here we will discuss both. Major features of the two approaches are shown in Fig. 2.4.

In the lattice-based approach (used in the codes microMegas [24] and TRIDIS [103]), a grid of computational points, i.e., a lattice, often with a simple cubic structure of spacing a, is predefined throughout the simulation cell. Based on the structure of this lattice, a finite set of dislocation orientations is then selected

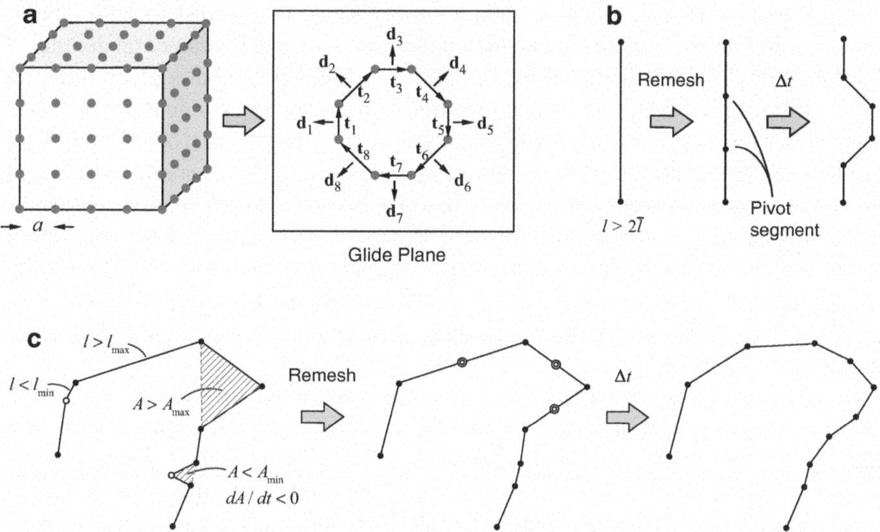

Fig. 2.4 Schematic depictions of (**a**,**b**) lattice-based and (**c**) node-based discretization. (**a**) The lattice grid used to define the segment directions $\mathbf{t}_i$ and movement directions $\mathbf{d}_i$. (**b**) Remeshing when a segment exceeds twice the average length $\bar{l}$, and response of pivot nodes after a time step Δt is taken. (**c**) Nodes are inserted (bullseye nodes) when $l > l_{\max}$ or $A > A_{\max}$, and removed (unfilled circles) when $l < l_{\min}$, or $A < A_{\min}$ with the area shrinking ($dA/dt < 0$)

and only these orientations are considered, as shown in Fig. 2.4a. These orientations define a unique set of straight line segments used to represent the dislocation lines (denoted as $\mathbf{t}_i$ in Fig. 2.4a). Dislocation motion is only considered in the direction orthogonal to each of these orientations (denoted as $\mathbf{d}_i$ in Fig. 2.4a). In this way, the segments are the degrees of freedom of the model. As the dislocation structure evolves, two different configurations of the dislocation lines are considered. The actual configuration is stored as the segments move continuously through space. When computing interaction forces and considering dislocation junctions, however, the actual configuration is projected onto the nearest set of lattice grid points in order to simplify the computations. Remeshing proceeds by dividing segments into smaller segments connected by "pivot segments" based on the user-specified average segment length, $\bar{l}$. The pivot segments initially have zero-length and extend along the direction set by the motion of their neighbors, allowing segments of new orientations to form, as depicted in Fig. 2.4b. In the lattice-based approach, the fidelity of the discretization is controlled by the spacing of the lattice grid, a, the number of line orientations allowed, and by the specified average dislocation segment length, $\bar{l}$.

With the node-based approach, dislocation lines are discretized according to a set of nodes and shape functions that connect the nodes, with the simplest case being linear shape functions that result in straight line segments. In this approach, any dislocation orientation is allowed and dislocation segments can move in any direction (consistent with their mobility law). In contrast to lattice-based discretization, in the node-based approach the nodes are the fundamental degrees of freedom. Only a single dislocation configuration is considered at a given time; the same configuration which is evolved in time is used for force calculations. Node-based codes have been written using linear segments (MDDP [64], NUMODIS [73], ParaDiS [5], PARANOID [89]) and cubic splines (PDD [35]) to connect the nodes. Given the greater versatility of the node-based approach, a larger set of remesh rules must be specified. For example, in ParaDiS two criteria are used for remeshing: segment lengths and the area enclosed by adjacent segments [9]. Both minimum (l_{min}, A_{min}) and maximum (l_{max}, A_{max}) values are specified for each, and nodes are added or removed to bring the dislocation structure into compliance with these ranges (see Fig. 2.4c).

In order to evolve the dislocation structure, we need to compute the forces acting on the segments or nodes. Generally, the forces per unit length discussed in Sect. 2.2.2.1 vary with position along the lines. To get the total force acting on node or segment i, we need to integrate the force along the line. In this respect, lattice-based and node-based discretization differ slightly. With lattice-based models, since the segments are the fundamental degrees of freedom, we need to calculate the total force acting on a segment with the line integral

$$\mathbf{f}_i = \int_{C_i} \mathbf{F}(s)dL(s) \tag{2.9}$$

where C_i denotes segment i. Note that a lower case $\mathbf{f}$ denotes a force, and an upper case $\mathbf{F}$ denotes a force per unit length. Node-based codes, on the other hand, require the total force acting on the nodes. This is determined in terms of the line shape function $N_i^j(s)$ which describes the contribution to node i from segment j as

$$\mathbf{f}_i^j = \int_{C_j} N_i^j(s)\mathbf{F}(s)dL(s). \tag{2.10}$$

For example, with linear segment j connecting nodes i and k, $N_i^j(s) = s$ where $s = 0$ at node k and $s = 1$ at node i. The total force on node i is then the sum of the contributions from each of the segments it is attached to:

$$\mathbf{f}_i = \sum_j \mathbf{f}_i^j. \tag{2.11}$$

These expressions are valid if the force per unit length acting along the line is known. However, in the case of the core force, determining the force per unit length is not very straightforward. Instead, it is easier to derive the force acting on a segment or node directly from the core energy per unit length expression, $E_c(\theta)$ [9]. Given $E_c(\theta)$, we can compute the total core energy E_{core} for a given discretized dislocation structure by summing the contribution from each segment, and then find the corresponding nodal or segment forces with

$$\mathbf{f}_i = -\frac{\partial E_{\text{core}}}{\partial \mathbf{r}_i}. \tag{2.12}$$

where $\mathbf{r}_i$ is the position vector of node or segment j.

Summing the Peach–Koehler and self-force contributions gives us the total force acting on a node or segment. However, mobility laws are usually written in terms of the force per unit length acting on the line. The force per unit length needed to evaluate the mobility law can be determined with

$$\mathbf{F}_i = \frac{\mathbf{f}_i}{\mathscr{L}_i}, \tag{2.13}$$

where $\mathscr{L}_i$ is a line length that depends on the discretization method. For the lattice-based approach, $\mathscr{L}_i$ is simply the length of segment i, $\mathscr{L}_i = l_i$. With node-based discretization, the following approximation[1] is commonly used: $\mathscr{L}_i = \sum_k l_{ik}/2$, where l_{ik} is the length of the segment connecting nodes i and k and the summation is over all nodes k connected to node i.

Now we have discretized the dislocation structure, and discussed the calculation of driving forces and subsequent velocity determination through the mobility law. Next we need to focus on evolving the positions of the nodes or segments, and the underlying dislocation structure they represent, in time.

2.2.2.4 Time Integration

As shown in Sect. 2.2.1, dislocation line motion is governed by a partial differential equation (PDE) in time (Eq. 2.4). After discretizing the dislocation lines, we can write this governing equation in terms of the motion of the nodes or segments, converting the PDE into a coupled system of N ordinary differential equations (ODEs). For example, in the nodal representation we have

$$\frac{d\mathbf{r}_i}{dt} = \mathbf{g}_i\left(\left\{\mathbf{r}_j\right\}, \sigma_{\text{ext}}, \ldots\right) \tag{2.14}$$

where $\mathbf{r}_i$ is the 3×1 position vector of node i and brackets denote the set of all nodes. In DD, we solve these ODEs using *time integration*, an approach where the solution is found over a series of sequential time steps. Many methods exist for time integrating coupled systems of ODEs, and in this section we discuss a few in the context of DD. In the following, for clarity we will assume $\left\{\mathbf{r}_j\right\}$ is the only argument of $\mathbf{g}(\cdot)$.

The simplest time integration scheme is the forward Euler method, which has the following form:

---

[1] A more rigorous definition can be written in terms of the line shape functions [9, 35].

$$\mathbf{r}_i^{k+1} = \mathbf{r}_i^k + \Delta t \, \mathbf{g}_i\left(\{\mathbf{r}_j^k\}\right). \tag{2.15}$$

Superscripts denote the time step number and Δt is the time step size. In this scheme we assume that the nodes maintain their current velocities over the duration of the time step, and update their positions accordingly. The forward Euler method is commonly used in DD simulations. One issue with this approach is that the error it introduces is unknown (without additional numerical methods). All time integration schemes introduce error and we must ensure this error does not overwhelm the solution. A simple method that provides an error estimate is the Heun method:

$$\mathbf{r}_{i,0}^{k+1} = \mathbf{r}_i^k + \Delta t \, \mathbf{g}_i\left(\{\mathbf{r}_j^k\}\right) \tag{2.16a}$$

$$\mathbf{r}_{i,l+1}^{k+1} = \mathbf{r}_i^k + \frac{\Delta t}{2}\left[\mathbf{g}_i\left(\{\mathbf{r}_j^k\}\right) + \mathbf{g}_i\left(\left\{\mathbf{r}_{j,l}^{k+1}\right\}\right)\right] \tag{2.16b}$$

$$e = \max_i \|\mathbf{r}_{i,l+1}^{k+1} - \mathbf{r}_{i,l}^{k+1}\|. \tag{2.16c}$$

Eq. (2.16a) is the forward Euler "predictor" and Eq. (2.16b) is the trapezoidal method "corrector." The corrector can be applied arbitrarily many times using a fixed-point iteration, with the second subscript denoting the iterate number, until the error estimate of Eq. (2.16c) falls below some user-specified tolerance. If the solution does not converge in a prespecified number of iterations, the time step must be reduced and the method applied anew. Note that in addition to providing an error estimate, the Heun method is globally second order accurate, meaning the solution converges as $\mathscr{O}(\Delta t^2)$, whereas the forward Euler method is only first order accurate, $\mathscr{O}(\Delta t)$. The Heun method is the default time integrator in ParaDiS.

Time integration turns out to be a challenging problem in DD, and is an active area of research. We defer discussion of more advanced topics, such as implicit time integration and subcycling, to Sect. 2.5.

2.2.2.5 Dislocation Collisions

When dislocation lines collide, they can react and form junctions or annihilate. The resulting junction formation and annihilation events can significantly influence the evolution of the dislocation structure. Hence, detecting and handling collision events reliably is important. To detect the collision of dislocation lines, a number of approaches have been developed. The simplest is a proximity-based algorithm, which assumes two lines have collided if they come within a user-defined minimum distance of each other. This approach can miss collisions, however, if dislocation lines are displaced too far in a time step. More advanced algorithms can safeguard against missing collisions [96]. Once a collision is detected, the appropriate topological changes must be made. The conservation of the Burgers vector must be invoked to determine the Burgers vectors of resulting segments. For instance, if two segments with opposite Burgers vectors collide they will annihilate with each other.

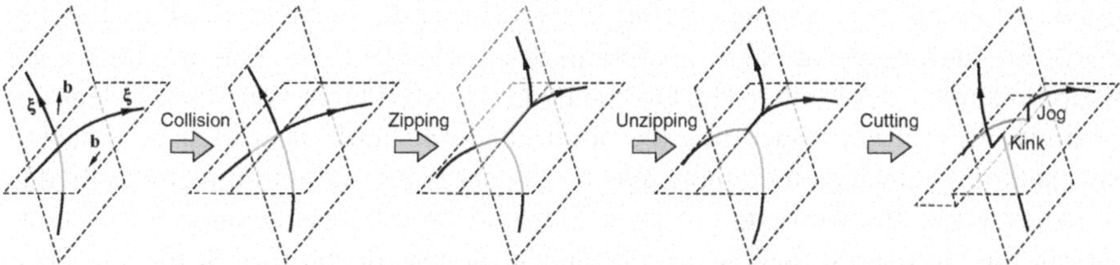

Fig. 2.5 Schematic showing the process of dislocation line collision, the zipping of a junction, subsequent unzipping, and then final dissolution after the dislocations cut each other. The cutting results in the formation of a jog and a kink

2.2.3 *Additional Aspects*

The fundamentals presented in Sect. 2.2.2 provide the basic toolset necessary to run a simple DD simulation. For example, the Frank–Read source simulations presented in Sect. 2.3.5.1 can be conducted using these methods. More advanced simulations require additional details, some of which are presented in this section.

2.2.3.1 Junctions and Dislocation Intersections

The discussion of dislocation collisions in Sect. 2.2.2.5 does not consider how to handle the formation and dissolution of dislocation junctions; we will elaborate these details here. When two dislocation lines moving in different planes collide, one of two things may occur. They may cut through each other and continue their motion, potentially producing Burgers-vector-sized steps on the lines known as jogs or kinks (depending on whether they are out of or within the glide planes, see Fig. 2.5). Or, they may zip together and react to annihilate or form a junction. Even if a junction does form, it may be ripped apart if a large enough force is applied, and the lines may cut each other and continue on as if the junction had never existed; this process is depicted in Fig. 2.5. Accurately capturing these behaviors is important because sessile (immobile) junctions (often referred to as locks) and dislocation intersections are thought to play vital roles in work hardening.

Considering this process in the context of DD, there are (at least) three different steps that need to be considered. First, the collision of dislocation lines needs to be detected, the result of which is a point junction between the two lines. The resulting point junction can then either zip together and form a proper junction, or split apart and possibly produce jogs and/or kinks. In some codes, the lines never formally react, and instead simply approach each other closely and align parallel to each other when forming a junction [24]. If the lines do formally react, the code must be able to detect whether the formation of a junction is favorable. This is typically done by applying an energy criterion to ensure that the system moves towards a state that maximizes its dissipation rate. Common examples include approximations based on line energy arguments [108], tests to see if the involved lines are moving apart [90], and the principle of maximum dissipation [9], which approximates the dissipation rate as the dot product of the nodal force with the velocity and seeks to maximize it. If it is decided that the point junction should instead split in such a way that lines cut each other, there may be an energy barrier inhibiting this split due to, for example, the formation of jogs. This barrier can be accounted for in terms of a splitting rate through the use of thermal activation theory (see Sect. 2.2.3.3), or athermally in terms of a junction strength dictating the minimum stress that must be applied for the split to occur. As an example for the latter scheme, Kubin et al. [54] have developed the following law to determine the strength of a junction:

$$\tau_{\mathrm{j}} = \frac{\beta \mu b}{l_{\mathrm{u}}} \tag{2.17}$$

where μ is the shear modulus, l_{u} is the length of the dislocation arms surrounding the junction, and β is a material constant that must be determined from experiments or atomistic simulations. If a cutting event like this does occur, the resulting jogs can influence the mobility of the dislocation lines [42]. However, most DD codes do not account for the presence of jogs.

2.2.3.2 Boundary Conditions

As with any initial-boundary value problem, the boundary conditions (BCs) need to be stated in order to have a well-defined problem. The specific form of the BCs is dictated by the geometry of interest. The types of BCs used in DD simulations can be categorized into three groups as shown in Fig. 2.6: (a) infinite BCs, (b) periodic BCs, and (c) heterogeneous BCs.

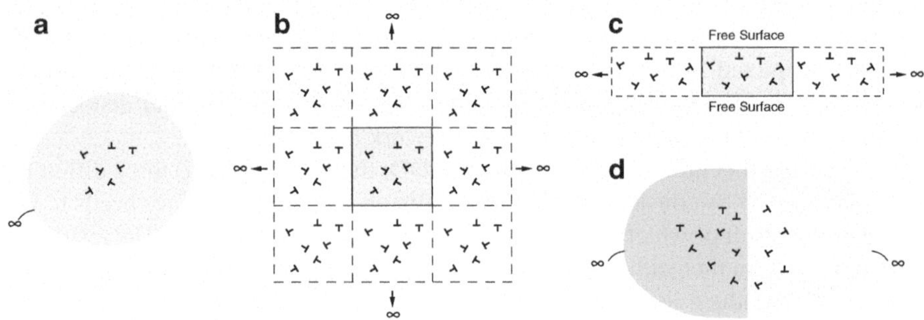

Fig. 2.6 Schematic depictions of different types of boundary conditions. (**a**) Infinite BCs, (**b**) periodic BCs, and heterogeneous BCs with (**c**) a free standing film with in-plane periodic BCs and (**d**) a bimaterial interface in an infinite medium. For simplicity, the dislocations are represented by the *perpendicular* symbol, even though the figure refers to 3D DD simulations

Infinite BCs (Fig. 2.6a) are the simplest, and correspond to the simulated dislocation lines being embedded in an infinite medium. Enforcement of an infinite BC in any coordinate direction requires simply that we allow dislocations to move an arbitrary distance along that axis. The stress expressions for dislocation lines (and even other defects) are known for an isotropic, homogeneous, infinite medium, so they may be implemented readily.

While infinite BCs provide a reasonable model for the behavior of dislocation lines far from free surfaces (i.e., in the bulk), it is computationally infeasible to keep track of all dislocation lines in an infinite medium that has a finite average dislocation density. This makes infinite BCs primarily useful for idealized test cases. Periodic BCs, in contrast, mimic an infinite medium while allowing for a nonzero average dislocation density. With periodic BCs, the simulation cell represents a so-called *supercell* which is repeated in all directions ad infinitum—Fig. 2.6b depicts this idea for a 2D geometry. The replicas surrounding the main simulation cell are called *images*. Periodic BCs provide a model for the simulation of bulk metals, where the material element being simulated is in the middle of a specimen many times larger than the cell. Any dislocation configuration or pattern, however, whose characteristic length scale is larger than the supercell cannot be captured with periodic BCs. To enforce periodic BCs, the total stress field due to every dislocation line in each of the infinite number of periodic images must be computed to determine the driving forces. In practice, only a finite number of images is considered, however, care must be taken to ensure the resulting stress field is well defined (due to conditional convergence [13, 55]). When a dislocation line crosses the supercell boundary, its next image over will enter the supercell from the opposing boundary. See [9] for a more detailed discussion of periodic boundary conditions.

The final type of boundary condition we will discuss applies to a much broader class of problems. In the case of a heterogeneous BC (Fig. 2.6c), some feature of the geometry breaks the homogeneity of the domain. Common examples are free surfaces, with geometries like cylinders, thin films, and half spaces, and bimaterial interfaces, as in the case of a layered material. As was discussed in Sect. 2.2.2.1, since analytic stress expressions generally only apply to an infinite, homogeneous medium, a corrective image stress field must be determined. Image stress solvers have been developed using the finite element method [100, 103, 108, 110], Fourier methods [32, 105, 107], and boundary element methods [26], as well as various other methods [29, 41, 49] to solve for the image field.

As a final note, we point out that these BCs can be combined. For instance, we may simulate a freestanding thin film [107] by employing periodic BCs in one or two coordinate directions and free surface BCs in the others (Fig. 2.6c).

2.2.3.3 Cross-Slip

Conservative dislocation motion occurs when no atomic diffusion is required and is termed dislocation glide. Nonconservative motion, on the other hand, requires the diffusion of vacancies and is referred to as dislocation climb. Assuming a dislocation only moves by conservative glide (i.e., no climb), it is confined to motion within the plane that contains both its Burgers vector and its line direction—this defines its glide plane. In the case of a screw dislocation, because the Burgers vector is parallel to the dislocation line, a unique glide plane cannot be defined. In principle, a screw dislocation can glide in any plane that contains its Burgers vector. Most of the time, however, dislocations prefer to glide along a few families of crystallographic planes that minimize their core energy, and this sets the slip systems for that metal. For instance, in FCC metals dislocations usually glide in $\{1\,1\,1\}$ planes, giving each screw dislocation two viable glide planes. The process of a screw dislocation changing from one glide plane to another is called *cross-slip*. Cross-slip is thought to be an important feature of dislocation motion, and in this section we will briefly outline the key aspects relevant to DD.

Cross-slip is known to be a *thermally activated* process [79]. This means that there is an energy barrier associated with its occurrence, and this barrier can be overcome by thermal fluctuations. The rate at which a thermally activated event occurs can be approximated with an Arrhenius-type relationship [51]:

$$R = \nu_0 \exp\left(-\frac{E_b}{k_\mathrm{B}T}\right) \tag{2.18}$$

where R is the rate in number of events per unit time, E_b is the energy barrier, k_B is Boltzmann's constant, T is the absolute temperature, and ν_0 is the attempt frequency. Often, the attempt frequency is approximated as $\nu_0 = \nu_D(L/L_0)$, where ν_D is the Debye frequency, L is the length of the dislocation segment, and L_0 is a reference length. Thus, in order to determine the cross-slip rate at a specified temperature, one needs to know the energy barrier and the attempt frequency. Atomistic simulations have commonly been used to determine these quantities, often finding that the energy barrier is sensitive to the local stress state (see Sect. 2.4.1).

Using thermal activation theory, cross-slip can be implemented in DD as follows. We test for cross-slip events once during each time step. We loop over all dislocation lines, looking for segments that are of screw character. If a screw segment is found, the energy barrier is calculated based on the local stress state at that segment, with which the cross-slip rate can be determined using Eq. (2.18). The cross-slip probability is then simply $R\Delta t$, where Δt is the time step size. We then select a random number ζ uniformly distributed in [0,1], and cross-slip occurs if $R\Delta t > \zeta$. The most difficult aspect of implementing this model is determining how the energy barrier depends on the local conditions (e.g., stress, local dislocation configuration, etc.).

2.2.3.4 2-Dimensional Dislocation Dynamics

As we have shown, fully three-dimensional dislocation dynamics simulations are complex and computationally expensive. Consequentially, many researchers have sought to develop dislocation dynamics in two-dimensions [3, 8, 18, 21, 37, 39,

72, 93, 100, 111]. In two-dimensional dislocation dynamics (2DDD), dislocations are assumed to be infinitely long and straight, so that they can be represented by point objects in the plane perpendicular to the dislocation line; this approximation greatly reduces the number of degrees of freedom and removes the need to track the complex topology present in three dimensions. 2DDD codes run much faster and can achieve much larger amounts of plastic deformation than 3D codes. However, these advantages are offset by the limited subset of problems that can be faithfully represented in two-dimensions (e.g., fatigue problems where dislocations are often long and straight). Because many physical phenomena are absent in the 2D picture, additional physics, such as sources for multiplication and obstacles [8, 16, 100], must be added. While many important contributions to DD have been made in a 2D setting, we will not elaborate on 2DDD further.

2.3 Running a DD Simulation

Over the past several decades, DD has been utilized to study a range of problems in crystal plasticity. While the specific details surrounding each of these simulations vary, they all share a number of basic ingredients. In this section we will briefly discuss each of these ingredients, and then provide several case studies.

2.3.1 Types of Simulations

DD simulations can be categorized into two groups: (1) small-scale—those interested in the interactions and behavior of one or a few dislocation lines and (2) large-scale—simulations examining the collective behavior of many dislocations. Examples of small-scale simulations include the simulation of intersecting dislocation lines, junction formation, and junction dissolution [58, 60]; the interaction of dislocations with precipitates [80] and solutes [17, 69]; and the interaction of dislocations with free surfaces [49, 98, 106]. Simulations of large-scale collective behavior generally involve simulating the stress–strain response of a material, with examples including work hardening in bulk metals [5, 11, 22, 96], the plasticity of micropillars [2, 19, 88, 106, 109], and plasticity during nanoindentation [30, 103]. Details below will be presented in terms of the two simulation types.

2.3.2 DD Codes

There are currently about a dozen 3D dislocation dynamics codes in use. Here we will briefly discuss some of their differences to aid the user in making a selection. See [28], [52], or [78] for additional reviews of DD codes.

As discussed in Sect. 2.2.2.3, DD codes can be categorized as either lattice-based or node-based. In practice, each of these discretization schemes has its own advantages and disadvantages in terms of accuracy, computing efficiency, simplicity, and flexibility. A strength of lattice-based simulations is that force calculations and tracking of dislocation intersections are simplified, since only a finite set of dislocation configurations (dictated by the lattice) are considered [24, 52].

With the node-based scheme, since dislocation segments can take any arbitrary orientation, dislocation lines tend to be smoother. In contrast, with lattice-based DD, the angles between neighboring segments remain unchanged regardless of how much the lattice or segment length is refined.

microMegas [24, 65] and TRIDIS [99, 103] are two examples of lattice-based DD codes dedicated to the 3D DD simulations of crystalline solids. microMegas, an open source code written mainly in Fortran, utilizes a base of eight line vectors per slip system, for describing dislocation lines in FCC, BCC, and HCP crystals, in addition to a few mineral materials. TRIDIS, suitable for the study of the mechanical response of FCC and BCC metals and alloys, is a parallel code that uses four line vectors per slip system and has been coupled to the finite element code CAST3M.

There are many node-based codes available and we briefly discuss a few. Parametric dislocation dynamics (PDD) is the only code which uses curved (cubic) dislocation segments [35, 77]; it was recently made open-source and renamed mechanics of defect evolution library (MODEL) [67, 77]. Multiscale dislocation dynamics plasticity (MDDP) [64] is a hybrid code coupling dislocation dynamics, continuum finite elements, and heat transfer models. Its DD code was originally named micro3d and was later implemented in MDDP. PARANOID [89] is a DD code suitable for DD simulations of thin films, strained layers, and bulk metals and semiconductors. Parallel dislocation simulator (ParaDiS) [5, 76] is an open-source, massively parallel DD code that has mobility laws implemented for FCC and BCC crystals incorporating glide and climb. NUMODIS [73] is a recently developed open-source, parallelized code, with features for simulations of polycrystals and polyphases.

2.3.3 Input Specification

Usually, DD simulations are controlled through two (or more) different input files. The *control file* specifies the parameters of the simulation. These include the material properties (elastic constants, drag coefficients, etc.), the loading conditions (strain rate, stress state, etc.), the numerical parameters (time step size, remeshing parameters, etc.), and output controls (e.g., what output to generate and how frequently). The *structure file* specifies the initial dislocation configuration and the geometry of the simulation cell or boundaries. This generally requires specifying where nodes are located, how segments connect the nodes, and what their Burgers vectors are. In the next section we will discuss how to select the necessary parameters and design a DD simulation.

2.3.4 *Designing a Simulation*

2.3.4.1 Initial Configuration

The initial dislocation configuration will be dependent upon the type of simulation. In small-scale settings, generally a few initially straight lines are used, and the dislocation character angle is often varied to see the different effects. The specific goals of the simulation will decide the initial geometry.

In large-scale simulations, initial configuration selection is more complex [71]. Usually, the initial configuration is intended to emulate a specific material state, for example, an annealed or cold-worked metal. The DD simulation would then predict the response of a material in such a state to the chosen loading. However, the full three-dimensional detail of dislocation structures in materials is generally not known; this means the initial configuration will have to be approximated somehow. Often, the following procedure is used. First, a simulation cell is populated with a chosen initial density of straight dislocation lines, usually randomly oriented and positioned. Then, the simulation cell is allowed to *relax*—equilibrate under zero imposed stress—until the dislocation structure reaches a meta-stable configuration. Once relaxed, the configuration may be used for further simulations.

2.3.4.2 Loads and Boundary Conditions

As with most solid mechanics simulations (and experiments), there are two common types of loadings in DD: *stress-controlled* and *strain-controlled*. Under stress-control, often referred to as creep loading, the applied stress is specified and the dislocation lines simply respond to the Peach-Kochler forces resulting from the applied stress and the stress fields of other dislocations. The stress state may be constant or vary in time.

Under strain-control, usually a strain rate tensor, $\dot{\epsilon}_{ij}$ is specified and the resulting stress state must be calculated as follows. The total strain at any time t is

$$\epsilon_{ij}^{\mathrm{tot}}(t) = \int_0^t \dot{\epsilon}_{ij}(\lambda)d\lambda. \qquad (2.19)$$

When $\dot{\epsilon}_{ij}$ is a constant, the result is simply $\epsilon_{ij}^{\mathrm{tot}}(t) = t\dot{\epsilon}_{ij}$. Using the procedure discussed in Sect. 2.3.4.3, the plastic strain due to the motion of the dislocation lines at time t, $\epsilon_{ij}^{\mathrm{p}}(t)$, can be determined. The elastic strain is then $\epsilon_{ij}^{\mathrm{el}}(t) = \epsilon_{ij}^{\mathrm{tot}}(t) - \epsilon_{ij}^{\mathrm{p}}(t)$ (assuming infinitesimal deformations), which is related to the stress through Hooke's law. For an isotropic linear elastic material with Lamé constants λ and μ (the shear modulus), they are related by

$$\sigma_{ij} = \lambda\bar{\epsilon}^{\mathrm{el}}\delta_{ij} + 2\mu\epsilon_{ij}^{\mathrm{el}} \qquad (2.20)$$

where δ_{ij} is the Kronecker delta and $\bar{\epsilon}^{\mathrm{el}} = \frac{1}{3}(\epsilon_{xx}^{\mathrm{el}} + \epsilon_{yy}^{\mathrm{el}} + \epsilon_{zz}^{\mathrm{el}})$ is the hydrostatic elastic strain. At each time step, the increments of total strain and plastic strain are computed, and then the stress state is updated according to Eq. (2.20).

The two loading conditions can also be combined. For instance, in the commonly used uniaxial tension loading condition, a normal strain rate is imposed along the loading direction while all other stress components are set to zero. In this case, assuming the imposed uniaxial strain rate is $\dot{\epsilon}_{xx}$, the externally applied stress state at any point in time is simply $\sigma_{xx} = E(t\dot{\epsilon}_{xx} - \epsilon_{xx}^{\mathrm{p}}(t))$, where E is the Young's modulus.

As discussed in Sect. 2.2.3.2, the boundary conditions will depend on the problem of interest. Periodic BCs are used to simulate bulk material response. Often infinite BCs are used when we are interested in the behavior of a few isolated dislocation lines. When running simulations under periodic boundary conditions, the size of the simulation cell is an important feature of the simulation; any dislocation structure whose length scale is larger than the simulation cell width cannot be accurately represented. Furthermore, if the cell is too small the interaction between a dislocation and its own periodic image can yield artificial behaviors.

2.3.4.3 Outputs

With DD, the positions of all the dislocation lines are known at each time step. This means that specific features of the dislocation structure can be extracted directly. For instance, we can determine how common a particular type of junction is or how predominant different line orientations are (e.g., edge versus screw). Often, it is useful to express features of the dislocation structure in terms of their density, ρ, the dislocation line length per unit volume (in units of [length]$^{-2}$). For example, a dislocation structure could be characterized in terms of the densities of the different slip systems. The density of a dislocation population can be computed by simply summing the length of all relevant segments and dividing by the simulation volume.

An important output for DD simulations is the plastic strain; it is needed for computing the stress state under strain-control (discussed in Sect. 2.3.4.2). In DD simulations, plastic deformation is produced by the motion of the dislocation lines. The area swept out by a dislocation segment in a time step is proportional to the plastic strain produced in the crystal according to the relation [4]

$$\delta\epsilon_{ij}^{\mathrm{p}} = \frac{b_i n_j + b_j n_i}{2\Omega}\delta A \tag{2.21}$$

where δA is the area swept out by the dislocation segment during its motion, Ω is the simulation volume, $\mathbf{b}$ is the burgers vector, and $\mathbf{n}$ is the slip plane normal. The total plastic strain produced in a time step is the summation of Eq. (2.21) over all dislocation segments.

2.3.4.4 Solution Convergence

As with any numerical simulation technique, it is important to ensure that the errors introduced by our discretizations in space and time are sufficiently small so that the solution converges. In DD, this means ensuring the time step and dislocation segments are small enough.

Two approaches have been used in DD to confirm that the time step size is adequately small. The first was discussed in Sect. 2.2.2.4, and involves approximating the truncation error of the time integrator, and selecting the time step size so that it falls below a user-specified tolerance. Another approach is to limit the time step size so that the dislocation structure does not change too much from step to step. This usually involves specifying a maximum displacement and/or rotation allowed for any dislocation segment during a time step, and limiting the time step so they are obeyed. While this approach does not directly control the error of the solution, it is commonly used and generally accepted.

Spatial discretization error is dictated by how well the discretized structure approximates the actual smooth structure of interest. The goal of the remeshing algorithms discussed in Sect. 2.2.2.3 is to provide a means for controlling the quality of the discretization. The remeshing algorithm operates according to the chosen remeshing parameters—the maximum and minimum segment lengths and areas. As these parameters are reduced, the discretization becomes more and more refined, and the discretization error is reduced. A refined structure is more accurate, but is also more computationally expensive. This is also true when choosing the shape of the dislocation segments. The cubic segments used in PDD better reproduce smooth dislocation structures, but at the cost of increased computational complexity. The user must decide where his or her simulation falls in the trade-off between speed and accuracy.

2.3.5 Example Simulations

Here we present three case studies showing the basics of running a DD simulation. First, we determine the activation stress of a Frank–Read source using the lattice-based code microMegas and the node-based code ParaDiS. Second, we examine the activation of a single-arm source in a micropillar using ParaDiS. Finally, we show results from a few simple work hardening simulations using ParaDiS. All simulations use the material properties for nickel at $T = 300\,\mathrm{K}$, which are given in Table 2.1, and the FCC mobility law presented in Sect. 2.2.2.2.

2.3.5.1 Case Study 1: Activation Stress of a Frank–Read Source

The Frank–Read source is a canonical case study in dislocation theory, showing how a single dislocation can multiply indefinitely by simply gliding in its slip plane under an applied shear stress [45]. A Frank–Read source can be modeled

Table 2.1 Parameters for nickel at $T = 300$ K used in DD simulations in all case studies

	microMegas			ParaDiS		
Property	Name	Value	Unit	Name	Value	Unit
Shear modulus (μ)	`ModuleG0`	76.0	GPa	`shearModulus`	76e9	Pa
Poisson's ratio (ν)	`DPOISS`	0.31	–	`pois`	0.31	–
Burgers vector (b)	`VecBurgers`	2.49	Å	`burgMag`	2.49e-10	m
Core energy parameter[a]		–		`Ecore`	6.05e9	Pa
Drag coefficient (B)	`Coef_visqueux`	1.61e-5	Pa s	`MobEdge`	62,112.0	(Pa s)$^{-1}$
or Mobility (M)				`MobScrew`	62,112.0	(Pa s)$^{-1}$
Core radius (r_c)		–		`rc`	1.0	b
Error tolerance		–		`rTol`	2.0	b
Reference scale	`Echelle`	6.75	–		–	
Time step size	`deltat0`	1e-12	s	Variable based on rTol		
Line tension type	`LINTEN`	4 (Mohles)			–	

Unspecified parameters were set to their default values. The mobility or drag coefficients are obtained from atomistic simulations [75]

[a]In ParaDiS, the core energy parameter `Ecore` controls the scaling of the core energy in the same way the $\mathcal{E}$ parameter does in Eq. (2.6) (hence it has units of Pa). The `Ecore` value used here leads to a core energy per unit length which scales as $\frac{\mu b_i^2}{4\pi}$, where b_i is the magnitude of the Burgers vector of segment i.

by considering a straight dislocation line of length L lying in its slip plane that is pinned at both ends. These pinning points could represent intersections with forest dislocations, impurities, obstacles, or any number of other pinning sites that occur in real metals. A force per unit length of τb (b is the magnitude of the Burgers vector) will be experienced by the dislocation line when a shear stress with magnitude τ is applied, which in turn causes the line to bow out. As τ increases, the radius of curvature decreases until the shear stress reaches τ_{act}, the activation stress. Figure 2.7a and e shows the configuration when $\tau = \tau_{\text{act}}$ from simulations in microMegas and ParaDiS, respectively. At stresses above τ_{act}, the line is able to bow around completely and partially annihilate with itself, as shown in (b,c) and (f,g). This process produces a new dislocation loop, shown in (d) and (h), which is free to continue expanding. The objective of this case study is to determine the activation stress for the Frank–Read source.

For these simulations, we choose a line direction of $[1\,\bar{1}\,2]$ and a Burgers vector of $\frac{b}{\sqrt{2}}[1\,1\,0]$, corresponding to an edge dislocation in an FCC metal with glide plane normal $(\bar{1}\,1\,1)$. The x-, y-, and z-axes of the coordinate system are along the $[1\,0\,0]$, $[0\,1\,0]$, and $[0\,0\,1]$ directions, respectively. The end nodes are flagged as immobile (velocities set to zero). Simulations were run under stress-control, with σ_{yy} applied to produce a resolved shear stress on the glide plane of τ, and all other stress components were set to zero. By slowly increasing the applied stress with increments of $\Delta\tau = 0.5$ MPa and monitoring for activation, we can determine the activation stress to within $\pm\Delta\tau/2$. To detect activation, we can simply watch for

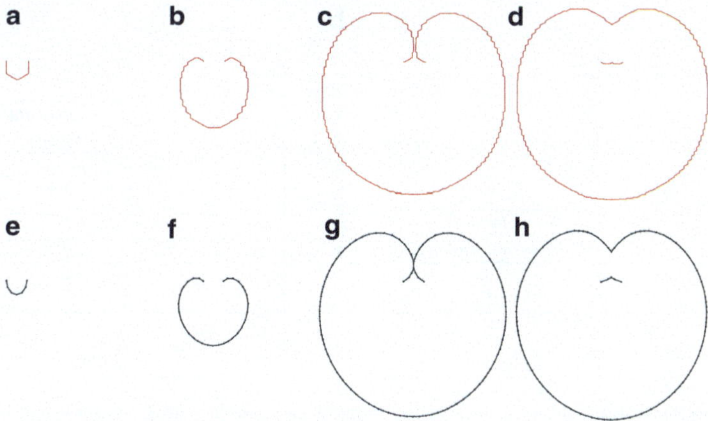

Fig. 2.7 Snapshots from Frank–Read source simulations using (**a**–**d**) microMegas with $L/\bar{l} = 2.4$ and (**e**–**h**) ParaDiS with $L/l_{max} = 2.4$, where $L = 0.596\,\mu$m (2400b) is the distance between the pinning points. ParaDiS graphics made with AtomEye [59]

whether activation occurs, or examine the plastic strain as a function of time—the plastic strain will plateau if the source is not activated. This is by no means the only possible approach for computing the activation stress, and strain-control or a combination of stress- and strain-control (like with microMegas deformation mode 6) could also be used.

The coarseness of the representation of the Frank–Read source will affect the activation stress. As the segment length is reduced, the activation stress should reach a converged value. Figure 2.8a shows the effect of the segment length on the activation stress with source length $L = 0.596\,\mu$m (2400b) in both codes. It is clear from the figure that the solution converges as smaller segments are used in both codes, and large segment lengths overestimate (in ParaDiS) or underestimate (in microMegas) the activation stress by up to 10 % in the parameter space considered here.

We can also study the effect of the source length, L, on the activation stress. Figure 2.8b shows the results for L ranging from 0.298 to 1.79 μm (1200b to 7200b) when $L/\bar{l} = L/l_{max} = 10$. The activation stresses estimated by the two codes differ by at most 2.2 % for all source lengths tested here. Also provided is the fit utilized by Foreman [31] of the form $\tau_{act} = [(A\mu b)/(4\pi L)]\ln(L/r_c)$ where r_c is the core radius, with $A = 1.2$, close to unity as Foreman found for an edge source.

2.3.5.2 Case Study 2: Spiral-Arm Source Activation in a Cylinder

In nanomaterials, the small specimen size allows spiral-arm or single-arm sources to operate. This type of source is similar to a Frank–Read source, except that only one end of the source is pinned in place; the rest of the source is free to rotate about

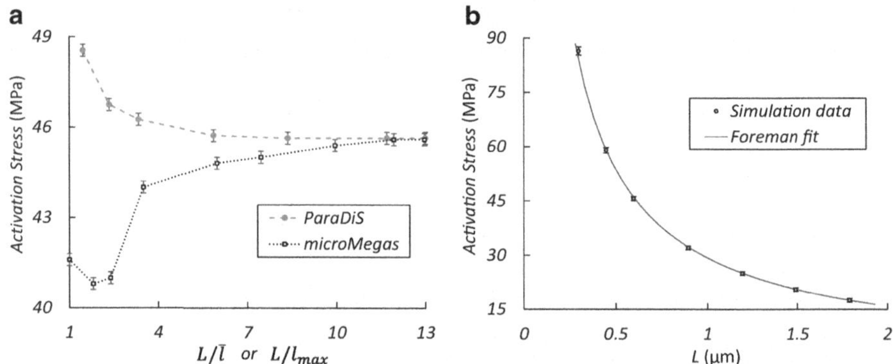

Fig. 2.8 (**a**) Activation stress τ_{act} of an edge Frank–Read source with $L = 0.596\,\mu\text{m}$ (2400b) as a function of the inverse of the segment length. The maximum segment length in ParaDiS and microMegas is, respectively, controlled by `maxSeg` (l_{max}) and `Ldis` ($\bar{l}$). Error bars show the inaccuracy caused by $\Delta\tau$, the stress increment used. (**b**) Activation stress τ_{act} as a function of the source length with $L/\bar{l} = L/l_{\text{max}} = 10$. Error bars include the differences between microMegas and ParaDiS as well. See text for explanation of Foreman fit

the pinning point under an applied stress. The result is a spiral-shaped dislocation line generating plastic strain with each revolution.

We here study the behavior of a single-arm source in a cylindrical specimen of radius R oriented along the [0 0 1] direction, imitating a source in a micropillar. For our source geometry, we choose a screw dislocation in an FCC metal with Burgers vector and line direction $[0\,1\,\bar{1}]$ and glide plane normal (1 1 1) that has a Lomer jog at its mid point. The Lomer jog is a section of dislocation with line direction $[0\,\bar{1}\,\bar{1}]$ which is out of the glide plane and treated as sessile in our simulation; in this way, the jog provides the pinning points for the source. Lomer jogs can form during plastic deformation when dislocations react and are often thought to act as immobile locks. We choose a jog height of $0.141\,\mu\text{m}$ (566b) for all simulations.

The simulation geometry is shown in Fig. 2.9. Initially, the arms are straight (Fig. 2.9a). Under an applied compressive stress σ_{zz}, the source will begin to rotate. Once again, the application of a stress greater than the activation stress σ_{act} is necessary for the source to activate and rotate freely about the jog. Figure 2.9b shows the configuration at activation when $R = 0.37\,\mu\text{m}$ (1500b). For this case study, we will again examine the activation stress, but now focusing on the effects of the free surface. The same procedure with stress-control taking steps of $\Delta\sigma = 0.5\,\text{MPa}$ is employed. We use ParaDiS to simulate the activation process, with a Fourier-based image stress solver [105]. A fast Fourier transform is used over a uniform grid on the surface of the cylinder to determine the image stress field. As with the discretization length, this grid spacing must be small enough to achieve a convergent solution. Periodic boundary conditions are used at either end of the cylinder, with a cell height of 6R, which gives us an approximately square $n \times n$ grid. The maximum segment length was set to $l_{\text{max}} = 2R/15$.

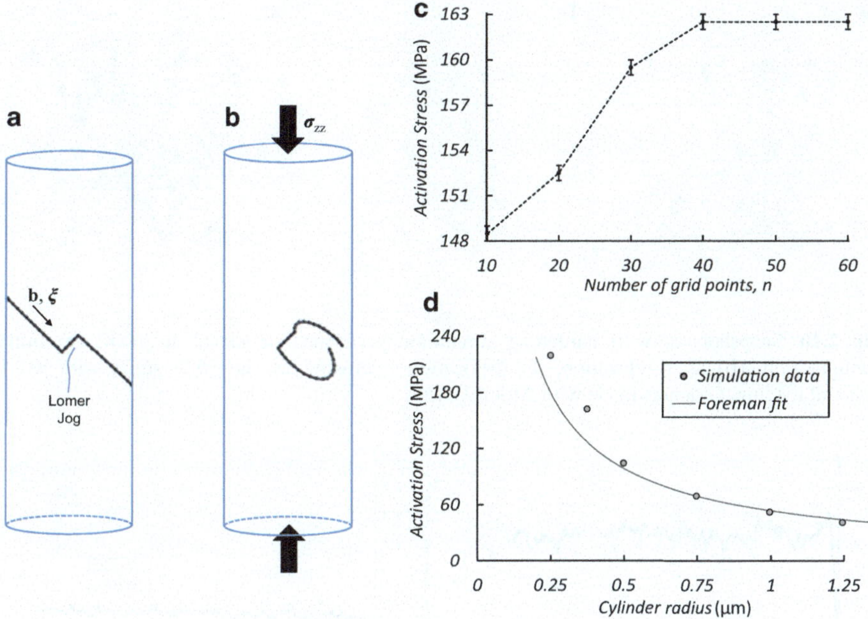

Fig. 2.9 Snapshots and results from single-arm source simulations using ParaDiS. (**a**) Initial configuration. (**b**) Configuration slightly below the activation stress when $R = 0.37\,\mu\text{m}$ (1500b). (**c**) Convergence of the activation stress as the number of grid points used for image stress calculation, n, is increased, with $R = 0.37\,\mu\text{m}$ (1500b). (**d**) Activation stress as a function of the cylinder radius. See text for the definition of Foreman fit. Graphics made with AtomEye [59]

First we examine the convergence behavior of the image stress solver. Figure 2.9c shows how the activation stress varies with the number of grid points. We see that about a 40×40 grid is required to achieve a converged result. Figure 2.9d demonstrates the dependence of the activation stress on the cylinder radius in the range $R = 0.124$–$1.24\,\mu\text{m}$ (500b–5000b) using $n = 50$ for the image stress calculation. At the larger radii, the activation stress again follows the Foreman behavior with $A = 2.15$ (using R in place of L), however, the smaller cylinders yield slightly higher values than the Foreman estimate.

2.3.5.3 Case Study 3: Bulk Plasticity Simulation

During plastic deformation the dislocation density tends to increase, causing the material to strengthen. This behavior is called *work-hardening* or *strain-hardening*. The study of work hardening is a key research area ripe for DD simulations. We close out this section with a few work hardening simulations.

For our work hardening simulations we use ParaDiS with a $10 \times 10 \times 10\,\mu\text{m}$ simulation cell, imposing periodic boundary conditions in all directions. No cross-

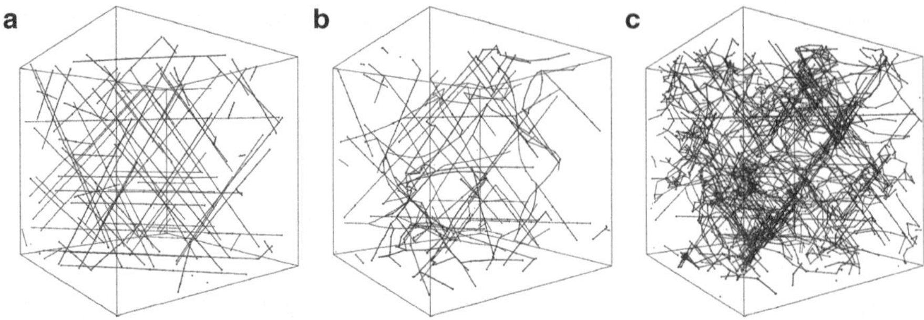

Fig. 2.10 Snapshots of work-hardening simulation performed on nickel at 300 K. (**a**) Initial configuration, (**b**) after relaxation, (**c**) dislocation microstructure at 0.5 % strain with [0 0 1] uniaxial loading. Graphics made with AtomEye [59]

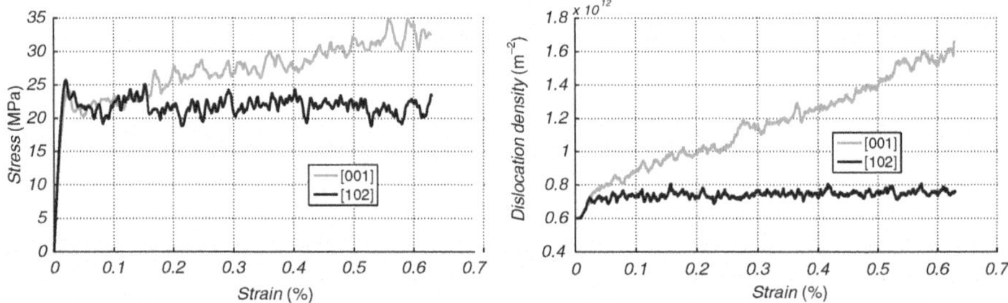

Fig. 2.11 Stress–strain and dislocation density-strain curves for the different loading directions

slip was allowed. The remesh parameter l_{max} (maxSeg) was set to 1.25 μm (5000*b*) (other remesh parameters were set to defaults). We start with 50 straight dislocation lines with a 60° character angle, random {1 1 1}-type glide plane, and random ⟨1 1 0⟩-type Burgers vector, as depicted in Fig. 2.10a, and then allow the system to relax under zero applied stress until it has reached the equilibrium configuration shown in Fig. 2.10b.

We study the response of the system under uniaxial tension with a constant strain rate of $10^3\,s^{-1}$ applied in the [0 0 1] and [1 0 2] directions. A recently developed subcycling-based time integrator was used [96], with simulations run for 40 and 7.2 h on a single CPU for [0 0 1] and [1 0 2] loading, respectively. The resulting dislocation configuration after a total strain of 0.5 % in the [0 0 1] direction is shown in Fig. 2.10c. Figure 2.11 shows the evolution of stress and dislocation density with respect to total strain. The initial yield strengths are similar for both loading directions. However for the [0 0 1] loading the crystal hardens with plastic strain as the dislocation density increases. In comparison, the flow stress and dislocation density remain relatively unchanged for the [1 0 2] loading.

2.4 Relation to Models at Other Length/Time Scales

Dislocation dynamics is just one of many tools that can be used to study the deformation behavior of materials. As this book demonstrates, these various models can be organized into a hierarchy that spans many orders of magnitude in both length and time scale. It is important to understand where a given model falls in this hierarchy, so that its connections to other models can be assessed. As discussed in the introduction, DD simulations are generally run at the length scale of about 0.1–10 μm and at time scales in the range of 1 μs to 1 ms, depending on the material. In this section, we will briefly discuss how DD relates to other material models, and examine a few examples of information propagation from one length/time scale to another.

2.4.1 Lower Scale Models

By its nature as a mesoscale modeling approach, DD requires numerous inputs that describe the physical behavior of dislocation lines. Elasticity and dislocation theory provide much of the information needed to define these models (e.g., Peach–Koehler forces, stress fields of dislocations, etc.). However, certain basic features of the behavior of dislocations are simply out of reach of these types of continuum models. A common example is the dislocation core. Many aspects of a dislocation's behavior are controlled by the structure at the dislocation core. Because the core is composed of a small number of atoms that are displaced far from their equilibrium positions, continuum models are often highly inaccurate. Where these models fail, experiments can be used to inform dislocation physics. However, it is usually challenging to extract information on individual dislocations from experiments.

Atomistic simulations, on the other hand, are well suited to informing DD models. Because the atomistic approach is closer to a "first principles" model, it can be used to study the fundamentals of dislocation physics. Atomistic simulations of one or a few dislocations can be conducted to study basic behaviors with different geometries, loading conditions, and temperature regimes, and this information can be included in the DD framework. Thus, we can think of DD as a model occupying the next larger length/time scale tier above atomistics. Common examples of the transfer of information from atomistic to DD include:

- Dislocation mobilities—This can be in the form of drag coefficients [36, 75, 83] or energy barriers [36, 38, 74, 83].
- Core energies—The core energy affects a number of features, including the core force (as discussed in Sect. 2.2.2.1). Core energy calculations have been carried out for a number of materials [10, 104, 114].
- Strength of junctions—In addition to using the scaling law discussed in Sect. 2.2.3.1, junction strengths can be calculated directly [10, 40].

- Cross-slip rate—Usually, this is calculated in the form of an energy barrier (as discussed in Sect. 2.2.3.3). Examples include the effects of different stress components [47, 56], intersection with forest dislocations [85], the presence of jogs [87, 102], and nucleation at the surface [86]. Many of these results have been incorporated in DD simulations [46].

2.4.2 *Higher Scale Models*

In the same way that MD can provide inputs for DD simulations, many researchers hope to use DD as a tool for informing higher length/time scale models. For example, a model residing at a larger length/time scale than DD is crystal plasticity (CP). In CP's continuum approach, constitutive laws are defined in terms of phenomenological models based on densities of different dislocation populations (e.g., forest and mobile dislocations). These dislocation densities are tracked at the continuum scale and dictate the loading response of each material element. DD can be used to develop the models which describe the relationship between dislocation densities, stress, and strain, thereby informing CP models.

An example of this transfer of information is the calculation of interaction coefficients in the Taylor hardening model. The generalized Taylor hardening law is commonly used in CP simulations, and states that the flow stress on slip system i is

$$\tau_i = \mu b \sqrt{\sum_j a_{ij}\rho_j} \tag{2.22}$$

where the summation is over all slip systems j, ρ_j is the dislocation density of slip system j, and a_{ij} is a matrix of *interaction coefficients* between the slip systems. The interaction coefficients can be determined using specialized DD simulations that target a specific pair of slip systems. These calculations have been performed for FCC metals [23, 61] and α-iron [81], and have been used to inform CP models [91].

2.4.3 *Concurrently Modeling Across Scales*

The approaches we have discussed so far involve passing information between modeling approaches using independently conducted simulations. However, it is also possible to transfer information between simulations as they both run concurrently. This approach may be useful in a number of settings. One example is if we are only interested in atomistic resolution over a small part of the domain, such as at the tip of a crack or beneath an indenter. Since atomistic resolution is not needed far from these regions where events such as dislocation nucleation are not occurring, we wish to represent the rest of the domain with a less expensive, higher scale model like DD. The atomistic and DD simulations would then be coupled at their mutual boundaries.

Such an approach has been implemented by Shilkrot et al. [92] in two-dimensions with the coupled atomistic and discrete dislocation (CADD) method for solving plasticity problems. In the CADD approach, the computational domain is divided up into atomistic and continuum regions; molecular dynamics is used in the atomistic region and 2D dislocation dynamics in the continuum domain [20]. For any concurrent modeling approach, the most challenging aspect is coupling the models at their shared domain boundaries. For instance, with CADD the code must detect when dislocations transmit between the domains. CADD has been used to study nanoindentation [66, 92, 93] as well as fracture and void growth [93].

2.5 Challenges and Current Research Topics

Here we will briefly list and introduce a few active research topics in the DD community. Some of the issues driving this research are purely mathematical or numerical in nature—for example, the fact that dislocation interactions cannot be calculated analytically in anisotropic elasticity. Other issues stem from the difficulty of accurately representing atomic-scale phenomena in a mesoscopic framework—for example, accounting for effects of the dislocation core structure. The following list is by no means comprehensive:

- *Time integration*—Efficiently time integrating the equations of motion in DD, i.e., taking a large time step with minimal computational expense, is a challenging but necessary task. Recent work examined implicit time integration methods [33, 43, 94] and time step subcycling [94, 96]. While larger time steps can be achieved with implicit methods, the additional computational cost makes performance gains less significant [33]. With subcycling, it has been shown that 100-fold speed-ups can be achieved [96].
- *Elastic anisotropy*—Most single crystals exhibit anisotropy in their elastic behavior, and yet most DD codes use isotropic elasticity to calculate the interactions between dislocation segments. This is because the analytic expressions for the stress fields of dislocations in anisotropic media are not known, and their numerical calculation is very expensive [112]. An approximate method was recently developed that utilizes spherical harmonics to estimate the interaction forces between dislocations [6]. With this approach, the computational cost can be adjusted according to the desired accuracy of the approximation.
- *Kinematics*—DD simulations are usually run under the assumption of infinitesimal deformations, so that the displacement field surrounding each dislocation is ignored. There are, however, instances where these displacements are known to be important. For instance, a symmetric tilt boundary can be thought of as a vertical array of edge dislocations; however, if the displacement fields of the dislocations are ignored then there is no tilt across the boundary. In addition to this effect, as dislocations move through a crystal, they alter the alignment of the crystallographic planes, i.e., they shift the connectivity of the planes of

atoms. This means two dislocations which are not coplanar initially may have their planes intersected by a series of dislocations which shift them onto the same plane [57]. This effect is related to the fact that when dislocations cut each other, jogs and/or kinks are produced. Incorporation of these effects in DD is challenging.

- *Core effects*—Some features of dislocation behavior are very sensitive to the nature of the core structure. These behaviors are challenging to capture in a framework that smears out all of these details into a simple line object. In some instances, certain features of the core can be included in the formulation presented above, for example, when constructing the mobility law or determining the stress dependence of the cross-slip rate. Sometimes explicit treatment of the core structure is important. For instance, FCC metals with low stacking fault energies have dislocations which are disassociated into Shockley partial dislocations that can be separated by 10s of nm. This can significantly influence the dislocation structures that develop. An approach for incorporating these effects in ParaDiS has been developed [63] .
- *Point defects*—Dislocations interact in a number of ways with point defects such as vacancies and solute atoms. These defects arise quite readily through material processing, alloying, and contamination, and give rise to many phenomena in dislocation physics. For example, solute atoms can accumulate on dislocations, forming so-called Cottrell atmospheres, which can slow down dislocation motion. Additionally, at high temperatures, dislocations are known to move out of their glide planes (climb) by consuming or producing vacancies. Some models have been developed to account for solutes [17, 69] and vacancy-driven climb[7, 70], however only a limited set of geometries have been considered.
- *Inclusions and precipitates*—The interactions of dislocations with inclusions and precipitates give rise to important phenomena such as precipitation hardening and kinematic hardening (Orowan looping). A number of researchers have conducted simulations examining the interaction of dislocations with a few precipitates in simplified settings [50, 68, 80, 84, 97], in addition to a few examples of large-scale simulations [44, 82, 101]. DD models describing the behavior of a dislocation as it cuts through a precipitate are still lacking.
- *Grain boundaries*—Most DD codes are only capable of simulating single crystals, whereas most structural materials are polycrystalline. The grain boundaries separating the individual grains of polycrystals can interact with dislocations in complex ways. Grain boundaries can both absorb and emit dislocations. A grain often experiences "misfit" stresses imposed by the surrounding grains during deformation, which can exert forces on dislocations. Dislocations can also transmit across grain boundaries, from one grain to another. As discussed in Chap. 11, DD simulations have been run with simplified grain and twin boundary models [27, 113], but a robust DD model for polycrystals still requires further development.

Acknowledgements We wish to thank Dr. Benoit Devincre for useful discussions. This work was supported by the US Department of Energy, Office of Basic Energy Sciences, Division of Materials Sciences and Engineering under Award No. DE-SC0010412. Sandia National Laboratories is a multi-program laboratory managed and operated by Sandia Corporation, a wholly owned subsidiary of Lockheed Martin Corporation, for the US Department of Energy's National Nuclear Security Administration under contract DE-AC04-94AL85000.

References

1. R. Abbaschian, L. Abbaschian, R.E. Reed-Hill, *Physical Metallurgy Principles* (Cengage Learning, Stamford, CT, 2009)
2. S. Akarapu, H.M. Zbib, D.F. Bahr, Analysis of heterogeneous deformation and dislocation dynamics in single crystal micropillars under compression. Int. J. Plast. **26**, 239–257 (2010)
3. R.J. Amodeo, N.M. Ghoniem, Dislocation dynamics. I. A proposed methodology for deformation micromechanics. Phys. Rev. B **41**(10), 6958–6967 (1990)
4. A.S. Argon, *Strengthening Mechanisms in Crystal Plasticity*. Oxford University Press, Oxford, 2008)
5. A. Arsenlis, W. Cai, M. Tang, M. Rhee, T. Oppelstrup, G. Hommes, T.G. Pierce, V.V. Bulatov, Enabling strain hardening simulations with dislocation dynamics. Model. Simul. Mater. Sci. Eng. **15**, 553 (2007)
6. S. Aubry, A. Arsenlis, Use of spherical harmonics for dislocation dynamics in anisotropic elastic media. Model. Simul. Mater. Sci. Eng. **21**, 065013 (2013)
7. B. Bakó, E. Clouet, L.M. Dupuy, M. Blétry, Dislocation dynamics simulations with climb: kinetics of dislocation loop coarsening controlled by bulk diffusion. Philos. Mag. **91**, 3173–3191 (2011)
8. A.A. Benzerga, Y. Bréchet, A. Needleman, E. van der Giessen, Incorporating three-dimensional mechanisms into two-dimensional dislocation dynamics. Model. Simul. Mater. Sci. Eng. **12**, 159–196 (2004)
9. V.V. Bulatov, W. Cai, *Computer Simulations of Dislocations* (Oxford University Press, Oxford, 2006)
10. V.V. Bulatov, F.F. Abraham, L.P. Kubin, B. Devincre, S. Yip, Connecting atomistic and mesoscale simulations of crystal plasticity. Nature **391**, 669–672 (1998)
11. V.V. Bulatov, L.L. Hsiung, M. Tang, A. Arsenlis, M.C. Bartelt, W. Cai, J.N. Florando, M. Hiratani, M. Rhee, G. Hommes, T.G. Pierce, T.D. de la Rubia, Dislocation multi-junctions and strain hardening. Nature **440**, 1174–1178 (2006)
12. W. Cai, V.V. Bulatov, Mobility laws in dislocation dynamics simulations. Mater. Sci. Eng. A **387**, 277–281 (2004)
13. W. Cai, V.V. Bulatov, J.P. Chang, J. Li, S. Yip, Periodic image effects in dislocation modeling. Philos. Mag. **83**, 539–567 (2003)
14. W. Cai, V.V. Bulatov, J. Chang, J. Li, S. Yip, Dislocation core effects on mobility, in *Dislocations in Solids*, ed. by F.R.N. Nabarro, J.P. Hirth, Vol. 12, Chap. 64 (Elsevier, Amsterdam, 2004), pp. 1–80
15. W. Cai, A. Arsenlis, C.R. Weinberger, V.V. Bulatov, A non-singular continuum theory of dislocations. J. Mech. Phys. Solids **54**, 561–587 (2006)
16. S.S. Chakravarthy, W.A. Curtin, Effect of source and obstacle strengths on yield stress: A discrete dislocation study. J. Mech. Phys. Solids **58**, 625–635 (2010)
17. Q. Chen, X.-Y. Liu, S.B. Biner, Solute and dislocation junction interactions. Acta Mater. **56**, 2937–2947 (2008)
18. H.H.M. Cleveringa, E. van der Giessen, A. Needleman, Comparison of discrete dislocation and continuum plasticity predictions for a composite material. Acta Mater. **45**(8), 3163–3179 (1997)
19. T. Crosby, G. Po, C. Erel, N. Ghoniem, The origin of strain avalanches in sub-micron plasticity of fcc metals. Acta Mater. **89**, 123–132 (2015)

20. W.A. Curtin, R.E. Miller, Atomistic/continuum coupling in computational materials science. Model. Simul. Mater. Sci. Eng. **11**, R33–R68 (2003)
21. V.S. Deshpande, A. Needleman, E. van der Giessen, Plasticity size effects in tension and compression of single crystals. J. Mech. Phys. Solids **53**, 2661–2691 (2005)
22. B. Devincre, L.P. Kubin, C. Lemarchand, R. Madec, Mesoscopic simulations of plastic deformation. Mater. Sci. Eng. A **309**, 211–219 (2001)
23. B. Devincre, L. Kubin, T. Hoc, Physical analyses of crystal plasticity by DD simulations. Scr. Mater. **54**, 741–746 (2006)
24. B. Devincre, R. Madec, G. Monnet, S. Queyreau, R. Gatti, L. Kubin, Modeling crystal plasticity with dislocation dynamics simulations: the 'microMegas' code, in *Mechanics of Nano-Objects*. (Presses de l'Ecole des Mines de Paris, Paris, 2011), pp. 81–100
25. G. deWit, J.S. Koehler, Interaction of dislocations with an applied stress in anisotropic crystals. Phys. Rev. **116**(5), 1113–1120 (1959)
26. J.A. El-Awady, S. Bulent Biner, N.M. Ghoneim, A self-consistent boundary element, parametric dislocation dynamics formulation of plastic flow in finite volumes. J. Mech. Phys. Solids **56**, 2019–2035 (2008)
27. H. Fan, S. Aubry, A. Arsenlis, J.A. El-Awady, The role of twinning deformation on the hardening response of polycrystalline magnesium from discrete dislocation dynamics simulations. Acta Mater. **92**, 126–139 (2015)
28. R.S. Fertig, S.P. Baker, Simulation of dislocations and strength in thin films: a review. Prog. Mater. Sci. **54**, 874–908 (2009)
29. M.C. Fivel, T.J. Gosling, G.R. Canova, Implementing image stresses in a 3D dislocation simulation. Model. Simul. Mater. Sci. Eng. **4**, 581–596 (1996)
30. M.C. Fivel, C.F. Robertson, G.R. Canova, L. Boulanger, Three-dimensional modeling of indent-induced plastic zone at a mesoscale. Acta Mater. **46**, 6183–6194 (1998)
31. A.J.E. Foreman, The bowing of a dislocation segment. Philos. Mag. **15**, 1011–1021 (1967)
32. S. Gao, M. Fivel, A. Ma, A. Hartmaier, Influence of misfit stresses on dislocation glide in single crystal superalloys: a three-dimensional discrete dislocation dynamics study. J. Mech. Phys. Solids **76**, 276–290 (2015)
33. D.J. Gardner, C.S. Woodward, D.R. Reynolds, G. Hommes, S. Aubry, A. Arsenlis, Implicit integration methods for dislocation dynamics. Model. Simul. Mater. Sci. Eng. **23**, 025006 (2015)
34. N.M. Ghoniem, R. Amodeo, Computer simulation of dislocation pattern formation. Solid State Phenom. **3** & **4**, 377 (1988)
35. N.M. Ghoniem, S.H. Tong, L.Z. Sun, Parametric dislocation dynamics: a thermodynamics-based approach to investigations of mesoscopic plastic deformation. Phys. Rev. B **61**, 913 (2000)
36. M.R. Gilbert, S. Queyreau, J. Marian, Stress and temperature dependence of screw dislocation mobility in α-Fe by molecular dynamics. Phys. Rev. B **84**, 174103 (2011)
37. D. Gómez-García, B. Devincre, L.P. Kubin, Dislocation patterns and the similitude principle: 2.5D mesoscale simulations. Phys. Rev. Lett. **96**, 125503 (2006)
38. P.A. Gordon, T. Neeraj, Y. Li, J. Li, Screw dislocation mobility in BCC metals: the role of the compact core on double-kink nucleation. Model. Simul. Mater. Sci. Eng. **18**, 085008 (2010)
39. P.J. Guruprasad, A.A. Benzerga, Size effects under homogeneous deformation of single crystals: A discrete dislocation analysis. J. Mech. Phys. Solids **56**, 132–156 (2008)
40. S.M. Hafez Haghighat, R. Schäublin, D. Raabe, Atomistic simulation of the $a_0 < 100 >$ binary junction formation and its unzipping in body-centered cubic iron. Acta Mater. **64**, 24–32 (2014)
41. A. Hartmaier, M.C. Fivel, G.R. Canova, P. Gumbsch, Image stresses in a free standing thin film. Model. Simul. Mater. Sci. Eng. **7**, 781–793 (1999)
42. J.P. Hirth, J. Lothe, *Theory of Dislocations*, 2nd edn. (Krieger Publishing Company, Malabar, FL, 1992)
43. J. Huang, N.M. Ghoniem, Accuracy and convergence of parametric dislocation dynamics. Model. Simul. Mater. Sci. Eng. **10**, 1–19 (2002)

44. M. Huang, L. Zhao, J. Tong, Discrete dislocation dynamics modelling of mechanical deformation of nickel-based single crystal superalloys. Int. J. Plast. **28**, 141–158 (2010)
45. D. Hull, D.J. Bacon, *Introduction to Dislocations*, 4th edn. (Butterworth Heinemann, Oxford, 2009)
46. A.M. Hussein, S.I. Rao, M.D. Uchic, D.M. Dimiduk, J.A. El-Awady, Microstructurally based cross-slip mechanisms and their effects on dislocation microstructure evolution in fcc crystals. Acta Mater. **85**, 180–190 (2015)
47. K. Kang, J. Yin, W. Cai, Stress dependence of cross slip energy barrier for face-centered cubic nickel. J. Mech. Phys. Solids **62**, 181–193 (2014)
48. Y. Kawasaki, T. Takeuchi, Cell structures in copper single crystals deformed in the [001] and [111] axes. Scr. Met. **14**, 183–188 (1980)
49. T.A. Khraishi, H.M. Zbib, Free-surface effects in 3D dislocation dynamics: formulation and modeling. ASME J. Eng. Mater. Technol. **124**(3), 342–351 (2002)
50. T.A. Khraishi, L. Yan, Y.L. Shen, Dynamic simulations of the interaction between dislocations and dilute particle concentrations in metal–matrix composites (MMCs). Int. J. Plast. **20**, 1039–1057 (2004)
51. U.F. Kocks, A.S. Argon, M.F. Ashby, Thermodynamics and kinetics of slip. Prog. Mater. Sci. **19**, 1–288 (1975)
52. L. Kubin, *Dislocations, Mesoscale Simulations and Plastic Flow*. (Oxford University Press, Oxford, 2013)
53. L.P. Kubin, G. Canova, M. Condat, B. Devincre, V. Pontikis, Y. Bréchet, Dislocation microstructures and plastic flow: a 3D simulation. Solid State Phenom. **23** & **24**, 455–472 (1992)
54. L.P. Kubin, B. Devincre, M. Tang, Mesoscopic modelling and simulation of plasticity in fcc and bcc crystals: dislocation intersections and mobility. J. Comput.-Aided Mater. Des. **5**, 31–54 (1998)
55. W.P. Kuykendall, W. Cai, Conditional convergence in 2-dimensional dislocation dynamics. Model. Simul. Mater. Sci. Eng. **21**, 055003 (2013)
56. W.P. Kuykendall, W. Cai, Effect of multiple stress components on the energy barrier of crossslip in nickel (2016, in preparation)
57. C. Laird, Chapter 27: fatigue, in *Physical Metallurgy*, ed. by R.W. Cahn, P. Haasen, 4th edn. (Elsevier Science, Amsterdam, 1996), pp. 2293–2397
58. S.W. Lee, S. Aubry, W.D. Nix, W. Cai, Dislocation junctions and jogs in a free-standing FCC thin film. Model. Simul. Mater. Sci. Eng. **19**, 025002 (2011)
59. J. Li, AtomEye: an efficient atomistic configuration viewer. Model. Simul. Mater. Sci. Eng. **11**, 173–177 (2003)
60. R. Madec, B. Devincre, L.P. Kubin, From dislocation junctions to forest hardening. Phys. Rev. Lett. **89**, 255508 (2002)
61. R. Madec, B. Devincre, L. Kubin, T. Hoc, D. Rodney, The role of collinear interaction in dislocation-induced hardening. Science **301**, 1879–1882 (2003)
62. J. Marian, A. Caro, Moving dislocations in disordered alloys: connecting continuum and discrete models with atomistic simulations. Phys. Rev. B **74**, 024113 (2006)
63. E. Martínez, J. Marian, A. Arsenlis, M. Victoria, J.M. Perlado, Atomistically informed dislocation dynamics in fcc crystals. J. Mech. Phys. Solids **56**, 869–895 (2008)
64. MDDP: Multiscale Dislocation Dynamics Plasticity code, http://www.cmms.wsu.edu/
65. microMegas (mM) code, http://zig.onera.fr/mm_home_page/
66. R.E. Miller, L.E. Shilkrot, W.A. Curtin, A coupled atomistics and discrete dislocation plasticity simulation of nanoindentation into single crystal thin films. Acta Mater. **52**, 271–284 (2004)
67. MODEL: Mechanics Of Defect Evolution Library, https://bitbucket.org/model/model/wiki/Home
68. V. Mohles, Dislocation dynamics simulations of particle strengthening. in *Continuum Scale Simulation of Engineering Materials: Fundamentals – Microstructures – Process Applications*, ed. by D. Raabe, F. Roters, F. Barlat, L.-Q. Chen (Wiley, Weinheim, 2004)

69. G. Monnet, B. Devincre, Solute friction and forest interaction. Philos. Mag. **86**, 1555–1565 (2006)
70. D. Mordehai, E. Clouet, M. Fivel, M. Verdier, Introducing dislocation climb by bulk diffusion in discrete dislocation dynamics. Philos. Mag. **88**, 899–925 (2008)
71. C. Motz, D. Weygand, J. Senger, P. Gumbsch, Initial dislocation structures in 3-D discrete dislocation dynamics and their influence on microscale plasticity. Acta Mater. **57**, 1744–1754 (2009)
72. A. Needleman, E. van der Giessen, Discrete dislocation and continuum descriptions of plastic flow. Mater. Sci. Eng. A **309–310**, 1–13 (2001)
73. NUMODIS: a Numerical Model for Dislocations, http://www.numodis.com/numodis/index.html
74. R.W. Nunes, J. Bennetto, D. Vanderbilt, Structure, barriers and relaxation mechanisms of kinks in the 90° partial dislocation in silicon. Phys. Rev. Lett. **77**, 1516–1519 (1996)
75. D.L. Olmsted, L.G. Hector Jr., W.A. Curtin, R.J. Clifton, Atomistic simulations of dislocation mobility in Al, Ni and Al/Mg alloys. Model. Simul. Mater. Sci. Eng. **13**, 371–388 (2005)
76. ParaDiS: Parallel Dislocation Simulator code, http://micro.stanford.edu/wiki/ParaDiS_Manuals
77. G. Po, N. Ghoniem, A variational formulation of constrained dislocation dynamics coupled with heat and vacancy diffusion. J. Mech. Phys. Solids **66**, 103–116 (2014)
78. G. Po, M.S. Mohamed, T. Crosby, C. Erel, A. El-Azab, N. Ghoniem, Recent progress in discrete dislocation dynamics and its applications to micro plasticity. J. Mat. **66**, 2108–2120 (2014)
79. W. Püschl, Models for dislocation cross-slip in close-packed crystal structures: a critical review. Prog. Mater. Sci. **47**, 415–461 (2002)
80. S. Queyreau, B. Devincre, Bauschinger effect in precipitation-strengthened materials: A dislocation dynamics investigation. Philos. Mag. Lett. **89**, 419–430 (2009)
81. S. Queyreau, G. Monnet, B. Devincre, Slip systems interactions in α-iron determined by dislocation dynamics simulations. Int. J. Plast. **25**, 361–377 (2009)
82. S. Queyreau, G. Monnet, B. Devincre, Orowan strengthening and forest hardening superposition examined by dislocation dynamics simulations. Acta Mater. **58**, 5586–5595 (2010)
83. S. Queyreau, J. Marian, M.R. Gilbert, B.D. Wirth, Edge dislocation mobilities in bcc Fe obtained by molecular dynamics. Phys. Rev. B **84**, 064106 (2011)
84. S. Rao, T.A. Parthasarathy, D.M. Dimiduk, P.M. Hazzledine, Discrete dislocation simulations of precipitation hardening in superalloys. Philos. Mag. **84**, 30, 3195–3215 (2004)
85. S. Rao, D.M. Dimiduk, J.A. El-Awady, T.A. Parthasarathy, M.D. Uchic, C. Woodward, Activated states for cross-slip at screw dislocation intersections in face-centered cubic nickel and copper via atomistic simulation. Acta Mater. **58**, 5547–5557 (2010)
86. S.I. Rao, D.M. Dimiduk, T.A. Parthasarathy, M.D. Uchic, C. Woodward, Atomistic simulations of surface cross-slip nucleation in face-centered cubic nickel and copper. Acta Mater. **61**, 2500 (2013)
87. S. Rao, D.M. Dimiduk, J.A. El-Awady, T.A. Parthasarathy, M.D. Uchic, C. Woodward, Screw dislocation cross slip at cross-slip plane jogs and screw dipole annihilation in FCC Cu and Ni investigated via atomistic simulations. Acta Mater. **101**, 10–15 (2015)
88. I. Ryu, W.D. Nix, W. Cai, Plasticity of bcc micropillars controlled by competition between dislocation multiplication and depletion. Acta Mater. **61**, 3233–3241 (2013)
89. K.W. Schwarz, Simulation of dislocations on the mesoscopic scale. I. Methods and examples. J. Appl. Phys. **85**, 108–119 (1999)
90. K.W. Schwarz, Local rules for approximating strong dislocation interactions in discrete dislocation dynamics. Model. Simul. Mater. Sci. Eng. **11**, 609–625 (2003)
91. P. Shanthraj, M.A. Zikry, Dislocation density evolution and interactions in crystalline materials. Acta Mater. **59**, 7695–7702 (2011)
92. L.E. Shilkrot, R.E. Miller, W.A. Curtin, Coupled atomistic and discrete dislocation plasticity. Phys. Rev. Lett. **89**(2) 025501 (2002)

93. L.E. Shilkrot, R.E. Miller, W.A. Curtin, Multiscale plasticity modeling: coupled atomistics and discrete dislocation mechanics. J. Mech. Phys. Solids **52**, 755–787 (2004)
94. R.B. Sills, W. Cai, Efficient time integration in dislocation dynamics. Model. Simul. Mater. Sci. Eng. **22**, 025003 (2014)
95. R.B. Sills, W. Cai, Solute drag on perfect and extended dislocations. Philos. Mag. **96**, 895–921 (2016)
96. R.B. Sills, A. Aghaei, W. Cai, Advanced time integration algorithms for dislocation dynamics simulations of work hardening. Model. Simul. Mater. Sci. Eng. **24**, 045019 (2016)
97. A. Takahashi, N.M. Ghoniem, A computational method for dislocation-precipitate interaction. J. Mech. Phys. Solids **56**, 1534–1553 (2008)
98. M. Tang, G. Xu, W. Cai, V.V. Bulatov, A hybrid method for computing forces on curved dislocations intersecting free surfaces in three-dimensional dislocation dynamics. Model. Simul. Mater. Sci. Eng. **14**, 1139–1151 (2006)
99. TRIDIS: the edge-screw 3D Discrete Dislocation Dynamics code, http://www.numodis.com/tridis/index.html
100. E. van der Giessen, A. Needleman, Discrete dislocation plasticity: a simple planar model. Model. Simul. Mater. Sci. Eng. **3**, 689–735 (1995)
101. A. Vattré, B. Devincre, A. Roos, Orientation dependence of plastic deformation in nickel-based single crystal superalloys: discrete-continuous model simulations. Acta Mater. **58**, 1938–1951 (2010)
102. T. Vegge, T. Rasmussen, T. Leffers, O.B. Pedersen, K.W. Jacobsen, Atomistic simulations of cross-slip of jogged screw dislocations in copper. Philos. Mag. Lett. **81**, 137 (2001)
103. M. Verdier, M. Fivel, I. Groma, Mesoscopic scale simulation of dislocation dynamics in fcc metals: Principles and applications. Model. Simul. Mater. Sci. Eng. **6**, 755 (1998)
104. G. Wang, A. Strachan, T. Cagin, W.A. Goddard III, Molecular dynamics simulations of 1/2 $a < 111 >$ screw dislocation in Ta. Mater. Sci. Eng. A **309–310**, 133–137 (2001)
105. C.R. Weinberger, W. Cai, Computing image stress in an elastic cylinder. J. Mech. Phys. Solids **55**, 2027–2054 (2007)
106. C.R. Weinberger, W. Cai, Surface-controlled dislocation multiplication in metal micropillars. Proc. Natl. Acad. Sci. **105**, 38, 14304–14307 (2008)
107. C.R. Weinberger, S. Aubry, S.W. Lee, W.D. Nix, W. Cai, Modelling dislocations in a free-standing thin film. Model. Simul. Mater. Sci. Eng. **17**, 075007 (2009)
108. D. Weygand, L.H. Friedman, E. van der Giessen, A. Needleman, Aspects of boundary-value problem solutions with three-dimensional dislocation dynamics. Model. Simul. Mater. Sci. Eng. **10**, 437–468 (2002)
109. D. Weygand, M. Poignant, P. Gumbsch, O. Kraft, Three-dimensional dislocation dynamics simulation of the influence of sample size on the stress–strain behavior of fcc single-crystalline pillars. Mater. Sci. Eng. A **483**, 188–190 (2008)
110. H. Yasin, H.M. Zbib, M.A. Khaleel, Size and boundary effects in discrete dislocation dynamics: coupling with continuum finite element. Mater. Sci. Eng. A **309–310**, 294–299 (2001)
111. S. Yefimov, I. Groma, E. van der Giessen, A comparison of a statistical-mechanics based plasticity model with discrete dislocation plasticity calculations. J. Mech. Phys. Solids **52**, 279–300 (2004)
112. J. Yin, D.M. Barnett, W. Cai, Efficient computation of forces on dislocation segments in anisotropic elasticity. Model. Simul. Mater. Sci. Eng. **18**, 045013 (2010)
113. C. Zhou, R. LeSar, Dislocation dynamics simulations of plasticity in polycrystalline thin films. Int. J. Plast. **30–31**, 185–201 (2012)
114. X.W. Zhou, R.B. Sills, D.K. Ward, R.A. Karnesky, Atomistic calculation of dislocation core energy in aluminum (2016 in preparation)

Chapter 3
Continuum Approximations

Joseph E. Bishop and Hojun Lim

3.1 Introduction

In continuum mechanics, instead of using discrete sums to assemble forces and assess equilibrium, as in molecular dynamics or dislocation dynamics, one instead uses the machinery of differential and integral calculus to cast Newton's second law in the form of a set of partial-differential equations, the solution of which gives the equilibrium configuration of the entire collection of discrete entities. While seemingly inapplicable to nanoscale structures, the use of continuum mechanics at the nanoscale is still a useful approximation with careful consideration of the assumptions inherent in the theory and with the inclusion of scale-dependent physical phenomena such as surface effects, microstructural effects, and nonlocal phenomena. The current literature on these generalized continuum theories is large and ever growing. Our goal in this chapter is to give a brief introduction to continuum mechanics with a focus on nanomechanics and how the classical theory can be modified to include phenomena such as surface effects and microstructural effects. Numerous references are given to more detailed expositions.

Section 3.2 begins with a brief review of classical continuum mechanics, an overview of micromorphic continuum formulations, and concludes with a discussion of continuum formulations that include an explicit surface stress. This section is also helpful in understanding the quasi-continuum method presented in Chap. 5

J.E. Bishop (✉)
Sandia National Laboratories, Engineering Sciences Center, Albuquerque, NM 87185, USA
e-mail: jebisho@sandia.gov

H. Lim
Sandia National Laboratories, Materials Science and Engineering Center, Albuquerque, NM 87185, USA
e-mail: hnlim@sandia.gov

C.R. Weinberger, G.J. Tucker (eds.), *Multiscale Materials Modeling for Nanomechanics*, Springer Series in Materials Science 245,
DOI 10.1007/978-3-319-33480-6_3

of this book. At the center of continuum mechanics is homogenization theory which provides a mathematically elegant and rigorous framework for replacing a discrete collection of interacting entities by an equivalent homogenous continuum with effective material properties. Furthermore, given the continuum approximation of the system, homogenization theory provides a method for recovering the solution of the original discrete or heterogeneous system. These concepts are discussed in Sect. 3.3. Continuum approaches to modeling crystal-plasticity are discussed in Sect. 3.4. These continuum crystal-plasticity models explicitly incorporate descriptions of the active slip systems and hardening phenomena at the crystal scale.

Errors in a continuum approximation to a discrete system are unavoidable whenever the introduced length scales are comparable to the length scale of the discrete system. Generalized continuum theories may be able to reproduce qualitatively correct physical phenomenon, such as a surface effect or optical-branch phonon dispersion curves, but the accuracy of the continuum theory must be judged with respect to the true behavior of the original discrete system. Two simple one-dimensional examples are given in the Appendix to demonstrate errors induced in a continuum approximation of a discrete system.

There are several excellent books available that include detailed discussions, examples, and tutorials for many of the concepts presented in this chapter. The book "Applied Mechanics of Solids" by Bower [19] gives a thorough yet accessible coverage of numerous topics in solid mechanics including continuum mechanics, constitutive modeling at the crystalline scale, and the finite-element method. The book "Nonlinear Continuum Mechanics for Finite Element Analysis" by Bonet and Woods [18] gives an introduction to tensor analysis, large deformation continuum mechanics, inelastic constitutive modeling, and nonlinear finite-element analysis. The book "Nonlinear Finite Elements for Continua and Structures" by Beyltschko et al. [15] thoroughly covers these topics and contains a chapter on dislocation density-based crystal-plasticity. The book "Crystal Plasticity Finite Element Methods in Materials Science and Engineering" by Roters et al. [127] gives an introduction to continuum mechanics, the finite-element method, homogenization, and presents numerous concepts in crystal-plasticity. The book "Practical Multiscaling" by Fish [50] covers numerous topics in homogenization and multiscale modeling.

3.2 Continuum Approximations

The continuum approximation is a mathematical idealization for modeling the collective response, or state, of discrete systems. The continuum approximation is extremely efficient. The very large number of degrees of freedom required to describe the complete state of a macroscale discrete system, for example, an Avogadro's number, is instead approximated using differential, integral, and functional calculus. Mathematically, the vector space of real numbers (also called the continuum) is used to represent the physical domain of the system. The mathematical continuum is continuous (no gaps) and infinitely divisible. Spatial fields, such as displacement and stress, are then defined on this continuum using familiar notions of a function. Finite differences and finite sums within the physical system are approximated with function derivatives and integrals, respectively. The approximation of a discrete system using differential and integral calculus is, operationally, the dual of approximating a continuous mathematical model using finite differences and finite sums; errors induced in the former are similar to the errors induced in the latter.

The classical continuum theory [90] does not contain an intrinsic length scale. One ramification for solid mechanics is that, for a given boundary-value problem, the stress and strain fields do not change when the physical dimensions change. The classic example from solid mechanics is that of a "hole in a plate." For the case of a homogeneous isotropic linear-elastic continuum, an infinite plate containing a circular hole subjected to far field uniaxial tension has a maximum stress around the hole that is exactly three times larger than the far field stress [146]. In the classical continuum theory, this stress ratio is independent of the size of the hole; a meter-sized hole for a geotechnical application gives the same stress concentration as a nanometer-sized hole in a nanotechnology application. This result is nonphysical. For a physical material, there is always an intrinsic length scale, for example, either the atomic spacing, size of a unit cell, or grain size for a polycrystalline material. When the size of the hole approaches these intrinsic length scales, the assumptions inherent in the continuum formulation are increasingly in error [40, 61, 101]. This error is clearly seen in the wave-propagation/phonon dispersion curves. The classical theory predicts no dispersion, while a discrete system will display a very complex dispersion response including both acoustical and optical branches [26, 27, 31].

Furthermore, classical continuum theory does not predict the existence of a surface effect. Surface effects, including edge and vertex effects, arise in atomic systems fundamentally due to the difference in the coordination number of atoms near the surface versus atoms in the interior and due to long-range atomic forces beyond nearest-neighbor interactions. These in turn lead to differences in the charge distributions, bond lengths, and bond angles near the surface versus the interior [67, 113, 114, 152]. Manifestations of the surface effect include surface tension in liquids as well as surface and interfacial stresses in solids [23, 57–60, 108, 130]. Surface effects can produce exotic physical behavior of both liquids and solids, such as capillarity, adsorption, and adhesion [24, 67]. Surface and interface stresses can also modify the local and far field deformations of nanoscale structures [40, 61, 96, 100, 131, 134]. Homogenization theory of periodic media also predicts the existence of a surface effect due to the difference in material confinement at the surface as compared to the interior [11, 39, 41]. This will be discussed in Sect. 3.3.

Several physically motivated generalizations to classical continuum theory have been proposed that introduce both a physical length scale and surface effects. These include the micromorphic theory of Mindlin [102] and Eringen [45, 48, 55, 140], the nonlocal theories [46, 47, 78] including peridynamics [136–138], and surface-stress formulations of Gurtin [57–60]. The micromorphic theory introduces additional microstructural degrees of freedom within a unit cell that result in a strain-gradient effect in the governing equilibrium equations. The classical theory is reproduced when the size of unit cell is reduced to zero, or in the long wave-length limit. Nonlocal theory has its foundations in the long-range interactions of interatomic forces. These theories posit that the constitutive response at a material point is dependent on the state in a nonlocal region around the point, unlike in classical theories. Strain-gradient theories can be viewed as special cases of the nonlocal formulations [3, 120, 124]. Homogenization theory of periodic media also predicts the existence of strain-gradient effects whenever the unit cell is finite [30, 139, 149, 154]. This will be discussed in Sect. 3.3.

It is important to keep in mind that these generalized continuum theories are still only approximations of the original discrete system. Errors in a continuum approximation to a discrete system are unavoidable whenever the introduced length scales are comparable to the length scale of the discrete system. The generalized theories may be able to reproduced qualitatively correct physical phenomenon, such as a surface effect or optical-branch phonon dispersion curves, but the accuracy of the continuum theory must be judged with respect to the true behavior of the original discrete system. For example, the surface effect may occur only over a few atomic spacings normal to the surface, while the governing equilibrium equations are still in the form of partial-differential equations. Examples are presented in the Appendix in order to demonstrate errors in the continuum approximation.

The classical theory of continuum mechanics is very briefly reviewed in Sect. 3.2.1. The micromorphic theories are briefly presented in Sect. 3.2.2. The surface-stress formulations are briefly presented in Sect. 3.2.3. The nonlocal theories are summarized in Sect. 3.2.4.

3.2.1 Classical Theory

There are many excellent texts on classical continuum mechanics [18, 19, 64, 83, 90]. In this section we give a very brief overview of the standard theory. We use primarily index notation, but also use vector notation for clarity when needed. Thus, x_i represents the three components ($i = 1, 2, 3$) of the vector $\mathbf{x}$. The summation convention of repeated indices within a product or quotient will also be used, for example, $x_i x_i = x_1 x_1 + x_2 x_2 + x_3 x_3$.

Consider the motion of a body $\mathscr{B}$ with interior domain Ω and boundary Γ subjected to a body force b_i per unit volume and applied surface tractions t_i. A Lagrangian description of the motion of $\mathscr{B}$ is used. The current position of a material point is given by x_i, and the original position is given by X_i. The displacement field is given by $u_i = x_i - X_i$. Since the spatial position of a material point is a function of it's original position $x_i(X_j)$, we can define the (material) derivative of the current position with respect to the original position, $\partial x_i / \partial X_j$. This vector derivative is called the deformation gradient, denoted by F_{ij}, and is one of the primary quantities used in continuum solid mechanics to describe the deformation of the body (kinematics). In terms of displacement,

$$F_{ij} = \partial u_i/\partial X_j + \delta_{ij} \; . \tag{3.1}$$

where δ_{ij} is the Kronecker delta with $\delta_{ij} = 1$ if $i = j$ and $\delta_{ij} = 0$ if $i \neq j$. From the deformation gradient, one can define measures of strain, for example, engineering strain, logarithmic strain, and Green strain [18]. For example, the Green strain tensor E_{ij} (also called the Lagrangian strain tensor) is defined as [18]

$$E_{ij} \doteq \frac{1}{2}\left(F_{ki}F_{kj} - \delta_{ij}\right) \; . \tag{3.2}$$

The physical interpretation of E_{ij} may be obtained by noting that it describes how the inner product between two infinitesimal material vectors, $d\mathbf{X}_1$ and $d\mathbf{X}_2$, change under deformation to $d\mathbf{x}_1$ and $d\mathbf{x}_2$, respectively,

$$\frac{1}{2}\left(d\mathbf{x}_1 \cdot d\mathbf{x}_2 - d\mathbf{X}_1 \cdot d\mathbf{X}_2\right) = d\mathbf{X}_1 \cdot \mathbf{E}\, d\mathbf{X}_2 \; . \tag{3.3}$$

For the special case of $d\mathbf{X}_1 = d\mathbf{X}_2 = d\mathbf{X}$, and consequently $d\mathbf{x}_1 = d\mathbf{x}_2 = d\mathbf{x}$, Eq. (3.3) becomes

$$\frac{1}{2}\left(dl^2 - dL^2\right) = d\mathbf{X} \cdot \mathbf{E}\, d\mathbf{X} \; , \tag{3.4}$$

where $(dl)^2 \doteq d\mathbf{x} \cdot d\mathbf{x}$ and $(dL)^2 \doteq d\mathbf{X} \cdot d\mathbf{X}$. Dividing by $(dL)^2$ gives the scalar version (uniaxial) of the Green strain,

$$\frac{dl^2 - dL^2}{2(dL)^2} = \frac{d\mathbf{X}}{dL} \cdot \mathbf{E} \frac{d\mathbf{X}}{dL} = \mathbf{N} \cdot \mathbf{E}\,\mathbf{N} = N_i E_{ij} N_j \; , \tag{3.5}$$

where $\mathbf{N}$ is a unit material vector in the direction of $d\mathbf{X}$. The full inner-product version given in Eq. (3.3) indicates that the strain tensor $\mathbf{E}$ contains both shear and normal strains, presented here in a large deformation formulation.

In order to describe the kinetics of the material, we need to derive both measures of stress and the constitutive relations between stress and strain. In the context of hyperelasticity, one posits a potential energy function W of the strain tensor,

$$W = W(\mathbf{E}) \; . \tag{3.6}$$

Certain restrictions are placed on W to obtain a constitutive model that is independent of the observer (objectivity) and to obtain a stable material. A stress tensor $\mathbf{S}$, called the second Piola–Kirchhoff stress, is defined to be work conjugate to the derivative of W with respect to the strain $\mathbf{E}$,

$$S_{ij} = \frac{\partial W}{\partial E_{ij}} \tag{3.7}$$

so that

$$dW = \mathbf{S} : d\mathbf{E} = S_{ij}\, dE_{ij} \; . \tag{3.8}$$

In many macroscale constitutive models, phenomenological formulations are developed for specific classes of materials such as those for plasticity and viscoplasticity (polycrystalline metals) [74, 83, 87, 111], hyperelasticity and viscoelasticity (polymers) [29], pressure dependent plasticity (porous materials) [5, 19]. These models use various internal-state variables to phenomenologically model physical effects such as dislocation slip, dislocation density, porosity, and damage. Through these constitutive models, the stress measure is related to the entire history of deformation. Constitutive models also exist at the single-crystal level [127]. Section 3.4 gives an overview of continuum crystal-plasticity modeling.

In the special case of small-deformation linear elasticity,

$$S_{ij} = C_{ijkl} E_{kl} \; , \tag{3.9}$$

where C_{ijkl} is the fourth-order elasticity stiffness tensor with $3^4 = 81$ components. For the special case of infinitesimal displacements, the various stress and strain measures are equivalent, and it is common to express the linear relation between stress and strain in Eq. (3.9) using the Cauchy stress σ_{ij}, or true stress (introduced below), and the infinitesimal strain tensor, ϵ_{ij},

$$\sigma_{ij} = C_{ijkl} \epsilon_{kl} \; , \tag{3.10}$$

where

$$\epsilon_{ij} \doteq \frac{1}{2}(\partial u_i / \partial x_j + \partial u_j / \partial x_i) \; . \tag{3.11}$$

In this special case, the potential energy function can be expressed as

$$W = \frac{1}{2} \epsilon_{ij} C_{ijkl} \epsilon_{kl} \; . \tag{3.12}$$

Since W is a scalar, it follows that $C_{ijkl} = C_{klij}$ (major symmetry) thus reducing the number of independent material constants to 36. Further, the symmetry of the tensors σ_{ij} and ϵ_{ij} require that $C_{ijkl} = C_{jikl}$ and $C_{ijkl} = C_{ijlk}$ (minor symmetry), respectively, thus reducing the number of independent elastic constants to 21 for a fully anisotropic crystal. Depending on the crystal symmetry, the number of independent elastic constants is further reduced: triclinic (13), orthotropic (9), ..., cubic (3). For the special case in which the material has no preferred orientation, an isotropic material, the number of independent elastic constants is 2 (e.g., the bulk and shear moduli, or Young's modulus and Poisson's ratio).

Physically, the energy stored in the material or body is related to the deformations of the atomic and molecular bonds. For nanoscale structures, this fact can be used to define W as a function of the history of $\mathbf{E}$ using the Cauchy–Born approximation [44, 135, 141, 142]. Interestingly, this approach can also reproduce surface effects [113, 114]. This will be described in more detail in Sect. 3.2.3.

Using Newton's second law, equilibrium of the body $\mathscr{B}$ may be expressed in terms of the divergence of the Cauchy stress tensor in the spatial frame [18],

$$\partial\sigma_{ij}/\partial x_j + b_i = \rho\ddot{u}_i\ , \tag{3.13}$$

where $\ddot{u}_i$ denotes the second derivative of u_i with respect to time, and ρ is the mass density. This set of equilibrium equations can also be cast in terms of other stress tensors, the choice of which is chosen for convenience, using relations such as $\sigma_{ij} = J^{-1}F_{ik}S_{kl}F_{jl}$, where $J = \det(F_{ij})$ [18].

The surface tractions, both normal and shear, are related to the Cauchy stress tensor by the relation

$$t_i = \sigma_{ij}n_j\ , \tag{3.14}$$

where n_j is the unit vector normal to the surface. This relation may be derived using Cauchy's tetrahedron [18], for example.

The field equations of classical continuum mechanics can be solved using a variety of numerical methods including the finite-element method [15, 18] and the boundary-element method [14]. Numerous commercial and research-based nonlinear finite-element solvers are available including Comsol [34], Abaqus [1], and Ansys [6].

3.2.2 Micromorphic Theories

The *micromorphic theories* of Eringen [45, 48, 55, 140] and Mindlin [102] introduce additional microstructural degrees of freedom within a unit cell or "micro-volume." The continuum is now thought of as a continuous collection of deformable cells or particles. *The size of this unit cell introduces a length scale into the continuum theory.* Each cell is free to deform and create a "micro-strain" in addition to the classical macro-strain. The effects of these additional microstructural degrees of freedom include strain-gradient effects, higher-order stresses, surface and size effects, and optical branches of the phonon dispersion curves [26, 27, 56, 102]. Malvern [90] integrates these generalized continuum formulations within his book on continuum mechanics.

Chen et al. [25, 28] give an overview of the hierarchy of generalized continua starting at the most general case with the micromorphic theory of Eringen [45, 48, 140]. With the assumption of infinitesimal linear-elastic deformations, this theory reduces to Mindlin's *microstructure theory* [102]. With the additional assumption that the unit cells are rigid, and thus ignoring internal motion within a cell, the theory reduces to the *micropolar theory* [48]. With the additional assumption that the orientations of the unit cells are fixed, the theory reduces to Cosserat theory [35].

With the additional assumption that the micro-motion is equal to the macro-motion, the theory reduces to the *couple-stress theory* [101, 105, 147, 148]. The classical continuum theory is reproduced when the size of unit cell is reduced to zero or in the long wave-length limit. In general, these formulations can be categorized as "strain-gradient" continuum theories [51, 103, 104, 119]. Various researchers have used these formulations to regularize softening and localization phenomena [92, 93, 150].

In Mindlin's microstructure theory [102], additional deformation modes, called micro-displacements and micro-deformations, are introduced as shown in Fig. 3.1. The macroscopic displacement field is given by $u_i = x_i - X_i$ as usual. A micro-displacement u'_i is defined as $u'_i = x'_i - X'_i$ where X'_i and x'_i are the reference and current positions of a material point within the micro-volume, referred to axes parallel to those of X_i and whose origin moves with displacement u_i (see Fig. 3.1). With the assumption of infinitesimal displacements for both the macro- and micro-displacements, $x_i \approx X_i$ and $x'_i \approx X'_i$. The displacement gradient within the micro-medium is given by

$$\psi_{ij} \doteq \partial u'_j / \partial x'_i \ , \tag{3.15}$$

where ψ_{ij} is called the *micro-deformation*. In Mindlin's theory, ψ_{ij} is taken to be constant within the micro-volume so that

$$u'_j = x'_i \psi_{ij} \ , \tag{3.16}$$

but ψ_{ij} is allowed to vary with position in the macroscale so that $\psi_{ij} = \psi_{ij}(\mathbf{x})$. The macro-gradient of the micro-deformation χ_{ijk} is given by

$$\chi_{ijk} \doteq \partial \psi_{jk} / \partial x_i \ . \tag{3.17}$$

The *relative deformation* γ_{ij} is defined as

$$\gamma_{ij} \doteq \partial u_j / \partial x_i - \partial u'_i / \partial x'_j = \partial u_j / \partial x_i - \psi_{ij} \ . \tag{3.18}$$

The usual infinitesimal strain tensor ϵ_{ij} is defined in Eq. (3.11). Note that while ϵ_{ij} is symmetric, γ_{ij} and χ_{ijk} are not in general symmetric or minor-symmetric, respectively.

The potential energy function per unit macro-volume is posited to be of the form

$$W = W(\epsilon_{ij}, \gamma_{ij}, \chi_{ijk}) \ . \tag{3.19}$$

The following stress tensors[1] are defined to be work conjugate to ϵ_{ij}, γ_{ij}, and χ_{ijk} [102],

[1]Note that the use of the symbols σ and τ is reversed from that of Mindlin [102] to be consistent with our use of σ for the Cauchy stress given in Sect. 3.2.1.

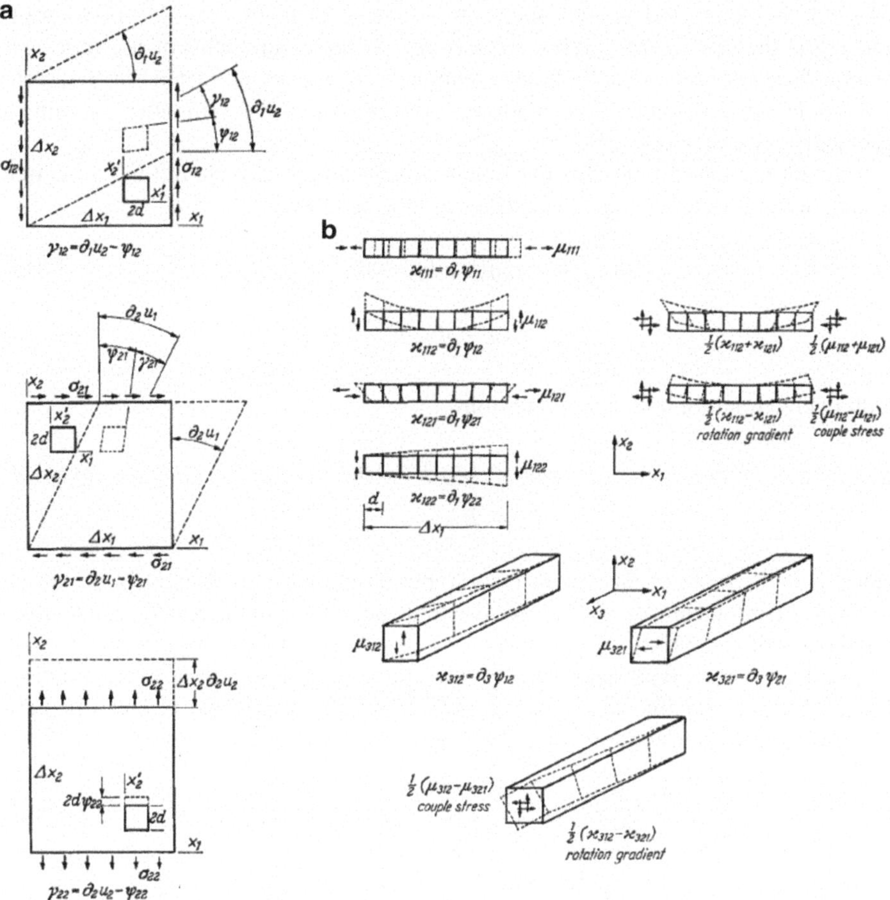

Fig. 3.1 Figures reproduced from Mindlin's original 1964 publication on his *microstructure* (micromorphic) theory [102]. (**a**) Example components of relative stress τ_{ij} (σ_{ij} in Mindlin's notation), displacement gradient $\partial u_i/\partial x_j$, micro-deformation ψ_{ij}, and relative deformation γ_{ij}. (**b**) Example components of the double stress μ_{ijk}, and the gradient of micro-deformation, χ_{ijk} (Reproduced with permission from [102])

$$\sigma_{ij} = \frac{\partial W}{\partial \epsilon_{ij}} \tag{3.20a}$$

$$\tau_{ij} = \frac{\partial W}{\partial \gamma_{ij}} \tag{3.20b}$$

$$\mu_{ijk} = \frac{\partial W}{\partial \chi_{ijk}} , \tag{3.20c}$$

where σ_{ij} is interpreted as the Cauchy stress tensor, τ_{ij} is the *relative stress* tensor, and μ_{ijk} is the *double stress* tensor. The twenty-seven components of μ_{ijk} represent *double-forces* per unit area. Example components of μ_{ijk} are shown in Fig. 3.1. Note that while σ_{ij} is symmetric, τ_{ij} and μ_{ijk} are not in general symmetric or minor-symmetric, respectively.

Developing expressions for the kinetic energy and using Hamilton's principle results in the following twelve equations of motion [102]:

$$\partial(\sigma_{ij} + \tau_{ij})/\partial x_i + b_j = \rho \ddot{u}_j \tag{3.21a}$$

$$\mu_{ijk},_i + \tau_{jk} + \Phi_{jk} = \frac{1}{3}\rho' d_{lj}^2 \ddot{\psi}_{lk} , \tag{3.21b}$$

with twelve traction boundary conditions,

$$t_j = (\sigma_{ij} + \tau_{ij})\, n_i \tag{3.22a}$$

$$T_{jk} = \mu_{ijk}\, n_i . \tag{3.22b}$$

Here, ρ' is the mass of the micro-material per unit macro-volume, d_{lk} is a unit-cell moment-of-inertia tensor, Φ_{jk} is a double-force per unit volume, and T_{jk} is a double-force per unit area [102].

Similar to the expression for W for the classical continuum formulation, Eq. (3.12), the potential energy of a microstructural continuum is taken to be a quadratic function of the forty-two variables ϵ_{ij}, γ_{ij}, and χ_{ijk},

$$W = \frac{1}{2}\epsilon_{ij}C_{ijkl}\epsilon_{kl} + \frac{1}{2}\gamma_{ij}B_{ijkl}\gamma_{kl} + \frac{1}{2}\chi_{ijk}A_{ijklmn}\chi_{lmn} \tag{3.23a}$$

$$+ \gamma_{ij}D_{ijklm}X_{klm} + \chi_{ijk}F_{ijklm}\epsilon_{lm} + \gamma_{ij}G_{ijkl}\epsilon_{kl} . \tag{3.23b}$$

The constitutive equations are then,

$$\sigma_{ij} = C_{ijpq}\epsilon_{pq} + G_{pqij}\gamma_{pq} + F_{pqrij}\chi_{pqr} \tag{3.24a}$$

$$\tau_{ij} = G_{ijpq}\epsilon_{pq} + B_{pqij}\gamma_{pq} + D_{ijpqr}\chi_{pqr} \tag{3.24b}$$

$$\mu_{ijk} = F_{ijkpq}\epsilon_{pq} + D_{pqijk}\gamma_{pq} + A_{ijkpqr}\chi_{pqr} . \tag{3.24c}$$

As noted by Mindlin [102], only 903 of these 1764 coefficients are independent due to the symmetry of ϵ_{ij} and the scalar property of W. Still, this number is vastly larger than 21 independent elastic constants for a fully anisotropic classical linear-elastic continuum described in Sect. 3.2.1. Even for an idealized isotropic microstructural medium, there are 18 independent coefficients [102].

A general three-dimensional finite-element implementation of the micromorphic continuum theory is ongoing within the Tahoe Development Project [143].

3.2.3 Surface Stress

For nanoscale structures, surface-to-volume ratios are relatively large, and surface effects can become significant [23, 40, 59, 134, 152]. Surface effects, including edge and vertex effects, arise in atomic systems fundamentally due to the difference in the coordination number of atoms near the surface versus atoms in the interior and due to long-range atomic forces beyond nearest-neighbor interactions. These in turn lead to differences in the charge distributions, bond lengths, and bond angles near the surface versus the interior [67].

Gurtin et al. [57–59] have developed a general continuum theory for elastic material surfaces and material interfaces. For solids, the concepts of surface tension, surface stress, and surface energy are distinctly different and cannot be used interchangeably. Only for liquids are all three the same [67]. (Chap. 17 of this book discusses surface-tension effects in thin liquid films.) For the special case of small deformations, these three quantities are related via the equation[2] [67],

$$\boldsymbol{\sigma}^{\mathrm{s}} = \sigma\,\mathbf{I}^{\mathrm{s}} + \frac{\partial \Gamma}{\partial \boldsymbol{\epsilon}^{\mathrm{s}}}\,, \tag{3.25}$$

where $\boldsymbol{\sigma}^{\mathrm{s}}$ is the surface-stress tensor, σ is the surface tension, $\mathbf{I}^{\mathrm{s}}$ is the identity tensor for surfaces, $\Gamma(\boldsymbol{\epsilon}^{\mathrm{s}})$ is the deformation-dependent surface energy, and $\boldsymbol{\epsilon}^{\mathrm{s}}$ is surface-strain tensor. Each tensor in this equation is defined in the surface manifold and has 2×2 components. For the idealized case of an isotropic material (surface and bulk) the following equilibrium and constitutive equations hold. For the bulk material, the equilibrium equations are the same as those given in Sect. 3.2.1, Eq. (3.13), but with the constitutive equations, Eq. (3.10), specialized for an isotropic material,

$$\operatorname{div}\boldsymbol{\sigma}^{\mathrm{b}} = 0 \tag{3.26a}$$

$$\boldsymbol{\sigma}^{\mathrm{b}} = 2\mu\boldsymbol{\epsilon}^{\mathrm{b}} + \lambda\,\mathrm{Tr}(\boldsymbol{\epsilon}^{\mathrm{b}})\,\mathbf{I}^{\mathrm{b}}\,, \tag{3.26b}$$

where $\boldsymbol{\sigma}^{\mathrm{b}}$ is the Cauchy stress in the bulk material, $\boldsymbol{\epsilon}^{\mathrm{b}}$ is the infinitesimal strain tensor in the bulk, $\mathrm{div}(\cdot)$ is the bulk divergence operator, λ and μ are the Lame' constants, $\mathrm{Tr}(\cdot)$ is the trace operator, and $\mathbf{I}^{\mathrm{b}}$ is the three-dimensional identity tensor. Equilibrium of the surface of the material, or interface between two materials, is given by [59, 134],

$$0 = [\boldsymbol{\sigma}^{\mathrm{b}} \cdot \mathbf{n}] + \operatorname{div}_{\mathrm{s}}\boldsymbol{\sigma}^{\mathrm{s}} \tag{3.27a}$$

$$\boldsymbol{\sigma}^{\mathrm{s}} = \sigma\,\mathbf{I}^{\mathrm{s}} + 2(\mu^{\mathrm{s}} - \sigma)\boldsymbol{\epsilon}^{\mathrm{s}} + (\lambda^{\mathrm{s}} + \sigma)\,\mathrm{Tr}(\boldsymbol{\epsilon}^{\mathrm{s}})\,\mathbf{I}^{\mathrm{s}}\,, \tag{3.27b}$$

[2]In this section, we use vector notation to simplify the representation of surface tensors.

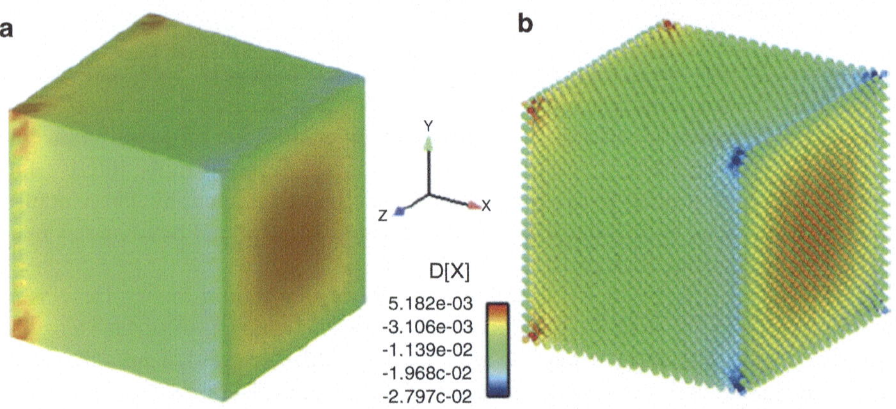

Fig. 3.2 The x-component of the displacement vector after the relaxation of a cube as calculated by (**a**) surface Cauchy–Born method and (**b**) molecular statics simulation (Reproduced with permission from [114])

where $\mathrm{div}_S(\cdot)$ is the surface divergence operator, λ^s and μ^s are the surface Lamé constants, and the square brackets in Eq. (3.27a) represent the jump in the quantity across the interface.

Determining the surface material constants in this continuum formulation is nontrivial [59, 67, 100] and somewhat ill-defined. (For example, how thick is the surface or interface?) At the nanoscale, when the surface effects are most prominent, the surface material constants are expected to depend on the orientation of the surface with respect to the underlying crystal lattice. Determining the material constants for all possible orientations of a surface is a daunting prospect. For these reasons, computational approaches based on the Cauchy–Born approximation have been pursued to model surface effects [113, 114]. In this approach, the surface and bulk strain energies are calculated directly using interatomic potentials, and the constitutive response is obtained using Eq. (3.7). The continuum displacement field is obtained by minimizing the total potential energy. An example from Park et al. [114] is shown in Fig. 3.2. This approach has the further advantage of incorporating edge and vertex effects that depend on the included angle of the intersecting surfaces and edges, respectively. This "quasi-continuum" approach to modeling nanoscale structures is described in detail in Chap. 5 of this book.

3.2.4 Nonlocal Theories

Nonlocal continuum theories were introduced by Kröner [78] in order to account for the long-range interactions present in atomic systems. These theories were later studied and further developed by Eringen [46, 47]. The relations between nonlocal theories and strain-gradient theories have been studied by several authors [2, 3, 120, 123, 124]. Nonlocal theories have been used to model nanoscale structures [7, 118, 151], and have been applied to macroscale systems in order to regularize strain-softening behavior [69, 120]. In Eringen's nonlocal theory, the nonlocal stress tensor $\boldsymbol{\sigma}(\mathbf{x})$ is related to the local (fictitious) stress tensor $\mathbf{s}(\mathbf{x})$ through an integral relation of the form

$$\boldsymbol{\sigma}(\mathbf{x}) = \int_V A(\mathbf{x}, \mathbf{x}')\mathbf{s}(\mathbf{x}')\, dV' \,, \tag{3.28}$$

where $A(\mathbf{x}, \mathbf{x}')$ is a scalar valued *attenuation function* or influence function, and $\mathbf{s}(\mathbf{x})$ is related to the local strain via the classical relation given by Eq. (3.10), $\mathbf{s}(\mathbf{x}) = \mathbb{C} : \boldsymbol{\epsilon}(\mathbf{x})$. The attenuation function is typically defined to be nonzero only within a certain radius R of the given point $\mathbf{x}$ so that $A(\mathbf{x}, \mathbf{x}') = 0$ when $||\mathbf{x}' - \mathbf{x}|| > R$. The attenuation function must be modified near the boundary of the domain to ensure certain consistency requirements in representing constant fields. This boundary effect in nonlocal models may be physically desired depending on the physical system being modeled.

An alternate nonlocal continuum formulation, called *peridynamics*, has been proposed by Silling [136–138]. In this formulation, the concept of strain is avoided all together in favor of generalized *bonds* connecting two disconnected points within a domain.

$$\int_V \mathbf{f}(\mathbf{u}' - \mathbf{u}, \mathbf{x}' - \mathbf{x})\, dV' + \mathbf{b} = \rho \ddot{\mathbf{u}} \,, \tag{3.29}$$

where $\mathbf{u}$ is the displacement vector, ρ is the mass density, $\mathbf{b}$ is the body force per unit volume, and $\mathbf{f}$ represents the *force density* per unit volume. This *bond-based* formulation has been generalized to the so-called *state-based* formulation [138] that allows for very general material behavior. Applications of peridynamics to the upscaling of molecular dynamics have been studied by Seleson et al. [125] with recent applications to nanomechanics [17, 42]. The peridynamic theory now has been extensively developed [49, 137]. Peridynamics has been implemented in the open-source massively parallel software LAMMPS [80, 115, 116].

3.3 Homogenization Theory

Homogenization is the mathematical process of replacing a heterogeneous material with a fictitious homogeneous material whose macroscopic response, in an energetic sense, is equal to that of the true heterogeneous material. The material properties of the fictitious homogeneous material are called the *effective* properties [65]. Homogenization is fundamental to continuum mechanics, since all materials are heterogeneous or discrete at some length scale and require homogenization in order to apply continuum principles at a higher length scale. There are several texts that cover homogenization theory for various physics and to various levels of mathematical rigor [16, 30, 50, 68, 95, 110, 117, 133]. The theory is well developed for materials with periodic microstructure using perturbation theory and the mathematical concepts of strong and weak convergence. The theory has been extended to random microstructures by Papanicolaou and Varadhan [112] using probabilistic definitions of convergence. Section 3.3.1 gives an overview of the two-scale asymptotic homogenization process for linear-elastic periodic microstructures. This mathematical approach is well developed, and leads to the concepts of higher-order stresses or "hyperstresses." Section 3.3.2 discusses the mathematical concepts of strong and weak convergence. Section 3.3.3 presents a three-dimensional homogenization example. The homogenization results of this example are compared to direct numerical simulations (DNS) of the heterogeneous microstructure. Section 3.3.4 discusses the concept of computational homogenization. Additionally, Sect. 3.3.4 discusses the fast Fourier transform (FFT) method for efficiently solving the unit-cell problem, in both the linear and nonlinear regimes.

Mean-field homogenization techniques such as the self-consistent method [106] do not attempt to directly model the exact field of a heterogeneous microstructure, but instead focus on modeling the response of a single inclusion within an approximately homogeneous continuum. Mean-field methods are not discussed here, but are covered in detail by Nemat-Nasser and Hori [110] along with other topics in the theory of composite materials.

3.3.1 Method of Two-Scale Asymptotic Expansion

Tran et al. [149] give a succinct yet detailed presentation of the two-scale asymptotic homogenization process for linear-elastic periodic microstructures. Their notation and presentation is followed here. Let L represent the length scale of the macrostructure. Let l represent the length scale of the unit cell. The ratio $\epsilon \doteq l/L$ of the two length scales is assumed to be less than 1, but not necessarily infinitesimally small. The existence of the small parameter ϵ suggests the use of perturbation theory [63]. A new "fast" oscillating variable is introduced as $\mathbf{y} \doteq \mathbf{x}/l$ which gives the position within an individual cell. The displacement field is now considered as a function of the two variables $\mathbf{x}$ and $\mathbf{y}$ so that $\mathbf{u} = \mathbf{u}(\mathbf{x}, \mathbf{y})$. The displacement field is expanded in a power series in ϵ,

$$\mathbf{u}(\mathbf{x}, \mathbf{y}) = \mathbf{u}^0(\mathbf{x}, \mathbf{y}) + \epsilon \mathbf{u}^1(\mathbf{x}, \mathbf{y}) + \epsilon^2 \mathbf{u}^2(\mathbf{x}, \mathbf{y}) + \cdots , \tag{3.30}$$

where the fields $\mathbf{u}^i(\mathbf{x}, \mathbf{y})$, $i = 1, 2, 3, \ldots$ are periodic with period l. This expansion is then substituted into Eqs. (3.10) and (3.13). By equating terms with the same power of ϵ, a hierarchy of cell problems for the quantities $\mathbf{u}^0(\mathbf{x}, \mathbf{y})$, $\mathbf{u}^1(\mathbf{x}, \mathbf{y})$, $\mathbf{u}^2(\mathbf{x}, \mathbf{y})$, etc., is obtained. It transpires that $\mathbf{u}^0(\mathbf{x}, \mathbf{y})$ is only a function of $\mathbf{x}$. With

$\mathbf{u}^0(\mathbf{x}, \mathbf{y}) \doteq \mathbf{U}(\mathbf{x})$, it can be shown that $\mathbf{U}(\mathbf{x}) = \frac{1}{V}\int_V \mathbf{u}(\mathbf{x}, \mathbf{y})\, dV$ where V is the volume of a unit cell. $\mathbf{U}(\mathbf{x})$ is then interpreted as the displacement of the centroid of the unit cell [149].

The total displacement field $\mathbf{u}(\mathbf{x}, \mathbf{y})$ can now be written as

$$\mathbf{u}(\mathbf{x}, \mathbf{y}) = \mathbf{U}(\mathbf{x}) + \epsilon \boldsymbol{\chi}^1(\mathbf{y}) : \mathbf{E}(\mathbf{x}) + \epsilon^2 \boldsymbol{\chi}^2(\mathbf{y}) : \cdot\, \mathbf{G}(\mathbf{x}) + \cdots \,, \tag{3.31}$$

where $\boldsymbol{\chi}^i(\mathbf{y})\,, i = 1, 2, 3, \ldots$ are called localization tensors and are obtained through the solution of the hierarchy of cell problems [149]. Here, $\mathbf{E}(\mathbf{x})$ is interpreted as the macroscopic strain tensor with

$$\mathbf{E}(\mathbf{x}) = \int_V \epsilon(\mathbf{x}, \mathbf{y})\, dV \,, \tag{3.32}$$

and $\mathbf{G}(\mathbf{x})$ is interpreted as the non-dimensional gradient of macroscopic strain, $\mathbf{G}(\mathbf{x}) = L\,\nabla\mathbf{E}(\mathbf{x})$. The additional terms in Eq. (3.31) consist of products of higher-order strain-gradients and localization tensors.

Tran et al. [149] were able to derive a generalization of the classical Hill–Mandel lemma,

$$\frac{1}{V}\int_V \boldsymbol{\sigma}(\mathbf{x}, \mathbf{y}) : \boldsymbol{\epsilon}(\mathbf{x}, \mathbf{y})\, dV = \Sigma(\mathbf{x}) : \mathbf{E}(\mathbf{x}) + \mathbf{T}(\mathbf{x}) : \cdot\, \nabla\mathbf{E}(\mathbf{x}) + \cdots \,, \tag{3.33}$$

where $\Sigma(\mathbf{x})$ is interpreted as the macroscopic stress, and $\mathbf{T}(\mathbf{x})$ is the "first hyper-stress." The additional terms in Eq. (3.33) consist of products of hyperstresses and higher-order strain gradients. It can be shown that

$$\frac{1}{V}\int_V \sigma_{ij}(\mathbf{x}, \mathbf{y})\, dV = \Sigma_{ij}(\mathbf{x}) - T_{ijk,k}(\mathbf{x}) + \cdots \,. \tag{3.34}$$

The macroscopic balance equation is given by [149]

$$\Sigma_{ij,j}(\mathbf{x}) - T_{ijk,jk}(\mathbf{x}) + \cdots + B_i = 0, \tag{3.35}$$

where $B_i = \frac{1}{V}\int_V b_i(\mathbf{x}, \mathbf{y})\, dV$ is the cell-averaged body force.

In the limit as $\epsilon \to 0$ or negligible strain gradients, $\nabla\mathbf{E}(\mathbf{x}) \to 0$, the hyperstresses are zero, and the classical homogenization results are recovered,

$$\frac{1}{V}\int_V \boldsymbol{\sigma}(\mathbf{x}, \mathbf{y}) : \boldsymbol{\epsilon}(\mathbf{x}, \mathbf{y})\, dV = \Sigma(\mathbf{x}) : \mathbf{E}(\mathbf{x}), \tag{3.36a}$$

$$\frac{1}{V}\int_V \sigma_{ij}(\mathbf{x}, \mathbf{y})\, dV = \Sigma_{ij}(\mathbf{x}), \tag{3.36b}$$

$$\Sigma_{ij,j}(\mathbf{x}) + B_i = 0 \,. \tag{3.36c}$$

The homogenized material properties relating the macroscopic stress Σ_{ij} to macroscopic strain E_{ij}, as well as relating the hyperstresses to the strain gradients (e.g., T_{ijk} to $E_{ij,k}$), are obtained from the unit-cell problems described previously. These are described by Tran et al. [149]. For first-order homogenization theory ($\epsilon \to 0$), the cell problem is given by

$$\sigma_{ij,j}(\mathbf{y}) = 0, \tag{3.37a}$$

$$\sigma_{ij}(\mathbf{y}) = C_{ijkl}(\mathbf{y})\epsilon_{kl}(\mathbf{y}) + p_{ij}(\mathbf{y}), \tag{3.37b}$$

$$\epsilon_{kl}(\mathbf{y}) = \frac{1}{2}\left(\partial u_k(\mathbf{y})/\partial y_l + \partial u_l(\mathbf{y})/\partial y_k\right), \tag{3.37c}$$

$$u_i(\mathbf{y}) \text{ periodic}, \quad \frac{1}{V}\int_V u_i(\mathbf{y})\, dV = 0. \tag{3.37d}$$

where $p_{ij}(\mathbf{y})$ is a polarization tensor given by $p_{ij}(\mathbf{y}) = C_{ijkl}(\mathbf{y})E_{kl}$, and E_{kl} is constant over the unit cell. By applying unit values of E_{kl} for each of the components, the homogenized stiffness tensor can be obtained. More complex polarization tensors and body forces arise for the higher-order cell problems. Because of the periodic boundary conditions, an efficient solution of this unit-cell problem can be obtained through the use of the FFT as described in Sect. 3.3.4. These results can be extended to the nonlinear regime using "computational homogenization" as described in Sect. 3.3.4.

As with atomic systems, the confinement of the unit cell at the surface of a body is different than a unit cell that is positioned within the bulk. In particular, at the surface the unit cell does not experience periodic boundary conditions. Thus, a surface effect will result when comparing unit-cell averages obtained from homogenization and unit-cell averages obtained from direct numerical calculations of the full heterogeneous material. This will be demonstrated in the example presented in Sect. 3.3.3.

3.3.2 Convergence: Strong and Weak

One of the main mathematical results of homogenization theory is that in the limit as $\epsilon \to 0$ the displacement solution of the macroscopic governing field equation containing the heterogeneous material converges *strongly* to the displacement solution of the macroscopic field equation containing the homogenized material [30]. Furthermore, in the limit as $\epsilon \to 0$ the strain (stress) field of the macroscopic field equation containing the heterogeneous material converges *weakly* to the strain (stress) field of the macroscopic field equation containing the homogenized material [30].

Recall that a sequence of functions (u_n), $n = 1, 2, 3, \ldots$, with $u_n \in L^2$, converges *strongly* to $u \in L^2$ if

$$\lim_{n\to\infty} \|u_n - u\| = 0\,, \tag{3.38}$$

and converges *weakly* if

$$\lim_{n\to\infty} \langle u_n, v\rangle = \langle u, v\rangle \quad \forall v \in L^2\,. \tag{3.39}$$

Here L^2 represents the space of square-integrable functions, $\langle u, v\rangle$ represents the L^2 inner product, and $\|u\| = \sqrt{\langle u, u\rangle}$ is the norm induced by the inner product.

For an example illustrating strong and weak convergence, consider a one-dimensional continuous bar of length L with variable elastic modulus, $E(x)$, given by

$$E(x) = E_0(1 + a\sin(2n\pi x/L)) \quad x \in [0, L] \quad n = 1, 2, 3, \tag{3.40}$$

where $a \in [0, 1)$ is a parameter that governs the degree of inhomogeneity. For this problem, the spatial period or unit cell is $l = L/n$ (*cf.* example problems given in the Appendix). For a one-dimensional bar, the governing equations given in Sect. 3.2.1 reduce to the following ordinary differential equation:

$$\frac{d}{dx}\left(E(x)\frac{du}{dx}\right) = 0, \tag{3.41}$$

where $u = u(x)$ is the uniaxial displacement along the bar. This equation may be integrated analytically using displacement boundary conditions $u(0) = 0$ and $u(L) = u_0$ to give the exact displacement field,

$$u(x) = \frac{u_0}{n\pi}\left[\pi\,\lfloor nx/L + 1/2\rfloor + \tan^{-1}\left(\frac{\tan\left(\frac{n\pi x}{L}\right) + a}{\sqrt{1-a^2}}\right) - \tan^{-1}\left(\frac{a}{\sqrt{1-a^2}}\right)\right] \tag{3.42}$$

with derivative (longitudinal strain) given by

$$\frac{du}{dx} = \frac{u_0}{L}\frac{\sqrt{1-a^2}}{1 + a\sin(2n\pi x/L)}\,. \tag{3.43}$$

In Eq. (3.42), the function $\lfloor\cdot\rfloor$ is the floor function, and the inverse tangent functions are understood to give principal values. These functions are plotted in Fig. 3.3 for $a = 0.9$ and $n = 5, 10, 20$. The displacement field of the homogenized beam, $u^0(x)$, is given simply by

$$u^0(x) = u_0\frac{x}{L}\,, \tag{3.44}$$

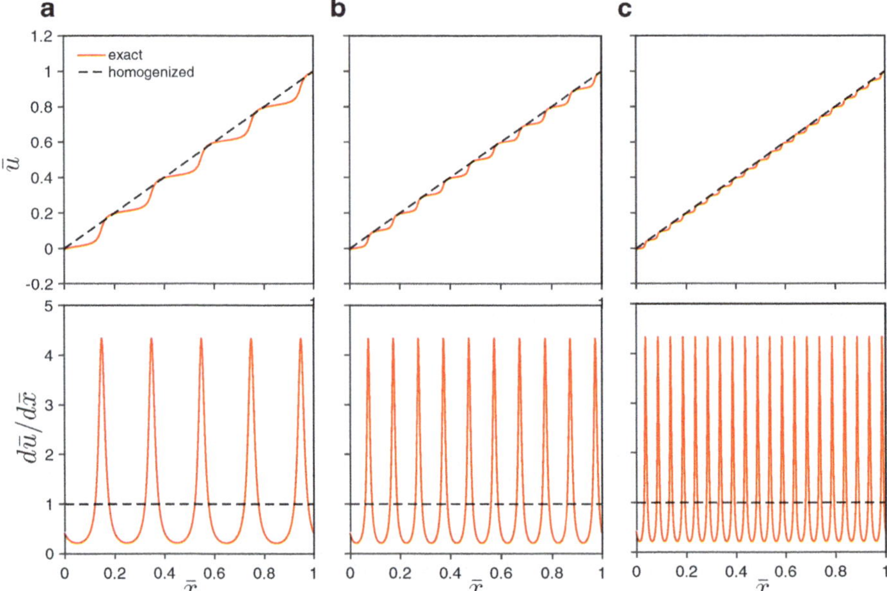

Fig. 3.3 Example of both strong (*top row*) and weak convergence (*bottom row*) using a one-dimensional continuous bar of length L with periodic elastic modulus given by Eq. (3.40). The exact displacement field is given by Eq. (3.42), and is shown here for the cases (**a**) $n = 5$, (**b**) $n = 10$, (**c**) $n = 20$ (*top row*), with $a = 0.9$ and $\bar{u} \doteq u/u_0, \bar{x} \doteq x/L$. This displacement converges *strongly* to the displacement field $\bar{u}(x) = \bar{x}$ as $n \to \infty$. However, the exact displacement derivative (strain), given by Eq. (3.43), converges only *weakly* to the derivative of the homogenized solution $d\bar{u}/d\bar{x} = 1$ as $n \to \infty$ (*bottom row*)

with constant derivative

$$\frac{du^0}{dx} = \frac{u_0}{L} . \tag{3.45}$$

These functions are also plotted in Fig. 3.3. It can be proven that $u(x)$ converges *strongly* to u^0 as $n \to \infty$, which is clear via inspection of Fig. 3.3. What is less clear, but still true, is that du/dx converges *weakly* to du^0/dx as $n \to \infty$.

It is instructive to verify the equivalency of the total energy stored using the two descriptions of the bar, direct and homogenized. For this one-dimensional example, the energy density given by Eq. (3.12) reduces to

$$W = \frac{1}{2}E(x)\left(\frac{du}{dx}\right)^2 , \tag{3.46}$$

where $E(x)$ is given by Eq. (3.40) and du/dx is given by Eq. (3.43). The total stored energy in the bar W_{total} is then

$$W_{\text{total}} = \int_0^L W\,dx = \frac{1}{2}E_0(1-a^2)\left(\frac{u_0}{L}\right)^2 \int_0^L \frac{1}{1+a\sin(2n\pi x/L)}dx$$
$$= \frac{1}{2}E_0\sqrt{1-a^2}\left(\frac{u_0}{L}\right)^2 L\,. \tag{3.47a}$$

For the homogenized bar, the effective modulus E^0 is simply the harmonic mean of the modulus within the unit cell[3] [30],

$$\frac{1}{E^0} = \frac{1}{l}\int_0^l \frac{1}{E(x)}dx = \frac{1}{l}\int_0^l \frac{1}{E_0(1+a\sin(2\pi x/l))}dx$$
$$= \frac{1}{E_0}\frac{1}{2\pi}\int_0^{2\pi}\frac{1}{1+a\sin(\xi)}d\xi = \frac{1}{E_0}\frac{1}{\sqrt{1-a^2}}\,. \tag{3.48a}$$

Thus,

$$E^0 = E_0\sqrt{1-a^2}\,. \tag{3.49}$$

The energy density of the homogenized bar W^0 is given by

$$W^0 = \frac{1}{2}E^0\left(\frac{du^0}{dx}\right)^2, \tag{3.50}$$

where E^0 is given by Eq. (3.49) and du^0/dx is given by Eq. (3.45). The total stored energy in the homogenized bar W^0_{total} is then

$$W^0_{\text{total}} = \int_0^L W^0\,dx = W^0 L = \frac{1}{2}E_0\sqrt{1-a^2}\left(\frac{u_0}{L}\right)^2 L\,, \tag{3.51}$$

which is identical to W_{total} given by Eq. (3.47a), as required.

3.3.3 *Homogenization Example*

For a homogenization example, consider the unit cell shown in Fig. 3.4a. This unit cell is a cube partitioned into nine subvolumes. In the center of the unit cell is a truncated octahedron. The remaining eight parts are the corners of the cube not contained in the truncated octahedron. We take each subvolume of the cube to be a crystal of stainless steel 304L (γ-Fe). The FCC crystal structure of γ-Fe (austenite) possesses cubic elastic symmetry with a relatively large anisotropy ratio A, with

[3]This simple relation does not hold in higher-dimensional problems.

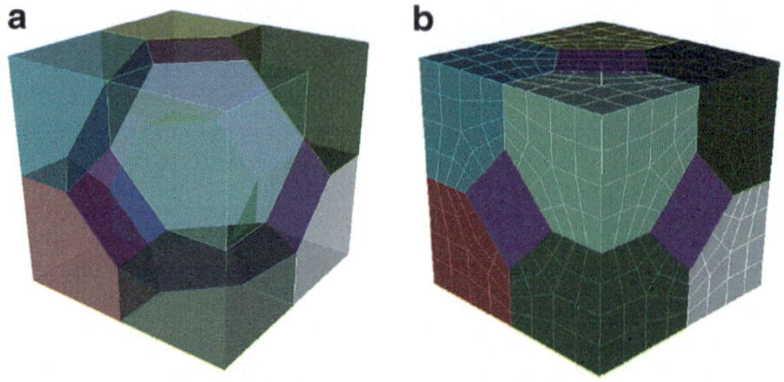

Fig. 3.4 (**a**) Example unit cell consisting of nine subvolumes. The center subvolume is a truncated octahedron. (**b**) Conformal hexahedral finite-element mesh of the unit cell

Table 3.1 Cubic elasticity constants for 304L stainless steel (γ-Fe) [81] (units are GPa)

Material	C_{11}	C_{12}	C_{44}	A
304L	204.6	137.7	126.2	3.8
α-Fe	231.4	134.7	116.4	2.4
Al	107.3	60.9	28.3	1.2
Cu	168.4	121.4	75.4	3.2

The anisotropy ratio $A = 2C_{44}/(C_{11} - C_{12})$ is also given. For an isotropic material, $A = 1$. Several other cubic metals are given for comparison [19]

$A \doteq 2C_{44}/(C_{11} - C_{12}) = 3.8$, where C_{11}, C_{12}, and C_{44} are the cubic elastic constants. For an isotropic material, $A = 1$. The values of these constants are given in Table 3.1 for austenite [81] along with several other cubic materials for comparison [19]. The relatively large anisotropy ratio of austenite should produce more pronounced surface and strain-gradient effects within a macroscale structure containing finite microstructure.

The orientation of each crystal (subvolume) within the unit cell is chosen such that the homogenized (first-order) material properties of the unit cell are orthotropic. For the truncated octahedron subvolume, the crystal aligns with the axes of the unit cell. For each corner subvolume, one axis of the crystal is in a $\langle 111\rangle$ direction. The second axis is in a $\langle 110\rangle$ direction.

The homogenized elastic constants (first-order) are obtained by solving the boundary-value problem defined in Eq. (3.37). The polyhedral algorithm within the Cubit meshing tool [36] was used to create a hexahedral mesh of each subvolume as shown in Fig. 3.4b. The finite-element mesh of the unit cell contained 832 hexahedral elements. Results were insensitive to further mesh refinement. The homogenized material properties are $E_1 = E_2 = 1.15737 \times 10^5$, $E_3 = 0.980940 \times 10^5$, $\mu_{12} = 5.01817 \times 10^4$, $\mu_{13} = \mu_{23} = 5.89234 \times 10^4$, $\nu_{21} = 0.26924$, $\nu_{31} = \nu_{32} = 0.38072$, where E_i, μ_i, ν_{ij}, $i, j = 1, 2, 3$ are the orthotropic Young's moduli, shear moduli, and Poisson's ratios, respectively. Note there is some additional symmetry present, since there are only six independent elastic constants whereas a fully orthotropic material has nine independent elastic constants. Additionally, recall that for an orthotropic material, $\nu_{12} = \frac{E_1}{E_2}\nu_{21}$, $\nu_{13} = \frac{E_1}{E_3}\nu_{31}$, and $\nu_{23} = \frac{E_2}{E_3}\nu_{32}$.

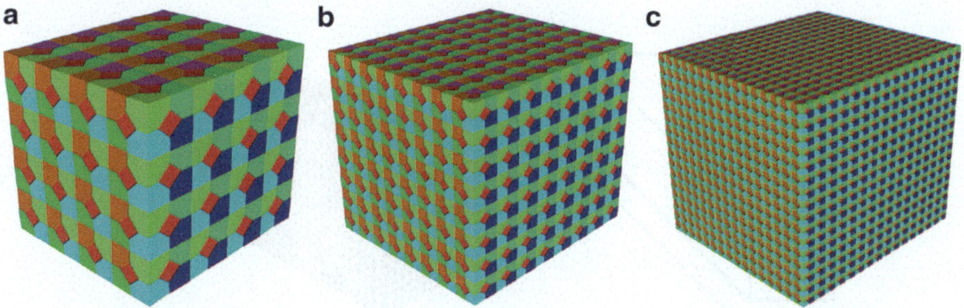

Fig. 3.5 Aggregates of the unit cell shown in Fig. 3.4: (**a**) $4 \times 4 \times 4$, (**b**) $8 \times 8 \times 8$, (**c**) $16 \times 16 \times 16$ (The color of each subvolume with the unit cell is arbitrary)

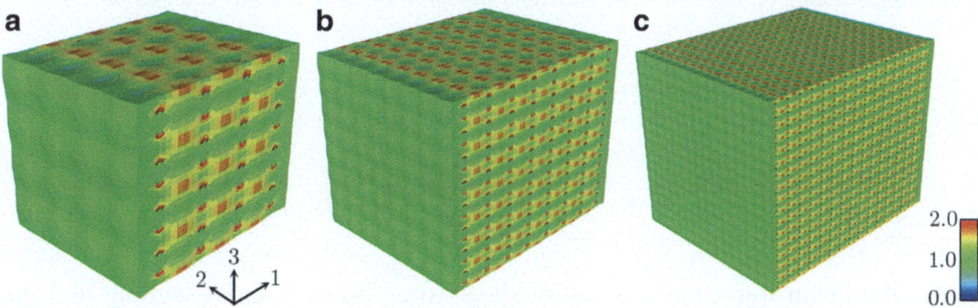

Fig. 3.6 Stress component σ_{11} resulting from the application of a uniform unit traction on the two surfaces normal to the [100] direction of the crystal structures shown in Fig. 3.5. (**a**) $4 \times 4 \times 4$, (**b**) $8 \times 8 \times 8$, (**c**) $16 \times 16 \times 16$. Note the apparent surface effects near the two planes on which the tractions are applied. For this simple loading case, the stress field resulting from the use of the homogenized material properties is $\sigma_{11} = 1$ with all other components identically zero. On the interior, the stress field is approximately periodic

Note that $E_1/E_3 = 1.17986$. Thus, the unit cell exhibits a significantly smaller (homogenized) anisotropy than the individual cubic crystals.

Now consider the $4 \times 4 \times 4$, $8 \times 8 \times 8$, and $16 \times 16 \times 16$ aggregates of unit cells as shown in Fig. 3.5. A uniform traction of unit value is applied to the two surfaces normal to the [100] direction. The resulting stress field is shown in Fig. 3.6 (component σ_{11}). For this simple loading case, the stress field resulting from the use of the homogenized material properties is $\sigma_{11} = 1$ with all other components identically zero. Note the apparent surface effects near the two planes on which the tractions are applied. Also, the surface effect does not disappear as the size of the unit cell gets smaller relative to the overall aggregate size. The stress field at the surface consistently deviates from that of the interior. In the interior, the stress field is approximately periodic.

In order to introduce macroscale strain gradients, consider the boundary-value problem of a linear-elastic prismatic beam of length L and rectangular cross-section of width $2a$ and height $2b$ as shown in Fig. 3.7a. The beam is fixed (weakly) on the end $x_3 = L$, and subjected to a transverse shear force F in the negative x_2-direction

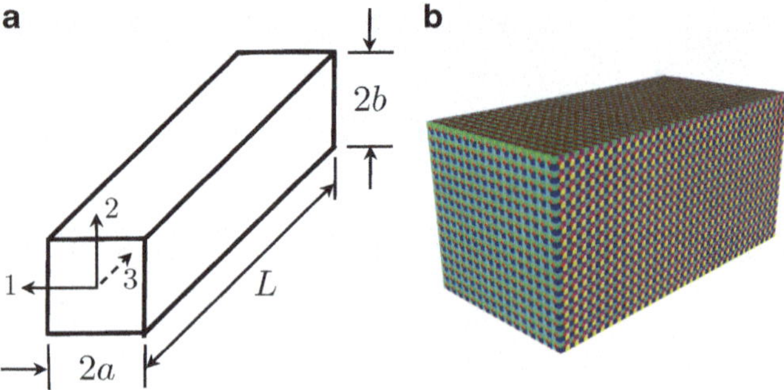

Fig. 3.7 (**a**) Prismatic beam of length L and rectangular cross-section of width $2a$ and height $2b$. (**b**) Beam consisting of $16 \times 16 \times 32$ unit cells shown in Fig. 3.4. For this case, $2a = 2b = 1$, $L = 2$

at the opposite end $x_3 = 0$. At any cross-section of the beam

$$\int_{-b}^{b}\int_{-a}^{a} \sigma_{32}\, dx_1\, dx_2 = F \quad \text{and} \quad \int_{-b}^{b}\int_{-a}^{a} \sigma_{33} x_2\, dx_1\, dx_2 = F x_3 \,, \tag{3.52}$$

so that the beam transmits a constant shear force on each cross-section, and the bending moment about the x_1-axis increases linearly from zero at $x_3 = 0$ to FL at $x_3 = L$. The transverse faces are traction free. For an isotropic material, Barber [12] gives the exact Cauchy stress field as

$$\sigma_{11} = \sigma_{21} = \sigma_{22} = 0 \tag{3.53a}$$

$$\sigma_{33} = \frac{F}{I} x_2 x_3 \tag{3.53b}$$

$$\sigma_{31} = \frac{F}{I}\frac{2a^2}{\pi^2}\frac{\nu}{1+\nu}\sum_{n=1}^{\infty}\frac{(-1)^n}{n^2}\sin(n\pi x_1/a)\frac{\sinh(n\pi x_2/a)}{\cosh(n\pi b/a)} \tag{3.53c}$$

$$\sigma_{32} = \frac{F}{I}\frac{b^2 - x_2^2}{2} + \frac{F}{I}\frac{\nu}{1+\nu}\left[\frac{3x_1^2 - a^2}{6} - \frac{2a^2}{\pi^2}\sum_{n=1}^{\infty}\frac{(-1)^n}{n^2}\cos(n\pi x_1/a)\frac{\cosh(n\pi x_2/a)}{\cosh(n\pi b/a)}\right] \tag{3.53d}$$

where ν is Poisson's ratio, and $I = 4ab^3/3$ is the second moment-of-area about the x_1-axis. An analysis of the derivation of this stress field shows that the solution also holds for the special case of an orthotropic material with the additional symmetries $E_1 = E_2$, $\mu_{13} = \mu_{23}$, and $\nu_{31} = \nu_{32}$. This is precisely the case for the unit cell considered in this example. The only modification required of Eq. (3.53) is to let $\nu \rightarrow \nu_{31} = \nu_{32}$.

Figure 3.7 shows a beam consisting of $16 \times 16 \times 32$ unit cells. For this case, $2a = 2b = 1$, $L = 2$, and $l/L = 1/16$. The total size of the finite-element mesh

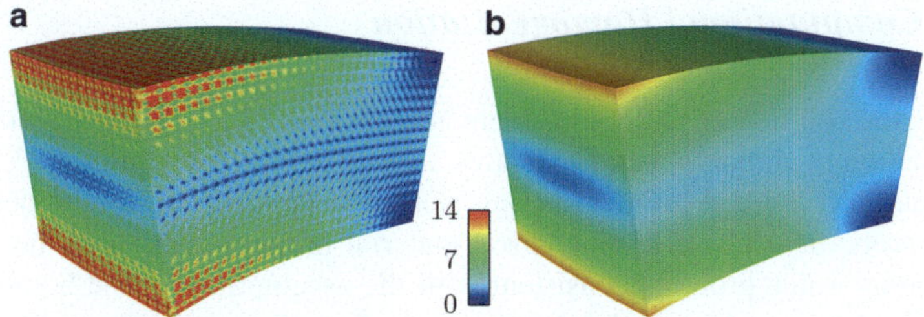

Fig. 3.8 (**a**) Von Mises stress field of the heterogeneous beam shown in Fig. 3.7. The beam is fixed (weakly) on the end $x_3 = L$, and subjected to a transverse shear force in the negative x_2-direction at the opposite end $x_3 = 0$. (**b**) Von Mises stress field of the same beam but instead using the homogenized material properties (The deformed shape is shown greatly magnified)

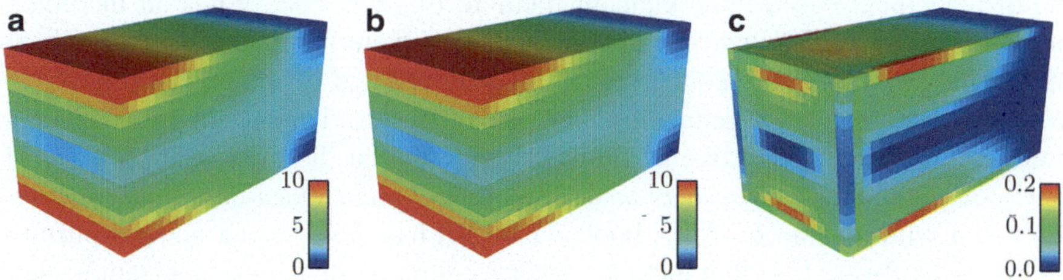

Fig. 3.9 (**a**) Magnitude of the cell-averaged Cauchy stress tensor for the heterogeneous beam. (**b**) Magnitude of the cell-averaged Cauchy stress tensor for the homogeneous beam. (**c**) Magnitude of the difference in the cell-averaged Cauchy stress tensors

is then $832 \times 8192 \approx 6.8$ million. Tractions are applied on the surfaces $x_3 = 0$ and $x_3 = L$ according to Eq. (3.53). The resulting Von Mises stress field is shown in Fig. 3.8a. The Von Mises stress field of the same beam, but instead using the homogenized unit-cell material properties, is shown in Fig. 3.8b.

To investigate the accuracy of homogenization solution for this example, we consider the average of the stress tensor over the unit cell according to Eq. (3.34) for both the DNS results and the homogenization results (H). Any difference can be attributed to either surface effects or hyperstresses. Figure 3.9a, b gives the Euclidean magnitude $||\langle\sigma_{ij}\rangle_V||$ of the cell-averaged Cauchy stress tensor for the both the DNS results and the homogenization results, respectively, where

$$\langle\sigma_{ij}\rangle_V \doteq \frac{1}{V}\int_V \sigma_{ij}(\mathbf{x}, \mathbf{y})\, dV \,, \tag{3.54}$$

and V represents the volume of the unit cell. Figure 3.9c shows the magnitude of the difference of the cell-averaged Cauchy stress, $||\langle\sigma_{ij}^{\rm DNS}\rangle_V - \langle\sigma_{ij}^{\rm H}\rangle_V||$. The difference is about 2 %.

3.3.4 Computational Homogenization

Typically, the solution of the cell problem given by Eq. (3.37) requires a numerical approximation in all but the simplest cases, as was demonstrated in Sect. 3.3.3. For linear materials, the cell problem is only solved once to obtain the homogenized material properties. These homogenized material properties can then be used in any boundary-value problem consistent with the assumptions of the homogenization. For inelastic problems, the deformation of the unit cell is typically history dependent, and thus cannot be homogenized *a priori* in general. Instead, the homogenization step must be performed concurrently during the simulation of the macroscale problem [75]. Within a finite-element framework, each macroscale finite element has a unit cell (or RVE) embedded within each integration point. For the displacement-based finite-element method, the host code sends an increment of deformation to the material model, and the material model sends back an increment of stress. This concurrent-multiscale process is sometimes referred to as "computational homogenization," or $(\mathrm{FE})^2$ (finite element squared), since a finite-element model is active at the macroscale and at the microscale (unit cell). This computational approach is analogous to the *local* quasi-continuum method discussed within Chap. 6 of this book. Fish [50] has developed a general-purpose multiscale software [94] that includes several computational homogenization capabilities.

For first-order homogenization, the main assumption is the presence of a scale separation between the microscale and macroscale, $l/L \ll 1$, where now the macroscopic length scale L should take into account strain gradients as well. The assumption of a scale separation and small strain gradients typically breaks down when the material localizes, such as when shear-banding and fracture occur. The range of applicability of computational homogenization can be extended by including strain-gradient effects (higher-order homogenization theory) [32, 33, 54, 76, 149].

While the unit-cell problem defined by Eq. (3.37) can be solved using standard numerical techniques, such as the finite-element method, a very efficient numerical algorithm has been developed by Moulinec and Suquet [107] based on the FFT. This approach transforms the set of partial-differential equations with periodic boundary conditions into a convolution-based integral equation that can be solved using fixed-point iteration in Fourier space. The FFT method for solving the unit-cell problem has proven effective for both linear and nonlinear computational homogenization [97, 98, 107]. Eisenlohr et al. [43] have compared the finite-element solution of the unit-cell problem to the FFT method for polycrystals in the large deformation regime. Massively parallel implementations of the FFT algorithm exist in three dimensions [122] as well as GPU implementations [99].

To set up the FFT-based fixed-point iteration algorithm, the concept of a linear homogeneous reference medium with stiffness tensor $\mathbf{C}_0$ is introduced, along with the polarization stress $\boldsymbol{\tau}$ defined by

$$\boldsymbol{\tau}(\mathbf{y}) = \boldsymbol{\sigma}(\mathbf{y}) - \mathbb{C}_0 : \boldsymbol{\epsilon}(\mathbf{y}) = (\mathbb{C}(\mathbf{y}) - \mathbb{C}_0) : \boldsymbol{\epsilon}(\mathbf{y}) \; . \tag{3.55}$$

Equation (3.37) can be cast in the form of the periodic Lippmann–Schwinger equation,

$$\boldsymbol{\epsilon}(\mathbf{y}) = -\int_V \Gamma_0(\mathbf{y},\mathbf{y}') : \boldsymbol{\tau}(\mathbf{y}')\, d\mathbf{y}' + \mathbf{E}, \tag{3.56}$$

where $\boldsymbol{\Gamma}_0(\mathbf{y},\mathbf{y}')$ is a fourth-order tensor related to the Green's function of the homogeneous reference medium [109], and here $\mathbf{E}$ represents a constant strain over the unit cell. Taking the Fourier transform of Eq. (3.56) turns the nonlocal convolution in real space into a local product in Fourier space,

$$\hat{\boldsymbol{\epsilon}}(\boldsymbol{\xi}) = -\widehat{\Gamma}_0(\boldsymbol{\xi}) : \hat{\boldsymbol{\tau}}(\boldsymbol{\xi}) \quad \text{for all } \boldsymbol{\xi} \neq \mathbf{0} \tag{3.57a}$$

$$\hat{\boldsymbol{\epsilon}}(\mathbf{0}) = \mathbf{E}\ , \tag{3.57b}$$

where $\boldsymbol{\xi}$ is the spatial frequency vector. In Fourier space, $\widehat{\Gamma}_0(\boldsymbol{\xi})$ is given explicitly by

$$\widehat{\Gamma}_{klij}(\boldsymbol{\xi}) = \frac{\delta_{ki}\xi_l\xi_j + \delta_{li}\xi_k\xi_j + \delta_{kj}\xi_l\xi_i + \delta_{lj}\xi_k\xi_i}{4\mu_o||\boldsymbol{\xi}||^2} - \frac{\lambda_o + \mu_o}{\mu_o(\lambda_o + 2\mu_o)}\frac{\xi_i\xi_j\xi_k\xi_l}{||\boldsymbol{\xi}||^4}\ , \tag{3.58}$$

where λ_0 and μ_0 are the Lame coefficients of the reference medium, and δ_{ij} is the Kronecker delta.

Since the polarization stress defined in Eq. (3.55) is a function of the unknown strain field $\boldsymbol{\epsilon}(\mathbf{x})$, an iterative solution of Eq. (3.57) can be obtained using the fixed-point scheme

$$\hat{\boldsymbol{\epsilon}}^{i+1}(\boldsymbol{\xi}) = -\widehat{\Gamma}_0(\boldsymbol{\xi}) : \hat{\boldsymbol{\tau}}^i(\boldsymbol{\xi}) \quad i = 0, 1, 2, \ldots\ . \tag{3.59}$$

The steps in this algorithm are shown in Fig. 3.10. The convergence test in the algorithm measures the deviation from equilibrium (divergence of stress equal to zero) within a given tolerance. Note that the algorithm does not involve the solution of a matrix equation. According to Banach's fixed-point theorem [77], the fixed-point iteration converges to a unique solution if the operator implicit in Eq. (3.56) is a contraction. This condition has been studied extensively, and can be assured under certain restrictions for λ_0 and μ_0 [98]. The number of iterations required for convergence scales with the ratio of maximum and minimum contrasts in the elastic moduli. For a nonlinear material, only step 6 needs to be modified to return the stress for the given value of strain. For a history-dependent material, an incremental version of the algorithm is used [43].

The development of numerical schemes that accelerate the basic FFT algorithm is an active area of research. Recent improvements include using a periodic Green's operator instead of one based on an infinite medium [20, 21, 70, 98, 156].

$$
\begin{array}{ll}
\text{Initialization:} & \boldsymbol{\varepsilon}^0(\mathbf{y}) = \mathbf{E}\,, \quad \text{for all } \mathbf{y} \in V \\
& \boldsymbol{\sigma}^0(\mathbf{y}) = \mathbb{C}(\mathbf{y}) : \boldsymbol{\varepsilon}^0(\mathbf{y})\,, \quad \text{for all } \mathbf{y} \in V \\
\\
\text{Iteration } i+1: & \text{with } \boldsymbol{\varepsilon}^i \text{ and } \boldsymbol{\sigma}^i \text{ known} \\
& 1. \quad \boldsymbol{\tau}^i(\mathbf{y}) = \boldsymbol{\sigma}^i(\mathbf{y}) - \mathbb{C}_0 : \boldsymbol{\varepsilon}^i(\mathbf{y}) \\
& 2. \quad \hat{\boldsymbol{\tau}}^i(\boldsymbol{\xi}) = \mathcal{F}(\boldsymbol{\tau}^i(\mathbf{y})) \\
& 3. \quad \text{convergence test} \\
& 4. \quad \hat{\boldsymbol{\varepsilon}}^{i+1}(\boldsymbol{\xi}) = -\widehat{\Gamma}_0(\boldsymbol{\xi}) : \hat{\boldsymbol{\tau}}^i(\boldsymbol{\xi}) \quad \text{for all } \boldsymbol{\xi} \neq \mathbf{0}\,, \quad \hat{\boldsymbol{\varepsilon}}^{i+1}(\mathbf{0}) = \mathbf{E} \\
& 5. \quad \hat{\boldsymbol{\varepsilon}}^{i+1}(\mathbf{x}) = \mathcal{F}^{-1}(\boldsymbol{\varepsilon}^{i+1}(\boldsymbol{\xi})) \\
& 6. \quad \boldsymbol{\sigma}^{i+1}(\mathbf{y}) = \mathbb{C}(\mathbf{y}) : \boldsymbol{\varepsilon}^{i+1}(\mathbf{y})
\end{array}
$$

Fig. 3.10 Iterative FFT algorithm for solving the homogenization unit-cell problem given by Eq. (3.37) [107]. $\mathscr{F}$ represents the FFT (in 3D), and $\mathscr{F}^{-1}$ represents the inverse FFT (in 3D). The convergence test measures the deviation from equilibrium

3.4 Crystal-Plasticity Models

In many macroscale constitutive models, phenomenological formulations are developed for specific classes of materials such as those for plasticity and viscoplasticity (polycrystalline metals) [74, 83, 87, 111], hyperelasticity and viscoelasticity (polymers) [29], and pressure dependent plasticity (porous materials) [5, 19]. These models use various internal-state variables to phenomenologically model physical effects such as dislocation slip, dislocation density, porosity, and damage. Through these constitutive models, the stress is related to the entire history of deformation. Constitutive models also exist that describe the deformation of the grain or single-crystal scale in metals by incorporating homogenization of discrete dislocation slip events [127]. These constitutive models are generally referred to as "crystal plasticity" (CP) models. Some CP models use dislocation densities explicitly as internal-state variables, and thus incorporate additional length scales [15]. In this section, crystal-plasticity models based on a continuum approximation will be discussed as one class of inelastic continuum model.

3.4.1 Background

Crystal-plasticity models are based on single-crystal plastic deformation via dislocation motion through a crystal lattice on specific slip systems. The applied stress is resolved onto predefined slip systems to accommodate plastic deformation of each grain or crystal. Figure 3.11 illustrates the Schmid law, $\tau = \sigma \cos\lambda \cos\phi$, that defines the relationship between the shear stress τ resolved on the slip plane and the uniaxial applied stress σ. Here, ϕ is the angle between the slip plane with normal $\mathbf{n}$ and the loading axis, and λ is the angle between the slip direction $\mathbf{m}$ and

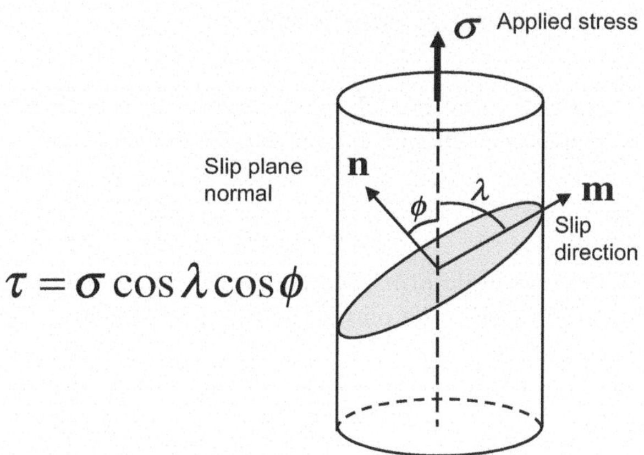

Fig. 3.11 Schematics of the Schmid law. The resolved shear stress τ is described by the applied stress σ, cosine of the angle ϕ between the slip-plane normal **n** and the loading axis, and cosine of the angle λ between the slip direction **m** and the loading axis

the loading axis. For the case of arbitrary loading, τ can be represented using a contraction between the Schmid tensor, $\mathbf{P} = 1/2(\mathbf{m} \otimes \mathbf{n} + \mathbf{n} \otimes \mathbf{m})$, and the applied stress tensor, $\boldsymbol{\sigma}$.

With some simplifying assumptions of the interactions between grains, single-crystal constitutive equations can be used to directly approximate the response of a polycrystal. Classical polycrystal plasticity models (texture models) assume simplified interactions between grains, e.g., the same stress state [132] or strain state [145] for each crystal. However, these models ignore intergranular compatibility and equilibrium, respectively. Recently, crystal-plasticity constitutive models implemented within nonlinear displacement-based finite-element (FEM) codes enforce compatibility strongly and equilibrium weakly. In addition to solving the equilibrium equations, the advantages of crystal-plasticity FEM (CP-FEM) include the ability to handle irregular shaped domains, complex boundary conditions, and multiple material phases. Furthermore, high performance computers enable simulations of polycrystalline domains on macroscopic length scales, i.e., having millions of grains. Polycrystalline simulations using CP-FEM can provide mechanical properties of the polycrystalline body with consideration of microstructural effects such as texture evolution (crystal orientations), multi-phases, defects (voids and dislocations), and grain morphology (shapes and sizes).

The basic assumptions used in conventional crystal-plasticity models are (1) the deformation gradient can be decomposed into elastic and plastic parts, and (2) plastic deformation occurs by dislocation slip on the predefined slip systems. Thus, conventional crystal-plasticity models ignore other deformation mechanisms such as cross slip, climb, dislocation twinning, grain-boundary sliding, and non-Schmid yield behavior. More sophisticated models have been developed to capture these effects [71, 86, 89]. In this section, the most widely adopted crystal-plasticity formulation by Peirce et al. [121] is outlined.

3.4.2 Model Formulations

Crystal-plasticity models generally adopt a multiplicative decomposition of the deformation gradient $\mathbf{F}$ into elastic $\mathbf{F}^{\mathrm{e}}$ and plastic $\mathbf{F}^{\mathrm{p}}$ parts [62, 82, 121, 126],

$$\mathbf{F} = \mathbf{F}^{\mathrm{e}}\,\mathbf{F}^{\mathrm{p}}\,. \tag{3.60}$$

As the crystal deforms, the lattice is rotated and elastically stretched and sheared according to $\mathbf{F}^{\mathrm{e}}$. The plastic deformation gradient $\mathbf{F}^{\mathrm{p}}$ is calculated from the crystal slip as [9]:

$$\mathbf{F}^{\mathrm{p}} = \mathbf{I} + \sum_{\alpha} \gamma^{\alpha}\mathbf{s}_0^{\alpha} \otimes \mathbf{n}_0^{\alpha}\,, \tag{3.61}$$

where γ^{α} is the amount of slip in α slip system, $\mathbf{s}_0^{\alpha}$ and $\mathbf{n}_0^{\alpha}$ are the unit vectors in the slip direction and slip-plane normal direction on the α slip system in the original configuration, respectively, and $\mathbf{I}$ is the identity tensor. The slip systems for different crystal structures, FCC and BCC, are given in Table 3.2.

The velocity gradient in the current configuration, $\mathbf{L} \doteq \partial\mathbf{v}/\partial\mathbf{x}$, can be written as

$$\mathbf{L} = \dot{\mathbf{F}}\,\mathbf{F}^{-1} = \mathbf{L}^{\mathrm{e}} + \mathbf{L}^{\mathrm{p}}\,, \tag{3.62}$$

Table 3.2 Slip systems for different crystal structures

α	Slip system	α	Slip system	α	Slip system	α	Slip system
Twelve FCC slip systems							
1	$(01\bar{1})[111]$	4	$(0\bar{1}\bar{1})[\bar{1}\bar{1}1]$	7	$(01\bar{1})[\bar{1}\bar{1}\bar{1}]$	10	$(0\bar{1}\bar{1})[11\bar{1}]$
2	$(\bar{1}01)[111]$	5	$(101)[\bar{1}\bar{1}1]$	8	$(\bar{1}01)[\bar{1}\bar{1}\bar{1}]$	11	$(101)[11\bar{1}]$
3	$(1\bar{1}0)[111]$	6	$(\bar{1}10)[\bar{1}\bar{1}1]$	9	$(1\bar{1}0)[\bar{1}\bar{1}\bar{1}]$	12	$(\bar{1}10)[11\bar{1}]$
Twenty four {110} *BCC slip systems*							
1	$(01\bar{1})[111]$	7	$(0\bar{1}\bar{1})[\bar{1}\bar{1}1]$	13	$(01\bar{1})[\bar{1}\bar{1}\bar{1}]$	19	$(0\bar{1}\bar{1})[11\bar{1}]$
2	$(\bar{1}01)[111]$	8	$(101)[\bar{1}\bar{1}1]$	14	$(\bar{1}01)[\bar{1}\bar{1}\bar{1}]$	20	$(101)[11\bar{1}]$
3	$(1\bar{1}0)[111]$	9	$(\bar{1}10)[\bar{1}\bar{1}1]$	15	$(1\bar{1}0)[\bar{1}\bar{1}\bar{1}]$	21	$(\bar{1}10)[11\bar{1}]$
4	$(\bar{1}0\bar{1})[\bar{1}11]$	10	$(10\bar{1})[1\bar{1}1]$	16	$(\bar{1}0\bar{1})[1\bar{1}\bar{1}]$	22	$(10\bar{1})[\bar{1}1\bar{1}]$
5	$(0\bar{1}1)[\bar{1}11]$	11	$(011)[1\bar{1}1]$	17	$(0\bar{1}1)[1\bar{1}\bar{1}]$	23	$(011)[\bar{1}1\bar{1}]$
6	$(110)[\bar{1}11]$	12	$(\bar{1}\bar{1}0)[1\bar{1}1]$	18	$(110)[1\bar{1}\bar{1}]$	24	$(\bar{1}\bar{1}0)[\bar{1}1\bar{1}]$
Twenty four {112} *BCC slip systems*							
1	$(11\bar{2})[111]$	7	$(\bar{1}\bar{1}\bar{2})[\bar{1}\bar{1}1]$	13	$(11\bar{2})[\bar{1}\bar{1}\bar{1}]$	19	$(\bar{1}\bar{1}\bar{2})[11\bar{1}]$
2	$(\bar{2}11)[111]$	8	$(2\bar{1}1)[\bar{1}\bar{1}1]$	14	$(\bar{2}11)[\bar{1}\bar{1}\bar{1}]$	20	$(2\bar{1}1)[11\bar{1}]$
3	$(1\bar{2}1)[111]$	9	$(\bar{1}21)[\bar{1}\bar{1}1]$	15	$(1\bar{2}1)[\bar{1}\bar{1}\bar{1}]$	21	$(\bar{1}21)[11\bar{1}]$
4	$(\bar{1}1\bar{2})[\bar{1}11]$	10	$(1\bar{1}\bar{2})[1\bar{1}1]$	16	$(\bar{1}1\bar{2})[1\bar{1}\bar{1}]$	22	$(1\bar{1}\bar{2})[\bar{1}1\bar{1}]$
5	$(\bar{1}\bar{2}1)[\bar{1}11]$	11	$(121)[1\bar{1}1]$	17	$(\bar{1}\bar{2}1)[1\bar{1}\bar{1}]$	23	$(121)[\bar{1}1\bar{1}]$
6	$(211)[\bar{1}11]$	12	$(\bar{2}\bar{1}1)[1\bar{1}1]$	18	$(211)[1\bar{1}\bar{1}]$	24	$(\bar{2}\bar{1}1)[\bar{1}1\bar{1}]$

where $\mathbf{L}^e$ and $\mathbf{L}^p$ are elastic and plastic parts of the velocity gradient, respectively, and are given by

$$\mathbf{L}^e = \dot{\mathbf{F}}^e (\mathbf{F}^e)^{-1} \quad \text{and} \quad \mathbf{L}^p = \mathbf{F}^e \dot{\mathbf{F}}^p (\mathbf{F}^p)^{-1} (\mathbf{F}^e)^{-1} . \tag{3.63}$$

The plastic part of the velocity gradient $\mathbf{L}^p$ is given as [121]

$$\mathbf{L}^p = \sum_{\alpha} \dot{\gamma}^{\alpha} \, \mathbf{s}_0^{\alpha} \otimes \mathbf{n}_0^{\alpha} , \tag{3.64}$$

where $\dot{\gamma}^{\alpha}$ is the slip rate on the α slip system.

The critical aspect of single-crystal constitutive equations is formulating how the slip rate is related to the applied stress. The most widely used form for a viscoplastic formulation is based on the power-law function [66, 126],

$$\dot{\gamma} = \dot{\gamma}_0 \left(\frac{\tau}{g} \right)^{1/m} \operatorname{sign}(\tau) , \tag{3.65}$$

where $\dot{\gamma}_0$ is the reference shear rate, m is the rate sensitivity, and g is the slip resistance. Alternatively, the slip rate can be modeled using a thermal activation form given by Ma et al. [89].

$$\dot{\gamma} = \dot{\gamma}_0 \exp\left[-\frac{Q}{kT} \left(1 - \frac{\tau}{g} \right) \right] \operatorname{sign}(\tau) , \tag{3.66}$$

where Q is the activation enthalpy, k is the Boltzmann constant, and T is temperature. This form is useful when the crystal slip exhibits large temperature dependence, e.g., for body-centered cubic (BCC) crystal structure.

The slip resistance g represents strain hardening. There are different models for g, e.g., isotropic hardening, slip-based hardening, or dislocation density-based hardening. Classical isotropic hardening models generally assume that all slip systems harden equally as a function of plastic strain [88]. Slip-based hardening models are also widely used in crystal-plasticity models, and have been used to successfully predict texture and anisotropic behaviors in polycrystals [72, 91]. In slip-based hardening models, the strain hardening term is related to the slip increment on all slip systems through the relation [8],

$$\dot{g}^{\alpha} = \sum_{\beta} h^{\alpha\beta} |\dot{\gamma}^{\alpha}|. \tag{3.67}$$

Here, $h^{\alpha\beta}$ is the hardening matrix that relates hardening on one slip system to other active slip systems. Different hardening matrices that account for anisotropy and dislocation–dislocation interactions have been proposed [13, 38, 52, 79]. The widely used form that successfully predicts stress–strain behavior of polycrystals is as follows [22]:

$$h^{\alpha\beta} = \mathbf{q}^{\alpha\beta} h_0 \left(1 - \frac{g^{\beta}}{g_s} \right)^{a} . \tag{3.68}$$

Here, h_0 is the initial hardening rate, g_s is the saturated flow stress, and a is the hardening exponent. $\mathbf{q}^{\alpha\beta}$ is a hardening matrix that determines the self to latent hardening ratios. The diagonal terms of $\mathbf{q}^{\alpha\beta}$, denoted by q_{self}, describe self-hardening while the non-diagonal terms denoted by q_{lat} describe the latent hardening effect. The values $q_{\text{self}} = 1$ and $q_{\text{lat}} = 1.4$ are commonly used in polycrystalline simulations [10, 91, 121].

The dislocation density-based Taylor hardening law is represented as follows [144]:

$$g = A\mu b \sqrt{\sum_{\beta=1} \rho^\beta}\,, \tag{3.69}$$

where A is a material constant, μ is the shear modulus, b is the Burger's vector (magnitude), and ρ^β is the dislocation density on slip system β. The evolution of dislocation density for the α slip system is obtained by a standard phenomenological equation [73],

$$\dot{\rho}^\alpha = \left(\kappa_1 \sqrt{\sum_{\beta=1}^{24} \rho^\beta} - \kappa_2 \rho^\alpha \right) \cdot |\dot{\gamma}^\alpha|\,, \tag{3.70}$$

where κ_1 and κ_2 are hardening parameters representing generation and annihilation of dislocations, respectively, and ultimately determine the shape of the stress–strain curve.

Other variations of crystal-plasticity models include different types of time integration schemes, e.g., implicit or explicit, as well as rate independent models. Crystal-plasticity constitutive models can be implemented within the standard finite-element formulation [127]. This is commonly done in commercial software such as ABAQUS [1] and ANSYS [6], as well as more specialized finite-element software such as ALBANY [4], DAMASK [37, 129], and ZEBULON [155].

3.4.3 CP and Nanomechanics

One of the strengths of the crystal-plasticity model is that it can accommodate microstructure in simulations and provides realistic length and time scales. Modern CP-FEM has been modified and applied to various micro- to grain-scale problems [128]:

- Texture evolution, plastic anisotropy
- Nonlocal formulations, grain boundary mechanics, grain size effects
- Metal forming, deep drawing, springback
- Surface roughening, ridging, roping.
- Damage and fracture, fatigue, void growth

- Micromechanics: nanoindentation, micropillar testing
- Creep, high temperature deformation, diffusion mechanisms
- Deformation twinning, martensitic transformation, shape memory, phase transformation, recrystallization

There are two main limitations in applying conventional CP models directly to nanoscale materials. First, as grain sizes are reduced to nanometer sizes in nanocrystalline materials, the assumption of a continuous dislocation density becomes increasingly suspect. Second, many experiments and lower length-scale simulation results show that conventional dislocation plasticity in non-nanocrystalline metals breaks down in nanoscale materials. For example, discrete dislocation grain-boundary interactions and grain boundary sliding become significant deformation mechanisms at this scale. Note that conventional CP models do not properly incorporate these effects. To overcome these limitations, advanced CP models have appeared in the literature [53, 84, 85, 153]. They either use quantized slip or incorporate grain boundary effects to capture deformation mechanisms at the nanoscale. More detailed description of the quantized crystal-plasticity (QCP) model is discussed in Chap. 13.

3.5 Conclusions

In this chapter, we have briefly reviewed continuum mechanics, homogenization theory, computational homogenization, and constitutive modeling including crystal-plasticity. Our review of continuum mechanics included an overview of micromorphic and nonlocal theories as well as continuum formulations that include an explicit surface stress. At the center of continuum mechanics is homogenization theory which provides a mathematically elegant and rigorous framework for replacing a discrete collection of interacting entities by an equivalent homogenous continuum with effective material properties. Numerous references to much more detailed expositions on these topics were provided throughout the chapter.

Errors in a continuum approximation to a discrete system are unavoidable whenever the introduced length scales are comparable to the length scale of the discrete system. Generalized continuum theories may be able to reproduce qualitatively correct physical phenomenon, such as a surface effect or optical-branch phonon dispersion curves, but the accuracy of the continuum theory must be judged with respect to the true behavior of the original discrete system.

Appendix

Example: Error in the Continuum Approximation

Discrete systems have an inherent length scale governed fundamentally by the interaction distance between entities. The accuracy of the continuum approximation depends critically on the size of the structure compared to the size of this intrinsic length scale. To illustrate the continuum approximation of a discrete system and its accuracy, consider a simple one-dimensional chain of N atoms of total initial length L with initial atomic spacing l subjected to a body force F per atom as shown in Fig. 3.12a. The initial position X_I of each atom is given by $X_I = l \cdot I$, $I = 0, 1, 2, 3, \ldots, N$ with $X_N = L$. For simplicity, we take the interatomic potential to be harmonic with spring constant K, pair-wise additive, with only nearest-neighbor interactions.

The forces between the atoms are linear with respect to the relative displacements so that

$$K(u_N - u_{N-1}) = F \tag{3.71a}$$

$$K(u_{N-1} - u_{N-2}) = 2F \tag{3.71b}$$

$$\vdots$$

$$K(u_I - u_{I-1}) = F(N - I + 1) \tag{3.71c}$$

$$\vdots$$

$$K(u_1 - u_0) = FN \tag{3.71d}$$

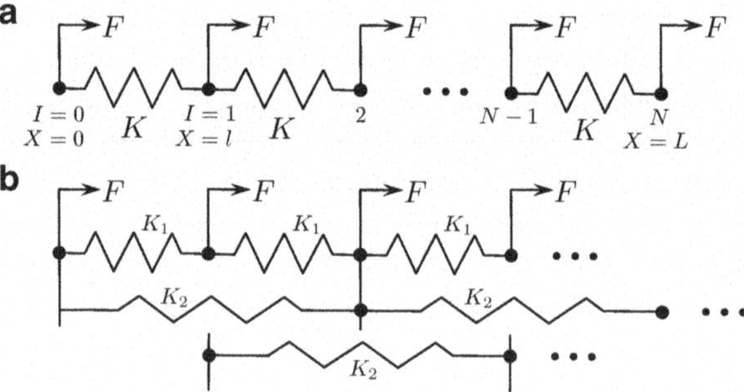

Fig. 3.12 (**a**) One-dimensional chain of atoms of total initial length L with initial atomic spacing l subjected to a body force F per atom. The initial position X_I of each atom is given by $X_I = l \cdot I$, $I = 0, 1, 2, 3, \ldots, N$ with $X_N = L$. The interatomic forces are governed by *local* interactions and a spring constant K. (**b**) One-dimensional chain of atoms with *nonlocal* interatomic forces. The nearest-neighbor spring constant is K_1, and next-nearest spring constant is K_2

where u_I is the displacement of atom I with $u_0 = 0$. Since this set of equations telescopes starting with $u_1 = FN/K$, we have

$$u_I = \frac{F}{K}\left[I \cdot N - \sum_{J=1}^{I}(J-1)\right] = \frac{F}{K}\left[I \cdot N - \frac{1}{2}I(I-1)\right]. \tag{3.72}$$

Since $N = L/l$ and $I = X_I/l$, we can write Eq. (3.72) as

$$u_I = \frac{(F/l)L^2}{(Kl)}\left[\frac{X_I}{L}\left(1 + \frac{l}{2L}\right) - \frac{1}{2}\left(\frac{X_I}{L}\right)^2\right]. \tag{3.73}$$

Note that the natural length-scale ratio for this problem is l/L.

To obtain the continuum version of this problem, we first use Eq. (3.13) reduced to the one-dimensional form, along with a linear-elastic constitutive model, $\sigma_{xx} = E du/dX$. This results in the following equation for the continuum displacement field in a one-dimensional bar,

$$\frac{d}{dX}\left(k\frac{du}{dX}\right) + f(X) = 0\,, \tag{3.74}$$

where f is the body force per unit length along the bar, and k is the cross-sectional stiffness. The solution to this equilibrium equation with the boundary condition $u(0) = 0$ is given by

$$u(X) = \frac{fL^2}{k}\left[\frac{X}{L} - \frac{1}{2}\left(\frac{X}{L}\right)^2\right]. \tag{3.75}$$

Note that there is no intrinsic length scale in this continuum solution. If we identify $k = K \cdot l$ and $F = f \cdot l$, then Eq. (3.73) converges to Eq. (3.75) in the limit as $l/L \to 0$. The absolute error in the continuum approximation is given by,

$$e_I \doteq |u(X_I) - u_I| = \frac{fL^2}{k}\left(\frac{X_i}{L}\right)\left(\frac{l}{2L}\right) \tag{3.76}$$

This error is proportional to the length-scale ratio l/L. Also, the error varies linearly along the chain, from $e_0 = 0$ at $X_0 = 0$ to a maximum value at $X_N = 0$. Thus, in the limit of infinitesimally small intrinsic length scale, the discrete solution converges to the continuum solution.

The normalized displacement solution for the discrete atom chain, given by Eq. (3.73) with $\bar{u} \doteq u/(FL^2/Kl^2)$, is shown in Fig. 3.13a as a function of normalized initial position, $\bar{X} \doteq X/L$, for several values of the length-scale ratio l/L. The continuum approximation for the displacement field, given by Eq. (3.75), is also shown. The error in the continuum approximation, given by Eq. (3.76) with $\bar{e} \doteq |\bar{u} - \bar{u}_I|$, is shown in Fig. 3.13b, and is seen to approach zero as $l/L \to 0$. Only for $l/L < 0.02$ ($N = 50$) is the maximum error less than 2 % of the peak displacement.

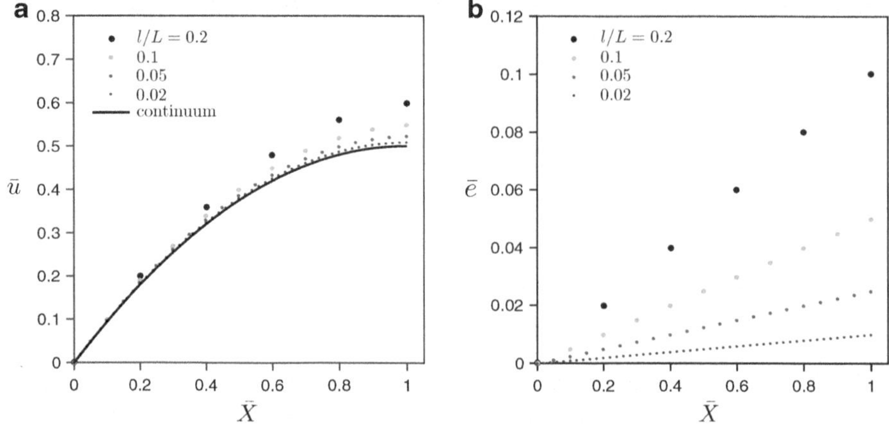

Fig. 3.13 (**a**) Normalized displacement $\bar{u}$ as a function of normalized initial position $\bar{X}$ of a one-dimensional chain of atoms subjected to a uniform body force and fixed at the end $\bar{X} = 0$ for various values of the ratio of atom spacing to total initial chain length, l/L. The continuum approximation is also shown. (**b**) Normalized error in the continuum approximation. Only for $l/L < 0.02$ ($N = 50$) is the maximum error less than 2 % of the peak displacement

Example: Absence of a Surface Effect in Classical Continuum Mechanics

Discrete systems can also display surface effects that are not present in classical continuum theories. To illustrate the continuum approximation of a discrete system and its accuracy, consider a simple one-dimensional chain of N atoms of total initial length L with initial atomic spacing l subjected to a body force F per atom as shown in Fig. 3.12b. The initial position X_I of each atom is given by $X_I = l \cdot I$, $I = 0, 1, 2, 3, \ldots, N$ with $X_N = L$. We take the interatomic potential to be harmonic, pair-wise additive, with both nearest-neighbor interactions with spring constant K_1, and nonlocal interactions with spring constant K_2. Note that the atoms at the end of the chain experience a distinctly different force environment than those atoms in the interior of the chain due to the number of interacting neighbors.

The forces between the atoms are linear with respect to the relative displacements so that

$$K_1(u_N - u_{N-1}) + K_2(u_N - u_{N-2}) = F \quad (3.77a)$$

$$-K_1(u_N - u_{N-1}) + K_1(u_{N-1} - u_{N-2}) + K_2(u_{N-1} - u_{N-3}) = F \quad (3.77b)$$

$$-K_2(u_N - u_{N-2}) - K_1(u_{N-1} - u_{N-2}) + K_1(u_{N-2} - u_{N-3}) + K_2(u_{N-2} - u_{N-4}) = F \quad (3.77c)$$

$$\vdots$$

$$-K_2(u_{I+2} - u_I) - K_1(u_{I+1} - u_I) + K_1(u_I - u_{I-1}) + K_2(u_I - u_{I-2}) = F \quad (3.77\text{d})$$

$$\vdots$$

$$-K_2(u_3 - u_1) - K_1(u_2 - u_1) + K_1(u_1 - u_0) = F \quad (3.77\text{e})$$

where u_I is the displacement of atom I with $u_0 = 0$. This system of equations results in a matrix equation $\mathbf{Ku} = \mathbf{F}$ where $\mathbf{K}$ is $N \times N$ banded matrix of the Toeplitz type and can be solved using standard methods.

The normalized displacement solution for the discrete atom chain with $\bar{u} \doteq u/(FL^2/K_1 l^2)$ is shown in Fig. 3.14a as a function of normalized initial position, $\bar{X} \doteq X/L$, for several values of the length-scale ratio l/L. For this example, we have chosen $K_2 = 0.5K_1$. In order to use the continuum approximation given by Eq. (3.75), we must first define an effective spring stiffness, K_{eff}. To this end, we isolate a unit cell of length $2l$ surrounding one *interior* atom. Within each cell, there are two K_1 springs acting in series thus contributing a value of $\frac{1}{2}K_1$ to K_{eff}. There is a full K_2 spring acting in parallel thus contributing a value of K_2 to K_{eff}. There are also two K_2 springs that effectively act in parallel to the unit cell, thus contributing a value of $\frac{1}{2}K_2 + \frac{1}{2}K_2$ to K_{eff}. Thus, $K_{\text{eff}} = \frac{1}{2}K_1 + 2K_2$, and $k \doteq K_{\text{eff}} \cdot (2l)$. Also, the effective force per unit length is $f = 2F/2l = F/l$. The continuum approximation

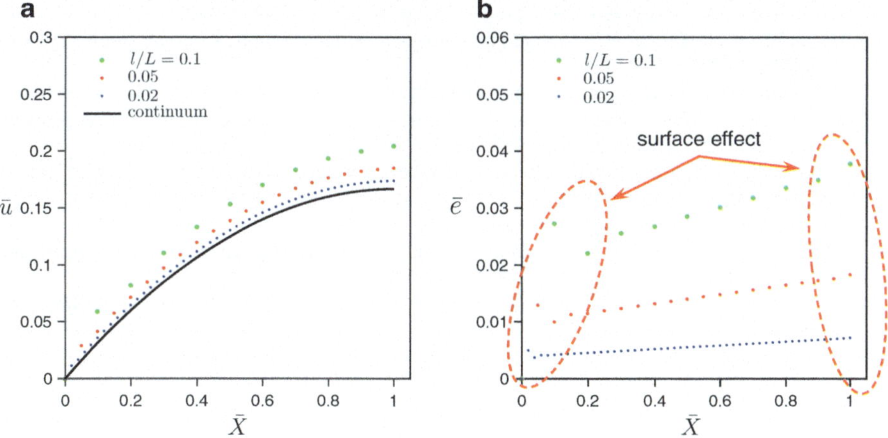

Fig. 3.14 Illustration of the surface effect for a one-dimensional discrete chain of atoms subjected to a body force F applied to each atom. The atoms are connected to their nearest neighbors by linear springs with spring constant K_1, and to their next-nearest neighbors by linear springs with spring constant K_2. The chain is fixed at $X = 0$. The total initial length of the chain is L, and the atomic spacing is l so that the number of atoms $N = L/l$. (**a**) Normalized displacement, $\bar{u} \doteq u/(FL^2/K_1 l^2)$, plotted as a function of the normalized initial position, $\bar{X} \doteq X/L$. Results are shown for $l/L = 0.1, 0.05, 0.02$ ($N = 10, 20, 50$, respectively) with $K_2 = 0.5K_1$. The continuum approximation, $l/L \to 0$, is also shown with $\bar{u} \doteq u/(fL^2/k)$ where $f \doteq F/l$, and $k \doteq K_1\, l$. (**b**) Error in the continuum approximation with $\bar{e} \doteq |\bar{u} - \bar{u}_i|$

for the displacement field is also shown in Fig. 3.14a. There is some noticeable surface effect on the atoms near the ends of the chain, particularly near $\bar{X} = 0$. This effect is more noticeable if we plot the error in the continuum approximation, $\bar{e} \doteq |\bar{u} - \bar{u}_I|$, as shown in Fig. 3.14b. Notice that the surface effect near $\bar{X} = 0$ affects several atoms. The surface effect is absent in the chosen continuum approximation.

Acknowledgements Sandia National Laboratories is a multi-program laboratory operated by Sandia Corporation, a wholly owned subsidiary of Lockheed Martin Corporation, for the US Department of Energy's National Nuclear Security Administration under contract DE-AC04-94AL85000.

References

1. ABAQUS, http://www.3ds.com/products-services/simulia/products/abaqus/ (2015)
2. E. Aifantis, Update on a class of gradient theories. Mech. Mater. **35**(3–6), 259–280 (2003)
3. E. Aifantis, On the gradient approach – relation to Eringen's nonlocal theory. Int. J. Eng. Sci. **49**(12), 1367–1377 (2011)
4. ALBANY, https://github.com/gahansen/Albany (2015)
5. A. Anandarajah, *Computational Methods in Elasticity and Plasticity: Solids and Porous Media* (Springer, New York, 2010)
6. ANSYS, http://www.ansys.com (2015)
7. B. Arash, Q. Wang, A review on the application of nonlocal elastic models in modeling of carbon nanotubes and graphenes. Comput. Mater. Sci. **51**(1), 303–313 (2012)
8. R.J. Asaro, Geometrical effects in the inhomogeneous deformation of ductile single crystals. Acta. Metall. **27**, 445 (1979)
9. R.J. Asaro, Micromechanics of crystals and polycrystals. Adv. Appl. Mech. **23**, 1–115 (1983)
10. R.J. Asaro, A. Needleman, Texture development and strain hardening in rate dependent polycrystals. Acta Metall. **33**, 923–953 (1985)
11. J. Auriault, G. Bonnet, Surface effects in composite materials: two simple examples. Int. J. Eng. Sci. **25**(3), 307–323 (1987)
12. J. Barber, *Elasticity* (Springer, New York, 2010)
13. J.L. Bassani, T.Y. Wu, Latent hardening in single crystals II. Analytical characterization and predictions. Proc. R. Soc. Lond. A. **435**, 21–41 (1991)
14. G. Beer, I. Smit, C. Duenser, *The Boundary Element Method with Programming* (Springer, Wien, 2008)
15. T. Belytschko, W. Liu, B. Moran, K. Elkhodary, *Nonlinear Finite Elements for Continua and Structures*, 2nd edn. (Wiley, London, 2014)
16. A. Bensoussan, J. Lions, G. Papanicolaou, *Asymptotic Analysis for Periodic Structures* (American Mathematical Society, Providence, 2011)
17. F. Bobaru, Influence of van der Waals forces on increasing the strength and toughness in dynamic fracture of nanofibre networks: a peridynamic approach. Model. Simul. Mater. Sci. Eng. **15**(5), 397–417 (2007)
18. J. Bonet, R. Wood, *Nonlinear Continuum Mechanics for Finite Element Analysis*, 2nd edn. (Cambridge University Press, Cambridge, 2008)
19. A. Bower, *Applied Mechanics of Solids* (CRC, New York, 2010)
20. S. Brisard, L. Dormieux, FFT-based methods for the mechanics of composites: a general variational framework. Comput. Mater. Sci. **49**(3), 663–671 (2010)
21. S. Brisard, L. Dormieux, Combining Galerkin approximation techniques with the principle of Hashin and Shtrikman to derive a new FFT-based numerical method for the homogenization of composites. Comput. Methods Appl. Mech. Eng. **217–220**, 197–212 (2012)

22. S.B. Brown, K.H. Kim, L. Anand, An internal variable constitutive model for hot working of metals. Int. J. Plast. **5**, 95–130 (1989)
23. H. Butt, M. Kappl, *Surface and Interfacial Forces*, 3rd edn. (Wiley-VCH, Weinheim, 2010)
24. H. Butt, K. Graf, M. Kappl, *Physics and Chemistry of Interfaces*, 3rd edn. (Wiley-VCH, Weinheim, 2013)
25. Y. Chen, J. Lee, Connecting molecular dynamics to micromorphic theory. (I). Instantaneous and averaged mechanical variables. Phys. A Stat. Mech. Appl. **322**, 359–376 (2003)
26. Y. Chen, J. Lee, Determining material constants in micromorphic theory through phonon dispersion relations. Int. J. Eng. Sci. **41**(8), 871–886 (2003)
27. Y. Chen, J. Lee, A. Eskandarian, Examining the physical foundation of continuum theories from the viewpoint of phonon dispersion relation. Int. J. Eng. Sci. **41**(1), 61–83 (2003)
28. Y. Chen, J. Lee, A. Eskandarian, Atomistic viewpoint of the applicability of microcontinuum theories. Int. J. Solids Struct. **41**(8), 2085–2097 (2004)
29. R. Christensen, *Theory of Viscoelasticity: An Introduction*, 2nd edn. (Academic, New York, 1982)
30. D. Cioranescu, P. Donato, *An Introduction to Homogenization* (Oxford University Press, Oxford, 1999)
31. A. Cleland, *Foundations of Nanomechanics: From Solid-State Theory to Device Applications* (Springer, New York, 2003)
32. E. Coenen, V. Kouznetsova, M. Geers, Enabling microstructure-based damage and localization analyses and upscaling. Model. Simul. Mater. Sci. Eng. **19**(7), 074,008 (2011)
33. E. Coenen, V. Kouznetsova, M. Geers, Novel boundary conditions for strain localization analyses in microstructural volume elements. Int. J. Numer. Methods Eng. **90**(1), 1–21 (2012)
34. COMSOL Multiphysics, http://www.comsol.com (2015)
35. E. Cosserat, *Theorie des Corps Deformable* (Hermann, Paris, 1909)
36. CUBIT Geometry and Meshing Toolkit, https://cubit.sandia.gov (2012)
37. DAMASK, http://damask.mpie.de (2015)
38. B. Devincre, L. Kubin, Scale transitions in crystal plasticity by dislocation dynamics simulations. C. R. Phys. **11**, 274–284 (2010)
39. F. Devries, H. Dumontet, G. Duvaut, F. Lene, Homogenization and damage for composite structures. Int. J. Numer. Methods Eng. **27**, 285–298 (1989)
40. R. Dingreville, J. Qu, M. Cherkaoui, Surface free energy and its effect on the elastic behavior of nano-sized particles, wires and films. J. Mech. Phys. Solids **53**, 1827–1854 (2005)
41. H. Dumontet, Study of a boundary layer problem in elastic composite materials. Math. Model. Numer. Anal. **20**(2), 265–286 (1987)
42. M. Duzzi, M. Zaccariotto, U. Galvanetto, Application of peridynamic theory to nanocomposite materials. Adv. Mater. Res. **1016**, 44–48 (2014)
43. P. Eisenlohr, M. Diehl, R. Lebenshohn, F. Roters, A spectral method solution to crystal elasto-viscoplasticity at finite strains. Int. J. Plast. **46**, 37–53 (2013)
44. J. Ericksen, On the Cauchy-Born rule. Math. Mech. Solids **13**, 199–200 (2008)
45. A. Eringen, *Microcontinuum Field Theories I: Foundations and Solids* (Springer, New York, 1999)
46. A. Eringen, *Nonlocal Continuum Field Theories* (Springer, New York, 2002)
47. A. Eringen, D. Edelen, On nonlocal elasticity. Int. J. Eng. Sci. **10**(3), 233–248 (1972)
48. A. Eringen, E. Suhubi, Nonlinear theory of simple microelastic solids – I. Int. J. Eng. Sci. **2**(2), 189–203 (1964)
49. E. Eringen, E. Oterkus, *Peridynamic Theory and its Applications* (Springer, New York, 2014)
50. J. Fish, *Practical Multiscaling* (Wiley, Chichester, 2014)
51. N. Fleck, J. Hutchinson, Strain gradient plasticity. Adv. Appl. Mech. **33**, 295–361 (1997)
52. P. Franciosi, A. Zaoui, Multislip in F.C.C crystals: a theoretical approach compared with experimental data. Acta Metall. **30**, 1627 (1982)
53. H. Fu, D.J. Benson, M.A. Meyers, Computational description of nanocrystalline deformation based on crystal plasticity. Acta Mater. **52**, 4413–4425 (2004)

54. M. Geers, V. Kouznetsova, W. Brekelmans, Multi-scale computational homogenization: trends and challenges. J. Comput. Appl. Math. **234**(7), 2175–2182 (2010)
55. P. Germain, The method of virtual power in continuum mechanics. part 2: microstructure. SIAM J. Appl. Math. **25**(3), 556–575 (1973)
56. S. Gonella, M. Greene, W. Liu, Characterization of heterogeneous solids via wave methods in computational microelasticity. J. Mech. Phys. Solids **59**, 959–974 (2011)
57. M. Gurtin, A. Murdoch, Addenda to our paper a continuum theory of elastic material surfaces. Arch. Ration. Mech. Anal. **59**(4), 389–390 (1975)
58. M. Gurtin, A. Murdoch, A continuum theory of elastic material surfaces. Arch. Ration. Mech. Anal. **57**(4), 291–323 (1975)
59. M. Gurtin, A. Murdoch, Surface stress in solids. Int. J. Solids Struct. **14**, 431–440 (1978)
60. M. Gurtin, J. Weissmuller, F. Larche, A general theory of curved deformable interfaces in solids at equilibrium. Philos. Mag. A **78**(5), 1093–1109 (1998)
61. L. He, Z. Li, Impact of surface stress on stress concentration. Int. J. Solids Struct. **43**, 6208–6219 (2006)
62. R. Hill, J.R. Rice, Constitutive analysis of elastic plastic crystals at arbitrary strain. J. Mech. Phys. Solids **20**, 401–413 (1972)
63. M. Holmes, *Introduction to Perturbation Methods*, 2nd edn. (Springer, New York, 2013)
64. G. Holzapfel, *Nonlinear Solid Mechanics: A Continuum Approach for Engineering* (Wiley, New York, 2000)
65. C. Huet, Application of variational concepts to size effects in elastic heterogeneous bodies. J. Mech. Phys. Solids **38**(6), 813–841 (1990)
66. J.W. Hutchinson, Bounds and self-consistent estimates for creep of polycrystalline materials. Proc. R. Soc. Lond. A **348**, 101–127 (1976)
67. H. Ibach, The role of surface stress in reconstruction, epitaxial growth and stabilization of mesoscopic structures. Surf. Sci. Rep. **29**, 193–263 (1997)
68. V. Jikov, S. Kozlov, O. Oleinik, *Homogenization of Differential Operators and Integral Functionals* (Springer, New York, 1994)
69. M. Jirásek, Nonlocal models for damage and fracture: comparison of approaches. Int. J. Solids Struct. **35**(31–32), 4133–4145 (1998)
70. M. Kabel, D. Merkert, M. Schneider, Use of composite voxels in FFT-based homogenization. Comput. Methods Appl. Mech. Eng. **294**, 168–188 (2015)
71. S.R. Kalidindi, Incorporation of deformation twinning in crystal plasticity models. J. Mech. Phys. Solids **46**, 267–290 (1998)
72. S.R. Kalidindi, C.A. Bronkhorst, L. Anand, Crystallographic texture evolution in bulk deformation processing of FCC metals. J. Mech. Phys. Solids **40**, 537 (1992)
73. U.F. Kocks, Laws for work-hardening and low-temperature creep. ASME J. Eng. Mater. Tech. **98**, 76–85 (1976)
74. U. Kocks, C. Tome, H. Wenk (eds.), *Texture and Anisotropy: Preferred Orientations in Polycrystals and Their Effect on Material Properties* (Cambridge University Press, New York, 1998)
75. V. Kouznetsova, W. Brekelmans, F. Baaijens, An approach to micro-macro modeling of heterogeneous materials. Comput. Mech. **27**(1), 37–48 (2001)
76. V. Kouznetsova, M. Geers, W. Brekelmans, Multi-scale constitutive modelling of heterogeneous materials with a gradient-enhanced computational homogenization scheme. Int. J. Numer. Methods Eng. **54**(8), 1235–1260 (2002)
77. E. Kreysig, *Introductory Functional Analysis with Applications* (Wiley, New York, 1978)
78. E. Kröner, Elasticity theory of materials with long range cohesive forces. Int. J. Solids Struct. **3**(5), 731–742 (1967)
79. L. Kubin, B. Devincre, T. Hoc, Modeling dislocation storage rates and mean free paths in face-centered cubic crystals. Acta Mater. **56**, 6040–6049 (2008)
80. LAMMPS, Molecular dynamics simulator. http://lammps.sandia.gov (2015)
81. H. Ledbetter, Monocrystal-polycrystal elastic constants of a stainless steel. Phys. Status Solidi A **85**(1), 89–96 (1984)

82. E.H. Lee, Elastic-plastic deformation at finite strains. Appl. Mech. **36**, 1–6 (1969)
83. J. Lemaitre, J. Chaboche, *Mechanics of Solid Materials* (Cambridge University Press, Cambridge, 1990)
84. L. Li, P.M. Anderson, M.G. Lee, E. Bitzek, P. Derlet, H.V. Swygenhoven, The stress-strain response of nanocrystalline metals: A quantized crystal plasticity approach. Acta Mater. **57**, 812–822 (2009)
85. L. Li, M.G. Lee, P.M. Anderson, Critical strengths for slip events in nanocrystalline metals: predictions of quantized crystal plasticity simulations. Metall. Mater. Trans. A Phys. Metall. Mater. Sci. **42**, 3875–3882 (2011)
86. H. Lim, C.R. Weinberger, C.C. Battaile, T.E. Buchheit, Application of generalized non-Schmid yield law to low temperature plasticity in BCC transition metals. Model. Simul. Mater. Sci. Eng. **21**, 045,015 (2013)
87. J. Lubliner, *Plasticity Theory* (Macmillan Publishing Company, New York, 1990)
88. P. Ludwik, *Element der Technologischen Mechanik* (Springer, New York, 1909)
89. A. Ma, F. Roters, D. Raabe, A dislocation density based constitutive law for BCC materials in crystal plasticity FEM. Comp. Mat. Sci. **39**, 91–95 (2007)
90. L. Malvern, *Introduction to the Mechanics of a Continuous Medium* (Prentice-Hall, Englewood Cliffs, 1969)
91. K.K. Mathur, P.R. Dawson, On modeling the development of crystallographic texture in bulk forming processes. Int. J. Plast. **5**, 67–94 (1989)
92. C. McVeigh, W. Liu, Linking microstructure and properties through a predictive multiresolution continuum. Comput. Methods Appl. Mech. Eng. **197**, 3268–3290 (2008)
93. C. McVeigh, W. Liu, Multiresolution continuum modeling of micro-void assisted dynamic adiabatic shear band propagation. J. Mech. Phys. Solids **58**, 187–205 (2010)
94. MDS: Multiscale Design Systems, http://multiscale.biz (2015)
95. C. Mei, B. Vernescu, *Homogenization Methods for Multiscale Mechanics*, 2nd edn. (World Scientific, New York, 2010)
96. C. Mi, D. Kouris, Nanoparticles under the influence of surface/interface elasticity. J. Mech. Mater. Struct. **1**(4), 763–791 (2006)
97. J. Michel, H. Moulinec, P. Suquet, Effective properties of composite materials with periodic microstructure: a computational approach. Comput. Methods Appl. Mech. Eng. **172**(1–4), 109–143 (1999)
98. J. Michel, H. Moulinec, P. Suquet, Comparison of three accelerated FFT-based schemes for computing the mechanical response of composite materials. Int. J. Numer. Methods Eng. **97**, 960–985 (2014)
99. B. Mihaila, M. Knezevic, A. Cardenas, Three orders of magnitude improved efficiency with high-performance spectral crystal plasticity on GPU platforms. Int. J. Numer. Methods Eng. (2014)
100. R. Miller, V. Shenoy, Size-dependent elastic properties of nanosized structural elements. Nanotechnology **11**, 139–147 (2000)
101. R. Mindlin, Influence of couple-stresses on stress concentrations. Exp. Mech. **3**(1), 1–7 (1963)
102. R. Mindlin, Micro-structure in linear elasticity. Arch. Ration. Mech. Anal. **16**(1), 51–78 (1964)
103. R. Mindlin, Second gradient of strain and surface-tension in linear elasticity. Int. J. Solids Struct. **1**, 417–438 (1965)
104. R. Mindlin, N. Eshel, On first strain-gradient theories in linear elasticity. Int. J. Solids Struct. **4**, 109–124 (1968)
105. R. Mindlin, H. Tiersten, Effects of couple-stresses in linear elasticity. Arch. Ration. Mech. Anal. **11**(1), 415–448 (1962)
106. T. Mori, K. Tanaka, Average stress in matrix and average elastic energy of materials with misfitting inclusions. Acta Metall. **21**(5), 571–574 (1973)
107. H. Moulinec, P. Suquet, A numerical method for computing the overall response of nonlinear composites with complex microstructure. Comput. Methods Appl. Mech. Eng. **157**(1–2), 69–94 (1998)

108. P. Muller, A. Saul, Elastic effects on surface physics. Surf. Sci. Rep. **54**, 157–258 (2004)
109. T. Mura, *Micromechanics of Defects in Solids* (Martinus Nijhoff Publishers, The Hague, 1982)
110. S. Nemat-Nasser, M. Hori, *Micromechanics: Overall Properties of Heterogeneous Materials*, 2nd edn. (Elsevier, Amsterdam, 1999)
111. A. Nowick, B. Berry, *Anelastic Relaxation in Crystalline Solids* (Academic, New York,1972)
112. G. Papanicolaou, S. Varadhan, Boundary value problems with rapidly oscillating random coefficients. Colloquia Math. Soc. J'anos Bolyai **27**, 835–873 (1979)
113. H. Park, P. Klein, A surface cauchy-born analysis of surface stress effects on metallic nanowires. Physical Review B **75**, 085,408:1–9 (2007)
114. H. Park, P. Klein, G. Wagner, A surface cauchy-born model for nanoscale materials. Int. J. Numer. Methods Eng. **68**, 1072–1095 (2006)
115. M. Parks, R. Lehoucq, S. Plimpton, S. Silling, Implementing peridynamics within a molecular dynamics code. Comput. Phys. Commun. **179**(11), 777–783 (2008)
116. M. Parks, S. Plimpton, R. Lehoucq, S. Silling, Peridynamics with LAMMPS: a user guide. Technical Report, SAND 2008-1035, Sandia National Laboratories (2008). http://www.sandia.gov/~mlparks
117. G. Pavliotis, A. Stuart, *Multiscale Methods: Averaging and Homogenization* (Springer, New York, 2008)
118. J. Peddieson, G. Buchanan, R. McNitt, Application of nonlocal continuum models to nanotechnology. Int. J. Eng. Sci. **41**(3–5), 305–312 (2003)
119. R. Peerlings, N. Fleck, Computational evaluation of strain gradient elasticity constants. Int. J. Multiscale Comput. Eng. **2**(4), 599–619 (2004)
120. R. Peerlings, M. Geers, R. de Borst, W. Brekelmans, A critical comparison of nonlocal and gradient-enhanced softening continua. Int. J. Solids Struct. **38**(44–45), 7723–7746 (2001)
121. D. Peirce, R.J. Asaro, A. Needleman, An analysis of nonuniform and localized deformation in ductile single crystals. Acta Metall. **30**, 1087–1119 (1982)
122. D. Pekurovsky, P3DFFT: A framework for parallel computations of Fourier transforms in three dimensions. SIAM J. Sci. Comput. **34**(4), C192–C209 (2012)
123. C. Polizzotto, Nonlocal elasticity and related variational principles. Int. J. Solids Struct. **38**(42–43), 7359–7380 (2001)
124. C. Polizzotto, Unified thermodynamic framework for nonlocal/gradient continuum theories. Eur. J. Mech. A. Solids **22**(5), 651–668 (2003)
125. P. Seleson, M. Parks, M. Gunzburger, R. Lehoucq, Peridynamics as an upscaling of molecular dynamics. Multiscale Model. Simul. **8**(1), 204–227 (2009)
126. J.R. Rice, Inelastic constitutive relations for solids, an internal-variable theory and its application to metal plasticity. J. Mech. Phys. Solids **19**, 443–455 (1971)
127. F. Roters, P. Eisenlohr, T. Bieler, D. Raabe, *Crystal Plasticity Finite Element Methods in Materials Science and Engineering* (Wiley-VCH, Berlin, 2010)
128. F. Roters, P. Eisenlohr, L. Hantcherli, D. Tjahjanto, T. Bieler, D. Raabe, Overview of constitutive laws, kinematics, homogenization and multiscale methods in crystal plasticity finite-element modeling: theory, experiments, applications. Acta Mater. **58**, 1152–1211 (2010)
129. F. Roters, P. Eisenlohr, C. Kords, D. Tjahjanto, M. Diehl, D. Raabe, DAMASK: the Düsseldorf Advanced MAterial Simulation Kit for studying crystal plasticity using an FE based or a spectral numerical solver. Proc. IUTAM **3**, 3–10 (2012)
130. A. Rusanov, Surface thermodynamics revisited. Surf. Sci. Rep. **58**, 111–239 (2005)
131. A. Rusanov, Surface thermodynamics of cracks. Surf. Sci. Rep. **67**, 117–140 (2012)
132. G. Sachs, Ableitung einer fliessbedingung. Z. Ver. Dtsch. Ing. **72**, 734–736 (1928)
133. E. Sanchez-Palencia, *Non-Homogeneous Media and Vibration Theory*. Lecture Notes in Physics, vol. 127 (Springer, New York, 1980)
134. P. Sharma, S. Ganti, Size-dependent Eshelby's tensor for embedded nano-inclusions incorporating surface/interface energies. J. Appl. Mech. **71**, 663–671 (2004)
135. V. Shenoy, R. Miller, E. Tadmor, D. Rodney, R. Phillips, M. Ortiz, An adaptive finite element approach to atomic-scale mechanics–the quasicontinuum method. J. Mech. Phys. Solids **47**, 611–642 (1999)

136. S. Silling, Reformulation of elasticity theory for discontinuities and long-range forces. J. Mech. Phys. Solids **48**(1), 175–201 (2000)
137. S. Silling, R. Lehoucq, Peridynamic theory of solid mechanics. Adv. Appl. Mech. **44**, 74–164 (2010)
138. S. Silling, M. Epton, O. Weckner, J. Xu, E. Askari, Peridynamic states and constitutive modeling. J. Elast. **88**(2), 151–184 (2007)
139. V. Smyshlyaev, K. Cherednichenko, On rigorous derivation of strain gradient effects in the overall behaviour of periodic heterogeneous media. J. Mech. Phys. Solids **48**, 1325–1357 (2000)
140. E. Suhubi, A. Eringen, Nonlinear theory of simple microelastic solids – II. Int. J. Eng. Sci. **2**(4), 389–404 (1964)
141. E. Tadmor, M. Ortiz, R. Phillips, Quasicontinuum analysis of defects in solids. Philos. Mag. A **73**(6), 1529–1563 (1996)
142. E. Tadmor, G. Smith, N. Bernstein, E. Kaxiras, Quasicontinuum analysis of defects in solids. Phys. Rev. B **59**(1), 235–245 (1999)
143. Tahoe Development Project, http://tahoe.sourceforge.net (2015)
144. G.I. Taylor, The mechanism of plastic deformation of crystals. Part I. Theoretical. Proc. R. Soc. A **165**, 362–387 (1934)
145. G.I. Taylor, Plastic strain in metals. J. Inst. Metals **62**, 307–324 (1938)
146. S. Timoshenko, J. Goodier, *Theory of Elasticity*, 3rd edn. (McGraw-Hill, New York, 1987)
147. R. Toupin, Elastic materials with couple-stresses. Arch. Ration. Mech. Anal. **11**(1), 385–414 (1962)
148. R. Toupin, Theories of elasticity with couple-stress. Arch. Ration. Mech. Anal. **17**(2), 85–112 (1964)
149. T. Tran, V. Monchiet, G. Bonnet, A micromechanics-based approach for the derivation of constitutive elastic coefficients of strain-gradient media. Int. J. Solids Struct. **49**(5), 783–792 (2012)
150. F. Vernerey, W. Liu, B. Moran, Multi-scale micromorphic theory for hierarchical materials. J. Mech. Phys. Solids **55**, 2603–2651 (2007)
151. Q. Wang, K. Liew, Application of nonlocal continuum mechanics to static analysis of micro- and nano-structures. Phys. Lett. A **363**(3), 236–242 (2007)
152. J. Wang, Z. Huang, H. Duan, S. Yu, X. Feng, G. Wang, W. Zhang, T. Wang, Surface stress effect in mechanics of nanostructured materials. Acta Mech. Solida Sin. **24**(1), 52–82 (2011)
153. D.H. Warner, J.F. Molinari, A semi-discrete and non-local crystal plasticity model for nanocrystalline metals. Scr. Mater. **54**, 1397–1402 (2006)
154. X. Yuan, Y. Tomita, T. Andou, A micromechanical approach of nonlocal modeling for media with periodic microstructures. Mech. Res. Commun. **35**(1–2), 126–133 (2008)
155. ZEBULON, http://www.nwnumerics.com/Zebulon/ (2015)
156. J. Zeman, J. Vondřejc, J. Novák, I. Marek, Accelerating a FFT-based solver for numerical homogenization of periodic media by conjugate gradients. J. Comput. Phys. **229**(21), 8065–8071 (2010)

Chapter 4
Density Functional Theory Methods for Computing and Predicting Mechanical Properties

Niranjan V. Ilawe, Marc N. Cercy Groulx, and Bryan M. Wong

4.1 Introduction

Over the past few decades, tremendous progress has been made in the development of computational methods for *predicting* the properties of materials. At the heart of this progress is density functional theory (DFT) [13, 17, 31, 39, 65], one of the most powerful and efficient computational modeling techniques for predicting electronic properties in chemistry, physics, and material science. Prior to the introduction of DFT in the 1960s [31, 39] the only obvious method for obtaining the electronic energies of materials required a direct solution of the many-body Schrödinger equation [62]. While the Schrödinger equation provides a rigorous path for predicting the electronic properties of *any* material system, analytical solutions for realistic systems having more than one interacting electron are out of reach. Moreover, since the Schrödinger equation is inherently a many-body formalism (3N spatial coordinates for N *strongly* interacting electrons), numerically accurate solutions of multi-electron systems are also impractical. Instead of the full 3N-dimensional Schrödinger equation, DFT recasts the electronic problem into a simpler yet mathematically equivalent 3-dimensional theory of non-interacting electrons (cf. Fig. 4.1). The exact form of this electron density, ρ $(= n(r))$, hinges on the mathematical form of the exchange-correlation functional, $E_{xc}[n(r)]$, which is crucial for providing accurate and efficient solutions to the many-body Schrödinger equation. Unfortunately, the exact form of the exchange-correlation functional is currently unknown, and all modern DFT functionals invoke various degrees of approximation. Starting with the pioneering work of Perdew and co-workers

N.V. Ilawe • M.N. Cercy Groulx • B.M. Wong (✉)
Department of Chemical & Environmental Engineering and Materials Science & Engineering, University of California, Riverside, CA 92521
e-mail: bryan.wong@ucr.edu

C.R. Weinberger, G.J. Tucker (eds.), *Multiscale Materials Modeling for Nanomechanics*, Springer Series in Materials Science 245,
DOI 10.1007/978-3-319-33480-6_4

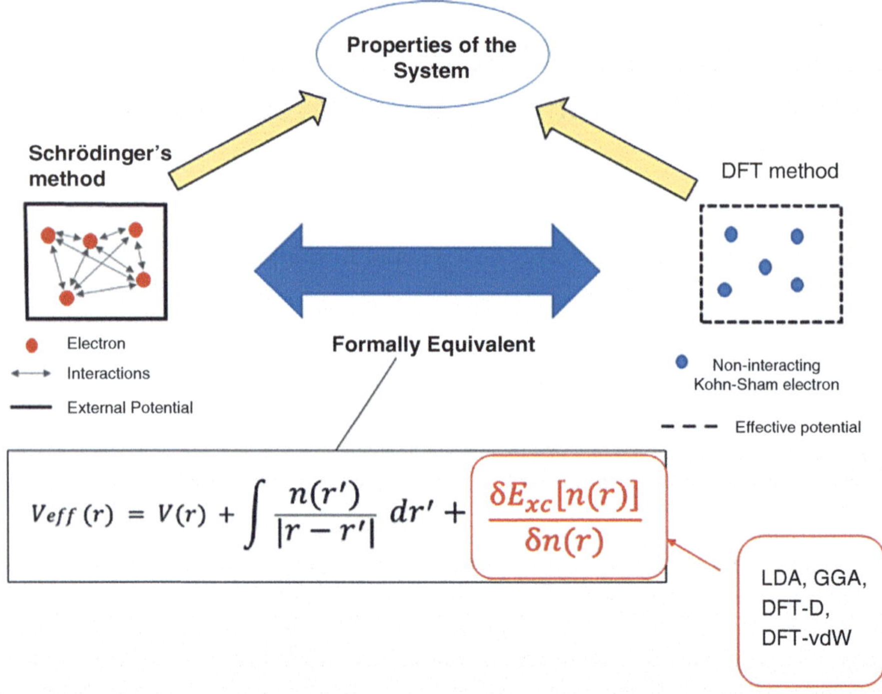

Fig. 4.1 All of the electronic properties of a system can be rigorously obtained by a direct solution of the 3N-dimensional Schrödinger equation (*left*). DFT (*right*) provides a more tractable (but formally equivalent) approach for solving the same many-body electron problem. Adapted from [46]

[3, 4, 51, 54], a hierarchy of approximate functionals [52] has been developed over the past three decades. The following sections in this book chapter focus on a specific set of four DFT families: the local density approximation (LDA) [39], generalized gradient approximation (GGA) [55], DFT-D [24], and DFT-vdW [30] methods. Our motivation in this chapter is not to provide a detailed, comprehensive review of exchange-correlation functionals; rather, we have chosen this specific family of functionals because of their direct relevance to mechanical and structural properties of materials currently studied by our group. In particular, to illustrate the use of these first-principles methods, we focus on two specific mechanical systems: palladium-hydride materials and spiropyran-based mechanochromic polymers, both of which pose unique challenges to electronic structure methods. To provide the necessary background for understanding these mechanical systems, we first give a brief summary of each of the four representative DFT methods while specifically highlighting both the advantages and disadvantages of these individual functional classes. Next, we explore how these four DFT methods can be utilized to predict important mechanical properties such as cohesive energies, maximum strength, stress–strain properties, and mechanochromic effects. We conclude this chapter with a detailed analysis of the different functional families and highlight the relevance of each method for predicting the diverse mechanical properties of these material systems.

4.2 Density Functional Theory

Hohenberg and Kohn [31] established that the total electron density, ρ, completely (and exactly) determines all of the ground-state properties of an N-electron system. Hence, knowledge of the electron density (ρ) implies knowledge of the wave function (up to a phase factor) and all other observables. The density functional approach can be succinctly summarized as follows:

$$\rho \Rightarrow v_{\text{ext}} \Rightarrow \Psi(\mathbf{r}_1, \mathbf{r}_2, \ldots, \mathbf{r}_N) \Rightarrow \text{everything!} \tag{4.1}$$

In other words, the electron density, ρ, uniquely determines the external potential, v_{ext}, which has a one-to-one correspondence with a unique wavefunction, Ψ, that allows us to access all known properties of the system. This, however, is not enough; for the theory to be self-contained we require a variational principle. Consider the multi-electron Schrödinger equation:

$$\left[\sum_{i=1}^{N}\left(-\frac{\hbar^2}{2m_e}\nabla_i^2 + v_{\text{ext}}(\mathbf{r}_i)\right) + \sum_{i<j}\frac{e^2}{\mathbf{r}_i - \mathbf{r}_j}\right]\Psi(\mathbf{r}_1, \mathbf{r}_2, \ldots, \mathbf{r}_N)$$
$$= E\Psi(\mathbf{r}_1, \mathbf{r}_2, \ldots, \mathbf{r}_N) \tag{4.2}$$

where the first term in the square brackets is the kinetic energy operator, and the third term is the inter-electronic repulsion. The first and the last term in the Hamiltonian, which do not involve the external potential, can be cast as a density functional for the total kinetic and Coulombic interaction energy:

$$F(\rho) = T(\rho) + V_{ee}(\rho) \tag{4.3}$$

A straightforward application of the variational principle gives

$$F(\rho') + \int v_{\text{ext}}\rho' \mathbf{dr} \geq F(\rho) + \int v_{\text{ext}}\rho \mathbf{dr} = E_0 \tag{4.4}$$

where ρ' is not the ρ corresponding to v_{ext}, but to some other external potential, and E_0 is the exact ground-state energy. Thus, the Hohenburg–Kohn variational principle permits a solution of the Schrödinger equation for non-degenerate ground states by variationally minimizing the functional $F(\rho)$. Though the functionals $T(\rho)$ and $V_{ee}(\rho)$ formally exist, an exact mathematical form for these density-dependent quantities is not currently known. Kohn and Sham [39] made critical progress in this area by mapping the full interacting system onto a fictitious non-interacting system where the electrons move in an effective "Kohn–Sham" single-particle potential. Within the Kohn–Sham approach, the kinetic energy for the non-interacting system retains a familiar form,

$$T_0(\rho) = -\frac{1}{2}\sum_i 2\int \Psi_i^*(\mathbf{r}_i)\nabla^2\Psi_i(\mathbf{r}_i)\,\mathrm{d}\mathbf{r} \tag{4.5}$$

where, for simplicity, we assume spin-neutral systems. The non-interacting Coulombic interaction energy, $J(\rho)$, in the Kohn–Sham approach was also approximated as the classical Coulomb self-energy,

$$J(\rho) = \frac{1}{2}\int\int \frac{\rho(\mathbf{r}_1)\rho(\mathbf{r}_2)}{\mathbf{r}_{12}}\,\mathrm{d}\mathbf{r}_1\,\mathrm{d}\mathbf{r}_2 \tag{4.6}$$

Kohn and Sham called the error made by these approximations, the exchange-correlation energy E_{xc}:

$$E_{\mathrm{xc}}(\rho) = T(\rho) + V_{\mathrm{xc}}(\rho) - T_0(\rho) - J(\rho) \tag{4.7}$$

The Kohn–Sham total energy functional hence becomes

$$E(\rho) = T_0(\rho) + \int v_{\mathrm{ext}}\rho\,\mathrm{d}\mathbf{r} + J(\rho) + E_{\mathrm{xc}}(\rho) \tag{4.8}$$

This decomposition is noteworthy since T_0 and J for the non-interacting system are given by known expressions, and the "unknown" functional, E_{xc}, is a relatively small part of the total energy. It is important to mention at this point that since T_0 and J depend explicitly on ρ, the Kohn–Sham equations must be solved iteratively to self-consistency. Several computational schemes exist for obtaining self-consistent solutions, ranging from simple under-relaxation techniques [42] to advanced extrapolation approaches [58]. In addition, Perdew and co-workers further showed that the exchange-correlation energy can further be decomposed as $E_{\mathrm{xc}} = E_x + E_c$, where E_x is due to the Pauli principle (exchange energy) and E_c is due to electron correlations. Given the exact exchange-correlation functional, one can, in principle, compute the exact density and total energy of any interacting electronic system. However, the accuracy and efficiency of such solutions hinges on approximations to the exchange-correlation functional, $E_{\mathrm{xc}}(\rho)$, which we discuss in the following section.

4.3 Exchange-Correlation Functionals

4.3.1 The Local Density Approximation

During the formative stages of DFT, the LDA was one of the first approaches for approximating the exchange-correlation energy in bulk solids [39]. Unlike many of the other functionals discussed in this chapter, the LDA approach is unique in that it only relies on the local value of the electron density in three-dimensional space (as opposed to derivatives or nonlocal integrals of the density). In order to provide practical solutions to the many-body problem, the LDA method approximates the system as a homogeneous electron gas (i.e., a jellium model). Within these approximations, the exchange-correlation energy ($E_{\mathrm{xc}} = E_x + E_c$) is relatively simple, with the exchange term, E_x, having an analytic solution, since the per-volume exchange energy of the homogenous electron gas in known exactly [27, 50]. Numerous different approximations exist for the correlation term, E_c, and several of them have been parameterized and fitted from quantum Monte Carlo calculations [8, 18]. Despite their simplicity, LDA methods are surprisingly accurate, particularly in systems characterized by a slow variation in electron density, such as homogenous solid metals. In these systems where the LDA method is applicable, the predicted bond lengths, bond angles, and vibrational frequencies are comparable to experimental values within a few percent. However, many realistic materials are strongly inhomogeneous, and the resulting parameterization in the LDA approach can lead to serious inaccuracies in predicted properties. For example, LDA methods tend to overestimate the cohesive energies and, therefore, bond strengths of most solids. In terms of electronic properties, the band gaps in semiconductors and insulators are seriously underestimated by LDA functionals. Over the years, subsequent improvements have been made to LDA methods in an attempt to correct some of these problems [53, 54, 69, 69]; however, substantial improvements require the implementation of additional variables (and constraints) to address the issues of inhomogeneous electron densities.

4.3.2 The Generalized Gradient Approximation

The GGA improves on many of the weaknesses in LDA by incorporating additional constraints and terms that account for the inhomogeneities of the electron density. Unlike the LDA approach, GGA methods utilize both the electron density and gradients of the density to approximate the true exchange-correlation potential. These additional constraints allow GGA methods to outperform LDA in predicting cohesive energies, bulk moduli, energy barriers, and structural energy differences of atoms. Moreover, the incorporation of gradients in the electron density partially corrects the energetic overbinding in LDA functionals, while also making improvements on both bond lengths and angles. Numerous approaches exist for incorporating gradients of the electron density to generate new GGA functionals. Perdew and co-workers strongly advocate an approach where GGA terms are constructed by only adding parameters that satisfy certain exact constraints of the exchange-correlation potential. A few of the most common functionals in this family are PW91 [56], PBE [55] and more recently RPBE [29], AM05 [45], and PBEsol [57]. In contrast, Becke and co-workers have popularized alternative approaches to constructing functionals by fitting unknown terms to experimental data and high-level calculated atomic and molecular properties. The BLYP [4] GGA falls into this family of exchange-correlation functionals and is one of the most commonly used GGA methods in quantum chemistry. It is important to mention that although GGA methods (in general) improve on many of the deficiencies inherent to the LDA approach, they still fail to account for van der Waals interactions [63] and accurate predictions of electronic bandgaps [63]. These shortcomings arise from the inability of GGA methods to account for nonlocal exchange effects as well as their failure to compensate for electron self-interaction effects. One approach to remedy these effects is to construct hybrid and/or range-separated functionals that mix a portion of nonlocal Hartree–Fock exchange with an appropriately modified GGA. These hybrid or range-separated functionals are seldom used for predicting mechanical properties of large systems and, consequently, are out of scope for this chapter on structural properties.

4.3.3 DFT-D

In order to accurately predict the structural and mechanical properties of large nanosystems and solid structures, it is necessary to go beyond conventional GGA approaches and explicitly account for van der Waals interactions [24]. Van der Waals or dispersion interactions are attractive forces due to electron correlation effects that arise from instantaneous fluctuating charge distributions between material surfaces [37]. While it is obvious that dispersion interactions play a significant role in materials characterized by noncovalent interactions (i.e., molecular aggregates [19, 33, 72]), graphene-based systems [32, 36], nanotube aggregates [70], and layered materials [71], we give several examples in Sect. 4.3 showing that dispersion interactions still play a dominant role in covalently bonded solid structures, particularly for predicting cohesive energies and maximum strength. A relatively simple way to modify GGA exchange-correlation functionals for dispersion interactions is to explicitly add the attractive C_6/R^6 van der Waals term for all atomic pairs. This pragmatic approach was first suggested by the Scoles group to correct Hartree–Fock energies for dispersion effects [30]. The same approach can also be used to correct DFT methods, and these corrections are collectively referred to as DFT-D methods in the scientific literature. There are several variants of DFT-D, and they are further classified according to updated derivations of the empirical dispersion coefficients [15, 20, 22]. For example, several of the next-generation DFT-D family of functionals are already implemented in software packages as DFT-D2 and

DFT-D3 methods. The DFT-D2 and DFT-D3 methods are based on the approaches suggested by Grimme [23, 24] and have already been successfully used in both molecular and solid-state applications. Within the DFT-D2 approach, an empirical atomic pairwise dispersion correction is added to the Kohn–Sham portion of the total energy (EKS-DFT) as

$$E_{\mathrm{DFT-D}} = E_{\mathrm{KS-DFT}} + E_{\mathrm{disp}} \tag{4.9}$$

where E_{disp} is given by

$$E_{\mathrm{disp}} = -s_6 \sum_{i=1}^{N_{\mathrm{at}}-1} \sum_{j=i+1}^{N_{\mathrm{at}}} \sum_{g} f_{\mathrm{damp}}(R_{ij,g}) \frac{C_{6,ij}}{R_{ij,g}^{6}} \tag{4.10}$$

The summation is over all atom pairs i and j, and over all $\mathbf{g}$ lattice vectors with the exclusion of the $i = j$ contribution when $\mathbf{g} = 0$ (this restriction prevents atomic self-interaction in the reference cell). The parameter is the dispersion coefficient for atom pairs i and j, calculated as the geometric mean of the atomic dispersion coefficients. The s_6 parameter is an empirical scaling factor specific to the adopted DFT method ($s_6 = 0.75$ for PBE), and $R_{ij,g}$ is the interatomic distance between atom i in the reference cell and j in the neighboring cell at distance . In order to avoid near-singularities for small interatomic distances, f_{damp} is a damping function that effectively re-scales interatomic forces to minimize interactions within the bonding distance $R_{ij,g}$.These additive approaches do not incur significant computational cost over conventional DFT calculations; however, they generally require the use of relatively large basis set to ensure accurate results. A significant improvement over the DFT-D2 method is the DFT-D3 scheme, which is characterized by higher accuracy, a broader range of applicability, and less empiricism compared to the DFT-D2 method. The D3 correction scheme by Grimme et al. [24] uses the following form of the dispersion correction

$$E_{\mathrm{disp}} = -\sum_{i=1}^{N_{\mathrm{at}}-1} \sum_{j=i+1}^{N_{\mathrm{at}}} \sum_{\mathbf{g}} f_{d,6}(R_{ij,g}) \frac{C_{6,ij}}{R_{ij,g}^{6}} + f_{d,8}(R_{ij,g}) \frac{C_{8,ij}}{R_{ij,g}^{8}} \tag{4.11}$$

where $f_{d,8}$ and are eighth-order damping functions and dispersion coefficients, respectively, for the additional repulsive potential. However, unlike the DFT-D2 method, the dispersion coefficients are geometry dependent and are adjusted as a function of the local geometry around atoms i and j. In the original DFT-D3 method, the $f_{d,6}$ and $f_{d,8}$ damping functions (and thus E_{disp}) were constructed to approach zero when $R_{ij} = 0$. A critical disadvantage of this zero-damping approach is that at small and medium distances, the atoms experience repulsive forces leading to even longer interatomic distances than those obtained without dispersion corrections [25]. As a practical solution for this counter-intuitive observation, Becke and Johnson [5, 35] proposed the DFT-D3(BJ) method which contains modified expressions for $f_{d,6}$ and

$f_{d,8}$ that lead to a constant contribution of E_{disp} to the total energy when $R_{ij} = 0$. The DFT-D3(BJ) method produces improved results for nonbonded distances/energies [25], and we use this variant of the D3 damping function throughout this work. Within the same family of empirically constructed dispersion methods is the DFT-TS method proposed by Tkatchenko and Scheffler [61, 66], which has also been previously applied to molecular solids [40]. While the expression for the dispersion correction in DFT-TS is identical to the DFT-D2 method (Eq. (4.10)), the important distinction is that the dispersion coefficients and damping function in DFT-TS are explicitly dependent on the charge-density. As such, the DFT-TS method takes into account van der Waals interactions due to the local chemical environment by directly calculating the polarizability, dispersion coefficients, and atomic radii from their free-atomic values [66].

4.3.4 vdW-DFT

As discussed in the previous section, the dispersion coefficients in conventional DFT-D methods are empirically derived and, therefore, limited in their accuracy and reliability for general systems. In order to generalize the DFT-D methods to account for the surrounding electronic environment, a natural approach is to explicitly incorporate van der Waals correlation effects directly into the exchange-correlation kernel. The vdW-DF methods proposed by Dion et al. [12] incorporate these dispersion forces by evaluating nonlocal double real space integrals of the electron density to account for these correlation effects. In order to reduce the computational effort to evaluate these nonlocal integrals, the numerical algorithms suggested by Roman-Perez and Soler [60] are often used to transform these real space integrals to reciprocal space. In addition to the original vdW-DF method by Dion et al. [12], there have been several additional improvements including a revised vdW-DF2 approach [41] as well as the "opt" family of functionals [38] (optPBE-vdW, optB88-vdW, and optB86b-vdW). In all of these nonlocal vdW functionals, the exchange-correlation energy, E_{xc}, takes the form

$$E_{\text{xc}} = E_x^{\text{GGA}} + E_c^{\text{LDA}} + E_c^{\text{nl}} \tag{4.12}$$

where E_x^{GGA} is the GGA exchange energy, E_c^{LDA} accounts for the local correlation energy within the LDA, and E_c^{nl} is the nonlocal correlation energy. An important distinguishing characteristic of these nonlocal vdW functionals (compared to the DFT-D approaches) is that they directly incorporate van der Waals effects explicitly in the exchange-correlation kernel as a nonlocal double real space integral of the electron density without assuming a priori an analytic form for the dispersion interaction. For additional information on these nonlocal-vdW correlation methods, the interested reader is encouraged to review the publications of Langreth and Lundqvist [12], as well as related work by the Michaelides group [38].

4.4 DFT Predictions of Mechanical Properties in Palladium-Hydride Systems

We now shift our focus to a few applications of the above-discussed DFT methods on practical calculations and predictions of mechanical properties. Within this section, we give a summary of our research on the mechanical properties of palladium-hydride materials. This particular material system was chosen to highlight the importance of carefully choosing appropriate DFT methods for accurately characterizing these systems [34]. In this context, over the past 7 years our group has focused on the development and use of both DFT and molecular dynamics methods for characterizing palladium-based materials. Our motivation stems from the widespread use of palladium in enabling technological advancements for both catalytic and hydrogen-storage applications. By far, the largest usage of palladium resides in catalytic converters within the auto industry and their use as catalytic materials in the agrochemical, pharmaceutical, and materials-based sectors [14, 67]. Within the hydrogen-storage and production industries, palladium is widely used in membrane reactors for the production of high-purity hydrogen [64], and palladium-based alloys are actively used in new fuel cell technologies as replacements for costly platinum-based alloys [1].

At room temperature, pure palladium reacts with both molecular and atomic hydrogen to form various palladium-hydride (PdH_x) stoichiometries [43]. As shown in Fig. 4.2, these metal-hydride materials are formed when hydrogen atoms become embedded into various interstitial sites within the metal lattice [28, 74]. Since the hydrogen absorption process in palladium is reversible [26], there is significant interest in using palladium-based alloys for large-scale storage of hydrogen and their isotopes. While there has been considerable work in examining the equilibrium hydrogen-storage properties of palladium, practical applications of these materials are still limited by their intrinsic dynamical embrittlement caused by the growth of gaseous bubbles in the lattice. In particular, several experimental and computational studies have shown that palladium-hydride materials containing large amounts of hydrogen are more susceptible to brittle fracture than their pure metallic counterparts [2]. In order to fully characterize these complex materials, it is necessary to carry out detailed predictions of both their static and dynamic structural properties to understand the limitations of first-principle-based methods for these processes. Specifically, we focus on the implications of using dispersion-corrected DFT methods for predicting the mechanical properties of various PdH_x materials to serve as guidance for experiment and also as benchmarks for constructing new Pd-H MD potentials.

Although most work on dispersion-corrected methods have focused on predicting properties of molecular systems at equilibrium (i.e., structures and energies at a global minimum), there has been significantly less work on understanding the performance of these methods for systems far from equilibrium. In particular, there is mounting evidence in the scientific literature demonstrating the importance of including non-equilibrium geometries and energies to construct new

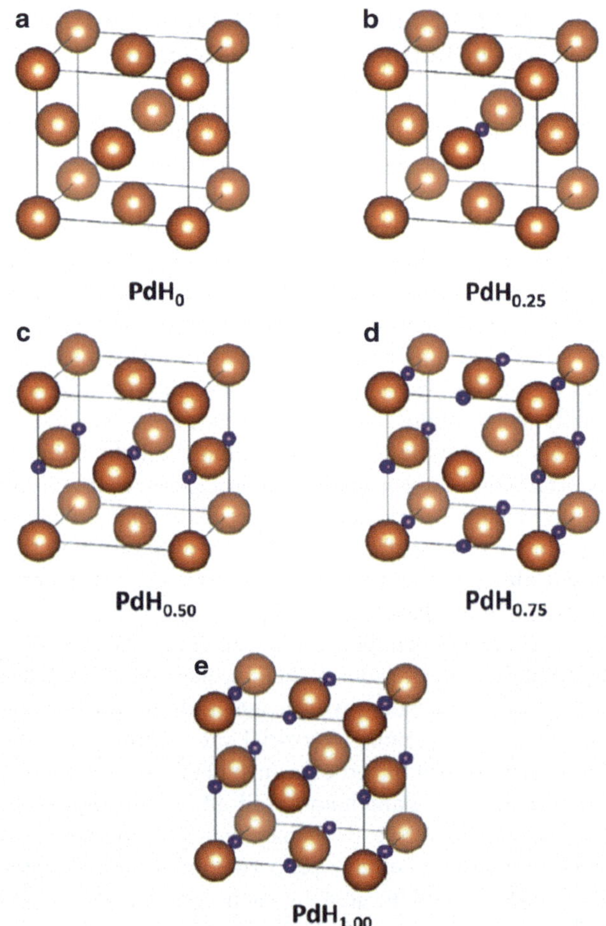

Fig. 4.2 Unit cells for various PdH_x stoichiometries where x = (**a**) 0, (**b**) 0.25, (**c**) 0.50, (**d**) 0.75, and (**e**) 1.00. The large orange spheres represent Pd atoms and the small purple spheres denote the placement of H atoms

dispersion-corrected DFT methods that are globally accurate [21, 44]. While there has been increased effort to incorporate these effects in molecular systems, we are not aware of any prior studies using dispersion-based DFT methods for bulk solids far from equilibrium (i.e., strained up to their maximum tensile strength). To bridge this knowledge gap, we carry out an extensive series of DFT calculations on both equilibrium (optimized geometries and cohesive energies) as well as non-equilibrium properties (stress–strain relationships and maximum tensile strength) in various PdH_x systems. We demonstrate that dispersion effects play a vital role in these complex solids and, most importantly, we show that some dispersion-corrected functionals can give spurious, *qualitatively incorrect* results for structural properties of solids far from equilibrium. We give a detailed analysis of both the static and dynamical properties predicted by the various DFT methods, and we discuss the implications of using these methods in developing new DFT parameterizations and fitted MD potentials for these materials.

4.4.1 Lattice Constants and Cohesive Energies

For all of the calculations in this work, we used a locally modified version of the VASP code for calculating the ideal tensile strength along arbitrary crystallographic directions (described further in Section 4.4.2). A planewave energy cutoff of 500 eV was used in conjunction with a dense 15 × 15 × 15 Monkhorst–Pack grid for sampling the Brillouin zone. All calculations utilized projected augmented wave pseudopotentials, and a Gaussian smearing scheme with a 0.2 eV width was used to describe the partial occupancies for all the Pd-based solids. Before proceeding to an analysis of non-equilibrium stress–strain properties in the various PdH_x systems, it is necessary to first characterize the static properties of these solids. While the static properties of pure Pd are well known, the lattice constants, bulk moduli, and cohesive energies of the other palladium-hydride systems are not well characterized. To serve as benchmarks for our survey of various DFT methods, we calculated the lattice constants and bulk moduli of pure Pd and compared their predicted values against experiment (corrected for zero-point energy effects) [10]. The predicted lattice constant and bulk moduli for Pd is presented in Table 4.1, and the relative errors between the predicted and experimental values are summarized in Fig. 4.3a.

As shown in Fig. 4.3a, LDA slightly underestimates the lattice constant for pure Pd. This error is not surprising, as bond lengths in LDA are typically underestimated due to the simplistic homogenous electron gas approximation within the LDA exchange-correlation potential. The second-generation vdW-DF2 functional by Langreth and co-workers significantly overestimates the lattice constant compared to all of the tested methods. This error in the vdW-DF2 functional is consistent with the previous study by Klimes et al. [38] which found that vdW-DF2 has average relative errors of 2.6 % in the lattice constants for various solids. The semilocal PBE functional also overestimates the lattice constant, although the error is surprisingly not as severe as the vdW-DF2 predictions. The PBE-D2, PBE-D3(BJ), PBE-TS, and the nonlocal optB86b-vdW functionals give the best agreement to experiment, with the PBE-D2 and PBE-D3(BJ) functionals only performing marginally better than PBE-TS and optB86b-vdW.

Table 4.1 Lattice constants and bulk moduli for pure Pd as predicted by various DFT methods

	Lattice constant (Å)		Bulk modulus (GPa)	
	Pure Pd	error (%)	Pure Pd	error (%)
LDA	3.839	0.92	223	14.6
PBE	3.946	1.84	167	14.4
PBE-D2	3.890	0.39	164	16.1
PBE-D3(BJ)	3.890	0.39	188	3.4
PBE-TS	3.917	1.07	183	6.4
vdW-DF2	4.095	5.68	119	38.9
optB86b-vdW	3.903	0.73	183	6.3

The experimental lattice constant and bulk modulus for Pd is 3.875 Å and 195 GPa, respectively [10]

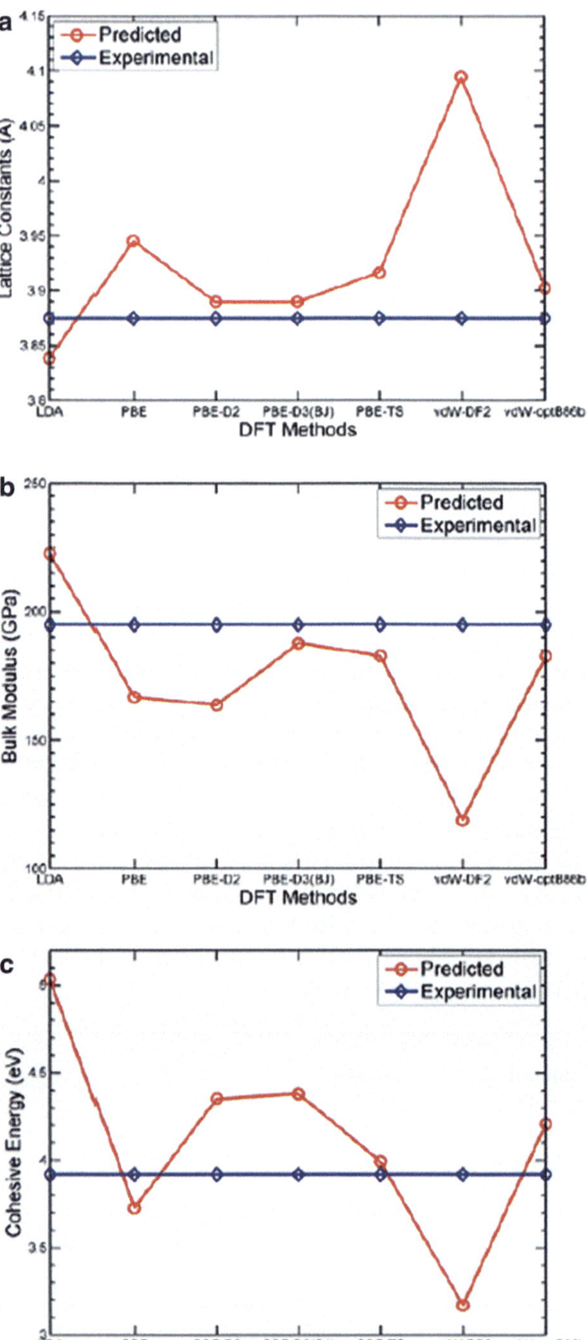

Fig. 4.3 (**a**) Lattice constants, (**b**) bulk moduli, and (**c**) cohesive energies of pure palladium predicted by various DFT methods

Table 4.2 Cohesive energies (in eV) for various PdH_x stoichiometries as predicted by various DFT methods

	PdH_0	$PdH_{0.25}$	$PdH_{0.50}$	$PdH_{0.75}$	$PdH_{1.00}$
LDA	5.036	4.606	4.328	4.130	3.974
PBE	3.728	3.459	3.284	3.156	3.051
PBE-D2	4.354	3.969	3.721	3.544	3.403
PBE-D3(BJ)	4.383	4.001	3.754	3.575	3.432
PBE-TS	3.993	3.689	3.491	3.349	3.233
vdW-DF2	3.169	2.976	2.849	2.754	2.674
optB86b-vdW	4.208	3.879	3.668	3.516	3.394

The experimental cohesive energy for pure PdH_0 is 3.92 eV [10]

To complement the lattice constant calculations described previously, we also calculated the cohesive energies (defined as the energy required for separating the condensed material into isolated free atoms) of all the various PdH_x systems using the expression:

$$E_{\text{cohesive}} = n_{\text{Pd}} E_{\text{Pd}} + n_{\text{H}} E_{\text{H}} - E_{\text{PdH}} \tag{4.13}$$

where, n_{Pd} is the number of palladium atoms in the PdH_x system, E_{Pd} is the energy of an isolated palladium atom, n_{H} is the number of hydrogen atoms in the PdH_x system, E_H is the energy of an isolated hydrogen atom, and E_{PdH_x} is the total electronic energy of the total PdH_x system. Based on its definition, the cohesive energy provides an indirect measure of the stability of the material (with larger positive values indicating more stability) relative to the atomic elements. The calculated cohesive energies are presented in Table 4.2, and the relative difference between the predicted and experimental values for pure Pd is summarized in Fig. 4.3b. The experimental cohesive energy for pure Pd, corrected for zero-point energy effects, is 3.92 eV as obtained from [10].

Figure 4.3b provides a complementary viewpoint of chemical bonding in comparison to the lattice constants plotted in Fig. 4.3a. In particular, Fig. 4.3b shows an opposite trend in the lattice constant compared to Fig. 4.3a. DFT methods that predict a smaller lattice constant also yield a higher cohesive energy (i.e., stronger bonding) and vice-versa. As such, the trends in the cohesive energy can be rationalized using the same arguments previously discussed for the lattice constant. For example, Fig. 4.3b shows that LDA significantly overestimates the cohesive energy for Pd, which is consistent with the small lattice constant predicted by LDA in Fig. 4.3a. Interestingly, the vdW-DF2 functional underestimates the cohesive energy with a similar magnitude as the overestimation by LDA. This error in the vdW-DF2 functional is also consistent with the previous analyses by Klimes et al. [38] which found that vdW-DF2 has quite large relative errors of −15.9 % in the atomization energies for various solids. These large errors have been observed in many other systems [9, 38, 59] and can be attributed to the steep behavior of the vdw-DF2 functional which at short distances is still too

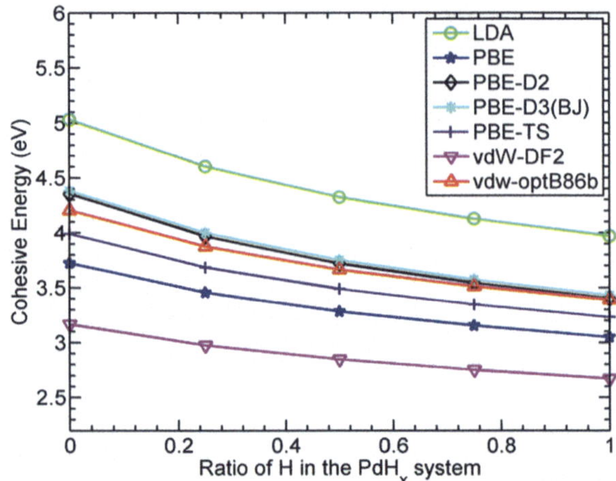

Fig. 4.4 Cohesive energy trends as a function of hydrogen percentage in the PdH_x system as predicted by various DFT methods

repulsive [47]. Both the PBE-D2 and PBE-D3(BJ) functionals overestimate the cohesive energy, although the error is not as severe as the LDA predictions. The PBE, PBE-TS, and the nonlocal vdW-optB86b functionals give the best agreement to experiment, with PBE slightly underestimating the cohesive energy compared to PBE-TS and vdW-optB86b. Finally, Fig. 4.4 summarizes the cohesive energy trends of the remaining PdH_x systems ($x = 0$, 0.25, 0.50, 0.75, and 1.00; cf. Fig. 4.2) for all the various DFT methods. Again, we have chosen these specific stoichiometries based on our previous molecular dynamics study of thermodynamically stable PdH_x configurations for these materials [28, 74]. While there are distinct differences between the magnitudes of the cohesive energies, it is important to mention that all DFT methods predict that the cohesive energy in the PdH_x systems decreases as the percentage of H in the metal-hydride lattice increases. These observations, coupled with our previous discussion of cohesive energies, corroborate the effects of embrittlement due to the presence of hydrogen in these materials.

4.4.2 Stress–Strain Relationships

With the static equilibrium properties now fully characterized for each of the DFT methods, we turn our attention to non-equilibrium stress–strain effects. For uniaxial tensile strain, we calculated the tensile stress, σ, using the expression

$$\sigma = \frac{1}{V(\epsilon)} \frac{\partial E}{\partial \epsilon} \tag{4.14}$$

where ϵ is the imposed tensile strain, E is the total electronic energy, and $V(\epsilon)$ is the volume at the imposed tensile strain ϵ. The stress–strain curves were obtained by first starting with the fully relaxed geometry and subsequently applying a fixed tensile stress by elongating the crystal along a specified loading axis (described further below). The VASP source code was modified to perform constrained relaxations in the directions perpendicular to the elongation axis to minimize the other 5 components of the stress tensor and to ensure uniaxial loading. This important modification allows us to (1) account for the Poisson ratio of the material by relaxing the lattice constants in the directions perpendicular to the applied stress and (2) simultaneously allow for relaxation of the atomic positions in the material. In order to obtain smooth plots of the tensile stress as a function of strain, we calculated these relaxed energies by varying ϵ from 0.00 to 0.60 in increments of 0.01 (i.e., 1 % strain). We thus obtained the electronic energy E as a function of ϵ which we then fitted to a piecewise polynomial form of a cubic spline interpolant. The first derivative of this piecewise polynomial was then evaluated numerically to obtain smooth values of the tensile strength as a function of ϵ, as required in Eq. (4.2). We carried out all of our tensile strength calculations by applying a uniaxial tensile strain along the 3 individual crystalline directions ([100], [110], and [111]) for each of the different PdH_x stoichiometries in Fig. 4.2. It is worth noting that while the tensile strength calculations along the [100] direction can be carried out using the primitive unit cells depicted in Fig. 4.2, the tensile strengths for the other [110] and [111] directions require the use of larger supercells. It is also worth mentioning that our present study on stress–strain properties for the various PdH_x stoichiometries comprises an immense number of DFT calculations, specifically, a total of 6,300 separate DFT relaxations (= 60 strain values $\times$ 5 different PdH_x stoichiometries $\times$ 7 different DFT methods $\times$ 3 crystalline directions). Figures 4.5 and 4.6 compare the stress–strain curves among the 5 different DFT methods for the [100] and [111] crystalline directions. We omit plots of the [110] stress–strain curve in the main text since the stress–strain properties along the [110] direction are known to exhibit qualitatively different characteristics. Specifically, it is well known in the literature that while bond breaking occurs for [100] and [111] strains in face-centered cubic metals, tension in the [110] direction leads to a phase transformation (Morris et al., 2003). As a result, for pure Pd this phase transformation results in a compressive stress in the pulling direction, leading to a counter-intuitive result. These complex phase transformations are tangential to our analyses of DFT methods, and we do not discuss the [110] stress–strain relationships further (plots of the [110] stress–strain curves, however, are included in the Supporting Information in [34] for completeness).

For the LDA, PBE, PBE-D3(BJ), PBE-TS, vdW-DF2, and optB86b-vdW methods only, the plots depicted in Figs. 4.5 and 4.6 are characteristic of stress–strain curves of most metallic solids [6]. Specifically at small strains, the stress–strain curves exhibit a nearly linear behavior up to a maximum stress; as the strain is increased even further past this point, a more gradual decrease in the stress is observed. This maximum along the stress–strain curve is known as the ideal tensile strength of the material [6], which is the last state just prior to the onset of instability in the crystal lattice. Mathematically, the ideal tensile strength also corresponds to the inflection point in the total energy-strain curve ($\partial^2 E/\partial\epsilon^2 = \partial\sigma/\partial\epsilon = 0$) and represents the maximum load that the material can sustain without undergoing the

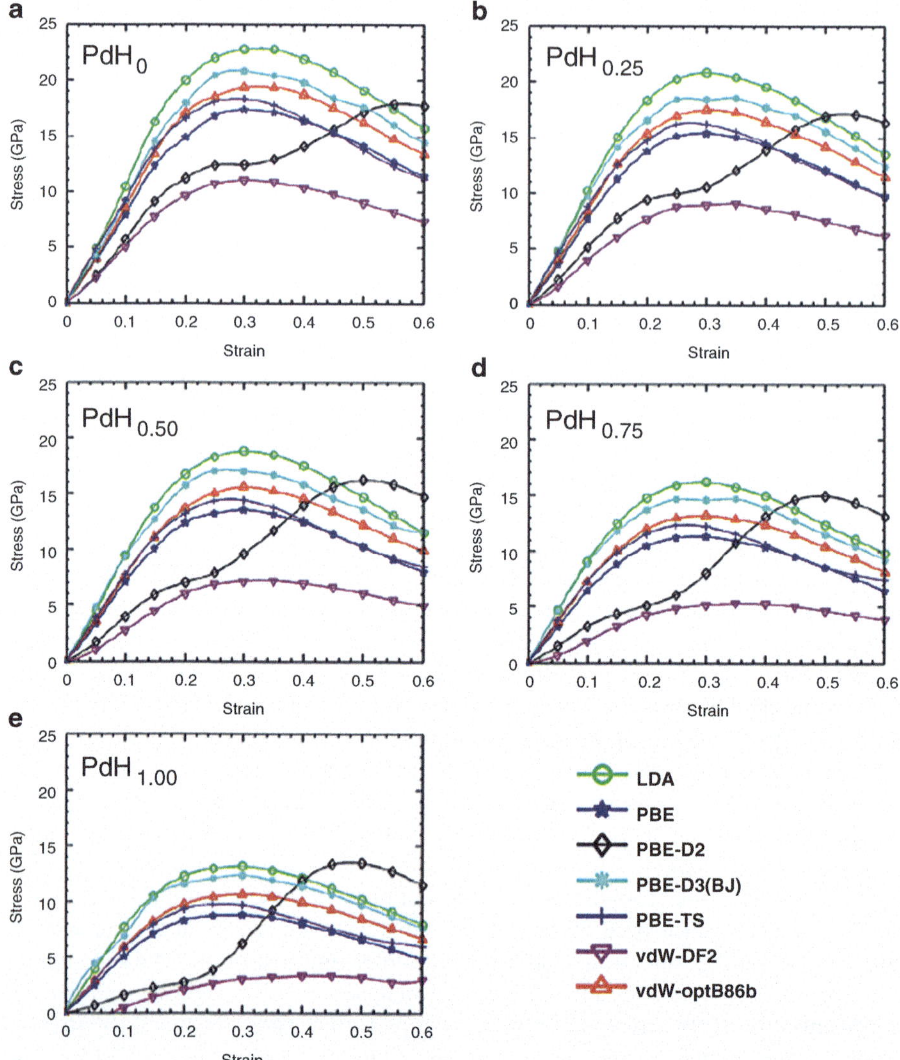

Fig. 4.5 Stress–strain curves in PdH_x for x = (**a**) 0, (**b**) 0.25, (**c**) 0.50, (**d**) 0.75, and (**e**) 1.00 along the [100] crystal axis

instability of necking, which leads inevitably to fracture. While the stress–strain curves for LDA, PBE, PBE-D3(BJ), PBE-TS, vdW-DF2, and optB86b-vdW appear reasonable, we find that the stress–strain relations predicted by the empirically constructed DFT-D2 method are extremely anomalous. Specifically, for both the [100] and [111] directions (as well as for all 5 PdH_x stoichiometries), the PBE-D2 functional predicts the ideal strength to occur at ϵ values that are nearly twice as

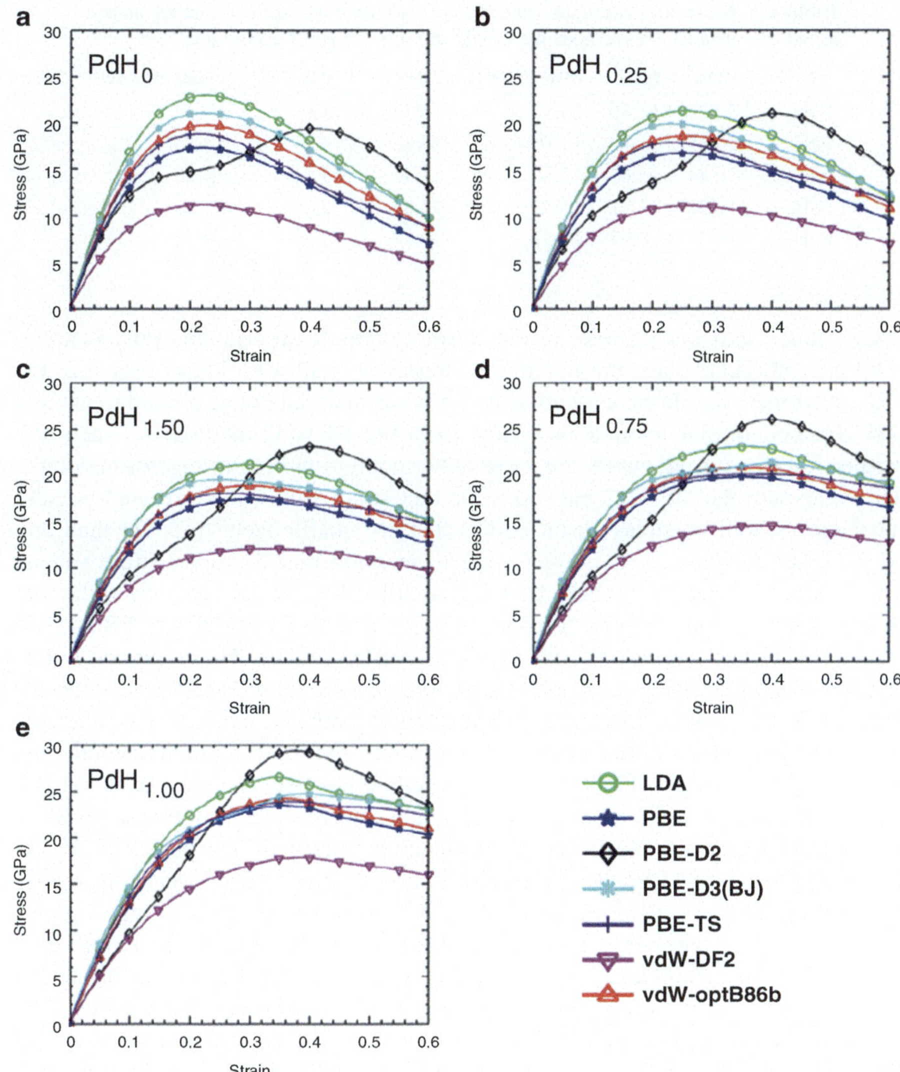

Fig. 4.6 Stress–strain curves in PdH_x for $x =$ (**a**) 0, (**b**) 0.25, (**c**) 0.50, (**d**) 0.75, and (**e**) 1.00 along the [111] crystal axis

large compared to the other DFT methods. Take, for example, all the [100] stress–strain curves shown in Fig. 4.5: the PBE-D2 method predicts the maximum stress to occur at $\epsilon \sim 0.5$ whereas all the other DFT methods produce a maximum near $\epsilon \sim 0.3$ instead. The [111] stress–strain curves in Fig. 4.6 show similar anomalous results: the PBE-D2 method yields a maximum stress at $\epsilon \sim 0.4$ instead of the $\epsilon \sim 0.25$ value predicted by all the other functionals. To put these anomalous

Table 4.3 Maximum strengths (in GPa) for various PdH_x stoichiometries as predicted by various DFT methods for tensile strain in the [100] crystal direction

	LDA	PBE	PBE-D3(BJ)	PBE-TS	vdW-DF2	optB86b-vdW
PdH_0	22.77	17.39	20.82	18.34	11.09	19.52
$PdH_{0.25}$	20.85	15.42	19.10	16.32	9.14	17.53
$PdH_{0.5}$	18.84	13.62	17.36	14.48	7.31	15.65
$PdH_{0.75}$	16.31	11.51	15.02	12.36	5.35	13.25
$PdH_{1.0}$	13.25	8.93	12.37	9.87	3.36	10.70

stress values into perspective, recent tensile strength experiments on palladium (Dillon et al. [11]) place the maximum stress at ϵ values no larger than 0.2. We also performed simple benchmark tests on other materials (and pseudopotentials) and obtained similar anomalous results from the DFT-D2 method. As such, the results described in this work are expected to apply to a broad range of materials, and we expect the DFT-D2 method to predict strain values that are unphysically too large as well as stress–strain curves that are qualitatively different than any of the other functionals. It is interesting to mention that for small strain values, $0.0 < \epsilon < 0.2$ (regimes where DFT-D2 is still valid), we observe the following trends in the computed stress for both Figs. 4.5 and 4.6: vdw-DF2 $\approx$ PBE-D2 $<$ PBE $\approx$ PBE-TS $\sim$ optB86b-vdW $<$ PBE-D3(BJ) $<$ LDA. These general trends can be rationalized using the same arguments discussed previously for the cohesive energies. For example, LDA overestimates bond strengths and cohesive energies and, therefore, predicts the largest stress among the DFT methods (within the $0.0 < \epsilon < 0.2$ range). The optB86b-vdW method lies between the PBE and LDA curves since the non-empirical vdW correlation effects added to the base optB86b functional corrects for the lower stress values predicted by the pure PBE GGA. The PBE-D3(BJ) and PBE-TS stress–strain curves yield slightly larger and smaller stresses, respectively, than the optB86b-vdW functional, which also reflects their trends in the cohesive energy. As mentioned previously, the second-generation vdW-DF2 functional significantly underestimates cohesive energies among the tested DFT methods and, therefore, predicts the lowest stress for small strain values. The maximum ideal tensile strengths are presented in Table 4.3 and summarized in Fig. 4.7. The ideal tensile strengths obtained from PBE-D2 are not tabulated since this functional produces unphysical results, as discussed previously. However, it is worth noting that the maximum ideal strength decreases as the proportion of hydrogen in the palladium system increases. These observations confirm recent experiments that show metal-hydrides are more prone to failure than their pure counterparts [11], and that their maximum tensile strength is inversely proportional to the amount of hydrogen present in the crystal lattice.

Returning to the anomalous stress–strain curves predicted by PBE-D2, it is somewhat surprising that even the simplest functionals (LDA and PBE) yield results that are much more realistic than the improved dispersion-corrected DFT-D2 method. We attribute these anomalous results to the (geometry-independent) dispersion

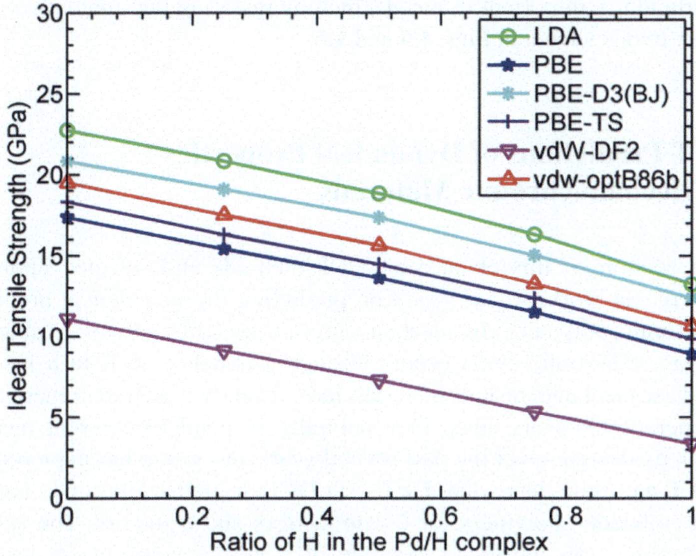

Fig. 4.7 Ideal tensile strengths along the [100] direction for the PdH_x system predicted with various DFT methods

coefficients and simple damping functions that are inherent to the empirical DFT-D2 method. To test these claims further, we also carried out DFT-D3(BJ) and DFT-TS single-point energies on top of the previously optimized DFT-D2 geometries. Interestingly, we found that both the DFT-D3(BJ) and DFT-TS single-point energy stress–strain curves were nearly identical to their corresponding relaxed-strain curves already shown in Figs. 4.5 and 4.6. Consequently, the anomalous DFT-D2 stress–strain curves are not the result of abnormal DFT-D2 structural optimizations or unusual deformations of the solid; rather, the electronic energies predicted by the empirical DFT-D2 method itself give strain energies that are unphysically too large. As the C_6 dispersion energies in DFT-D2 are constructed from simple estimates of atomic ionization potentials and static dipole polarizabilities, this simplistic approach is expected to fail in much more complex environments in solids, particularly configurations that are far from equilibrium. It is interesting to note that the atomic values for C_6 in DFT-D2, DFT-D3, and DFT-TS are 428, 609, and 158 a.u., respectively, whereas the bulk values for C_6 in DFT-D3 and DFT-TS are correspondingly 266 and 174 a.u. (the atomic and bulk values for C_6 are the same in DFT-D2). Moreover, while it is well known that DFT-D2 has a tendency to overestimate the binding energies of isolated molecules in the gas phase [7], we are not aware of any previous studies on highly non-equilibrium properties of bulk solids. In contrast, the PBE-D3(BJ), PBE-TS, and optB86b-vdW functionals take into account van der Waals effects in a less empirical fashion than the pairwise interactions in DFT-D2. All of the former functionals account for screening effects that are particularly important in metals, thereby restoring the qualitatively correct stress–strain trends shown in Figs. 4.5 and 4.6.

4.5 DFT Predictions of Dynamical Properties in Mechanochromic Materials

In the last section of this chapter on DFT methods and selected applications, we present recent work by our group on predicting the mechanical properties of mechanochromic polymers. In mechanochromic systems, an observable optical response (i.e., a chromic shift) occurs through a dynamic shift in a mechanical property. These mechanochromic materials have recently garnered immense interest as next-generation sensors since they naturally respond to external mechanical deformation by design. Over the past several years, our group has explored the use of both DFT and time-dependent DFT (TD-DFT) to understand and characterize spiropyran-embedded polymers for use in sensors and actuators. The spiropyran polymer system is an intriguing example of a mechanochromic material since the photochromic response in these systems results from a dynamic coupling between photoexcitation and mechanical deformation. Upon exposure to UV light, the spiropyran undergoes a reversible photochemical rearrangement that leads to a dramatic color change in the material. Similarly, this color change can also be triggered by mechanical deformation of the spiropyran-embedded polymer, which is shown dramatically in Fig. 4.8. First-principles DFT calculations play an important role in characterizing these systems since they provide mechanistic insight into properties that can be further fine-tuned to enhance the mechanochromic response of these materials. We give a brief discussion of the computational methods that our group has utilized for these mechanochromic materials, and we conclude with a summary of properties and trends obtained by our analysis.

Fig. 4.8 Photo series displaying mechanochromism in a spiropyran-embedded polymer as it is deformed to failure. From [49]

In order to predict and understand the dynamic properties of this spiropyran system, both DFT and TD-DFT must be utilized to capture the coupled mechanical and optical properties of this system. Because the applied stress on poly (ϵ-caprolactone) creates an optical transition in the material (cf. Fig. 4.8), it is necessary to choose a DFT functional that accurately captures this optical process. To this end, all DFT calculations were performed using a range-separated LC-BLYP functional (discussed briefly in Section 4.3.2) which incorporates an exact $-1/r$ exchange energy dependence at large inter-electronic distances [49]. First, to simulate an applied stress, the initial equilibrium geometry for a closed spiropyran unit tethered with pendant poly(ϵ-caprolactone) polymer chains was optimized with DFT. Starting with this equilibrium geometry, an applied external stress was obtained by gradually increasing the distance between the two terminal methyl groups (denoted by arrows in Fig. 4.9a) in small increments of 0.1 Å up to a final molecular elongation of 60 %. Throughout this entire procedure, all of the other unconstrained internal coordinates were fully optimized in order to minimize the total strain energy. The resulting energy curve is depicted in Fig. 4.9b, which shows a monotonic increase in energy as the polymer is stretched until an abrupt transition occurs at 39 % elongation. At this transition point, the C-O spiro bond suddenly ruptures, and the energy sharply decreases from 2.7 to 1.6 eV as the spiropyran has now relaxed into the open charge-separated zwitterionic merocyanine form. It is remarkable to point out that our DFT calculations naturally predict the C-O spiro bond to be the weakest in the entire spiropyran polymer, even though we have not constrained this particular bond-energy at all in our first-principles calculations. After the weak C-O spiro bond has broken, the energy for elongations greater than 39 % is associated with stretching the strong molecular bonds along the open zwitterionic backbone.

Next, in parallel with the force-constrained DFT optimizations, we also carried out excited-state calculations to understand the photochromic properties of the nanoribbon-spiropyran polymer. As described in our previous work, reversible photoswitching of a spiropyran-embedded polymer is experimentally realizable since light of different wavelengths can initiate both the forward and reverse isomerizations in these materials. In particular, real-time monitoring of a spiropyran-polymer film gives a dramatic demonstration of a reversible cyclic response to light irradiation [49]. In order to evaluate the photoresponse in this system, we carried out several time-dependent DFT calculations at each of the force-constrained polymer geometries. In Fig. 4.9c, we plot the excitation wavelength having the strongest absorption maximum along the mechanical deformation reaction path. For molecular elongations less than 39 %, Fig. 4.9c shows that it only absorbs in the UV with a maximum absorbance at 250 nm. However, upon cleavage of the weak C-O spiro bond, the stretched spiropyran structure shows a strong $S_0 \rightarrow S_1$ absorption maximum (450 nm) that is significantly red-shifted due to the larger conjugated π-electron system, confirming previous experimental observations of a strong photochromic shift in the polymer. To further characterize this dramatic change in electronic character, we also performed new calculations of the electric dipole moment in Fig. 4.9d along the mechanical deformation reaction path. In the

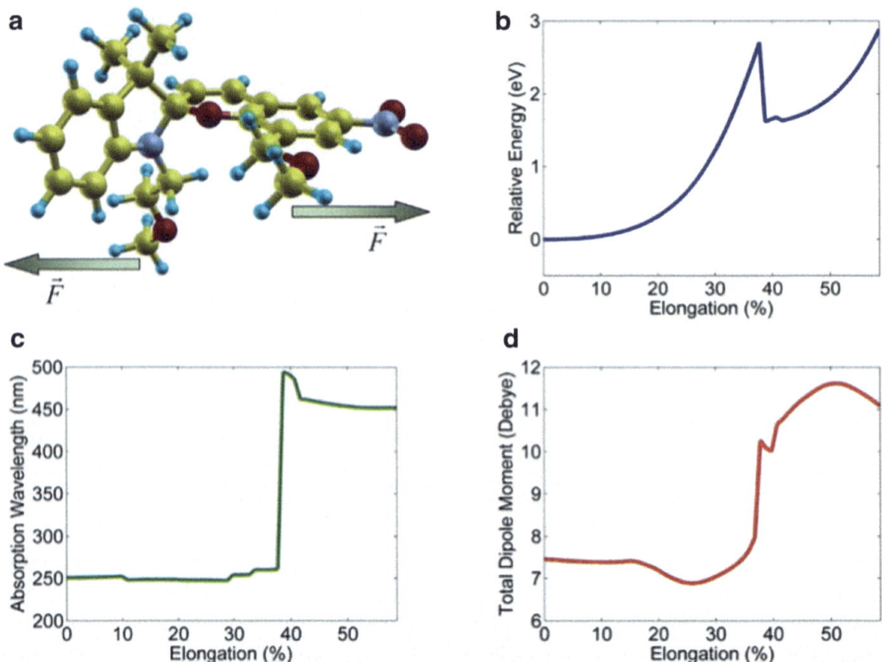

Fig. 4.9 (**a**) Molecular structure of a closed spiropyran unit tethered with pendant poly(ϵ-caprolactone) polymer chains. An external mechanical stress was applied across this polymer unit by gradually increasing the distance between the methyl units (denoted by *arrows*) while optimizing all other internal coordinates. (**b**) Force-constrained potential energy curve for mechanochemical switching of the spiropyran system. A sharp transition between the spiropyran and merocyanine forms occurs at 39 % elongation length. (**c**) Absorption wavelength for photochromic switching of spiropyran. The unstretched spiropyran form absorbs in the UV (< 39 % elongation), while the stretched system strongly absorbs in the visible (> 39 % elongation). (**d**) Electric dipole moment as a function of molecular elongation. When the spiropyran polymer is stretched, the dipole moment nearly doubles as the deformation proceeds towards complete mechanical failure of the material

unstretched-spiropyran region, the calculated dipole moment has a value of 7.5 D that begins to change discontinuously in the same energy region where the C-O spiro bond is broken and where the polymer begins to strongly absorb at 450 nm. Beyond this transition point, the dipole moment almost doubles to a value of 11.5 D due to increasing polarization of the C-O bond as the polymer is stretched. Our first-principles calculations give detailed mechanistic insight into this evolution of the dipole moment, and they further predict the mechanical deformation of the system to occur at 39 % elongation, or at the 250 nm/450 nm crossover point.

4.6 Discussions and Conclusions

In summary, we have given a brief overview of various representative DFT methods as well as presented practical applications for predicting various mechanical properties. While DFT offers a powerful capability for modeling these diverse properties, the examples presented in this chapter highlight the extra care that must be taken in choosing the appropriate functional for each application. In our first example on palladium-hydride materials, we first carried out a series of calculations on static properties, including lattice constants and cohesive energies, using representative functionals within the LDA, GGA, DFT-D, and nonlocal vdW DFT families. For these simple static properties, we find that dispersion interactions are not negligible and still play a significant role even in these covalently bonded solid structures. Specifically, both the empirically constructed PBE-D2, PBE-D3(BJ), and PBE-TS methods and the nonlocal optB86-vdW dispersion-corrected functional give the best agreement to the experimental lattice constants and cohesive energies. However, the testing of DFT methods on static properties only gives limited information for solid structures near equilibrium, and a more representative analysis requires additional benchmarks on properties far from equilibrium. To further characterize and thoroughly test the various DFT methods at these extreme conditions, we also carried out extensive analyses of non-equilibrium properties including stress–strain relationships and maximum tensile strengths. Among all the tested functionals, we surprisingly found that the empirically constructed DFT-D2 method gives stress–strain relationships that are extremely anomalous. Specifically, the PBE-D2 functional predicts the ideal strength to occur at ϵ values that are nearly twice as large compared to the other DFT methods. To the best of our knowledge, this present study is the first to benchmark dispersion-corrected DFT methods on highly non-equilibrium properties of bulk solids and also the first to report the anomalous tensile strength results produced by DFT-D2 methods.

Our findings have important ramifications for both method development and future parameterizations of structural properties in solids: (1) on a practical note, we find that the next-generation dispersion-corrected methods (i.e., DFT-D3 and DFT-TS) and the nonlocal optB86b-vdW method yield accurate results for both equilibrium and non-equilibrium properties. While the DFT-D2 method gives accurate results for lattice constants, bulk moduli, and cohesive energies, the stress–strain curves predicted by DFT-D2 are extremely anomalous and physically incorrect (the opposite is true for the nonlocal vdW-DF2 functional which gives reasonable results for stress–strain properties but poor results for lattice constants, bulk moduli, and cohesive energies). As such, we recommend usage of the PBE-D3(BJ), PBE-TS, or optB86-vdW functionals for accurately computing both equilibrium and non-equilibrium properties in bulk solids; (2) for future development of new dispersion-corrected DFT functionals, we advocate the inclusion of simple tensile strength benchmarks to ensure that anomalous stress–strain results are not produced. There has already been prior work to include non-equilibrium molecular geometries (in vacuum) to construct new dispersion-corrected DFT methods; however, the use of non-equilibrium solid-state structures has not been utilized. These non-equilibrium tensile strength tests require relatively small unit cells, and a simple plot of the stress–strain curve can be easily used as a diagnostic for testing the fidelity of new dispersion-corrected functionals; and (3) most importantly, one should proceed with caution in using DFT-D2 (and other coarse-grained parameterizations obtained from DFT-D2) for computing material properties in extreme environments, particularly for bulk solids. While the newer dispersion-corrected DFT-D3, DFT-TS, and nonlocal vdW methods are now commonly used for molecular interactions, the older DFT-D2 method is still widely employed for bulk solids, surfaces, and condensed phase systems (a cursory keyword search for "DFT-D2" vs. "DFT-D3" in the titles and abstracts of materials science journals in the Thomson Reuters Web of Science yields 41 papers for DFT-D2 and only 12 papers for DFT-D3, during the years 2010–2015). Moreover, there are still ongoing efforts in parameterizing new molecular dynamics force fields and equation-of-state models from DFT-D2 benchmarks [16, 68, 73]. While these new force fields and models will give reliable results for properties near equilibrium, one should proceed with caution in using DFT-D2-parameterized methods for properties far from equilibrium (i.e., in extreme stress–strain environments and irreversible phase transformations). For materials under extreme stress/strain conditions, particularly for bulk solids as demonstrated in this work, DFT-D2-based force fields may need to be re-examined or re-parameterized.

In our final example on using DFT methods to predict mechanical properties, we presented an application on mechanochromic systems. Using both force-constrained DFT optimizations and time-dependent DFT, we show that first-principles calculations give detailed mechanistic insight into these mechanically activated processes. Moreover, when compared to experiment, these calculations also give accurate predictions of mechanical deformation in these complex systems. Looking forward, it would be extremely interesting to postulate how one can further optimize this hybrid system in future mechanochromic applications. As mentioned previously, while our DFT calculations were motivated from experimentally available spiropyran materials, it is very likely that further modifications of the spiropyran unit would lead to even larger mechanochromic effects. In particular, further chemical functionalization via other strong electron donor or acceptor groups of the spiropyran would further modulate the optical and mechanical couplings in this system. We are currently using both DFT and time-dependent DFT to exploring these options in parallel with experimental efforts to understand their effect on the mechanical properties of spiropyran-based polymers. As a result, we anticipate that the mechanochromic effects presented here are applicable to opto-mechanical properties in other nanostructures, and first-principles DFT calculations can provide further insight into these effects in other mechanically coupled systems of increasing complexity.

References

1. E. Antolini, Palladium in fuel cell catalysis. Energy Environ. Sci. **2**(9), 915–931 (2009)
2. A. Barnoush, H. Vehoff, Recent developments in the study of hydrogen embrittlement: hydrogen effect on dislocation nucleation. Acta Math. **58**(16), 5274–5285 (2010)
3. A.D. Becke, A new mixing of Hartree–Fock and local density-functional theories. J. Chem. Phys. **98**(2), 1372–1377 (1993)
4. A.D. Becke, Density-functional exchange-energy approximation with correct asymptotic behavior. Phys. Rev. A **38**(6), 3098 (1988)
5. A.D. Becke, E.R. Johnson, A density-functional model of the dispersion interaction. J. Chem. Phys. **123**(15), 154,101 (2005)
6. F.P. Beer, E. Johnston, J. DeWolf, D. Mazurek, *Mechanics of Materials*, 5th edn. (McGraw-Hill, New York, 2011)
7. J. Brndiar, I. Stich, van der Waals interaction energies of small fragments of P, As, Sb, S, Se, and Te: comparison of complete basis set limit CCSD (T) and DFT with approximate dispersion. J. Chem. Theory Comput. **8**(7), 2301–2309 (2012)
8. D.M. Ceperley, B. Alder, Ground state of the electron gas by a stochastic method. Phys. Rev. Lett. **45**(7), 566 (1980)
9. S.D. Chakarova-Käck, E. Schröder, B.I. Lundqvist, D.C. Langreth, Application of van der Waals density functional to an extended system: Adsorption of benzene and naphthalene on graphite. Phys. Rev. Lett. **96**(14), 146,107 (2006)
10. G.I. Csonka, J.P. Perdew, A. Ruzsinszky, P.H. Philipsen, S. Lebègue, J. Paier, O.A. Vydrov, J.G. Ángyán, Assessing the performance of recent density functionals for bulk solids. Phys. Rev. B **79**(15), 155,107 (2009)
11. E. Dillon, G. Jimenez, A. Davie, J. Bulak, S. Nesbit, A. Craft, Factors influencing the tensile strength, hardness, and ductility of hydrogen-cycled palladium. Mater. Sci. Eng. A **524**(1), 89–97 (2009)
12. M. Dion, H. Rydberg, E. Schröder, D.C. Langreth, B.I. Lundqvist, Van der Waals density functional for general geometries. Phys. Rev. Lett. **92**(24), 246,401 (2004)
13. P.A. Dirac, Note on exchange phenomena in the Thomas Atom. in, *Mathematical Proceedings of the Cambridge Philosophical Society*, vol. 26, pp. 376–385 (Cambridge University Press, Cambridge, 1930)
14. C. Drahl, Palladium's hidden talent. Chem. Eng. News **86**(35), 53–56 (2008)
15. M. Elstner, P. Hobza, T. Frauenheim, S. Suhai, E. Kaxiras, Hydrogen bonding and stacking interactions of nucleic acid base pairs: a density-functional-theory based treatment. J. Chem. Phys. **114**(12), 5149–5155 (2001)
16. H. Fang, P. Kamakoti, J. Zang, S. Cundy, C. Paur, P.I. Ravikovitch, D.S. Sholl, Prediction of co2 adsorption properties in zeolites using force fields derived from periodic dispersion-corrected DFT calculations. J. Phys. Chem. C **116**(19), 10,692–10,701 (2012)
17. E. Fermi, A statistical method for determining some properties of the atoms and its application to the theory of the periodic table of elements. Z. Phys. **48**, 73–79 (1928)
18. W. Foulkes, L. Mitas, R. Needs, G. Rajagopal, Quantum Monte Carlo simulations of solids. Rev. Mod. Phys. **73**(1), 33 (2001)
19. M.R. Golder, B.M. Wong, R. Jasti, Photophysical and theoretical investigations of the [8] cycloparaphenylene radical cation and its charge-resonance dimer. Chem. Sci. **4**(11), 4285–4291 (2013)
20. A. Goursot, T. Mineva, R. Kevorkyants, D. Talbi, Interaction between n-alkane chains: applicability of the empirically corrected density functional theory for van der Waals complexes. J. Chem. Theory Comput. **3**(3), 755–763 (2007)
21. L. Gráfová, M. Pitonak, J. Rezac, P. Hobza, Comparative study of selected wave function and density functional methods for noncovalent interaction energy calculations using the extended s22 data set. J. Chem. Theory Comput. **6**(8), 2365–2376 (2010)

22. S. Grimme, Accurate description of van der Waals complexes by density functional theory including empirical corrections. J. Comput. Chem. **25**(12), 1463–1473 (2004)
23. S. Grimme, Semiempirical GGA-type density functional constructed with a long-range dispersion correction. J. Comput. Chem. **27**(15), 1787–1799 (2006)
24. S. Grimme, J. Antony, S. Ehrlich, H. Krieg, A consistent and accurate ab initio parametrization of density functional dispersion correction (DFT-d) for the 94 elements h-pu. J. Chem. Phys. **132**(15), 154,104 (2010)
25. S. Grimme, S. Ehrlich, L. Goerigk, Effect of the damping function in dispersion corrected density functional theory. J. Comput. Chem. **32**(7), 1456–1465 (2011)
26. W. Grochala, P.P. Edwards, Thermal decomposition of the non-interstitial hydrides for the storage and production of hydrogen. Chem. Rev. **104**(3), 1283–1316 (2004)
27. E.K. Gross, R.M. Dreizler, *Density Functional Theory*, vol. 337 (Springer Science & Business Media, New York, 2013)
28. L.M. Hale, B.M. Wong, J.A. Zimmerman, X. Zhou, Atomistic potentials for palladium–silver hydrides. Model. Simul. Mater. Sci. Eng. **21**(4), 045,005 (2013)
29. B. Hammer, L.B. Hansen, J.K. Nørskov, Improved adsorption energetics within density-functional theory using revised Perdew-Burke-Ernzerhof functionals. Phys. Rev. B **59**(11), 7413 (1999)
30. J. Hepburn, G. Scoles, R. Penco, A simple but reliable method for the prediction of intermolecular potentials. Chem. Phys. Lett. **36**(4), 451–456 (1975)
31. P. Hohenberg, W. Kohn, Inhomogeneous electron gas. Phys. Rev. **136**(3B), B864 (1964)
32. C. Huang, M. Kim, B.M. Wong, N.S. Safron, M.S. Arnold, P. Gopalan, Raman enhancement of a dipolar molecule on graphene. J. Phys. Chem. C **118**(4), 2077–2084 (2014)
33. N.V. Ilawe, A.E. Raeber, R. Schweitzer-Stenner, S.E. Toal, B.M. Wong, Assessing backbone solvation effects in the conformational propensities of amino acid residues in unfolded peptides. Phys. Chem. Chem. Phys. **17**(38), 24,917–24,924 (2015)
34. N.V. Ilawe, J.A. Zimmerman, B.M. Wong, Breaking badly: DFT-d2 gives sizeable errors for tensile strengths in palladium-hydride solids. J. Chem. Theory Comput. **11**(11), 5426–5435 (2015)
35. E.R. Johnson, A.D. Becke A post-Hartree–Fock model of intermolecular interactions. J. Chem. Phys. **123**(2), 024,101 (2005)
36. P.S. Johnson, C. Huang, M. Kim, N.S. Safron, M.S. Arnold, B.M. Wong, P. Gopalan, F. Himpsel, Orientation of a monolayer of dipolar molecules on graphene from X-ray absorption spectroscopy. Langmuir **30**(9), 2559–2565 (2014)
37. P. Jurečka, J. Černý, P. Hobza, D.R. Salahub, Density functional theory augmented with an empirical dispersion term. interaction energies and geometries of 80 noncovalent complexes compared with ab initio quantum mechanics calculations. J. Comput. Chem. **28**(2), 555–569 (2007)
38. J. Klimeš, D.R. Bowler, A. Michaelides, Van der Waals density functionals applied to solids. Phys. Rev. B **83**(19), 195,131 (2011)
39. W. Kohn, L.J. Sham, Self-consistent equations including exchange and correlation effects. Phys. Rev. **140**(4A), A1133 (1965)
40. L. Kronik, A. Tkatchenko, Understanding molecular crystals with dispersion-inclusive density functional theory: pairwise corrections and beyond. Acc. Chem. Res. **47**(11), 3208–3216 (2014)
41. K. Lee, É.D. Murray, L. Kong, B.I. Lundqvist, D.C. Langreth, Higher-accuracy van der Waals density functional. Phys. Rev. B **82**(8), 081,101 (2010)
42. A.W. Long, B.M. Wong, Pamela: an open-source software package for calculating nonlocal exact exchange effects on electron gases in core-shell nanowires. AIP Adv. **2**(3), 032,173 (2012)
43. F. Manchester, A. San-Martin, J. Pitre, The H-Pd (hydrogen-palladium) system. J. Phase Equilib. **15**(1), 62–83 (1994)
44. N. Marom, A. Tkatchenko, M. Rossi, V.V. Gobre, O. Hod, M. Scheffler, L. Kronik, Dispersion interactions with density-functional theory: benchmarking semiempirical and interatomic pairwise corrected density functionals. J. Chem. Theory Comput. **7**(12), 3944–3951 (2011)

45. A.E. Mattsson, R. Armiento, Implementing and testing the am05 spin density functional. Phys. Rev. B **79**(15), 155,101 (2009)
46. A.E. Mattsson, P.A. Schultz, M.P. Desjarlais, T.R. Mattsson, K. Leung, Designing meaningful density functional theory calculations in materials science—a primer. Model. Simul. Mater. Sci. Eng. **13**(1), R1 (2004)
47. É.D. Murray, K. Lee, D.C. Langreth, Investigation of exchange energy density functional accuracy for interacting molecules. J. Chem. Theory Comput. **5**(10), 2754–2762 (2009)
48. J.W. Morris, D.M. Clatterbuck, D.C. Chrzan, C.R. Krenn, W. Luo, M.L. Cohen, Elastic stability and the limits of strength, in *Thermec'2003*, Pts 1–5, 426–4, 4429–4434 (2003)
49. G. O'Bryan, B.M. Wong, J.R. McElhanon, Stress sensing in polycaprolactone films via an embedded photochromic compound. ACS Appl. Mater. Interfaces **2**(6), 1594–1600 (2010)
50. R.G. Parr, W. Yang, *Density-Functional Theory of Atoms and Molecules*, ed. by R. Breslow. International Series of Monographs on Chemistry, vol. 16 (Oxford University Press, New York, 1989), pp. 160–180
51. J.P. Perdew, Density-functional approximation for the correlation energy of the inhomogeneous electron gas. Phys. Rev. B **33**, 8822 (1986)
52. J.P. Perdew, K. Schmidt, Jacob's ladder of density functional approximations for the exchange-correlation energy. AIP Conf. Proc. **577**, 1–20 (2001)
53. J.P. Perdew, Y. Wang, Accurate and simple analytic representation of the electron-gas correlation-energy. Phys. Rev. B **45**, 13,244–13,249 (1992)
54. J.P. Perdew, A. Zunger, Self-interaction correction to density-functional approximations for many-electron systems. Phys. Rev. B **23**, 5048–5079 (1981)
55. J.P. Perdew, K. Burke, M. Ernzerhof, Generalized gradient approximation made simple. Phys. Rev. Lett. **77**(18), 3865 (1996)
56. J.P. Perdew, J. Chevary, S. Vosko, K.A. Jackson, M.R. Pederson, D. Singh, C. Fiolhais, Atoms, molecules, solids, and surfaces: Applications of the generalized gradient approximation for exchange and correlation. Phys. Rev. B **46**(11), 6671 (1992)
57. J.P. Perdew, A. Ruzsinszky, G.I. Csonka, O.A. Vydrov, G.E. Scuseria, L.A. Constantin, X. Zhou, K. Burke, Restoring the density-gradient expansion for exchange in solids and surfaces. Phys. Rev. Lett. **100**, 136406 (2008)
58. P. Pulay, Convergence acceleration of iterative sequences. the case of SCF iteration. Chem. Phys. Lett. **73**, 393–398 (1980)
59. A. Puzder, M. Dion, D.C. Langreth, Binding energies in benzene dimers: nonlocal density functional calculations. J. Chem. Phys. **124**, 164,105 (2006)
60. G. Roman-Perez, J.M. Soler, Efficient implementation of a van der Waals density functional: application to double-wall carbon nanotubes. Phys. Rev. Lett. **103**, 096,102 (2009)
61. V.G. Ruiz, W. Liu, E. Zojer, M. Scheffler, A. Tkatchenko, Density-functional theory with screened van der Waals interactions for the modeling of hybrid inorganic-organic systems. Phys. Rev. Lett. **108**, 146,103 (2012)
62. E. Schrödinger, An undulatory theory of the mechanics of atoms and molecules. Phys. Rev. **28**, 1049–1070 (1926)
63. L.J. Sham, M. Schloter, Density-functional theory of the energy gap. Phys. Rev. Lett. **51** (1888)
64. J. Shu, B.P.A. Grandjean, A.V. Neste, S. Kaliaguine, Catalytic palladium-based membrane reactors: a review. Can. J. Chem. Eng. **69**, 1036–1060 (1991)
65. L.H. Thomas, The production of characteristic X-rays by electronic impact. Proc. Camb. Philos. Soc. **23**, 829–831 (1927)
66. A. Tkatchenko, M. Scheffler, Accurate molecular van der Waals interactions from ground-state electron density and free-atom reference data. Phys. Rev. Lett. **102**, 073,005 (2009)
67. J. Tsuji, *Palladium Reagents and Catalysts: New Perspectives for the 21st Century* (Wiley, New York, 2004)
68. T. Tuttle, W. Thiel, Omx-D: semiempirical methods with orthogonalization and dispersion corrections. Implementation and biochemical application. Phys. Chem. Chem. Phys. **10**, 2159–66 (2008)
69. S.H. Vosko, L. Wilk, Influence of an improved local-spin-density correlation-energy functional on the cohesive energy of alkali-metals. Phys. Rev. B **22**, 3812–3815 (1980)

70. B.M. Wong, S.H. Ye, Self-assembled cyclic oligothiophene nanotubes: electronic properties from a dispersion-corrected hybrid functional. Phys. Rev. B **84.** (2011)
71. B.M. Wong, S.H. Ye, G. O'Bryan, Reversible, opto-mechanically induced spin-switching in a nanoribbon-spiropyran hybrid material. Nanoscale **4**, 1321–1327 (2012)
72. J.L. Xia, M.R. Golder, M.E. Foster, B.M. Wong, R. Jasti, Synthesis, characterization, and computational studies of cycloparaphenylene dimers. J. Am. Chem. Soc. **134**, 19,709–19,715 (2012)
73. J. Zang, S. Nair, D.S. Sholl, Prediction of water adsorption in copper-based metal organic frameworks using force fields derived from dispersion-corrected DFT calculations. J. Phys. Chem. C **117**, 7519–7525 (2013)
74. X.W. Zhou, J.A. Zimmerman, B.M. Wong, J.J. Hoytt, An embedded-atom method interatomic potential for Pd-H alloys. J. Mater. Res. **23**, 704–718 (2008)

Chapter 5
The Quasicontinuum Method: Theory and Applications

Dennis M. Kochmann and Jeffrey S. Amelang

5.1 Introduction

The state of the art in material modeling offers highly accurate methods for each individual scale, from density functional theory (DFT) and molecular dynamics (MD) at the lower scales all the way up to continuum theories and associated computational tools for the macroscale and structural applications. Unfortunately, a wide gap exists due to a lack of models applicable at the intermediate scales (sometimes referred to as *mesoscales*). Here, the continuum hypothesis fails because the discreteness of the atomic crystal becomes apparent, e.g., through the emergence of size effects. At the same time, atomistic techniques tend to incur prohibitively high computational expenses when reaching the scales of hundreds of nanometers or microns. Mastering this gap between atomistics and the continuum is the key to understanding a long list of diverse open problems. These include the mechanical response of nanoporous or nanostructured (i.e., nanocrystalline or nanotwinned) metals, the effective mechanical properties of nanometer- and micron-sized structures, devices and engineered (meta)materials, further the underlying mechanisms leading to inelasticity and material failure, or heat and mass transfer in nanoscale materials systems. Overall, there is urgent need for techniques that bridge across length and time scales in order to accurately describe, to thoroughly understand, and to reliably predict the mechanics and physics of solids (Fig. 5.1).

D.M. Kochmann (✉)
California Institute of Technology, Graduate Aerospace Laboratories, 1200 E California Blvd MC 105-50, Pasadena, CA 91125, USA
e-mail: kochmann@caltech.edu

J.S. Amelang
Harvey Mudd College, 301 Platt Blvd, Claremont, CA 91711, USA
e-mail: jamelang@caltech.edu

C.R. Weinberger, G.J. Tucker (eds.), *Multiscale Materials Modeling for Nanomechanics*, Springer Series in Materials Science 245,
DOI 10.1007/978-3-319-33480-6_5

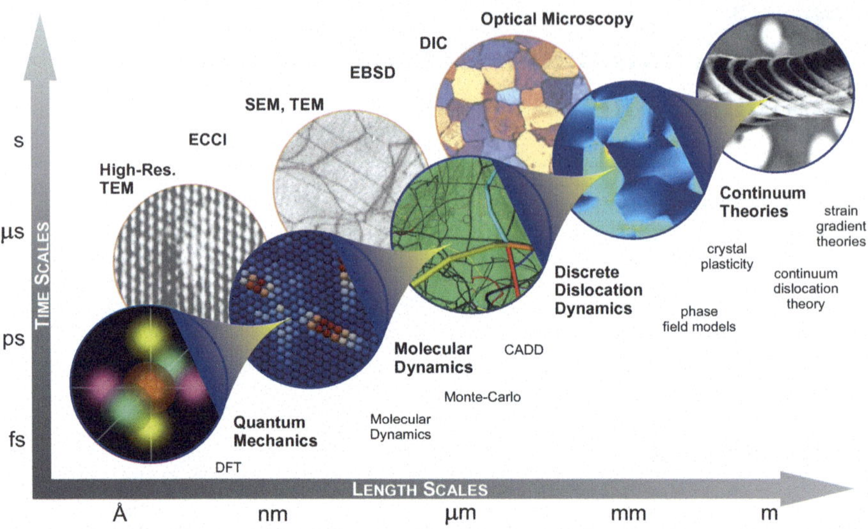

Fig. 5.1 Bridging across scales in crystalline solids: from the electronic structure all the way up to the macroscale (including some of the prominent modeling and experimental techniques). Abbreviations stand for Coupled Atomistic/Discrete-Dislocation (CADD), Density Functional Theory (DFT), Digital Image Correlation (DIC), Electron Back-Scatter Diffraction (EBSD), Electron Channelling Contrast Imaging (ECCI), and Scanning/Transmission Electron Microscopy (SEM/TEM)

Over the decades, various methodologies have been developed to bridge across scales. On the one hand, *hierarchical* schemes are the method of choice when a clear separation of scales can be assumed. In this case, homogenization techniques can extract the effective constitutive response at the macroscale from representative simulations at the lower scales. Examples include multiple-level finite-element (FE) analysis or FE^n [30, 56, 70], as well as the homogenization of atomistic ensembles to be used at the material-point level in macroscale FE simulations [15, 16, 62, 73]. By contrast, *concurrent scale-coupling* methods avoid the aforementioned separation of scales and instead integrate different constitutive descriptions into a single-scale model by spatially separating domains treated, e.g., by first-principles, molecular dynamics, discrete defect mechanics, and continuum theories. Prominent examples comprise Coupled Atomistic/Discrete-Dislocation (CADD) models [17, 59, 74] and AtoDis [13], furthermore the Bridging Domain Method (BDM) [12] and Bridging Scale Decomposition [50], as well as Macroscopic, Atomistic, AB Initio Dynamics (MAAD) [2, 14] which couples several scales. In such methods, a key challenge arises from the necessity to pass information across interfaces between different model domains. To pick out one example, CADD is an elegant methodology which embeds a small MD domain into a larger region treated by a Discrete Dislocation (DD) description. Here, the passing of lattice defects across the interface between the two domains is a major challenge. In order to circumvent such difficulties, *coarse-graining techniques* apply the same lower-scale constitutive description to the entire model but scale up in space and/or in time. Classical examples include Coarse-Grained MD (CGMD) [69] as well as the quasicontinuum (QC) method [82]. We note that, while upscaling in time and in space are equally important, the primary focus here will be on spatial coarse-graining, which is the main achievement of the QC approximation. Upscaling of atomistic simulations in the time domain has been investigated in the context of MD simulations, see, e.g., [41, 89, 90], and can be added to a spatial coarse-graining scheme such as the QC method. A particular QC formulation (applicable to finite temperature) to study, e.g., long-term mass and heat transfer phenomena was introduced recently [6, 88, 91]. The QC approximation relies upon the crystalline structure and is ideally suited to carry out zero-temperature calculations, as presented in the following. A variety of finite-temperature QC extensions exist, see, e.g., [21, 27, 34, 45, 54, 69, 72, 85, 86], which can be applied on top of the presented spatial coarse-graining strategies.

Spatial coarse-graining reduces the number of degrees of freedom by introducing geometric constraints, thereby making the lower-scale accuracy efficiently available for larger-scale simulations [37, 38, 78, 82]. Coarse-graining strategies offer a number of advantages over domain-coupling methods: (1) the model is solely based on the lower-scale constitutive laws and hence comes with superior accuracy (in contrast to coupling methods, there is no need for a separate and oftentimes empirical constitutive law in the continuum region); (2) the transition from fully resolved to coarse-grained regions is seamless (no approximate hand-shake region is required between different domains in general); (3) depending on the chosen formulation, model adaptation techniques can efficiently reduce computational complexity by tying full resolution to those regions where it is indeed required (such as in the vicinity of lattice defects or cracks and voids).

The *quasicontinuum method* was introduced to bridge from atomistics to the continuum by applying finite element interpolation schemes to atoms in a crystal lattice [58, 65, 82]. This is achieved by three integral components: *geometric constraints* (which interpolate lattice site positions from the positions of a reduced set of representative atoms), *summation rules* or *sampling rules* (which avoid the computation of thermodynamic quantities from the full atomistic ensemble), and *model adaptation* schemes (which localize atomistic resolution and thereby efficiently minimize the total number of degrees of freedom). To date, numerous QC flavors have been developed which mainly differ in the choice of how the aforementioned three aspects are realized. Despite their many differences, all QC-based techniques share as a common basis the interpolation of atomic positions from a set of representative atoms as the primary coarse-graining tool.

The family of QC methods has been applied to a wide range of problems of scientific and technological interest such as studies of the interactions of lattice defects in fcc and bcc metals. Typical problems include nanoindentation [48, 51, 83], interactions of lattice defects [35, 65, 76] or of defects with nanosized cracks [36, 68, 80] and nanovoids [6, 66, 98], or with individual interfaces [47, 97]. Beyond metals, the developed modeling techniques can be extended, e.g., to ferroelectric materials and ionic crystals via electrostatic interactions [20] and multi-lattice approaches [1, 22], further to non-equilibrium thermodynamics [45, 54, 69, 72, 88], also to structural mechanics [7, 9] and two-dimensional materials [64, 92], and in principle to any material system with crystalline order as long as a suitable position-dependent interatomic potential is available.

In this chapter, we will review the common theoretical basis of the family of QC methods, followed by the description of particular QC schemes and available techniques. We then proceed to highlight past and recent applications and extensions of the QC technique to study the mechanics and physics of solids and structures, and we conclude by summarizing some open questions and challenges to be addressed in the future or subject to ongoing research. Since such a book chapter cannot give a full account of all related research, we focus on—in our view—important QC advances and contributions and apologize if we have overlooked specific lines of research. For a list of related publications, the interested reader is also referred to the *qcmethod.org* website maintained by Tadmor and Miller [81].

5.2 The Quasicontinuum Method

The rich family of QC methods is united by the underlying concepts of (1) significantly reducing the number of degrees of freedom by suitable interpolation schemes, and (2) approximating the thermodynamic quantities of interest by summation or sampling rules. Let us briefly outline those concepts before summarizing particular applications and extensions of the QC method.

5.2.1 Representative Atoms and the Quasicontinuum Approximation

In classical mechanics, an atomistic ensemble containing N atoms is uniquely described by how their positions $\mathbf{q} = \{\mathbf{q}_1, \ldots, \mathbf{q}_N\}$ and momenta $\mathbf{p} = \{\mathbf{p}_1, \ldots, \mathbf{p}_N\}$ with $\mathbf{p}_i = m_i \dot{\mathbf{q}}_i$ evolve with time t. Here and in the following, m_i represents the mass of atom i, and dots denote material time derivatives. Then, the ensemble's total Hamiltonian $\mathscr{H}$ is given by

$$\mathscr{H}(\mathbf{q}, \mathbf{p}) = \sum_{i=1}^{N} \frac{|\mathbf{p}_i|^2}{2\,m_i} + V(\mathbf{q}), \tag{5.1}$$

where the first term accounts for the kinetic energy and the second term represents the potential energy of the ensemble with V denoting a suitable atomic interaction potential which we assume to depend on atomic positions only. The time evolution of the system is governed by Hamilton's equations, which yield Newton's equations of motion for all atoms $i = 1, \ldots, N$,

$$m_i \ddot{\mathbf{q}}_i = \mathbf{f}_i(\mathbf{q}) = -\frac{\partial V}{\partial \mathbf{q}_i}(\mathbf{q}), \tag{5.2}$$

where $\mathbf{f}_i(\mathbf{q})$ represents the total (net) force acting on atom i. We note that V can, in principle, depend on both positions $\mathbf{p}$ and momenta $\mathbf{q}$ (the latter may be necessary, e.g., when using the quasiharmonic and other approximations for quasistatic finite-temperature formulations without time discretization). For conciseness, we here restrict ourselves to position-dependent potentials only; the extension to include momenta is straightforward. As one of the cornerstones of the QC method, we limit our analysis to crystalline solids in which the ground state atomic positions coincide with sites of a regular lattice described by a set of Bravais vectors that span the discrete periodic array of lattice sites. In most materials (including metals, ceramics, and organic materials), the interaction potential usually allows for an additive decomposition, i.e.

$$V(\mathbf{q}) = \sum_{i=1}^{N} E_i(\mathbf{q}) \tag{5.3}$$

with E_i being the energy of atom i. Some authors in the scientific literature have preferred to distinguish between external and internal forces by introducing external forces $\mathbf{f}_{i,\text{ext}}$ on all atoms $i = 1, \ldots, N$ so that

$$\mathbf{f}_i(\mathbf{q}) = \mathbf{f}_{i,\text{ext}}(\mathbf{q}_i) - \frac{\partial V}{\partial \mathbf{q}_i}(\mathbf{q}). \tag{5.4}$$

Ideally (and in any reasonable physical system), such external forces are conservative and derive from an external potential $V_{\text{ext}}(\mathbf{q})$ and we may combine internal and external forces into a single potential, which is tacitly assumed in the following. Examples of conservative external forces include gravitation, long-range Coulombic interactions, or multi-body interactions such as during contact. For computational convenience, the latter is oftentimes realized by introducing artificial potentials, see, e.g., [40] for an effective external potential for spherical indenters.

Due to limitations of computational resources, it is generally not feasible to apply the above framework to systems that are sufficiently large to simulate long-range elastic effects, even for short-range interatomic potentials. However, except in the vicinity of flaws and lattice defects such as cracks and dislocations, respectively, the local environment of each atom in a crystal lattice is almost identical up to rigid body motion. Therefore, the *QC approximation* replaces the full atomistic ensemble by a reduced set of $N_h \ll N$ representative atoms (often referred to as *repatoms* for short). This process is shown schematically in Fig. 5.2.

Let us denote the repatom positions by $\mathbf{x}(t) = \{\mathbf{x}_1(t), \ldots, \mathbf{x}_{N_h}(t)\}$. The approximate current positions $\mathbf{q}_i^h$ and momenta $\mathbf{p}_i^h$ of all atoms $i = 1, \ldots, N$ are now obtained from interpolation, i.e. we have

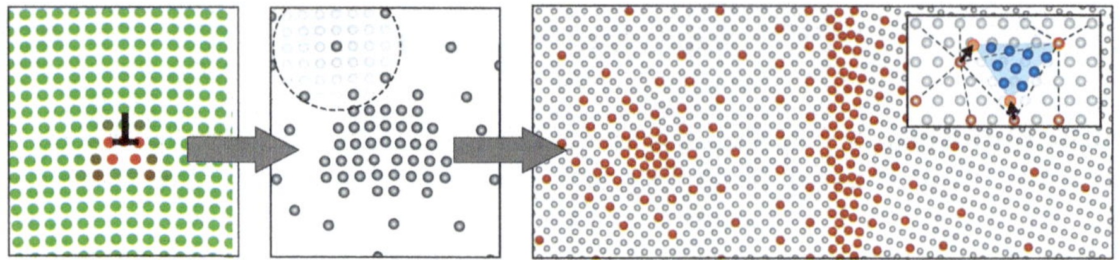

Fig. 5.2 Illustration of the QC methodology: identification of atoms of interest (e.g., high-centrosymmetry lattice sites near a dislocation core), reduction to the set of repatoms (coarsening the full atomic lattice by choosing a small set of representative atomic sites), and the interpolation of atomic positions (*blue* in the small *inset*) from repatom positions (*red*) with an example repatom distribution around a dislocation and a grain boundary in two dimensions

$$\mathbf{q}_i \approx \mathbf{q}_i^h = \sum_{a=1}^{N_h} N_a(\mathbf{X}_i)\, \mathbf{x}_a \tag{5.5}$$

and consequently

$$\mathbf{p}_i \approx \mathbf{p}_i^h = m_i\, \dot{\mathbf{q}}_i^h = m_i \sum_{a=1}^{N_h} N_a(\mathbf{X}_i)\, \dot{\mathbf{x}}_a. \tag{5.6}$$

$N_a(\mathbf{X}_i)$ is the shape function of repatom a evaluated at the position $\mathbf{X}_i$ of lattice site i in the undeformed (reference) configuration (analogously to shape functions commonly used in the finite element method). As an essential feature, one usually requires this coarse-graining scheme to locally recover the exact atomic ensemble when all atoms are turned into repatoms; i.e., if every lattice site is a repatom, then we should recover full atomistics exactly. In other words, in the atomistic limit we require $\mathbf{q}_i = \mathbf{x}_i$. This in turn implies that shape functions should be chosen to satisfy the Kronecker property, i.e., $N_a(\mathbf{X}_b) = \delta_{ab}$ for all $1 \leq a, b \leq N_h$ with δ_{ij} denoting Kronecker's delta ($\delta_{ij} = 1$ if $i = j$ and 0 otherwise). This is the case, e.g., for Lagrange interpolation functions (including affine interpolation), which ensure that shape functions are 1 at their respective node and 0 at all other nodes.

By borrowing concepts from the finite element method, the above geometric constraints within the original QC method [42, 71, 82] made use of an affine interpolation on a Delaunay-triangulated mesh. More recent versions have explored higher-order polynomial shape functions [46, 63] as well as meshless interpolations using smoothed-particle approaches [94, 96] or local maximum-entropy shape functions [43].

The introduction of the geometric constraints in (5.5) has reduced the total number of independent degrees of freedom from $d \times N$ in d dimensions to $d \times N_h$. Therefore, the approximate Hamiltonian $\mathscr{H}^h$ of the coarse-grained system, now involving approximate atomic positions $\mathbf{q}^h = \{\mathbf{q}_1^h, \ldots, \mathbf{q}_N^h\}$ and momenta $\mathbf{p}^h = \{\mathbf{p}_1^h, \ldots, \mathbf{p}_N^h\}$, only depends on the positions and momenta of the repatoms through (5.5):

$$\mathscr{H}^h(\mathbf{x},\dot{\mathbf{x}}) = \sum_{i=1}^{N} \frac{|\mathbf{p}_i^h|^2}{2m_i} + V(\mathbf{q}^h). \tag{5.7}$$

Instead of solving for the positions and momenta of all N lattice sites, the QC approximation allows us to update only the positions and momenta of the N_h repatoms, which requires to compute forces on repatoms. These are obtained from the potential energy by differentiation, which yields the net force on repatom k:

$$\mathbf{F}_k(\mathbf{x}) = -\frac{\partial V(\mathbf{q}^h)}{\partial \mathbf{x}_k} = \sum_{j=1}^{N} \mathbf{f}_j^h(\mathbf{q}^h)\, N_k(\mathbf{X}_j) \tag{5.8}$$

with

$$\mathbf{f}_j^h(\mathbf{q}^h) = -\frac{\partial V(\mathbf{q}^h)}{\partial \mathbf{q}_j^h} = -\sum_{i=1}^{N} \frac{\partial E_i(\mathbf{q}^h)}{\partial \mathbf{q}_j^h}, \tag{5.9}$$

the total force acting on atom j. It is important to note that, in an infinite Bravais lattice (i.e., in a defect-free single-crystal in the absence of external loading or free surfaces), the forces on all atoms vanish, i.e. we have $\mathbf{f}_i^h(\mathbf{q}^h) = \mathbf{0}$, so that the net forces on all repatoms vanish as well ($\mathbf{F}_k = \mathbf{0}$).

5.2.2 *Summation Rules and Spurious Force Artifacts*

Summation rules have become an integral ingredient of all QC methods (even though they are sometimes not referred to as such). Although the above introduction of repatoms has reduced the total number of degrees of freedom significantly from $d \times N$ to $d \times N^h$, the calculation of repatom forces still requires computing the forces between all N atoms and their on average N_b neighbors located within each atom's radius of interaction. These $\mathscr{O}(N \times N_b)$ operations become computationally not feasible for realistically sized systems. We note that, in principle, summations in (5.8) can be reduced to the support of the shape functions (i.e., to regions in which $N_k(\mathbf{X}_j) \neq 0$) but even this is prohibitively expensive in regions of dilute repatom concentrations. Therefore, *summation rules* or *sampling rules* have been introduced to approximate the thermodynamic quantities of interest of the full atomistic ensemble by those of a small set of carefully chosen lattice sites (in the following referred to as *sampling atoms*), comparable to quadrature rules commonly found in the finite element method.

QC summation rules have either approximated the energy of the system (so-called *energy-based QC*) [5, 28] or the forces experienced by the repatoms (*force-based QC*) [42]. Within each of those two categories, multiple versions have been proposed. Here, we will first describe what is known as *fully nonlocal* energy-based and force-based QC formulations and then extract the original *local/nonlocal* QC method as a special case of energy-based QC.

5.2.2.1 Energy-Based QC

In energy-based nonlocal QC formulations, the total Hamiltonian is approximated by a weighted sum over a carefully selected set of *sampling atoms* (which do not have to coincide with the repatoms introduced above). To this end, we replace the sum over index i in (5.3) or (5.8) by a weighted sum over N_s carefully-chosen sampling atoms, see, e.g., [5, 28, 29]. Consequently, the total potential energy is approximated by

$$V(\mathbf{q}^h) = \sum_{i=1}^{N} E_i(\mathbf{q}^h) \quad \approx \quad \widetilde{V}(\mathbf{q}^h) = \sum_{\alpha=1}^{N_s} w_\alpha \, E_\alpha(\mathbf{q}^h), \tag{5.10}$$

where w_α is the weight of sampling atom α. Physically, w_α denotes the number of lattice sites represented by sampling atom α.

The force experienced by repatom k is obtained by differentiation in analogy to (5.8):

$$\widetilde{\mathbf{F}}_k(\mathbf{x}) = -\frac{\partial \widetilde{V}(\mathbf{q}^h)}{\partial \mathbf{x}_k} = -\sum_{\alpha=1}^{N_s} w_\alpha \sum_{j=1}^{N} \frac{\partial E_\alpha(\mathbf{q}^h)}{\partial \mathbf{q}_j^h} N_k(\mathbf{X}_j). \tag{5.11}$$

Now, repatom force calculation has been reduced to $\mathcal{O}(N_s \times N_b)$ operations (accounting for the fact that the sum over j above effectively only involves N_b neighbors within the potential's cutoff radius). The selection of sampling atoms and the calculation of sampling atom weights aim for a compromise between maximum accuracy (ideally $N_s = N$ and $w_\alpha = 1$) and maximum efficiency (requiring $N_s \ll N$ and $w_\alpha \gg 1$).

In order to seamlessly bridge from atomistics to the continuum without differentiating between atomistic and coarse-grained regions, the above summation rules can be interpreted as a fully *nonlocal QC* approximation which treats the entire simulation domain in the same fashion. Summation rules now differ by the choice of (1) sampling atom locations and (2) sampling atom weights. Successful examples of summation rules introduced previously include *node-based cluster summation* [42] with sampling atoms located in clusters around repatoms, and *quadrature-type summation* [33] with sampling atoms chosen nearest to Gaussian quadrature points with or without the repatoms included as sampling atoms. Furthermore, *element-based summation rules* have been introduced in the nonlocal QC context [33, 95] and have demonstrated superior accuracy over traditional cluster-based summation schemes [37]. Recently, a *central summation rule* was proposed in [11] as an ad-hoc compromise between local and quadrature summation rules, which is similar in

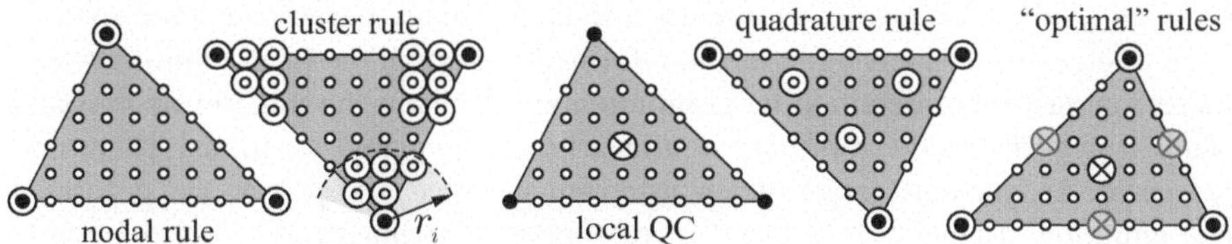

Fig. 5.3 Illustration of popular summation rules with *small open circles* denoting lattice sites, *solid circles* representing repatoms, and *large open circles* are sampling atoms (*crossed circles* denote sampling atoms whose neighborhoods undergo affine deformations). Optimal summation rules [5] are shown of first and second order (the *gray* sampling atoms only exist in the second-order rule)

spirit and contained as a special case within the optimal summation rules recently introduced in [5]. Figure 5.3 illustrates some of the most popular summation rules.

Many summation rules were introduced in an ad-hoc manner to mitigate particular QC deficiencies (e.g., cluster rules were introduced to remove zero-energy modes in force-based QC [42]) or by borrowing schemes from related models (e.g., finite element quadrature rules [33]). The recent *optimal sampling rules* [5] followed from optimization and ensure small or vanishing force artifacts in large elements while seamlessly bridging to full atomistics. Their new second-order scheme also promises superior accuracy when modeling free surfaces and associated size effects at the nanoscale [4].

5.2.2.2 Force Artifacts

Summation rules on the energy level not only reduce the computational complexity but they also give rise to *force artifacts*, see, e.g., [5, 28] for reviews within the energy-based context. By rewriting (5.11) without neighborhood truncations as

$$\widetilde{\mathbf{F}}_k(\mathbf{x}) = -\sum_{\alpha=1}^{N_s} w_\alpha \sum_{j=1}^{N} \frac{\partial E_\alpha(\mathbf{q}^h)}{\partial \mathbf{q}_j^h} N_k(\mathbf{X}_j) = \sum_{j=1}^{N} \left(-\sum_{\alpha=1}^{N_s} w_\alpha \frac{\partial E_\alpha(\mathbf{q}^h)}{\partial \mathbf{q}_j^h} \right) N_k(\mathbf{X}_j), \tag{5.12}$$

and comparing with (5.8), we see that the term in parentheses in (5.12) is not equal to force $\mathbf{f}_j^h$ and hence does not vanish in general even if $\mathbf{f}_j^h = \mathbf{0}$ for all atoms (unless we choose all atoms to be sampling atoms, i.e., unless $N_s = N$ and $w_\alpha = 1$). As a consequence, the QC representation with energy-based summation rules shows what is known as *residual forces* in the undeformed ground state which are non-physical.

In uniform QC meshes (i.e. uniform repatom spacings, uniform sampling atom distribution, and a regular mesh), these residual forces indeed disappear due to symmetry; specifically, the sum in parentheses in (5.12) cancels pairwise when carrying out the full sum. Residual forces hence appear only in spatially non-uniform meshes, see also [37] for a discussion. In general, one should differentiate between residual and spurious forces [5, 28]. Here, we use the term *residual force* to denote force artifacts arising in the undeformed configuration. In an infinite crystal, these can easily be identified by computing all forces in the undeformed ground state. However, force errors vary nonlinearly with repatom positions. Therefore, we speak of *spurious forces* when referring to force artifacts in the deformed configuration. As shown, e.g., in [5], correcting for residual forces by subtracting those as dead loads does not correct for spurious forces and can lead to even larger errors in simulations. Indeed, in many scenarios this dead-load correction is not easily possible since the undeformed ground state may contain, e.g., free surfaces. In this case, the computed forces in the undeformed ground state contain both force artifacts and physical forces. Thus, without running a fully atomistic calculation for comparison, it is impossible to differentiate between residual force artifacts and physical forces arising, e.g., due to surface relaxation [4]. Even though residual and spurious forces are conservative in the energy-based scheme, they are non-physical and can drive the coarse-grained system into incorrect equilibrium states. Special correction schemes have been devised which effectively reduce or remove the impact of force artifacts, see, e.g., the *ghost force correction* techniques in [71, 75]. However, such schemes require a special algorithmic treatment at the interface between full resolution and coarse-grained regions, which may create computational difficulties and, most importantly, requires the notion of an interface.

Based on mathematical analyses, further energy-based schemes have been developed, which blend atomistics and coarse-grained descriptions by a special formulation of the potential energy or the repatom forces. Here, an interfacial domain between full resolution and coarse regions is introduced, in which the thermodynamic quantities of interest are taken as weighted averages of the exact atomistic and the approximate coarse-grained ones [49, 53]. Such a formulation offers clear advantages, including the avoidance of force artifacts. It also requires the definition of an interface region, which may be disadvantageous in a fully nonlocal QC formulation with automatic model adaption since no notion of such interfaces may exist strictly [5].

5.2.2.3 Force-Based QC

In order to avoid force artifacts, *force-based summation rules* were introduced in [42]. These do not approximate the Hamiltonian but the repatom forces explicitly. Ergo, the summation rule is applied directly to the repatom forces in (5.8), viz.

$$\mathbf{F}_k(\mathbf{x}) = \sum_{i=1}^{N} \mathbf{f}_i^h(\mathbf{q}^h)\, N_k(\mathbf{X}_i) \quad \approx \quad \widetilde{\mathbf{F}}_k(\mathbf{x}) = \sum_{\alpha=1}^{N_s} w_\alpha\, \mathbf{f}_\alpha^h(\mathbf{q}^h)\, N_k(\mathbf{X}_\alpha). \tag{5.13}$$

As a consequence, this force-based formulation does not produce residual forces because $\mathbf{f}_\alpha^h(\mathbf{q}^h) = \mathbf{0}$ in an undeformed infinite crystal. The same holds true for an infinite crystal that is affinely deformed: for reasons of symmetry the repatom forces cancel, and spurious force artifacts are effectively suppressed.

Yet, this approximation gives rise to a new problem: the force-based method is non-conservative; hence, there is no potential from which repatom forces derive. This has a number of drawbacks, as discussed, e.g., in [28, 37, 57]. In quasistatic problems, the non-conservative framework may lead to slow numerical convergence, cause numerical instability, or converge to non-physical equilibrium states, see, e.g., the analyses in [24–26, 52, 61]. Furthermore, for dynamic or finite-temperature scenarios, a QC approximation using force-based summation rules cannot be used to simulate systems in the microcanonical ensemble (where the system's energy is to be conserved; other ensembles may, of course, be used). Finally, repatom masses, required for dynamic simulations, are not uniquely defined because there is no effective kinetic energy potential when using force-based summation rules (see also Sect. 5.2.2.5 below). In contrast, energy-based summation rules lead to conservative forces and to strictly symmetric stiffness matrices with only the six admissible zero eigenvalues [28]. Moreover, the proper convergence of energy-based QC techniques was shown recently for harmonic lattices [29].

5.2.2.4 Local/Nonlocal QC

The original QC method of [82] uses affine interpolation within elements and may be regarded as a special energy-based QC scheme, in which atomistic and coarse-grained regions are spatially separated. In the atomistic region and its immediate vicinity, the full atomistic description is applied and one solves the discrete (static) equations of motion for each atom and approximations thereof near the interface (*nonlocal QC*). In the coarse-grained region, a particular element-based summation rule is employed (*local QC*): for each element, the energy is approximated by assuming an affine deformation of all atomic neighborhoods within the element, so that one Cauchy Born-type sampling atom per element is sufficient and the total potential energy is the weighted sum over all such element energies. This concept has been applied successfully to a myriad of examples, primarily in two dimensions (or in 2.5D by assuming periodicity along the third dimension). Since the element-based summation rule in the nonlocal coarse region produces no force artifacts [5], force artifacts here only appear at the interface between atomistic and coarse-grained regions and have traditionally been called *ghost forces* [58, 71]. For a comprehensive comparison of this technique with other atomistic-to-continuum-coupling techniques, see [57]. Figure 5.4 schematically illustrates the concepts of local/nonlocal and fully nonlocal QC.

5.2.2.5 Repatom Masses

Besides repatom forces, energy-based summation rules provide consistent repatom masses, as mentioned above. When approximated by the QC interpolation scheme,

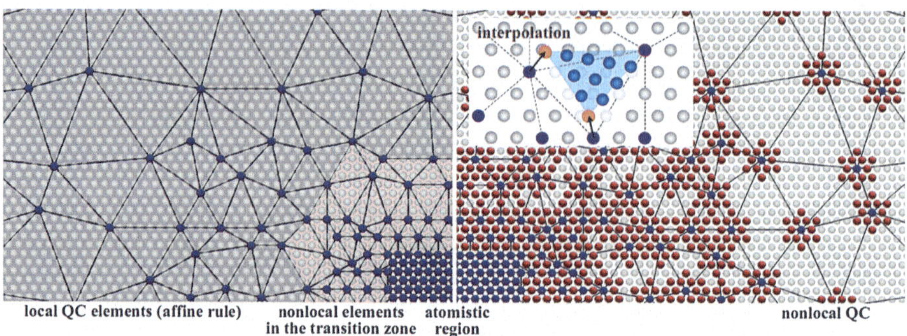

Fig. 5.4 Schematic view of the local/nonlocal (*left*) and fully nonlocal (*right*) QC formulations. The shown nonlocal scheme uses a node-based cluster summation rule. The *inset* on the right illustrates the affine interpolation of atomic positions from repatoms, which is the same in both formulations

the total kinetic energy of an atomic ensemble becomes

$$\begin{aligned}\frac{1}{2}\sum_{i=1}^{N} m_i\,(\dot{\mathbf{q}}_i^h)^2 &= \frac{1}{2}\sum_{i=1}^{N} m_i\left|\sum_{a=1}^{N_h} N_a(X_i)\,\dot{\mathbf{x}}_a\right|^2\\ &= \frac{1}{2}\sum_{a=1}^{N_h}\sum_{c=1}^{N_h}\dot{\mathbf{x}}_a\cdot\left(\sum_{i=1}^{N} m_i\,N_a(X_i)\,N_c(X_i)\right)\dot{\mathbf{x}}_c,\end{aligned} \tag{5.14}$$

where the term in parentheses may be interpreted as the components M_{ac}^h of a consistent mass matrix $\mathbf{M}^h$. In avoidance of solving a global system during time stepping, one often resorts to mass lumping in order to diagonalize $\mathbf{M}^h$. Applying the energy-based summation rule to the total Hamiltonian (and thus to the kinetic energy) yields a consistent approximation of the kinetic energy with mass matrix components

$$M_{ac}^h = \sum_{i=1}^{N} m_i\,N_a(X_i)\,N_c(X_i) \quad\approx\quad \widetilde{M}_{ac}^h = \sum_{b=1}^{N_s} w_b\,m_b\,N_a(X_b)\,N_c(X_b). \tag{5.15}$$

In case of mass matrix lumping, the equations of motion of all repatoms now become

$$\tilde{m}_k^h\,\ddot{\mathbf{x}}_k = \widetilde{\mathbf{F}}_k(\mathbf{x}), \tag{5.16}$$

which can be solved by (explicit or implicit) finite difference schemes. The calculation of repatom masses is of importance not only for dynamic QC simulations, but it is also relevant in quasistatic finite-temperature QC formulations.

5.2.2.6 Example: Embedded Atom Method

The QC approximation outlined above is sufficiently general to model the performance of most types of crystalline solids, as long as their potential energy can be represented as a function of atomic positions. Let us exemplify the general concept by a frequently used family of interatomic potentials for metals (which has also been used most frequently in conjunction with the QC method). The embedded atom method (EAM) [19] defines the interatomic potential energy of a collection of N atoms by

$$E_i(\mathbf{q}) = \frac{1}{2}\sum_{j\neq i}\Phi(r_{ij}) + \mathscr{F}(\rho_i), \qquad \rho_i = \sum_{j\neq i} f(r_{ij}). \tag{5.17}$$

The pair potential $\Phi(r_{ij})$ represents the energy due to electrostatic interactions between atom i and its neighbor j, whose distance r_{ij} is the norm of the distance vector $\mathbf{r}_{ij} = \mathbf{q}_i - \mathbf{q}_j$. ρ_i denotes an effective electron density sensed by atom i due to its neighboring atoms. $f(r_{ij})$ is the electron density at site i due to atom j as a function of their distance r_{ij}. $\mathscr{F}(\rho_i)$ accounts for the energy release upon embedding atom i into the local electron density ρ_i. From (5.2), the exact force $\mathbf{f}_k$ acting on atom k is obtained by differentiation, viz.

$$\mathbf{f}_k(\mathbf{q}) = -\sum_{i=1}^{N}\frac{\partial E_i(\mathbf{q})}{\partial \mathbf{q}_k} = -\sum_{j\in n_I(k)}^{N}\left[\Phi'(r_{kj}) + \{\mathscr{F}'(\rho_k) + \mathscr{F}'(\rho_j)\}f'(r_{kj})\right]\frac{\mathbf{r}_{kj}}{r_{kj}}. \tag{5.18}$$

Since most (non-ionic) potentials are short-range (in particular those for metals), one can efficiently truncate the above summations to include only neighboring atoms within the radius of interaction ($n_I(i)$ denotes the set of neighbors within the sphere of interaction of atom i). Introducing the QC approximation along with an energy-based summation rule now specifies the approximate repatom force as

$$\widetilde{\mathbf{F}}_k(\mathbf{x}) = -\sum_{\alpha=1}^{N_s} w_\alpha \sum_{j\in n_I(\alpha)}\left[\frac{1}{2}\Phi'(r^h_{\alpha j}) + \mathscr{F}'(\rho^h_\alpha)f'(r^h_{\alpha j})\right]\frac{\mathbf{r}^h_{\alpha j}}{r^h_{\alpha j}}\left[N_k(\mathbf{X}_\alpha) - N_k(\mathbf{X}_j)\right], \tag{5.19}$$

which completes the coarse-grained description required for QC simulations. Note that, as discussed above, repatom forces (5.19) are prone to produce residual and spurious force artifacts in non-uniform QC meshes.

5.2.2.7 Adaptive Refinement

One of the general strengths of the QC method is its suitability for adaptive model refinement. This is particularly useful when there is no a-priori knowledge about where within a simulation domain atomistic resolution will be required during the course of a simulation. For example, when studying defect mechanisms near a crack tip or around a pre-existing lattice defect, it may be sufficient to restrict full atomistic resolution to regions in the immediate vicinity of those microstructural features and efficiently coarsen away from those. However, when investigating many-defect interactions or when the exact crack path during ductile failure is unknown, it is beneficial to make use of automatic model adaptation. By locally refining elements down to the full atomistic limit, full resolution can effectively be tied to evolving defects in an efficient manner. In the fully nonlocal QC method [5], this transition is truly seamless as there is no conceptual differentiation between atomistic and coarse domains. In the local/nonlocal QC method as well as in blended QC formulations the transition is equally possible but may require a special algorithmic treatment.

Mesh refinement requires two central ingredients: a criterion for refinement and a geometric refinement algorithm. The former can be realized on the element level, e.g., by checking invariants of the deformation gradient within each element and comparing those to a threshold for refinement [42]. Alternatively, one can introduce criteria based on repatoms or sampling atoms, e.g., by checking the energy or the centrosymmetry of repatoms or sampling atoms and comparing those to refinement thresholds. The most common geometric tool is element bisection; however, a large variety of tools exist and these are generally tied to the specifically chosen interpolation scheme.

Unlike in the finite element method, mesh adaptation within the QC method is challenging since element vertices, at least in the most common implementations, are restricted to sites of the underlying Bravais lattice (this ensures that full atomistics is recovered upon ultimate refinement). The resulting constrained mesh generation and adaptation has been discussed in detail, e.g. in [3]. While mesh refinement is technically challenging but conceptually straightforward, mesh coarsening presents a bigger challenge. Meshless formulations [43, 94, 96] appear promising but have not reached sufficient maturity for large-scale simulations.

5.2.3 Features and Extensions

The QC methods described above have been applied to a variety of scenarios which go beyond the traditional problem of quasistatic equilibration at zero-temperature. In particular, QC extensions have been proposed in order to describe phase transformations and deformation twinning, finite temperature, atomic-level heat and mass transfer, or ionic interactions such as in ferroelectrics. In addition, the basic concept of the QC method has been applied to coarse-grain discrete mechanical systems beyond atomistics. Here, we give a brief (non-exhaustive) summary of such special QC features and extensions.

5.2.3.1 Multilattices

The original (local) QC approximation applies an affine interpolation to all lattice site positions within each element in the coarse domain, which results in affine neighborhood changes within elements (the same applies to the optimal summation rules within fully nonlocal QC). By using Cauchy–Born (CB) kinematics, one can account for relative shifts within the crystalline unit cells such as those occurring during domain switching in ferroelectrics or during solid-state phase transitions and deformation twinning. The CB rule is thus used in coarse-grained regions to relate atomic motion to continuum deformation gradients. In order to avoid failure of the CB kinematics, such a multilattice QC formulation has been augmented with a phonon stability analysis which detects instability and identifies the minimum required periodic cell size for subsequent simulations. This augmented approach has been referred to as *Cascading Cauchy–Born Kinematics* [23, 77] and has been applied, among others, to ferroelectrics [84] and shape-memory alloys [23].

5.2.3.2 Finite Temperature and Dynamics

Going from zero to finite temperature within the QC framework is a challenge that has resulted in various approximate descriptions but has not been resolved entirely. In contrast to full atomistics, the physical behavior of continua is generally governed by thermodynamic state variables such as temperature and by thermodynamic quantities such as the free energy. Therefore, many finite-temperature QC frameworks have constructed an effective, temperature-dependent free energy potential; for example, see [27, 34, 45, 54, 72, 85, 86]. Dynamic simulations of finite-temperature and non-equilibrium processes raise additional questions. Here, we discuss a few representative finite-temperature QC formulations (often referred to as *hotQC* techniques); for reviews, see also [44, 85, 87, 88] and the references therein, to name but a few.

Within the local/nonlocal QC framework, finite-temperature has often been enforced in different ways in the atomistic and the continuum regions, see, e.g., [85]. While in the atomistic region a thermostat can be used to maintain a constant temperature, an effective, temperature-dependent free energy is formulated in the coarse domain, e.g., based on the *quasiharmonic approximation* [85]. For quasistatic equilibrium calculations, the dynamic atomistic ensemble is then embedded into the static continuum description of the coarse-grained model. Alternatively, in dynamic hotQC simulations of this type, all repatoms (including those in both regions) evolve dynamically. The quasiharmonic approximation was shown to be well suited to describe, e.g., the volumetric lattice expansion at moderate temperatures (where the phonon-based model serves as a leading-order approximation). Drawbacks include the inaccuracy observed at elevated temperature levels. Computational drawbacks in terms of instability or technical difficulties may arise from the necessity to compute up to third-order derivatives of the interatomic potentials. These must hence be sufficiently smooth but often derive from piece-wise polynomials or tabulated values in reality.

Temperature and *dynamics* are intimately tied within any framework based on atomistics. In principle, the zero-temperature QC schemes described in previous sections can easily be expanded into dynamic calculations by accounting for inertial effects and solving the dynamic equations of motion instead of their static counterparts. In non-uniform QC meshes, however, this inevitably leads to the reflection and refraction of waves at mesh interfaces and within non-uniform regions. While such numerical artifacts may be of limited concern within, e.g., the finite element method, they become problematic in the present atomistic scenario, since atomic vibrations imply heat. For example, when high-frequency, short-wavelength signals cannot propagate into the coarse region, the fully refined region will trap high-frequency motion, which leads to artificial heating of the atomistic region and thus to non-physical simulation artifacts. Dynamic finite-temperature QC techniques hence face the challenge of removing such artifacts.

Any dynamic atomistic simulation must cope with the small time steps on the order of femtoseconds, which are required for numerical stability of the explicit finite difference schemes. This, of course, also applies to QC. In general, techniques developed for the acceleration of MD can also be applied to the above QC framework. One such example, *hyperQC* [41] borrows concepts of the original hyperdynamics method [89] and accelerates atomistic calculations by energetically favoring rare events through the modification of the potential energy landscape. This approach was reported as promising in simple benchmark calculations [41].

As a further example, finite-temperature QC has been realized by the aid of *Langevin thermostats* [54, 87]. Here, the QC approximation is applied within the framework of dissipative Lagrangian mechanics with a viscous term that expends the thermal energy introduced by a Langevin thermostat through a random force at the repatom level. In a nutshell, the repatoms can be pictured as an ensemble of nodes suspended in a viscous medium which represents the neglected degrees of freedom. The effect of this medium is approximated by frictional drag on the repatoms as well as random fluctuations from the thermal motion of solvent particles. As a consequence, high-frequency modes (i.e., phonons) not transmitted across mesh interfaces are dampened out by the imposed thermostat in order to sample stable canonical ensemble trajectories. This method is anharmonic and was used to study non-equilibrium, thermally activated processes. Unfortunately, the reduced phonon spectra in the coarse-grained domains result in an underestimation of thermal properties such as thermal expansion. For a thorough discussion, see [87].

In contrast to the above finite-temperature QC formulations which interpret continuum thermodynamic quantities as atomistic time averages, one may alternatively assume ergodicity and take advantage of averages in phase space. This forms the basis of the *maximum-entropy* hotQC formulation [44, 45, 87]. By recourse to mean-field theory and statistical mechanics, this hotQC approach is based on the assumption of maximizing the atomistic ensemble's entropy. To this end, repatoms are equipped with an additional degree of freedom which describes their mean vibrational frequency and which must be solved for in addition to the repatom positions at a given constant temperature of the system. This approach seamlessly bridges across scales and does not conceptually differentiate between atomistic and continuum domains. As described in the following section, this concept has also been extended to account for heat and mass transfer [66, 88].

5.2.3.3 Heat and Mass Transfer

The aforementioned maximum-entropy framework has been applied to study not only finite-temperature equilibrium mechanics but has also been extended to describe non-equilibrium heat and mass transfer in coarse-grained crystals. As explained above, the motion of each atom can be decomposed into high-frequency oscillations (i.e., heat) and its mean path (slow, long-term motion). The principle of maximum entropy along with variational mean-field theory provides the tools to apply this concept to the QC method by providing governing equations for the repatom positions as well as their vibrational frequencies [45, 87, 88]. In addition, the same framework is suitable to describe multi-species systems by equipping repatoms with additional degrees of freedom that represent the local chemical composition. For an atomistic ensemble, this yields an effective total Hamiltonian [66, 87, 88]

$$\begin{aligned}\mathscr{H}_{\mathrm{eff}}(\overline{\mathbf{q}},\overline{\mathbf{p}},\boldsymbol{\omega}) = \sum_{i=1}^{N}\Bigg\{ & \frac{|\overline{\mathbf{p}}_i|^2}{2m_i} + \langle V_i\rangle(\overline{\mathbf{q}},\overline{\mathbf{p}},\boldsymbol{\omega},\mathbf{x}) \\ & +3k_{\mathrm{B}}\theta_i\left[\frac{3}{2}N + \log\left(\frac{\hbar\,\omega_i}{k_{\mathrm{B}}\theta_i}\right) - 1 + \sum_{k}^{n} x_{ik}\log x_{ik}\right]\Bigg\},\end{aligned} \tag{5.20}$$

where $\langle\cdot\rangle$ denotes the phase-space average to be computed, e.g., by numerical quadrature; the associated probability distribution and partition function are available in closed form [88]. N, k_{B}, and $\hbar$ are the total number of atoms, Boltzmann's and Planck's constants, respectively. ω_i represents the vibrational frequency of atom i, and θ_i is its absolute temperature. In addition to obtaining Hamilton's equations of motion for the mean positions $\overline{\mathbf{q}}(t)$ and momenta $\overline{\mathbf{p}}(t)$ of all N atoms, minimization of the above Hamiltonian also yields the vibrational frequencies ω_i as functions of temperature. In order to describe n different species, x_{ik} denotes the effective molar fraction of species k at atomic site i (i.e., for full atomistics it is either 1 or 0, in the coarse-grained context it will assume any value $x_{ik} \in [0; 1]$). m_i is an effective atomic mass. In addition to interpolating (mean) atomic positions and momenta as in the traditional QC method, one now equips repatoms with vibrational frequencies and molar fractions as independent degrees of freedom to be interpolated: $(\overline{\mathbf{q}},\overline{\mathbf{p}},\boldsymbol{\omega},\mathbf{x})$. While governing equations for atomic positions follow directly, the modeling of heat and mass transfer requires (empirical) relations for atomic-level mass and heat exchange to complement the above framework [88].

We note that the evaluation of the above Hamiltonian requires effective interatomic potentials for many-body ensembles with known average molar fractions x_{ik}. Possible choices include empirical EAM potentials [31, 39] and related lower-bound approximations [88] as well as DFT-informed EAM potentials [67, 93].

5.2.3.4 Ionic Crystals

Ionic interactions, such as those arising, e.g., in solid electrolytes or in complex oxide ferroelectric crystals, present an additional challenge for the QC method because atomic interactions cease to be short-range. The QC summation or sampling rules take advantage of short-range interatomic potentials which admit the local evaluation of thermodynamic quantities of interest, ideally at the element level. When long-range interactions gain importance, that concept no longer applies and new effective summation schemes must be developed such as those reported in [55].

5.2.3.5 Beyond Atomistics

The QC approximation is a powerful tool whose application is not necessarily restricted to atomistic lattices. In general, any periodic array of interacting nodes can be coarsened by a continuum description which interpolates the full set of nodal positions from a small set of representative nodes. Therefore, since its original development for coarse-grained atomistics, the QC method has been extended and applied to various other fields.

As an example at even lower length scales, coarse-grained formulations of *Density Functional Theory* have been reported [32, 79], which considerably speed up electronic calculations. While the key concepts here appertain to the particular formulation of quantum mechanics based on electron densities, the coarse-graining of the set of atomic nuclei may be achieved by recourse to the QC method.

As a further example at larger scales, the QC method can also be applied at the meso- and macroscales in order to coarse-grain periodic structures such as truss or fiber networks. Here, interatomic potentials are replaced by the discrete interactions of truss members or fibers, and the collective response of large networks is again approximated by the selection of representative nodes and suitable interpolation schemes. This methodology was applied to truss lattices [7, 9, 10] and electronic textiles [8]. As a complication, nodal interactions in those examples are not necessarily elastic, so that internal variables must be introduced to account for path dependence in the nodal interactions and dissipative potentials can be introduced based on the virtual power theorem or variational constitutive updates. Figure 5.5 shows a schematic of the coarse-graining of truss lattices as well as the simulated failure of a coarse-grained periodic truss structure loaded in three-point bending. Detailed results will be shown in Sect. 5.3.3.

5.2.4 Codes

A variety of QC codes are currently being used around the world, most of which are owned and maintained by individual research groups. The most prominent and freely available simulation software is the Fortran-based code developed

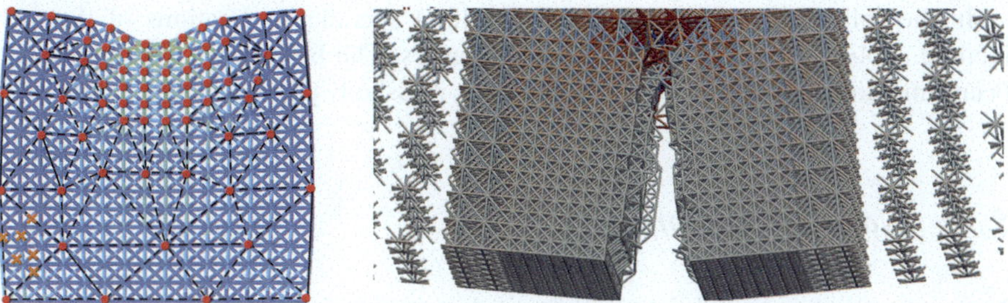

Fig. 5.5 Schematic view of a coarse-grained truss network undergoing indentation (*left*; shown are truss members, nodes, representative nodes, and example sampling truss members within one element) and the simulation result of a truss lattice fracturing during three-point bending (*right*, color-coded by the axial stress within truss members which are modeled as elastic-plastic)

and maintained by Tadmor et al. [81]. This framework is currently restricted to 2.5D simulations and has been used successfully for numerous studies of defect interactions and interface mechanics at the nanoscale [81]. The code is available through the website [81], which also includes a list of references to examples and applications. High-performance codes for 3D QC simulations exist primarily within academic research groups; these include, among others, those of M. Ortiz, P. Ariza, I. Romero et al. (hotQC), J. Knap, J. Marian, and others (force-based QC), E. Tadmor et al. (the 3D extension of [81]), and the authors (fully nonlocal 2D and 3D QC). This is, of course, a non-exhaustive summary of current codes.

5.3 Applications

It is a difficult task to summarize all applications of the QC method to date; for a thorough overview, the interested reader is referred to the QC references found in [81]. Without claiming completeness, the following topics have attracted significant interest in the QC community (for references, see [81]):

- size effects and deformation mechanisms during nanoindentation,
- nanoscale contact and friction, nanoscale scratching and cutting processes,
- interactions between dislocations and grain boundaries (GBs), GB mechanisms,
- phase transformations and deformation twinning, shape-memory alloys,
- fracture and damage in fcc/bcc metals and the brittle-to-ductile transition,
- void and cavity growth at zero and finite temperature,
- deformation and failure mechanisms in single-, bi-, and polycrystals, in particular the mechanics of nanocrystalline solids,
- crystalline sheets and rods, including graphene and carbon nanotubes,
- ferroelectrics and polarization switching,
- electronic textile, fiber networks and truss structures.

In the following sections, we will highlight specific applications of the QC method applied to fcc metals and truss structures. The benchmark examples have been simulated by the authors' fully nonlocal, massively parallel 3D QC code.

5.3.1 Nanoindentation

Nanoindentation is one of the most classical examples that has been studied by the QC method since its inception due to the relatively simple problem setup and the rich inelastic deformation mechanisms that can be observed. Oftentimes, the indenter is conveniently modeled by an external potential [40], which avoids the handling of contact and the presence of two materials.

As an illustrative example of the nonlocal QC method, Fig. 5.6 shows the results of a virtual indentation test into single-crystalline pure copper with a spherical indenter of radius 5 nm up to a maximum indentation depth of 3 nm.

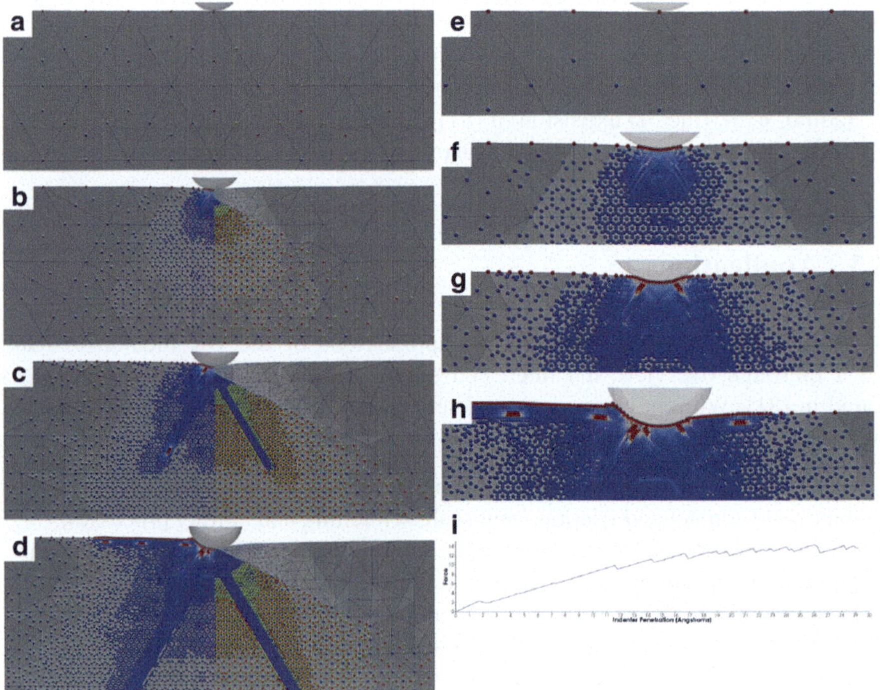

Fig. 5.6 Nanoindentation into a Cu single-crystal: (**a**)–(**d**) show the distribution of repatoms in *blue* (*bottom left panels*) as well as color-coded by centrosymmetry (*top left panels*), the sampling atoms of the first- and second-order summation rules of [5] (shown in *red* and *green*, respectively, in the *bottom right panels*), and the mesh (*top right panels*). (**e**)–(**h**) are zooms of (**a**)–(**d**) illustrating the sampling atom centrosymmetry. (**i**) shows the force-indentation curve (force in eV/A)

All nanoindentation simulations were performed at an indentation rate of $7.8 \cdot 10^8$ at zero temperature, using an extended Finnis–Sinclair potential [18]. For ease of visualization, this example is restricted to two dimensions. Of course, this scenario could easily be simulated by MD but we deliberately present simple scenarios first for purposes of detail visualization. The four panels in Fig. 5.6a–d illustrate the distribution of repatoms and sampling atoms along with the QC mesh. Automatic mesh adaptation leads to local refinement underneath the indenter and around lattice defects. The latter are visualized by plotting the centrosymmetry parameter [40] of sampling atoms (which agree with lattice sites in the fully resolved regions).

Figure 5.7 shows analogous results obtained from a pyramidal indenter penetrating into the same Cu single-crystal. As before, full atomistic resolution is restricted to those regions where it is indeed required: underneath the indenter as well as in the vicinity of lattice defects. Figure 5.7c, d shows the spreading of dislocations into the crystal after emission from the indenter, resulting in full resolution in the wake of the dislocations. The remainder of the simulation domain remains coarse-grained, thus allowing for efficient simulations having significantly fewer degrees of freedom than full atomistic calculations. As before, the distribution of repatoms and sampling atoms is shown along with the QC mesh.

The curves of load vs. indentation depth for both cases of spherical and pyramidal indenters in two dimensions are summarized in Fig. 5.8. As can be expected from experiments, results demonstrate broad hysteresis loops stemming from incipient plasticity underneath the indenters. Data also show pronounced size effects in case of the spherical indenter as well as clear geometrical effects for the (self-similar) pyramidal indenters.

Figures 5.9 and 5.10 show 3D nanoindentation simulations with spherical and pyramidal indenters, respectively. Both graphics show results for Cu single-crystals modeled by an EAM potential [18]. The spherical indenter has a radius of 40 nm and results are shown up to an indentation depth of 3 nm; the pyramidal indenter has an angle of 65.3° against the vertical axis and a maximum indentation depth of 5 nm. Via the centrosymmetry parameter, lattice defects have been identified and highlighted (for better visibility, only those atoms are shown which contribute to lattice defects as identified by a higher centrosymmetry parameter).

As shown in these examples, the QC method efficiently reduces computational complexity by assigning full atomistic resolution where it is indeed required (underneath the indenter and near defects). While full resolution provides locally the same accuracy of molecular statics or dynamics, the QC approximation efficiently coarse-grains the remainder of the model domain, thereby allowing for simulations of significantly larger sample sizes. Similarly, by placing grain or twin boundaries as well as cavities and pre-existing defects in the crystal underneath the indenter, the QC method has been used to study defect interactions (again, the interested reader is referred to [81] for a full list of references).

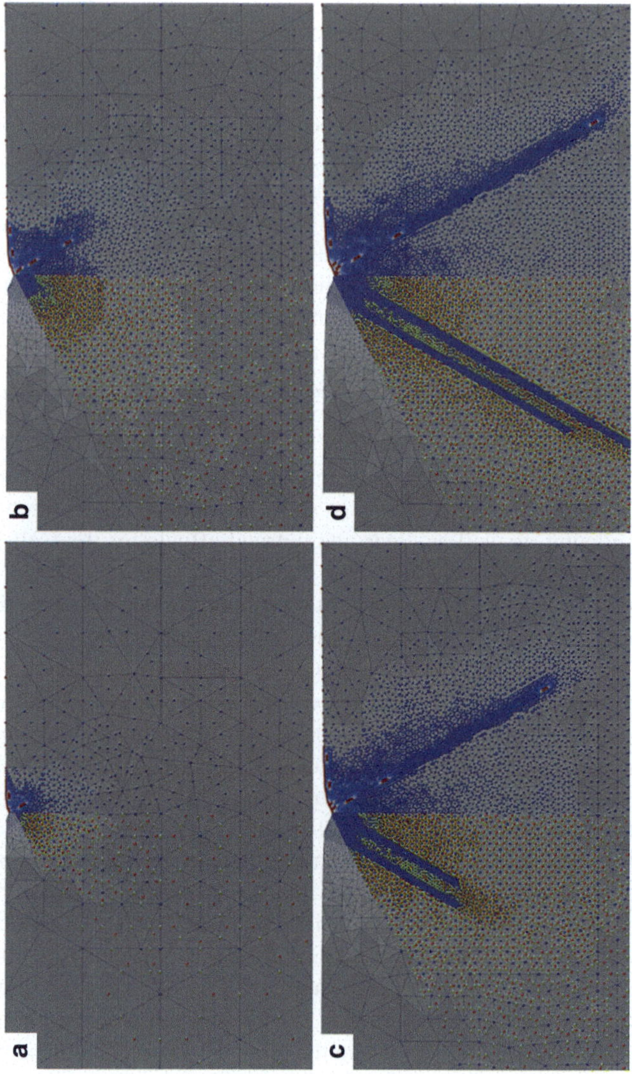

Fig. 5.7 Nanoindentation into a Cu single-crystal with a conical indenter, showing the distribution of repatoms, sampling atoms, the QC mesh, and the local centrosymmetry. Colors and panel partitioning are identical to those of Fig. 5.6a–d

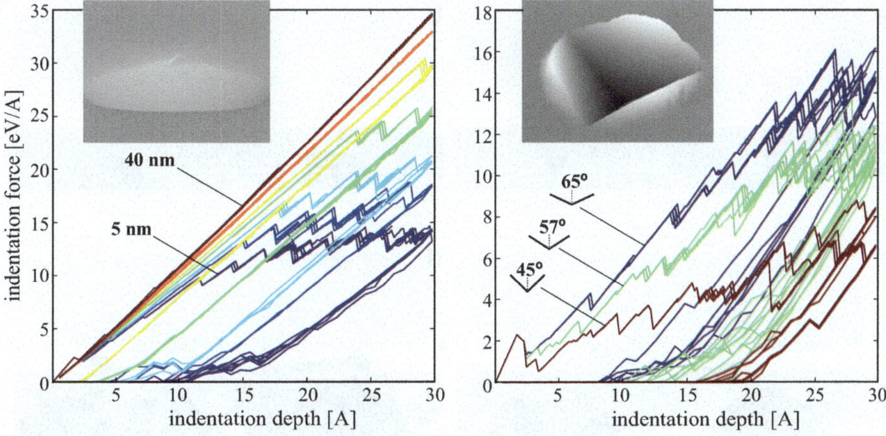

Fig. 5.8 Load vs. indentation depth for spherical indenters (of radii 5, 7.5, 10, 15, 20, 30, and 40 nm) and pyramidal indenters (of different pyramidal angles) in 2D for a (100) single-crystalline Cu, sample

Fig. 5.9 QC Simulation of 3D nanoindentation with a spherical indenter (radius 40 nm) into single-crystalline (100) Cu up to a penetration depth of about 3 nm, for details see [5]

5.3.2 *Surface Effects*

Surface effects play an important role in the mechanics of nanoscale structures and devices. At those small scales, the abundance of free surfaces and the associated high surface-to-volume ratios give rise to—both elastic and plastic—size effects. Using the QC approximation near free surfaces introduces errors because (1) the associated interpolation of atomic positions may prevent surface relaxation and (2) as a consequence of the chosen summation rule, atoms below the surface may be represented by those on the surface and vice versa. The fully nonlocal QC method combined with the second-order summation rules of [5] reduces the errors near free surfaces and therefore has been utilized extensively to model deformation mechanisms near free surfaces, see, e.g., [4].

As an example, consider a thin single-crystalline plate (thickness 12 nm) with a cylindrical hole, which is being pulled uniaxially at remote locations far away from the hole. Near the hole, full atomistic resolution is required to capture defect mechanisms, but MD simulations are too expensive to model large plates. Common MD solutions focus on a small plate instead and either apply periodic boundary conditions (which introduces artifacts due to void interactions) or apply the remote

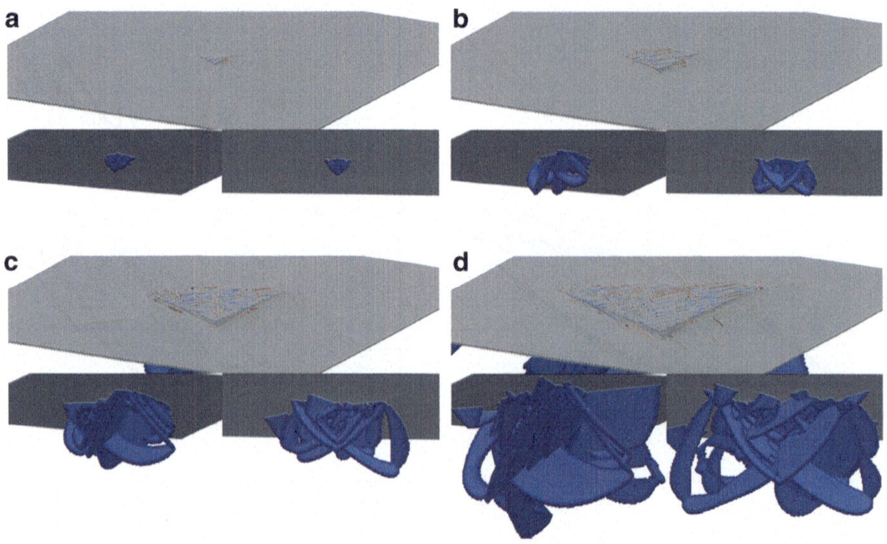

Fig. 5.10 QC simulation of 3D nanoindentation with a pyramidal indenter (angle of 65.3° against the vertical axis) into single-crystalline (100) Cu up to a penetration depth of about 5 nm

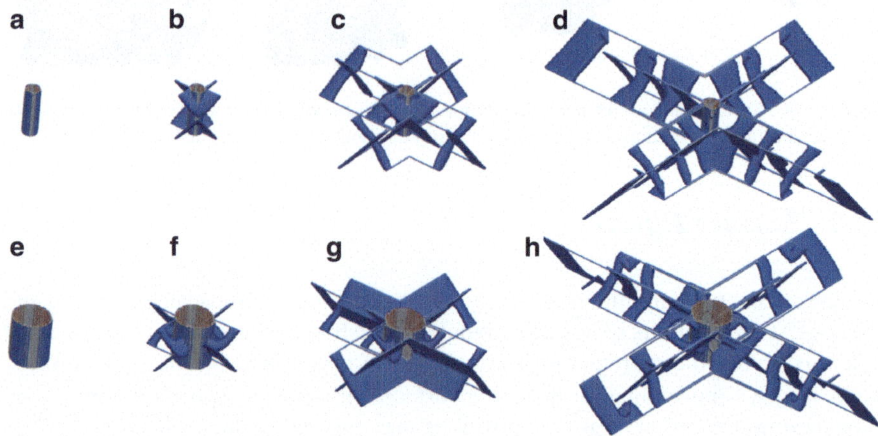

Fig. 5.11 QC simulation of dislocations emitted from cylindrical holes in thin films of single-crystalline Cu (hole radii are 1.5 and 4.5 nm in the top and bottom graphics, respectively) at strains of (**a**) 1.40 %, (**b**) 1.63 %, (**c**) 1.88 %, (**d**) 2.13 % and (**e**) 1.33 %, (**f**) 1.50 %, (**g**) 1.73 %, (**h**) 1.93 %

boundary conditions directly (which also introduces artifacts due to edge effects). Here, we use full atomistic resolution near the hole and efficiently coarsen the remaining simulation domain. By using the second-order summation rule of [4], the coarsening has no noticeable impact on the representation of the free surface. Figure 5.11 illustrates the microstructural evolution for hole radii of 1.5 and 4.5 nm.

Dislocations are emitted from the hole at a critical strain level (which increases as the hole radius decreases); the particular dislocation structures are constrained by the free surfaces at the top and bottom.

5.3.3 *Truss Networks*

As discussed above, the QC method applies not only to atomic lattices but can also form the basis for the coarse-graining of discrete structural lattices, as long as a thermodynamic potential is available which depends on nodal degrees of freedom. For elastic loading, the potential follows from the strain energy stored within structural components. When individual truss members are loaded beyond the elastic limit, extensions of classical QC are required in order to introduce internal variables and account for such effects as plastic flow, or truss member contact and friction. For such scenarios, effective potentials can be defined, e.g., by using the virtual power theorem [7, 9, 10]. Alternatively, variational constitutive updates [60] can be exploited to introduce effective incremental potentials, as will be shown here for a truss network. For simplicity, we restrict our study to elastic-plastic bars which undergo only nonlinear stretching deformation and large rotations (but do not deform in bending).

Let us briefly review the constitutive model for elastic-plastic trusses used in the following example. Consider a truss structure consisting of two-node bars having undeformed lengths L_{ij} and deformed lengths $r_{ij} = |\mathbf{r}_{ij}| = |\mathbf{q}_i - \mathbf{q}_j|$ with nodal positions $\mathbf{q}_i$ $(i = 1, \ldots, N)$. The total strain of each bar is given by $\varepsilon = (r_{ij} - L_{ij})/L_{ij}$. By using variational constitutive updates [60], we can introduce an effective potential for elastic-plastic bars. To this end, we choose as history variables ε_p^n (the plastic strain) and e_p^n (the accumulated plastic strain) for each bar, and we update these history variables at each converged load step n as in classical computational plasticity. In linearized strains, the elastic response of each bar is characterized by $\sigma = E(\varepsilon - \varepsilon_p)$ with axial stress σ and Young's modulus E. Further, assume that the full elastic-plastic stress-strain response is piecewise linear; i.e., under monotonic loading, the bar will first yield at a stress τ_0 followed by linear hardening with a slope of $2EH/(E + 2H)$ with some hardening modulus H. Thus, for monotonic loading, we assume

$$\sigma = \begin{cases} E\,\varepsilon, & \text{if} \quad \varepsilon \leq \tau_0/E, \\ \tau_0 + 2H/(E + 2H)(E\varepsilon - \tau_0), & \text{if} \quad \varepsilon > \tau_0/E, \end{cases} \tag{5.21}$$

and the analysis for non-monotonic loading can be carried out analogously. This model can be cast into an effective potential using variational constitutive updates as follows.

Let us discretize the time response into time steps Δt and define the stress and strain at time $t^n = n \cdot \Delta t$ as σ^n and ε^n (the internal variables follow analogously).

After each load increment, we must determine the new internal variables

$$\varepsilon_p^{n+1} = \varepsilon_p^n + \Delta\varepsilon_p, \qquad e_p^{n+1} = e_p^n + |\Delta\varepsilon_p|. \tag{5.22}$$

Assume an elastic-plastic effective energy density is given by

$$W(\varepsilon^{n+1}, \Delta\varepsilon_p; \varepsilon_p^n, e_p^n) = \frac{E}{2}\left[\varepsilon^{n+1} - (\varepsilon_p^n + \Delta\varepsilon_p)\right]^2 + H\left(e_p^n + |\Delta\varepsilon_p|\right)^2 + \Delta t\, \tau_0 \left|\frac{\Delta\varepsilon_p}{\Delta t}\right|, \tag{5.23}$$

where the first term represents the elastic energy at the new load step $n+1$, the second term represents linear plastic hardening with hardening modulus H, and the final term defines dissipation in a rate-independent manner (τ_0 is again the yield stress, and $\Delta t = t_{n+1} - t_n > 0$ is taken as a constant time increment). The potential energy of a bar with cross-section A and length L_{ij} is now the above energy density multiplied by the bar volume AL_{ij}. Using the theory of [60], an effective potential energy of the bar can be defined as

$$V\left(\mathbf{q}_i^{n+1}, \mathbf{q}_j^{n+1}; \varepsilon_p^n, e_p^n\right) = AL_{ij} \cdot \inf_{\Delta\varepsilon_p}\left\{W\left(\frac{r_{ij}^{n+1} - L_{ij}}{L_{ij}}, \Delta\varepsilon_p; \varepsilon_p^n, e_p^n\right)\right\}. \tag{5.24}$$

One can easily show that this potential recovers the above elastic-plastic bar model as follows. Minimization can be carried out analytically for this simple case. In particular, the stationarity condition yields the kinetic rule of plastic flow:

$$\frac{\partial W}{\partial \Delta\varepsilon_p} = -E\left(\varepsilon^{n+1} - \varepsilon_p^n - \Delta\varepsilon_p\right) + 2H\left(e_p^n + |\Delta\varepsilon_p|\right) + \tau_0 \operatorname{sign} \Delta\varepsilon_p = 0. \tag{5.25}$$

Consider first the case of $\Delta\varepsilon_p > 0$, which leads to

$$\Delta\varepsilon_p = \frac{E\left(\varepsilon^{n+1} - \varepsilon_p^n\right) - (\tau_0 + 2He_p^n)}{E + 2H} > 0. \tag{5.26}$$

Analogously, for $\Delta\varepsilon_p < 0$ we have

$$\Delta\varepsilon_p = \frac{E\left(\varepsilon^{n+1} - \varepsilon_p^n\right) + (\tau_0 + 2He_p^n)}{E + 2H} < 0. \tag{5.27}$$

If none of the two inequalities is satisfied, the bar deforms elastically and $\Delta\varepsilon_p = 0$. The stress in the bar connecting nodes i and j at the new step t^n is thus given by

$$\sigma_{ij}^{n+1} = E\left(\varepsilon^n - \left[\varepsilon_{p,ij}^n + \Delta\varepsilon_{p,ij}\right]\right), \tag{5.28}$$

and the corresponding axial bar force is $f_{ij} = A\sigma_{ij}$. Note that for monotonic loading this recovers (5.21). In other words, the above elastic-plastic model can be cast into the effective potential (5.24). Finally, a failure criterion can be included by defining a critical stress σ_{cr} or $\varepsilon_{\mathrm{cr}}$ and removing bars whose stress or strain reaches the maximum allowable value.

The total Hamiltonian of the truss structure with nodal positions $\mathbf{q} = \{\mathbf{q}_1, \dots, \mathbf{q}_N\}$ and momenta $\mathbf{p} = \{\mathbf{p}_1, \dots, \mathbf{p}_N\}$ becomes

$$\mathscr{H}(\mathbf{q}, \mathbf{p}) = \sum_{i=1}^{N} \frac{|\mathbf{p}_i|^2}{2\, m_i} + \sum_{i=1}^{N} \sum_{j \in \mathscr{C}(i)} V\left(\mathbf{q}_i^{n+1}, \mathbf{q}_j^{n+1}; \varepsilon_{p,ij}^{n}, e_{p,ij}^{n}\right), \tag{5.29}$$

where we assumed that the mass of each bar is lumped to its nodes, so m_i represents the total mass of node i. $\mathscr{C}(i)$ denotes the star of each node, i.e. the set of all adjacent nodes connected to node i through a bar. As the structure of (5.29) is identical to that of atomistics, cf. (5.1), the QC method can be applied in the very same manner as described above for atomistic ensembles, including summation/sampling rules as well as adaptive remeshing. We note that the sums in (5.29) only involve nearest-neighbor interactions, which is why no force artifacts are expected from node-based summation rules as well as from those of [5], even in non-uniform meshes (as long as elements are sufficiently large).

As an example of this approach, we show simulation results for a three-point bending test of a periodic truss lattice with an initial notch. For small numbers of truss members, such simulations can efficiently be run with full resolution (i.e., modeling each truss member as a bar element). Advances in additive manufacturing over the past decade have continuously pushed the frontiers of fabricating micron- and nanometer-sized truss structures. This has resulted in periodic and hierarchical truss lattices with an unprecedented architectural design space of (meta)material properties. As a consequence, modeling techniques are required that can efficiently predict deformation and failure mechanisms in truss structures containing millions or billions of individual truss members. The 3D three-point bending scenario of Fig. 5.12 is one such example, where the QC method reduces computational complexity by efficient coarse-graining of the bar network (bars are assumed to be monolithic and slender, undergoing the stretching-dominated elastic-plastic model described above with a failure strain of 10 %). Loads are applied by three cylindrical indenter potentials. To account for manufacturing imperfections and to induce local failure, we vary the elastic modulus of all truss members by a Gaussian distribution (with a standard deviation of 50 % of the nominal truss stiffness). Figure 5.12 compares the simulated truss deformation and stress distribution before and after crack propagation. (a) and (b) demonstrate schematically the coarse-grained nodes, whereas (c) through (f) illustrate the advancing crack front in the fully resolved region. Like in the atomistic examples in the previous sections, the QC method allows us to efficiently model large periodic systems with locally full resolution where needed.

Of course, this is only an instructive example. Truss members may be subject to flexural deformation, nodes may contribute to deformation mechanisms, and during truss compression one commonly observes buckling and densification accompanied

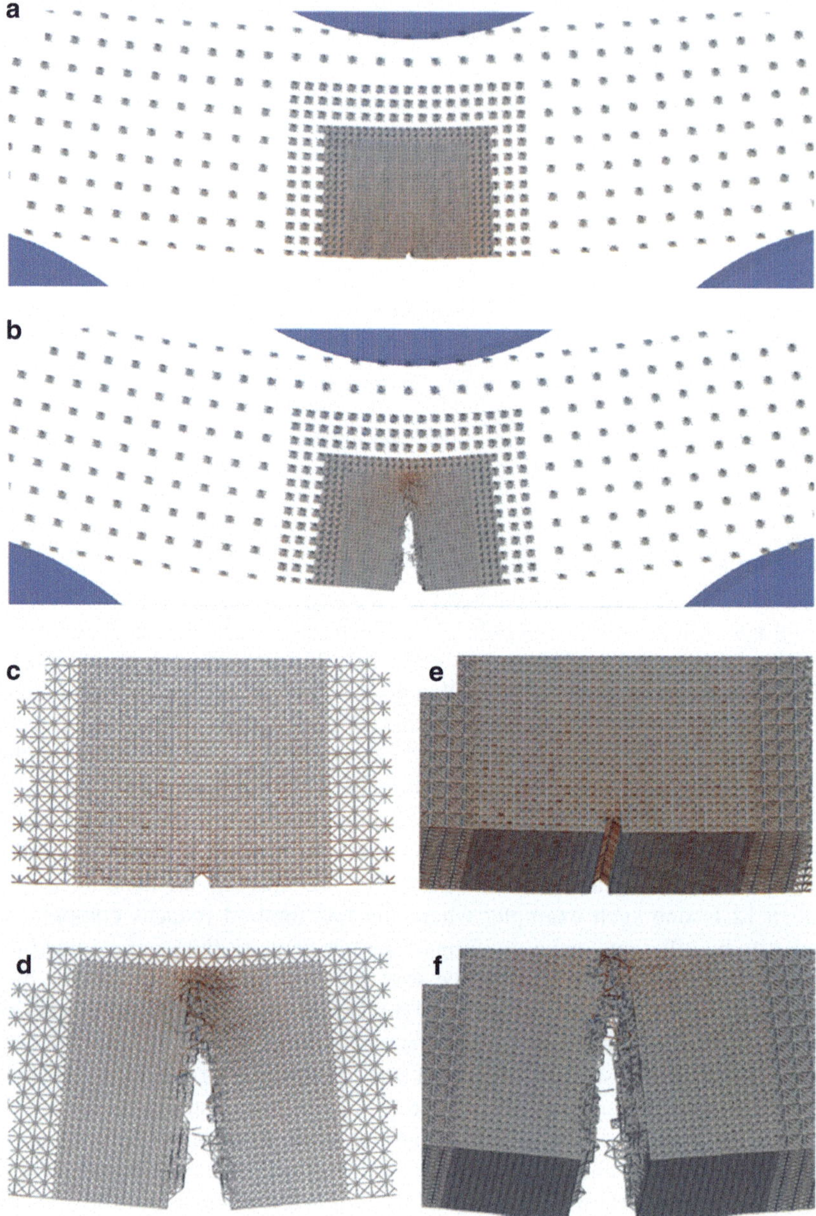

Fig. 5.12 Failure of a periodic 3D truss network under three-point bending; color-code indicates stresses within truss members (in order to account for imperfections and stress concentrations; stresses scaled to maximum). (**a**) and (**b**) visualize the representative nodes in the coarse-grained QC description (full resolution around the notch and coarsening away from the notch; the three indenters are shown as *blue balls*). (**c**/**e**) and (**d**/**f**) show, respectively, the deformed structure with its stress distribution before and after the crack has advanced

by truss member contact and friction. Model extensions for such effects have been proposed recently and are subject to ongoing research, see, e.g., [7, 9, 10].

5.4 Summary and Open Challenges

We have reviewed the fundamental concepts as well as a variety of extensions of the traditional quasicontinuum (QC) method for concurrent scale-coupling simulations with a focus on coarse-grained atomistics. This technique is solely based on interatomic potentials without the need for further empirical or phenomenological constitutive relations. Coarse-graining is achieved through the selection of representative atoms (and the interpolation of atomic positions from the set of repatoms), the introduction of sampling or summation rules (to approximate thermodynamic quantities based on a small set of sampling atoms and associated weights), and model adaptation schemes (to restrict full atomistic resolution to those regions where it is indeed beneficial). For each of those three QC pillars, a multitude of variations have been proposed and implemented. Important extensions of the method include multilattices, finite temperature, mass and heat transfer, long-range and dissipative interactions, to name but a few of those discussed in this chapter. It is important to note that the focus here has been on the QC method as a computational tool to facilitate scale-bridging simulations; as such, we have emphasized modeling concepts and applications rather than the mathematical proofs of convergence or stability (the interested reader is referred to the rich literature in this field of research, see, e.g., [29] and the references therein as well as the website [81]). Among all existing QC flavors, we have chosen a fully nonlocal QC formulation for our simulation examples to demonstrate its application to crystal plasticity and periodic truss networks.

Although the QC method was established almost two decades ago, many challenges have remained and new ones have aroused. These include, among others:

- *Adaptive coarsening*: adaptive model refinement may be geometrically challenging in practice but it is conceptually sound, ultimately turning all lattice sites into repatoms and sampling atoms and thus recovering (locally) molecular dynamics or statics. Model coarsening, by contrast, is challenging conceptually and a key open problem in many QC research codes due to the Lagrangian formulation. Overcoming this difficulty will allow to efficiently coarsen the model by removing full resolution (e.g., in the wake of a traveling lattice defect which leaves behind a perfect crystal that is now fully resolved). Without adaptive model coarsening, simulations will accumulate atomistic resolution and ultimately turn large portions of the model domain into full MD, thus producing prohibitive computational expenses.

- *Large-scale simulations*: the number of existing massively parallel, distributed-memory QC implementations is small for a variety of reasons, especially in three dimensions. This limits the size of domains that can be simulated.
- *Dynamics* is a perpetual problem of the QC method. Non-uniform meshes result in wave reflections and wave refraction, which affects phonon motion and thereby corrupts heat propagation within the solid. For these reasons, dynamic QC simulations call for new approaches. Several finite-temperature QC formulations have aimed at overcoming some of the associated problems, but those generally come with phenomenological or simplifying assumptions.
- *Mass and heat transfer* is intimately tied to the long-term dynamics of a (coarse-grained) atomistic ensemble. Recent progress [88] has enabled the computational treatment via statistical mechanics combined with mean-field theory, yet such approaches require constitutive relations for heat and mass transfer at the atomic scale or effective transport equations, and they have not been explored widely.
- *Potentials*: every atomistic simulation stands and falls by the accuracy and reliability of its interatomic potentials. Unlike MD, QC may require conceptual extensions when it comes to, e.g., long-range interactions or interaction potentials that not only depend on the positions of the atomic nuclei.
- *Surfaces* play a crucial role in small-scale structures and devices and most QC variants do not properly account for free surfaces (misrepresenting or not at all accounting for surface relaxation). Improved surface representations are hence an open area of research, see, e.g., [4] for a recent discussion.
- *Structures*: rather recently, the QC approximation has been applied to truss and fiber networks where new challenges arise due to the complex interaction mechanisms (including, e.g., plasticity, failure, or contact). This branch of the QC family is still in its early stages with many promising applications.
- *Imperfections*: The QC method assumes perfect periodicity of the underlying (crystal or truss) lattice. When geometric imperfections play a dominant role (such as in many nano- and microscale truss networks), new extensions may be required to account for random or systematic variations within the arrangement of representative nodes, all the way to non-periodic or irregular systems.

Of course, this can only serve as a short excerpt of the long list of open challenges associated with the family of QC methods and as an open playground for those interested in scale-bridging simulations.

Acknowledgements The authors gratefully acknowledge support from the National Science Foundation (NSF) under grant number CMMI-123436.

References

1. A. Abdulle, P. Lin, A.V. Shapeev, Numerical methods for multilattices. Multiscale Model. Simul. **10**(3), 696–726 (2012)
2. F.F. Abraham, J.Q. Broughton, N. Bernstein, E. Kaxiras, Spanning the continuum to quantum length scales in a dynamic simulation of brittle fracture. Europhys. Lett. **44**(6), 783 (1998)
3. J.S. Amelang, D.M. Kochmann, Surface effects in nanoscale structures investigated by a fully-nonlocal energy-based quasicontinuum method, Mechanics of Materials **90**, 166–184 (2015)
4. J.S. Amelang, G.N. Venturini, D.M. Kochmann, Summation rules for a fully-nonlocal energy-based quasicontinuum method, J. Mech. Phys. Solids **82**, 378–413 (2015)
5. J.S. Amelang, A fully-nonlocal energy-based formulation and high-performance realization of 842 the quasicontinuum method. PhD thesis, California Institute of Technology (2015)
6. M. Ariza, I. Romero, M. Ponga, M. Ortiz, Hotqc simulation of nanovoid growth under tension in copper. Int. J. Fract. **174**, 75–85 (2012)
7. L.A.A. Beex, R.H.J. Peerlings, M.G.D. Geers, A quasicontinuum methodology for multiscale analyses of discrete microstructural models. Int. J. Numer. Methods Eng. **87**(7), 701–718 (2011)
8. L.A.A. Beex, C.W. Verberne, R.H.J. Peerlings, Experimental identification of a lattice model for woven fabrics: application to electronic textile. Compos. Part A Appl Sci. Manuf. **48**, 82–92 (2013)
9. L.A.A. Beex, R.H.J. Peerlings, M.G.D. Geers, A multiscale quasicontinuum method for dissipative lattice models and discrete networks. J. Mech. Phys. Solids **64**, 154–169 (2014)
10. L.A.A. Beex, R.H.J. Peerlings, M.G.D. Geers, A multiscale quasicontinuum method for lattice models with bond failure and fiber sliding. Comput. Methods Appl. Mech. Eng. **269**, 108–122 (2014)
11. L.A.A. Beex, R.H.J. Peerlings, M.G.D. Geers, Central summation in the quasicontinuum method. J. Mech. Phys. Solids **70**, 242–261 (2014)
12. T. Belytschko, S.P. Xiao, Coupling methods for continuum model with molecular model. Int. J. Multiscale Comput. Eng. **1**(1), 115–126 (2003)
13. S. Brinckmann, D.K. Mahajan, A. Hartmaier, A scheme to combine molecular dynamics and dislocation dynamics. Model. Simul. Mater. Sci. Eng. **20**(4), 045001 (2012)
14. J.Q. Broughton, F.F. Abraham, N. Bernstein, E. Kaxiras, Concurrent coupling of length scales: methodology and application. Phys. Rev. B **60**, 2391–2403 (1999)
15. P.W. Chung, Computational method for atomistic homogenization of nanopatterned point defect structures. Int. J. Numer. Methods Eng. **60**(4), 833–859 (2004)
16. J.D. Clayton, P.W. Chung, An atomistic-to-continuum framework for nonlinear crystal mechanics based on asymptotic homogenization. J. Mech. Phys. Solids **54**(8), 1604–1639 (2006)
17. W.A. Curtin, R.E. Miller, Atomistic/continuum coupling in computational materials science. Model. Simul. Mater. Sci. Eng. **11**(3), R33 (2003)
18. X.D. Dai, Y. Kong, J.H. Li, B.X. Liu, Extended finnin-sinclair potential for bcc and fcc metals and alloys. J. Phys. Condens. Matter **18**, 4527–4542 (2006)
19. M.S. Daw, M.I. Baskes, Embedded-atom method: derivation and application to impurities, surfaces, and other defects in metals. Phys. Rev. B **29**, 6443–6453 (1984)
20. K. Dayal, J. Marshall, A multiscale atomistic-to-continuum method for charged defects in electronic materials, in *Presented as the Society of Engineering Science 2011 Annual Technical Conference*, Evanston (2011)
21. D.J. Diestler, Z.B. Wu, X.C. Zeng, An extension of the quasicontinuum treatment of multiscale solid systems to nonzero temperature. J. Chem. Phys. **121**, 9279–9282 (2004)
22. M. Dobson, R.S. Elliott, M. Luskin, E.B. Tadmor, A multilattice quasicontinuum for phase transforming materials: cascading cauchy born kinematics. J. Comput. Aided Mater. Des. **14**, 219–237 (2007)

23. M. Dobson, R.S. Elliott, M. Luskin, E.B. Tadmor, A multilattice quasicontinuum for phase transforming materials: cascading cauchy born kinematics. J. Comput. Aided Mater. Des. **14**(1), 219–237 (2007)
24. M. Dobson, M. Luskin, C. Ortner, Accuracy of quasicontinuum approximations near instabilities. J. Mech. Phys. Solids **58**, 1741–1757 (2010)
25. M. Dobson, M. Luskin, C. Ortner, Sharp stability estimates for the force-based quasicontinuum approximation of homogeneous tensile deformation. Multiscale Model. Simul. **8**, 782–802 (2010)
26. M. Dobson, M. Luskin, C. Ortner, Stability, instability, and error of the force-based quasicontinuum approximation. Arch. Ration. Mech. Anal. **197**, 179–202 (2010)
27. L.M. Dupuy, E.B. Tadmor, R.E. Miller, R. Phillips, Finite-temperature quasicontinuum: molecular dynamics without all the atoms. Phys. Rev. Lett. **95**, 060202 (2005)
28. B. Eidel, A. Stukowski, A variational formulation of the quasicontinuum method based on energy sampling in clusters. J. Mech. Phys. Solids **57**, 87–108 (2009)
29. M.I. Espanol, D.M. Kochmann, S. Conti, M. Ortiz, A gamma-convergence analysis of the quasicontinuum method. SIAM Multiscale Model. Simul. **11**, 766–794 (2013)
30. F. Feyel, J.-L. Chaboche, Fe2 multiscale approach for modelling the elastoviscoplastic behaviour of long fibre sic/ti composite materials. Comput. Methods Appl. Mech. Eng. **183**(3–4), 309–330 (2000)
31. S.M. Foiles, M.I. Baskes, M.S. Daw, Embedded-atom-method functions for the fcc metals Cu, Ag, Au, Ni, Pd, Pt, and their alloys. Phys. Rev. B **33**, 7983–7991 (1986)
32. V. Gavini, K. Bhattacharya, M. Ortiz, Quasi-continuum orbital-free density-functional theory: a route to multi-million atom non-periodic DFT calculation. J. Mech. Phys. Solids **55**(4), 697–718 (2007)
33. M. Gunzburger, Y. Zhang, A quadrature-rule type approximation to the quasi-continuum method. Multiscale Model. Simul. **8**(2), 571–590 (2010)
34. S. Hai, E.B. Tadmor, Deformation twinning at aluminum crack tips. Acta Mater. **51**, 117–131 (2003)
35. K. Hardikar, V. Shenoy, R. Phillips, Reconciliation of atomic-level and continuum notions concerning the interaction of dislocations and obstacles. J. Mech. Phys. Solids **49**(9), 1951–1967 (2001)
36. L. Huai-Bao, L. Jun-Wan, N. Yu-Shan, M. Ji-Fa, W. Hong-Sheng, Multiscale analysis of defect initiation on the atomistic crack tip in body-centered-cubic metal Ta. Acta Phys. Sin. **60**(10), 106101 (2011)
37. M. Iyer, V. Gavini, A field theoretical approach to the quasi-continuum method. J. Mech. Phys. Sol. **59**(8), 1506–1535 (2011)
38. S. Izvekov, G.A. Voth, A multiscale coarse-graining method for biomolecular systems. J. Phys. Chem. B **109**(7), 2469–2473 (2005)
39. R.A. Johnson, Alloy models with the embedded-atom method. Phys. Rev. B **39**, 12554–12559 (1989)
40. C.L. Kelchner, S.J. Plimpton, J.C. Hamilton, Dislocation nucleation and defect structure during surface indentation. Phys. Rev. B **58**, 11085–11088 (1998)
41. W.K. Kim, M. Luskin, D. Perez, A.F. Voter, E.B. Tadmor, Hyper-qc: an accelerated finite-temperature quasicontinuum method using hyperdynamics. J. Mech. Phys. Solids **63**, 94–112 (2014)
42. J. Knap, M. Ortiz, An analysis of the quasicontinuum method. J. Mech. Phys. Solids **49**(9), 1899–1923 (2001)
43. D.M. Kochmann, G.N. Venturini, A meshless quasicontinuum method based on local maximum-entropy interpolation. Model. Simul. Mater. Sci. Eng. **22**, 034007 (2014)
44. Y. Kulkarni, Coarse-graining of atomistic description at finite temperature. PhD thesis, California Institute of Technology (2007)
45. Y. Kulkarni, J. Knap, M. Ortiz, A variational approach to coarse graining of equilibrium and non-equilibrium atomistic description at finite temperature. J. Mech. Phys. Solids **56**, 1417–1449 (2008)

46. S. Kwon, Y. Lee, J.Y. Park, D. Sohn, J.H. Lim, S. Im, An efficient three-dimensional adaptive quasicontinuum method using variable-node elements. J. Comput. Phys. **228**(13), 4789–4810 (2009)
47. J. Li, J. Mei, Y. Ni, H. Lu, W. Jiang, Two-dimensional quasicontinuum analysis of the strengthening and weakening effect of cu/ag interface on nanoindentation. J. Appl. Phys. **108**(5), 054309 (2010)
48. J. Li, H. Lu, Y. Ni, J. Mei, Quasicontinuum study the influence of misfit dislocation interactions on nanoindentation. Comput. Mater. Sci. **50**(11), 3162–3170 (2011)
49. X.H. Li, M. Luskin, C. Ortner, A.V. Shapeev, Theory-based benchmarking of the blended force-based quasicontinuum method. Comput. Methods Appl. Mech. Eng. **268**, 763–781 (2014)
50. W.K. Liu, H.S. Park, D. Qian, E.G. Karpov, H. Kadowaki, G.J. Wagner, Bridging scale methods for nanomechanics and materials. Comput. Methods Appl. Mech. Eng. **195**(13–16), 1407–1421 (2006)
51. H. Lu, Y. Ni, J. Mei, J. Li, H. Wang, Anisotropic plastic deformation beneath surface step during nanoindentation of fcc al by multiscale analysis. Comput. Mater. Sci. **58**, 192–200 (2012)
52. M. Luskin, C. Ortner, An analysis of node-based cluster summation rules in the quasicontinuum method. SIAM J. Numer. Anal. **47**(4), 3070–3086 (2009)
53. M. Luskin, C. Ortner, B. Van Koten, Formulation and optimization of the energy-based blended quasicontinuum method. Comput. Methods Appl. Mech. Eng. **253**, 160–168 (2013)
54. J. Marian, G. Venturini, B.L. Hansen, J. Knap, M. Ortiz, G.H. Campbell, Finite-temperature extension of the quasicontinuum method using langevin dynamics: entropy losses and analysis of errors. Model. Simul. Mater. Sci. Eng. **18**(1), 015003 (2010)
55. J. Marshall, K. Dayal, Atomistic-to-continuum multiscale modeling with long-range electrostatic interactions in ionic solids. J. Mech. Phys. Solids **62**, 137–162 (2014). Sixtieth anniversary issue in honor of Professor Rodney Hill
56. C. Miehe, J. Schröder, J. Schotte, Computational homogenization analysis in finite plasticity simulation of texture development in polycrystalline materials. Comput. Methods Appl. Mech. Eng. **171**, 387–418 (1999)
57. R.E. Miller, E.B. Tadmor, A unified framework and performance benchmark of fourteen multiscale atomistic/continuum coupling methods. Model. Simul. Mater. Sci. Eng. **17**, 053001 (2009)
58. R.E Miller, M. Ortiz, R. Phillips, V. Shenoy, E.B. Tadmor, Quasicontinuum models of fracture and plasticity. Eng. Fract. Mech. **61**, 427–444 (1998)
59. A.K. Nair, D.H. Warner, R.G. Hennig, W.A. Curtin, Coupling quantum and continuum scales to predict crack tip dislocation nucleation. Scr. Mater. **63**(12), 1212–1215 (2010)
60. M. Ortiz, L. Stainier, The variational formulation of viscoplastic constitutive updates. Comput. Methods Appl. Mech. Eng. **171**(3–4), 419–444 (1999)
61. C. Ortner, A priori and a posteriori analysis of the quasinonlocal quasicontinuum method in 1d. Math. Comput. **80**(275), 1265–1285 (2011)
62. H.S. Park, E.G. Karpov, W.K. Liu, P.A. Klein, The bridging scale for two-dimensional atomistic/continuum coupling. Philos. Mag. **85**(1), 79–113 (2005)
63. J.Y. Park, S. Im, Adaptive nonlocal quasicontinuum for deformations of curved crystalline structures. Phys. Rev. B **77**, 184109 (2008)
64. J.Y. Park, C.-H. Park, J.-S. Park, K.-J. Kong, H. Chang, S. Im, Multiscale computations for carbon nanotubes based on a hybrid qm/qc (quantum mechanical and quasicontinuum) approach. J. Mech. Phys. Solids **58**(2), 86–102 (2010)
65. R. Phillips, D. Rodney, V. Shenoy, E. Tadmor, M. Ortiz, Hierarchical models of plasticity: dislocation nucleation and interaction. Model. Simul. Mater. Sci. Eng. **7**, 769–780 (1999)
66. M. Ponga, M. Ortiz, M.P. Ariza, Finite-temperature non-equilibrium quasi-continuum analysis of nanovoid growth in copper at low and high strain rates. Mech. Mater. **90**, 253–267 (2015)
67. A. Ramasubramaniam, M. Itakura, E.A. Carter, Interatomic potentials for hydrogen in α-iron based on density functional theory. Phys. Rev. B **79**, 174101 (2009)

68. I. Ringdalen Vatne, E. Ostby, C. Thaulow, Multiscale simulations of mixed-mode fracture in bcc-fe. Model. Simul. Mater. Sci. Eng. **19**(8), 085006 (2011)
69. R.E. Rudd, J.Q. Broughton, Coarse-grained molecular dynamics: nonlinear finite elements and finite temperature. Phys. Rev. B **72**, 144104 (2005)
70. J. Schröder, Homogenisierungsmethoden der nichtlinearen Kontinuumsmechanik unter Beachtung von Stabilitätsproblemen. Habilitation thesis, Universität Stuttgart (2000)
71. V.B. Shenoy, R. Miller, E.B. Tadmor, R. Phillips, M. Ortiz, Quasicontinuum models of interfacial structure and deformation. Phys. Rev. Lett. **80**, 742–745 (1998)
72. V.B. Shenoy, V. Shenoy, R. Phillips, Finite temperature quasicontinuum methods. Mater. Res. Soc. Symp. Proc. **538**, 465–471 (1999)
73. M.S. Shephard, C. Picu, D.K. Datta, Composite grid atomistic continuum method: an adaptive approach to bridge continuum with atomistic analysis. Int. J. Multiscale Comput. Eng. **2**(3) (2004)
74. L.E. Shilkrot, R.E. Miller, W.A. Curtin, Multiscale plasticity modeling: coupled atomistics and discrete dislocation mechanics. J. Mech. Phys. Solids **52**, 755–787 (2004)
75. T. Shimokawa, J.J. Mortensen, J. Schiøtz, K.W. Jacobsen, Matching conditions in the quasicontinuum method: removal of the error introduced at the interface between the coarse-grained and fully atomistic region. Phys. Rev. B **69**, 214104 (2004)
76. T. Shimokawa, T. Kinari, S. Shintaku, Interaction mechanism between edge dislocations and asymmetrical tilt grain boundaries investigated via quasicontinuum simulations. Phys. Rev. B **75**, 144108 (2007)
77. V. Sorkin, R.S. Elliott, E.B. Tadmor, A local quasicontinuum method for 3d multilattice crystalline materials: application to shape-memory alloys. Model. Simul. Mater. Sci. Eng. **22**(5), 055001 (2014)
78. P. Suryanarayana, Coarse-graining Kohn-Sham density functional theory. PhD thesis, California Institute of Technology (2011)
79. P. Suryanarayana, K. Bhattacharya, M. Ortiz, Coarse-graining Kohn-Sham Density Functional Theory. J. Mech. Phys. Solids **61**(1), 38–60 (2013)
80. E.B. Tadmor, S. Hai, A Peierls criterion for the onset of deformation twinning at a crack tip. J. Mech. Phys. Solids **51**(5), 765–793 (2003)
81. E.B. Tadmor, R.E. Miller, http://qcmethod.org/ (2016)
82. E.B. Tadmor, M. Ortiz, R. Phillips, Quasicontinuum analysis of defects in solids. Philos. Mag. A **73**(6), 1529–1563 (1996)
83. E.B. Tadmor, R. Miller, R. Philipps, M. Ortiz, Nanoindentation and incipient plasticity. J. Mater. Res. **14**, 2233–2250 (1999)
84. E.B. Tadmor, U.V. Waghmare, G.S. Smith, E. Kaxiras, Polarization switching in PbTiO3: an ab initio finite element simulation. Acta Mater. **50**(11), 2989–3002 (2002)
85. E.B. Tadmor, F. Legoll, W. Kim, L. Dupuy, R. Miller, Finite-temperature quasi-continuum. Appl. Mech. Rev. **65**(1), 010803 (2013)
86. Z. Tang, H. Zhao, G. Li, N.R. Aluru, Finite-temperature quasicontinuum method for multiscale analysis of silicon nanostructures. Phys. Rev. B **74**, 064110 (2006)
87. G.N. Venturini, Topics in multiscale modeling of metals and metallic alloys. PhD thesis, California Institute of Technology (2011)
88. G.N. Venturini, K. Wang, I. Romero, M.P. Ariza, M. Ortiz, Atomistic long-term simulation of heat and mass transport. J. Mech. Phys. Solids **73**, 242–268 (2014)
89. A.F. Voter, A method for accelerating the molecular dynamics simulation of infrequent events. J. Chem. Phys. **106**(11), 4665–4677 (1997)
90. A.F. Voter, F. Montalenti, T.C. Germann, Extending the time scale in atomistic simulation of materials. Annu. Rev. Mater. Res. **32**(1), 321–346 (2002)
91. K.G. Wang, M. Ortiz, M.P. Ariza, Long-term atomistic simulation of hydrogen diffusion in metals. Int. J. Hydrog. Energy **40**(15), 5353–5358 (2015)
92. X. Wang, X. Guo, Quasi-continuum model for the finite deformation of single-layer graphene sheets based on the temperature-related higher order cauchy-born rule. J. Comput. Theor. Nanosci. **10**(1), 154–164 (2013)

93. L. Ward, A. Agrawal, K.M. Flores, W. Windl, Rapid production of accurate embedded-atom method potentials for metal alloys. ArXiv e-prints (2012)
94. S. Xiao, W. Yang, A temperature-related homogenization technique and its implementation in the meshfree particle method for nanoscale simulations. Int. J. Numer. Methods Eng. **69**(10), 2099–2125 (2007)
95. Q. Yang, E. Biyikli, A.C. To, Multiresolution molecular mechanics: statics. Comput. Methods Appl. Mech. Eng. **258**, 26–38 (2013)
96. W. Yang, S. Xiao, The applications of meshfree particle methods at the nanoscale, in *Computational Science – ICCS 2005*, ed. by V.S. Sunderam, G.D. Albada, P.M.A. Sloot, J. Dongarra. Lecture Notes in Computer Science, vol. 3516 (Springer, Berlin/Heidelberg, 2005), pp. 284–291
97. H. Yoshihiko, Y. Nobuhiro, S.V. Dmitriev, K. Masanori, T. Shingo, Large scale atomistic simulation of cu/al2o3 interface via quasicontinuum analysis. J. Jpn. Inst. Metals **69**(1), 90–95 (2005)
98. W. Yu, S. Shen, Initial dislocation topologies of nanoindentation into copper film with a nanocavity. Eng. Fract. Mech. **77**(16), 3329–3340 (2010)

Chapter 6
A Review of Enhanced Sampling Approaches for Accelerated Molecular Dynamics

Pratyush Tiwary and Axel van de Walle

6.1 Introduction

Molecular dynamics (MD) simulations have been a tool of widespread use over the last several decades, with applications in materials science, chemistry, biology, geology, and several other fields [1]. These simulations provide a direct insight into the temporal evolution of molecular systems along with a full atomistic resolution. Especially with the advent of easily available parallel computing resources, one can now deal with fairly large system sizes—often well into the range of a few million atoms [2], which is nearly the length scale relevant for several practical nanoscience applications. However, in spite of such massively parallel resources, MD still has an analogous and often debilitating timescale problem. MD is restricted to integration timesteps of a few femtoseconds, which can be partially mitigated by multiple timestep algorithms [3]. However, it is not yet routinely feasible to go into the millisecond regime and beyond for any system with more than a few thousand atoms. Unlike space, time is sequential, and this makes the timescale problem less amenable to raw parallelization.

Both the cause and several proposed solutions of the timescale problem are embedded in the observation that for many interesting systems, the energy landscape is characterized by numerous metastable states separated by large kinetic barriers. Crossing these barriers to visit new states becomes a rare event, relative to the mandatory few femtosecond integration timestep of MD. As such, a large number

P. Tiwary (✉)
Department of Chemistry, Columbia University, New York, NY, USA
e-mail: pt2399@columbia.edu

A. van de Walle
School of Engineering, Brown University, Providence, RI, USA
e-mail: avdw@brown.edu

C.R. Weinberger, G.J. Tucker (eds.), *Multiscale Materials Modeling for Nanomechanics*, Springer Series in Materials Science 245,
DOI 10.1007/978-3-319-33480-6_6

of so-called *enhanced sampling* approaches have been proposed over the last two decades, that allow one to accelerate the dynamics of the system and access much longer timescales [4–6].

In this review, we will describe the theoretical underpinnings and practical implementations of several such methods. We begin by providing a broad summary of the theoretical prerequisites for the methods (Sect. 6.2). In Sect. 6.3, we will describe the methods themselves—how they work, why they work, and notes of caution. For each method, Sect. 6.3 also gives examples of typical applications and enlists the available software resources for using these. Due to practical length constraints, we have naturally not been able to cover all available methods. This is not because they are not important or useful, and we have tried to at least mention these methods and provide related references for the interested reader. This review is restricted to acceleration of *classical* MD, wherein there are no electronic degrees of freedom. An analogous and even more severe timescale problem pertains to ab initio MD methods, but that is beyond the scope of this review. We do, however, mention in the respective applications (Section 6.3) if a particular method can be used for the acceleration of ab initio MD.

A natural question that might be asked at this point is why another review? Indeed there are several excellent reviews for enhanced sampling approaches for accelerated molecular dynamics, for instance, see [4, 5, 7, 8]. This review should complement these references, and be unique due to the following reasons:

1. This review describes diverse methods developed from the vantages of potential energy surface, free energy surface, and other varied viewpoints. These viewpoints have developed independently and in different scientific communities. Having access to them in one review should be of use from a general intellectual purpose, as well as for interdisciplinary applications where a creative use of more than one method might be called for.
2. Any enhanced sampling method essentially involves one or more theoretical approximations. A key intention of the authors of this review is to highlight the various such approximations underlying the different methods in a compact, accessible, and theoretically sound manner. For this reason, a decent fraction of this review summarizes relevant theoretical concepts in a self-contained manner.
3. Last but not the least, this review also provides a handy list of applications and available software resources.

Through these we hope that the reader will have a range of methods at his or her disposal, and armed with a good theoretical understanding of their individual pros and cons, pick the methods most suited and relevant for the application at hand.

6.2 Theoretical Prerequisites

The challenge of accelerating molecular dynamics has been addressed over the last few decades by building on a wide base of theoretical ideas, primarily coming from equilibrium statistical mechanics [9, 10]. In this section we will broadly cover these various theoretical prerequisites, occasionally leaning more on the side of giving intuition to the reader than on being rigorous. References such as [8, 11] are suggested for more rigorous descriptions.

6.2.1 Rare Events, Separation of Timescales, and Markovianity

Throughout this review we will consider a system comprising N atoms in $2dN$-dimensional phase space, with dimensionality $d = 1$ to 3, and N anywhere in between 1 to a few millions. Starting from a given configuration, MD simulation involves evolving the system under numerical integration of Newton's laws of motion at a temperature T or equivalently inverse temperature $\beta = \frac{1}{k_B T}$, under the action of a given classical force-field or interatomic potential. The temperature is enforced in practice by the presence of a thermostat. In addition to constant (within fluctuations) temperature, the simulation could be under constant volume V or constant pressure P conditions. These are, respectively, referred to as NVT and NPT ensembles. The methods in this review are applicable to both of these, and possibly other ensembles as well.

Our interest lies in processes that typically occur at timescales much longer than the vibration period of individual atoms, the latter typically equaling a few femtoseconds. This is to be contrasted with the timescale of the process one is interested in studying, say nucleation of a dislocation or coalescence of vacancies, which is typically much slower than even microseconds or milliseconds. While each degree of freedom of the system constantly undergoes fluctuations with an average thermal energy of $\frac{1}{2}k_B T$, only very rarely do the fluctuations get concentrated in the specific modes required for the particular event to happen. Thus, an event that might be routinely seen and studied in the experimentalist's laboratory becomes in practice a *rare event* for the molecular dynamics simulator [4].

Closely related to the notion of rare events are the twin concepts of timescale separation and Markovianity.

By timescale separation, we mean that there exists a spectrum of timescales which can be clearly demarcated into distinct non-overlapping regimes. In more intuitive terms, timescale separation means that the system is undergoing motion on a landscape characterized by deep basins (Fig. 6.1), and decorrelation of dynamical variables in each basin occurs much faster compared to the typical basin escape times. Thus, there is a molecular relaxation timescale τ_{mol}, a typical basin escape time τ_e, and one has $\tau_{\text{mol}} \ll \tau_e$.

By Markovianity, we mean that before the system leaves a stable basin, it has lost memory of how it got in there in the first place. This is reasonable because systems of many interacting particles exhibit strong sensitivity to initial conditions [12, 13]. As a result, imperceptible changes in initial conditions prior to and soon after entering the basin (e.g., due to numerical noise in the thermostatting) quickly lead to diverging trajectories within the basin, thus making the system's precise state, within the basin, effectively random after the molecular relaxation time has elapsed.

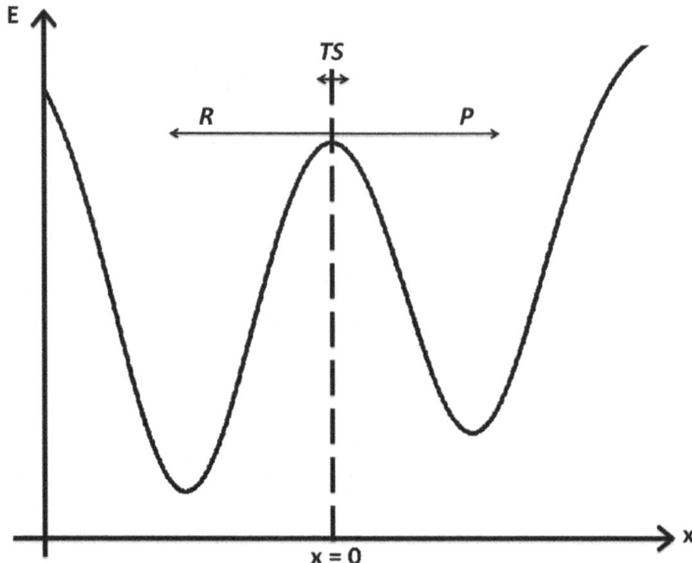

Fig. 6.1 A schematic 1-d energy landscape where the x-axis denotes some reaction coordinate and y-axis is the energy. R, TS, and P stand for reactant, transition state, and product, respectively. Here the depth of either basin is much larger than k_BT or the typical fluctuation in the energy associated with the coordinate x. As such, moving from basin R to P becomes a rare event, and only very occasionally the system visits the TS region

As such, if one maps the system's trajectory into a list of states it visits, then it makes sense to talk of a unique state-to-state timescale that can be characterized solely by two inputs: the identities of the state being exit and of the state being entered. Most, if not all, of the methods discussed in this review exploit this timescale separation and Markovianity in one way or the other. Checking whether these two are actually followed also provides an excellent criterion to assess reliability of the enhanced sampling method at hand.

6.2.2 Potential and Free Energy Surfaces

In the previous subsection we mentioned the notion of a landscape or a surface characterized by deep basins. Now we will make this notion slightly more rigorous.

One way to characterize this landscape could be using coordinates R from the full $3N$-dimensional configuration space (i.e., all atomic coordinates) and looking at the total potential energy $U(R)$ of the system. A constant energy surface in this space is known as a *potential energy surface* (PES). Such surfaces have been extensively and successfully used for relatively small systems, and especially at rather low temperatures [14]. At sufficiently low temperatures and small system sizes, accelerating molecular dynamics becomes closely related to identifying relevant saddle points on the PES that the system must cross while moving from one basin to another. Numerous effective techniques exist for this purpose [15–17], some of which we will tangentially mention in Sect. 6.3.1.

Often, one does not have the luxury of working with small system sizes, and then it becomes prohibitively expensive computationally to characterize or work with the PES. For high-dimensional systems, the PES contains far too many saddle points, and many of them are irrelevant to the dynamics one is interested in studying [18]. Or, it could be that the simulation temperature is high enough that it is more appropriate and effective to describe the large number of states visited via an entropic description rather than an exhaustive enumeration of states that a PES description would require [14]. One can still work around the PES, however, separate approximations (see HTST [4, 19] in Sect. 6.2.4, for example) are then needed to calculate the appropriate entropic corrections.

Instead of focusing on PES, one alternative strategy then (but not the only strategy) is to look at a low-dimensional *free energy surface* (FES), defined as a function of a small number of the so-called *collective variables* $s = \{s_1, \ldots, s_k\}$ where $k \ll N$. These collective variables (CVs) represent "interesting" degrees of freedom or "reaction coordinates". In general, they may be more than simple linear projections—they can be complex nonlinear function of all atomic coordinates [20]. In terms of the potential energy $U(R)$, the free energy $F(s)$ is defined by:

$$F(s) = -\beta^{-1} \ln \int dR\delta(s - s(R))e^{-\beta U(R)} \tag{6.1}$$

This definition differs from the conventional Helmholtz free energy [9] solely by the term $\delta(s - s(R))$, which picks out the region of phase space associated with a specific value of s of the collective variable. Integrating without the term $\delta(s - s(R))$ in Eq. (6.1) would naturally yield the total Helmholtz free energy of the system.

In analogy with the PES description, one now looks at basins in the FES, which is low-dimensional by construction and thus easier to handle. Furthermore, since the free energy as per Eq. (6.1) has temperature built in, one expects to be able to deal with entropic effects in an explicit way. Many methods on calculating and using FES to accelerate dynamics will be discussed in Sect. 6.3.

However, in computer simulations as in life, there is no free lunch [1]. The reduction in dimensionality by switching from PES to FES is closely linked to a judicious choice of a small number of effective collective variables, and often also to a priori knowledge of all possible existing and relevant deep stable basins in the system. Take note that knowledge of stable basins is in general a weaker requirement than knowledge of escape pathways. We will revisit in Sect. 6.3.2.2 as to what makes the choice of a CV good or bad, and how does one identify good CVs.

6.2.3 *Transition States and Transition State Ensemble*

Now we have at our disposal the notion of a landscape, either in potential energy or free energy space, comprising deep stable basins separated by high barriers or transition states (TS) [14, 18, 21]. In this section, we will clarify what exactly do we mean by a transition state.

On a PES, a *transition state* is synonymous to a saddle point on the $3N$-dimensional surface. That is, points at which the quantity $U(R)$ is at a maximum with respect to small displacements in few select k number of directions, and minimum for all others (see Fig. 6.2). Such a saddle point is called as a k-th order saddle point. At such a point, the Hessian matrix H of the potential energy $U(R)$, given by $H_{ij} = \frac{\partial^2 U}{\partial R_i \partial R_j}$, will have k negative eigenvalues [14]. These saddle points correspond to the lowest energy reaction path that connects two stable states. Very efficient and widely used techniques are available for identifying any order saddle points in high-dimensional PES [15–17].

On a FES, in principle one could again define the TS as a saddle point on the low-dimensional surface defined in terms of CVs. However, one faces at least two problems in proceeding this way. Unlike $U(R)$, $F(s(R))$ as defined in Eq. (6.1) is not known a priori, and specialized techniques need to be used to get $F(s(R))$ to begin with, before calculating the saddle points (though there are recent exceptions, see, e.g., [22, 23]). This significantly reduces the appeal of using saddle points to characterize the TS on a FES.

The second problem is more generic, and pertains to the sufficiency of using saddle points to characterize the TS on either the PES or FES. The very reason we are interested in knowing the TS is because it marks the bottleneck in configuration

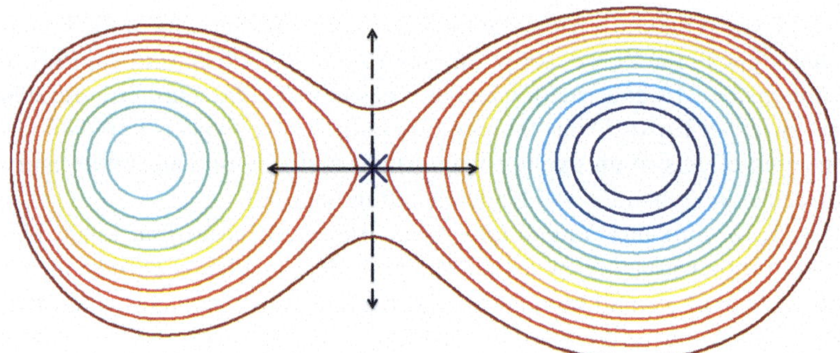

Fig. 6.2 A schematic 2-d energy landscape (energy increases from *blue* to *red*) projected in two-dimensions. The saddle point is marked by an *asterisk*. Note that at the saddle point, movement in the direction indicated by the *solid line* either way would lead to the energy decreasing, while movement in the direction indicated by the *dashed line* either way would lead to the energy increasing. Since there is only one direction in which the energy *decreases*, this is called a 1st order saddle point

space that the rare, successful trajectories must squeeze through while escaping a stable basin. These are the so-called reactive trajectories. While it is true that reactive trajectories will indeed pass through a region near the saddle point, at any finite temperature there will be a distribution of trajectories that *do not* go through the actual saddle but around it [21], especially when the saddle is defined in terms of low-dimensional collective variables. Conflating the concept of a saddle point with the TS could thus have non-trivial errors in how one makes use of any TS based theory (Sect. 6.2.4). Approximations such as harmonic transition state theory (TST) [4, 19] which we describe in Sect. 6.2.4 attempt to negate this error by considering a harmonic region around the saddle, and in this way introduce a distribution of transition states.

Another alternative to thinking in terms of saddle points is to directly work with the "transition state ensemble" (TSE) [18, 21]. Instead of taking a static 0 temperature view, one now defines this ensemble as the collection of points from which the system has equal probability of going into the reactant or the product basin, if one was to launch from these points a large number of MD trajectories with randomized momenta at temperature T. The saddle point is just one member of this more general ensemble.

6.2.4 Rate Constants Through Transition State Theory

In majority of the methods considered in the review, a central concept is the use of rate constants calculated through TST or variants thereof. We will now explain the basics of this theory, referring the interested reader to reviews such as [11, 24] for more rigorous and detailed expositions.

Again consider the two states R and P (standing without loss of generality for reactant and product) marked in Fig. 6.1, along with a transition state region TS separating them. Note that in view of the discussion in previous Sect. 6.2.3, the region TS has a finite volume associated with it. Also note the unidimensional reaction coordinate x that we use to demarcate various regions, which takes the value 0 when in TS. TST [11, 24, 25] assumes that there exists a steady-state probability of finding the system in any of the locations R, P, or TS. That is, if one was to run a long MD trajectory thermostatted at temperature T, then one could define numbers Z_R, Z_P, and Z_{TS} such that log Z_R, log Z_P, and log Z_{TS} are proportional to the probabilities of finding the system in those respective regions. Here, for any region Ω, Z_Ω is the partition function for the system confined to that region defined as $Z_\Omega = \int_\Omega dRe^{-\beta U(R)}$.

TST then states that the rate k of escaping from R to P is proportional to the relative flux crossing through the TS region per unit time [4, 11]:

$$k_{\mathrm{TST}} = \omega\kappa \frac{Z_{\mathrm{TS}}}{Z_R} \tag{6.2}$$

Here ω is a normalization constant detailed in [11, 24]. κ is the so-called transmission coefficient [11, 24], accounting for the fact that not every trajectory that crosses over from R to P through TS is necessarily a reactive trajectory. That is, a certain fraction of such trajectories might actually quickly recross the TS and then settle back in the region R. If one assumes that the mass m of the reaction coordinate (see [11] for a rigorous definition) is constant over the region TS, then Eq.(6.2) reduces to a simple average:

$$k = \kappa \sqrt{\frac{2}{\pi m \beta}} \frac{\langle \delta(x) e^{-\beta U(x)} \rangle_R}{\langle e^{-\beta U(x)} \rangle_R} \tag{6.3}$$

where δ is the Dirac delta function that is summed only when $x = 0$ or equivalently x is in TS (Fig. 6.1). Angular brackets with subscript R denote averaging only when the system is in R. This amounts to calculating the probability of the system being in the region TS relative to the total probability of being in the region comprising both R and TS, multiplied by the recrossing correction κ. There exist ways to explicitly calculate κ [26, 27], but most often researchers either assume it to be 1, or assume that it depends only on the properties of the TS. Equation (6.3) will be the most general form of TST that we will use in this review.

To end this section, we introduce common simplifications of TST that some (but not all) of the methods use. First we set $\kappa = 1$. Then, we assume that the TS is a 1-st order saddle point (see Sect. 6.2.3), and that the basin minimum in region R or P can be well described with a harmonic approximation. Thus the Hessian introduced earlier will have (a) in the saddle region, $3N-1$ positive and 1 negative eigenvalues, and (b) in the basin minima, $3N$ positive eigenvalues. Collectively we call these as $\{\nu_i^{TS}\}$ and $\{\nu_i^R\}$, respectively, with corresponding dimensionalities $3N - 1$ and $3N$. We denote the potential energy difference between the minimum and the saddle point as E_a, the so-called activation energy. Then the *harmonic* TST (HTST) rate is given by [4, 19]:

$$k_{\mathrm{HTST}} = \frac{\prod_{i=1}^{3N} \nu_i^R}{\prod_{i=1}^{3N-1} \nu_i^{\mathrm{TS}}} e^{-\beta E_a} \tag{6.4}$$

Some of the methods reviewed in Sec. 6.3 rely heavily on the use of Eq. (6.4).

6.3 Selected Methods: Theory, Applications, and Software Resources

Now that we have covered the key underlying theoretical concepts, in this section we will get to the heart of the matter, i.e., the enhanced sampling methods themselves. Given the obvious importance of the timescale problem in fairly diverse fields, it is no surprise that over the last nearly three decades a whole plethora of methods have been proposed [4, 5, 7, 8]. This often leads to confusion for a prospective user of accelerated molecular dynamics, who might find themselves wondering as to which method to pick from the very many available. One of the objectives of this section is to create a brief manual of sorts where the user can browse through various available methods and the respective underlying theoretical assumptions, and ascertain which is the method best suited for the application at hand. We would like to point out that there are yet more methods available in addition to the ones described here [23, 28–32]. We hope that a basic understanding of the theory behind the ones discussed here will make the user well equipped to critically ascertain the advantages and shortcomings of any other methods they come across.

There is no one best way to classify these methods. In fact, there is no real need to even classify the methods, except that a classification might help the user find his coordinates better when he comes across a new method not discussed in this review. Here we choose to categorize them on the basis of whether the method is

1. based around the PES, and involves calculation of saddle points/transition states a priori or "on the fly"
2. based around the FES, and thus requires a priori identification of suitable low-dimensional collective variables
3. neither of the above two. While the above two classes invoke TST directly or indirectly, this third class comprises methods that completely bypass the use of TST in any form.

For every method, we outline the key underlying idea, and a broad sense of the assumptions one should be willing to make if using the particular method. For each method, we also provide a brief overview of its applications, restricting our attention to nanomaterials and nanomechanics. We also enlist for each method the software resources available in public domain that we are aware of. Please note that while we have tried our best to be inclusive, we could still have missed on some methods and applications, as is always a possibility in a review of this breadth but with space constraints.

6.3.1 Potential Energy Surface Based Methods

6.3.1.1 Hyperdynamics

Theory The hyperdynamics method [33, 34] is a popular and well-established method that offers an elegant and practical way to increase the rate of infrequent events. It consists of adding a potential energy bias that makes the potential wells, in which the system normally remains trapped for extended periods, less deep. A timescale correction is also evaluated in terms of the bias potential. The hyperdynamics method, especially with the advent of a variety of easy to implement biasing forms [35–37], has seen several compelling applications over the past years (see section Applications).

In this approach, the potential energy $V(R)$ is modified by a pre-determined bias potential $\Delta V(R) \geq 0$, where R as usual denotes the full set of microscopic coordinates in $3N$-dimensional configuration space. This bias potential should ideally be such that the deep basins in the PES are lifted but the saddle points are left untouched. Assuming that such an ideal bias potential can be constructed, then the system is more likely to escape from deep basins without getting trapped. Furthermore, since the saddle points have been left untouched, the relative escape rates for the various escape mechanisms from any states are left untouched. As such, the system moves correctly from one stable state to another at an accelerated pace, but preserving the distribution of the state-to-state sequence. By making use of Eq. (6.2), and again, *only if* the bias has not corrupted the TS, the accelerated time can be quantified and easily computed through the following relation [33, 34]:

$$t_{\text{accelerated}} = \Sigma_{j=1}^{n} \Delta t_{\text{MD}} e^{\beta \Delta V(R(t_j))} \tag{6.5}$$

where n is the number of MD steps (each of duration Δt_{MD}) that have been carried out till a certain point. Often, one uses the term *acceleration factor* or *boost* to denote the ratio of the net accelerated time and MD time. Thus under the application of bias, a time evolution of Δt_{MD} in the MD routine is corrected using the simple relation above, and the user advances the *true time* by $\Delta t_{\text{MD}} e^{\beta \Delta V(R(t))}$. This estimate of $t_{\text{accelerated}}$ unfortunately can sometimes be noisy in practice, especially if the bias ΔV is too strong, because it is obtained as the average of an exponential term which can vary by orders of magnitude [35].

Through the years a range of bias potentials have been designed to meet the requirements for hyperdynamics: namely, no bias in the TS region and high boost without much computational overhead in calculating the bias. In its original form [33, 34] , hyperdynamics involved constructing a Hessian based bias potential, with some clever techniques to replace an expensive full Hessian calculation with a simple optimization problem (reminiscent of the one later used as a key step of the well-known dimer method [17], a technique for saddle point search described in Sect. 6.3.1.3). Independently from hyperdynamics, Grubmueller proposed the "conformational flooding" approach where multi-variate Gaussians depending on local population density were used to form the bias potential inside the stable basins [38]. Several groups also proposed using the potential energy itself as a parameter to define the bias potential, i.e., the bias $\Delta V(r) = \Delta V\{U(r)\}$ where $U(r)$ is the system's potential energy [36, 37]. This specific implementation has been given the somewhat confusing name of accelerated molecular dynamics (aMD) [36]. When it comes to obtaining accelerated timescales, many of these bias potentials have, however, been limited in applicability to relatively low-dimensional systems (less than around a few hundred atoms). This has primarily been due to the difficulties of designing a bias potential that gives a high boost yet meets the stringent requirements of going to zero in transition states [4].

Currently, one of the most popular and robust biasing forms is that of the bond-boost method, originally proposed by Fichthorn and co-workers [35]. This is a good bias potential for systems in which the dynamics is governed by bond breaking and forming events. In such systems, transitions between states will be correlated with large distortions of one or more bonds. This forms the central inspiration for the bond-boost method. The bias potential is controlled by the maximal fractional change ϵ among all bond lengths in the system, and the bias essentially smoothly goes to 0 as soon as ϵ exceeds a cut-off ϵ_c which is to be set by the user.

Most applications of hyperdynamics so far have been restricted to the range of up to a thousand atoms or so. Roughly put, the reason for this is that as the system size increases, it becomes increasingly difficult to design a bias potential that gives high boost, yet also leaves transition states unperturbed [4, 39]. Very recently, a local hyperdynamics method has been proposed in an effort to overcome system size limitations [40]. In this method, the force on each atom is obtained by differentiating a local bias energy that depends only on the coordinates of atoms within a small range of this atom. Intuitively, this scheme breaks down a large system into a large number of small systems, each of which can then be sped up significantly. A unique aspect of this approach is that different parts of the system must continuously adjust the magnitudes of their boost potentials to ensure that all parts share the same boost factor on average through the use of the so-called boostostat. This, in turn, requires the boost factor estimates to be well converged. Another recently proposed variation on hyperdynamics that could ascend length scale issues involves coupling it with the popular quasicontinuum (QC) method. This coupled implementation has been given the name of "hyper-QC" [41].

One practical concern that can arise in using hyperdynamics is when the system is externally driven—say by the application of a tensile load or a magnetic field. If under the application of this external time-dependent perturbation the PES and the saddle themselves are getting affected, then the bias potential must be designed to take this into account. Another obvious concern in using hyperdynamics type methods for driven systems is that the timescale evolution per Eq. (6.5) is choppy, and strictly speaking correct only in the very long time limit. As such, one has to be very careful in application of a driving force that in turn depends on a correct evaluation of the accelerated time.

Later in this section we will also see how using the popular technique metadynamics [5, 42, 43] one can switch focus to the FES instead of PES, and design hyperdynamics-inspired bias potentials for systems where the stable states are known a priori.

Applications Hyperdynamics so far has been quite successful in modeling accelerated dynamics of solid state metallic systems [7]. These include bulk and surface diffusion, adatom ripening [44], desorption of organic molecules from graphitic substrates [45], dislocation nucleation [46], and heteroepitaxial growth [47]. A key issue in hyperdynamics is the choice of bias potential, and evidently most of the applications of hyperdynamics to solid state metallic systems have been through the bond-boost potential [35].

Software Resources Most applications of hyperdynamics have been performed through patches not publicly released. However, hyperdynamics can apparently now be performed through the recently developed EON package [48], developed and maintained by the Henkelman group at UT Austin. In principle, hyperdynamics can also be performed by using a fixed bias set up through the plugin software PLUMED, which can be patched with a large number of MD routines such as LAMMPS, GROMACS, NAMD, etc. [49, 50]. At the time of writing of this review, the group of Art Voter and collaborators at Los Alamos was about to release a public version of software named AMDF implementing bond-boost hyperdynamics [7].

6.3.1.2 Temperature Accelerated Dynamics

Theory This is a method that involves performing dynamics at an elevated temperature to estimate dynamics at a specified lower temperature of interest. This approach does not require a bias potential as in hyperdynamics, but it makes the additional approximation of harmonic TST [51] (see Eq. (6.4)). The central idea is to let the system evolve at a higher temperature, causing transitions to occur more rapidly and, through this, derive information about the evolution of the system at the lower temperature of interest.

To obtain the correct transition rates at the lower temperature of interest, the rates obtained at high temperature need to be extrapolated back to the lower temperature. This is done by assuming that the rates follow an Arrhenius-type dependence on the temperature given by HTST through Eq. (6.4). However, in addition to having assumed the validity of HTST, one has to be careful that the most commonly occurring transition at high temperatures might not be the preferred transition at the actual low temperature of interest, since rates for different events do not change by the same factor when changing the temperature. As such it is not sufficient to get one escape event at high temperature. One needs to get several of these, and build a whole catalog of prospective transitions that could happen. After sufficiently many escape events at higher temperature are identified, the algorithm involves calculating the saddle point energy and frequency (assuming first order saddles only) for each escape event [16], to go into Eq. (6.4). This knowledge is needed for the correct extrapolation of the rates. Once the rates have been mapped into a collection of much longer estimated times in the low temperature system, the event with the shortest escape time at the low temperature is taken to be the correction transition from the current basin, and the clock is correspondingly advanced.

In order to estimate the energy barrier and saddle frequency for each event, the temperature accelerated dynamics (TAD) method typically invokes the double-ended saddle point method called "Nudged Elastic Band (NEB)" [16], which needs both the starting and ending states between which the saddle point is to be found. Since the high temperature escape event gives information on both the starting and the ending states, this is not a problem. The NEB replica runs can efficiently be carried out on parallel processors. Recent developments [52, 53] have also made it possible to use TAD in principle for much larger systems than previously possible.

We would like to point out two possible concerns, that while avoidable in many cases, one should always be aware of while using TAD:

1. one needs to be careful in deciding when "sufficiently many" events have been observed at the higher temperature. Reference [4] provides a formal numerical test for this, and we refer the interested reader to the same.
2. anharmonicity in the transition states, especially if a very high temperature is used for acceleration purposes, could lead to an uncontrollable and unquantifiable error in the timescales obtained from the use of Eq. (6.4).

Applications TAD has seen several convincing applications over the years. TAD has been demonstrated to be very effective for studying grain boundary assisted annealing of defects in radiation damage [54]. An MD/TAD combined procedure has been used to model thin-film growth of Ag and Cu on Ag(100) [55]. Other applications for which TAD has proven effective include crystal growth [56], defect diffusion on oxide surfaces [57], the diffusion of interstitial clusters in Si [58], and defect diffusion in plutonium [59].

Software Resources The software LAMMPS provides a functionality for performing TAD simulations [60], though it is apparently still under development. A code implementing TAD can be directly obtained upon request from the authors of the method [51]. TAD is also implemented in the DLPOLY code [61].

6.3.1.3 Kinetic Monte Carlo and on-the-Fly KMC

Theory The kinetic Monte Carlo (or KMC) method [4, 62] is a fairly simple and useful technique that generates a sequence of configurations and the times at which transitions occur between these configurations. It relies on all of the following assumptions:

1. One knows all the possible stable states that a system can exist in.
2. For each of these states, one knows all possible exit pathways and their respective rates.
3. All these processes are independent and memoryless (i.e., the escape times follow independent Poisson distributions).

Given these assumptions, KMC generates a state-to-state sequence and an associated timescale that has the same statistical distribution as the real system. The algorithm can be roughly summarized as follows [62].

One starts in a certain state, which for KMC means a set of atomic positions, typically constrained to be on a lattice throughout the course of the simulation (though recent developments make it possible to do KMC without lattice constraints [39, 63]). Velocities do not play any role in KMC. One has at his disposal a list of possible transition events from this state, out of which any one escape path is chosen randomly. The key is in how to implement this random selection of escape path. In a nutshell, each path is weighted by its rate constant (which is known a priori); thus, the probability of selecting one path out of the very many available paths is simply proportional to its respective rate constant. The system is then moved to the state dictated by the chosen path. Right at this point, the system's clock is incremented in a way consistent with the average time for escape from the just-exited state, with the averaging carried over all possible exit paths. Thus note that the time increment has nothing to do with which path was actually selected. The process is then repeated for the new state.

Typically the rates for all escape events are computed before starting the simulation, with the use of HTST, along with either classical interatomic potentials or first-principles calculations. This way one builds a catalog of states and rates. Strictly speaking, one could get the catalog of rates without invoking TST but that is at least not common practice, mainly due to the potentially large size of the catalog of possible transitions.

The most important restriction of this approach is that one has to know all possible escape events and rates in advance. The key charm of the KMC protocol is that if the rate catalog is complete, the state-to-state dynamical evolution predicted by KMC is exact. However, if the rate catalog is incomplete, there is no known way to quantify the error the incompleteness could have induced in the dynamics. In KMC as well one has to be careful with driven systems, where the PES, hence the transition states and consequently the rate catalog could be changing as a function of time.

The reliance on a priori knowledge of transitions can become rather prohibitive in many cases, such as highly disordered solids or biomolecular systems. On this note, several workers have been recently making efforts to design algorithms that can predict transition pathways *on-the-fly* as the simulation progresses, such as the adaptive KMC method and the activation relaxation technique [39, 63–65]. In these methods, one builds the rate catalog as the simulation progresses. Several groups have proposed variants of this idea. One universal feature is the use of some sort of single-ended saddle point search method such as the dimer method [17] (as contrasted to NEB [16] which is a double-ended method that we mentioned in Sect. 6.3.1.2). Thus, in any given state, one uses saddle point search techniques of the type that do not require the final state to be known [15, 17]. These searches can be carried out independently on parallel computers, which is one of the main appeals of such adaptive methods. Through these searches one builds a catalog of saddle points and the states that can be reached through them. As the simulation progresses this catalog becomes increasingly accurate. However, at any given point there is no guarantee that all relevant saddles have been found. If the search is systematically missing any of the saddles, there is no known way to quantify that. Nevertheless, this can be a very powerful way of exploring the configuration space if not much is known a priori about the system.

Applications Plain KMC has been applied to an immense number of problems, practically in all aspects of nanomaterials [62]. Adaptive KMC methods as well are seeing an increasingly large number of applications—most of which are not directly tractable through normal KMC. A class of examples would be disordered materials such as amorphous semiconductors, glasses, and polymers where it is not possible to construct a rate table beforehand [66–69]. Even here though, one cannot be hundred percent sure that all possible transitions have been identified. It is also worth noting that adaptive KMC type methods have been successfully applied for accelerating the timescales of not just classical but also ab initio calculations [70].

Software Resources KMC itself generally does not require more than a few lines of code, and thus is mostly done through private implementations. The hard part is generation of the rate table, for which one has to use classical or ab initio transition state methods. A good resource for these is the VASP TST module maintained by the Henkelman group. One can also use the SPPARKS (Stochastic Parallel PARticle Kinetic Simulator) package maintained by Sandia Labs, which is a parallel Monte Carlo code for on-lattice and off-lattice models. Adaptive or on-the-fly KMC methods are, however, far more serious, since they are coupled with classical MD or ab initio calculations. These can be performed through the EON package (for the adaptive KMC flavor) [48], or through the software packages available at the Mousseau group website (for the kinetic activation relaxation technique flavor).

6.3.1.4 κ-Dynamics

Theory κ-dynamics is a recently proposed accelerated dynamics method [71] that involves following forward and backward trajectories from a hypersurface separating a reactant state from its associated product states. The method proceeds by re-initiating trajectories from the hypersurface until one is found that leads to a product without recrossing back to the reactant state. The simulation is then resumed, considering the newly visited product state as the reactant state. In spite of its name, the method actually does not involve an explicit calculation of the transmission coefficient κ in Eq. (6.2). Instead, the success rate of the transition is estimated numerically by repeated simulated trajectories which do not need to be initiated from the transition surface intersecting the saddle point of the transition. The method, however, relies on its user to properly specify an appropriate hypersurface that can demarcate reactant, product, and transition state regions as well as to specify a numerical scheme to generate Boltzmann-distributed starting points on this hypersurface. It is also somewhat agnostic regarding how the boost factor or the actual timescale is to be computed. One benefit of κ-dynamics is that correct state-to-state trajectories themselves could be obtained at a computational cost much lower than obtaining the associated timescales.

Applications κ-dynamics was found to work well in study of FCC adatom ripening and island growth [71].

6.3.1.5 SISYPHUS

Theory SISYPHUS (stochastic iterations to strengthen yield of path hopping over upper states) is a recent mixed Monte Carlo–MD method for accessing extended timescales in atomistic simulations [72, 73]. The key idea is that each time the system enters a stable basin, it will be well thermalized and decorrelated before it leaves this basin and commits itself to another basin. Thus, instead of following the dynamics precisely inside the basins, quicker biased Monte Carlo (MC) runs, with biasing implemented through compensating repulsive potentials as function of some collective variable, are used to decorrelate and thermalize the system inside the basins and enhance the probability of escape.

More precisely, SISYPHUS proceeds by separating the configuration space into basins, and transition regions between the basins, based on some general collective variable Ψ. The stable basins are defined as regions with $\Psi \leq \Psi_c$ where Ψ_c is a user-defined cut-off. Once the system enters the stable basins, then instead of doing unbiased MD the algorithm switches to a biased MC run (called type A) that quickly thermalizes the system within that basin. Once thermalized, the system is brought back to the periphery of the basin, and launched outwards with a randomly assigned velocity at the temperature T of interest [72, 73]. The transition regions and any region outside the deep basins (i.e., states with $\Psi > \Psi_c$) are treated via traditional MD.

In parallel to the decorrelation MC, a second MC routine (called type B) that uses an efficient adiabatic switching based scheme estimates the time the true unperturbed system would have spent inside the basins. Many such routines can be launched on independent processors to come up with a quick estimate of the time-correction, thus making efficient use of loosely coupled parallel computers. The timescale correction is added to the molecular dynamics clock, and so one obtains an accurate accelerated timescale.

SISYPHUS handles the issue of the transmission coefficient κ is a way that is akin to the parallel replica [74] (Sect. 6.3.3.1) or to the κ-dynamics method [71]. When the system enters a basin ($\Psi < \Psi_c$), the algorithm does not immediately switch from MD to MC. Instead, one waits until the system's *correlation time* t_c has elapsed before starting MC. This represents an effective way to ensure that local equilibrium has been reached within the basin, so that TST (Eq. (6.2)) is only used when it indeed applies. If the system exits the new product basin within the correlation time, TST is not invoked, thus allowing a proper representation of "ballistic" (instead of thermally activated) events. This scheme also automatically guards against recrossing events. That is, any MD trajectories that come back to the reactant region before t_c has elapsed and stay therein for an amount exceeding t_c, do not lead to either type of MC routine being launched.

SISYPHUS offers various advantages in terms of (1) excellent use of parallel resources, (2) avoiding reliance on harmonic TST, and (3) avoiding the need to enumerate all possible transition events. However, there is some sensitivity to the choice of the variable Ψ and its cut-off value Ψ_c. As in any other accelerated dynamics method, the reconstructed timescale should ideally be verified by ascertaining insensitivity of the accelerated dynamics to the details of the acceleration—in this case possibly by redoing parts of the simulation for different Ψ_c values. For a poor choice of Ψ , or for too aggressive of a choice of Ψ_c even for a carefully chosen Ψ , many sub-basins could coalesce into one basin, and the calculation of the timescale correction through the second MC routine, while correct in the long time limit, would be extremely slow in converging.

For solid state systems, Tiwary and van de Walle have proposed a modified bond-distortion function for Ψ, which they defined as a p-norm over the magnitudes of all the first neighbor bond distortions in the system [73]. For $p \to \infty$, this becomes the bond-boost function we saw earlier in Sect. 6.3.1.1 on Hyperdynamics [35]. By modulating the value of p, one obtains flexibility to switch between few bonds distorting very sharply ($p \gg 1$) or several bonds collectively undergoing a small but coordinated distortion ($p < 8 \ldots 12$).

Applications SISYPHUS has been successfully applied to a range of problems in nanomaterials and nanomechanics. For instance, in [72] the authors used SISYPHUS to study source-controlled plasticity and deformation behavior in Au nanopillars at strain rates of 10^4/s and lower. SISYPHUS has also been used to study vacancy diffusion in BCC Fe and Ta at a wide range of temperatures, and for the study of adatom island ripening in FCC Al [73].

Software Resources A software package implementing SISYPHUS is available at http://alum.mit.edu/www/avdw/sisyphus.html.

6.3.2 *Free Energy Surface Based Methods*

6.3.2.1 Umbrella Sampling

Theory Umbrella sampling [75] might very well be the *grand daddy* of some of the methods considered in this review. While it is primarily aimed at recovering the underlying FES as a function of some reaction coordinate, it can also be used to obtain timescales. It was first suggested by Torrie and Valleau [75] in 1977, and is essentially an instance of the *importance sampling* technique heavily used by statisticians [76].

The focus here as well is on systems in which high energy barriers (much larger than $k_B T$) separate two or more regions of configuration space. Standard unbiased MD will then suffer from poor sampling, as the barriers will rarely be crossed. In umbrella sampling, the standard Boltzmann weighting for the sampling is replaced by a potential chosen to partially negate the influence of the energy barriers present. This artificial weight $w(s)$ is chosen as a function of specific 1-d or 2-d collective variables s, that force the system to sample a particular range of s values. Average values for any thermodynamic property A calculated from a biased umbrella sampling run performed in this manner can be transformed into unbiased values by applying the formula:

$$\langle A \rangle = \frac{\langle A/w \rangle_w}{\langle 1/w \rangle_w} \tag{6.6}$$

where the subscript w denotes sampling under the non-Boltzmann biased distribution.

A typical umbrella sampling simulation actually comprises a series of harmonic restraints along different s-values each in turn forcing the system to sample separate but overlapping regions of the s-space. Each simulation window can then be treated with Eq. (6.6) to give the unbiased distribution for that window. It, however, remains to combine the unbiased distributions from the various windows. For this, one uses the weighted histogram analysis method [77] that patches the probability distributions from the various windows in a self-consistent manner minimizing the overall error.

Once one has the FES through this procedure, then there are two ways to get the rates:

1. identify the transition state on the FES, and then use Eq. (6.2) to calculate the rate [11, 78, 79]. Often this will involve the laborious calculation of the transmission coefficient κ. In principle, any sub-optimal choice of the TS can be corrected by a corresponding calculation of κ. But for a poor choice of the TS or if the TS is intrinsically very broad, trajectories can recross the TS several times, which can make $\kappa \ll 1$ and hard to converge. Note that this will give an estimate of the rate constant, but not actual dynamical trajectories themselves.
2. calculate the position-dependent diffusion coefficient along the chosen collective variable, and then perform Brownian dynamics along this coordinate (see for eg. [80]). We will not cover the details of how to calculate position-dependent diffusion constant or performing Brownian dynamics in this review, and refer the interested reader to articles such as [80–82] instead.

Applications Umbrella sampling has seen a large number of applications for calculating static FESs, and subsequently, rate constants. For example, Ryu et al used it to study the entropic effects in homogeneous dislocation nucleation, and source-controlled plasticity in nanowires [79]. Buehler, van Duin, and other groups have used it to study the nanomechanics of day-to-day materials such as hair, feather, wool, etc., as well as novel materials such as nanotubes [83, 84].

Software Resources Umbrella sampling can be easily used through the plugin software PLUMED [49], which can be interfaced with a large number of MD routines such as LAMMPS, GROMACS, NAMD, etc. PLUMED also allows direct use of pre-designed as well as custom-made collective variables.

6.3.2.2 Metadynamics

Theory Metadynamics is a well-established method for exploring complex FESs by constructing a time-dependent bias potential. One first identifies a small subset of relevant collective variables (CVs) [5, 42, 43, 85, 86]. To enhance the sampling of regions of CV space that are rarely visited, a memory dependent bias potential is constructed through the simulation as a function of these CVs. This bias is typically in the form of repulsive Gaussians added wherever the system visits in the CV space. Thus slowly the system starts to avoid the places where it has already visited. This leads to a gradual enhancement of the fluctuations in the CVs, through which the system is discouraged from getting trapped in the low free energy basins. At the end of a metadynamics run the probability distribution of any observable, whether biased directly or not can be computed through a reweighting procedure [86]. This easy reweighting functionality is one of the many features of metadynamics that has made it a very popular method for calculation of FESs.

Recently, Tiwary and co-workers extended the scope of metadynamics by showing how to extract unbiased rates from biased ones with minimal extra computational burden [87]—and without explicitly calculating transition states or performing Brownian dynamics on the FES. Their key idea was to use metadynamics as a tool to construct on-the-fly the ideal bias potential needed for the validity of Eq. (6.5). For this, they made the following two key assumptions:

1. the process being investigated is characterized by movements from one stable state to another via dynamical bottlenecks that are rarely but quickly crossed.
2. while there is no need to know beforehand the precise nature or location of these bottlenecks, one should have CVs that can distinguish between all stable basins of relevance. Note that this is much less stringent a requirement than having to know the true *reaction coordinate* [88], as the latter needs information not just about stable states but also the pathway connecting them.

Under these two key assumptions, simply by making the bias deposition slower than the time spent in dynamical bottlenecks, one can then keep the bottlenecks bias-free throughout the course of the metadynamics run. This preserves the unbiased sequence of state-to-state transitions and allows one to access the acceleration of transition rates through Eq. (6.5). Note that this so-called *infrequent metadynamics* procedure will work only for transitions that are rare but fast. Otherwise the bias deposition might have to be made so infrequent that while the method still remains correct, one will not have any computational gain.

In a successive work, a way was proposed to assess the reliability of the two assumptions above [89]. This relies on the fact that the escape times from a long-lived metastable state obey time-homogeneous Poisson statistics [90]. A statistical analysis based on the Kolmogorov-Smirnov (KS) test can quantitatively assess how precisely these assumptions are met [89]. Any deviations from a time-homogeneous Poisson fit signal that the reconstructed kinetics cannot be trusted. This can happen because either the bias deposition was too frequent, leading to gradual corruption of the transition state region, or the chosen CVs cannot resolve between some stable basins.

The complexity of the problem is now shifted to point (2) above in the list of assumptions: identification of CVs that distinguish between stable states. In practice, one can start with a trial choice of CV, use it to explore the FES, and if the corresponding transition statistics is not Poisson, then simply iterate upon improving the CVs and repeating the simulations until a good Poisson fit is obtained.

Applications Metadynamics has seen a large number of applications for calculating static FESs. In the context of nanomechanics, of special note is the work done by Buehler and co-workers using metadynamics to study bioinspired and natural mechanical systems—such as spider silk, biological glue secreted by marine animals, effect of carbon dioxide on ice crystals, and several other applications [91, 92]. The very recent extension to obtain kinetic rate constants from metadynamics has already been used to study stress-induced denaturation bubbles in DNA [93], self-assembly and disassembly of nanostructures in the presence of molecular solvent [94, 95], and liquid droplet nucleation rates.

Software Resources Metadynamics can be easily used through the plugin software PLUMED [49], described in the Umbrella sampling section. The VASP [96] electronic structure code also implements a flavor of metadynamics that allows it to be used in ab initio MD.

6.3.3 Few Other Methods That Need Neither PES nor FES

6.3.3.1 Parallel Replica Dynamics

Theory This is probably the simplest of all accelerated dynamics methods with minimal theoretical assumptions [74]. The key idea is that for a memoryless rare event, running N copies of the same simulation, each starting from a properly defined decorrelated starting state, makes it N times more likely to see an event within a given time window than if one was to run only one simulation. We will see shortly why the rare event should be memoryless.

One runs several replicas of the system in parallel while waiting for transition from one stable state (which could be a deep free energy / potential energy basin) to another to happen [74]. Each replica runs on a different processor. Once a transition is recorded on some processor **P**, all the processors are stopped, and the run is restarted by creating replicas of the state one obtained after transition on the processor **P**. Before restarting the simulations, the replicas are decorrelated for a small duration t_c, which adds to the computational overhead of the process. Of course, unlike the other methods discussed herein, this method only provides a reduction in wall clock time but not a reduction in overall computational requirements.

Now suppose that one was running parallel replica with N total processors, and that the first transition event is seen when each of these processors individually has covered molecular dynamics time equaling t. As derived in [74], if the rare event is memoryless, i.e., if its time distribution follows Poisson statistics, then the "accelerated time" is given by tN. Furthermore, again as shown in [74], if the individual processors have different speeds and have covered different times, then the accelerated time will be given by the sum of the individual molecular dynamics times attained on the various processors.

Theoretically this algorithm achieves an enhancement in the timescale that is linearly proportional to the number of replicas used. There are, however, problems in using this technique for large systems (more than a few hundred atoms) where transitions start becoming too frequent, thereby leading to prohibitively large overhead time spent in decorrelation of the system. Another technical challenge is in defining a "stable state" and detecting transitions out of these.

However, it was recently demonstrated on a sound mathematical framework, that even for arbitrarily defined stable states, the principle of parallel replica dynamics of adding up simulation times is still correct, as long as the system before exit from any state is decorrelated in it [90]. Thus, strictly speaking for any stable state definition, there exists a decorrelation time t_c, such that parallel replica dynamics is correct if (1) one decorrelates for time $> t_c$ in each replica, and (2) any trajectory that leaves the stable state within t_c is simply restarted. Some rigorous bounds have been provided on t_c, and one hopes to see this leading to further popularity of the method.

Applications Parallel replica dynamics has been successfully applied to a very large number of difficult problems pertaining to the field of nanomechanics and nanomaterials. These include the diffusion of H_2 in crystalline C_{60} [97], the pyrolysis of hexadecane [98], the diffusion of defects in plutonium [99], the transformation of voids into stacking fault tetrahedra in FCC metals [100], the stretching of carbon nanotubes [101], grain boundary sliding in Cu [102], the fracture process of metals [103], and several other applications [7].

Software Resources The softwares LAMMPS [60] or EON [48] can be used to provides parallel replica dynamics calculations.

6.3.3.2 Transition Path Sampling

Theory Transition path sampling (TPS) concerns the problem of calculating transition rates between two well-defined states at a temperature of interest [18, 104]. It does not assume TST in any form, nor does it even assume that the transition event is Poissonian.

TPS begins by considering a system with two known stable states A and B. One needs to have order parameters that demarcate the two states, and a trial path that connects the two states. Normally, this trial path is generated by running the dynamics at an elevated temperature. Starting from this trial pathway, one performs a Monte Carlo random walk in the *space of paths*. Given the initial path, TPS generates new pathways by perturbing that path and then creating a new one. As in any Monte Carlo scheme, the new pathway is accepted or rejected with a certain probability. By iterating this procedure, one then generates an ensemble of transition paths. All relevant information including the reaction mechanism, the transition states, and the associated rate constants can then be extracted from the ensemble [18].

The rate constants from TPS can be very accurate; however, the method itself can be somewhat computationally demanding especially if (a) one has more than 2 stable states, or (b) there are multiple pathways joining the 2 states, such that the pathways themselves are separated by high barriers. In the latter case, the typical path generation algorithm of TPS might be inefficient, or be sensitive to the choice of the initial trial path, but improvements on this are available. Methods such as transition interface sampling [105], which we will not discuss here, have also been proposed that build up on TPS to improve the rate calculation.

Applications TPS has been used to study phase transformations in nanomaterials under the effect of pressure, temperature, or some other driving force [106]. It can be especially useful for nucleation studies [107, 108].

Software Resources The software LibTPS can be used to perform TPS calculations.

6.4 Summary and Outlook

While MD simulations have now without doubt become ubiquitous across disciplines, they continue to suffer from timescale limitations. However, over the last few years a variety of enhanced sampling approaches have come up that mitigate this timescale problem by providing access to accelerated dynamics and long timescales often well beyond the microsecond and millisecond regimes. While it is clear that there have been several striking applications of such approaches for accelerated dynamics, one still would not call these mainstream. Most of these approaches involve some sort of theoretical assumptions, and their use invariably requires the user to be very cautious and aware of the assumptions—more so than the caution needed in the use of an ordinary unbiased MD code. Thus, as is the case for most simulation techniques, an effective and correct use of enhanced sampling is contingent upon a good understanding and awareness of the involved theoretical principles and approximations.

In this review, we have tried to consolidate the theoretical underpinnings, methodological details, and typical applications of many of these enhanced sampling approaches. Through this, we have attempted to create a fairly self-contained introductory manual, where a prospective user with a specific application at hand, can understand and judiciously select from various available approaches. Furthermore, by presenting together a range of approaches originally developed in different sub-disciplines, this review might also serve as a useful vantage point for a method developer interested in developing new, more generic methods for tackling problems in interdisciplinary fields, especially at the confluence of nano- and biomaterials. The field of enhanced sampling approaches for accelerated molecular dynamics is now starting to come of age, and the authors hope that this review will facilitate future applications and developments.

Acknowledgements The authors thank Pablo Piaggi for a careful reading of the manuscript. PT would like to acknowledge numerous discussions and arguments on the subject over the years with Bruce Berne, Michele Parrinello, and Art Voter.

References

1. D. Frenkel, B. Smit, *Understanding Molecular Simulation: From Algorithms to Applications*, vol. 1 (Academic, New York, 2001)
2. F.F. Abraham, R. Walkup, H. Gao, M. Duchaineau, T.D. De La Rubia, M. Seager, Simulating materials failure by using up to one billion atoms and the world's fastest computer: Brittle fracture. Proc. Natl. Acad. Sci. **99**(9), 5777–5782 (2002)
3. M. Tuckerman, B.J. Berne, G.J. Martyna, Reversible multiple time scale molecular dynamics. J. Chem. Phys. **97**(3), 1990–2001 (1992)
4. A.F. Voter, F. Montalenti, T.C. Germann, Extending the time scale in atomistic simulation of materials. Ann. Rev. Mater. Res. **32**(1), 321–346 (2002)
5. A. Barducci, M. Bonomi, M. Parrinello, metadynamics. Wiley Interdiscip. Rev. Comput. Mol. Sci. **1**(5), 826–843 (2011)
6. A. van de Walle, Simulations provide a rare look at real melting. Science **346**(6210), 704–705 (2014)
7. D. Perez, B.P. Uberuaga, Y. Shim, J.G. Amar, A.F. Voter, Accelerated molecular dynamics methods: introduction and recent developments. Annu. Rep. Comput. Chem. **5**, 79–98 (2009)
8. E. Weinan, *Principles of Multiscale Modeling* (Cambridge University Press, Cambridge, 2011)
9. D. Chandler, *Introduction to Modern Statistical Mechanics*, vol. 1 (Oxford University Press, Oxford, 1987)
10. D.A. McQuarrie, *Statistical Thermodynamics* (Harper Collins Publishers, New York, 1973)
11. B.J. Berne, M. Borkovec, J.E. Straub, Classical and modern methods in reaction rate theory. J. Phys. Chem. **92**(13), 3711–3725 (1988)
12. B.J. Berne, N. De Leon, R. Rosenberg, Isomerization dynamics and the transition to chaos. J. Phys. Chem. **86**(12), 2166–2177 (1982)
13. J.-P. Eckmann, D. Ruelle, Ergodic theory of chaos and strange attractors. Rev. Mod. Phys. **57**(3), 617 (1985)
14. D. Wales, *Energy Landscapes: Applications to Clusters, Biomolecules and Glasses* (Cambridge University Press, Cambridge, 2003)
15. D. Sheppard, R. Terrell, G. Henkelman, Optimization methods for finding minimum energy paths. J. Chem. Phys. **128**(13), 134106 (2008)
16. G. Henkelman, B.P. Uberuaga, H. Jónsson, A climbing image nudged elastic band method for finding saddle points and minimum energy paths. J. Chem. Phys. **113**(22), 9901–9904 (2000)
17. G. Henkelman, H. Jónsson, A dimer method for finding saddle points on high dimensional potential surfaces using only first derivatives. J. Chem. Phys. **111**(15), 7010–7022 (1999)
18. P.G. Bolhuis, D. Chandler, C. Dellago, P.L. Geissler, Transition path sampling: throwing ropes over rough mountain passes, in the dark. Ann. Rev. Phys. Chem. **53**(1), 291–318 (2002)
19. G.H. Vineyard, Frequency factors and isotope effects in solid state rate processes. J. Phys. Chem. Solids **3**(1), 121–127 (1957)
20. G.A. Tribello, M. Ceriotti, M. Parrinello, Using sketch-map coordinates to analyze and bias molecular dynamics simulations. Proc. Natl. Acad. Sci. **109**(14), 5196–5201 (2012)
21. P.G. Bolhuis, C. Dellago, D. Chandler, Reaction coordinates of biomolecular isomerization. Proc. Natl. Acad. Sci. **97**(11), 5877–5882 (2000)
22. A. Samanta, E. Weinan, Atomistic simulations of rare events using gentlest ascent dynamics. J. Chem. Phys. **136**(12), 124104 (2012)
23. W. Ren, E. Vanden-Eijnden, Finite temperature string method for the study of rare events. J. Phys. Chem. B **109**(14), 6688–6693 (2005)
24. P. Hänggi, P. Talkner, M. Borkovec, Reaction-rate theory: fifty years after Kramers. Rev. Mod. Phys. **62**(2), 251 (1990)
25. H. Eyring, The activated complex in chemical reactions. J. Chem. Phys. **3**(2), 107–115 (1935)
26. R.G. Mullen, J.-E. Shea, B. Peters, Transmission coefficients, committors, and solvent coordinates in ion-pair dissociation. J. Chem. Theory Comput. **10**(2), 659–667 (2014)

27. R.F. Grote, J.T. Hynes, The stable states picture of chemical reactions. ii. rate constants for condensed and gas phase reaction models. J. Chem. Phys. **73**(6), 2715–2732 (1980)
28. A.K. Faradjian, R. Elber, Computing time scales from reaction coordinates by milestoning. J. Chem. Phys. **120**(23), 10880–10889 (2004)
29. T.T. Lau, A. Kushima, S. Yip, Atomistic simulation of creep in a nanocrystal. Phys. Rev. Lett. **104**(17), 175501 (2010)
30. J.-H. Prinz, H. Wu, M. Sarich, B. Keller, M. Senne, M. Held, J.D. Chodera, C. Schütte, F. Noé, Markov models of molecular kinetics: Generation and validation. J. Chem. Phys. **134**(17), 174105 (2011)
31. J.E. Straub, B.J. Berne, A rapid method for determining rate constants by molecular dynamics. J. Chem. Phys. **83**(3), 1138–1139 (1985)
32. A. Samanta, M.E. Tuckerman, T.-Q. Yu, E. Weinan, Microscopic mechanisms of equilibrium melting of a solid. Science **346**(6210), 729–732 (2014)
33. A.F. Voter, Hyperdynamics: accelerated molecular dynamics of infrequent events. Phys Rev Lett **78**, 3908–3911 (1997)
34. A.F. Voter, A method for accelerating the molecular dynamics simulation of infrequent events. J. Chem. Phys. **106**(11), 4665–4677 (1997)
35. R.A. Miron, K.A. Fichthorn, Accelerated molecular dynamics with the bond-boost method. J. Chem. Phys. **119**(12), 6210–6216 (2003)
36. D. Hamelberg, J. Mongan, J.A. McCammon, Accelerated molecular dynamics: a promising and efficient simulation method for biomolecules. J. Chem. Phys. **120**(24), 11919–11929 (2004)
37. M. Steiner, P.-A. Genilloud, J. Wilkins, Simple bias potential for boosting molecular dynamics with the hyperdynamics scheme. Phys. Rev. B **57**(170, 10236 (1998)
38. H. Grubmüller, Predicting slow structural transitions in macromolecular systems: conformational flooding. Phys. Rev. E **52**, 2893–2906 (1995)
39. G. Henkelman, H. Jónsson, Long time scale kinetic Monte Carlo simulations without lattice approximation and predefined event table. J. Chem. Phys. **115**(21), 9657–9666 (2001)
40. S.Y. Kim, D. Perez, A.F. Voter, Local hyperdynamics. J. Chem. Phys. **139**(14), 144110 (2013)
41. W.K. Kim, M. Luskin, D. Perez, A. Voter, E. Tadmor, Hyper-qc: an accelerated finite-temperature quasicontinuum method using hyperdynamics. JJ. Mech. Phys. Solids **63**, 94–112 (2014)
42. A. Laio, M. Parrinello, Escaping free-energy minima. Proc. Natl. Acad. Sci. **99**(20), 12562–12566 (2002)
43. A. Barducci, G. Bussi, M. Parrinello, Well-tempered metadynamics: a smoothly converging and tunable free-energy method. Phys. Rev. Lett. **100**(2), 020603–020606 (2008)
44. Y. Lin, K.A. Fichthorn, Accelerated molecular dynamics study of the GaAs (001) β 2 (2× 4)/c (2× 8) surface. Phys. Rev. B **86**(16), 165303 (2012)
45. K.E. Becker, M.H. Mignogna, K.A. Fichthorn, Accelerated molecular dynamics of temperature-programed desorption. Phys. Rev. lett. **102**(4), 046101 (2009)
46. S. Hara, J. Li, Adaptive strain-boost hyperdynamics simulations of stress-driven atomic processes. Phys. Rev. B **82**(18), 184114 (2010)
47. R.A. Miron, K.A. Fichthorn, Heteroepitaxial growth of co/ cu (001): an accelerated molecular dynamics simulation study. Phys. Rev. B **72**(3), 035415 (2005)
48. S.T. Chill, M. Welborn, R. Terrell, L. Zhang, J.-C. Berthet, A. Pedersen, H. Jonsson, G. Henkelman, Eon: software for long time simulations of atomic scale systems. Model. Simul. Mater. Sci. Eng. **22**(5), 055002 (2014)
49. M. Bonomi, D. Branduardi, G. Bussi, C. Camilloni, D. Provasi, P. Raiteri, D. Donadio, F. Marinelli, F. Pietrucci, R.A. Broglia et al., Plumed: a portable plugin for free-energy calculations with molecular dynamics. Comp. Phys. Comm. **180**(10), 1961–1972 (2009)
50. G.A. Tribello, M. Bonomi, D. Branduardi, C. Camilloni, G. Bussi, Plumed 2: new feathers for an old bird. Comput. Phys. Commun. **185**(2), 604–613 (2014)
51. M.R. So, A.F. Voter et al., Temperature-accelerated dynamics for simulation of infrequent events. J. Chem. Phys. **112**(21), 9599–9606 (2000)

52. Y. Shim, J.G. Amar, B.P. Uberuaga, A.F. Voter, Reaching extended length scales and time scales in atomistic simulations via spatially parallel temperature-accelerated dynamics. Phys. Rev. B **76**, 205439 (2007)
53. V. Bochenkov, N. Suetin, S. Shankar, Extended temperature-accelerated dynamics: enabling long-time full-scale modeling of large rare-event systems. J. Chem. Phys. **141**(9), 094105 (2014)
54. X.-M. Bai, A.F. Voter, R.G. Hoagland, M. Nastasi, B.P. Uberuaga, Efficient annealing of radiation damage near grain boundaries via interstitial emission. Science **327**(5973), 1631–1634 (2010)
55. J.A. Sprague, F. Montalenti, B.P. Uberuaga, J.D. Kress, A.F. Voter, Simulation of growth of cu on ag(001) at experimental deposition rates. Phys. Rev. B **66**, 205415 (2002)
56. F. Montalenti, M. Sørensen, A. Voter, Closing the gap between experiment and theory: crystal growth by temperature accelerated dynamics. Phys. Rev. Lett. **87**(12), 126101 (2001)
57. B. Uberuaga, R. Smith, A. Cleave, F. Montalenti, G. Henkelman, R. Grimes, A. Voter, K. Sickafus, Structure and mobility of defects formed from collision cascades in MgO. Phys. Rev. Lett. **92**(11), 115505 (2004)
58. M. Cogoni, B. Uberuaga, A. Voter, L. Colombo, Diffusion of small self-interstitial clusters in silicon: temperature-accelerated tight-binding molecular dynamics simulations. Phys. Rev. B **71**(12), 121203 (2005)
59. B.P. Uberuaga, S.M. Valone, M. Baskes, Accelerated dynamics study of vacancy mobility in-plutonium. J. Alloys Compd. **444**, 314–319 (2007)
60. S. Plimpton, P. Crozier, A. Thompson, *Lammps-Large-Scale Atomic/Molecular Massively Parallel Simulator*, vol. 18 (Sandia National Laboratories, Albuquerque, 2007)
61. W. Smith, C. Yong, P. Rodger, Dl_poly: application to molecular simulation. Mol. Simul. **28**(5), 385–471 (2002)
62. A.F. Voter, Introduction to the kinetic monte carlo method, in *Radiation Effects in Solids* (Springer, Berlin, 2007), pp. 1–23
63. L.K. Béland, P. Brommer, F. El-Mellouhi, J.-F. Joly, N. Mousseau, Kinetic activation-relaxation technique. Phys. Rev. E **84**(4), 046704 (2011)
64. N. Mousseau, G. Barkema, Traveling through potential energy landscapes of disordered materials: the activation-relaxation technique. Phys. Rev. E **57**(2), 2419 (1998)
65. H. Xu, Y.N. Osetsky, R.E. Stoller, Self-evolving atomistic kinetic monte carlo: fundamentals and applications. J. Phys. Condens. Matter **24**(37), 375402 (2012)
66. J.-F. Joly, L.K. Béland, P. Brommer, N. Mousseau, Contribution of vacancies to relaxation in amorphous materials: a kinetic activation-relaxation technique study. Phys. Rev. B **87**(14), 144204 (2013)
67. L.K. Béland, N. Mousseau, Long-time relaxation of ion-bombarded silicon studied with the kinetic activation-relaxation technique: microscopic description of slow aging in a disordered system. Phys. Rev. B **88**(21), 214201 (2013)
68. P. Brommer, L.K. Béland, J.-F. Joly, N. Mousseau, Understanding long-time vacancy aggregation in iron: a kinetic activation-relaxation technique study. Phys. Rev. B **90**(13), 134109 (2014)
69. H. Kallel, N. Mousseau, F. Schiettekatte, Evolution of the potential-energy surface of amorphous silicon. Phys. Rev. letters **105**(4), 045503 (2010)
70. L. Xu, G. Henkelman, Adaptive kinetic monte carlo for first-principles accelerated dynamics. J. Chem. Phys. **129**(11), 114104 (2008)
71. C.-Y. Lu, D.E. Makarov, G. Henkelman, Communication: κ-dynamics—an exact method for accelerating rare event classical molecular dynamics. J. Chem. Phys. **133**(20), 201101 (2010)
72. P. Tiwary, A. van de Walle, Hybrid deterministic and stochastic approach for efficient atomistic simulations at long time scale s. Phys. Rev. B **84**, 100301–100304 (2011)
73. P. Tiwary, A. van de Walle, Accelerated molecular dynamics through stochastic iterations and collective variable based basin identification. Phys. Rev. B **87**, 094304–094307 (2013)
74. A.F. Voter, Parallel replica method for dynamics of infrequent events. Phys. Rev. B **57**(22), R13985 (1998)

75. G.M. Torrie, J.P. Valleau, Nonphysical sampling distributions in monte carlo free-energy estimation: umbrella sampling. J. Comput. Phys. **23**(2), 187–199 (1977)
76. D.P. Landau, K. Binder, *A Guide to Monte Carlo Simulations in Statistical Physics* (Cambridge University Press, Cambridge, 2014)
77. J.M. Rosenbergl, The weighted histogram analysis method for free-energy calculations on biomolecules. i. the method. J. Comput. Chem. **13**(8), 1011–1021 (1992)
78. S.H. Northrup, M.R. Pear, C.-Y. Lee, J.A. McCammon, M. Karplus, Dynamical theory of activated processes in globular proteins. Proc. Natl. Acad. Sci. **79**(13), 4035–4039 (1982)
79. S. Ryu, K. Kang, W. Cai, Entropic effect on the rate of dislocation nucleation. Proc. Natl. Acad. Sci. **108**(13), 5174–5178 (2011)
80. J.A. Morrone, J. Li, B.J. Berne, Interplay between hydrodynamics and the free energy surface in the assembly of nanoscale hydrophobes. J. Phys. Chem. B **116**(1), 378–389 (2011)
81. J.E. Straub, B.J. Berne, B. Roux, Spatial dependence of time-dependent friction for pair diffusion in a simple fluid. J. Chem. Phys. **93**(9), 6804–6812 (1990)
82. G. Hummer, Position-dependent diffusion coefficients and free energies from Bayesian analysis of equilibrium and replica molecular dynamics simulations. New J. Phys. **7**(1), 34 (2005)
83. S. Keten, C.-C. Chou, A.C. van Duin, M.J. Buehler, Tunable nanomechanics of protein disulfide bonds in redox microenvironments. J. Mech. Behav. Biomed. Mater. **5**(1), 32–40 (2012)
84. R. Vijayaraj, S. Van Damme, P. Bultinck, V. Subramanian, Molecular dynamics and umbrella sampling study of stabilizing factors in cyclic peptide-based nanotubes. J. Phys. Chem. B **116**(33), 9922–9933 (2012)
85. J.F. Dama, M. Parrinello, G.A. Voth, Well-tempered metadynamics converges asymptotically. Phys. Rev. Lett. **112**(24), 240602–240605 (2014)
86. P. Tiwary, M. Parrinello, A time-independent free energy estimator for metadynamics. J. Phys. Chem. B **119**(3), 736–742 (2015). doi:10.1021/jp504920s
87. P. Tiwary, M. Parrinello, From metadynamics to dynamics. Phys. Rev. Lett. **111**, 230602–230606 (2013)
88. R.B. Best, G. Hummer, Reaction coordinates and rates from transition paths. Proc. Natl. Acad. Sci. U. S. A. **102**(19), 6732–6737 (2005)
89. M. Salvalaglio, P. Tiwary, M. Parrinello, Assessing the reliability of the dynamics reconstructed from metadynamics. J. Chem. Theory Comput. **10**(4), 1420–1425 (2014)
90. T. Lelièvre, *Two Mathematical Tools to Analyze Metastable Stochastic Processes* (Springer, Berlin, 2013), pp. 791–810
91. G. Gronau, Z. Qin, M.J. Buehler, Effect of sodium chloride on the structure and stability of spider silk's n-terminal protein domain. Biomater. Sci. **1**(3), 276–284 (2013)
92. D. Lau, K. Broderick, M.J. Buehler, O. Büyüköztürk, A robust nanoscale experimental quantification of fracture energy in a bilayer material system. Proc. Natl. Acad. Sci. **111**(33), 11990–11995 (2014)
93. F. Sicard, N. Destainville, M. Manghi, DNA denaturation bubbles: free-energy landscape and nucleation/closure rates. J. Chem. Phys. **142**(3), 034903 (2015)
94. P. Tiwary, V. Limongelli, M. Salvalaglio, M. Parrinello, Kinetics of protein-ligand unbinding: predicting pathways, rates, and rate-limiting steps. Proc. Natl. Acad. Sci. **112**(5), E386–E391 (2015)
95. P. Tiwary, J. Mondal, J. Morrone, B. Berne, Understanding the influence of water and steric effects on the kinetics of cavity-ligand unbinding (2015). Proc. Natl. Acad. Sci. **112**(39), 12015–12019 (2015). doi:10.1073/pnas.1516652112
96. G. Kresse, J. Furthmüller, Software VASP, Vienna (1999). Phys. Rev. B **54**(11), 169 (1996)
97. B.P. Uberuaga, A.F. Voter, K.K. Sieber, D.S. Sholl, Mechanisms and rates of interstitial h 2 diffusion in crystalline c 60. Phys. Rev. Lett. **91**(10), 105901 (2003)
98. O. Kum, B.M. Dickson, S.J. Stuart, B.P. Uberuaga, A.F. Voter, Parallel replica dynamics with a heterogeneous distribution of barriers: application to n-hexadecane pyrolysis. J. Chem. Phys. **121**(20), 9808–9819 (2004)

99. B.P.Uberuaga, S.M. Valone, M. Baskes, Accelerated dynamics study of vacancy mobility in-plutonium. J. Alloys Compd. **444**, 314–319 (2007)
100. B. Uberuaga, R. Hoagland, A. Voter, S. Valone, Direct transformation of vacancy voids to stacking fault tetrahedra. Phys. Rev. Lett. **99**(13), 135501 (2007)
101. B.P. Uberuaga, S.J. Stuart, A.F. Voter, Parallel replica dynamics for driven systems: derivation and application to strained nanotubes. Phys. Rev. B **75**(1), 014301 (2007)
102. Y. Mishin, A. Suzuki, B. Uberuaga, A. Voter, Stick-slip behavior of grain boundaries studied by accelerated molecular dynamics. Phys. Rev. B **75**(22), 224101 (2007)
103. K. Baker, D. Warner, Extended timescale atomistic modeling of crack tip behavior in aluminum. Model. Simul. Mater. Sci. Eng. **20**(6), 065005 (2012)
104. T. Dumitrica, *Trends in Computational Nanomechanics: Transcending Length and Time Scales*, vol. 9 (Springer Science & Business Media, New York, 2010)
105. T.S. Van Erp, P.G. Bolhuis, Elaborating transition interface sampling methods. J. Comput. Phys. **205**(1), 157–181 (2005)
106. M. Grünwald, E. Rabani, C. Dellago, Mechanisms of the wurtzite to rocksalt transformation in CdSe nanocrystals. Phys. Rev. lett. **96**(25), 255701 (2006)
107. M. Gruünwald, C. Dellago, Nucleation and growth in structural transformations of nanocrystals. Nano lett. **9**(5), 2099–2102 (2009)
108. D. Moroni, P.R. Ten Wolde, P.G. Bolhuis, Interplay between structure and size in a critical crystal nucleus. Phys. Rev. Lett. **94**(23), 235703 (2005)

Chapter 7
Principles of Coarse-Graining and Coupling Using the Atom-to-Continuum Method

Reese E. Jones, Jeremy Templeton, and Jonathan Zimmerman

7.1 Introduction

Molecular dynamics (MD) and finite element (FE) simulation are both powerful, widely applied methods in their own right. MD enables the study of the atomic motion that underlies material deformation and failure mechanisms. It has been a tool in understanding phenomena such as diffusion, energy transport, and fracture at a fundamental level. In contrast, FE simulation of continuum processes uses preconceived knowledge about mechanisms, in the form of constitutive models, to predict the response of structures and devices that span from microns to meters. Just as each method possesses unique strengths, they also have their limits. Typically, MD cannot be used to simulate even microscopic devices due to computational inefficiencies of resolving atomic motion at that scale, whereas FE relies heavily on phenomenological constitutive models that may not encompass all the mechanisms needed for predictive simulation.

To provide a tool to simulate and design nanoscale devices and nanostructured materials, we have developed a suite of numerical methods that bring together the best aspects of MD and FE. This coupling of methods enables us to capture multiscale phenomena [1–3], reduce atomic data to connect with continuum theory [4–7], simulate large systems with atomic detail and statistical characterization [8, 9], and add physics not intrinsic to MD [10]. In general, the methods rely on the fact that atomistic behavior is asymptotic to continuum, which is the basis for Green–Kubo methods [3] for example, and that the scales at which the two representations of materials become consistent are surprisingly short and small [11, 12].

R.E. Jones (✉) • J. Templeton • J. Zimmerman
Sandia National Laboratories, Livermore, CA, USA
e-mail: rjones@sandia.gov; jatempl@sandia.gov; jzimmer@sandia.gov

C.R. Weinberger, G.J. Tucker (eds.), *Multiscale Materials Modeling for Nanomechanics*, Springer Series in Materials Science 245,
DOI 10.1007/978-3-319-33480-6_7

In this chapter, we review our approaches for both coarse-graining (the averaging of MD-level quantities to inform models used within FE) and coupling (concurrent simulation at the MD and FE levels such that information is exchanged through interface conditions) that have been implemented in the Atom-to-Continuum (AtC) user module available with the widely used, large-scale molecular simulation code LAMMPS [13]. As we will discuss, coupling generally involves four ingredients: (a) consistent upscaling/coarse-graining of atomic data, (b) multiscale governing balances connecting atomic and continuum representations, (c) consistent continuum constitutive models, and (d) atomistic control schemes to effect the influence of the continuum on the atomistic representation.

In the next section, Sect. 7.2, we will develop a means of coarse-graining of atomic data into continuum fields that is consistent with accepted conservation laws. Then, we will revisit (a) and cover (b)–(d) in Sect. 7.3. Finally, in Sect. 7.4 we will show examples of both coarse-graining and coupling that demonstrate the utility and versatility of these techniques. We refer the reader to Fig. 7.1 for a guide to the notation and basic geometry used in this chapter.

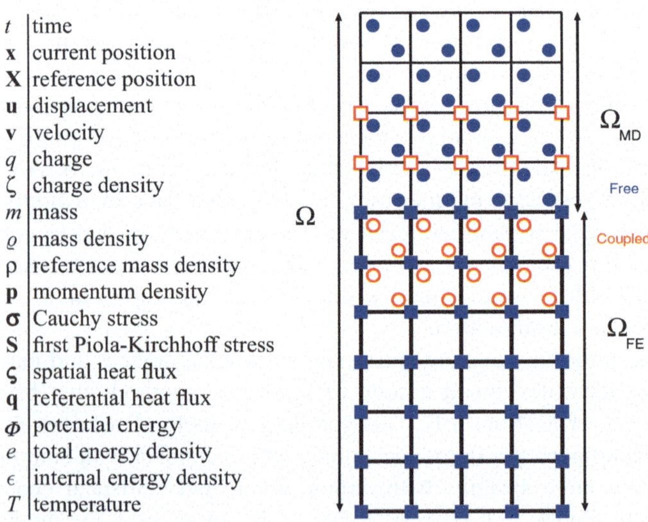

Fig. 7.1 Notation used for physical properties and fields within this chapter. Atomic quantities are indexed with Greek α subscripts and nodal/continuum fields with Latin I subscripts and unless explicitly noted α ranges over the whole set of atoms $\mathscr{A}$ and I ranges over all nodes $\mathscr{N}$. The schematic shows the finite element Ω_{FE} and molecular dynamic Ω_{MD} regions comprising the complete system $\Omega = \Omega_{\mathrm{FE}} \cup \Omega_{\mathrm{MD}}$. The nodes and atoms that are *free* evolve according to their own constitutive models reside in Ω_{FE} and Ω_{MD}, respectively, and those that are *coupled* to the other paradigm reside in Ω_{MD} and Ω_{FE}, respectively

7.2 Coarse Graining

Coarse-graining is the averaging of atomic-scale quantities in order to estimate fields that have well-defined physical meaning at larger scales. These continuum fields can be used to interpret atomistic data and construct constitutive models that guide FE simulations, or to provide interface conditions for concurrent MD/FE simulations. The development of expressions to calculate continuum fields from pointwise atomistic information dates back to the late nineteenth century, when Clausius [14] and Maxwell [15] developed the virial theorem (VT) to define the stress applied to the bounding surface of a fixed volume containing interacting particles at finite temperature. In 1950, Irving and Kirkwood [16] derived expressions for local measures of stress and heat flux from microscopic/atomic densities for mass, momentum and energy and the associated continuum balance equations.

Subsequent to these foundations, there have been many efforts to improve on atomic-based definitions for stress [4, 17–38] and heat flux [35, 38]. We direct the reader to [4, 35–37] for more complete discussions of these derivations. Notable among these efforts is the work by Hardy and colleagues [21, 39, 40] which replaced the Dirac delta employed by Irving and Kirkwood with a more computationally amenable smooth, finite localization function to establish a self-consistent manner of distributing discrete atomic contributions to thermomechanical fields. Hardy's original formulation is based on an Eulerian/spatial representation where localization volumes are essentially control volumes fixed in space that matter occupies at a particular time. Hence, as was the case for the expressions by Irving and Kirkwood [16], Hardy's expressions for (Cauchy) stress, $\boldsymbol{\sigma}$, and heat flux, $\boldsymbol{\varsigma}$, contain both potential and kinetic terms.

A Lagrangian/material frame representation affords an alternative approach particularly suited to solids and tracking *material* motion from a continuum perspective. In this case, the first Piola–Kirchhoff stress tensor $\mathbf{S}$, the amount of current force exerted on a unit area as measured in the reference configuration, is the relevant stress measure. The material frame heat flux, $\mathbf{q}$, has a similar definition and has units of energy per time per unit reference area. Expressions to calculate $\mathbf{S}$ and $\mathbf{q}$ have been developed by Andia et al. [29–32], and more recently by Zimmerman et al. [36] who used a Hardy-like formalism.

In this section, we present a generalized weighted-residual formulation for coarse-graining atomic data. We derive consistent fields from minimizing the L^2 norm of the difference between an atomic-based description of a continuum field variable and a representation using nodal variables and FE basis functions. We compare this approach with expressions developed by Irving and Kirkwood, Hardy and similar efforts found in the literature, and elucidate their connections as well as their differences. Lastly, although becoming prominent in recent research [41, 42] we do not cover the uncertainty quantification (UQ) aspects of parameter and property estimation in this chapter nor give an extensive treatment of measuring or coupling to the scale-dependent fluctuations intrinsic in MD [43].

7.2.1 Formulation

As in [44], we take a least squares statement:

$$\min_{\varrho_I} \int_{\Omega} \|\varrho^* - \sum_I N_I \varrho_I\|^2 \, \mathrm{d}V \Longleftrightarrow \sum_J \left[\int_{\Omega} N_I N_J \, \mathrm{d}V \right] \varrho_J = \int_{\Omega} N_I \varrho \, \mathrm{d}V \tag{7.1}$$

as starting point to relate, in this case, the mass density ϱ^* to its approximation $\sum_I N_I(\mathbf{x})\varrho_I(t)$ with a basis $\{N_I(\mathbf{x})\}$ covering region $\Omega \equiv \Omega_{\mathrm{MD}}$ filled with atoms. This is a sub-case of the more general scenario shown in Fig. 7.1.

As in Irving and Kirkwood's seminal work [16], we form microscopic densities, e.g.,

$$\varrho^*(\mathbf{x}, t) = \sum_{\alpha} m_\alpha \delta(\mathbf{x} - \mathbf{x}_\alpha(t)) \tag{7.2}$$

in terms of atomic quantities, here m_α is the mass of atom α, and δ is the Dirac delta operator. Hence Eq. (7.1) becomes

$$\sum_J \underbrace{\left[\int_{\Omega} N_I N_J \, \mathrm{d}V \right]}_{M_{IJ}} \varrho_J = \sum_{\alpha} N_I(\mathbf{x}_\alpha) m_\alpha = \sum_{\alpha} N_{I\alpha} m_\alpha \tag{7.3}$$

where $N_{I\alpha} = N_I(\mathbf{x}_\alpha)$ is the basis evaluated at atomic positions. The solution of Eq. (7.3) is the projection

$$\varrho_I = \sum_{J,\alpha} M_{IJ}^{-1} N_{J\alpha} m_\alpha = \sum_{\alpha} \Delta_{I\alpha} m_\alpha, \tag{7.4}$$

given the mass matrix M_{IJ}, and introduce the localization function $\Delta_I = \sum_J M_{IJ}^{-1} N_J$, with $\Delta_{I\alpha} = \Delta_I(\mathbf{x}_\alpha)$, which has units of inverse volume [45]. We can reduce this projection to a *restriction*

$$\varrho_I \approx \sum_{\alpha} \frac{1}{V_I} N_{I\alpha} m_\alpha = \sum_{\alpha} \Delta_{I\alpha} m_\alpha \tag{7.5}$$

by row-sum lumping the mass matrix $M_{IJ} \approx \sum_J M_{IJ} = \int_{\Omega} N_I \, \mathrm{d}V = V_I$ using the partition of unity property $\sum_I N_I = 1$ of the basis N_I, which localizes the influence of atomic data on specific nodes. In this case, $\Delta_{I\alpha} = N_{I\alpha}/V_I$. In a similar fashion, we can take the localization function to be a moving least squares (MLS) kernel [46] like in Hardy's work [21, 39, 47] so that $\Delta_{I\alpha} = \Delta(\mathbf{x}_I - \mathbf{x}_\alpha)$ with the normalization $\int_{\Omega} \Delta \, \mathrm{d}V = 1$, or a reproducing kernel [48] which is polynomially complete. Using either a projection (7.4), a restriction (7.5) or a moving least squares estimate to obtain nodal values, the continuum field is then interpolated using the basis

$$\varrho(\mathbf{x}, t) = \sum_I N_I(\mathbf{x}) \varrho_I(t) = \sum_{I,\alpha} N_I(\mathbf{x}) \Delta_I(\mathbf{x}_\alpha(t)) m_\alpha. \tag{7.6}$$

Note that N_I will always denote the chosen basis but the particular form of the localization Δ_I will depend on the appropriate mass matrix, which we will assume is implicit in the context of particular Δ_I.

Two characteristic sizes have been introduced into our formulation: (a) the nodal spacing of the mesh used (i.e., the scale of the basis $N_I \sim \sqrt[3]{V_I}$), which dictates the spatial refinement of our coarse-grained field, and (b) the size of the region over which averaging is performed at each node (i.e., the scale of Δ_I). These two length-scales are independent, but typically the averaging region is taken to be commensurate with the mesh spacing for convenience. This practice becomes problematic in certain cases. For example, the anticipation of large gradients may require a fine mesh; however, too fine a mesh will result in nodal averages that depend on only a few atoms. Conversely, too coarse a mesh will result in averages that are limited in their spatial variation. Alternatively, one can separate these two length-scales, using a larger size for atomic averaging and a smaller one for mesh definition,[1] as we demonstrate in the example presented in Sect. 7.4.1. Instead of a completely empirical approach, the asymptotic analysis of Ulz et al. can be employed to balance smoothness and resolution of the estimate [49].

Taking the (partial) time derivative of Eq. (7.1)

$$\sum_J \left[\int_\Omega N_I N_J \, \mathrm{d}V \right] \frac{\partial}{\partial t} \varrho_J = -\sum_\alpha \nabla_{\mathbf{x}} N_{I\alpha} \cdot m_\alpha \mathbf{v}_\alpha \tag{7.7}$$

using the short-hand

$$\frac{\partial}{\partial t} N_{I\alpha}(t) = \frac{\partial}{\partial t} N_I(\mathbf{x}_I - \mathbf{x}_\alpha(t)) = -\nabla_{\mathbf{x}} N_{I\alpha} \cdot \mathbf{v}_\alpha \tag{7.8}$$

for the application of the necessary chain rule, we see the estimates are consistent with the usual continuum Eulerian/spatial mass balance

$$\frac{\partial}{\partial t} \varrho_I + \nabla_{\mathbf{x}} \cdot \mathbf{p}_I = \dot{\varrho}_I + \varrho_I \nabla_{\mathbf{x}} \cdot \mathbf{v}_I = 0 \tag{7.9}$$

at the nodes. Here we introduce the material time derivative $\dot{\varrho} = \frac{\partial}{\partial t}\varrho + \nabla_{\mathbf{x}} \varrho \cdot \mathbf{v}$ of ϱ and the linear momentum density field $\mathbf{p} = \rho \mathbf{v}$. This approach gives an explicit sense of scale at which the estimated fields are consistent with the appropriate balance in contrast with MLS/Hardy approach [21] where at any point

$$\frac{\partial \varrho}{\partial t} = \frac{\partial}{\partial t} \sum_\alpha \Delta(\mathbf{x} - \mathbf{x}(t)) m_\alpha = -\sum_\alpha \nabla_{\mathbf{x}} \Delta(\mathbf{x} - \mathbf{x}(t)) \cdot m_\alpha \mathbf{v}_\alpha = -\nabla_{\mathbf{x}} \cdot \mathbf{p} \tag{7.10}$$

[1] This choice has similarities with the Bubnov–Galerkin weighted residual where the weight space and primary field have different bases.

The Lagrangian mass balance is trivially satisfied by

$$\rho = \sum_I N_I(\mathbf{X})\rho_I = \sum_{I,\alpha} N_I \Delta_I(\mathbf{X}_\alpha) m_\alpha \tag{7.11}$$

since $\dot{\rho}_I = 0$.

A further generalization is to introduce time averaging by revisiting the microscopic density Eq. (7.2)

$$\varrho^*(\mathbf{x}, t) = \sum_\alpha \langle m_\alpha \delta(\mathbf{x} - \mathbf{x}_\alpha(t)) \rangle \tag{7.12}$$

using a causal time-filter

$$f(t) \equiv \int_{-\infty}^{t} f(s) w(t-s)\, \mathrm{d}s. \tag{7.13}$$

The kernel $w(t)$ must asymptote to zero as $t \to -\infty$ sufficiently fast for the integral to converge, which, in conjunction with the properties of the convolution operator (7.13),[2] results in this time-average having the commutation property

$$\frac{\partial}{\partial t} \langle f \rangle = \left\langle \frac{\partial}{\partial t} f \right\rangle. \tag{7.14}$$

Thus, the time derivative of Eq. (7.1) becomes

$$\begin{aligned} \sum_J \left[\int_\Omega N_I N_J \, \mathrm{d}V \right] \frac{\partial}{\partial t} \varrho_J &= \frac{\partial}{\partial t} \left\langle \sum_\alpha N_{I\alpha} m_\alpha \right\rangle = \left\langle \sum_\alpha \frac{\partial}{\partial t} N_{I\alpha} m_\alpha \right\rangle \\ &= -\left\langle \sum_\alpha \nabla_{\mathbf{x}} N_{I\alpha} \cdot m_\alpha \mathbf{v}_\alpha \right\rangle = -\nabla_{\mathbf{x}} \cdot \left\langle \sum_\alpha N_{I\alpha} m_\alpha \mathbf{v}_\alpha \right\rangle \end{aligned} \tag{7.15}$$

and hence the time averaged definition of ϱ also satisfies the (weak) mass balance. Given the extent of the kernel into the past, which allows the filter to be invertible, means of initializing the causal filter with consistent initial conditions at some finite

[2]Using Leibniz's rule and $w(t) = 0 \forall t > 0$:

$$\begin{aligned} \frac{\partial}{\partial t} \langle f \rangle &= \frac{\partial}{\partial t} \int_{-\infty}^{t} f(s) w(t-s)\, \mathrm{d}s = -\frac{\partial}{\partial t} \int_{-\infty}^{t} f(t-s) w(s)\, \mathrm{d}s \\ &= -\int_{-\infty}^{t} \frac{\partial}{\partial t} f(t-s) w(s)\, \mathrm{d}s - \varrho(0) w(t) = -\int_{-\infty}^{t} \frac{\partial}{\partial t} f(t-s) w(s)\, \mathrm{d}s = \left\langle \frac{\partial}{\partial t} f \right\rangle \end{aligned}$$

for $t > 0$.

time are necessary [9]. With an exponential filter, $w(t) = \frac{1}{\tau}\exp\left(\frac{t}{\tau}\right)$, the ordinary differential equation (ODE)

$$\frac{\mathrm{d}}{\mathrm{d}t}\langle f\rangle = \frac{1}{\tau}\left(f - \langle f\rangle\right) \tag{7.16}$$

can be used to update the filtered value $\langle f\rangle$. Similarly the ODE

$$\frac{\mathrm{d}}{\mathrm{d}t}\langle\!\langle f\rangle\!\rangle = \frac{1}{\tau}\left((f - \langle f\rangle)^2 - \langle\!\langle f\rangle\!\rangle\right) \tag{7.17}$$

can be used to apply a variance estimator $\langle\!\langle f\rangle\!\rangle$ with this particular kernel.

7.2.2 Atomic Data

Now that we have examined the example of mass density we can develop other consistent field estimators. For example, the expression for linear momentum density akin to Eq. (7.6),

$$\mathbf{p}(\mathbf{x},t) = \sum_{I,\alpha} N_I(\mathbf{x})\Delta_{I\alpha} m_\alpha \mathbf{v}_\alpha(t), \tag{7.18}$$

can be used together with the spatial version of the momentum balance,

$$\frac{\partial}{\partial t}\mathbf{p} = \nabla_{\mathbf{x}} \cdot \left(\boldsymbol{\sigma} - \frac{1}{\varrho}\mathbf{p}\otimes\mathbf{p}\right), \tag{7.19}$$

to derive the expression for the Cauchy stress $\boldsymbol{\sigma}$. Starting from the left-hand side of Eq. (7.19), at the nodes we obtain

$$\begin{aligned}\frac{\partial}{\partial t}\mathbf{p}_I &= \sum_{I,\alpha}\frac{\partial}{\partial t}\left(\Delta_{I\alpha} m_\alpha \mathbf{v}_\alpha\right) = \sum_{I,\alpha}\left(\Delta_{I\alpha}\mathbf{f}_\alpha + m_\alpha \mathbf{v}_\alpha \frac{\partial \Delta_{I\alpha}}{\partial t}\right)\\ &= \sum_{I,\alpha,\beta}\Delta_{I\alpha}\mathbf{f}_{\alpha\beta} - \sum_{I,\alpha} m_\alpha \nabla_{\mathbf{x}}\Delta_{I\alpha}\cdot\mathbf{v}_\alpha\otimes\mathbf{v}_\alpha\\ &= \frac{1}{2}\sum_{I,\alpha,\beta}\left(\Delta_{I\alpha}\mathbf{f}_{\alpha\beta} - \Delta_{I\beta}\mathbf{f}_{\alpha\beta}\right) - \nabla_{\mathbf{x}}\cdot\sum_{I,\alpha} m_\alpha\mathbf{v}_\alpha\otimes\mathbf{v}_\alpha\Delta_{I\alpha},\end{aligned} \tag{7.20}$$

where $\mathbf{f}_{\alpha\beta}$ is the portion of the total force on atom α due to atom β, such that $\mathbf{f}_\alpha = \sum_\beta \mathbf{f}_{\alpha\beta}$. As in Hardy's work [21, 47], we introduce the bond function

$$B_{I\alpha\beta} \equiv \int_0^1 \Delta_I(\lambda\mathbf{x}_{\alpha\beta} + \mathbf{x}_\beta - \mathbf{x}_I)\mathrm{d}\lambda, \tag{7.21}$$

such that

$$-\nabla_{\mathbf{x}} B_{I\alpha\beta} \cdot \mathbf{x}_{\alpha\beta} = \Delta_{I\alpha} - \Delta_{I\beta}, \tag{7.22}$$

where $\mathbf{x}_{\alpha\beta} \equiv \mathbf{x}_\alpha - \mathbf{x}_\beta$. By combining Eqs. (7.19), (7.20), and (7.22), we arrive at an expression for the Cauchy stress field,

$$\boldsymbol{\sigma}(\mathbf{x}, t) = -\frac{1}{2} \sum_{I,\alpha,\beta} \mathbf{f}_{\alpha\beta} \otimes \mathbf{x}_{\alpha\beta} B_{I\alpha\beta} N_I(\mathbf{x}) - \sum_{I,\alpha} m_\alpha \mathbf{w}_\alpha \otimes \mathbf{w}_\alpha \Delta_{I\alpha} N_I(\mathbf{x}), \tag{7.23}$$

where we define a relative atomic velocity

$$\mathbf{w}_\alpha = \mathbf{v}_\alpha - \sum_I \Delta_{I\alpha} \mathbf{v}_I \approx \mathbf{v}_\alpha - \frac{1}{V_I} \sum_I N_{I\alpha} \mathbf{v}_I. \tag{7.24}$$

As in [45], the projection form of Eq. (7.24) decouples the large-scale kinetic energy from the fine-scale, whereas, for the restriction form, this decomposition is only approximate.

A similar manipulation of the material-frame momentum density expression,

$$\mathbf{p}(\mathbf{X}, t) = \sum_{I,\alpha} N_I(\mathbf{X}) m_\alpha \mathbf{v}_\alpha \Delta_{I\alpha}, \tag{7.25}$$

where $\Delta_{I\alpha} = \Delta_I(\mathbf{X}_\alpha)$, within a material-frame momentum balance

$$\dot{\mathbf{p}} = \nabla_{\mathbf{X}} \cdot \mathbf{S}, \tag{7.26}$$

produces an expression for the first Piola–Kirchhoff stress,

$$\mathbf{S}(\mathbf{X}, t) = -\frac{1}{2} \sum_{I,\alpha,\beta} \mathbf{f}_{\alpha\beta}(t) \otimes \mathbf{X}_{\alpha\beta} B_{I\alpha\beta} N_I(\mathbf{X}). \tag{7.27}$$

A similar exercise can be done with the balance of energy, as shown in [35] and [36], by starting with a definition of nodal energy density,

$$\varrho_I e_I = \sum_\alpha \left(\frac{1}{2m_\alpha} \mathbf{p}_\alpha \cdot \mathbf{p}_\alpha + \phi_\alpha \right) \Delta_{I\alpha}, \tag{7.28}$$

where $\varrho e(\mathbf{x}, t) = \sum_I N_I(\mathbf{x}) \varrho_I(t) e_I(t)$. Here, we have partitioned the total potential energy Φ into separate contributions from each atom, ϕ_α, such that $\Phi = \sum_\alpha \phi_\alpha$. (For pair potentials ϕ_α is simply $\phi_\alpha = \frac{1}{2} \sum_\beta \phi_{\alpha\beta}$.) We require a direct relation between $\mathbf{f}_{\alpha\beta}$ and these individual atomic energies, specifically: $\mathbf{f}_{\alpha\beta} \equiv -\left\{ \frac{\partial \phi_\alpha}{\partial r_{\alpha\beta}} + \frac{\partial \phi_\beta}{\partial r_{\alpha\beta}} \right\} \frac{\mathbf{x}_{\alpha\beta}}{r_{\alpha\beta}}$, where $r_{\alpha\beta} = ||\mathbf{x}_{\alpha\beta}||$. Also, although energy density is a primary field, i.e. is a conserved quantity associated with a balance law, we approximate ρe to conform to the more common convention of a per mass energy density. Using these expressions and relations within the Eulerian energy balance,

$$\varrho \frac{\partial e}{\partial t} = \nabla_{\mathbf{x}} \cdot (\boldsymbol{\sigma} \cdot \mathbf{v} - \varrho e \mathbf{v} - \boldsymbol{\varsigma}) \tag{7.29}$$

enables derivation of an expression for the heat flux[3] ς,

$$\varsigma = -\sum_{I,\alpha,\beta} \left(\frac{\partial \phi_\beta}{\partial r_{\alpha\beta}} \frac{\mathbf{x}_{\alpha\beta}}{r_{\alpha\beta}} \cdot \mathbf{w}_\alpha \right) \mathbf{x}_{\alpha\beta} B_{I\alpha\beta} N_I(\mathbf{x}) + \sum_{I,\alpha} \left(\frac{1}{2} m_\alpha \mathbf{w}_\alpha \cdot \mathbf{w}_\alpha + \phi_\alpha \right) \Delta_{I\alpha} N_I(\mathbf{x}). \tag{7.30}$$

For the Lagrangian energy balance,

$$\rho \dot{e} = \mathbf{S} : \dot{\mathbf{F}} - \nabla_{\mathbf{X}} \cdot \mathbf{q}, \tag{7.31}$$

the resulting reference frame heat flux is

$$\mathbf{q} = -\sum_{I,\alpha,\beta} \left(\frac{\partial \phi_\beta}{\partial r_{\alpha\beta}} \frac{\mathbf{x}_{\alpha\beta}}{r_{\alpha\beta}} \cdot \mathbf{w}_\alpha \right) \mathbf{X}_{\alpha\beta} B_{I\alpha\beta} N_I(\mathbf{X}) \, . \tag{7.32}$$

Given the fundamental definitions of mass ρ and momentum density $\mathbf{p}$, we can define the velocity field $\mathbf{v}$ such that $\varrho \mathbf{v} = \mathbf{p}$. In our L^2 formalism:

$$\int_\Omega N_I \rho \mathbf{v} \, \mathrm{d}V = \int_\Omega N_I \mathbf{p} \, \mathrm{d}V \Longrightarrow \tag{7.33}$$

$$\int_\Omega N_I \sum_{J,\alpha} m_\alpha \delta(\mathbf{x} - \mathbf{x}_\alpha) N_J \, \mathrm{d}V \, \mathbf{v}_J = \int_\Omega N_I(\mathbf{x}) \sum_\alpha m_\alpha \mathbf{v} \delta(\mathbf{x} - \mathbf{x}_\alpha) \, \mathrm{d}V \Longrightarrow$$

$$\sum_{J,\alpha} [N_{I\alpha} m_\alpha N_{J\alpha}] \, \mathbf{v}_J = \sum_\alpha N_{I\alpha} m_\alpha \mathbf{v};$$

however, we choose to simply take the nodal velocities to be: $\mathbf{v}_I \equiv \mathbf{p}_I / \varrho_I$, and interpolate these values with the basis N_I. The corresponding displacement for node I is defined as

$$\mathbf{u}_I = \frac{1}{\varrho_I} \sum_{I\alpha} \Delta_{I\alpha} m_\alpha \mathbf{u}_\alpha = \frac{1}{\varrho_I} \sum_{I\alpha} \Delta_{I\alpha} m_\alpha \left(\mathbf{x}_\alpha - \mathbf{X}_\alpha \right) \tag{7.34}$$

[3]The term $\frac{\partial \phi_\beta}{\partial r_{\alpha\beta}} \frac{\mathbf{x}_{\alpha\beta}}{r_{\alpha\beta}}$ cannot be replaced with $-\mathbf{f}_{\alpha\beta}$, as is often done in the literature. This issue was examined by Admal and Tadmor [38], who determined that doing so also requires replacing $\mathbf{w}_\alpha$ with the average relative velocities of both atoms α and β, and modifying the internal energy density with an extra term that involves the difference between the velocities of the two atoms.

given reference configuration $\{\mathbf{X}_\alpha\}$. While this relation is exact for a Lagrangian analysis (as $\dot{\Delta}_{I\alpha} = 0$), with an Eulerian description an additional term involving $\mathbf{v}_\alpha$ breaks the correspondence between the displacement and its time-derivative, velocity. The displacement gradient $\mathbf{H} = \nabla_{\mathbf{X}}\mathbf{u}$, directly related to the deformation gradient $\mathbf{F} \equiv \nabla_{\mathbf{X}}\mathbf{x} = \mathbf{I} + \mathbf{H}$, can be derived via the basis

$$\mathbf{H} = \sum_I \mathbf{u}_I \otimes \nabla_{\mathbf{X}} N_I. \tag{7.35}$$

We can also define an expression for temperature T using the kinetic definition based on the principle of equipartition of energy [50, Sect. 6.4]:

$$3k_B T = \frac{2}{N}\sum_{\alpha=1}^{N} \langle k'_\alpha \rangle \tag{7.36}$$

i.e., an ensemble's internal energy equals twice its average fluctuating kinetic energy k'_α.[4] Along the lines of the derivation for the velocity $\mathbf{v}_I$, Eq. (7.33), we use Eq. (7.36) to obtain a temperature field

$$\sum_{J,\alpha} [N_{I\alpha} 3k_B N_{J\alpha}]\, T_J = \sum_\alpha N_{I\alpha} \underbrace{m_\alpha \mathbf{w}_\alpha \cdot \mathbf{w}_\alpha}_{2k'_\alpha}. \tag{7.37}$$

which we restrict to

$$T_I = \frac{1}{3k_B \sum_\alpha N_{I\alpha}} \sum_\alpha N_{I\alpha} m_\alpha \mathbf{w}_\alpha \cdot \mathbf{w}_\alpha = \frac{1}{3k_B}\sum_\alpha \Delta_{I\alpha} m_\alpha \mathbf{w}_\alpha \cdot \mathbf{w}_\alpha. \tag{7.38}$$

Our coarse-graining methodology can also be applied to diffusion/ionic conduction phenomena [3]. We start with the per-species mass density field akin to Eq. (7.6):

$$\varrho_I^{(a)} = \sum_{\alpha \in \mathscr{A}^{(a)}} m_\alpha \Delta_{I\alpha} \tag{7.39}$$

and associated flux

$$\mathbf{J}_I^{(a)} = \sum_{\alpha \in \mathscr{A}^{(a)}} m_\alpha \mathbf{v}_\alpha \Delta_{I\alpha} \tag{7.40}$$

[4]In statistical mechanics, temperature is defined in terms of the amount of phase space a system visits. We need to employ 'the "local equilibrium" and ergodic assumptions' in order to make the temperature a field variable and feasible to compute. We assume that the strict definition and ours coincide in the limit of large averaging volumes and long averaging times.

where $\mathscr{A}^{(a)}$ is the group of atoms that are of species a. Then the electrical current density $\mathbf{I}_0$ is the flux of ionic charge q_α

$$\begin{aligned}\mathbf{I}_0 &= \sum_{J,\alpha} q_\alpha \mathbf{v}_\alpha \Delta_{J\alpha} N_J = \sum_a z_a \mathbf{J}_a \\ &= \varrho z \mathbf{v} + \underbrace{\sum_a z^{(a)} \mathbf{J}^{(a)}}_{\mathbf{I}}\end{aligned} \tag{7.41}$$

split into convective, $\varrho z\mathbf{v}$, and diffusive, $\mathbf{I}$, components. Here $z^{(a)} = q^{(a)}/m^{(a)}$ is the valence of species a and

$$\varrho z = \sum_a \varrho^{(a)} z^{(a)} = \sum_{J,\alpha} q_\alpha \Delta_{J\alpha} N_J \tag{7.42}$$

is the total charge density. The diffusive ionic flux $\mathbf{I}$ satisfies the conservation equation

$$\varrho \dot{z} + \nabla_{\mathbf{x}} \cdot \mathbf{I} = 0 \tag{7.43}$$

by virtue of each of the species satisfying their respective mass conservation equations at the nodes and the fact that $z^{(a)}$ is constant:

$$\begin{aligned}\varrho \dot{z}_I &= \frac{\mathrm{d}}{\mathrm{d}t} \sum_\alpha q_\alpha \Delta_{I\alpha} = -\sum_\alpha q_\alpha \nabla_{\mathbf{x}} \Delta_{I\alpha} \cdot (\mathbf{v}_\alpha - \mathbf{v}) \\ &= -\nabla_{\mathbf{x}} \cdot \left[\sum_\alpha q_\alpha (\mathbf{v}_\alpha - \mathbf{v}) \Delta_{I\alpha} \right] = -\nabla_{\mathbf{x}} \cdot \mathbf{I}\end{aligned} \tag{7.44}$$

It is important to note that the material time derivative of the charge density $\dot{z}$ is with respect to the barycentric velocity $\mathbf{v}$ of the fluid [3].

7.2.3 *Molecular Data*

One can extend this same coarse-graining procedure to properties of molecular materials. As one such example, we consider electrical charge and the continuum quantities of electric field $\mathbf{E}$, electric displacement, $\mathbf{D}$, and polarization vector, $\mathbf{P}$. As discussed in [51], polarization quantifies the density of dipole moments in a dielectric material. Dipole moments can either be induced, as in non-polar molecules, or permanent, in polar molecules such as water. Here, we briefly develop expressions for $\mathbf{E}$, $\mathbf{D}$, and $\mathbf{P}$ using our upscaling formalism.

The (microscopic) electrostatic balance considering only atomic point charges is

$$\epsilon_0 \nabla_{\mathbf{x}} \cdot \mathbf{E} = \rho z^* \,, \tag{7.45}$$

where ϵ_0 is the vacuum permittivity and ρz is the microscopic charge density. The total charge density, ρz, is given by the summation of both free charges, q_α, over the group $\mathscr{A}_f$ of ions and charges on atoms, $q_{\beta m}$, belonging to molecules m:

$$\varrho z^* = \sum_\alpha q_\alpha \delta(\mathbf{x} - \mathbf{x}_\alpha) = \sum_{\alpha \in \mathscr{A}_f} q_\alpha \delta(\mathbf{x} - \mathbf{x}_\alpha) + \sum_{m,\beta} q_{\beta m} \delta(\mathbf{x} - (\mathbf{x}_m + \mathbf{x}_{\beta m})). \tag{7.46}$$

Following our coarse-graining methodology and employing a Taylor series expansion for the molecular charges around their (center-of-mass) coordinates $\mathbf{x}_m$

$$\begin{aligned} \epsilon_0 \int_\Omega N_I \nabla_{\mathbf{x}} \mathbf{E} \, \mathrm{d}V = & \sum_{\alpha \in \mathscr{A}_f} N_{I\alpha} q_\alpha + \sum_{m,\beta \in \mathscr{M}_n} N_{Im} q_{\beta m} \\ & + \nabla_{\mathbf{x}} \cdot \underbrace{\left[\sum_{m,\beta \in \mathscr{M}_n} \left(-\mathbf{x}_{\beta m} N_{Im} + \frac{1}{2!} \mathbf{x}_{\beta m} \otimes \mathbf{x}_{\beta m} \cdot \nabla_{\mathbf{x}} N_{Im} + \ldots \right) q_{\beta m} \right]}_{\mathbf{P}} . \end{aligned} \tag{7.47}$$

where we recognize the polarization vector $\mathbf{P}$ contains contributions from molecular dipole, $\sum_{\beta \in \mathscr{M}_m} q_{\beta m} \mathbf{x}_{\beta m}$; quadrupole, $\frac{1}{2!} \sum_{\beta \in \mathscr{M}_m} q_{\beta m} \mathbf{x}_{\beta m} \otimes \mathbf{x}_{\beta m}$; and higher moments. An integral form of the macroscopic Maxwell's equation result in $\nabla_{\mathbf{x}} \cdot \mathbf{D} = \zeta$, after defining $\mathbf{D} = \epsilon_0 \mathbf{E} + \mathbf{P}$, the electric displacement vector, and the coarse-grained charge related to the macroscopic free-charge density [51]

$$\zeta_I = \sum_{\alpha \in \mathscr{A}_f} N_{I\alpha} q_\alpha + \sum_{m,\beta \in \mathscr{M}_n} N_{Im} q_{\beta m} \,. \tag{7.48}$$

For molecules that are charge neutral, ζ_I reduces to $\sum_{I\alpha \in \mathscr{A}_f} N_{I\alpha} q_\alpha$.

Also, as noted in [51], the accuracy of $\mathbf{P}$ depends on the order of the polynomial used for the coarse-graining function. For example, if one uses a constant coarse-graining function, then only the dipole moments can be recovered. Likewise, if the basis is linear in $\mathbf{x}$, then the quadrupole moments may be computed. Lastly, the same methodology can be used to extract other quantities; for instance, by replacing the charge q_α in the expression for $\mathbf{P}$ in Eq. (7.47) with the mass m_α the moments of inertia can be extracted.

7.3 Coupling

As mentioned in the introduction, Sect. 7.1, in this section we will develop the basis for atom-to-continuum coupling simulation: (a) consistent upscaling/coarse-graining of the atomic state, (b) multiscale governing balances, (c) consistent continuum surrogate models, and (d) atomistic control schemes. In Sect. 7.2 we have covered the first ingredient in detail which stands as useful tool in it own right. In this section we will discuss the remaining components of a fully coupled multiscale algorithm where $\Omega_{\mathrm{FE}} \neq \emptyset$ and there is an interface, $\partial\Omega_{\mathrm{FE}} \cap \partial\Omega_{\mathrm{MD}}$, between the FE and MD regions.

Before developing the coupling methodology, we provide a brief overview of atomistic-to-continuum multiscale methods. This field is particularly rich in mechanical coupling schemes, and the interested reader is referred to the review article by Miller and Tadmor [52] for a more detailed description of the many strategies as well as references. Apparently the idea of using a coupled finite element model of material to alleviate the computational burden of computing atomic trajectories in regions that are expected to behave in a continuum fashion goes back to the early work of Kohlhoff et al. [53, 54]. Kohlhoff's application to fracture became one of the primary motivating examples for MD/FE mechanical coupling. Motivated by the same application, Tadmor, Ortiz et al. developed one of the most long-lived coupling algorithms: the quasicontinuum (QC) method [55]. Unlike in a strict domain-decomposition, QC takes the particles in the computation to transition between atoms driven by the interatomic potential and finite element nodes obeying a corresponding Cauchy–Born rule [56, 57]. In this method where primarily atomic domains transition to continuum, the particles become more widely spaced to reduce the burden of resolving all the atoms. Broughton et al. [58] derived a three-method algorithm based on hand-shaking regions, localized domains in which information is exchanged between different models which overlap in them, like in Kohlhoff's scheme, which combined tight-binding, MD, and FE. For each scale of exchange, a modified energy functional was derived incorporating contributions from both components. Another popular method is the bridging scale method of Wagner and Liu [45]. Their approach employs a hybrid Lagrangian incorporating the energy from both the atomic and continuum. Galerkin projection is used to partition the resulting forces and stresses between the two systems. Klein and Zimmerman [59] replaced standard finite elements with MLS and reproducing kernel bases for the continuum fields and to enable the multiscale information propagation, but also introduced a Cauchy–Born surrogate model for the continuum corrected near the boundary.

Since a continuum is an incomplete representation of an atomic system, other coupling methodologies use uncertainty quantification to exchange information between the two domains. In some cases, the goal is finding optimally consistent parameters for continuum closures using MD data [3, 60, 61]. In other cases, new closure model forms have been identified using MD samples to estimate a stochastic representation. This strategy has been employed in both off-line [62] and on-line [63] modes based on concurrency of the MD and continuum simulations.

7.3.1 Atomic Regulators

To effect the information transfer from continuum region to the atomistic system necessary for concurrent coupling, we have developed a variety of control/regulation strategies akin to the isokinetic thermostat [64] applied to a field of target values instead of a single system temperature. Both our method and the isokinetic thermostat are based on Gauss's principle of least constraint (GLC). In our application, GLC takes the form

$$\min_{\mathbf{f}_\alpha} \max_{\lambda_I} \underbrace{\left(\frac{1}{2} \sum_\alpha \|\mathbf{f}_\alpha - \mathbf{f}^*_\alpha\|^2 - \sum_I \lambda_I \dot{g}_I \right)}_{J}, \tag{7.49}$$

where $\mathbf{f}^*_\alpha \equiv -\partial_{\mathbf{x}_\alpha} \Phi$ are the unconstrained forces on atoms, g_I are the constraints dependent on atomic data, and λ_I are the associated Lagrange multipliers. It is clear from Eq. (7.49) that the principle is designed so that the Lagrange multipliers do the least work on the system necessary to enforce, the constraints. Also, note that the derivative of the constraint $\dot{g}_I$ is enforced not the constraint itself g_I. Our constraints g_I can be flux balances or field matching conditions analogous to Neumann or Dirichlet interface conditions and take the general form

$$g_I = \sum_\alpha N_{I\alpha} a_\alpha - A_I = 0, \tag{7.50}$$

where $a_\alpha = a(\mathbf{x}_\alpha, \mathbf{v}_\alpha)$ is a phase function corresponding to the nodal/continuum quantity A_I.

The (first order) optimality conditions arising from the extremization of the functional J in Eq. (7.49) recover the derivative of the constraint:

$$\partial_{\lambda_I} J = \dot{g}_I = \sum_\alpha N_{I\alpha} \dot{a}_\alpha - \dot{A}_I = \sum_\alpha N_{I\alpha} \left(\partial_{\mathbf{x}_\alpha} a_\alpha \cdot \mathbf{v}_\alpha + \partial_{\mathbf{v}_\alpha} a_\alpha \cdot \dot{\mathbf{v}}_\alpha \right) - \dot{A}_I = 0, \tag{7.51}$$

assuming a Lagrangian description ($\dot{N}_{I\alpha} = 0$), and the condition:

$$\partial_{\mathbf{f}_\alpha} J = \mathbf{f}_\alpha + \partial_{\mathbf{x}_\alpha} \Phi - \sum_I \lambda_I N_{I\alpha} \, \partial_{\mathbf{f}_\alpha} \dot{a}_\alpha = \mathbf{f}_\alpha + \partial_{\mathbf{x}_\alpha} \Phi - \sum_I \lambda_I N_{I\alpha} \, m_\alpha \partial_{\mathbf{v}_\alpha} a_\alpha = \mathbf{0}, \tag{7.52}$$

where we have used $m_\alpha \partial_{\mathbf{f}_\alpha} = \partial_{\dot{\mathbf{v}}_\alpha}$ from Newton's law: $m_\alpha \dot{\mathbf{v}}_\alpha = \mathbf{f}_\alpha$. We can rearrange Eq. (7.52) into an augmented form of Newton's law

$$m_\alpha \dot{\mathbf{v}}_\alpha = \mathbf{f}_\alpha = -\partial_{\mathbf{x}_\alpha} \Phi + \underbrace{\sum_I \lambda_I N_{I\alpha} \, m_\alpha \partial_{\mathbf{v}_\alpha} a_\alpha}_{\mathbf{f}^\lambda_\alpha} \tag{7.53}$$

which when we substitute it into the constraint, Eq. (7.51), gives a means of solving for λ_I:

$$\sum_{J,\alpha} \left[N_{I\alpha} \left(m_\alpha \|\partial_{\mathbf{v}_\alpha} a_\alpha\|^2\right) N_{J\alpha}\right] \lambda_J = \dot{A}_I - \sum_\alpha N_{I\alpha} \left(\partial_{\mathbf{x}_\alpha} a_\alpha \cdot \mathbf{v}_\alpha - \partial_{\mathbf{v}_\alpha} a_\alpha \cdot \partial_{\mathbf{x}_\alpha} \Phi\right). \tag{7.54}$$

Now that we have a solution for λ_I, it is worth re-examining the structure of the problem. With the constraints g_I posed at nodes of the finite element mesh, the term $\sum_I \lambda_I N_{I\alpha}$ is the interpolation of the nodal field λ at the location of atom α. This is the conduit for propagation of continuum information in the form of a flux balance or field consistency to the atoms. Meanwhile, $\partial_{\mathbf{v}_\alpha} a_\alpha$ is strictly an atomic quantity and, as will be shown in detail, is related to conserved quantities. Hence, the correction $\mathbf{f}_\alpha^\lambda$ to the total force on the atom $\mathbf{f}_\alpha$ is a mixed-scale term incorporating aspects of both the small and large scales present in the problem and will introduce a correlation at the atomic level on the length-scale of the element size.

In our coupling schemes there are two instances of the nature of the constraint which are mutually exclusive at any particular node, like classical Dirichlet and Neumann conditions. For the first case of using the GLC framework to effect consistency between atomic data and a continuum field, we recognize the constraint

$$\underbrace{\sum_\alpha N_{I\alpha} \underbrace{m_\alpha \dot{\mathbf{v}}_\alpha}_{\dot{a}_\alpha} = \sum_J M_{IJ}^A \dot{\mathbf{p}}_J}_{\dot{A}_I}, \tag{7.55}$$

as an expression for the dynamics for a coarse-grained quantity, e.g. momentum. In general we apply M_{IJ}^A, the mass matrix over the atomic domain associated with quantity A, to obviate the need for applying a projection to the atomic information, i.e., the left-hand side of Eq. (7.55). On the other hand, in the case of enforcing a flux balance on the interface of the FE and MD regions, we tie the rate of change of a conserved quantity to the normal component of the associated flux, for instance

$$\sum_\alpha N_{I\alpha} \mathbf{f}_\alpha = \int_{\partial\Omega_{\mathrm{FE}}} N_I \boldsymbol{\sigma} \cdot \mathrm{d}\mathbf{A}. \tag{7.56}$$

This form is derivable from partitioning the global balance of the appropriate flux, see [10, 44] for more details.

These atomistic analogs of Dirichlet and Neumann boundary conditions can be applied simultaneously on disjoint sets of nodes. However, a consequence of our least squares formulation is that λ is, in general, non-zero throughout the domain. As a result, λ peaks at the MD-FE interface where the constraints are initially violated by uncorrected velocities and decays in an oscillatory fashion further into the MD domain. If this behavior is not desired, it is possible to localize λ near the MD boundary, but the method of localization is constraint-dependent. For the case of field-based constraints, Eq. (7.55), nodal λ_I's are only defined at nodes where the constraint is imposed. The cost of this change is that the effect of the GLC affects the atomic domain an element away from the MD/FE boundary. Localization of the flux-based GLC approaches results in a similar effect, but results from row-sum lumping the matrix in Eq. (7.54) such that the global conservation balance is still respected [9]. Constraints involving time filtering can also be posed [44], and the formulation presented here is compatible with complex geometries, albeit at the cost of extra computational expense when localization is used [9].

Lastly, given that GLC derived regulators only control the derivative of the desired constraint it is necessary to set up initial conditions consistent with constraint. Unlike in the traditional FE framework, MD simulations typically need to be conditioned from some unlikely/unphysical state, e.g. a perfect lattice, to a representative sample of the desired ensemble. We find it is expedient to use a version of the velocity rescaling algorithm sometimes used as a thermostat to prepare an initial state for the subsequent dynamics controlled by GLC regulation. With

$$\sum_{\alpha} N_{I\alpha} a_{\alpha} = A_I, \tag{7.57}$$

we can pick a_α to be either the momentum $m_\alpha \mathbf{v}_\alpha$ or the kinetic energy $\frac{1}{2} m_\alpha \mathbf{v}_\alpha \cdot \mathbf{v}_\alpha$. To affect the change in a_α through $\mathbf{v}^*_\alpha = \sqrt[p]{s_\alpha}\mathbf{v}_\alpha$ where p is the power of $\mathbf{v}_\alpha$ in a_α, we construct the scaling field s_I interpolated to the atoms $s_\alpha = \sum_I N_{I\alpha} s_I$, which has the solution

$$\sum_{J,\alpha} \left[N_{I\alpha} a_\alpha N_{J\alpha} \right] s_J = A_I. \tag{7.58}$$

7.3.2 *Mechanical Coupling*

We now turn to the specific problem of mechanical coupling. In this case, the corresponding conserved quantities are atomic momentum $m_\alpha \mathbf{v}_\alpha$ and continuum momentum density $\mathbf{p} = \rho \mathbf{v}$. Using a Lagrangian description and the least-squares formalism introduced in Sect. 7.2.1:

$$\sum_J \left[\int_\Omega N_I \rho N_J \, \mathrm{d}V \right] \mathbf{v}_J = \int_{\Omega_{\mathrm{FE}}} N_I \rho \mathbf{v} \, \mathrm{d}V + \sum_\alpha N_{I\alpha} m_\alpha \mathbf{v}_\alpha \,, \tag{7.59}$$

we can solve for the velocity $\mathbf{v}$, instead of the momentum density $\mathbf{p}$, since the mass density field ρ is known and constant. A consistent decomposition of the left-hand side integral results in

$$\sum_J \underbrace{\left[\int_{\Omega_{\mathrm{FE}}} N_I \rho N_J \, \mathrm{d}V + \sum_\alpha N_{I\alpha} m_\alpha N_{J\alpha} \right]}_{M^V_{IJ}} \mathbf{v}_J = \int_{\Omega_{\mathrm{FE}}} N_I \rho \mathbf{v} \, \mathrm{d}V + \sum_\alpha N_{I\alpha} m_\alpha \mathbf{v}_\alpha \tag{7.60}$$

using an atomic definition of mass density $\rho_\alpha = \frac{m_\alpha}{V_\alpha}$ based on a consistent atomic volume/quadrature weight V_α such that $\int_{\Omega_{\mathrm{MD}}} \mathrm{d}V = \sum_\alpha V_\alpha$. In the case of Eulerian frame equations, the shape function $N_{I\alpha}$ is now a function of time which reproduces the advective/convective fluxes; however, the atomic volume remains unknown but can be approximated by representing the atomic volume as a prolongation of a FE field consistent with $\int_{\Omega_{\mathrm{MD}}} \mathrm{d}V = \sum_\alpha N_{I\alpha} V_I$ without modifying the governing equations. Hence, the dynamical equation governing the continuum velocity is

$$\sum_J \left[\int_{\Omega_{\mathrm{FE}}} N_I \rho N_J \,\mathrm{d}V + \sum_\alpha N_{I\alpha} m_\alpha N_{J\alpha} \right] \dot{\mathbf{v}}_J = \int_{\Omega_{\mathrm{FE}}} N_I \nabla_{\mathbf{X}} \cdot \mathbf{S} \,\mathrm{d}V + \sum_\alpha N_{I\alpha} \mathbf{f}_\alpha \quad (7.61)$$

after substituting the balance of linear momentum, Eq. (7.26), for the nodes, and Newton's law for the atoms. Before moving on to the influence of atomic control forces, several observations can be made. First, Eq. (7.61) is entirely consistent with either the finite element momentum equation or coarse-grained atomic momentum, Eq. (7.18), in the event only one type of region is present. Second, information propagates from the atomic system to the finite element region through the coarse-grained atomic force (the complementary flow of information was described in the previous section, Sect. 7.3.1).

To perform a coupled simulation, a final ingredient is required: a surrogate model for the interatomic-potential [65, 66] in the continuum region. Closures such as this are necessary in all continuum mechanics to account for the physics associated with the missing degrees of freedom and atomic interactions; in the present context they are assumed to be accurate for a limited regime of the possible atomic motion, e.g. nearly homogeneous deformation. It is important to note that the degree of consistency between the continuum closure and the true contributions from the atoms will impact the accuracy and sometimes the stability of the method. The Cauchy–Born model derived directly from the interatomic potential and the lattice [67, 68] and thermal models inferred from MD [61, 62] are examples of surrogate models particularly suited to MD/FE coupling.

For illustration of the present case, we introduce an elastic surrogate model which assumes a linear relationship between the stress tensor, $\mathbf{S}$, and (infinitesimal) strain:

$$\mathbf{S}(\mathbf{X}) \approx \mathbb{C} : \nabla_{\mathbf{X}} \mathbf{u}(\mathbf{x}) \approx \mathbb{C} : \sum_I \mathbf{u}_I \nabla_{\mathbf{X}} N_I, \quad (7.62)$$

with the fourth order elasticity tensor $\mathbb{C}$. After substituting this constitutive relationship into (7.61) and integrating by parts, we arrive at

$$\sum_J \left[\int_{\Omega_{\mathrm{FE}}} N_I \rho N_J \,\mathrm{d}V + \sum_\alpha N_{I\alpha} m_\alpha N_{J\alpha} \right] \dot{\mathbf{v}}_J = -\sum_J \left[\int_{\Omega_{\mathrm{FE}}} \nabla_{\mathbf{X}} N_I \,\mathbb{C} : \nabla_{\mathbf{X}} N_J \right] \mathrm{d}V \,\mathbf{u}_J + \int_{\partial\Omega_{\mathrm{FE}}} N_I \mathbf{S} \cdot \mathrm{d}\mathbf{A} + \sum_\alpha N_{I\alpha} \mathbf{f}_\alpha, \quad (7.63)$$

which is completely prescribed except for the stress at the interface between the atomic and continuum domains, which we will address presently.

In order to couple the entire system, the continuum state must also influence the atoms. We accomplish this by using the constraint formalisms outlined in Sect. 7.3.1. The MD and FE sub-systems can be coupled either strongly by constraining the atomic forces based on the continuum velocity or weakly by using the continuum stress. Strong coupling uses the constraint:

$$\sum_{\alpha} N_{I\alpha} \mathbf{f}_{\alpha} = \sum_{J,\alpha} [N_{I\alpha} m_{\alpha} N_{J\alpha}] \dot{\mathbf{v}}_J. \tag{7.64}$$

Then Eq. (7.54) becomes

$$\sum_{J,\alpha} [N_{I\alpha} N_{J\alpha}] \boldsymbol{\lambda}_J = \sum_{J,\alpha} [N_{I\alpha} m_{\alpha} N_{J\alpha}] \dot{\mathbf{v}}_J + \sum_{\alpha} N_{I\alpha} \partial_{\mathbf{x}_\alpha} \Phi. \tag{7.65}$$

Similarly, weak coupling is derived using conservation of momentum for the total system,

$$\sum_{\alpha} \mathbf{f}_{\alpha} = -\int_{\partial\Omega_{\mathrm{FE}}} \mathbf{S} \cdot \mathrm{d}\mathbf{A}, \tag{7.66}$$

which after partitioning using the basis functions N_I implies

$$\sum_{\alpha} N_I \mathbf{f}_{\alpha}^{\lambda} = -\int_{\partial\Omega_{\mathrm{FE}}} N_I \mathbf{S} \cdot d\mathbf{A}. \tag{7.67}$$

The equation for $\boldsymbol{\lambda}$ is then

$$\sum_{J,\alpha} [N_{I\alpha} N_{J\alpha}] \boldsymbol{\lambda}_J = -\int_{\partial\Omega_{\mathrm{FE}}} N_I \mathbf{S} \cdot \mathrm{d}\mathbf{A}. \tag{7.68}$$

The last step in the derivation involves eliminating the unknown boundary flux in Eq. (7.63) by equating it with the force arising from the constraint, resulting in

$$\begin{aligned} &\sum_J \left[\int_{\Omega_{\mathrm{FE}}} N_I \rho N_J \,\mathrm{d}V + \sum_{\alpha} N_{I\alpha} m_{\alpha} N_{J\alpha} \right] \dot{\mathbf{v}}_J \\ &\quad = -\sum_J \left[\int_{\Omega_{\mathrm{FE}}} \nabla_{\mathbf{X}} N_I \,\mathbb{C} : \nabla_{\mathbf{X}} N_J \,\mathrm{d}V \right] \mathbf{u}_J - \sum_{\alpha} N_{I\alpha} \partial_{\mathbf{x}_\alpha} \Phi. \end{aligned} \tag{7.69}$$

In either case, the atomic dynamics result from

$$m_{\alpha} \dot{\mathbf{v}}_{\alpha} = -\partial_{\mathbf{x}_\alpha} \Phi + \sum_I N_{I\alpha} \lambda_I. \tag{7.70}$$

In many practical calculations, a layer of ghost atoms outside Ω_{MD} is used to exert forces to keep the unconstrained atoms within Ω_{MD}, refer to Fig. 7.1. The locations of the ghost atoms can be tied to the continuum displacement field, but in this case the forces the ghost atoms exert on the other atoms must be incorporated into the momentum conservation constraint. An alternative is simply to set the boundary stress based on the forces exerted by the ghost atoms, which also conserves momentum.

In fact there are a variety of means of coupling the atomistic and continuum motions and forces, but not all are inherently stable [69]. In particular, coupling the continuum to the atomistic flux, for instance the virial in mechanical coupling, generally leads to instabilities. Another issue that has received considerable attention [45] is how to handle the waves that are supported in the atomic region but cannot be transmitted to the continuum region due to the inherent mismatch in the dispersion characteristics of the two representations. Typically they are selectively damped out of the system [70] which violates overall energy conservation. This is another aspect of the incompleteness of the finite element representation.

7.3.3 Thermal Coupling

We now examine thermal coupling. With the definition of atomic temperature, Eq. (7.37), in hand, the least squares minimization procedure from Sect. 7.2.1 can be used to construct a finite element temperature field:

$$\sum_J \left[\int_{\Omega} N_I \rho c N_J \, \mathrm{d}V \right] T_J = \int_{\Omega_{\mathrm{FE}}} N_I \rho c T \, \mathrm{d}V + \sum_\alpha N_{I\alpha} m_\alpha \mathbf{w}_\alpha \cdot \mathbf{w}_\alpha \tag{7.71}$$

where the total fluctuating energy consistent with Eq. (7.37) appears on the right-hand side. The expression is a form of conservation of energy in absence of mechanical work, and hence we can reduce $\mathbf{w}_\alpha$ to $\mathbf{v}_\alpha$ and use a Lagrangian description of the material. Similar to the partition of the right-hand side into continuum and atomic domains, the left-hand side can be decomposed as

$$\sum_J \underbrace{\left[\int_{\Omega_{\mathrm{FE}}} N_I \rho c N_J \, \mathrm{d}V + 3k_B \sum_\alpha N_{I\alpha} N_{J\alpha} \right]}_{M^T_{IJ}} T_J = \int_{\Omega_{\mathrm{FE}}} N_I \rho c T \, \mathrm{d}V + \sum_\alpha N_{I\alpha} m_\alpha \mathbf{v}_\alpha \cdot \mathbf{v}_\alpha \tag{7.72}$$

using the Dulong–Petit law [71, Chap. 22] for the heat capacity

$$\rho c = \frac{3k_B}{V_\alpha} \tag{7.73}$$

of a classical solid material in conjunction with atomic quadrature weights V_α. For a system without a continuum region, this equation reduces to a projection of the temperature as in Eq. (7.37).

To reach the final form of the governing equation, we take the time derivative of Eq. (7.72):

$$\sum_J \left[\int_{\Omega_{\mathrm{FE}}} N_I \rho c N_J \, \mathrm{d}V + 3k_B \sum_\alpha N_{I\alpha} N_{J\alpha} \right] \dot{T}_J = \int_{\Omega_{\mathrm{FE}}} N_I \nabla \cdot \mathbf{q} \, \mathrm{d}V + 2 \sum_\alpha N_{I\alpha} \mathbf{v}_\alpha \cdot \mathbf{f}_\alpha . \tag{7.74}$$

Into this balance for the finite element temperature field we substitute Fourier's law for the referential heat flux

$$\mathbf{q}(\mathbf{X}) \approx -\boldsymbol{\kappa} \nabla_{\mathbf{X}} T = -\boldsymbol{\kappa} \sum_I T_I \nabla_{\mathbf{X}} N_I , \tag{7.75}$$

as a constitutive relationship for the heat flux $\mathbf{q}$ where $\boldsymbol{\kappa}$ is the thermal conductivity tensor. (Note that Fourier's law is not always an accurate surrogate model at small scales, see, e.g., [61].) Integrating the continuum right-hand side term of Eq. (7.74) by parts completes the derivation of the multiscale balance:

$$\begin{aligned} \sum_J \left[\int_{\Omega_{\mathrm{FE}}} N_I \rho c N_J \, \mathrm{d}V + 3k_B \sum_\alpha N_{I\alpha} N_{J\alpha} \right] \dot{T}_J &= \sum_J \left[\int_{\Omega_{\mathrm{FE}}} \nabla N_I \cdot \boldsymbol{\kappa} \nabla N_J \, \mathrm{d}V \right] T_J \\ &\quad + \int_{\partial\Omega_{\mathrm{FE}}} N_I \mathbf{q} \cdot \mathrm{d}\mathbf{A} + 2 \sum_\alpha N_{I\alpha} \mathbf{v}_\alpha \cdot \mathbf{f}_\alpha . \end{aligned} \tag{7.76}$$

Now coupling can be imposed via constraints on the atomic force in the framework in Sect. 7.3.1. To maintain the consistency of the coarse-grained atomic temperature and the finite element temperature, their time derivatives are constrained to match using Eq. (7.50):

$$2 \sum_\alpha N_{I\alpha} \mathbf{v}_\alpha \cdot \mathbf{f}_\alpha = 3k_B \sum_{J,\alpha} N_{I\alpha} N_{J\alpha} \dot{T}_J . \tag{7.77}$$

In this case, the equation for the Lagrange multipliers, Eq. (7.54), becomes

$$\sum_{J,\alpha} \left[N_{I\alpha} \mathbf{v}_\alpha \cdot \mathbf{v}_\alpha N_{J\alpha} \right] \lambda_J = \frac{3k_B}{2} \sum_{J,\alpha} \left[N_{I\alpha} N_{J\alpha} \right] \dot{T}_J + \sum_\alpha N_{I\alpha} \partial_{\mathbf{x}_\alpha} \Phi \cdot \mathbf{v}_\alpha . \tag{7.78}$$

To conserve energy in the exchange between the continuum and atomic domains, the finite element boundary flux

$$\int_{\partial\Omega_{\mathrm{FE}}} N_I \mathbf{q} \cdot \mathrm{d}\mathbf{A} = -2 \sum_\alpha N_{I\alpha} \mathbf{v}_\alpha \cdot \mathbf{f}_\alpha^\lambda \tag{7.79}$$

is equated to the power due to the constraint force $\mathbf{f}_\alpha^\lambda$ from GLC. The governing equations for the FE and MD sub-systems are

$$\sum_J \left[\int_{\Omega_{\mathrm{FE}}} N_I \rho c N_J \, \mathrm{d}V + 3k_B \sum_\alpha N_{I\alpha} N_{J\alpha} \right] \dot{T}_J \tag{7.80}$$

$$= \sum_J \left[\int_{\Omega_{\mathrm{FE}}} \nabla N_I \cdot \boldsymbol{\kappa} \nabla N_J \, \mathrm{d}V \right] T_J - 2 \sum_\alpha N_{I\alpha} \mathbf{v}_\alpha \cdot \partial_{\mathbf{x}_\alpha} \Phi,$$

$$m_\alpha \dot{\mathbf{v}}_\alpha = -\partial_{\mathbf{x}_\alpha} \Phi + 2\mathbf{v}_\alpha \sum_I N_{I\alpha} \lambda_I. \tag{7.81}$$

When flux-based coupling is desired, conservation of energy is used to derive the constraint:

$$\partial_{\mathbf{x}_\alpha} \Phi \cdot \mathbf{v}_\alpha + \mathbf{f}_\alpha \cdot \mathbf{v}_\alpha = \mathbf{f}_\alpha^\lambda \cdot \mathbf{v}_\alpha = -\sum_I \int_{\partial\Omega_{\mathrm{FE}}} N_I \mathbf{q} \cdot \mathrm{d}\mathbf{A}, \tag{7.82}$$

resulting in the governing equation for the Lagrange multipliers:

$$\sum_{J,\alpha} \left[N_{I\alpha} N_{J\alpha} \mathbf{v}_\alpha \right] \lambda_J = -\int_{\partial\Omega_{\mathrm{FE}}} N_I \mathbf{q} \cdot \mathrm{d}\mathbf{A}, \tag{7.83}$$

when partitioned akin to Eq. (7.56), see [44]. However, in this case the flux is overcompensated for by our equipartition assumption, $3k_B T = m_\alpha \langle \mathbf{v}_\alpha \cdot \mathbf{v}_\alpha \rangle$, so the governing equations are

$$\sum_J \left[\int_{\Omega_{\mathrm{FE}}} N_I \rho c N_J \, \mathrm{d}V + 3k_B \sum_\alpha N_{I\alpha} N_{J\alpha} \right] \dot{T}_J \tag{7.84}$$

$$= \sum_J \left[\int_{\Omega_{\mathrm{FE}}} \nabla N_I \cdot \boldsymbol{\kappa} \nabla N_J \, \mathrm{d}V \right] T_J - 2 \sum_\alpha N_{I\alpha} \partial_{\mathbf{x}_\alpha} \Phi \cdot \mathbf{v}_\alpha + \sum_\alpha N_{I\alpha} \mathbf{v}_\alpha \cdot \mathbf{f}_\alpha^\lambda,$$

$$m_\alpha \dot{\mathbf{v}}_\alpha = -\partial_{\mathbf{x}_\alpha} \Phi + \mathbf{v}_\alpha \sum_I N_{I\alpha} \lambda_I. \tag{7.85}$$

Lastly, the required finite element heat fluxes can be computed using face-based quadrature schemes if the domain boundary aligns with the finite element faces, or using an approximate L^2 projection of the heat flux [44]. As a final comment, the thermostat force in the augmented Newton's equation (7.85) is of the recognizable velocity drag form common to many thermostats including the Langevin thermostat [72] but without the random force from the Mori–Zwanzig formalism.

7.3.4 Thermomechanical Coupling

The primary difference in developing a framework for thermomechanical coupling from the separate mechanical and thermal coupling we developed in the previous two sections is that the atomic momentum and temperature are no longer independent. The multiscale momentum balance (7.61) for the velocity is unchanged, but to derive the equation for temperature, we must start with the rate of change of the total energy:

$$\sum_J \left[\int_{\Omega} N_I \rho N_J \, \mathrm{d}V \right] \dot{e}_J = \int_{\Omega_{\mathrm{FE}}} N_I \left(\rho \dot{\varepsilon} + \rho \mathbf{v} \cdot \dot{\mathbf{v}} \right) \mathrm{d}V + \sum_\alpha N_{I\alpha} \left(m_\alpha \dot{\mathbf{v}}_\alpha + \partial_{\mathbf{x}_\alpha} \Phi \right) \cdot \mathbf{v}_\alpha \,. \tag{7.86}$$

We assume the total energy density e is the sum of internal energy density ε, which has thermal and elastic components, and coarse-scale kinetic energy[5]

$$\rho \dot{e} = \rho \dot{\epsilon} + \rho \mathbf{v} \cdot \dot{\mathbf{v}} = \rho c \dot{T} + \mathbf{S} \cdot \nabla_{\mathbf{X}} \mathbf{v} + \rho \mathbf{v} \cdot \dot{\mathbf{v}} \tag{7.87}$$

We can reduce Eq. (7.86) to

$$\begin{aligned}
\sum_J \int_{\Omega} N_I \rho N_J \, \mathrm{d}V \, \dot{\varepsilon}_J &= \int_{\Omega_{\mathrm{FE}}} N_I \rho \dot{\varepsilon} \, \mathrm{d}V - \int_{\Omega_{\mathrm{MD}}} N_I \rho \mathbf{v} \cdot \dot{\mathbf{v}} \, \mathrm{d}V \\
&\quad + \sum_\alpha N_{I\alpha} \left(m_\alpha \dot{\mathbf{v}}_\alpha + \partial_{\mathbf{x}_\alpha} \Phi \right) \cdot \mathbf{v}_\alpha \qquad (7.88) \\
&= \int_{\Omega_{\mathrm{FE}}} N_I \rho \dot{\varepsilon} \, \mathrm{d}V \\
&\quad + \sum_\alpha N_{I\alpha} \underbrace{\left(m_\alpha \mathbf{v}_\alpha \cdot \dot{\mathbf{v}}_\alpha - \sum_J N_{J\alpha} \rho_J \mathbf{v}_J \cdot \dot{\mathbf{v}}_J V_\alpha + \partial_{\mathbf{x}_\alpha} \Phi \cdot \mathbf{v}_\alpha \right)}_{\dot{k}'_\alpha}
\end{aligned}$$

using the atomic quadrature based on weights V_α and making the particular definition of the fluctuating kinetic energy k'_α. Note that our use of inexact projections/restrictions leads to k'_α not being identified with our previous definition of the thermal energy $\frac{1}{2} m_\alpha \mathbf{w}_\alpha \cdot \mathbf{w}_\alpha$ from Sect. 7.2. Using Eq. (7.61), the coupled multi-scale equations for the nodal velocities, $\mathbf{v}_I$ and temperatures, T_I are

$$M^V_{IJ} \dot{\mathbf{v}}_J = - \int_{\Omega^{\mathrm{FE}}} \nabla_{\mathbf{X}} N_I \cdot \mathbf{S} \, \mathrm{d}V + \int_{\partial \Omega_{\mathrm{FE}}} N_I \mathbf{S} \cdot \mathrm{d}\mathbf{A} + \sum_\alpha N_{I\alpha} \mathbf{f}_\alpha , \tag{7.89}$$

[5]This a common assumption that neglects, for example, interactions between the thermal and the coarse-scale mechanical energy related to thermal expansion at the macro-scale.

$$M_{IJ}^T \dot{T}_J = -\int_{\Omega_{\rm FE}} \nabla_{\mathbf{X}} N_I \cdot \mathbf{q}\, \mathrm{d}V + \int_{\partial\Omega_{\rm FE}} N_I \mathbf{q} \cdot \mathrm{d}\mathbf{A} + \sum_\alpha N_{I\alpha} \left(\dot{k}'_\alpha + \partial_{\mathbf{x}_\alpha} \Phi \cdot \mathbf{v}_\alpha \right). \tag{7.90}$$

As in the previous two sections, we need surrogate models for the first Piola–Kirchhoff stress $\mathbf{S}$ and referential heat flux $\mathbf{q}$. A good model for the former is the Cauchy–Born model based on a quasi-harmonic free energy detailed in [68].

To form the appropriate constraints, we begin with the global conservation of momentum and (total) energy:

$$\frac{\mathrm{d}}{\mathrm{d}t} \left(\sum_\alpha m_\alpha \mathbf{v}_\alpha + \int_{\Omega_{\rm FE}} \mathbf{p}\, \mathrm{d}V \right) = \sum_\alpha \mathbf{f}_\alpha + \int_{\partial\Omega_{\rm FE}} \mathbf{S}\, \mathrm{d}\mathbf{A} = \int_{\partial\Omega} \mathbf{S}\, \mathrm{d}\mathbf{A} \tag{7.91}$$

$$\frac{\mathrm{d}}{\mathrm{d}t} \left(\sum_\alpha m_\alpha e_\alpha + \int_{\Omega_{\rm FE}} \rho e\, \mathrm{d}V \right) = \sum_\alpha (\mathbf{f}_\alpha + \partial_{\mathbf{x}_\alpha} \Phi) \cdot \mathbf{v}_\alpha + \int_{\partial\Omega_{\rm FE}} (\mathbf{v} \cdot \mathbf{S} - \mathbf{q}) \cdot \mathrm{d}\mathbf{A} = \int_{\partial\Omega} (\mathbf{v} \cdot \mathbf{S} - \mathbf{q}) \cdot \mathrm{d}\mathbf{A} \tag{7.92}$$

which we reduce and partition to form:

$$\mathbf{g}_I^V = \sum_\alpha N_{I\alpha} \mathbf{f}_\alpha - \int_{\partial\Omega \backslash \partial\Omega_{\rm FE}} N_I \mathbf{S}\, \mathrm{d}\mathbf{A} = \mathbf{0} \tag{7.93}$$

$$g_I^T = \sum_\alpha N_{I\alpha} (\mathbf{f}_\alpha + \partial_{\mathbf{x}_\alpha} \Phi) \cdot \mathbf{v}_\alpha - \int_{\partial\Omega \backslash \partial\Omega_{\rm FE}} N_I (\mathbf{v} \cdot \mathbf{S} - \mathbf{q}) \cdot \mathrm{d}\mathbf{A} = 0 \tag{7.94}$$

As previously discussed, the augmented force in the MD component is $\mathbf{f}_\alpha = -\partial_{\mathbf{x}_\alpha} \Phi + \mathbf{f}_\alpha^\lambda$ where in this case $\mathbf{f}_\alpha^\lambda$ takes the form

$$\mathbf{f}_\alpha^\lambda = -\sum_I N_I \left(\boldsymbol{\lambda}_I^V + \lambda_I^T \mathbf{v}_\alpha \right), \tag{7.95}$$

with a vector $\boldsymbol{\lambda}_I^V$ and scalar λ_I^T Lagrange multipliers similar to system-wide momentum and temperature control found in [73]. This modified force is substituted into each constraint to obtain the following block symmetric system of equations for $\boldsymbol{\lambda}_I^V$ and λ_I^T:

$$\sum_{J,\alpha} [N_{I\alpha} N_{J\alpha}] \boldsymbol{\lambda}_I^V + \sum_{J,\alpha} [N_{I\alpha} \mathbf{v}_\alpha N_{J\alpha}] \lambda_J^T = -\sum_\alpha N_{I\alpha} \partial_{\mathbf{x}_\alpha} \Phi - \int_{\partial\Omega_{\rm MD}} N_I \mathbf{S}\, \mathrm{d}\mathbf{A} \tag{7.96}$$

$$\sum_{J,\alpha} [N_{I\alpha} \mathbf{v}_\alpha N_{J\alpha}] \cdot \boldsymbol{\lambda}_I^V + \sum_{J,\alpha} [N_{I\alpha} \mathbf{v}_\alpha \cdot \mathbf{v}_\alpha N_{J\alpha}] \lambda_J^T = -\int_{\partial\Omega \backslash \partial\Omega_{\rm FE}} N_I (\mathbf{v} \cdot \mathbf{S} - \mathbf{q}) \cdot \mathrm{d}\mathbf{A} \tag{7.97}$$

7.4 Examples

In this section we demonstrate the utility of the methods we have developed with selected applications of the coarse-graining, Sect. 7.2, and coupling, Sect. 7.3, methodologies.

7.4.1 Inclusion

With this example we demonstrate the utility and versatility of our coarse-graining methods by examining the simple case of insertion of an oversized inclusion within a constrained lattice. Figure 7.2a shows the face-centered cubic (fcc) gold lattice 20×20 unit cells on a side and 3 unit cells thick in the out-of-plane direction (4800 atoms, lattice parameter of 4.08 Å). Periodic boundary conditions are used in all directions. We employ a Lennard-Jones [74, 75] pair potential smoothly truncated at distance of about 5.46 Å. We expand the center region of 4×4 unit cells by 0.5 % and hold the atoms in this inclusion region fixed, while the outer material is allowed to relax via energy minimization.

Figure 7.2a shows the atomic displacement magnitudes, with the largest displacements occurring at the inclusion corners. Figure 7.2b shows the coarse-graining of these displacements onto a 10×10 element mesh that overlays the atomic system. Here, we use the mesh's own interpolation function as the coarse-graining operator. The resulting displacement field has similar features to the atomic one, albeit with a noticeably lower peak magnitude.

Figures 7.2c–e illustrates the use of a localization kernel at each node to coarse-grain the atomic displacements, where the kernel differs from the mesh interpolation function N_I. Here, we use a quartic polynomial that depends on the radial distance from the node's position, with its maximum value at the node and smoothly reaching zero at a distance just over the cutoff of the potential, 6 Å (as recommended in [4]). Comparing Fig. 7.2b and Fig. 7.2c, which use the same 10×10 mesh, we observe that the field coarseness remains the same, but the peak value is somewhat higher for the kernel-averaged system. Figures 7.2d, e reveal how use of a successively refined mesh (20×20 and 40×40, respectively) improves the resolution and fidelity of the coarse-grained field.

This comparison shows the advantage of using independent localization kernels for coarse-graining—in fact the kernel size could vary throughout the domain and be tied to atomic density or field gradients. Use of kernels enables the calculation of robust averages and continuous fields that correspond to the local continuum limits [49] for an arbitrary degree of mesh resolution. In contrast, exclusive use of the mesh's interpolation function would produce either a field with converged values but coarse features (for a mesh with large element size), or one with fine features but spatially fluctuating, non-converged values (for a mesh with small element size).

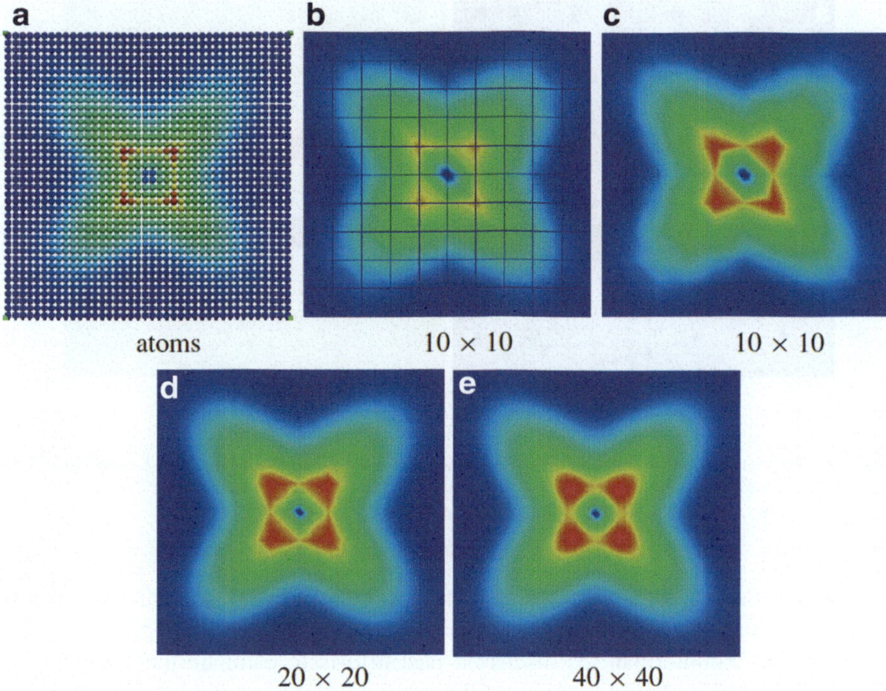

Fig. 7.2 Displacement magnitude, $\|\mathbf{u}\|$, for insertion of an oversized inclusion in fcc gold. (**a**) Atoms colored by atomic displacements. (**b**) Coarse-graining on a 10×10 mesh using interpolation functions. (**c**)–(**e**) Coarse-graining using localization kernels on 10×10, 20×20 and 40×40 meshes. Here, $\|\mathbf{u}\|$ ranges from 0 (*blue*) to 0.05 Å for the atomic system, or 0.037 Å for the coarse-grained systems (*red*)

We also used our coarse-graining methods to calculate the continuum stress fields for this inclusion problem. Figure 7.3 shows the $\boldsymbol{\sigma}_{11}$ and $\boldsymbol{\sigma}_{12}$ fields using the finely resolved 40 × 40 mesh with localization kernels. The resulting fields are smoothly varying and appear to be consistent with expectations of continuum mechanics.

7.4.2 J-Integral

Coarse-graining methods enable the use of continuum fields to estimate other metrics defined within continuum theory, e.g. configurational forces. Based on the seminal work by Eshelby [76] and developed in the context of fracture mechanics by Rice [77], the J-integral is a path independent contour or surface integral (in two- or three-dimensions, respectively) that measures the energetic driving force on a defect. The J-integral is commonly used in numerical simulations of continuum

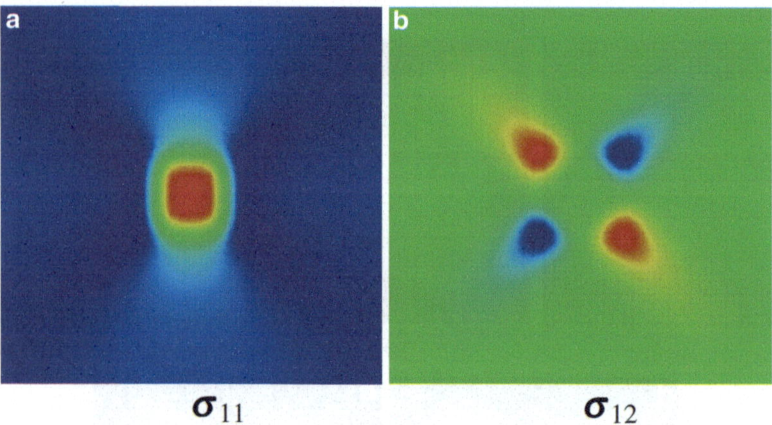

Fig. 7.3 (**a**) σ_{11} and (**b**) σ_{12} for insertion of an oversized inclusion in fcc gold. Coarse-graining is performed using localization kernels on a 40×40 mesh. Stress values range from -0.59 to 3.71 GPa for σ_{11} and from -0.85 to 0.85 GPa for σ_{12} (*blue-to-red*)

mechanical deformation, such as the finite element method, to indicate when a critical loading state has been achieved that will result in crack propagation.

Jones and Zimmerman [5] discussed past efforts to estimate the J-integral at the atomic scale, and proposed use of the coarse-graining methods covered in this chapter as a means to ensure consistency with linear elastic fracture mechanics (LEFM), and to preserve the path independence of the J-integral. In that work, the J-integral expression for a isothermal, equilibrium material is given as

$$\mathbf{J} = \int_{\partial\Omega} \left(W\mathbf{I} - \mathbf{H}^T\mathbf{S} \right) \, \mathrm{d}\mathbf{A}, \tag{7.98}$$

where W is the material frame internal energy density, and $\mathbf{H}$ and $\mathbf{S}$ have already been defined as the displacement gradient and 1st Piola–Kirchhoff stress fields, respectively. In [5], it was shown that W and $\mathbf{S}$ exhibited thermodynamic consistency (i.e., $\mathbf{S} = \partial W / \partial \mathbf{F}$), thereby ensuring that the J-integral around a closed region with a smooth motion is zero and consequently that $\mathbf{J}$ is path independent for arbitrary contours around regions that contain singularities such as crack tips.

Figure 7.4a shows the S_{22} stress field for a single crack in a cylinder composed of the same LJ gold used in the inclusion example. Here, displacements are imposed on atoms within an outer annulus of the cylinder in accordance with the LEFM solution for a mode I loaded crack tip. Details about this simulation are given in [5]. We note that this stress field contains the same characteristic pattern as predicted by the LEFM solution. Figure 7.4b plots the J-integral for a square contour that encircles the crack tip (shown clearly in Fig. 7.4a) as a function of the square of the applied stress intensity factor, K_I. In LEFM theory, J_1 for this configuration should be proportional to K_I^2, as shown by the straight, black line in the figure. We see

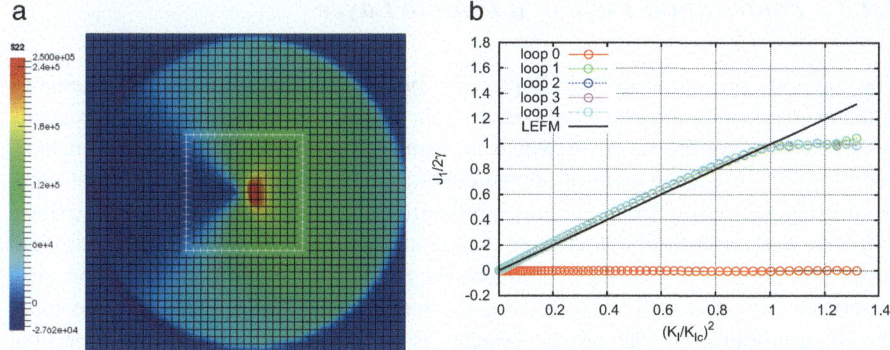

Fig. 7.4 (**a**) The S_{22} stress (units in *bars*) from a coarse-grained estimate. A contour loop and the FE interpolation grid are also shown. (**b**) The calculated J-integral for the single crack configuration showing path independence. J-integral values are normalized by twice the surface energy of the Lennard-Jones system and the loading parameter K_I by the corresponding critical value K_{Ic}

that for a set of concentric loops of varying size, our estimation follows this trend up to the point when crack propagation begins, $J_1 = 2\gamma$, thereby confirming path independence of the J-integral. Discrepancies from exact linearity may be due to the anisotropic and non-linear aspects of the LJ potential, as well as the non-ideal aspects of the crack face geometry. We also observe that for a loop that does not enclose the crack tip singularity (loop 0), $J_1 = 0$, as expected.

In [6], this approach was expanded to treat systems at a finite (i.e., non-zero) temperature. In this case, the J-integral is given by the expression

$$\mathbf{J} = \int_{\partial\Omega} \left(\Psi\mathbf{I} - \mathbf{F}^T\mathbf{S}\right) \, \mathrm{d}\mathbf{A}, \tag{7.99}$$

where Ψ is the Helmholtz free energy density, a function of both deformation gradient $\mathbf{F}$ and temperature T. At finite temperature, Ψ has contributions from both internal (strain) energy and entropy [68]. In [6], this entropic term was determined by a local harmonic (LH) approximation that requires calculation of a simplified dynamical matrix that neglects coupling between atoms. Details on this method are given in [6], along with an analysis that shows the LH approximation to closely correspond with a more exact calculation of free energy using thermodynamic integration up to significant temperatures ($\lesssim$ 400 K) for substantial amounts of uniaxial and volumetric strains. Application of Eq. (7.99) to the single crack tip geometry showed that although the J_1 dependence on applied stress intensity factor changes slightly due to thermal stresses that arise in the heated system, path independence at a given temperature is maintained [6].

7.4.3 Polarization Field of a Double Layer

The methods presented in this chapter can be used to investigate inhomogeneous and anisotropic phenomena which occur at the atomic scale. An interesting example is the electric double layer, in which an ionic solution covers a charged surface, resulting in a screening layer of oppositely charged ions attracted from the solution. Double layers are important in many applications, ranging from electrokinetic flows in micro- and nano-fluidic devices to energy storage devices including batteries and super-capacitors. Despite their ubiquity, they are still poorly understood because experimentally it is difficult to resolve the length-scales over which they develop, and the configuration for realistic systems is too complex for a purely theoretical treatment. However, molecular dynamics studies [2, 78–80] have provided insights into double layer structure, and the coarse-graining theories for atoms and molecules enable a deeper understanding of the important physics.

In this example, we model a box of salt water using molecular dynamics periodic in two directions and constrained by a uniform force field in the third to mimic a nano-channel geometry. A uniform electric field is also applied in this direction to account for a potential drop across the channel. When this happens, a structured layer of solvent and solute particles formed, followed slightly further from the wall by a diffuse layer in which the solute remains at elevated concentrations but no significant structure is present. Figure 7.5 shows this structure at the boundary. Coarse-graining the atoms enables a continuum density field to be post-processed, as shown in Fig. 7.6. The densities are important because they enable the calculation of the electric field and electric potential throughout the domain. These quantities allow us to determine the structure of the condensed layer, how far it extends away from the boundary, and how much charge the double layer can store.

We can understand the double layer structure more deeply by quantifying its electrical properties [51] using the methods outlined in Sect. 7.2.3. Molecular coarse-graining elucidates the degree to which the solvent is aligned with the electric field. Figure 7.7 shows the strong polarization present in the structured layer to be more than an order of magnitude greater than the bulk values. More importantly, the polarization is beyond the level at which a constant relative permittivity is a useful description, implying that the physics cannot be understood without appeal to the atomic nature of matter. However, resolving the boundary region allows us

Fig. 7.5 Structure of ions and molecules near the boundary. *Grey* denotes oxygen, *red* hydrogen, and *green* counter-ions. Reproduced from [79] with permission

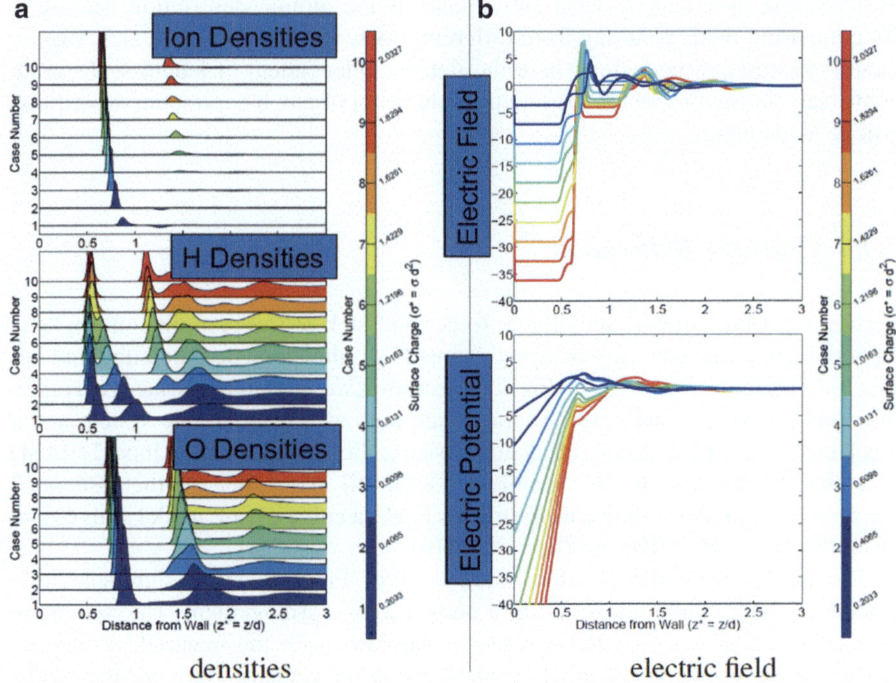

Fig. 7.6 Fields obtained from nanochannel simulations: (**a**) densities associated with different types of atoms using atomic coarse-graining, and (**b**) the electric field and electric potential from Gauss' law using the coarse-grained charge density. Reproduced from [79] with permission

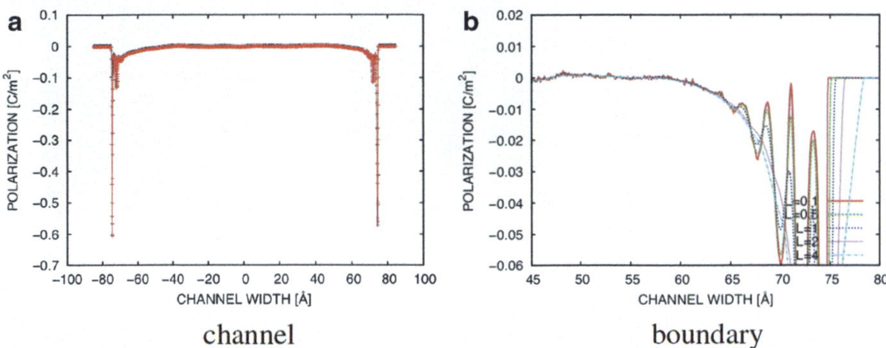

Fig. 7.7 Polarization profile in a nanochannel: throughout the entire channel (**a**) and near the boundary (**b**). Reproduced from [51] with permission

to determine over exactly what sub-domain is the atomic description necessary. By comparing fields resulting from different coarse-graining length-scales, we can identify continuum behavior where the field is independent of length-scale. And, conversely, in regions where the length-scale is important, a continuum description will be inadequate.

7.4.4 Surface Relaxation

Surface relaxation due to interatomic forces and lack of full coordination shells is a phenomenon that can only be approximated in finite element simulation and not in a fully predictive manner. In this demonstration we model the surface relaxation of a cube of Ni nominally 38.72 Å on a side both as a fully atomic system and a mechanically coupled atomic/FE system. We use the embedded atom model (EAM) potential [81–83] for the interatomic forces in Ni and either (a) the associated Cauchy–Born model for the elastic stresses or (b) a cubic elastic model with $C_{11} = 261$ GPa, $C_{12} = 151$ GPa, and $C_{44} = 132$ GPa.

The results from the all atom simulation in Fig. 7.8a show significant displacement as the cube relaxes from a perfect lattice arrangement. The maximum displacements $|u_1|, |u_2| \approx 0.2$ Å occur at the corners of the material. As can be seen from Fig. 7.8b,c the coupled simulation with the Cauchy–Born and the significantly less computationally intensive cubic elastic surrogate models,[6] respectively,

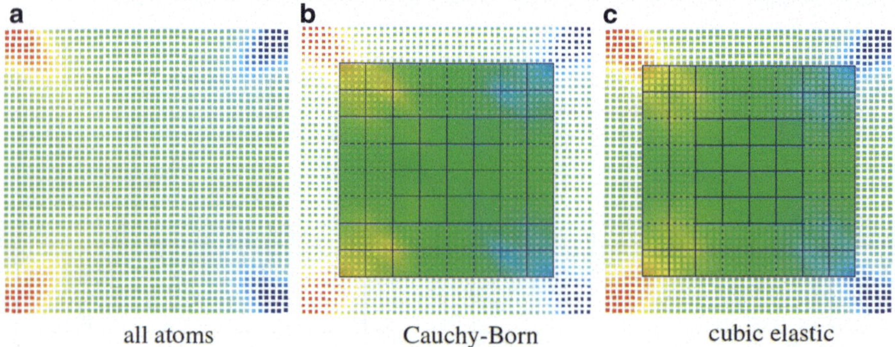

Fig. 7.8 Displacement in the x-direction: (**a**) all atoms, (**b**) CB coupled, (**c**) cubic elastic coupled. Range ± 0.2 Å

[6]The cost of the molecular statics solution is proportional to the number of atoms times the number of neighbors per atoms, while the cost of the Cauchy–Born solution is proportional to the number of elements times the number of integration points per element times the number of neighbors per atoms, while the cost of the Cauchy-Born cubic elastic solution is proportional to merely the product of the number of elements and the number of integration points per element.

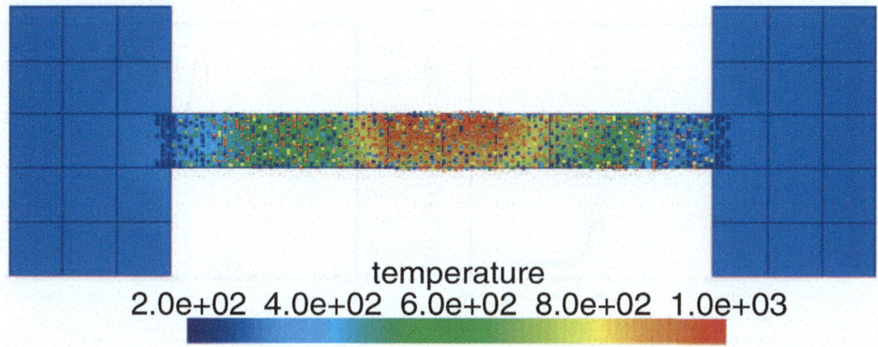

Fig. 7.9 Metallic CNT embedded in an FE mesh showing phonon temperature near the beginning of the heating phase, $t = 10\,\text{ps}$. Reproduced from [10] with permission

reproduce the all atom displacement field with considerable fidelity. We attribute this result to the fact that the coupled scheme preserves momentum and that in this zero temperature simulation the field of displacements is smooth on the scale of the mesh.

7.4.5 Laser Heating of a Carbon Nanotube

In this example a metallic (8,8) armchair carbon nanotube (CNT), 12.6 nm long, is suspended by embedding its ends in solid graphite, see Fig. 7.9 and [10], and heated with a shaped laser pulse directly heating the CNT's electron gas. We use the Tersoff potential [84, 85] to model the CNT. The graphite substrate is modeled with a continuum with the same thermal properties as the CNT. The exposed surface of the reservoirs and the tube are insulated by the air so that no heat crosses those boundaries and the remaining surfaces of the reservoirs are fixed at a constant temperature of 300 K. The electronic heat capacity has a temperature dependency: $c_e = \gamma\theta_e$ with $\gamma = 1.5\,\text{J/m}^3\,\text{K}^2$, and the electronic heat conductivity $k_e = L\sigma\theta_e$ is estimated with the Franz–Wiedemann law with $L\sigma = 2.443 \times 10^{-3}\,\text{W/m}\,\text{K}^2$. In addition to temperature dependence of the electron heat capacity and conductivity, the measured form of the electron–phonon exchange for CNTs [86] is highly non-linear in temperature, $g = h(\theta_e - \theta_p)^5$ with $h = 3.7 \times 10^4\,\text{W/m}^3\,\text{K}^5$. Also, the fact that the CNT lattice is not space filling creates no algorithmic difficulties since its contribution directly and fully determines the phonon temperature in the regions where there are atoms. In these elements all the effects of the phonon constitutive model are removed. To offset the larger fluctuations associated with basis functions with few atoms in their support we employ a time-filter, with characteristic time-scale $\tau = 0.01$ ps.

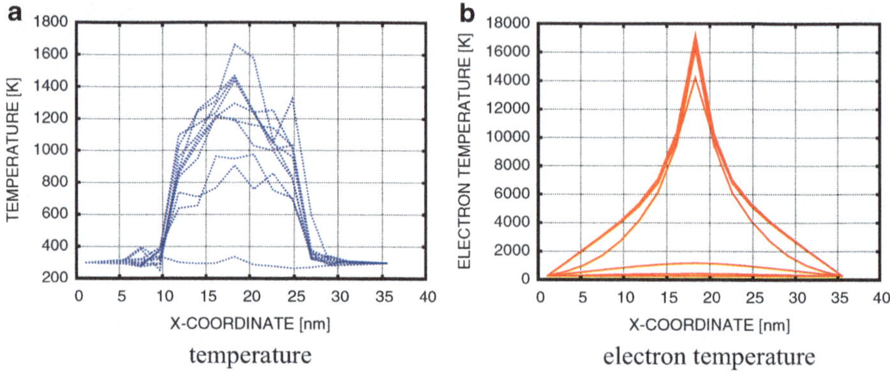

Fig. 7.10 Sequence of temperature (**a**) and electron temperature (**b**) profiles along the axis of the CNT. Reproduced from [10] with permission

The electron system of the CNT interacts with a focused radiation source that has a power input of $1.6 \times 10^{-12} \exp(-(x_1^2 + x_2^2)/(0.1\,[\mathrm{nm}])^2)\,\mathrm{W/m^3}$. We turn on this localized source for 50 ps and then allow the system to relax. The sequence of temperature profiles along the axis of the tube in Fig. 7.10 shows very localized electron and relatively diffuse phonon temperatures in correspondence with their diffusivities. These profiles through the axis of the CNT extend into FE regions without atoms; in the reservoir regions, we see a distinct change in slope due to the reservoirs' higher thermal mass, especially for the phonons. As the experiments [87] demonstrate, we expect mixed ballistic/diffusive transport in the CNT, which is modeled entirely by the MD. This mixed harmonic/enharmonic transport must transition to purely diffusive heat flux at the CNT-reservoir boundary, given the nature of the coupling. The large scale oscillations that start to become apparent at about $t = 40$ ps in Fig. 7.11 indicate that the input energy to electrons eventually excites a strong fundamental mode resonance [88, 89] which can be directly observed in Fig. 7.12.

7.5 Conclusion

We have presented the framework for the ATC methods available in LAMMPS together with illustrative examples. In contrast to other coupling methods, ATC has the distinct advantage of treating the full possibilities of thermal transport at the nanoscale since its inception [45] and provides a generalized framework of consistent multiscale balances, coarse graining, and control schemes that are applicable to a wide range of multiphysics problems. In its present state of development it is particularly suited to warm, slow processes like: predicting the growth of large nanostructures via vapor deposition, simulating the steep gradation

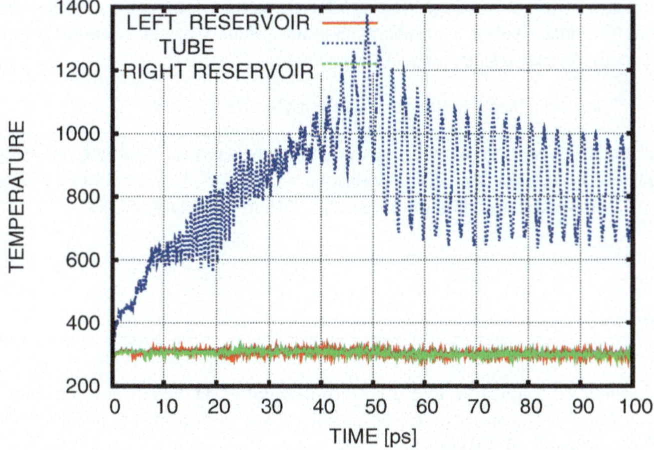

Fig. 7.11 Evolution of average temperatures of the two explicitly modeled reservoirs and the CNT. Reproduced from [10] with permission

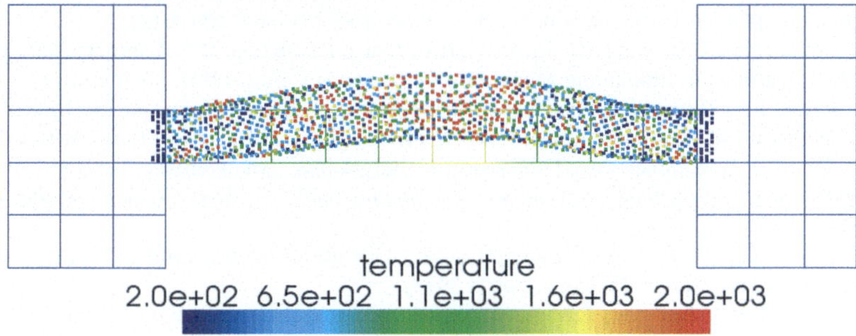

Fig. 7.12 Fundamental mode excited by focused irradiation. The atoms and the mesh are both colored by the phonon temperature. Reproduced from [10] with permission

from an electrical double layer to a bulk fluid, and modeling the deformation of polycrystalline materials with complex grain boundary structure. In general, the method is appropriate for systems with characteristic sizes reaching micrometers containing large regions of regular behavior interacting through structures requiring atomic detail. We have laid the groundwork of treating the ubiquitous problem of spurious wave reflection in the shock and dynamic regimes [70] in the ATC framework but this topic has also been an area of intense development by other researchers; see, for example, the review by Miller and Tadmor [90] and the seminal paper by Wagner and Liu [45].

The scope of the theory presented and the particular version of the methodology have not been presented elsewhere and we hope this chapter serves as a concise and coherent overview of the work we have done in the upscaling and coupling arenas. Full-fledged fluid coupling is notably absent from our exposition mainly due to the complexity introduced by treating open systems.

Acknowledgements Thanks to Greg Wagner (Northwestern University), Mike Parks (Sandia), Steve Plimpton (Sandia), Aidan Thompson (Sandia), Paul Crozier (Sandia) for assistance in developing the methods that reside in the ATC package:

http://lammps.sandia.gov/doc/fix_atc.html

Sandia is a multiprogram laboratory managed and operated by Sandia Corporation, a wholly owned subsidiary of Lockheed Martin Corporation, for the U.S. Department of Energy's National Nuclear Security Administration under contract No. DE-AC04-94AL85000.

References

1. J.A. Templeton, R.E. Jones, J.W. Lee, J.A. Zimmerman, B.M. Wong, A long-range electric field solver for molecular dynamics based on atomistic-to-continuum modeling. J. Chem. Theory Comput. **7**(6), 1736–1749 (2011)
2. J.W. Lee, J.A. Templeton, K.K. Mandadapu, J.A. Zimmerman, Comparison of molecular and primitive solvent models for electrical double layers in nanochannels. J. Chem. Theory Comput. **9**, 3051–3061 (2013)
3. R.E. Jones, D.K. Ward, J.A. Templeton, Spatial resolution of the electrical conductance of ionic fluids using a Green-Kubo method. J. Chem. Phys. **141**(18), 184110 (2014)
4. J.A. Zimmerman, E.B. Webb III, J.J. Hoyt, R.E. Jones, P.A. Klein, D.J. Bammann, Calculation of stress in atomistic simulation. Model. Simul. Mater. Sci. Eng. **12**(4), S319 (2004)
5. R.E. Jones, J.A. Zimmerman, The construction and application of an atomistic J-integral via hardy estimates of continuum fields. J. Mech. Phys. Solids **58**(9), 1318–1337 (2010)
6. R.E. Jones, J.A. Zimmerman, J. Oswald, T. Belytschko, An atomistic J-integral at finite temperature based on hardy estimates of continuum fields. J. Phys. Condens. Matter **23**(1), 015002 (2011)
7. J.A. Zimmerman, R.E. Jones, The application of an atomistic J-integral to a ductile crack. J. Phys. Condens. Matter **25**(15), 155402 (2013)
8. R.E. Jones, J.A. Templeton, T.W. Rebold, Simulated real-time detection of a small molecule on a carbon nanotube cantilever. J. Comput. Theor. Nanosci. **8**(8), 1364–1384 (2011)
9. J.A. Templeton, R.E. Jones, G.J. Wagner, Application of a field-based method to spatially varying thermal transport problems in molecular dynamics. Model. Simul. Mater. Sci. Eng. **18**(8), 085007 (2010)
10. R.E. Jones, J.A. Templeton, G.J. Wagner, D. Olmsted, N.A. Modine, Electron transport enhanced molecular dynamics for metals and semi-metals. Int. J. Numer. Methods Eng. **83**(8–9), 940–967 (2010)
11. M.H. Ulz, K.K. Mandadapu, P. Papadopoulos, On the estimation of spatial averaging volume for determining stress using atomistic methods. Model. Simul. Mater. Sci. Eng. **21**(1), 15010–15024 (2013)
12. G.I. Barenblatt, *Scaling, Self-Similarity, and Intermediate Asymptotics: Dimensional Analysis and Intermediate Asymptotics*, vol. 14 (Cambridge University Press, Cambridge, 1996)
13. LAMMPS : Large-scale Atom/Molecular Massively Parallel Simulator, Sandia National Laboratories (2015), http://lammps.sandia.gov
14. R.J.E. Clausius, On a mechanical theorem applicable to heat. Philos. Mag. **40**, 122–127 (1870)
15. J.C. Maxwell, On reciprocal figures, frames and diagrams of forces. Trans. R. Soc. Edinb. **XXVI**, 1–43 (1870)
16. J.H. Irving, J.G. Kirkwood, The statistical mechanical theory of transport processes. IV. the equations of hydrodynamics. J. Chem. Phys. **18**(6), 817–829 (1950)
17. W. Noll, Die herleitung der grundgleichungen der thermomechanik der kontinua aus der statistischen mechanik. J. Ration. Mech. Anal. **4**(5), 627–646 (1955)

18. R.B. Lehoucq, A.V. Lilienfeld-Toal, Translation of Walter Noll's "Derivation of the fundamental equations of continuum thermodynamics from statistical mechanics". J. Elast. **100**(1–2), 5–24 (2010)
19. D.H. Tsai, The virial theorem and stress calculation in molecular dynamics. J. Chem. Phys. **70**, 1375–1382 (1979)
20. P. Schofield, J.R. Henderson, Statistical mechanics of inhomogeneous fluids. Proc. R. Soc. Lond. A **379**, 231–240 (1982)
21. R.J. Hardy, Formulas for determining local properties in molecular-dynamics simulations: shock waves. J. Chem. Phys. **76**(1), 622–628 (1982)
22. J.F. Lutsko, Stress and elastic constants in anisotropic solids: molecular dynamics techniques. J. Appl. Phys. **64**(3), 1152–1154 (1988)
23. J.S. Rowlinson, B. Widom, *Molecular Theory of Capillarity* (Clarendon Press, Oxford, 1989)
24. K.S. Cheung, S. Yip, Atomic-level stress in an inhomogeneous system. J. Appl. Phys. **70**(10), 5688–5690 (1991)
25. J. Cormier, J.M. Rickman, T.J. Delph, Stress calculation in atomistic simulations of perfect and imperfect solids. J. Appl. Phys. **89**(1), 99–104 (2001)
26. M. Zhou, D.L. McDowell, Equivalent continuum for dynamically deforming atomistic particle systems. Philos. Mag. A **82**, 2547–2574 (2002)
27. M. Zhou, A new look at the atomic level virial stress: on continuum-molecular system equivalence. Proc. R. Soc. Lond. Ser. A **459**, 2347–2392 (2003)
28. A.I. Murdoch, On the microscopic interpretation of stress and couple stress. J. Elast. **71**, 105–131 (2003)
29. F. Costanzo, G.L. Gray, P.C. Andia, On the notion of average mechanical properties in md simulation via homogenization. Model. Simul. Mater. Sci. Eng. **12**, S333–S345 (2004)
30. F. Costanzo, G.L. Gray, P.C. Andia, On the definitions of effective stress and deformation gradient for use in md: Hill's macro-homogeneity and the virial theorem. Int. J. Eng. Sci. **43**, 533–555 (2005)
31. P.C. Andia, F. Costanzo, G.L. Gray, A lagrangian-based continuum homogenization approach applicable to molecular dynamics simulation. Int. J. Solids Struct. **42**, 6409–6432 (2005)
32. P.C. Andia, F. Costanzo, G.L. Gray, A classical mechanics approach to the determination of the stress-strain response of particle systems. Model. Simul. Mater. Sci. Eng. **14**, 741–757 (2006)
33. M. Zhou, Thermomechanical continuum representation of atomistic deformation at arbitrary size scales. Proc. R. Soc. Lond. Ser. A **461**, 3437–3472 (2006)
34. A.I. Murdoch, A critique of atomistic definitions of the stress tensor. J. Elast. **88**, 113–140 (2007)
35. E.B. Webb III, J.A. Zimmerman, S.C. Seel, Reconsideration of continuum thermomechanical quantities in atomic scale simulations. Math. Mech. Solids **13**, 221–266 (2008)
36. J.A. Zimmerman, R.E. Jones, J.A. Templeton, A material frame approach for evaluating continuum variables in atomistic simulations. J. Comput. Phys. **229**(6), 2364–2389 (2010)
37. N.C. Admal, E.B. Tadmor, A unified interpretation of stress in molecular systems. J. Elast. **100**, 63–143 (2010)
38. N.C. Admal, E.B. Tadmor, Stress and heat flux for arbitrary multibody potentials: a unified framework. J. Chem. Phys. **134**, 184106 (2011)
39. R.J. Hardy, A.M. Karo, Stress and energy flux in the vicinity of a shock front, in *Shock Compression of Condensed Matter. Proceedings of the American Physical Society Topical Conference* (Elsevier, Amsterdam, 1990), pp. 161–164
40. R.J. Hardy, S. Root, D.R. Swanson, Continuum properties from molecular simulations, in *12th International Conference of the American-Physical-Society-Topical-Group on Shock Compression of Condensed Matter. AIP Conference Proceedings*, vol. 620, Pt. 1 (American Institute of Physics, Melville, 2002), pp. 363–366
41. F. Rizzi, R.E. Jones, B.J. Debusschere, O.M. Knio, Uncertainty quantification in md simulations of concentration driven ionic flow through a silica nanopore. I. Sensitivity to physical parameters of the pore. J. Chem. Phys. **138**(19), 194104 (2013)
42. F. Rizzi, R.E. Jones, B.J. Debusschere, O.M. Knio, Uncertainty quantification in MD simulations of concentration driven ionic flow through a silica nanopore. II. Uncertain potential parameters. J. Chem. Phys. **138**(19), 194105 (2013)

43. A. Donev, J.B. Bell, A.L. Garcia, B.J. Alder, A hybrid particle-continuum method for hydrodynamics of complex fluids. Multiscale Model. Simul. **8**(3), 871–911 (2010)
44. G.J. Wagner, R.E. Jones, J.A. Templeton, M.L. Parks, An atomistic-to-continuum coupling method for heat transfer in solids. Comput. Methods Appl. Mech. Eng. **197**(41), 3351–3365 (2008)
45. G.J. Wagner, W.K. Liu, Coupling of atomistic and continuum simulations using a bridging scale decomposition. J. Comput. Phys. **190**(1), 249–274 (2003)
46. P. Lancaster, K. Salkauskas, Surfaces generated by moving least squares methods. Math. Comput. **37**, 141–158 (1981)
47. S. Root, R.J. Hardy, D.R. Swanson, Continuum predictions from molecular dynamics simulations: shock waves. J. Chem. Phys. **118**(7), 3161–3165 (2003)
48. W.K. Liu, S. Jun, Y.F. Zhang, Reproducing kernel particle methods. Int. J. Numer. Methods Fluids **20**(8–9), 1081–1106 (1995)
49. M.H. Ulz, S.J. Moran, Optimal kernel shape and bandwidth for atomistic support of continuum stress. Model. Simul. Mater. Sci. Eng. **21**(8), 085017 (2013)
50. K. Huang, *Statistical Mechanics*, 2nd edn. (Wiley, New York, 1987)
51. K.K. Mandadapu, J.A. Templeton, J.W. Lee, Polarization as a field variable from molecular dynamics simulations. J. Chem. Phys. **139**, 054115 (2013)
52. R.E. Miller, E.B. Tadmor, A unified framework and performance benchmark of fourteen multiscale atomistic/continuum coupling methods. Model. Simul. Mater. Sci. Eng. **17**(5), 053001 (2009)
53. S. Kohlhoff, S. Schmauder, A new method for coupled elastic-atomistic modelling, in *Atomistic Simulation of Materials* (Springer, New York, 1989), pp. 411–418
54. S. Kohlhoff, P. Gumbsch, H.F. Fischmeister, Crack propagation in bcc crystals studied with a combined finite-element and atomistic model. Philos. Mag. A **64**(4), 851–878 (1991)
55. E.B. Tadmor, M. Ortiz, R. Phillips, Quasicontinuum analysis of defects in solids. Philos. Mag. A **73**(6), 1529–1563 (1996)
56. E. Weinan, P. Ming, Cauchy–Born rule and the stability of crystalline solids: static problems. Arch. Ration. Mech. Anal. **183**(2), 241–297 (2007)
57. J.L. Ericksen, On the Cauchy–Born rule. Math. Mech. Solids **13**(3–4), 199–220 (2008)
58. J.Q. Broughton, F.F. Abraham, N. Bernstein, E. Kaxiras, Concurrent coupling of length scales: methodology and application. Phys. Rev. B **60**(4), 2391 (1999)
59. P.A. Klein, J.A. Zimmerman, Coupled atomistic–continuum simulations using arbitrary overlapping domains. J. Comput. Phys. **213**(1), 86–116 (2006)
60. F. Rizzi, M. Salloum, Y.M. Marzouk, R.G. Xu, M.L. Falk, T.P. Weihs, G. Fritz, O.M. Knio, Bayesian inference of atomic diffusivity in a binary Ni/Al system based on molecular dynamics. SIAM Multi. Model. Simul. **9**, 486–512 (2011)
61. M. Salloum, J. Templeton, Inference and uncertainty propagation of atomistically-informed continuum constitutive laws, part 1: Bayesian inference of fixed model forms. Int. J. Uncertain Quantif. **4**(2), 151–170 (2014)
62. M. Salloum, J. Templeton, Inference and uncertainty propagation of atomistically-informed continuum constitutive laws, part 2: genereralized continuum models based on gaussian processes. Intl. J. Uncertain. Quantif. **4**(2), 171–184 (2014)
63. M. Salloum, K. Sargsyan, R. Jones, B. Debusschere, H.N. Najm, H. Adalsteinsson, A stochastic multiscale coupling scheme to account for sampling noise in atomistic-to-continuum simulations. Multiscale Model. Simul. **10**(2), 550–584 (2012)
64. D.J. Evans, W.G. Hoover, B.H. Failor, B. Moran, A.J.C. Ladd, Nonequilibrium molecular dynamics via gauss's principle of least constraint. Phys. Rev. A **28**(2), 1016 (1983)
65. F. Rizzi, H.N. Najm, B.J. Debusschere, K. Sargsyan, M. Salloum, H. Adalsteinsson, O.M. Knio, Uncertainty quantification in md simulations. part I: forward propagation. Multiscale Model. Simul. **10**(4), 1428 (2012)
66. F. Rizzi, H.N. Najm, B.J. Debusschere, K. Sargsyan, M. Salloum, H. Adalsteinsson, O.M. Knio, Uncertainty quantification in md simulations. part II: Bayesian inference of force-field parameters. Multiscale Model. Simul. **10**(4), 1460 (2012)

67. P. Steinmann, A. Elizondo, R. Sunykm, Studies of validity of the Cauchy-Born rule by direct comparison of continuum and atomistic modelling. Model. Simul. Mater. Sci. Eng. **15**, S271–S281 (2007)
68. C.J. Kimmer, R.E. Jones, Continuum constitutive models from analytical free energies. J. Phys. Condens. Matter **19**(32), 326207 (2007)
69. W. Ren, Analytical and numerical study of coupled atomistic-continuum methods for fluids. J. Comput. Phys. **227**(2), 1353–1371 (2007)
70. R.E. Jones, C.J. Kimmer, Efficient non-reflecting boundary condition constructed via optimization of damped layers. Phys. Rev. B **81**(9), 094301 (2010)
71. N.W. Ashcroft, N.D. Mermin, *Solid State Physics* (Brooks-Cole, Belmont, 1976)
72. T. Schneider, E. Stoll, Molecular-dynamics study of a three-dimensional one-component model for distortive phase transitions. Phys. Rev. B **17**, 1302–1322 (1978)
73. T. Ikeshoji, B. Hafskjold, Non-equilibrium molecular dynamics calculation of heat conduction in liquid and through liquid-gas interface. Mol. Phys. **81**(2), 251–261 (1994)
74. J.E. Lennard-Jones, The determination of molecular fields I. From the variation of the viscosity of a gas with temperature. Proc. R. Soc. Lond. **106A**, 441 (1924)
75. J.E. Lennard-Jones, The determination of molecular fields II. From the equation of state of a gas. Proc. R. Soc. Lond. **106A**, 463 (1924)
76. J.D. Eshelby, The force on an elastic singularity. Philos. Trans. R. Soc. Lond. A **244**(877), 87–112 (1951)
77. J.R. Rice, A path independent integral and approximate analysis of strain concentration by notches and cracks. J. Appl. Mech. **35**(2), 379–386 (1968)
78. R. Qiao, N.R. Aluru, Ion concentrations and velocity profiles in nanochannel electroosmotic flows. J. Chem. Phys. **118**(10), 4692–4701 (2003)
79. J.W. Lee, R.H. Nilson, J.A. Templeton, S.K. Griffiths, A. Kung, B.M. Wong, Comparison of molecular dynamics with classical density functional and poisson–boltzmann theories of the electric double layer in nanochannels. J. Chem. Theory Comput. **8**(6), 2012–2022 (2012)
80. J.W. Lee, A. Mani, J.A. Templeton, Atomistic and molecular effects in electric double layers at high surface charges. Langmuir **31**(27), 7496–7502 (2015)
81. M.S. Daw, M.I. Baskes, Semiempirical, quantum mechanical calculation of hydrogen embrittlement in metals. Phys. Rev. Lett. **50**(17), 1285 (1983)
82. M.S. Daw, M.I. Baskes, Embedded-atom method: derivation and application to impurities, surfaces, and other defects in metals. Phys. Rev. B **29**(12), 6443 (1984)
83. S.M. Foiles, M.I. Baskes, M.S. Daw, Embedded-atom-method functions for the fcc metals Cu, Ag, Au, Ni, Pd, Pt, and their alloys. Phys. Rev. B **33**(12), 7983 (1986)
84. J. Tersoff, New empirical-approach for the structure and energy of covalent systems. Phys. Rev. B **37**(12), 6991–7000 (1988)
85. J. Tersoff, Modeling solid-state chemistry: interatomic potentials for multicomponent systems. Phys. Rev. B **39**, R5566–R5568 (1989)
86. T. Hertel, R. Fasel, G. Moos, Charge-carrier dynamics in single-wall carbon nanotube bundles: a time-domain study. Appl. Phys. A **75**(4), 449–465 (2002)
87. J. Wang, J.-S. Wang, Carbon nanotube thermal transport: ballistic to diffusive. Appl. Phys. Lett. **88**(11), 111909 (2006)
88. E.H. Feng, R.E. Jones, Equilibrium thermal vibrations of carbon nanotubes. Phys. Rev. B **81**(12), 125436 (2010)
89. E.H. Feng, R.E. Jones, Carbon nanotube cantilevers for next-generation sensors. Phys. Rev. B **83**(19), 195412 (2011)
90. R.E. Miller, E.B. Tadmor, A unified framework and performance benchmark of fourteen multiscale atomistic/continuum coupling methods. Model. Simul. Mater. Sci. Eng. **17**(5), 053001 (2009)

Chapter 8
Concurrent Atomistic-Continuum Simulation of Defects in Polyatomic Ionic Materials

Shengfeng Yang and Youping Chen

8.1 Introduction

The dynamic interactions between defects in materials, such as dislocations, cracks, and grain boundaries (GBs), play a crucial role in determining the properties of the materials, including ductility, strength, toughness, and hardness [1, 2]. The dynamic behavior of defects generally involves a wide range of length scales. For example, not only does the dislocation-GB interaction depend on the structural details of the GBs on the atomic-scale [3, 4] but the long-range fields of defects can also have a significant effect on the interaction. To study the mechanisms of interaction between defects in general and the interaction between defects and GBs in polycrystalline materials in particular, concurrent multiscale methods are advantageous over fully atomistic methods, as the size of the computational model can increase through coarse-graining the regions away from critical regions such as GBs. As a result, the long-range effect of defects can be considered, while the fine details of the structure and the deformation of the critical regions can be retained.

In the past two decades, a large body of literature in science and engineering has focused on the development of various multiscale computational tools [5–9], with the hope of expanding the atomistic simulation-based predictive capability from nanometers to microns. However, despite two decades of intense research efforts, a general successful multiscale methodology has not been established. In most existing concurrent multiscale methods, classical molecular dynamics (MD) and the linear elasticity-based finite element (FE) method are directly combined. Given that the linear elasticity of continua is the long wavelength limit of acoustic

S. Yang (✉) • Y. Chen
Department of Mechanical and Aerospace Engineering, University of Florida, Gainesville, FL 32611, USA
e-mail: saint628@ufl.edu; ypchen2@ufl.edu

C.R. Weinberger, G.J. Tucker (eds.), *Multiscale Materials Modeling for Nanomechanics*, Springer Series in Materials Science 245,
DOI 10.1007/978-3-319-33480-6_8

vibrations of lattice, the simulation of the dynamic behavior of polyatomic materials, i.e., materials that have two or more atoms in the primitive unit cell, has been a challenge for these combined MD/FE methods. Another challenge for the existing multiscale method is to pass defects from the atomistic to the continuum region. For example, to model polycrystalline materials, dislocations and cracks nucleated from the atomic-scale GBs where MD is used need to be smoothly propagating into the far field, where FE is usually employed for the crystalline grains. Such MD/FE interfaces prevent defect-induced discontinuities passing from the atomic to the FE region if no special numerical treatment is included.

To meet these challenges, a concurrent multiscale representation of general crystalline materials [10–14] has been formulated and a *concurrent atomistic-continuum* (CAC) method based on the formulation has been developed [15–25]. In CAC, a crystalline material is described in terms of continuously distributed lattice cells, while, within each lattice cell, a group of discrete atoms is embedded. Such a two-scale description of crystalline materials makes CAC applicable for the modeling and simulation of polyatomic materials with concurrently coupled atomistic and continuum descriptions even in the coarse-grained domain. In addition, building a continuum field formulation of a complete set of balance equations with atomistic information in the formulation ensures that both the atomistic and the continuum domains are governed by the same set of governing equations. Such a unique feature enables the coarse-grained simulation of defect dynamics in the continuum domain as well as the smooth passage of defects from the atomic-scale to the coarse scale.

This chapter reviews the fundamentals of the CAC method and demonstrates the method through simulations of the dynamics of defects in single-crystal, bi-crystal, and polycrystalline strontium titanate ($SrTiO_3$). The chapter is organized as follows. In Sect. 8.2, the CAC formulation is reviewed. In Sect. 8.3, the finite element implementation of the formulation is described. Sect. 8.4 presents three case studies on the nucleation and evolution of defects in $SrTiO_3$ (STO). A summary of the chapter is given in Sect. 8.5.

8.2 The Concurrent Atomistic-Continuum Formulation

8.2.1 Existing Formalism for Deriving Hydrodynamics Balance Equations from Molecular Variables

Irving and Kirkwood made the pioneering contribution in linking molecular and continuum descriptions of matter [26]. In their classical paper on the equations of hydrodynamics, they derived conservation equations for mass, linear momentum, and energy, using Newton's mechanics in terms of ensemble averages. Later, Hardy [27] derived formulas for continuum quantities and conservation laws for mass, momentum, and energy in terms of instantaneous quantities. The fundamental equation in their formulations is the link between microscopic quantities in phase space and the local continuum densities at a point $\mathbf{z}$ in physical space. In Hardy's formulation this link is defined as

$$\mathbf{f}(\mathbf{z},t) = \sum_{k=1}^{n} \mathbf{F}(\mathbf{r},\mathbf{p})\,\delta\left(\mathbf{R}^k - \mathbf{z}\right) \tag{8.1}$$

where $\mathbf{r} = \left\{\mathbf{R}^k, k = 1,2,\ldots,n\right\}$ and $\mathbf{p} = \left\{m^k\mathbf{V}^k, k = 1,2,\ldots,n\right\}$ are, respectively, the positions and momenta of particles, with m^k, $\mathbf{R}^k$, and $\mathbf{V}^k$ being the mass, the position, and the velocity of the kth particle in the system, respectively; $\mathbf{F}(\mathbf{r},\mathbf{p})$ is a dynamic function in the phase space, and $\mathbf{f}(\mathbf{z},t)$ is the corresponding continuum description of the local density in the physical space. The localization function $\delta\left(\mathbf{R}^k - \mathbf{z}\right)$ is any non-negative function that has a peak value at $\mathbf{R}^k = \mathbf{z}$ and becomes zero as $\left|\mathbf{R}^k - \mathbf{z}\right|$ is larger, with $\int_V \delta\left(\mathbf{R}^k - \mathbf{z}\right)d\mathbf{z} = 1$. Based on Eq. (8.1), the local densities of mass, momentum, and energy can be defined as

$$\rho = \sum_{k=1}^{n} m^k\delta\left(\mathbf{R}^k - \mathbf{z}\right), \tag{8.2}$$

$$\rho\mathbf{v} = \sum_{k=1}^{n} m^k\mathbf{V}^k\delta\left(\mathbf{R}^k - \mathbf{z}\right), \tag{8.3}$$

$$\rho e = \sum_{k=1}^{n}\left[\frac{1}{2}m^k\left(\mathbf{V}^k\right)^2 + U^k\right]\delta\left(\mathbf{R}^k - \mathbf{z}\right), \tag{8.4}$$

where U^k is the potential energy of the kth particle, and $\mathbf{v}$ is the velocity field. As an exact consequence of Newton's laws, the conservation equations for the mass, the momentum, and the energy can be found to be

$$\frac{\partial\rho}{\partial t} + \nabla_z\cdot(\rho\mathbf{v}) = 0, \tag{8.5}$$

$$\frac{\partial(\rho\mathbf{v})}{\partial t} = \nabla_z\cdot(\mathbf{t} - \rho\mathbf{v}\otimes\mathbf{v}) + \mathbf{f}, \tag{8.6}$$

$$\frac{\partial(\rho e)}{\partial t} = \nabla_z\cdot(\mathbf{q} + \mathbf{t}\cdot\mathbf{v} - \mathbf{v}\rho e) + \mathbf{f}\cdot\mathbf{v}, \tag{8.7}$$

where $\mathbf{t}$, $\mathbf{f}$, and $\mathbf{q}$ are the stress tensor, the external force vector, and the heat flux vector, respectively.

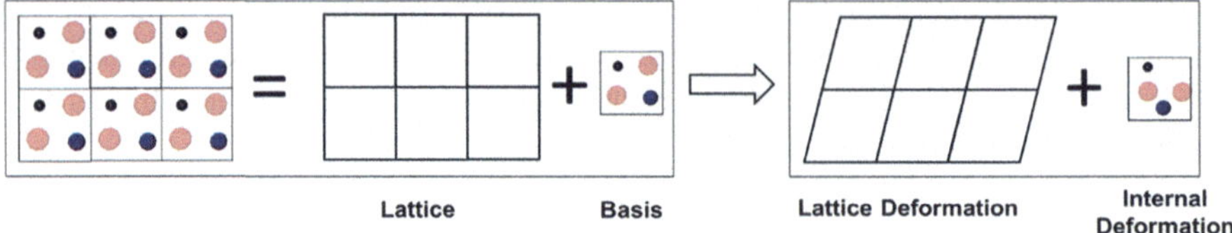

Fig. 8.1 Atomistic view of crystal structures: (**a**) Crystal structure = lattice + basis (atomic structural units); (**b**) Atomic displacement = lattice displacement + internal displacement relative to the lattice

8.2.2 The Concurrent Multiscale Formalism

Kirkwood's *statistical mechanical theory of transport processes* links the macroscopic description of thermodynamic quantities to the molecular variables, i.e., the mass, the position, and the momentum of the molecules in a molecular system with the internal motion within a molecule being ignored [26]. In the first paper of a series of 14 papers on statistical mechanical theory of transport processes, Kirkwood envisioned the extension of his formulation to molecules possessing internal degrees of freedom (DOF) [28]. Note that the idea of internal DOF can also be applied to crystalline systems. In Solid State Physics, the structure of all crystals is described in terms of continuously distributed lattice cells, but with a group of atoms located in each lattice cell [29] (Fig. 8.1); correspondingly, the atomic displacements in a general lattice can be expressed as a sum of the displacements of the lattice and the subscale internal displacements of the atoms within the lattice cells [30].

Motivated by Kirkwood's idea of molecule having internal deformation, and also by the two-level structural description of materials in Micromorphic Theory [31–40], Chen et al. [10, 12] have developed a description of crystalline materials as a continuous collection of lattice cells, while with a group of atoms being embedded within each lattice cell. This concurrent two-level materials description leads to a new link between a continuously distributed local density function $\mathbf{f}(\mathbf{x}, \mathbf{y}^{\alpha}, t)$ in the physical space and a phase space dynamical function $\mathbf{F}(\mathbf{R}^{k\lambda}, \mathbf{V}^{k\lambda})$, with

$$\mathbf{f}(\mathbf{x}, \mathbf{y}^{\alpha}, t) = \sum_{k=1}^{n} \sum_{\lambda=1}^{\nu} \mathbf{F}\left(\mathbf{R}^{k\lambda}, \mathbf{V}^{k\lambda}\right) \delta\left(\mathbf{R}^{k} - \mathbf{x}\right) \tilde{\delta}\left(\Delta\mathbf{r}^{k\lambda} - \mathbf{y}^{\alpha}\right), \tag{8.8}$$

where $\mathbf{R}^k$ is the position vector of the mass center of the kth unit cell, $\Delta\mathbf{r}^{k\lambda}$ the internal position of the λth atom within the kth unit cell, $\mathbf{R}^{k\lambda} = \mathbf{R}^k + \Delta\mathbf{r}^{k\lambda}$, $\mathbf{V}^{k\lambda} = \dot{\mathbf{R}}^{k\lambda}$; $\mathbf{x}$ is the physical space coordinate vector and $\mathbf{y}^{\alpha}$ ($\alpha = 1, 2, \ldots \nu$, in which ν denotes the number of atoms within a unit cell) is an internal variable describing the internal position of atom α located within the lattice point at $\mathbf{x}$. The localization functions satisfy

$$\int_{V(\mathbf{x})} \delta\left(\mathbf{R}^k - \mathbf{x}\right) d^3\mathbf{x} = 1 \quad (k = 1, 2, 3, \ldots n), \tag{8.9}$$

and

$$\tilde{\delta}\left(\Delta\mathbf{r}^{k\lambda} - \mathbf{y}^\alpha\right) = \int_{V(\mathbf{y}^\alpha)} \left[\delta\left(\Delta\mathbf{r}^{k\lambda} - \mathbf{y}\right)\right] d\mathbf{y} = \begin{cases} 1 & \text{if} \quad \Delta\mathbf{r}^{k\lambda} = \mathbf{y}^\alpha \text{ or } \lambda = \alpha \\ 0 & \text{if} \quad \Delta\mathbf{r}^{k\lambda} \neq \mathbf{y}^\alpha \text{ or } \lambda \neq \alpha \end{cases} \tag{8.10}$$

With the identities:

$$\begin{aligned} \nabla_{\mathbf{R}^k}\delta\left(\mathbf{R}^k - \mathbf{x}\right) &= -\nabla_{\mathbf{x}}\delta\left(\mathbf{R}^k - \mathbf{x}\right) \\ \nabla_{\Delta\mathbf{r}^{k\lambda}}\tilde{\delta}\left(\Delta\mathbf{r}^{k\lambda} - \mathbf{y}^\alpha\right) &= -\nabla_{\mathbf{y}^\alpha}\tilde{\delta}\left(\Delta\mathbf{r}^{k\lambda} - \mathbf{y}^\alpha\right), \end{aligned} \tag{8.11}$$

the time evolution of a local density defined in Eq. (8.8) can be expressed as

$$\begin{aligned} \frac{\partial \mathbf{f}(\mathbf{x},\mathbf{y}^\alpha,t)}{\partial t}\Big|_{\mathbf{x},\mathbf{y}^\alpha} = &\sum_{k=1}^{n}\sum_{\xi=1}^{\upsilon} \dot{\mathbf{F}}(\mathbf{r},\mathbf{p})\,\delta\left(\mathbf{R}^k - \mathbf{x}\right)\tilde{\delta}\left(\Delta\mathbf{r}^{k\lambda} - \mathbf{y}^\alpha\right) \\ &-\nabla_{\mathbf{x}}\cdot\sum_{k=1}^{n}\sum_{\lambda=1}^{\upsilon} \mathbf{V}^k \otimes \mathbf{F}(\mathbf{r},\mathbf{p})\,\delta\left(\mathbf{R}^k - \mathbf{x}\right)\tilde{\delta}\left(\Delta\mathbf{r}^{k\lambda} - \mathbf{y}^\alpha\right) \\ &-\nabla_{\mathbf{y}^\alpha}\cdot\sum_{k=1}^{n}\sum_{\lambda=1}^{\upsilon} \Delta\mathbf{v}^{k\alpha} \otimes \mathbf{F}(\mathbf{r},\mathbf{p})\,\delta\left(\mathbf{R}^k - \mathbf{x}\right)\tilde{\delta}\left(\Delta\mathbf{r}^{k\lambda} - \mathbf{y}^\alpha\right), \end{aligned} \tag{8.12}$$

where $\mathbf{V}^k = \dot{\mathbf{R}}^k$ and $\Delta\mathbf{v}^{k\lambda} = \Delta\dot{\mathbf{r}}^{k\lambda}$. When $\mathbf{f}(\mathbf{x},\mathbf{y}^\alpha,t)$ is the local density of a conserved quantity, Eq. (8.12) represents the corresponding conservation equation.

8.2.3 The Multiscale Representation of Balance Laws

The new link defined in Eq. (8.8) leads to a new definition of the local density of the mass, the momentum, and the energy

$$\rho^\alpha = \sum_{k=1}^{n}\sum_{\lambda=1}^{\nu} m^\lambda \delta\left(\mathbf{R}^k - \mathbf{x}\right)\tilde{\delta}\left(\Delta\mathbf{r}^{k\lambda} - \mathbf{y}^\alpha\right), \tag{8.13}$$

$$\rho^\alpha\left(\mathbf{v} + \Delta\mathbf{v}^\alpha\right) = \sum_{k=1}^{n}\sum_{\lambda=1}^{\nu} m^\lambda \mathbf{V}^{k\lambda}\delta\left(\mathbf{R}^k - \mathbf{x}\right)\tilde{\delta}\left(\Delta\mathbf{r}^{k\lambda} - \mathbf{y}^\alpha\right), \tag{8.14}$$

$$\rho^\alpha e^\alpha = \sum_{k=1}^{n}\sum_{\lambda=1}^{\upsilon}\left[\frac{1}{2}m^\lambda\left(\mathbf{V}^{k\lambda}\right)^2 + U^{k\lambda}\right]\delta\left(\mathbf{R}^k - \mathbf{x}\right)\tilde{\delta}\left(\Delta\mathbf{r}^{k\lambda} - \mathbf{y}^\alpha\right). \tag{8.15}$$

Following from Eq. (8.12), the time evolution laws for the conserved quantities, mass, linear momentum, and energy, can be, respectively, derived as

$$\begin{aligned}\frac{\partial\rho^\alpha}{\partial t} &= -\nabla_{\mathbf{x}}\cdot\sum_{k=1}^{n}\sum_{\lambda=1}^{\upsilon} m^\alpha\mathbf{V}^k\delta\left(\mathbf{R}^k - \mathbf{x}\right)\tilde{\delta}\left(\Delta\mathbf{r}^{k\lambda} - \mathbf{y}^\alpha\right)\\ &\quad - \nabla_{\mathbf{y}^\alpha}\cdot\lambda\sum_{k=1}^{l}\sum_{\lambda=1}^{\upsilon} m^\alpha\Delta\mathbf{v}^{k\lambda}\delta\left(\mathbf{R}^k - \mathbf{x}\right)\tilde{\delta}\left(\Delta\mathbf{r}^{k\lambda} - \mathbf{y}^\alpha\right),\end{aligned} \tag{8.16}$$

$$\begin{aligned}\frac{\partial(\rho^\alpha(\mathbf{v}+\Delta\mathbf{v}^\alpha))}{\partial t} &= \sum_{k=1}^{n}\sum_{\lambda=1}^{\upsilon} m^\lambda\dot{\mathbf{V}}^{k\lambda}\delta\left(\mathbf{R}^k - \mathbf{x}\right)\tilde{\delta}\left(\Delta\mathbf{r}^{k\lambda} - \mathbf{y}^\alpha\right)\\ &- \nabla_{\mathbf{x}}\cdot\left(\sum_{k=1}^{n}\sum_{\lambda=1}^{\upsilon}\tilde{\mathbf{V}}^k\otimes m^\lambda\tilde{\mathbf{V}}^{k\lambda}\delta\left(\mathbf{R}^k - \mathbf{x}\right)\tilde{\delta}\left(\Delta\mathbf{r}^{k\lambda} - \mathbf{y}^\alpha\right) + \mathbf{v}\otimes\rho\left(\mathbf{v}+\Delta\mathbf{v}^\alpha\right)\right)\\ &-\nabla_{\mathbf{y}^\alpha}\cdot\left(\sum_{k=1}^{n}\sum_{\lambda=1}^{\upsilon}\Delta\tilde{\mathbf{v}}^{k\lambda}\otimes m^\lambda\tilde{\mathbf{V}}^{k\lambda}\delta\left(\mathbf{R}^k-\mathbf{x}\right)\tilde{\delta}\left(\Delta\mathbf{r}^{k\lambda}-\mathbf{y}^\alpha\right)+\Delta\mathbf{v}^\alpha\otimes\rho\left(\mathbf{v}+\Delta\mathbf{v}^\alpha\right)\right)\end{aligned} \tag{8.17}$$

$$\begin{aligned}\frac{\partial(\rho^\alpha e^\alpha)}{\partial t} &= \sum_{k=1}^{n}\sum_{\lambda=1}^{\upsilon}\left(m^\lambda\mathbf{V}^{k\lambda}\dot{\mathbf{V}}^{k\lambda} + \dot{U}^{k\lambda}\right)\delta\left(\mathbf{R}^k - \mathbf{x}\right)\tilde{\delta}\left(\Delta\mathbf{r}^{k\lambda} - \mathbf{y}^\alpha\right) + \mathbf{f}^\alpha\cdot\left(\mathbf{v}+\Delta\mathbf{v}^\alpha\right)\\ &- \nabla_{\mathbf{x}}\cdot\left(\sum_{k=1}^{n}\sum_{\lambda=1}^{\upsilon}\tilde{\mathbf{V}}^k\left[\frac{1}{2}m^\lambda\left(\mathbf{V}^{k\lambda}\right)^2 + U^{k\lambda}\right]\delta\left(\mathbf{R}^k - \mathbf{x}\right)\tilde{\delta}\left(\Delta\mathbf{r}^{k\lambda} - \mathbf{y}^\alpha\right) + \mathbf{v}\rho^\alpha e^\alpha\right)\\ &-\nabla_{\mathbf{y}^\alpha}\cdot\left(\sum_{k=1}^{n}\sum_{\lambda=1}^{\upsilon}\Delta\tilde{\mathbf{v}}^{k\lambda}\left[\frac{1}{2}m^\lambda\left(\mathbf{V}^{k\lambda}\right)^2+U^{k\lambda}\right]\delta\left(\mathbf{R}^k-\mathbf{x}\right)\tilde{\delta}\left(\Delta\mathbf{r}^{k\lambda}-\mathbf{y}^\alpha\right)+\Delta\mathbf{v}^\alpha\rho^\alpha e^\alpha\right)\end{aligned} \tag{8.18}$$

Note that for polyatomic crystalline materials, the momentum flux and the heat flux can be decomposed into components due to homogeneous lattice deformation (denoted as $\mathbf{t}^\alpha$ and $\mathbf{q}^\alpha$, respectively) and that due to rearrangements of atoms within lattice cells (denoted as $\boldsymbol{\tau}^\alpha$ and $\mathbf{j}^\alpha$, respectively). It follows

$$\begin{aligned}\mathbf{t}^\alpha &= -\sum_{k=1}^{n}\tilde{\mathbf{V}}^k\otimes\left(m^\alpha\tilde{\mathbf{V}}^{k\alpha}\right)\delta\left(\mathbf{R}^k - \mathbf{x}\right)\tilde{\delta}\left(\Delta\mathbf{r}^{k\alpha} - \mathbf{y}^\alpha\right)\\ &\quad -\frac{1}{2}\sum_{k,l=1}^{n}\sum_{\lambda,\eta=1}^{\upsilon}\left(\mathbf{R}^k-\mathbf{R}^l\right)\otimes\overset{\frown}{\mathbf{F}}^{k\lambda}A\left(k,\lambda,l,\eta,\mathbf{x},\mathbf{y}^\alpha\right),\end{aligned} \tag{8.19}$$

$$\boldsymbol{\tau}^{\alpha} = -\sum_{k=1}^{n} \Delta\tilde{\mathbf{v}}^{k\alpha} \otimes \left(m^{\alpha}\tilde{\mathbf{V}}^{k\alpha}\right)\delta\left(\mathbf{R}^{k}-\mathbf{x}\right)\tilde{\delta}\left(\Delta\mathbf{r}^{k\alpha}-\mathbf{y}^{\alpha}\right) \\ -\frac{1}{2}\sum_{k,l=1}^{n}\sum_{\lambda,\eta=1}^{\nu}\left(\Delta\mathbf{r}^{k\lambda}-\Delta\mathbf{r}^{l\eta}\right)\otimes\overset{\frown k\lambda}{\mathbf{F}} A\left(k,\lambda,l,\eta,\mathbf{x},\mathbf{y}^{\alpha}\right), \tag{8.20}$$

$$\mathbf{q}^{\alpha} = -\sum_{k=1}^{n} \tilde{\mathbf{V}}^{k}\left[\frac{1}{2}m^{\alpha}\left(\tilde{\mathbf{V}}^{k\alpha}\right)^{2}+U^{k\alpha}\right]\delta\left(\mathbf{R}^{k}-\mathbf{x}\right)\tilde{\delta}\left(\Delta\mathbf{r}^{k\alpha}-\mathbf{y}^{\alpha}\right) \\ -\frac{1}{2}\sum_{k,l=1}^{n}\sum_{\lambda,\eta=1}^{\nu}\left(\mathbf{R}^{k}-\mathbf{R}^{l}\right)\tilde{\mathbf{V}}^{k\lambda}\cdot\overset{\frown k\lambda}{\mathbf{F}} A\left(k,\lambda,l,\eta,\mathbf{x},\mathbf{y}^{\alpha}\right), \tag{8.21}$$

$$\mathbf{j}^{\alpha} = -\sum_{k=1}^{n} \Delta\tilde{\mathbf{v}}^{k\alpha}\left[\frac{1}{2}m^{\alpha}\left(\tilde{\mathbf{V}}^{k\alpha}\right)^{2}+U^{k\alpha}\right]\delta\left(\mathbf{R}^{k}-\mathbf{x}\right)\tilde{\delta}\left(\Delta\mathbf{r}^{k\alpha}-\mathbf{y}^{\alpha}\right) \\ -\frac{1}{2}\sum_{k,l=1}^{n}\sum_{\lambda,\eta=1}^{\nu}\left(\Delta\mathbf{r}^{k\lambda}-\Delta\mathbf{r}^{l\eta}\right)\tilde{\mathbf{V}}^{k\lambda}\cdot\overset{\frown k\lambda}{\mathbf{F}} A\left(k,\lambda,l,\eta,\mathbf{x},\mathbf{y}^{\alpha}\right), \tag{8.22}$$

where $\tilde{\mathbf{V}}^{k\alpha} = \mathbf{V}^{k\alpha} - \mathbf{v} - \Delta\mathbf{v}^{\alpha}$, $\tilde{\mathbf{V}}^{k} = \mathbf{V}^{k} - \mathbf{v}$, $\Delta\tilde{\mathbf{V}}^{k\alpha} = \Delta\mathbf{v}^{k\alpha} - \Delta\mathbf{v}^{\alpha}$, $\overset{\frown k\lambda}{\mathbf{F}} = -\frac{1}{2}\sum_{l,m}\sum_{\eta,\varsigma}\frac{\partial\left(U_{k\lambda}^{l\eta}+U_{m\varsigma}^{l\eta}\right)}{\partial\mathbf{R}^{k\lambda}}$, $U_{k\lambda}^{l\eta}$ is the interatomic potential for systems involving three-body interaction forces, and

$$A\left(k,\lambda,l,\eta,\mathbf{x},\mathbf{y}^{\alpha}\right) \equiv \int_{0}^{1}\left[\delta\left(\mathbf{R}^{l}\left(1-\lambda\right)+\mathbf{R}^{k}\lambda-\mathbf{x}\right)\tilde{\delta}\left(\Delta\mathbf{r}^{l\eta}\left(1-\lambda\right)-\Delta\mathbf{r}^{k\lambda}\lambda-\mathbf{y}^{\alpha}\right)\right]d\lambda, \tag{8.23}$$

is a line function that connecting two particle $k\lambda$ and $l\eta$ and passing through point $\mathbf{x}+\mathbf{y}^{\alpha}$.

Substituting the momentum flux, i.e., stress tensor, and the heat flux vectors into Eqs. (8.16), (8.17), and (8.18), we obtain the complete field representation of the balance laws for a general atomistic system:

$$\frac{\partial\rho^{\alpha}}{\partial t} = -\nabla_{\mathbf{x}}\cdot\left(\rho^{\alpha}\mathbf{v}\right)-\nabla_{\mathbf{y}^{\alpha}}\cdot\left(\rho^{\alpha}\Delta\mathbf{v}^{\alpha}\right), \tag{8.24}$$

$$\frac{\partial\rho^{\alpha}(\mathbf{v}+\Delta\mathbf{v}^{\alpha})}{\partial t} = \nabla_{\mathbf{x}}\cdot\left(\mathbf{t}^{\alpha}-\rho^{\alpha}\mathbf{v}\otimes\left(\mathbf{v}+\Delta\mathbf{v}^{\alpha}\right)\right) \\ +\nabla_{\mathbf{y}^{\alpha}}\cdot\left(\boldsymbol{\tau}^{\alpha}-\rho^{\alpha}\Delta\mathbf{v}^{\alpha}\otimes\left(\mathbf{v}+\Delta\mathbf{v}^{\alpha}\right)\right)+\mathbf{f}^{\alpha}, \tag{8.25}$$

$$\begin{aligned}\frac{\partial}{\partial t}\left(\rho^\alpha e^\alpha\right) = \nabla_{\mathbf{x}} \cdot \left(\mathbf{q}^\alpha + \mathbf{t}^\alpha \cdot \left(\mathbf{v} + \Delta\mathbf{v}^\alpha\right) - \mathbf{v}\rho^\alpha e^\alpha\right) \\ + \nabla_{\mathbf{y}^\alpha} \cdot \left(\mathbf{j}^\alpha + \boldsymbol{\tau}^\alpha \cdot \left(\mathbf{v} + \Delta\mathbf{v}^\alpha\right) - \Delta\mathbf{v}^\alpha \rho^\alpha e^\alpha\right) + \mathbf{f}^\alpha \cdot \left(\mathbf{v} + \Delta\mathbf{v}^\alpha\right),\end{aligned} \tag{8.26}$$

where ρ^α, $\rho^\alpha\left(\mathbf{v} + \Delta\mathbf{v}^\alpha\right)$, and $\rho^\alpha e^\alpha$ are the local density of mass, momentum, and energy, respectively; $\mathbf{v}$ is the homogeneous velocity field; $\mathbf{v} + \Delta\mathbf{v}^\alpha = \rho^\alpha\left(\mathbf{v} + \Delta\mathbf{v}^\alpha\right)/\rho^\alpha$ is the atomic-level velocity; and $\mathbf{f}^\alpha$ is the external force.

Expressing the material time derivative of tensor $\mathbf{S}^\alpha = \mathbf{F}\left(\mathbf{x}, \mathbf{y}^\alpha, t\right)$ as

$$\frac{d\mathbf{S}^\alpha}{dt} \equiv \frac{\partial \mathbf{S}^\alpha}{\partial t} + \left(\mathbf{v} \cdot \nabla_{\mathbf{x}}\right)\mathbf{S}^\alpha + \left(\Delta\mathbf{v}^\alpha \cdot \nabla_{\mathbf{y}^\alpha}\right)\mathbf{S}^\alpha, \tag{8.27}$$

we can rewrite the balance equation of linear momentum, i.e., Eq. (8.25), in a different form:

$$\rho^\alpha \frac{d}{dt}\left(\mathbf{v} + \Delta\mathbf{v}^\alpha\right) = \nabla_{\mathbf{x}}\mathbf{t}^\alpha + \nabla_{\mathbf{y}^\alpha}\boldsymbol{\tau}^\alpha + \mathbf{f}^\alpha. \tag{8.28}$$

8.3 Numerical Implementation

Using the classical definitions of the internal force density [11] and the general definition of kinetic temperature [17], the balance equation of linear momentum, Eq. (8.28), can be further rewritten as:

$$\rho^\alpha \frac{d}{dt}\left(\mathbf{v} + \Delta\mathbf{v}^\alpha\right) = \mathbf{f}^\alpha_{\text{int}} + \mathbf{f}^\alpha_{\text{ext}} - \frac{\gamma^\alpha k_B}{\Delta V}\nabla_{\mathbf{x}}T, \tag{8.29}$$

or

$$\rho^\alpha \ddot{\mathbf{u}}^\alpha(\mathbf{x}) = \mathbf{f}^\alpha_{\text{int}}(\mathbf{x}) + \mathbf{f}^\alpha_{\text{ext}} - \mathbf{f}^\alpha_T, \tag{8.30}$$

where

$$\rho^\alpha(\mathbf{x}, t) = m^\alpha / \Delta V(\mathbf{x}). \tag{8.31}$$

Here, m^α is the mass of αth atom; $\Delta V(\mathbf{x})$ is the volume of the unit cell at $\mathbf{x}$; $\mathbf{u}^\alpha(\mathbf{x})$ is the displacement vector of the αth atom within the unit cell at $\mathbf{x}$; k_B is the Boltzmann constant; $\gamma^\alpha = \frac{m^\alpha}{M\Delta V}$; $M = \sum_{\alpha=1}^{\nu} m^\alpha$ is the sum of the masses of atoms in a unit cell; $\mathbf{f}^\alpha_{ext} = \mathbf{f}^\alpha$ represents the external force field acting on the αth atom, and the internal force density

$$\mathbf{f}^\alpha_{\text{int}}(\mathbf{x}) = \int_{\Omega(\mathbf{s})} \sum_{\beta=1}^{\nu} \mathbf{f}\left(\mathbf{u}^\alpha(\mathbf{x}) - \mathbf{u}^\beta(\mathbf{s})\right) ds \tag{8.32}$$

is a nonlocal and nonlinear function of the relative displacements. This function of the internal force field and its related parameters can be obtained from fitting to quantum mechanical calculations or to experimental measurements. They can also be derived from empirical interatomic potentials.

It should be noted that in both conventional dynamic finite element method and the state-of-the-art molecular dynamics method, the dominant computational cost is the calculation of the internal force. In particular, the calculation of the internal force in ionic materials is most expensive, since ionic interactions involve the long-range Coulomb force. Efficient computational methods for calculating Coulomb force, such as the Ewald summation method and the Particle–Particle Particle–Mesh method, have been developed based on periodic boundary condition. As a result, they are only suitable for periodical models. To provide a computationally efficient and highly accurate method for non-periodic models, Fennell et al. [41] extended the shifted-form method to mimic the long-ranged ordering. This is realized by including the distance depending damping function. The idea is based on a damped and cutoff-neutralized Coulomb sum proposed by Wolf et al. [42]. This method (called Wolf method) can drastically reduce the range of direct electrostatic interactions. At the same time, it makes sure that a violent step in the force and the potential energy will be avoided at the cutoff. In CAC, the Wolf method is implemented to calculate the long-range Coulomb force.

It is noticed from Eqs. (8.29) or (8.30) that there is a temperature gradient term in the balance equation of linear momentum. For systems with a given temperature field, i.e., a homogeneous temperature field or a steady state temperature field with a constant temperature gradient, Eq. (8.30) can serve as the governing equation for the evolution of the displacement field in space and time. In particular, for systems with a given homogeneous (constant) temperature field, the temperature term only acts as a surface traction on the boundaries and can only account for the effect of thermal expansion or contraction. Note that the value of the Boltzmann constant k_B ($1.3806488 \times 10^{-23}$ $\mathrm{m^2\ kgs^{-2}\ K^{-1}}$) is extremely small. Thus, in such situation, the temperature effect is negligible. On the other hand, for coupled thermal-mechanical problems, i.e., when there is heat generation during plastic deformation or when there are thermally activated dislocations, both the linear momentum equation and the energy equation then need to be solved for both the displacement field and the local temperature. The focus of this chapter is mechanical problem only. Therefore, we only need to solve the balance equation of linear momentum.

To numerically solve the governing equation, Eq. (8.30), we discretize a crystalline solid at lattice level into finite elements (shown in Fig. 8.2), with each finite element (FE) node corresponding to a primitive unit cell. Following the standard procedure of the finite element method, we approximate the displacement field using a conventional FE shape function. It follows

$$\widehat{\mathbf{u}}^{\alpha}(\mathbf{x}) = \mathbf{\Phi}_{\lambda}(\mathbf{x})\,\mathbf{U}_{\lambda}{}^{\alpha}, \quad \lambda = 1, 2, \ldots n, \tag{8.33}$$

where $\hat{\mathbf{u}}^{\alpha}(\mathbf{x})$ is the approximate displacement field with an element, $\mathbf{\Phi}_{\lambda}(\mathbf{x})$ is the shape function, and $\mathbf{U}_{\lambda}^{\alpha}$ is the nodal displacement vector of the αth atom at the λth

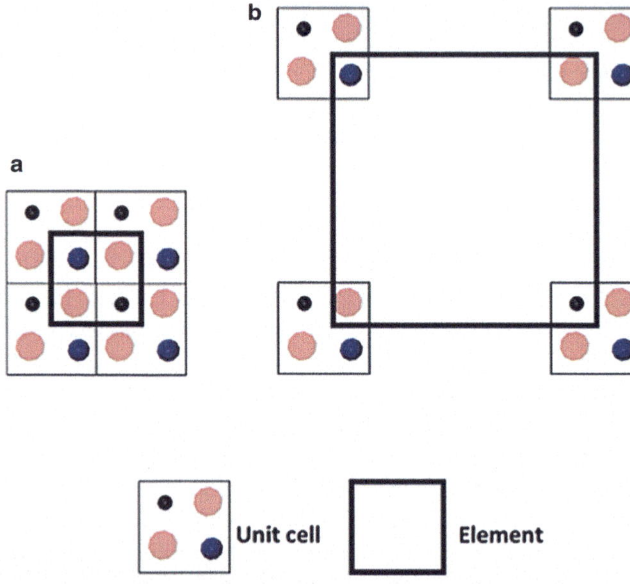

Fig. 8.2 Element with different sizes: (**a**) smallest element; (**b**) coarse element (reprinted from [48]. Copyright (2015) by Taylor & Francis Group)

element node. The weak form of Eq. (8.30) can be expressed as,

$$\int_{\Omega(\mathbf{x})} \rho^{\alpha} \mathbf{\Phi}_{\eta}(\mathbf{x}) \mathbf{\Phi}_{\lambda}(\mathbf{x})\, d\mathbf{x} \ddot{\mathbf{U}}_{\lambda}^{\alpha} = \int_{\Omega(\mathbf{x})} \mathbf{\Phi}_{\eta}(\mathbf{x}) \mathbf{f}_{\text{int}}^{\alpha}(\mathbf{x})\, d\mathbf{x} + \int_{\Omega(\mathbf{x})} \mathbf{\Phi}_{\eta}(\mathbf{x}) \left(\mathbf{f}_{\text{ext}}^{\alpha} + \mathbf{f}_{\text{T}}^{\alpha}\right) d\mathbf{x}, \tag{8.34}$$

where

$$\int_{\Omega(\mathbf{x})} \mathbf{\Phi}_{\eta}(\mathbf{x}) \mathbf{f}_{\text{int}}^{\alpha}(\mathbf{x})\, d\mathbf{x} = \int_{\Omega(\mathbf{x})} \mathbf{\Phi}_{\eta}(\mathbf{x}) \int_{\Omega(\mathbf{y})} \sum_{\beta=1}^{\upsilon} \mathbf{f}\left[\mathbf{\Phi}_{\lambda}(\mathbf{x}) \mathbf{U}_{\lambda}{}^{\alpha} - \mathbf{\Phi}_{\lambda}(\mathbf{y}) \mathbf{U}_{\lambda}{}^{\beta}\right] d\mathbf{y} d\mathbf{x}.$$

Equation (8.34) can be written in a matrix form as

$$\mathbf{M}^{\alpha} \ddot{\mathbf{U}}^{\alpha} = \mathbf{F}_{\text{int}}^{\alpha} + \mathbf{F}_{(\text{ext}+T)}^{\alpha} \tag{8.35}$$

in which

$$\mathbf{M}^{\alpha} = \int_{\Omega(\mathbf{x})} \rho^{\alpha} \mathbf{\Phi}_{\eta}(\mathbf{x}) \mathbf{\Phi}_{\lambda}(\mathbf{x})\, d\mathbf{x}, \tag{8.36}$$

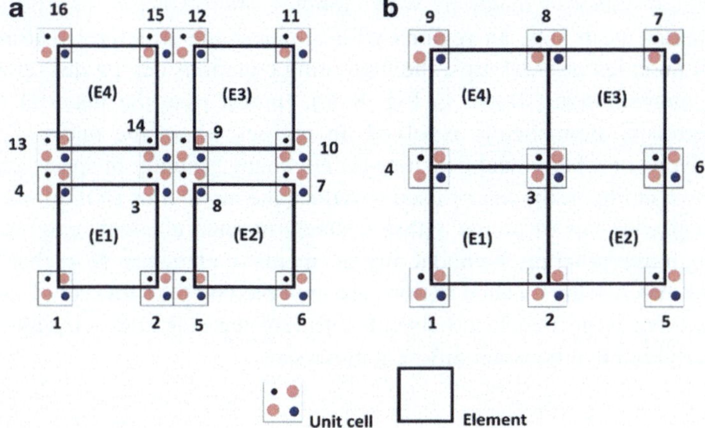

Fig. 8.3 Four elements without and with connectivity: (**a**) elements without connectivity; (**b**) elements with connectivity

$$\mathbf{F}^{\alpha}_{\text{int}} = \int_{\Omega(\mathbf{x})} \mathbf{\Phi}_{\eta}(\mathbf{x}) \int_{\Omega(\mathbf{y})} \sum_{\beta=1}^{\upsilon} \mathbf{f}\left[\mathbf{\Phi}_{\lambda}(\mathbf{x})\,\mathbf{U}_{\lambda}{}^{\alpha} - \mathbf{\Phi}_{\lambda}(\mathbf{y})\,\mathbf{U}_{\lambda}{}^{\beta}\right] d\mathbf{y}d\mathbf{x}, \tag{8.37}$$

$$\mathbf{F}^{\alpha}_{(\text{ext}+T)} = \int_{\Omega(\mathbf{x})} \mathbf{\Phi}_{\eta}(\mathbf{x}) \left(\mathbf{f}^{\alpha}_{\text{ext}} + \mathbf{f}^{\alpha}_{T}\right) d\mathbf{x}. \tag{8.38}$$

The values of the integrals can be numerically computed through Gauss quadrature following the procedure of standard FE method. The discretized governing equation, i.e., Equation (8.35), is the ordinary differential equations of the second order in time, and can be solved using the central-difference algorithm.

We would like to mention again that in CAC, the interatomic potential is the only constitutive relation. This is different from classic continuum mechanics-based or generalized continuum mechanics-based methods or many existing mesoscale simulation methods. The CAC simulation tool can simulate any crystalline materials and the deformation processes as long as the interatomic potential that describes the atomic interaction in the material is available. Moreover, because the internal force–displacement relationship is nonlocal, the connectivity between finite elements is not required in CAC. Figure 8.3 shows the comparison between elements with connectivity and those without connectivity. Without the need of connectivity between elements, nucleation and propagation of cracks and/or dislocations can be simulated through separation and sliding between elements, as direct consequences of the governing equations and the microstructure of the materials. No additional rules or empirical parameters are needed for modeling and simulation of the initiation and propagation of defects or of the interactions between defects.

Since the governing equations are constructed in terms of local densities, different meshes can be used in regions of different structural features. Multiple morphological scales in materials with complex microstructure can be modeled using different mesh size. In regions with atomic-scale structural features, such as grain boundaries or crack tips, the governing equations can be discretized with the finest element size (shown in Fig. 8.2a). In that case the material behavior of that region is atomistically resolved. In regions where the material deforms collaboratively, or where local densities do not change rapidly in space, coarse FE mesh (shown in Fig. 8.2b) can be used to reduce the number of DOF of the system, and the displacements of atoms within coarsely meshed elements can be mapped back through interpolating the nodal displacements of elements. Note that both the atomic and the coarsely meshed regions are governed by the same set of governing equations, there is no interface between different theoretical descriptions for the same material but that between different mesh sizes.

8.4 Computational Models and Simulation Results

Perovskite oxides are a very special class of functional materials that have a broad application. Strontium titanate ($SrTiO_3$, STO) is one of the most important perovskite oxides. It is one of the most commonly used substrates, has applications in dielectric and optical devices, and is a potentially important thermoelectric material [43]. A fundamental understanding of the dynamics of defects in STO is crucial, as defects can significantly affect functionality [44, 45]. In this section, we present three case studies to illustrate and to demonstrate the CAC method through simulations and predictions of the dynamic behavior of GBs, cracks, and dislocations in STO [46–48].

To test the efficiency and the accuracy of the CAC method with respect to fully atomistic molecular dynamics (MD) simulation, the open-source software LAMMPS [49] is used for the corresponding MD simulations. The rigid-ion potential for STO developed by Thomas et al. [50] is employed to describe the force field in both the CAC and MD models. This potential includes the short-range pair interactions represented by the Born–Mayer model and long-range Coulomb interactions.

8.4.1 Case I: Prediction of the Equilibrium Grain Boundary Structures

Grain boundaries are planar defects that connect abutting grains with various crystallographic orientations. The ratio between the number of lattice points in the unit cell of a *coincident site lattice* (CSL) and the number of lattice points in a unit cell of the generating lattice is called Sigma ($\sum$), which describes the degree of fit between two neighboring grains. To describe a GB geometrically, there are eight

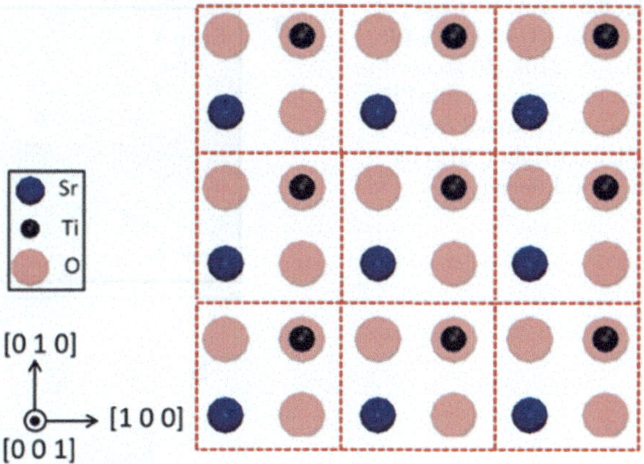

Fig. 8.4 Illustration of unit cells of STO viewed from [001] (reprinted from [47]. Copyright (2015) by the Royal Society)

DOFs, among which three DOFs are used to define the normal of a GB plane, one to define the rotation angle, one to define the twist angle between the lattices, and three to define the rigid-body translation states that determine the relative position of the two grains.

The CAC method is applied in this case to obtain equilibrium GB structures of STO. The primitive unit cell of the STO contains five atoms (one chemical formula) and has a lattice constant of 3.905 Å. We employ the cubic eight-node linear solid element to mimic the shape of the cubic primitive unit cell of the STO. An illustration of the cubic primitive unit cell is shown in Fig. 8.4.

To simulate the GB-defect interaction by CAC, it is necessary for the method to reproduce the equilibrium or stable GB structure as a result of the interaction of atoms in the material. To obtain a stable GB structure, we first construct an initial CAC bi-crystal model of STO. The model shown in Fig. 8.5 contains the $\sum 5$ GB with the GB plane (210) and is constructed by rotating *Grain A* by 26.6° counterclockwise and *Grain B* by 26.6° clockwise around the [001] axis. The GB region is meshed with atomic resolution, while regions away from the GB are meshed with low resolution, with each finite element containing 512 unit cells. The equilibrium structures of the bi-crystals are then obtained through CAC simulation of the models under various translation states but without an applied external force field. Thereafter, the GB energy is calculated for each specified microscopic translation state [51].

The equilibrium GB structure of $\sum 5$ (210) predicted by CAC is presented in Fig. 8.6, in which the unit structures around the GB plane are outlined with solid lines. The equilibrium GB structures in STO tilt $\sum 5$ (210) GB have been obtained by DFT calculations as well as by measurements from experiments such as high angular annular dark field scanning transmission electron microscopy (HAADF-STEM) by Lee et al. [52]. A comparison between CAC and existing results shows that the GB microstructures, including both the shape of the GB structural units and the arrangement of ions within each unit predicted by CAC are in excellent

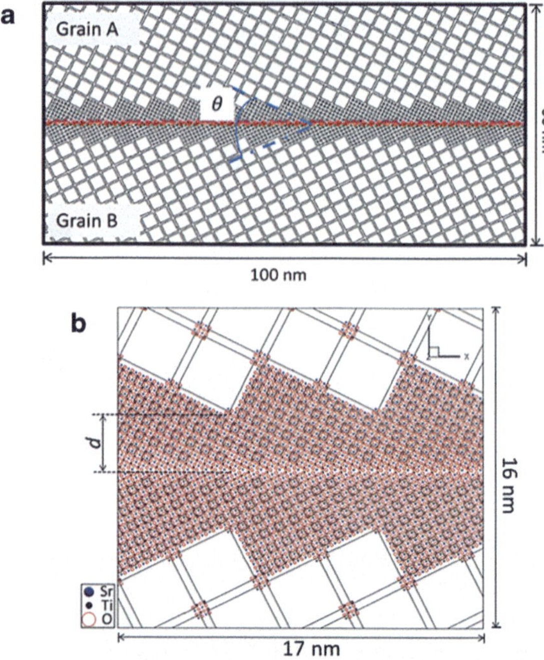

Fig. 8.5 (**a**) CAC model of bi-crystal STO with the tilt GB ∑5 (210); (**b**) zoomed-in view of coarse and fine FE elements (reprinted from [47]. Copyright (2015) by The Royal Society)

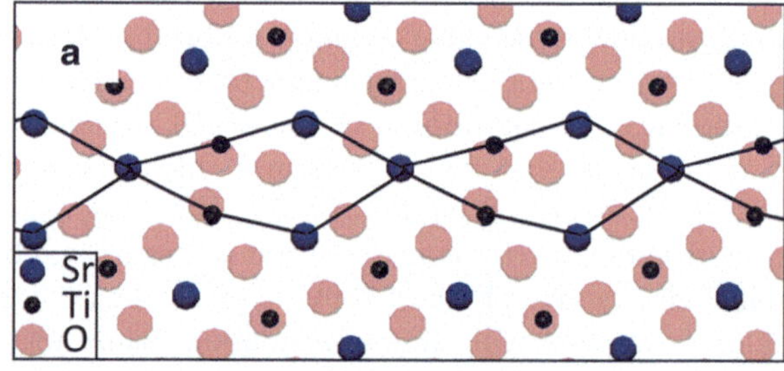

Fig. 8.6 The GB structure for ∑5 (210) with a translation state of (0.0, 0.5, 1.5 Å) from the CAC result (reprinted from [47]. Copyright (2015) by The Royal Society)

agreement with those obtained by DFT and HAADF-STEM experiments [52]. The predicted equilibrium structures for two other GBs, ∑5 (310) and ∑13 (510), are presented in Figs. 8.7a, b, respectively. These two GB structures are also found to agree with those from DFT calculations as well as the experimental results obtained by Imaeda et al. [53] and Lee et al. [52].

Furthermore, the variation in GB energy with different translation states is calculated and compared with the results obtained from the DFT calculation by Lee et al. [52]. The comparison shows that the CAC result qualitatively agrees well with the DFT result, but the overall magnitude of GB energy from the CAC simulation is slightly larger than that from the DFT calculation. The difference in the overall

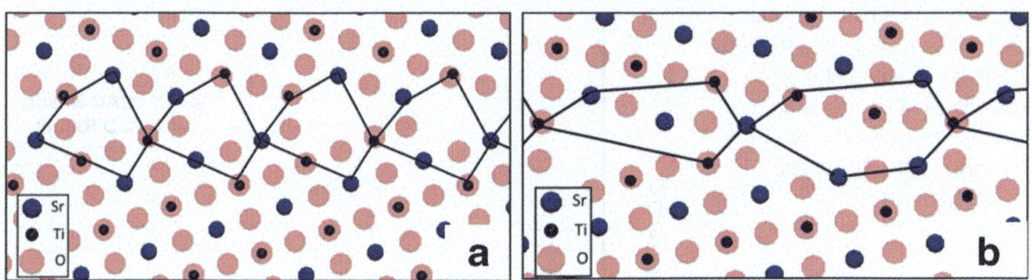

Fig. 8.7 GB structures from CAC results for (**a**) $\sum 5$ (310) with a translation state of (4.5, 0.5, 0.5 Å) and (**b**) $\sum 13$ (510) with a translation state of (5.0, 1.0, 0.0 Å) (reprinted from [47]. Copyright (2015) by The Royal Society)

magnitude of GB energy between the CAC and DFT results is mainly caused by the employment of the empirical rigid-ion model potential [50, 54, 55].

To quantify the effect of the size of the atomic-resolution region on the GB structure and energies predicted by CAC, we have constructed ten CAC models with different sizes for the atomically resolved GB regions. The size of the atomically resolved GB region is defined by the smallest distance between the GB and the atomic-continuum interface, denoted as d in Fig. 8.5b. The CAC and MD results for GB energies as a function of d are compared in Fig. 8.8. We have also compared the entire displacement field obtained by each CAC model with the atomic displacements obtained by MD. The magnitude of the displacement discrepancy between the CAC and MD results is plotted in Fig. 8.9. In this figure, we present the differences between the CAC and MD displacement fields for the models with different d. According to the comparisons of both the displacement fields plotted in Fig. 8.9 and the GB energies presented in Fig. 8.8, it is concluded that, with the increasing d, the CAC simulation results are closer to the corresponding MD results and eventually become identical. When d is larger than 12 Å, CAC simulation can accurately reproduce the structures and energies of GBs. To reproduce the GB structure with reasonable accuracy, d is suggested to be larger than 12 Å in the CAC simulations involving GBs in the following two cases.

In summary, we have successfully demonstrated in this case that the CAC method is applicable in predicting microstructures and energies of the grain boundaries in polyatomic ionic materials by comparing the CAC results with existing DFT calculations and experiments. In addition, the effect of the size of the atomic-resolution region on the GB structure and energies predicted by the CAC method is quantified.

8.4.2 Case II: Crack Initiation and Propagation in STO

It is well known that STO is extremely brittle. It is a highly ionic crystal with very low fracture toughness. For ionic materials, Schultz et al. [56] proposed the *crystal*

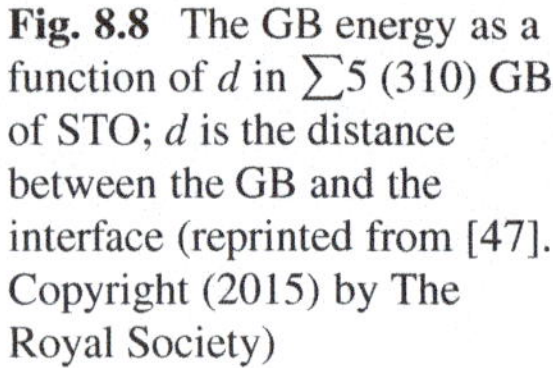

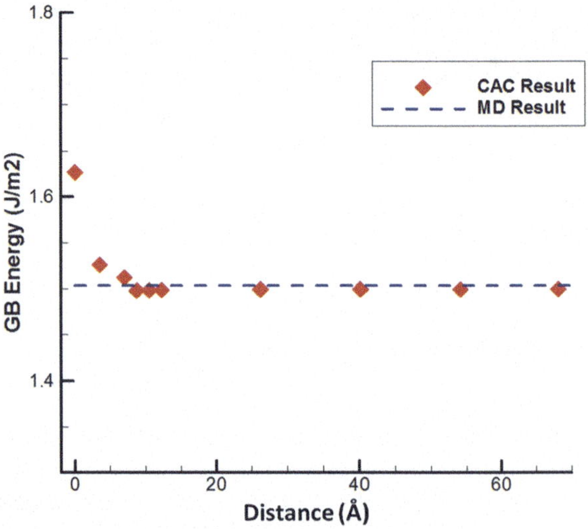

Fig. 8.8 The GB energy as a function of d in $\sum 5$ (310) GB of STO; d is the distance between the GB and the interface (reprinted from [47]. Copyright (2015) by The Royal Society)

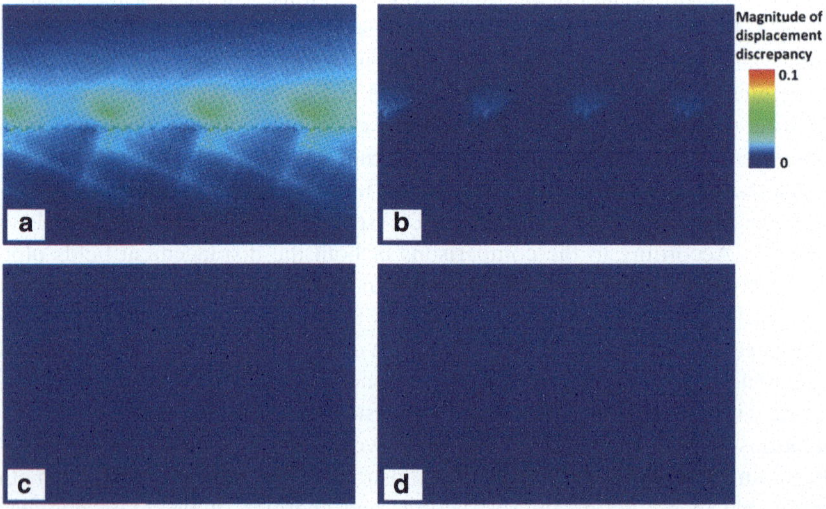

Fig. 8.9 The differences between the CAC and MD displacement fields for the models with different d (the distance between the GB and the continuum-atomic interface): (**a**) 7.0 Å; (**b**) 8.73 Å; (**c**) 12.22 Å; (**d**) 26.2 Å (reprinted from [47]. Copyright (2015) by The Royal Society)

plane neutrality condition, which suggests that cleavage occurs only at neutral planes in highly ionic crystal structures. According to this neutrality condition criterion, the {1 0 0} planes that bound the primitive unit cells in STO are the cleavage planes because both of the possible terminating planes, SrO and TiO_2, are electrically neutral. The objective of this case is to study crack initiation and propagation in STO.

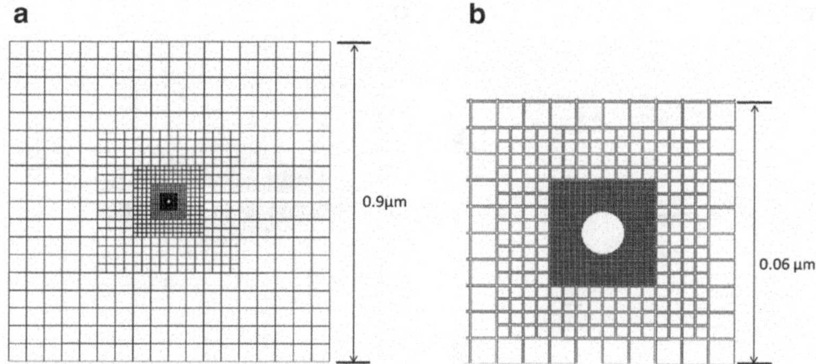

Fig. 8.10 CAC model with a single void at the center: (**a**) the whole model; (**b**) a zoomed-in view of the center region

8.4.2.1 Single-Crystal STO

A large-scale CAC simulation is performed on a single-crystal STO model under tensile loading. To allow cracks to initiate and propagate at inter-element boundaries, the eight-node linear cubic solid element shown in Fig. 8.4 is employed in this work to mimic the shape of the cubic primitive unit cell of the STO perovskite structure, with all the element surfaces being the cleavage planes, i.e., $\{1\,0\,0\}$ planes. In Fig. 8.10, we present the CAC computer model, which has the dimensions of 0.9 μm by 0.9 μm by 0.781 nm. A void is initially introduced into the CAC model at the center to simulate the initiation of cracks from the void and the subsequent propagation into the far regions. The computer model is coarsely meshed with five different element sizes that have 128, 64, 32, 16, and 8 unit cells in each edge of the element, respectively. Only 2000 elements are used in the CAC model for the STO specimen, which contains about ~55 million atoms. After relaxation, a tensile loading is applied in the CAC model along the vertical direction.

In Fig. 8.11, we plot the higher-order displacement field in the vertical direction for better visualization of the propagating crack. The displacement in the vertical direction v can be decomposed into two parts: the displacement v_c caused by the rigid-body motion and constant strain and the higher-order displacement v_h. In this model, the displacement v_c is much larger than v_h. Furthermore, the displacement v_c varies linearly with the vertical coordinates, so v_c is the same around the horizontal crack in this model. Therefore, to visualize the propagating crack more clearly, we plot the high-order displacement field in Fig. 8.11, which is obtained by subtracting v_c from the total displacement v in the vertical direction, i.e., $v_h = v - v_c$. The zoomed-in view in Fig. 8.11b shows that the crack nucleates from the atomistically resolved region around the void. Thereafter, the crack propagates into the continuum region meshed with coarse elements. Without any numerical treatment or additional rules, the cracks can smoothly propagate across the interfaces between the atomistic

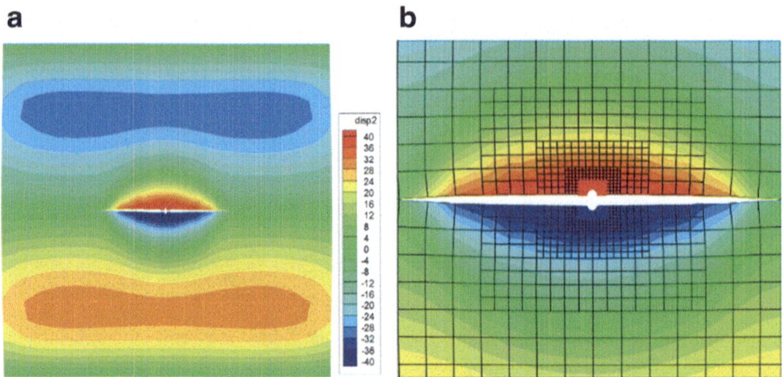

Fig. 8.11 A contour plot of the higher-order displacement field in the CAC model: (**a**) a view of the whole specimen without element edges; (**b**) a zoomed-in view with element edges for the voided region

and the continuum regions, as well as the interfaces between regions meshed with non-uniform elements.

To compare the efficiency of the CAC method with the MD method, we compare the number of equations to be solved or the number of mathematic operations needed in these two methods. In each time step, the ratio of the number of the solved governing equations in CAC to that in MD scales as (N^{-3}), where N is the number of unit cells in each edge of the coarse-grained element in CAC. To compare the computing time in each time step of the simulation in CAC and MD, we compare the number of mathematic operations needed in CAC and MD, respectively. The ratio of the number of mathematic operations in CAC to that in MD also scales as $O(N^{-3})$.

Both the ratio of the number of the solved governing equations and the ratio of the number of mathematic operations decrease with increasing N. In other words, the CAC shall become more efficient when the element size (N) increases. For example, in the single-crystal model with the dimensions of 0.9 μm × 0.9 μm × 0.781 nm (55 million atoms), only 2000 elements are used in the CAC model in total. The ratio of the number of solved equations in CAC to that in MD is 0.1408 %. The ratio of the number of mathematic operations in CAC to that in MD is 0.468 %.

8.4.2.2 Bi-Crystal STO

The bi-crystal configuration allows for the study of specific GB structures and the GB structure–property relationship and hence can be used to identify the mechanisms of GB-defect interaction and the effect on the dynamic fracture behavior. CAC is used here to simulate the dynamic processes of crack interaction with GBs in bi-crystal STO.

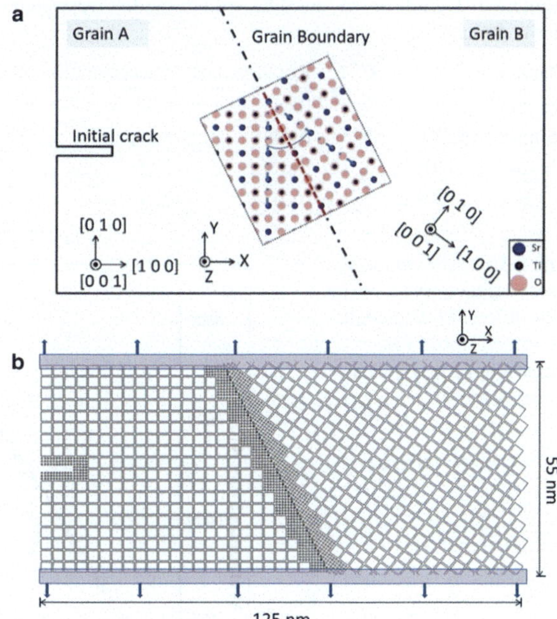

Fig. 8.12 A bi-crystal model of STO: (**a**) schematic illustration of two neighboring grains; (**b**) the CAC model with FE mesh (reprinted from [47]. Copyright (2015) by The Royal Society)

A notched bi-crystal with two neighboring grains that have different crystal orientations is schematically illustrated in Fig. 8.12a. The [100], [010], and [001] directions of *Grain A* are parallel to the X, Y, and Z directions in the global coordinate system, respectively. The bi-crystal model contains a $\sum 5$ (210) GB with the initial translation state of (0.0, 0.5, 1.5 Å). The corresponding CAC model for the study of crack-GB interaction is presented in Fig. 8.12b, in which the finest mesh, i.e., atomic resolution, is used in regions near the predefined crack and the GB, while a uniform coarse mesh with each element containing 512 unit cells is used in the regions away from the crack and the GB. After relaxation, a tensile loading is applied in the CAC model along the vertical direction.

Four different submicron-sized CAC bi-crystal STO models are constructed. The four models with different misorientation angles are constructed by rotating Grain B while keeping Grain A fixed. Specific translation states are then applied between the two grains in the models. Details of these models are listed in Table 8.1. The first two models contain the $\sum 5$ (210) and $\sum 5$ (310) GBs, respectively. The last two models have the same $\sum 13$ (510) GB structure but different translation states, corresponding to the stable and metastable GB structures, respectively. Note that both stable and metastable $\sum 13$ (510) GB structures are observed in experiments and DFT calculations [52].

In Fig. 8.13, we present the variation in the average stress with average strain for the four bi-crystal models under tensile loading. The stress–strain curves for the four different models overlap at the initial loading stage before the tensile stress

Table 8.1 Details of the four different GB models (reprinted from [47]. Copyright (2015) by The Royal Society)

Models	CSL	Rotation axis	Misorientation angle (°)	GB plane	Translation (Å)
Sigma 5 (210)	$\sum 5$	[001]	53.2	(210)	(0.0, 0.5, 1.5)
Sigma 5 (310)	$\sum 5$	[001]	36.88	(310)	(4.5, 0.5, 0.5)
Sigma 13 (510)-A	$\sum 13$	[001]	22.6	(510)	(5.0, 1.0, 0.0)
Sigma 13 (510)-B	$\sum 13$	[001]	22.6	(510)	(10.5, 1.0, 0.0)

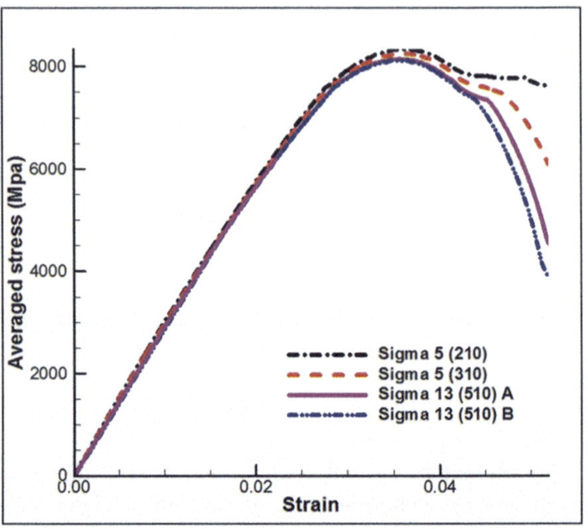

Fig. 8.13 Average tensile stress–strain curves for the four different GB models (reprinted from [47]. Copyright (2015) by The Royal Society)

reaches the maximum. When the strain is 0.042, we can clearly see a change in the stress–strain curve's slope, which is caused by the interaction between the propagating crack and the GB. Furthermore, the slope change is different among the four models because of different interaction mechanisms. When the applied strain reaches 0.0525, the crack almost propagates across the whole specimen. We observe that the corresponding stresses, i.e., the fracture strength, decrease in the order of $\sum 5$ (210), $\sum 5$ (310), $\sum 13$ (510)-A, and $\sum 13$ (510)-B, with $\sum 5$ (210) having the highest fracture strength. It has been observed that the fracture strength of $\sum 5$ (210) is larger than that of $\sum 5$ (310) in the micro-tensile tests of Ni-20Cr [57]. This experimental result is consistent with our CAC predictions: the specimens with larger misorientation angles between the two grains will exhibit higher fracture strength.

Different interaction mechanisms are identified by investigating the crack-GB interactions in the four models. In Fig. 8.14, we present the atomic configurations near GBs for the four models when the loading strain reaches 0.0452. Figure 8.14a shows that the propagating crack in the $\sum 5$ (210) model is arrested by the GB, and then many voids are formed along the GB. The formed voids then weaken the material and consequently lead to inter-granular fracture between the two

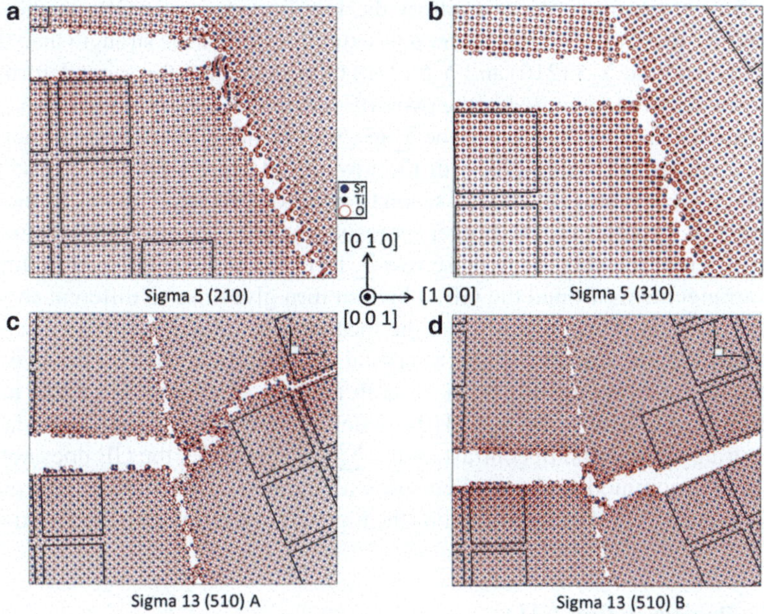

Fig. 8.14 Snapshots of CAC results showing the crack-GB interaction at the applied strain of 4.52 %. (**a**) Sigma 5 (210). (**b**) Sigma 5 (310). (**c**) Sigma 13 (510) A. (**d**) Sigma 13 (510) B (reprinted from [47]. Copyright (2015) by The Royal Society)

neighboring grains. In Fig. 8.14b, we observe that the propagating crack in the $\sum 5$ (310) model is also stopped by the GB, and then these two adjacent grains slide with each other along the GB. After tracing the movement of these two structural units during the deformation process in the model, we found that the deformation mechanism transitioned from crack propagation to GB sliding. An intra-granular fracture in the adjacent grain is observed in the $\sum 13$ (510)-A model, as shown in Fig. 8.14c, and we see that the crack has passed through the GB and propagated into the neighboring grain. Although the crack path in the second grain has a zigzag pattern, the overall crack path is still in the $\{1\ 0\ 0\}$ cleavage planes. This again demonstrates the feasibility of the CAC method in passing cracks from an atomically resolved region into a coarse-grained region without the need for any ad-hoc interfaces or any additional criterion. Figure 8.14d shows the similar intra-granular fracture pattern in the $\sum 13$ (510)-B model, in which only the translation state is different from that in the $\sum 13$ (510)-A model. The difference between the results from the two models is the location where intra-granular fracture occurs.

Further analysis indicates that the identified crack-GB interaction mechanisms can be used to explain the observed difference in the GB tensile strengths of the four bi-crystal models. As observed in the two $\sum 13$ (510) models, the crack propagates straight into the neighboring grain without much GB resistance; after the crack has passed through the GB, the intra-granular fracture in the adjacent grain is the deformation mechanism. As a result of the low impedance of GBs to the crack propagations, these two $\sum 13$ (510) models exhibit lower tensile strength than that of the models with the $\sum 5$ (210) and $\sum 5$ (310) GBs. Although the same deformation mechanism has been identified in the two different models, the $\sum 13$ (510)-A model exhibits higher strength than that of the $\sum 13$ (510)-B model. The reason is that $\sum 13$ (510)-A contains the stable GB with the lowest GB energy, while $\sum 13$ (510)-B contains the metastable GB with the second-lowest GB energy. Obviously, the more stable the GB structure is, the higher its resistance to crack propagations will be. It is the different translations in these two $\sum 13$ (510) models that lead to different atomic arrangements around the GB, which in turn gives rise to different strengths of the two models. This indicates that the local atomic arrangement can also affect the overall fracture strength of the bi-crystal. In contrast to that in the two $\sum 13$ (510) models, the propagating crack in both the $\sum 5$ (310) and $\sum 5$ (210) models is stopped by the GB. In the $\sum 5$ (310) model, the GB slides to accommodate the applied strain along GBs. In contrast, in the $\sum 5$ (210) model, the GB does not slide as easily to accommodate the applied strain along the GB for further deformation. Therefore, the $\sum 5$ (210) model exhibits the highest strength among the four models.

8.4.2.3 Polycrystalline STO

To simulate the fracture behavior of polycrystalline polyatomic ionic materials, we perform CAC simulations of a polycrystalline STO specimen under indentation. The computer model contains four columnar hexagonal grains, has the dimensions of $727 \times 607 \times 7.8$ Å, and has a nominal grain size of 346.4 Å. The regions near GBs in the CAC model are discretized into the smallest finite elements, i.e., with atomic resolution; while other regions of the model are discretized into coarsely meshed finite elements, with each element containing 512 unit cells. As mentioned regarding Case I, to reproduce the GB structures with a reasonable level of accuracy, the size of the atomic-resolution region is ensured to be larger than 12 Å in this CAC model.

The atoms located in GB regions are examined to ensure that the distance between any two atoms is smaller than a critical value. The value of the critical distance between two atoms depends on the species of the two atoms. In this work, the sum of the actual crystal ionic radii of two ions is used as the critical distance between the two ions. The crystal ionic radius for ions in strontium titanate is as follows: $r_{Ti} = 0.745$ Å; $r_{Sr} = 1.32$ Å; and $r_O = 1.26$ Å. For example, the critical distance between a strontium ion and oxygen ion is $D_{Sr-O} = r_{Sr} + r_O = 1.32 + 1.26 = 2.58$ Å. After relaxation, a spherical indenter shown in Fig. 8.15 is used to apply the indentation loading on both CAC and the corresponding atomistically resolved MD models.

In Fig. 8.16, we compare the CAC and MD simulation results of the force–displacement responses. Figure 8.16 shows that the slope of the curves and the maximum force values reproduced by the CAC simulation are similar to those from MD. In Fig. 8.17, we present the out-of-plane normal stress obtained by CAC and

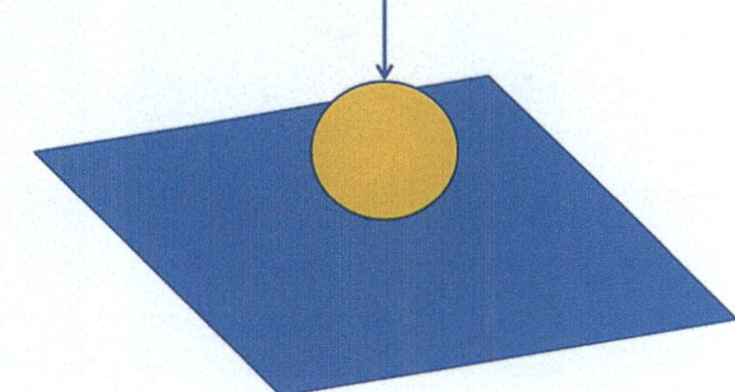

Fig. 8.15 Illustration of a polycrystalline STO specimen under indentation (reprinted from [48]. Copyright (2015) by Taylor & Francis Group)

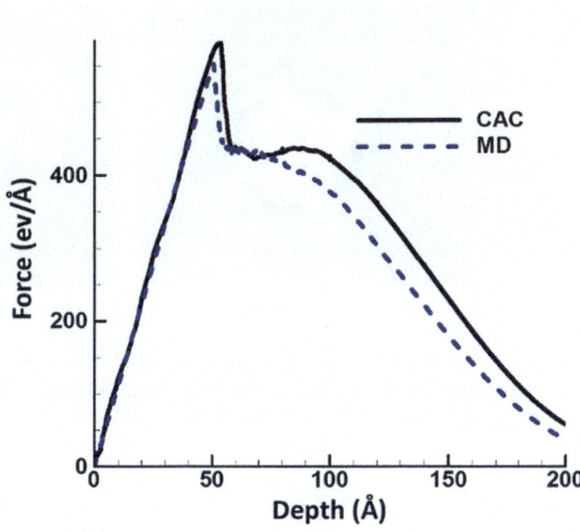

Fig. 8.16 Force–displacement curves from CAC and MD simulations of polycrystalline STO specimens under indentation (reprinted from [48]. Copyright (2015) by Taylor & Francis Group)

MD simulations at two different indenter displacements, respectively. Figure 8.17 shows that the cracks nucleate in the center region of the specimen and propagate along four different directions. Note that all the cracks nucleate and propagate on the {1 0 0} cleavage planes. These planes are electrically neutral planes in STO. A comparison between the CAC and MD simulation results shows that the CAC results are in good agreement with the MD results in many respects, including the number of cracks, the directions of cracks, and the stress distribution around the cracks. More importantly, the indentation simulation results presented in Figs. 8.16 and 8.17 demonstrate that (1) CAC allows the spontaneous nucleation of cracks from coarse-scale finite element regions and that (2) CAC allows the propagation of cracks in both the atomic regions and the coarsely meshed finite element regions, without the need for special constitutive relations to dictate the initiation and propagation of cracks or additional numerical treatments to facilitate the passage of cracks. Although the number of DOF has been greatly reduced by coarse-graining the atomistic system, the CAC method has successfully reproduced the initiation and propagation of cracks in polycrystalline STO such that they are comparable with MD results.

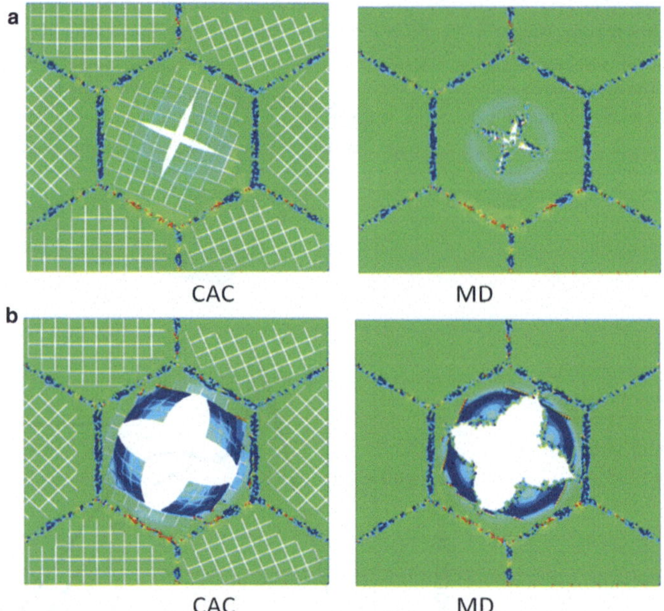

Fig. 8.17 Contour plots of stresses from CAC and MD simulations at different displacements (i.e., U) of the indenter with the radius of 140 Å (unit: GPa) (**a**) $U = 60$ Å. (**b**) $U = 150$ Å (reprinted from [48]. Copyright (2015) by Taylor & Francis Group)

In summary, the CAC method has been used in this case to study the dynamic processes of crack initiation and propagation in single-crystal, bi-crystal, and polycrystalline STO. The simulation results show that CAC can allow the spontaneous nucleation of cracks as well as the smooth propagation of cracks across the interfaces between the atomistic and the continuum regions and that between regions meshed with non-uniform elements. The efficiency of the CAC method has been discussed by comparing it with the MD method. Different interaction mechanisms are identified by investigating the crack-GB interactions in four different GB models.

8.4.3 Case III: Dislocation Nucleation and Migration in STO

Dislocations have been experimentally observed in STO during compression and nanoindentation tests [58]. In ionic materials, the existing criterion suggests that slip systems are those that do not involve bringing ions of the same sign close to each other during the dislocation glide. This criterion has been demonstrated to work very well in perovskite-structured materials such as $BaTiO_3$, $CaTiO_3$ [59], and $KZnF_3$ [60]. In these materials, the slip system with the $\{1\ -1\ 0\}$ slip planes and the

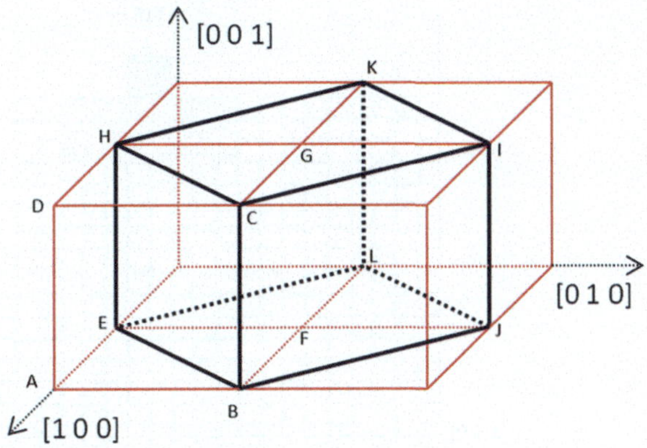

Fig. 8.18 The tetragonal element for STO. There are four cubic primitive unit cells in the figure, with the ABCDEFGH cube being that of the cell. The BCIJEHKL block (bounded with a *black bold line*) is the tetragonal element (reprinted from [46]. Copyright (2013) by Elsevier)

⟨1 1 0⟩ slip directions, denoted as ⟨1 1 0⟩ {1 −1 0}, is one of the easiest glide systems that prevent cations (or anions) from being brought close to each other during the dislocation process. The objective of this case is to study the dislocation nucleation and migration in STO.

8.4.3.1 Single-Crystal STO

In Fig. 8.18, we present an eight-node linear tetragonal finite element for simulation of the dynamics of dislocations. All the element surfaces are the {1 1 0} slip planes of STO. The cubic cell bounded with red thin lines contains four primitive cubic unit cells, each of which contains five atoms. The element bounded with black lines is the tetragonal element that has two primitive unit cells and contains ten atoms. This tetragonal element is the smallest possible finite element that exposes all the available {1 1 0} slip planes on the element surfaces for STO crystal under the plane strain condition [61]. To simulate the behavior of dislocations, two indenters with different shapes, rectangular and cylindrical, are applied in the fully relaxed CAC model. The computer model is shown in Fig. 8.19.

In Fig. 8.20, we present the variation of indentation force with indentation depth from CAC and MD simulations with a rectangular indenter. The force-depth curves from the CAC and MD simulations show that CAC agrees very well with those from MD in the elastic deformation stage. Each load drop, i.e., the pop-in in the force-depth curves corresponds to the nucleation of partial dislocations. In the plastic deformation stage, slight differences in the sequence and magnitude of the pop-ins between the force-depth curves of CAC and MD are observed in Fig. 8.20. By examining the atomic-scale dislocation activities associated with the pop-ins in the force-depth curves it is found that the differences between the force-depth response of CAC and that of MD are caused by different dislocation nucleation sequences in the CAC and MD simulations.

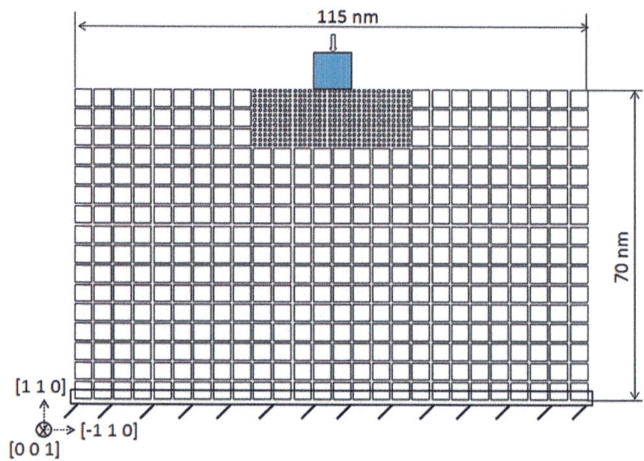

Fig. 8.19 The CAC model with a rectangular indenter (reprinted from [46]. Copyright (2013) by Elsevier)

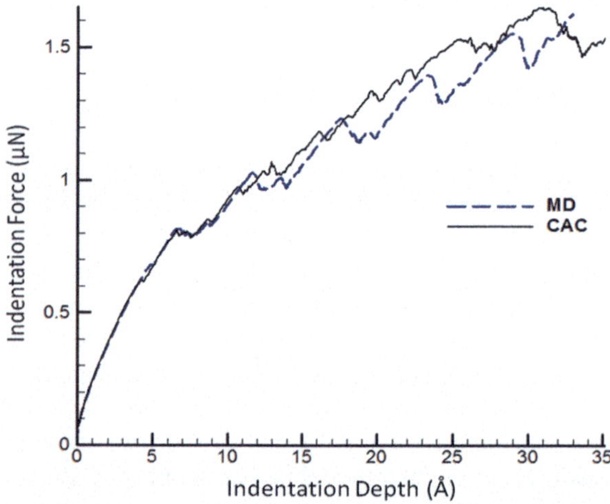

Fig. 8.20 Force-depth curves from CAC and MD simulations with a rectangular indenter (reprinted from [46]. Copyright (2013) by Elsevier)

Fig. 8.21 plots the centrosymmetry parameter of atoms in both the CAC and MD models with a rectangular indenter, in which only the atoms with a large centrosymmetry parameter are shown in the snapshots for better visualization. The plot of the centrosymmetry parameter [62] in Fig. 8.21 shows that the dissociation of $\langle 1\ 1\ 0 \rangle$ dislocations occurs on the $\{1\ -1\ 0\}$ slip planes. This is consistent with experimental observations [63, 64] and the first-principles calculations [65]. The stacking fault width is estimated to be 33.13 Å (Fig. 8.21), which is comparable to the recent experimental measurement [64]. The interface between the atomistic and continuum region is indicated by the red dashed line in Fig. 8.21. The results clearly show that the dislocation partials transmit smoothly from the atomic region to the coarse-grained region without any reflections.

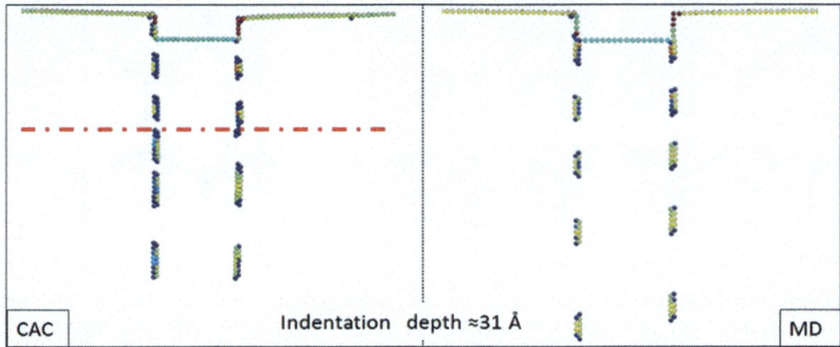

Fig. 8.21 A snapshot of the dislocation structure from CAC (*left column*) and MD (*right column*) simulations with a rectangular indenter. The dislocation partial is denoted by *blue* atoms, while atoms within stacking faults are shown in *yellow*. The *red dashed line* in CAC indicates the interface of the finest and coarse meshed domain (reprinted from [46]. Copyright (2013) by Elsevier)

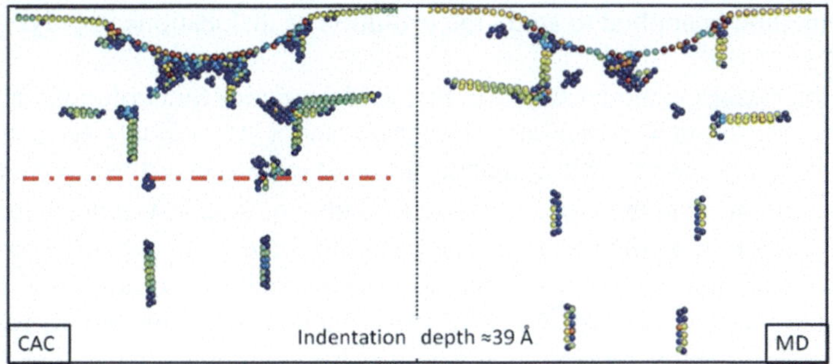

Fig. 8.22 Snapshots of the dislocation structure in CAC (*left column*) and MD simulations (*right column*) with a cylindrical indenter. The *red dashed line* in CAC indicates the interface between the atomic- and the coarse-grained domain (reprinted from [46]. Copyright (2013) by Elsevier)

In Fig. 8.22, wepresent the centrosymmetry parameter of atoms near dislocations and stacking faults in CAC and MD simulations with a cylindrical indenter. Figure 8.22 shows that the first dislocation nucleates underneath the indenter in the slip plane, i.e., the {1 1 0} planes, even though the resolved shear force is larger in the {1 0 0} planes in this model. We see that, under the cylindrical indenter, the maximum shear stress does not appear in the {1 1 0} planes, yet the dislocations still only nucleate and propagate in the {1 1 0} planes. The simulation result thus verifies the criterion for the easy glide system mentioned at the beginning of this case. Using the proportional relationship between the hardness and temperature [66], we estimate the value of the hardness at room temperature to be 9.4 GPa. Experimental measurements of the hardness of a macroscopic single-crystal STO at room temperature have been reported to be 9.5 GPa [67] and 11.5 GPa [68]. Therefore, it is reasonable to believe that the CAC predictions of the hardness of STO are comparable to existing experimental measurements.

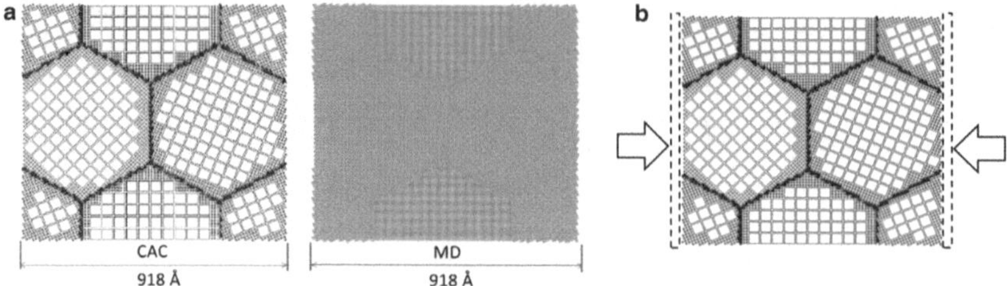

Fig. 8.23 Illustration of the polycrystalline model under compression: (**a**) CAC and MD models; (**b**) the compressive loading condition (reprinted from [48]. Copyright (2015) by Taylor & Francis Group)

8.4.3.2 Polycrystalline STO

The polycrystalline model is more realistic and more complex than the single-crystal model in studying the dynamics of dislocations because it considers the existence of GBs and triple junctions and interactions between dislocations and GBs. Here, the CAC method is applied to study the evolution of dislocations in polycrystalline STO.

The polycrystalline model shown in Fig. 8.23a contains four columnar hexagonal grains with different orientations. Both MD and CAC models have the same dimensions of $918 \times 833 \times 7.8$ Å, and the nominal grain size is 476.4 Å. In the CAC model, the regions near the GBs are discretized into the smallest finite elements, i.e., atomistic resolution, while the grain interior regions are modeled with coarse finite elements. As mentioned in Case I, the size of the atomic-resolution region in this CAC model is larger than 12 Å to reproduce the GB structures with a reasonable level of accuracy. After relaxation, a compressive loading, shown in Fig. 8.23b, is applied on the CAC and MD models.

In Fig. 8.24, we present the variation in the average stress with the average strain in the horizontal direction for both CAC and MD simulations. The CAC simulation reproduces both an elastic modulus and a maximum stress that are comparable to those from the MD simulations. In Fig. 8.25, we present the contour plots of stress distributions in both the CAC and the corresponding MD models. The figure shows that the stress in the GB regions is higher than the stress in other regions. Both compressive and tensile stresses are observed in regions near GBs. Similar stress distributions have also been observed in the full MD simulations of polycrystalline copper by Farkas [69]. The dislocation activities in polycrystalline STO materials are visualized by the centrosymmetry parameter and atoms with a color other than blue to indicate defects (e.g., dislocations or stacking faults), GBs, or surfaces.

Figure 8.26 plots the centrosymmetry parameter of all atoms in both the CAC and MD models. This figure shows that dislocations nucleated from the triple junctions and the GBs. With the increasing applied strain, more dislocations nucleate and migrate in the grains. Comparisons between the results from the MD and CAC simulations (Figs. 8.24, 8.25, and 8.26) show that CAC reproduces the atomic-level

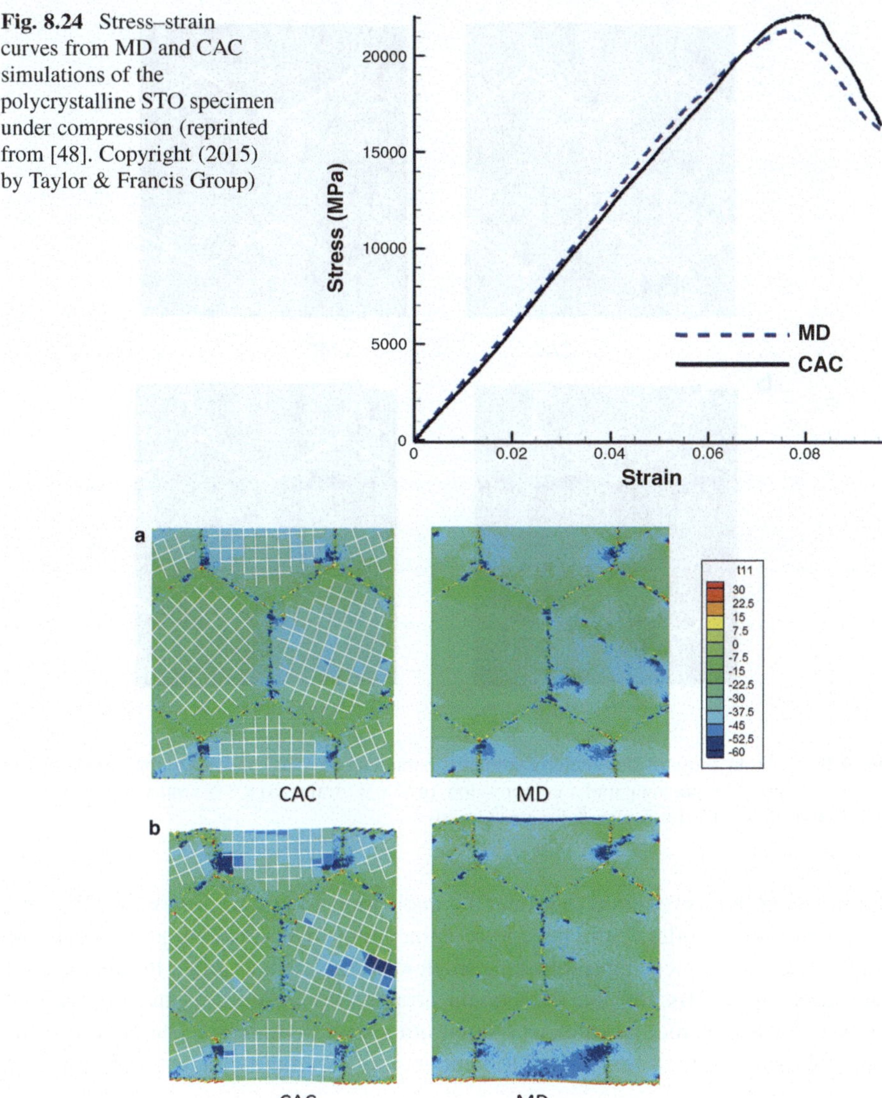

Fig. 8.24 Stress–strain curves from MD and CAC simulations of the polycrystalline STO specimen under compression (reprinted from [48]. Copyright (2015) by Taylor & Francis Group)

Fig. 8.25 Contour plots of stresses from CAC and MD simulations of the polycrystalline STO specimen under compression. (**a**) 7.6 % strain. (**b**) 8.7 % strain (reprinted from [48]. Copyright (2015) by Taylor & Francis Group)

simulation results with noticeable similarity, including the constitutive responses, the stress distribution, and the evolution of dislocation microstructures.

Note that the simulation results of polycrystalline STO show that dislocations can be nucleated in both the GBs and grain interiors and that many dislocations may propagate across the entire specimen in polycrystalline materials. Thus, if an

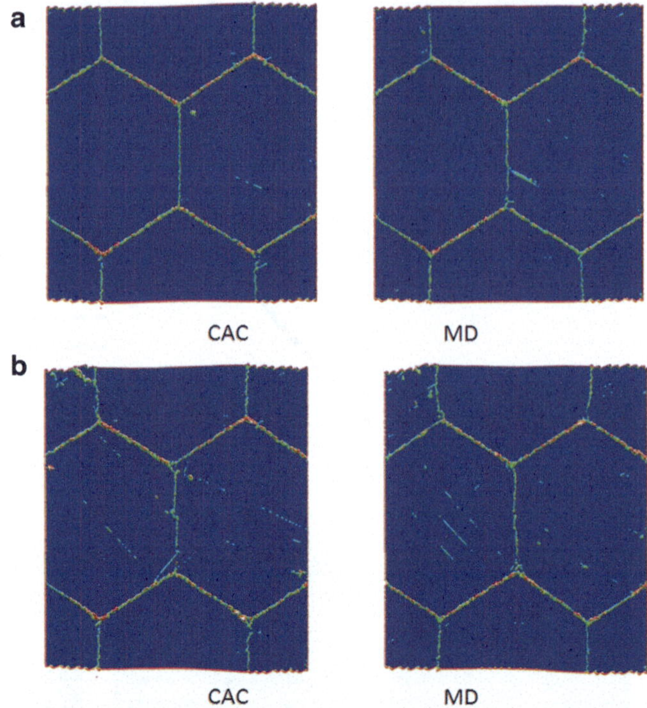

Fig. 8.26 Contour plots of the centerosymmetry parameter in CAC and MD simulations of the polycrystalline STO specimen under compression. (**a**) 7.6 % strain. (**b**) 8.7 % strain (reprinted from [48]. Copyright (2015) by Taylor & Francis Group)

atomic resolution is required to model the nucleation and migration of dislocations, almost the entire model would need to be refined into an atomistic model. In contrast to other existing concurrent multiscale methods, CAC can simulate the dislocation nucleation from GBs and the subsequent propagation into grain interiors without the need for supplemental rules or special numerical treatments. The need for this capability for a multiscale method to simulate polycrystalline materials is justified in this simulation.

Three CAC models with different grain sizes are constructed to further study the dynamics of dislocations in polycrystalline STO. These three models have the same dimensions of 187 nm $\times$ 162 nm $\times$ 0.78 nm. Models N4, N16, and N36 contain 4, 16, and 36 complete grains, respectively. Thus, the nominal grain sizes in these three models are 93.6 nm, 46.8 nm, and 30.9 nm, respectively. The compressive loading is applied on both the left and right sides of the CAC models.

Figure 8.27 plots the distribution of the stress along the horizontal direction for these three models with the applied strain of 8.5 %. In Fig. 8.28, wepresent the contour plots of the centrosymmetry parameter in the models under the same strain. These two figures show that several dislocations have nucleated from the

Fig. 8.27 Contour plots of the centerosymmetry parameter in models with different grain sizes at 8.5 % strain. (**a**) N4. (**b**) N16. (**c**) N36 (reprinted from [48]. Copyright (2015) by Taylor & Francis Group)

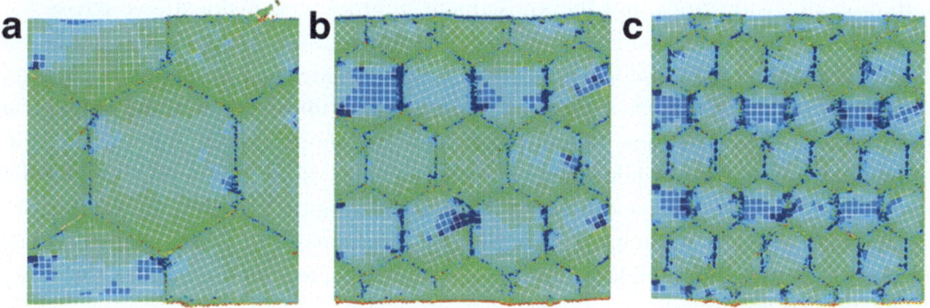

Fig. 8.28 Contour plots of stresses in polycrystalline models with different grain sizes at 8.5 % strain. (**a**) N4. (**b**) N16. (**c**) N36 (reprinted from [48]. Copyright (2015) by Taylor & Francis Group)

triple junctions, grain boundaries, and free surfaces. Dislocations propagate across the entire specimens, and more severe plastic deformations are observed near grain boundaries in the three polycrystalline models. In Fig. 8.27, we observe highly localized stresses in the regions near the dislocations and grain boundaries. Figure 8.28 shows that the distribution of dislocations becomes more uniform when the nominal grain size in the model decreases. The results of Model N4, shown in Fig. 8.28a, indicate that the migration of a few dislocations and GB sliding are found between the center grain and the upper-left grain. Figure 8.28b shows that many dislocations are migrating in the grains of Model N16. These dislocations eventually pile up on the GBs; thereafter, cracks initiate at the locations of the pile-ups. Figure 8.28c shows that dislocations are piling up near the GBs in Model N36. Consequently, the deformation is more pronounced in regions near the triple junctions.

The CAC simulation results provide insights into the mechanisms of plastic deformations in polycrystalline STO under compressive loadings. First, the emission of dislocations from GBs, triple junctions, and free surfaces is observed. Dislocations emit consecutively from the same location where large plastic deformation occurs. Second, grain boundary sliding is observed in the regions subjected to large shear deformations. For example, the deformation configuration shown in Fig. 8.28a exhibits an obvious grain boundary sliding between the upper-left grain and the grain in the center of the specimen. Third, the simulation results show that the angle between the slip planes and the GB planes plays a key role in determining how the GBs affect the dislocation motion. In general, dislocations pile up near the triple junctions and the grain boundaries, as shown in Fig. 8.28b. However, if the orientation of the slip plane is such that the slip plane lies almost parallel to the GB, the dislocations do not pile up at the GB but rather move within the GB. Finally, pile-ups of dislocations are formed at the grain boundaries, leading to the nucleation of cracks and the migration of several dislocations across GBs.

We have applied the CAC method in this case to study the evolution of dislocation in single-crystal STO as well as polycrystalline STO. Through the simulation of single-crystal STO, we have demonstrated that CAC enables the spontaneous initiation of dislocations and the subsequent migration of dislocations across the interfaces between the atomistic and the continuum regions. Through the simulation of polycrystalline STO, we show that CAC can simulate the dislocation nucleation from GBs, the subsequent propagation into grain interiors, and the migration across the entire specimen without the need for supplemental rules or special numerical treatments. In addition, simulation results show that CAC can predict reasonable stacking fault width and STO nanoindentation hardness values that are comparable to the experimental data. Several mechanisms of plastic deformations in polycrystalline STO under compressive loadings are identified from the CAC simulation results.

8.5 Summary

This chapter reviews a CAC method for simulations of defect dynamics in polyatomic ionic materials. The CAC method combines a unified atomistic and continuum formulation of balance laws and a modified finite element method. Three case studies are presented to demonstrate the capabilities of the CAC method in simulating the dynamics of defects, including cracks, dislocations, and GBs, in polyatomic ionic materials. The features of the CAC method are summarized as follows:

1. The CAC method is able to simulate the defects in polyatomic ionic materials. The CAC method stands apart from other multiscale methods in its ability to solve both the deformation of lattice cells and the internal deformation within each lattice cell. In this chapter we demonstrate the necessity to include the internal DOF of atoms from the three sublattices in STO into the methodology in order to resolve the core structures of dislocations and cracks, and describe the actual processes of defect evolution in ionic materials.

2. The CAC method can allow the passage of defects across atomistic-continuum interfaces without special numerical treatments. The simulation of the propagation of cracks and dislocations in STO demonstrates that the CAC allows the smooth propagation of defects across the interfaces between the atomistic and the continuum regions.
3. The CAC simulations of bi-crystal STO demonstrates that the CAC can reproduce the equilibrium structures and energies of grain boundaries in polyatomic ionic materials, in good agreement with those obtained from existing experiments and DFT calculations.
4. For the first time, the concurrent multiscale method is successfully used to study complex behaviors of multiple defects in polycrystalline materials. The simulations of polycrystalline STO reveal that the CAC can reproduce the nucleation of defects such as cracks and dislocations, the migration of defects across the entire specimen, and the interplay between cracks, dislocations, and GBs with high accuracy by comparing to the corresponding fully atomistic MD simulations.

Several mechanisms for the evolution of defects in STO have been identified from the simulation results. The findings are summarized as follows:

1. The CAC simulations have confirmed the criteria for cleavage planes and slip planes and have reproduced the open GB structures in STO. Cracks propagate along the $\{1\ 0\ 0\}$ cleavage planes; dislocations dissociate and migrate along the $\{1\ 1\ 0\}$ slip planes in STO.
2. Different crack-GB interaction mechanisms (GB sliding, inter- and intra-granular fracture) are observed in bi-crystal STO with different GB structures. The GB resistance to crack propagations significantly contributes to the fracture toughness of the materials, and the results imply that the GB resistance to crack propagation is determined by the misorientation angles, as well as the atomic-level details of the GB structures.
3. In polycrystalline STO, dislocations initiate from GBs and triple junctions. The angle between the slip planes and the GB planes plays a key role in determining how the GBs affect the dislocation motion. Pile-ups of dislocations are formed at the grain boundaries, leading to the nucleation of cracks and the migration of several dislocations across GBs.

Acknowledgments This material is based upon the work supported by Department of Energy under award number DOE/DE-SC0006539. The CAC computer code in its present form is a culmination of developments supported in part by National Science Foundation under Award Numbers CMMI-1233113 and CMMI-1129976. This chapter draws on material previously published in [46–48].

References

1. Y. Mishin, M. Asta, J. Li, Atomistic modeling of interfaces and their impact on microstructure and properties. Acta Mater. **58**, 1117–1151 (2010)
2. J.P. Hirth, The influence of grain boundaries on mechanical properties. Metall. Trans. A Phys. Metall. Mater. Sci. **3**, 3047–3067 (1972)
3. Y. Cheng, M. Mrovec, P. Gumbsch, Atomistic simulations of interactions between the ½ ⟨111⟩ edge dislocation and symmetric tilt grain boundaries in tungsten. Philos. Mag. **88**, 547–560 (2008)
4. Z.H. Jin, P. Gumbsch, K. Albe, E. Ma, K. Lu, H. Gleiter, H. Hahn, Interactions between non-screw lattice dislocations and coherent twin boundaries in face-centered cubic metals. Acta Mater. **56**, 1126–1135 (2008)
5. Y. Chen, J. Zimmerman, A. Krivtsov, D.L. McDowell, Assessment of atomistic coarse-graining methods. Int. J. Eng. Sci. **49**, 1337–1349 (2011)
6. M. Dewald, W.A. Curtin, Multiscale modeling of dislocation/grain boundary interactions: I. Edge dislocations impinging on $\sum 11$ (1 1 3) tilt boundary in Al. Model. Simul. Mater. Sci. Eng. **15**, S193–S215 (2007)
7. M. Dewald, W.A. Curtin, Multiscale modeling of dislocation/grain boundary interactions. II. Screw dislocations impinging on tilt boundaries in Al. Philos. Mag. **87**, 4615–4641 (2007)
8. M. Dewald, W.A. Curtin, Multiscale modeling of dislocation/grain-boundary interactions: III. 60° dislocations impinging on $\sum 3$, $\sum 9$ and $\sum 11$ tilt boundaries in Al. Model. Simul. Mater. Sci. Eng. **19**, 055002 (2011)
9. J. Marshall, K. Dayal, Atomistic-to-continuum multiscale modeling with long-range electrostatic interactions in ionic solids. J. Mech. Phys. Solids **62**, 137–162 (2014)
10. Y. Chen, J.D. Lee, Atomistic formulation of a multiscale theory for nano/micro physics. Philos. Mag. **85**, 4095–4126 (2005)
11. Y. Chen, Local stress and heat flux in atomistic systems involving three-body forces. J. Chem. Phys. **124**, 054113 (2006)
12. Y. Chen, Reformulation of microscopic balance equations for multiscale materials modeling. J. Chem. Phys. **130**, 134706 (2009)
13. Y. Chen, J.D. Lee, L. Xiong, Stresses and strains at nano/micro scales. J. Mech. Mater. Struct. **1**(4), 705–723 (2006)
14. Y. Chen, J.D. Lee, Conservation laws at nano/micro scales. J. Mech. Mater. Struct. **1**(4), 681–704 (2006)
15. Q. Deng, L. Xiong, Y. Chen, Coarse-graining atomistic dynamics of brittle fracture by finite element method. Int. J. Plast. **26**, 1402–1414 (2010)
16. Q. Deng, Y. Chen, A coarse-grained atomistic method for 3D dynamic fracture simulation. Int. J. Multiscale Comput. Eng. **11**, 227–237 (2013)
17. L. Xiong, G. Tucker, D.L. McDowell, Y. Chen, Coarse-grained atomistic simulation of dislocations. J. Mech. Phys. Solids **59**, 160–177 (2011)
18. L. Xiong, Q. Deng, G. Tucker, D.L. McDowell, Y. Chen, Coarse-grained atomistic simulations of dislocations in Al, Ni and Cu crystals. Int. J. Plast. **38**, 86–101 (2012)
19. L. Xiong, S. Xu, D.L. McDowell, Y. Chen, Concurrent atomistic–continuum simulations of dislocation–void interactions in fcc crystals. Int. J. Plast. **65**, 33–42 (2015)
20. L. Xiong, D.L. McDowell, Y. Chen, Sub-THz Phonon drag on dislocations by coarse-grained atomistic simulations. Int. J. Plast. **55**, 268–278 (2014)
21. L. Xiong, X. Chen, N. Zhang, D.L. McDowell, Y. Chen, Prediction of phonon properties of 1D polyatomic systems using concurrent atomistic–continuum simulation. Arch. Appl. Mech. **84**, 1665–1675 (2014)
22. L. Xiong, Q. Deng, G. Tucker, D.L. McDowell, Y. Chen, A concurrent scheme for passing dislocations from atomistic to continuum domains. Acta Mater. **60**, 899–913 (2012)
23. L. Xiong, Y. Chen, Coarse-grained simulations of single-crystal silicon. Modell. Simul. Mater. Sci. Eng. **17**, 035002 (2009)

24. L. Xiong, Y. Chen, Multiscale modeling and simulation of single-crystal MgO through an atomistic field theory. Int. J. Solids Struct. **46**, 1448–1455 (2009)
25. S. Xu, R. Che, L. Xiong, Y. Chen, D.L. McDowell, A quasistatic implementation of the concurrent atomistic-continuum method for FCC crystals. Int. J. Plast. **72**, 91–126 (2015)
26. J.H. Irving, J.G. Kirkwood, The statistical mechanical theory of transport processes. IV. The equations of hydrodynamics. J. Chem. Phys. **18**, 817–829 (1950)
27. R.J. Hardy, Formulas for determining local properties in molecular dynamics simulations: shock waves. J. Chem. Phys. **76**, 622–628 (1982)
28. J.G. Kirkwood, The statistical mechanical theory of transport processes I. General theory. J. Chem. Phys. **14**, 180–201 (1946)
29. C. Kittel, *Introduction to Solid State Physics* (John Wiley & Sons Inc, New York, 1956)
30. M. Born, K. Huang, *Dynamical Theory of Crystal Lattices* (Oxford University Press, London, 2014)
31. Y. Chen, J.D. Lee, L. Xiong, A generalized continuum theory and its relation to micromorphic theory. J. Eng. Mech. **135**, 149–155 (2009)
32. Y. Chen, J.D. Lee, A. Eskandarian, Atomistic viewpoint of the applicability of microcontinuum theories. Int. J. Solids Struct. **41**, 2085–2097 (2004)
33. Y. Chen, J.D. Lee, Connecting molecular dynamics to micromorphic theory. (I). Instantaneous and averaged mechanical variables. Phys. A Stat. Mech. Appl. **322**, 359–376 (2003)
34. Y. Chen, J.D. Lee, Connecting molecular dynamics to micromorphic theory. (II). Balance laws. Phys. A Stat. Mech. Appl. **322**, 377–392 (2003)
35. Y. Chen, J.D. Lee, Multiscale modeling of polycrystalline silicon. Int. J. Eng. Sci. **42**, 987–1000 (2004)
36. Y. Chen, J.D. Lee, A. Eskandarian, Micropolar theory and its applications to mesoscopic and microscopic problems. Comput. Model. Eng. Sci. **5**, 35–43 (2004)
37. Y. Chen, J.D. Lee, Determining material constants in micromorphic theory through phonon dispersion relations. Int. J. Eng. Sci. **41**, 871–886 (2003)
38. Y. Chen, J.D. Lee, A. Eskandarian, Examining the physical foundation of continuum theories from the viewpoint of phonon dispersion relation. Int. J. Eng. Sci. **41**, 61–83 (2003)
39. A.C. Eringen, *Mechanics of Micromorphic Continua* (Springer, Heidelberg, 1958)
40. X. Zeng, Y. Chen, J.D. Lee, Determining material constants in nonlocal micromorphic theory through phonon dispersion relations. Int. J. Eng. Sci. **44**, 1334–1345 (2006)
41. C.J. Fennell, J.D. Gezelter, Is the Ewald summation still necessary? Pairwise alternatives to the accepted standard for long-range electrostatics. J. Chem. Phys. **124**, 234104 (2006)
42. D. Wolf, P. Keblinski, S.R. Phillpot, J. Eggebrecht, Exact method for the simulation of Coulombic systems by spherically truncated, pairwise r summation. J. Chem. Phys. **110**, 8254–8282 (1999)
43. H. Ohta, Thermoelectrics based on strontium titanate. Mater. Today **10**, 44–49 (2007)
44. C.L. Canedy, H. Li, S.P. Alpay, L. Salamanca-Riba, A.L. Roytburd, R. Ramesh, Dielectric properties in heteroepitaxial Ba0.6Sr0.4TiO3 thin films: effect of internal stresses and dislocation-type defects. Appl. Phys. Lett. **77**, 1695–1697 (2000)
45. M.W. Chu, I. Szafraniak, R. Scholz, C. Harnagea, D. Hesse et al., Impact of misfit dislocations on the polarization instability of epitaxial nanostructured ferroelectric perovskites. Nat. Mater. **3**, 87–90 (2004)
46. S. Yang, L. Xiong, Q. Deng, Y. Chen, Concurrent atomistic and continuum simulation of strontium titanate. Acta Mater. **61**, 89–102 (2013)
47. S. Yang, Y. Chen, Concurrent atomistic and continuum simulation of bi-crystal strontium titanate with tilt grain boundary. Proc. Math. Phys. Eng. Sci. **471**, 20140758 (2009)
48. S. Yang, N. Zhang, Y. Chen, Concurrent atomistic and continuum simulation of polycrystalline strontium titanate. Philos. Mag. **95**, 2697–2716 (2015)
49. S. Plimpton, Fast parallel algorithms for short-range molecular dynamics. J. Comput. Phys. **117**, 1–19 (1995)
50. B. Thomas, N. Marks, B.D. Begg, Developing pair potentials for simulating radiation damage in complex oxides. Nucl. Instrum. Methods Phys. Res. Sect. B **228**, 288–292 (2005)

51. D. Spearot, D. McDowell, Atomistic modeling of grain boundaries and dislocation processes in metallic polycrystalline materials. J. Eng. Mater. Technol. **131**, 041204 (2009)
52. H.S. Lee, T. Mizoguchi, J. Mistui et al., Defect energetics in $SrTiO_3$ symmetric tilt grain boundaries. Phys. Rev. B **83**, 104110 (2011)
53. M. Imaeda, T. Mizoguchi, Y. Sato et al., Atomic structure, electronic structure, and defect energetics in [001](310) $\sum 5$ grain boundaries of $SrTiO_3$ and $BaTiO_3$. Phys. Rev. B **78**, 245320 (2008)
54. N. Benedeck, A.S. Chua, C. Elsässer, A. Sutton, M. Finnis, Interatomic potentials for strontium titanate: an assessment of their transferability and comparison with density functional theory. Phys. Rev. B **78**, 064110 (2008)
55. B. Thomas, N. Marks, B.D. Begg, Defects and threshold displacement energies in $SrTiO_3$ perovskite using atomistic computer simulations. Nucl. Inst. Methods Phys. Res. B **254**, 211–218 (2007)
56. R.A. Schultz, M.C. Jensen, R.C. Bradt, Single crystal cleavage of brittle materials. Int. J. Fract. **65**, 291–312 (1994)
57. A. Suzuki, M.F.X. Gigliotti, P.R. Subramanian, Novel technique for evaluating grain boundary fracture strength in metallic materials. Scr. Mater. **64**, 1063–1066 (2011)
58. K. Yang, N. Ho, H. Lu, Deformation microstructure in (001) Single crystal strontium titanate by Vickers indentation. J. Am. Ceram. Soc. **92**, 2345–2353 (2009)
59. N. Doukhan, J.C. Doukhan, Dislocations in perovskites $BaTiO_3$ and $CaTiO_3$. Phys. Chem. Miner. **13**, 403–410 (1986)
60. J.P. Poirier, J. Peyronneau, J.Y. Gesland, G. Brebec, Viscosity and conductivity of the lower mantle; an experimental study on a MgSiO3 perovskite analogue, KZnF3. Phys. Earth Planet. Inter. **32**, 273–287 (1983)
61. V.B. Shenoy, R. Philips, E.B. Tadmor, Nucleation of dislocations beneath a plane strain indenter. J. Mech. Phys. Solids **48**, 649–673 (2000)
62. J. Li, AtomEye: an efficient atomistic configuration viewer. Model. Simul. Mater. Sci. Eng. **11**, 173 (2003)
63. K. Matsunaga, S. Li, C. Iwamoto, T. Yamamoto, Y. Ikuhara, In situ observation of crack propagation in magnesium oxide ceramics. Nanotechnology **15**, S376–S381 (2004)
64. M. Castillo-Rodriguez, W. Sigle, Dislocation dissociation and stacking-fault energy calculation in strontium titanate. Scr. Mater. **62**, 270–273 (2010)
65. P. Hirel, P. Marton, M. Mrovec, C. Elsässer, Theoretical investigation of {110} generalized stacking faults and their relation to dislocation behavior in perovskite oxides. Acta Mater. **58**, 6072–6079 (2010)
66. W. Sigle, C. Sarbu, D. Brunner, M. Ruhle, Dislocations in plastically deformed $SrTiO_3$. Philos. Mag. **86**, 4809–4821 (2006)
67. O. Bernard, M. Andrieux, S. Poissonnet, A.M. Huntz, Mechanical behaviour of ferroelectric films on perovskite substrate. J. Eur. Ceram. Soc. **24**, 763 (2004)
68. P. Paufler, B. Bergk, M. Reibold, A. Belger, N. Patzke, D. Meyer, Why is $SrTiO_3$ much stronger at nanometer than at centimeter scale? Solid State Sci. **8**, 782–792 (2006)
69. D. Farkas, Atomistic simulations of metallic microstructures. Curr. Opin. Solid State Mater. Sci. **17**, 284–297 (2013)

Chapter 9
Continuum Metrics for Atomistic Simulation Analysis

Garritt J. Tucker, Dan Foley, and Jacob Gruber

9.1 Introduction

Over the past two decades, computational methods have been increasingly utilized to study materials and their structure–property relationships. One of the most commonly used computational methods to explore materials and other physical sciences at the nanoscale is atomistic simulations (AS). Two of the more common AS methods are molecular dynamics (MD) and molecular statics (MS)—see the fundamental chapter on these methods for more information. The common feature of all AS methods is that the trajectory of every atom in the simulation is explicitly solved for and tracked. Traditionally, AS results lead to impressive colored images of materials and structures (including various defects, cracks, interfaces, etc.), along with fundamental insight into the material behavior with evolutionary plots/fits of energy, phases, structural distributions, and forces. AS have become valuable tools for nearly all Materials Science programs to explore and understand the myriad of fundamental processes and transformations within materials. AS, such as MD, have provided enormous insight into the structure–property relationships of materials, especially as it pertains to the mechanics and physical mechanisms at the nanoscale, including dislocations, grain boundaries, surfaces, and the interplay of more complex topologies and compositions. Despite the continued improvement and accuracy of atomistic modeling that is occurring, *the impact or value of these tools will ultimately be linked to extracting key information from the simulations.*

Therefore, this chapter aims to provide an overview of methods used to analyze AS data, and introduce the reader to a new and emerging set of kinematic metrics

---

G.J. Tucker (✉) • D. Foley • J. Gruber
Department of Materials Science and Engineering, Drexel University, Philadelphia, PA 19104, USA
e-mail: gtucker@coe.drexel.edu

C.R. Weinberger, G.J. Tucker (eds.), *Multiscale Materials Modeling for Nanomechanics*, Springer Series in Materials Science 245,
DOI 10.1007/978-3-319-33480-6_9

that can be utilized to extract potentially more useful information from AS (and possibly be used as part of a multiscale modeling framework) [6–10]. A significant value of these new metrics is that they are based on continuum mechanics theory, but formulated within an atomistic framework. In other words, their mathematical treatment originates from the continuum treatment of materials mechanics (as material points), but has been adapted to a non-local point-wise (or atomistic) system. By originating these metrics from continuum mechanics, they naturally possess a number of advantages as compared to geometric-based approaches, as will be detailed later. In addition, the non-local consideration of the calculated kinematics can easily be extended by including additional nearest neighbor shells around each atom in the calculations (assuming proper weighting on additional atomic neighbors).

After providing an overview of the metrics and their formulation, this chapter contains a set of implementation instructions for the reader to successfully utilize the metrics. Following this, the value of these new metrics in analyzing AS data will be illustrated by applying them to a sample of important deformation mechanics in nanostructured materials. At the conclusion of this chapter, there will be conclusions, along with potential extensions of these metrics, and an overview of the available codes and software for these metrics.

9.2 Mathematical Formulations

In this section, we provide an overview of the mathematical formulations and use of metrics implemented to analyze AS. In the first subsection, we briefly discuss four of the traditional metrics or quantities used, and then discuss the newly developed microscale continuum-based kinematic metrics for AS analysis in the second subsection. Overall, there are two distinct classes of atomistic metrics covered in this chapter: (1) *instantaneous* or *static* and (2) *non-static* or *path-dependent*. For (1), calculations are based solely on the instantaneous atomic configuration of interest, while for (2), the user must prescribe a reference configuration. The advantage of (2) is that the "history" or "path" of the atomic neighborhood can be quantitatively tracked using a reference configuration, while no "history" or "path" is included in the calculation of metrics that fall within type (1).

9.2.1 *Traditional Metrics for Atomistic Analysis*

There are four particular metrics that we highlight in this section that are commonly utilized for analysis of nanomechanical studies using AS. Specifically, we discuss potential energy, centrosymmetry, the common neighbor analysis method [1, 4], and slip vector [11]. The first three are *instantaneous* or *static* metrics, while the fourth is *non-static* or *path-dependent*. *Instantaneous* or *static* metrics are those atomic quantities which only require a single atomic snapshot to characterize the constituent atoms. In other words, there is no information in this metrics of past positions or structures. Such metrics are very useful for determining structures and identifying defects. As their name implies *instantaneous* metrics characterize atoms based on their *instantaneous* position relative to their neighbors. Another commonly used metric for atomistic simulation data analysis is slip vector. The atomic slip vector estimates the relative slip in the surrounding atomic neighborhood, as compared to a prescribed reference configuration. For example, slip vector can help to uncover regions that have been traversed by dislocations in a crystalline lattice, or the atomic shuffling/rearrangement that occurs in a grain boundary due to relaxation or some external stimuli, such as mechanical loading. Additional descriptions and a more detailed comparison of these four traditional metrics are provided in this subsection.

9.2.1.1 Potential Energy

The interatomic energy describes the potential energy of an atom, based on its local atomic neighborhood and chosen interatomic potential. For each atom (α) in the simulation, there is a defined cut-off radius that defines the atomic neighborhood around it. Each atom located within the cut-off radius is considered a neighbor atom (β) of atom α. Each atom β contributes to the potential energy of atom α, and the potential energy, or energy, of atom α can be used to visualize and determine defected structures within AS. Atoms with higher energy will be located within defects, such as dislocations, grain boundaries, and surfaces. To determine the energy of atom α, a functional form of the interatomic potential must first be determined and summed over all β neighbors. See the fundamental chapter on AS for an overview of common interatomic potential functionals used in solid materials.

9.2.1.2 Centrosymmetry

In a centrosymmetric crystal (such as face-centered cubic), an atom α and its opposing nearest neighbors β will exist along a straight line through their atomic centers. Any deviation from this centrosymmetric configuration will be captured by numerical differences in the centrosymmetry value. More specifically, centrosymmetry is defined by Eq. (9.1).

$$p_{CSP} = \sum_{i=1}^{N/2} |r_i + r_{i+N/2}|^2 \tag{9.1}$$

According to this definition, atoms in a centrosymmetric crystalline system will have a centrosymmetry of zero, while defected atoms in that same crystal will have a centrosymmetry greater than zero. Therefore, the centrosymmetry metric is particularly useful in identifying defected atoms and distinguishing between various defected states. This includes atoms in a strained state and those adjacent to defects and interfaces. Despite its useful application in identifying defected atoms, centrosymmetry is also sensitive to deviations from centrosymmetry due to thermal fluctuations (e.g., at high temperature), Therefore, care must be taken when interpreting the results from centrosymmetry when substantial thermal energy is present in the system.

9.2.1.3 Common Neighbor Analysis Method

Another commonly used metric is common neighbor analysis (CNA). A particular advantage of CNA is that it can be leveraged to distinguish between the various crystalline phases or structures and is less sensitive to thermal vibrations than centrosymmetry. CNA has multiple components, the first of which is the number of neighbors within a nearest neighbor cut-off distance. The nearest neighbor cut-off radius is usually determined to be the mean of the first and second nearest neighbor distances. The remaining components of CNA are the three characteristic numbers defined as n_{cn}, the number of neighbors shared by a central atom and its bonded neighbor, n_b, the total number of bonds between those common neighbors, and n_{lcb}, the longest chain of bonds connecting the common neighbors. The combination of the number of nearest neighbors and the bond numbers makes CNA a unique identifier for crystalline atoms. Table 9.1 provides CNA data of the FCC, BCC, and HCP systems.

9.2.1.4 Atomic Slip Vector

Slip vector provides a measure of deformation experienced between a central atom α and its nearest neighbors β. Zimmerman et al. [11] defined slip vector as outlined in Eq. (9.2)

$$s^{\alpha} = -\frac{1}{n_s}\sum_{\beta\neq\alpha}^{n}(x^{\alpha\beta} + X^{\alpha\beta}) \tag{9.2}$$

Table 9.1 This table shows the CNA data for common crystal structures. CNA data is arranged as $n_{neigh} \times (n_{cn}n_bn_{lcb})$

FCC	BCC	HCP
12 × (421)	8 × (666)	6 × (421)
	6 × (444)	6 × (422)

In Eq. (9.2), s^{α} is the slip vector of atom α, n is the number of nearest neighbors β, n_s is the number of slipped neighbors, and $x^{\alpha\beta}$ and $X^{\alpha\beta}$ are the interatomic vectors between atoms α and β in the current and reference configurations, respectively. Since the slip vector measures local deformation relative to a reference configuration, it is very useful for identifying defects. One of the primary advantages of the slip vector is that it can be used to estimate the Burgers vector of a dislocation, thereby providing a means of distinguishing a perfect dislocation from a partial dislocation in FCC metals. It must be noted that since slip vector relies on a reference configuration, we consider slip vector as a *non-static* or *path-dependent* metric.

9.2.1.5 Comparison of Metrics

When investigating deformation and nanoscale mechanics from AS data, the ability to effectively visualize results is often pivotal to success. The metrics discussed so far in this chapter provide a convenient means for visualizing atoms in a way that highlights various structures and configurations. The images in Fig. 9.1a–d show a nanocrystalline Cu simulation, visualized with OVITO [3], where atoms are color coded according to energy (Fig. 9.1a), centrosymmetry (Fig. 9.1b), CNA (Fig. 9.1c), and slip vector magnitude (Fig. 9.1d). In each image, different defects common to crystalline materials are highlighted with colored arrows.

The images in Fig. 9.1a–d demonstrate how each metric can be used to identify grain boundaries and defects. Atomic potential energy (Fig. 9.1a) shows grain boundary atoms and atoms associated with a defect as being green, indicating a higher energy than the bulk crystalline atoms. However, atomic energy does not completely capture the stacking fault present in the simulation. Centrosymmetry improves on energy in identifying the stacking faults and grain boundaries, but provides little means of distinguishing various defects and crystalline structures from each other. CNA successfully identifies the stacking fault as HCP phase, even highlighting the bounding partial dislocation at the termination end of the stacking fault in red. Like centrosymmetry and CNA, slip vector successfully identify grain boundaries and the stacking fault. However, unlike the previous metrics, slip vectors also show differences in the grain boundaries present and slip planes traversed by a perfect dislocation.

9.2.2 Continuum-Based Kinematic Metrics for Atomistic Analysis

In this section, a mathematical overview is provided for a set of new continuum-based kinematic metrics that can be implemented within AS. Instead of relying on material points, these formulations rely on the exact positions of the atoms.

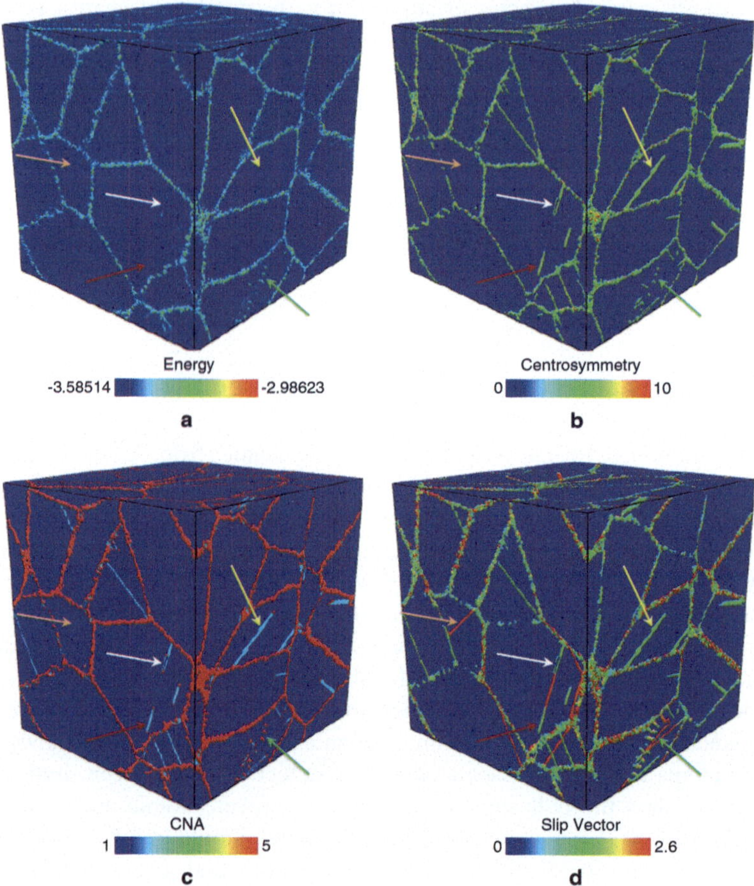

Fig. 9.1 A comparison of coloring atoms according to (**a**) energy, (**b**) centrosymmetry, (**c**) CNA, and (**d**) slip vector for a nanocrystalline copper structure. In each image, different features/mechanisms have been highlighted using colored arrows: (*yellow*) stacking fault, (*red*) extended dislocation, (*white*) twin boundaries, (*orange*) glide plane that has been traversed by a perfect dislocation

As mentioned previously, there are many distinct advantages to these metrics over geometry-based approaches for kinematic and deformation quantification, and there are a few points to highlight. First, the mathematical foundations originate from continuum mechanics with no presumed ideal shape or structure of an atomic neighborhood. Second, the reference state is defined explicitly by the user and is entirely flexible. Third, the effective volume of the calculation can be easily varied by including additional neighbors, or by increasing the cut-off radius. Finally, these kinematic tools can be applied to a variety of materials, even systems that lack a crystalline or ordered structure, as later described in the Conclusions and Outlook.

9.2.2.1 Deformation Gradient

From continuum mechanics, a reference configuration is mapped to a current or deformed configuration by the deformation mapping **F** as defined by

$$\mathbf{F} = \frac{\partial \mathbf{x}}{\partial \mathbf{X}} \tag{9.3}$$

In this chapter, reference configuration quantities will be noted by upper case symbols and current configuration quantities will be noted by lowercase symbols. In the equation above, **x** is the vector in the current configuration between two material points, and **X** is the vector between those same two material points in the reference configuration. However, as stated above, since material points do not explicitly exist in AS, the atomic positions and interatomic distances are used instead for this mapping. Using the interatomic separation distance, $(x^{\alpha\beta})$, the deformation mapping of atom α and a neighboring atom β can be written as

$$(x^{\alpha\beta})_i = F_{iI}(X^{\alpha\beta})_I \tag{9.4}$$

where the subscripts i and I are summed over each dimension. This relationship certainly holds for each α-β pair, but cannot accurately capture **F** for all atomic neighbors. Therefore, to obtain a more accurate estimation of **F**, the formulation relies on averaging within a specified finite domain that will encompass numerous neighbor atoms β. This is especially important in atomistic systems where the deformation in the neighborhood around each atom α is not homogeneous, as illustrated in Fig. 9.2.

To approximate the solution, the error is minimized over all neighbors within that cut-off radius and is performed by rearranging the above equation into

$$(x^{\alpha\beta})_i - F_{iI}(X^{\alpha\beta})_I = 0 \tag{9.5}$$

and summing the squared error over all neighbors gives

$$\Phi_i^{\alpha} = \sum_{\beta=1}^{N} \left((x^{\alpha\beta})_i - F_{iI}(X^{\alpha\beta})_I\right)^2 \tag{9.6}$$

Then, by some choice of F_{iI}^{α} for every i and I

$$\frac{\partial \Phi_i^{\alpha}}{\partial F_{iI}^{\alpha}} = 0 \tag{9.7}$$

As described in [10], this approach will lead to an estimated atomic deformation gradient tensor for each atom α in the system based on its neighboring atoms β from the minimization of the least squares summation. If we assume that there are

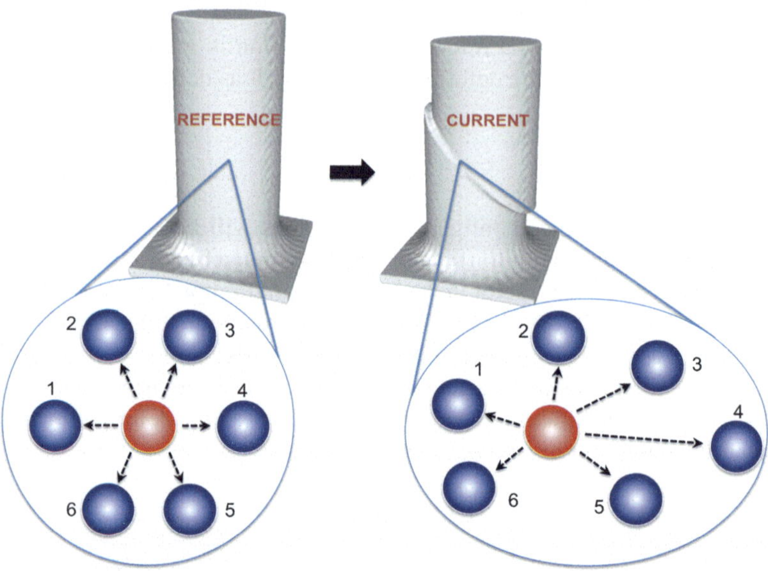

Fig. 9.2 An illustration showing the tracking of the local atomic neighborhood from the reference configuration to the current configuration about an atom α (*red*), and how the interatomic vectors between α and each neighboring atom β can change during deformation

N number of β atomic neighbors, the atomic deformation gradient tensor for atom α ($\mathbf{F}^{\alpha}$) is then reduced to

$$\left(\sum_{\beta=1}^{N}(x^{\alpha\beta})_i(X^{\alpha\beta})_M\right)_{iM} = F^{\alpha}_{iI}\left(\sum_{\beta=1}^{N}(X^{\alpha\beta})_I(X^{\alpha\beta})_M\right)_{IM} \tag{9.8}$$

If we let

$$(\omega^{\alpha})_{iM} = \sum_{\beta=1}^{N}(x^{\alpha\beta})_i(X^{\alpha\beta})_M \tag{9.9}$$

and

$$(\eta^{\alpha})_{IM} = \sum_{\beta=1}^{N}(X^{\alpha\beta})_I(X^{\alpha\beta})_M \tag{9.10}$$

we can simplify our expression for the atomic deformation gradient tensor ($\mathbf{F}^{\alpha}$) to be

$$(\omega^{\alpha})_{iM} = (F^{\alpha})_{iI}(\eta^{\alpha})_{IM} \tag{9.11}$$

This formulation for the atomic deformation gradient tensor was also shown to satisfy the compatibility relation $curl(\mathbf{F}) = 0$ in the interior of a deformed crystal by Zimmerman et al. [10]. Note the dependence of properly calculating the atomic deformation gradient on interatomic distances in both the current and reference configurations. In addition, the reference neighbor list defines the summation over β. Following from the multiplicative decomposition of **F** into the rotation tensor **R** and stretch tensor **U**, additional kinematic metrics can now be obtained. For further details on **F**, its decomposition, and its use in analyzing grain boundary deformation, the reader is encouraged to consult [6–9].

9.2.2.2 Microrotation

To extend the usefulness and capabilities of these kinematic metrics, we build upon the decomposition of the atomic deformation gradient tensor (**F**) into the rotation tensor (**R**) and stretch tensor (**U**) by formulating a few additional kinematic quantities. First, the rotation tensor (**R**) and its transpose ($\mathbf{R}^T$) are used to produce the skew-symmetric part of **R** as ($\mathbf{R}_{skew}$) according to

$$\mathbf{R}_{skew} = \frac{1}{2}\left(\mathbf{R} - \mathbf{R}^T\right) \tag{9.12}$$

Then, an axial representation of $\mathbf{R}_{skew}$, the microrotation vector ($\boldsymbol{\phi}$), is defined using the permutation tensor (ϵ_{ijk}) as

$$\phi_k = -\frac{1}{2}\epsilon_{ijk}\left(R_{skew}\right)_{ij} \tag{9.13}$$

Using the same atomic configuration shown in Fig. 9.1, but where atoms are now colored according to the magnitude of the calculated microrotation vector ($\boldsymbol{\phi}$), a more detailed kinematic understanding emerges (see Fig. 9.3), as compared to the traditional metrics shown in Fig. 9.1.

9.2.2.3 Strain

Over the years, there have been many attempts to accurately quantify strain within an atomistic framework. In our approach, we once again leverage the mathematical origins of strain from continuum mechanics. Specifically, we use our formulation for the atomic deformation gradient **F** to estimate the Green-Lagrange Strain tensor **E** as defined by

$$\mathbf{E} = \frac{1}{2}\left[\mathbf{F}^T\mathbf{F} - \mathbf{I}\right] \tag{9.14}$$

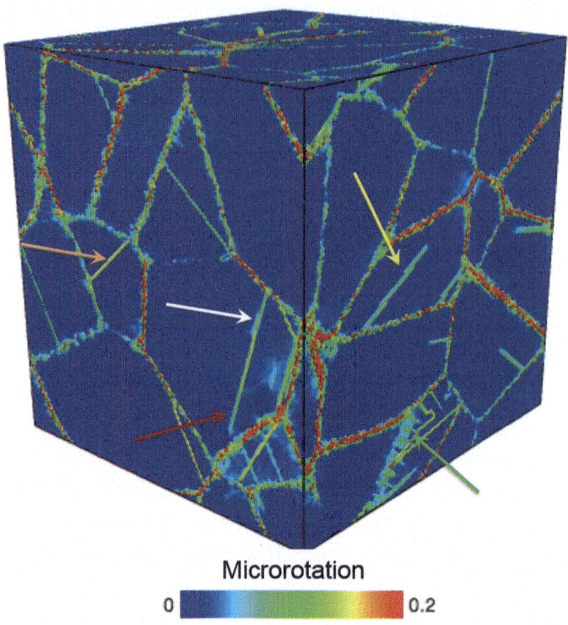

Fig. 9.3 Atoms are colored according to the magnitude of their microrotation vector, and the same colored arrows from Fig. 9.1 are included to highlight those particular locations within the microstructure. Comparing this images with those in Fig. 9.1, the path-dependent kinematic information from slip vector is captured, along with additional information about lattice elastic strain near defects and grain boundaries. In addition, both dislocation and grain boundary deformation are highlighted using microrotation

Then, we can approximate atomic volumetric strain (e—dilatation for small strain) from the first invariant of **E** as

$$e = tr(\mathbf{E}) \tag{9.15}$$

9.2.2.4 Velocity Gradient

Each of the previous kinematic metrics (i.e., **F**, **R**, $\boldsymbol{\phi}$, and **E**) relied on a prescribed reference state for their formulation. The next couple of metrics do not have that requirement. Instead, these metrics will only use the current or instantaneous configurational properties. Utilizing the instantaneous atomic velocities, the velocity gradient **L** is formulated using a similar approach employed for **F**, and is defined as

$$\left(\sum_{\beta=1}^{N}(v^{\alpha\beta})_i(x^{\alpha\beta})_m\right)_{im} = L^{\alpha}_{ik}\left(\sum_{\beta=1}^{N}(x^{\alpha\beta})_k(x^{\alpha\beta})_m\right)_{km} \tag{9.16}$$

where

$$(\rho^{\alpha})_{im} = \sum_{\beta=1}^{N}(v^{\alpha\beta})_i(x^{\alpha\beta})_m \tag{9.17}$$

and

$$(\tau^{\alpha})_{km} = \sum_{\beta=1}^{N} (x^{\alpha\beta})_k (x^{\alpha\beta})_m \tag{9.18}$$

simplifies our expression for **L** to be

$$(\rho^{\alpha})_{im} = (L^{\alpha})_{ik} (\tau^{\alpha})_{km} \tag{9.19}$$

It is important to reiterate that this definition is a function of the current configuration properties only (i.e., atomic velocities and neighbors), and therefore provides insight into current deformation behavior of the simulation, as shown later.

9.2.2.5 Vorticity

Following continuum mechanics theory, the vorticity (or spin) tensor **W** is a function of both **L** and its transpose ($\mathbf{L}^T$) as defined by

$$\mathbf{W} = \frac{1}{2} \left(\mathbf{L} - \mathbf{L}^T \right) \tag{9.20}$$

Then, in a similar approach used for the microrotation vector, the vorticity vector, **w**, is calculated as the dual vector of **W**,

$$w_k = -\frac{1}{2} \epsilon_{ijk} \left(W \right)_{ij} \tag{9.21}$$

9.3 Methodology and Implementation

In this section, we provide an overview of the steps/procedure on how to implement the metrics outlined above to analyze the atomic output files from an AS. It is assumed that the reader fully understands their chosen approach to produce AS data. For our purposes, we output a series of atomic files (i.e., dump files) that contain information on the simulation cell and atoms (in a LAMMPS dump file format). Each dump file is produced for a specific timestep in the simulation, and data for each atom is output into the dump file (e.g., position coordinates, CNA value (see above), centrosymmetry, potential energy, the atomic stress tensor, and sometimes the atomic velocity vector, if vorticity is needed). However, for many of the metrics outlined in this chapter, the user will only need the position coordinates of each atom in the dump files.

As many of the continuum kinematic metrics detailed in the last section require a reference configuration, the user will need to specify this state. In our examples, we choose the undeformed (i.e., zero strain) configuration for our reference state, because our examples are inducing strain into the simulation cell. Properly defining the reference state is important as the atomic neighbor list for each atom is calculated from this state. Therefore, if we are studying the uniaxial strain of a metallic nanowire, we will choose the undeformed nanowire configuration at zero strain for our reference state, and calculate the neighbor list for each atom from this atomistic configuration. Then, the continuum kinematic metrics can be calculated for each atom in subsequent "deformed" states. The general approach for implementation is as follows:

- Collect all simulation data containing atomic positions at various "times" or system states throughout the simulation.
- Determine the appropriate reference configuration or state for the project goals, and the defined cut-off radius (we use the first neighbor shell).
- Implement a nearest neighbor routine on the reference configuration for every atom, to store the reference neighbor list for each atom.
- In the current (or deformed) configuration file, loop through every atom of interest and perform the metric calculations as outlined above for its neighbor list obtained from the reference file. It is critical that the identities of the atoms be preserved through the simulation to correctly calculate the interatomic distances between atom α and its neighbors, where those neighbors in the current state are the same atoms from the reference neighbor list.
- Store or output the tensorial data to a file for future visualizations or further analysis. Note: if additional neighbor shells are included, proper weighting should be considered [8].

In the following sections, we cover a few examples that highlight some advantages and distinctions of leveraging these metrics over other methods, and their utility in uncovering the fundamental mechanisms and kinematics during materials deformation.

9.4 Dislocations in Nanostructured Materials

The deformation and mechanics in nanostructured materials is complex. A variety of mechanisms or strain accommodation processes have been stated to operate during deformation. One of the more important mechanisms or defects in crystalline materials is the dislocation. As dislocations are the main carrier of plastic deformation in crystalline materials, understanding the activation, motion, and role during the straining of a crystal is critical for materials mechanics and modeling. Thus, in this section we illustrate the utility of the kinematic metrics in identifying and quantifying the role of dislocations in regard to nanomechanics.

An AS of the uniaxial compression of nanocrystalline copper was performed, where snapshots of the atomic structure were output. The nanostructure was studied around the yield point and in one grain we found a dislocation being emitted from a triple junction. In Fig. 9.4, a series of snapshots are shown of this event in the

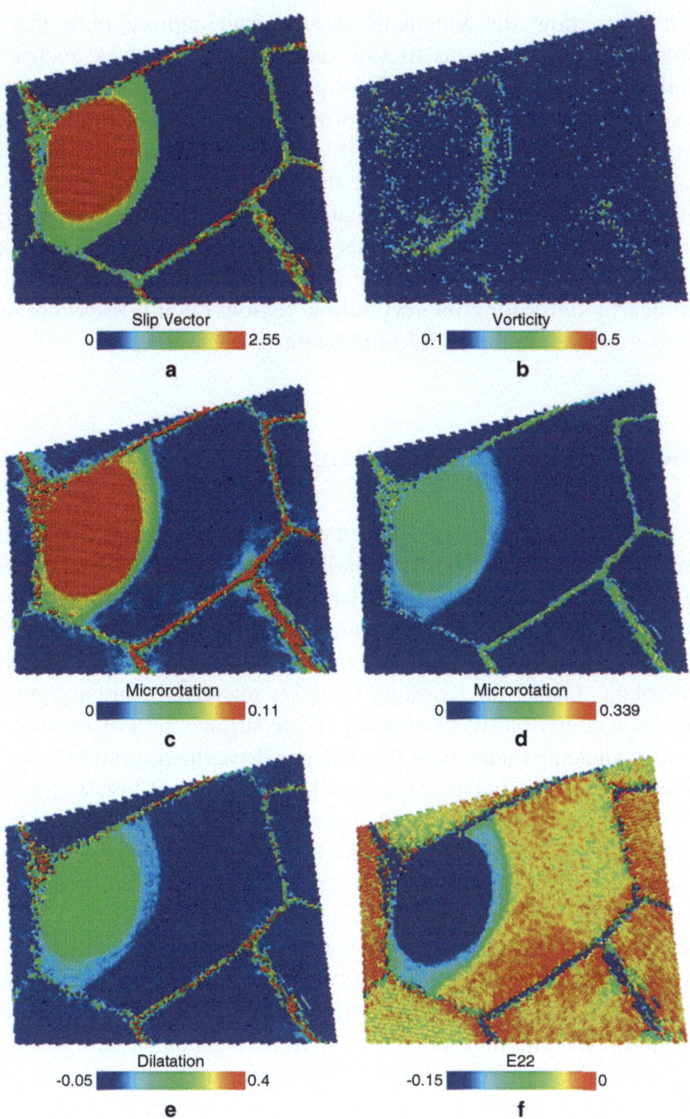

Fig. 9.4 Dislocation emission from a triple junction in nanocrystalline copper with atoms colored according to (**a**) slip vector, (**b**) vorticity—first nearest neighbors, (**c**-**d**) microrotation, (**e**) dilatation, and (**f**) the yy-component of the strain tensor

nanostructure, where atoms are colored according to different metrics to highlight their various advantages. It is important to note that this atomic configuration is also shown in a previous chapter on atomistic methods where atoms are colored according to potential energy, centrosymmetry, and the CNA method in that chapter.

Here, we show how the kinematic metrics can capture both the structure and deformation path, analogous to slip vector. However, these metrics contain additional information in regard to the deformation, rotation, and strain of the atomic neighborhood. In addition, as vorticity is formulated from the instantaneous velocity distribution, the current deformation activity within the microstructure can be easily identified and extracted using this metric (e.g., dislocation position and velocity). The microrotation metric provides qualitatively similar results to the slip vector, where regions of various slip magnitudes have unique microrotation magnitudes as well. Finally, note that microrotation also captures regions within the lattice (adjacent to nearby deforming intercrystalline regions) that possess curvature (see Fig. 9.4c), as referenced to the undeformed state.

9.5 Coherent Twin Boundary Migration

As the previous example illustrates, the rotation of an atomic field can be captured from the microrotation metric. Twin boundaries are known to play a fundamental and critical role in the deformation and mechanics of many metallic materials. One important mechanism of twin boundaries is migration. Due to the slip of a twinning partial dislocation adjacent to a twin boundary, the boundary will migrate normal to its plane. Often, the twinning partial is nucleated from a grain boundary at the termination edge of a twin boundary, a free surface, or a dislocation loop can homogeneously nucleate on the twin boundary to drive the migration process. In the current example, we construct an Al nanopillar (as shown in Fig. 9.2) that contains an inclined twin boundary. Molecular dynamics are then employed to indent the top surface of the nanopillar while holding the bottom surface fixed. This results in a twinning partial being nucleated from the free surface and gliding along the twin boundary leading to twin boundary migration.

In Fig. 9.5, a series of images are shown that illustrate how the inclined coherent twin boundary in the nanopillar migrates normal to its plane through the slip of twinning partial dislocations along the twin boundary plane. We colored the atoms according to three specific metrics to highlight the mechanism and show the value of our microrotation metric over slip vector in capturing twinned volumes [6]. The twin boundary migrates normal to its plane due to the slip of a twinning partial dislocation. As this processes continues, the boundary migrates further leading to a surface step in the nanopillar. When the atoms are colored according to slip vector (Fig. 9.5c), the region that has been twinned shows a value of zero; however, the same twinned region shows non-zero microrotation (Fig. 9.5d). Therefore, when twin boundary migration is active, calculating atomic microrotation can be leveraged to isolate regions or volumes within the structure that have been traversed by a migrating twin boundary [6].

In regard to nanomechanics, the ability of separating atomic regions due to their deformation or rotation is important because we are often interested in resolving the contributions of various mechanisms at the nanoscale to the total strain of the

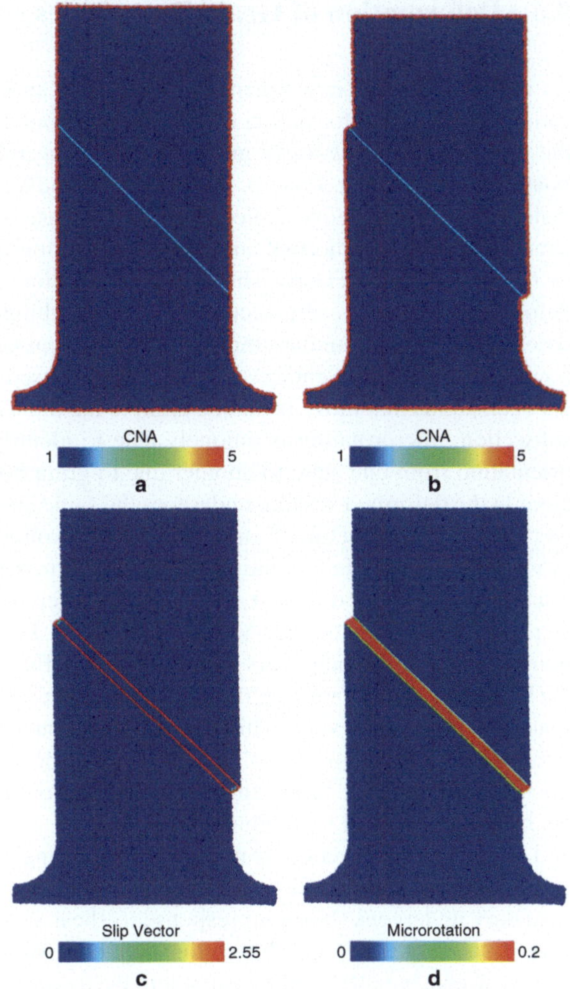

Fig. 9.5 Twin boundary migration under uniaxial compression shown by a series of slices through an Al nanopillar. (**a**) The initial structure with atoms colored according to CNA method. The deformed configuration after twin boundary migration with atoms colored according to (**b**) CNA, (**c**) slip vector, and (**d**) microrotation

material of structure. In the present example, a total compressive strain of 0.034 has been imposed on the nanopillar and no other mechanisms (lattice dislocation migration) were observed besides twin boundary migration. In other words, we are now able to quantify the role of the twin migration mechanism in accommodating strain in these nanopillars with inclined interfaces. This is true for other structures and deformation mechanisms in nanostructured materials as well, such as grain boundary plasticity.

9.6 Deformation of Grain Boundaries

As mentioned, these new kinematic metrics are an invaluable tool for the identification of discrete deformation events at grain boundaries (or interfaces). Consider the migration of a low-angle grain boundary, specifically the ⟨881⟩ symmetric tilt boundary in Cu (Fig. 9.6a–c). This grain boundary is composed of an array of dislocations which move collectively in response to an externally applied shear strain through both a normal and translation component. Coloring atoms according to CNA (Fig. 9.6a) clearly shows the dislocation cores at both the initial and deformed states, while slip vector (Fig. 9.6b) highlights the paths of the dislocation cores during grain boundary migration. The magnitude of the microrotation vector, however, identifies the entire volume that is traversed by the migrating grain boundary (i.e., atoms that move from one grain to the other). As explained in the previous subsection, microrotation is uniquely able to identify changes in crystallographic orientation from one state to another due to grain boundary migration. Since each atom in the deformed volume undergoes the same crystallographic orientation shift, the microrotation vector of each atom in the volume is roughly equivalent. The presence of this spatially contiguous, uniform microrotation field, bounded by two grain boundaries at different positions in different states represents the unique signature of a grain boundary migration event. For example, compare the grain boundary migration behaviors of that shown in Fig. 9.5 with that in Fig. 9.6. With the low-angle grain boundary, the migration of the dislocation cores govern the grain boundary migration event, while for the twin boundary, the migration of twinning partials along the boundary plane are responsible for the migration. By using the microrotation metric, these subtle, but very fundamental, mechanistic differences are both captured and highlighted.

Like migration events, grain boundary sliding events also possess a unique signature when examined using kinematic metrics. The ⟨221⟩ Cu symmetric tilt boundary undergoes sliding in response to shear strain (Fig. 9.6d–g). Visualization using CNA (Fig. 9.6d) and slip vector (Fig. 9.6e) shows only that the grain boundary has deformed, but offers little insight into the nature of deformation. That is, discrete sliding events and atomic reordering events are not easily distinguishable using these metrics. While the magnitude of the microrotation (Fig. 9.6f) is similarly murky, the component of the microrotation vector corresponding to rotation in the shear plane (Fig. 9.6g) shows relatively consistent values with a non-zero average, while components out of plane average to 0, indicative of random restructuring/reordering events. While less straightforward than migration events, kinematic metrics still provide valuable information that allows sliding and restructuring events to be distinguished. Kinematic metrics therefore offer the unique possibility of the fine characterization of the deformation behavior of interfaces.

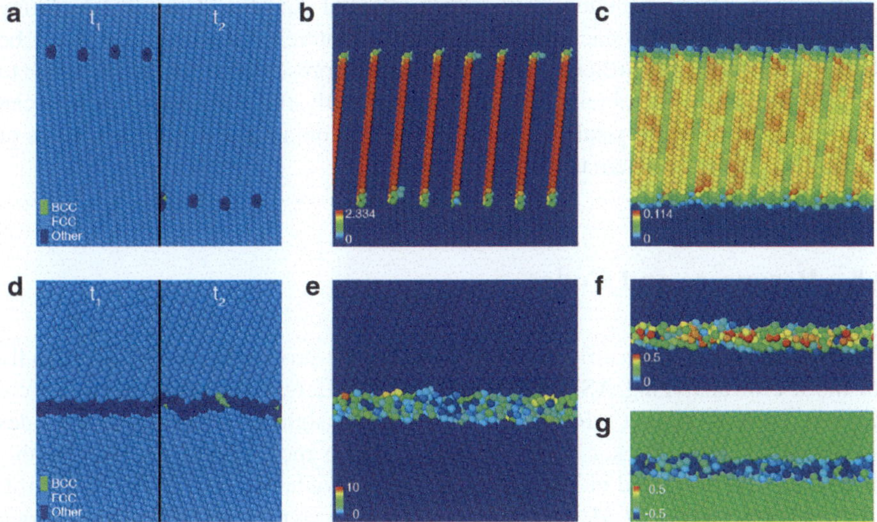

Fig. 9.6 Migration: Atomic visualizations of (**a**) CNA at two timesteps, (**b**) the slip vector magnitude and (**c**) microrotation magnitude computed between these timesteps. **Sliding**: Visualization of (**d**) CNA at two timesteps, (**e**) slip vector magnitude, (**f**) microrotation magnitude, and (**g**) microrotation in the sliding plane computed between the two timesteps

9.7 Conclusions and Outlook

This chapter outlines the development, implementation, and usefulness of a new set of microscale kinematic metrics applied to AS derived from continuum mechanics theory. Furthermore, specific examples were provided to illustrate the utility of these metrics in the visualization, interpretation, and analysis of AS results of deformation within nanostructured materials. By leveraging these new metrics, we have shown how unique insight into the nanoscale deformation and mechanics can be obtained and estimated—including the quantification of deformation, rotation, and vorticity fields. In addition, we presented similar analysis using traditionally used metrics to demonstrate how these new metrics can provide both complimentary and additional understanding.

The metrics overviewed in this chapter should serve as a first-step in the development of additional metrics to assist in extracting more useful and quantitative data from AS. There are numerous additional continuum-based metrics that can potentially be formulated as well as other approximations of deformation and rotation. We encourage researchers to use these metrics as a starting point to further advance the field of AS analysis.

As the formulation and development of these metrics is founded on a point-based approach, these same tools can be used for other material systems and simulation data as well. The formulation does not assume any reference geometry. The user has full control over the chosen reference state and cut-off radius for calculating the nearest neighbors to each point (or atom) of interest. However, care must be taken, as the unique identifier of each point must be preserved from the reference to current configuration to obtain accurate results. Finally, these kinematic metrics can also be applied to non-crystalline systems as well to obtain non-local estimations of the neighboring field kinematics.

9.8 Resources and Codes

This chapter outlines the utilization of multiple post-processing codes, specifically for the use of analyzing AS data. A number of different kinematic metrics were employed here to extract information, trends, and insight from the AS data. Codes developed by Prof. Tucker [5–9] containing these metrics and outlined in this chapter will be formatted in Python and be freely available on the website of Prof. Tucker's *Computational Materials Science and Design (CMSD)* research group [http://extreme.materials.drexel.edu]. In addition, example LAMMPS scripts and atomistic structures will also be freely available for download.

Acknowledgements The atomistic simulations in the chapter were performed using LAMMPS (http://lammps.sandia.gov) [2] and the visualizations were created using OVITO [3]. GT would like to acknowledge the support and collaborations with Prof. David L. McDowell and Dr. Jonathan Zimmerman. This material is based on work partially supported by the National Science Foundation under Grant No. DMR-1410970 (GT and DF), and through a GAANN fellowship (JG). Work reported here was run on hardware supported by Drexel's University Research Computing Facility.

References

1. D. Faken, H. Jonsson, Systematic analysis of local atomic structure combined with 3d computer graphics. Comput. Mater. Sci. **2**(2), 279–286 (1994)
2. S. Plimpton, Fast parallel algorithms for short-range molecular dynamics. J. Comput. Phys. **117**(1), 1–19 (1995)
3. A. Stukowski, Visualization and analysis of atomistic simulation data with ovito - the open visualization tool. Model. Simul. Mater. Sci. Eng. **18**, 015012 (2010)
4. H. Tsuzuki, P.S. Branicio, J.P. Rino, Structural characterization of deformed crystals by analysis of common atomic neighborhood. Comput. Phys. Commun. **177**(6), 518–523 (2007)
5. G.J. Tucker, S.M. Foiles, Molecular dynamics simulations of rate-dependent grain growth during the surface indentation of nanocrystalline nickel. Mater. Sci. Eng. A **571**(1), 207–214 (2013)
6. G.J. Tucker, S.M. Foiles, Quantifying the influence of twin boundaries on the deformation of nanocrystalline copper using atomistic simulations. Int. J. Plast. **65**, 191–205 (2015)
7. G.J. Tucker, S. Tiwari, J.A. Zimmerman, D.L. McDowell, Investigating the deformation of nanocrystalline copper with microscale kinematic metrics and molecular dynamics. J. Mech. Phys. Solids **60**(3), 471–486 (2011)
8. G.J. Tucker, J.A. Zimmerman, D.L. McDowell, Shear deformation kinematics of bicrystalline grain boundaries in atomistic simulations. Model. Simul. Mater. Sci. Eng. **18**(1), 015002 (2010)

9. G.J. Tucker, J.A. Zimmerman, D.L. McDowell, Continuum metrics for deformation and microrotation from atomistic simulations: Application to grain boundaries. Int. J. Eng. Sci. **49**(12), 1424–1434 (2011)
10. J.A. Zimmerman, D.J. Bammann, H. Gao, Deformation gradients for continuum mechanical analysis of atomistic simulations. Int. J. Solids Struct. **46**(2), 238–253 (2009)
11. J.A. Zimmerman, C.L. Kelchner, P.A. Klein, J.C. Hamilton, S.M. Foiles, Surface step effects on nanoindentation. Phys. Rev. Lett. **87**(16), 165507 (2001)

Chapter 10
Visualization and Analysis Strategies for Atomistic Simulations

Alexander Stukowski

10.1 Introduction

Molecular dynamics (MD) and other atomistic modeling techniques are powerful and established research tools that enable us to study processes at the nanoscale with full atomic resolution. An important aspect of such simulation studies is the meaningful visualization of atomic configurations and trajectories, often contributing a lot to the interpretation of numerical results. Visualization and analysis methods are to computational materials scientists what imaging and microscopy techniques are to experimentalists: key tools to reveal processes and mechanisms occurring in materials, which ultimately help us understand materials behavior. Section 10.2 in this chapter will look at specialized data visualization and analysis tools that have been developed for this purpose.

Furthermore, it is worth noting that MD models typically do not prescribe the structures formed by atoms nor do they explicitly keep track of them. Atoms rather dynamically move and arrange according to the acting forces, and the configurational space accessible to a system is subject to few restrictions. This freedom necessitates a careful analysis of the simulation results. Often the precise identification of phases and defects formed by atoms is essential to understand or quantify materials processes. It is therefore necessary to explicitly recover or reconstruct such local information from MD models and identify atomic structures using suitable computational methods [21]. Section 10.3 of this chapter will present several analysis techniques that address this issue and other common challenges in the field of atomistic modeling of materials mechanics.

A. Stukowski (✉)
Technische Universität Darmstadt, Germany
e-mail: stukowski@mm.tu-darmstadt.de

C.R. Weinberger, G.J. Tucker (eds.), *Multiscale Materials Modeling for Nanomechanics*, Springer Series in Materials Science 245,
DOI 10.1007/978-3-319-33480-6_10

10.2 Visualization and Analysis of Atomistic Simulations

In the past decades, a considerable number of software packages have been developed for the purpose of visualizing and analyzing the output of molecular dynamics simulations. Most of these packages have non-commercial or academic origins and were developed by their authors to solve a specific problem. Unfortunately, this often means that the work on visualizations tools is discontinued after a few years, and they quickly become incompatible with modern computer hardware and simulation codes. A few packages, however, have survived (and grown) over the years, and are actively developed by a group of professional developers or by contributors from the user community.

These software packages often target specific audiences and research fields. For example, the popular software *Visual Molecular Dynamics* (VMD) is a very powerful molecular visualization program for displaying, animating, and analyzing large biomolecular systems such as proteins, nucleic acids, etc. In contrast, the software *Visualization for Electronic and Structural Analysis* (VESTA) is designed to visualize structure models and electron densities typically obtained from electronic structure calculations. In nanomechanics research, MD simulations based on classical potentials are common which can involve models with millions of atoms. Two established visualization programs are available that are designed to handle such datasets. AtomEye is an older visualization code developed since 2003, which provides the essential visualization capabilities for working with condensed-matter system. However, at the time of this writing, the most recent program update had been released in 2012. This certainly has contributed to the fact that in more recent times the software *Open Visualization Tool* (OVITO), which will be described in the following section, has become a more frequently used visualization tool in the atomistic modeling community.

10.2.1 OVITO: The Open Visualization Tool

OVITO is a professional software package developed by the author of this chapter for visualizing and analyzing results from molecular dynamics and other atomistic simulation models. The software is available free of charge under an open source license and runs on all major operating systems. In-depth information on the software and a complete list of features can be found on the official website (www.ovito.org). OVITO is being continuously extended since 2007 and has become a widely used research tool in the computational materials science field and related disciplines. It combines state-of-the-art data visualization capabilities with an intuitive user interface and a large set of analysis functions, some of which will be introduced in more detail in this chapter.

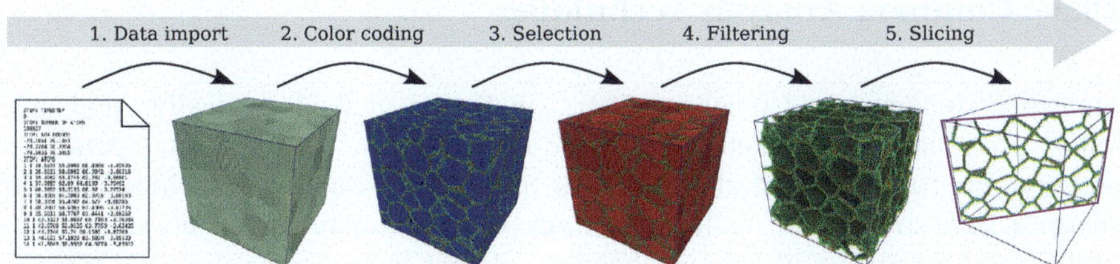

Fig. 10.1 Illustration of OVITO's data pipeline concept. Data manipulation, filtering, and analysis functions are arranged in a sequence by the user. The final output of the pipeline is displayed on screen or exported to a file

The initial focus of OVITO had been on medium to large classical molecular dynamics simulations of materials and condensed-matter systems, and it still provides an extensive toolkit of analysis functions that aim at nanomechanics studies. Nowadays, however, OVITO is also being used for the visualization and analysis of many other types of particle datasets.

Like other visualization programs, OVITO provides various tools for selecting, filtering, coloring, or transforming atoms, bonds, or other types of data. A powerful and unique feature of OVITO, however, which sets it apart from most other molecular visualization programs, is its concept of a non-destructive workflow [19]. The program's data pipeline architecture (Fig. 10.1) makes it possible for the user to retroactively change, delete, or amend applied editing actions. The program keeps a copy of the original simulation data in memory and, in addition, the full list of manipulation steps performed by the user during the program session. This list, which is editable by the user, allows changing parameters of existing data operations, removing operations, or even inserting new operations at arbitrary positions in the modification sequence. Upon any such change, OVITO immediately recalculates the effect of all data modifier functions that are part of the current pipeline and updates the real-time viewports to display the results. This type of non-destructive workflow can also be found in modern photo editing, 3D computer graphics, or scientific data analysis programs from other fields (e.g., PARAVIEW). In OVITO this approach allows the user to combine various basic data operations (e.g., for selecting, filtering, transforming, or analyzing particles) to accomplish complex analysis and visualization tasks. It can even serve as a graphical programming tool for setting up automated data analysis pipelines for batch-processing a series of atomistic simulation outputs.

Finally, OVITO itself provides a Python-based scripting interface in addition to the graphical user interface. This programming interface allows users to extend OVITO with new data filtering and analysis functions, to automate tasks, and to interface with other popular frameworks such as the *Atomistic Simulation Environment* (ASE) or custom tool chains.

10.3 Common Analysis Techniques

Molecular dynamics and other atomistic simulation methods typically yield two kinds of output: global quantities like the energy of the system, its temperature, pressure, etc., and per-particle quantities like atom positions, velocities, atomic energies, local stresses, etc.—all as functions of simulation time. In many simulation studies, and in the case of nanomechanics problems in particular, the latter type of simulation output contains the most valuable information. The purpose and strength of atomistic simulation models lie in the capability to reveal atomic-level processes and mechanisms that are not observable or investigable by other means. However, often the key to new insights is buried in the raw and overly verbose simulation data generated by such models. The right analysis techniques must be applied to extract the essential and insightful information.

The following sections will give a short introduction to several important analysis methods for atomistic simulations of nanomechanics problems. Some are commonly used techniques, which have been around for many years, while others are cutting-edge computational analysis tools, which emerged from recent research efforts. All techniques described here have in common that they are readily available in OVITO and can be used out of the box.

10.3.1 Displacement Vectors

Any deformation or transition taking place in a material must be accompanied by some motion of the atoms that constitute the material. Thus, if we want to identify and understand the underlying process or mechanism, we have to visualize and analyze the motion of atoms. A very rudimentary, but often also very illuminating approach is to consider the atomic displacements vectors (Fig. 10.2a):

$$\mathbf{u}_i = \mathbf{x}_i(t_1) - \mathbf{x}_i(t_0). \tag{10.1}$$

Here, $\mathbf{u}_i$ denotes the displacement vector of the i-th atom in the system, $\mathbf{x}_i(t)$ is the trajectory function of that atom computed by MD, and t_0 and t_1 are simulation times before and after the transition or deformation event. The magnitude of the displacement vector, $|\mathbf{u}_i|$, can be used to filter out atoms that remained at rest and reveal those affected by a transition or deformation process.

In OVITO you can calculate and visualize the displacement vectors as shown in Fig. 10.2c by inserting an instance of the *Displacement Vector* modifier into the modification pipeline. OVITO will ask for a simulation file that contains the reference coordinates of the atoms. Some precaution is necessary when the storage orders of atoms are not the same in the reference and the deformed simulation snapshots. Some MD codes will reorder atoms as needed during the course of a parallel simulation. In this case it is important to write out the unique atom IDs, which help OVITO to link an atom in the deformed configuration with its counterpart in the reference configuration.

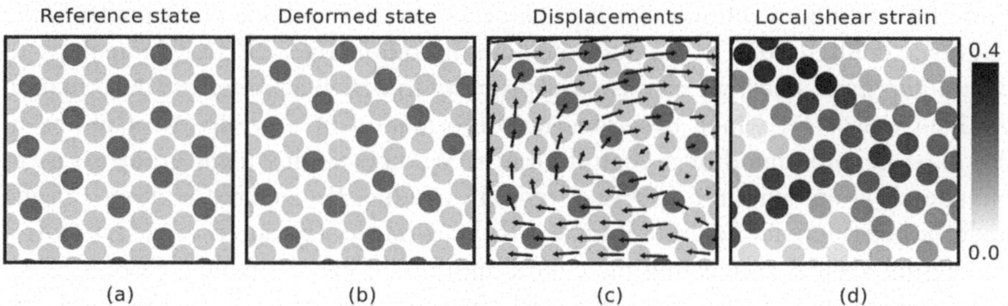

Fig. 10.2 Atomic displacement vectors and atomic shear strain values computed from the reference state and the deformed state of the system

Special attention must also be paid when determining the atomic displacements in systems that are subject to periodic boundary conditions. Typically, MD codes output the *wrapped* coordinates of atoms. That means $\mathbf{x}_i(t)$ is always inside the primary image of the simulation box, and when an atom leaves the box through a periodic boundary, it reappears on the opposite side of the box. Even though the physical trajectory is continuous, the trajectory written by the MD code is not, resulting in a wrong and extremely large displacement vector for atoms that have crossed a periodic boundary. To avoid this, it is a good idea to let the MD code output *unwrapped* atomic coordinates if this option is available. In cases where it is not, OVITO provides a solution that works in many cases: If a computed displacement vector spans more than half of the simulation box in a periodic direction, then a complete cell vector is added to or subtracted from the displacement vector such that it becomes shorter. This approach is related to the so-called *minimum image convention* for periodic systems, and it relies on the assumption that an atom never moves a distance larger than half of the box size.

10.3.2 Atomic Deformation Gradient and Strain Tensors

The displacement vectors discussed in the previous section directly visualize the motion of atoms in space. In many cases, however, it rather is the *relative* motion of atoms that is of interest and not the movement of the body as a whole, because only non-uniform atomic motion indicates a change in the material. It therefore makes sense to analyze how displacements vary within a local group of atoms by looking at the spatial derivative of the displacement field. From continuum mechanics we know that the deformation gradient $\mathbf{F}$ is a measure of the spatial variation of the displacement field $\mathbf{u}(\mathbf{X})$ at a material point $\mathbf{X}$:

$$\mathbf{F}(\mathbf{X}) = \nabla\mathbf{u}(\mathbf{X}) + \mathbf{I}. \tag{10.2}$$

From this definition follows that $\mathbf{F}$ transforms infinitesimal line elements from the reference configuration to the deformed configuration, $\mathrm{d}\mathbf{x} = \mathbf{F}\,\mathrm{d}\mathbf{X}$.[1] In analogy to continuum mechanics, we can use this property to define an *atomistic* deformation gradient tensor $\mathbf{F}_i$, which describes the deformation of a group $\mathscr{C}_i$ of atoms in the local neighborhood of a central atom i. We choose this local group so as to include all atoms within a spherical region of radius R_c centered at atom i in the undeformed reference configuration. Let $\mathbf{R}_{ij} = \mathbf{x}_j(t_0) - \mathbf{x}_i(t_0)$ and $\mathbf{r}_{ij} = \mathbf{x}_j(t_1) - \mathbf{x}_i(t_1)$ denote the vectors connecting the central atom with one of its neighbors, $j \in \mathscr{C}_i$, in the undeformed and in the deformed configuration, respectively. We seek the local deformation gradient tensor that transforms all interatomic vectors from the undeformed configuration to their deformed counterparts:

$$\mathbf{r}_{ij} = \mathbf{F}_i \mathbf{R}_{ij}, \qquad \forall j \in \mathscr{C}_i. \tag{10.3}$$

Note that, for such a linear transformation to exist, the deformation needs to be uniform, i.e., the local group has to undergo an affine deformation. Generally this is not the case. We therefore have to define the atomistic deformation gradient in a least-squares sense to approximate the local deformation. In other words we seek the $\mathbf{F}_i$ that minimizes

$$\sum_{j \in \mathscr{C}_i} \left| \mathbf{F}_i \mathbf{R}_{ij} - \mathbf{r}_{ij} \right|^2. \tag{10.4}$$

It can be shown that the closed-form solution to this linear least-squares minimization problem [17] is given by

$$\mathbf{F}_i = \mathbf{W}_i \cdot \mathbf{V}_i^{-1}, \qquad \text{with}$$
$$\mathbf{W}_i \equiv \sum_{j \in \mathscr{C}_i} \mathbf{R}_{ij} \mathbf{r}_{ij}^{\mathrm{T}} \quad \text{and} \quad \mathbf{V}_i \equiv \sum_{j \in \mathscr{C}_i} \mathbf{R}_{ij} \mathbf{R}_{ij}^{\mathrm{T}}. \tag{10.5}$$

This atomistic deformation gradient represents an average over a spherical volume with a finite radius R_c. When we increase the radius to include additional neighbor atoms, it leads to a smoothening effect. Typically, R_c is chosen such that it includes the first shell of neighbors of the central atom. At the very least it must be large enough to include three non-coplanar neighbors, otherwise the minimization problem is underdetermined and the matrix $\mathbf{V}_i$ becomes singular.

Once $\mathbf{F}_i$ is known, we can calculate the corresponding symmetric Green–Lagrangian strain tensor on a per-atom basis:

$$\mathbf{E}_i = \frac{1}{2}(\mathbf{F}_i^{\mathrm{T}} \mathbf{F}_i - \mathbf{I}). \tag{10.6}$$

[1] We assume here that all vectors are column vectors.

Plastic deformation, for example, due to dislocation glide, is typically associated with the generation of localized shear deformation at the atomic scale (Fig. 10.2d). A good measure for local shear deformation, irrespective of its direction, is the *von Mises* shear strain tensor invariant [17], which can be computed from the six components of the atomic strain tensor:

$$\gamma_i = \sqrt{E_{xy}^2 + E_{xz}^2 + E_{yz}^2 + \frac{(E_{xx} - E_{yy})^2 + (E_{yy} - E_{zz})^2 + (E_{zz} - E_{xx})^2}{6}}. \qquad (10.7)$$

The shear strain measure above is commonly used to visualize regions that undergo deformation in crystalline or amorphous materials. The first example, shown in Fig. 10.3, is a dataset from an MD simulation study [16] of a nanocrystalline PdAu alloy under tensile deformation. The atomistic model contains 54 grains with random orientations and an average size of 15 nm. The imposed external uniaxial strain leads to localized plastic deformation of the microstructure, which can be made visible with the atomic strain tensor method. The strain map in Fig. 10.3b reveals that the material has deformed via sliding and shuffling processes localized in the grain boundaries and, simultaneously, via conventional crystal slip produced by transient dislocations nucleated and absorbed at grain boundaries.

Figure 10.4 shows an example from a simulation study which investigates the formation of shear bands in CuZr metallic glass nanowires. As discussed

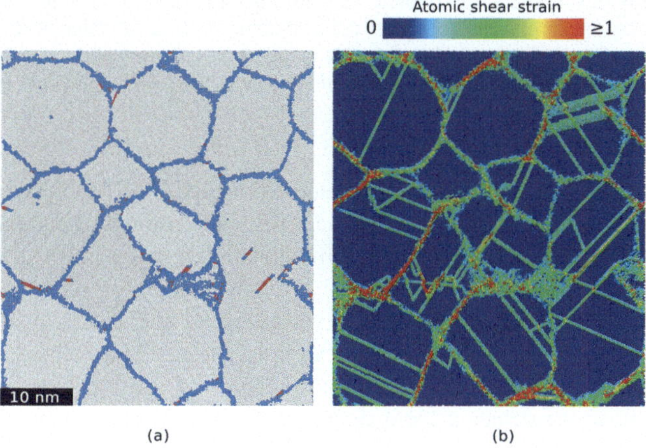

Fig. 10.3 MD simulation of a nanocrystalline Pd structure under uniaxial straining along the vertical axis [16]. (**a**) The common neighbor analysis method (see Sect. 10.3.3) has been used to mark atoms that belong to grain boundaries and other crystal defects such as dislocation cores. The cross-section shows that grain interiors are almost free of dislocation defects at this stage of the deformation ($\varepsilon = 10\,\%$). (**b**) The atomic-level shear strain calculated from Eq. (10.7), however, reveals that transient dislocations have traveled through many of the grains and left behind slip traces. Furthermore, significant strain concentration in the grain boundaries indicates that grain boundary sliding contributed as an additional deformation mechanism

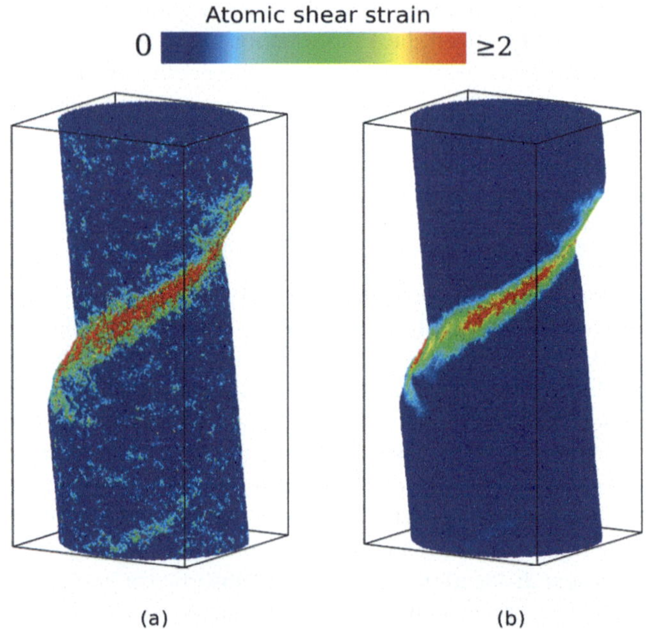

Fig. 10.4 MD simulation of a cylindrical nanowire made from a CuZr metallic glass that deforms under tensile load via formation of a shear band. The wire diameter is 30 nm and the model consists of 2.6 million atoms. Colors show the atomic-level shear strain, which is calculated from Eq. (10.7). The results depend on the value of the cutoff radius parameter R_c. (**a**) A value of $R_c = 4\,\text{Å}$, which includes only the first shell of neighbors, reveals small shear transformation zones at the surface of the wire. (**b**) A larger radius ($R_c = 16\,\text{Å}$) leads to a smoothing effect and only extended shear deformation zones remain visible

above, the per-atom strain tensors calculated from Eqs. (10.5) and (10.6) reflect the deformation of a local atom group, whose size is determined by the parameter R_c. Depending on our choice of this size parameter, small features in the strain field will be resolved or averaged out.

Finally, it is worth mentioning that variations to the classical atomic strain tensor method have been proposed. For example, the residual defined in Eq. (10.4) can be used as a measure for the non-affine part of the local deformation [7]. As such it allows to detect a local deviation from a homogeneous deformation, which is characteristic for atomic-level shuffling events in amorphous materials. Furthermore, more advanced formulations of the atomic deformation gradient exist, which weight the contribution of neighbor atoms depending on their distance from the central atom. This is in contrast to the classical formulation presented here, in which all neighbors enter with the same uniform weight. Finally, a practical approach has been proposed to decompose the atomic-level deformation gradient and strain tensors into elastic and plastic contributions [25], something that is only possible in crystalline structures, and which helps in quantifying the contributions of different plasticity mechanisms.

10.3.3 *Local Structure Identification*

In many instances it is necessary to determine the local, atomic-level structure of simulated materials, e.g., to identify defects in a crystal lattice, to detect phase transitions, or to study other structural changes that occur during a simulation. The goal is to get a precise understanding of which atoms are associated with which phases, and which are associated with defects. Typically, order parameters are used for this purpose, which can characterize the local arrangement of atoms. Each atom in the system is assigned a value (integer, real, or a vector of several numbers), which is computed from the positions of the atoms in the local neighborhood. Thus, the purpose of the order parameter is to distill the local structure into a single number or vector, which is constructed to be both informative and computationally tractable [12, 14, 20].

10.3.3.1 Centrosymmetry Parameter

A typical example for such order parameters, designed to detect defects in crystal lattices, is the *centrosymmetry parameter (CSP)* [11]. It quantifies the local loss of centrosymmetry at an atomic site, which is characteristic for crystal defects in simple centrosymmetric lattices (mainly FCC and BCC). The CSP value of an atom having N nearest neighbors ($N = 12$ for FCC, $N = 8$ for BCC) is defined as

$$p_{\mathrm{CSP}} = \sum_{i=1}^{N/2} \left| \mathbf{r}_i + \mathbf{r}_{i+N/2} \right|^2 \tag{10.8}$$

where $\mathbf{r}_i$ and $\mathbf{r}_{i+N/2}$ are vectors from the central atom to a pair of opposite neighbors. For lattice sites in an ideal centrosymmetric crystal, all pairs of neighbor vectors in Eq. (10.8) cancel, and hence the resulting order parameter value is zero. Atomic sites in a defect region, in contrast, typically have a defective, non-centrosymmetric neighborhood. Then the CSP becomes positive. Using a suitable threshold, which allows for small perturbations such as thermal displacements, the CSP can be used to extract atoms that form crystal defects.

10.3.3.2 Common Neighbor Analysis

A limitation of continuous order parameters like the CSP or the atomic potential energy is that degeneracies can arise in describing neighborhoods that are structurally distinct but which map to the same order parameter value. Discrete order parameters can help to overcome this limitation [13]. A popular example for such discrete classification methods is the *common neighbor analysis* (CNA) [6, 10].

The CNA is a relatively simple and efficient classification scheme to characterize the topology of local bond networks in the neighborhood of atoms and to identify simple crystalline structures. Note that in this context the term "bond" does not refer to physical bonds but rather to the geometric distance between pairs of atoms: Two atoms are considered bonded if their separation is less than a specified maximum distance (R_c). For densely packed structures (FCC and HCP) this cutoff is typically chosen to be halfway between the first and the second neighbor distance. That is for an FCC crystal,

$$R_c(\text{fcc}) = \frac{1}{2}\left(\sqrt{1/2}+1\right) a_{\text{fcc}} \simeq 0.854\, a_{\text{fcc}}, \tag{10.9}$$

where a_{fcc} is the equilibrium lattice parameter. Positioning the cutoff halfway between the two neighbor shells ensures that nearest neighbors are considered bonded while second nearest are not, even if the positions are perturbed. Given a central atom we can find its neighbors within the radius R_c, which, for an atom in a perfect FCC crystal, should be exactly 12. Next, the same cutoff is used to determine which of these 12 neighbors are in range of each other, thereby defining a local network of bonds between the 12 neighbors. For each neighbor, a triplet of indices ($n_{cn}\, n_b\, n_{lc}$) is computed, with

n_{cn}	Number of common neighbors of the central atom and its neighbor.
n_b	Number of bonds between these common neighbors.
n_{lc}	Number of bonds in the longest continuous chain of bonds between the common neighbors.

Figure 10.5 shows that for each of the 12 nearest neighbors of an FCC atom this triplet turns out to be the same: (4 2 1). In other words, every neighbor has 4 bonds to other neighbors of the central atom, there are 2 bonds connecting some of these 4 common neighbors with each other, and the longest chain these bonds form has length 1.

The CNA bond indices allow to quickly identify atoms with different structural environments. While FCC atoms have 12 neighbor bonds of (4 2 1)-type, HCP atoms have 6 neighbors of (4 2 2)-type and 6 neighbors of (4 2 1)-type. To identify BCC atoms, second nearest neighbors must be included in the cutoff radius, because the nearest neighbors alone would have no common neighbors. The cutoff radius is therefore placed between the second and third neighbor shell of the BCC lattice:

$$R_c(\text{bcc}) = \frac{1}{2}\left(1+\sqrt{2}\right) a_{\text{bcc}} \simeq 1.207\, a_{\text{bcc}}. \tag{10.10}$$

When using this cutoff radius, atoms in perfect BCC environments possess eight first neighbors of (6 6 6)-type and six second neighbors of (4 4 4)-type.

The CNA is often employed as a filtering technique in simulations of crystalline materials. Figure 10.6 shows a typical example of an MD simulation of a nanocrystalline FCC palladium structure. Here, the CNA serves an effective tool to filter out all atoms in perfect FCC environments and to reveal crystal defects such as grain boundaries, dislocations, and stacking faults. Intrinsic stacking faults and coherent twin boundaries parallel to $\{111\}$ crystal planes consist of atoms having a HCP-like local coordination. This makes them particularly easy to identify with the help of the CNA.

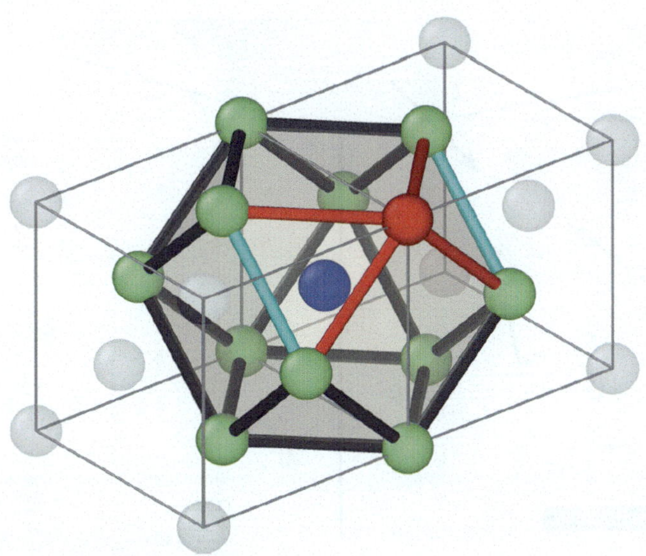

Fig. 10.5 Calculation of CNA indices in the FCC structure. The picture shows two cubic FCC cells. The central atom (*blue*) has 12 nearest neighbors (*green, one marked red*). The marked atom shares *4* common neighbors (indicated by *red bonds*) with the central atom. There exist *2* bonds between these four common neighbors (*cyan bonds*). These bonds do not form any chain. Thus, the maximum chain length is *1*. In summary, the CNA index vector computed for the bond from the central atom to the red neighbor is (4 2 1)

10.3.3.3 Variations of the Common Neighbor Analysis Method

The only input parameter to the CNA is the cutoff radius R_c. As explained in the previous section, it must be chosen according to the crystal structure at hand. But how do we choose it in simulations that involve multiple crystal phases or if phase transitions take place? To address this problem, an adaptive version of the CNA [20] has been proposed, which picks the cutoff radius R_c automatically and individually for every atom. The method starts by guessing an optimal cutoff radius R_c(fcc) from the distance of the 12 nearest neighbor of a central atom. That means the algorithm builds a sorted neighbor list first, takes the 12 leading entries, and uses their distances to determine R_c(fcc), the optimal cutoff for an FCC crystal. Based on this local cutoff, the CNA indices are computed to find a possible match with the FCC signature. If this succeeds, we are done. If not, the algorithm continues by testing other known crystal structures, re-computing the corresponding optimal local cutoff radius for each of them.

Note that the CNA is not directly suitable for materials with a diamond structure, because nearest neighbor pairs of atoms do not possess any common neighbors in this structure, and taking into account second nearest neighbors, like it is done for BCC, does not work well, because the second and third neighbor shells in

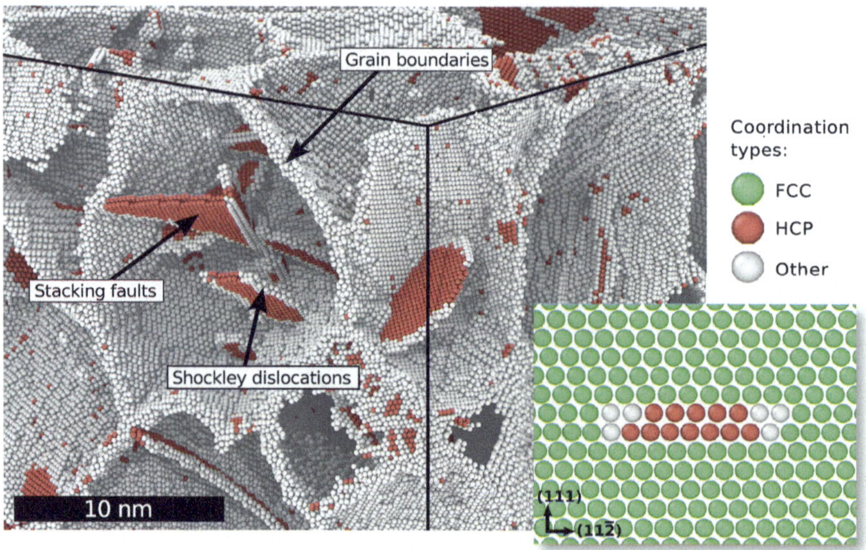

Fig. 10.6 Visualization of a simulated nanocrystalline Pd microstructure [24]. FCC atoms as identified by the CNA have been removed to reveal the crystal defects. The inset shows a cross-section of a $1/2\,\langle 110\rangle$ dislocation splitting into a pair of $1/6\,\langle 112\rangle$ Shockley partial dislocations, which are connected by an intrinsic stacking fault

the diamond lattice are not well separated. Hence, even a small elastic strain or thermal displacements easily disturb the computed CNA fingerprints, rendering the method unreliable. For this reason, a modified CNA method for diamond structures [27] has been proposed, which exploits the fact that the cubic diamond structure consists of two interleaved FCC lattices. The modified algorithm simply skips the four nearest neighbors of every atom and computes the CNA indices on the basis of their respective neighbors, which are the 12 second nearest neighbors of the central atom. These atoms form a regular FCC lattice, which can be reliably identified. The trick also works for hexagonal diamond structures, where the second nearest neighbors are arranged on a HCP sublattice.

10.3.4 Geometric Surface Reconstruction

In classical atomistic simulations the atoms that constitute a solid are modeled as point-like objects. Consequently, the simulated solid has no volume and no surface if it is described as such a set of zero-dimensional points in space. This is in contrast to our perception of solids and the continuum descriptions typically used at larger length scales, where solids always have a well-defined volume and boundaries. To bridge this gap between the atomistic model and the continuum world, a special type of transformation method is needed. Without it, we would be unable to properly quantify important properties of atomistic systems like their solid volume, surface area, surface orientation and curvature, or porosity.

The aim is to reconstruct the geometric boundaries of a set of points in space. The result needs to be a two-dimensional, closed, and oriented manifold that divides space into an inner (=solid) and an outer (=open) region. If we want to define the surface of an atomistic solid, an indispensable ingredient is some sort of length scale parameter. How else should we discern the empty space in between atomic points of a dense solid from actual pores or small cracks? A common solution to this problem is the concept of a *probe sphere* [18]. The open region is here defined as those parts of space that are accessible to the virtual probe sphere without touching any of the atomic points. Note that this includes cavities inside the solid as long as they can accommodate the virtual sphere. The radius of the probe sphere is the length scale parameter that needs to be specified by the user. It determines how many details and small features of the solid's geometric shape are resolved by the method.

The algorithm which actually constructs the boundary surface [21], and which is implemented in OVITO, is based on the *α-shape method* of Edelsbrunner and Mücke [3]. It starts with the Delaunay tetrahedrization of the input point set (Fig. 10.7a). From the resulting tetrahedra, which fill the convex hull of the point set, all elements are removed whose circumsphere does not fit into the virtual probe sphere (Fig. 10.7b). The remaining tetrahedra form the solid region. Now the closed surface mesh can be extracted, which consists of the triangular faces of the tessellation that divide the solid from the open region (Fig. 10.7c).

The resulting surface mesh reflects the atomic steps that are typically present on the surfaces of a solid (Fig. 10.8b). This may lead to an overestimation of the macroscopic surface area. Therefore, if such detailed surface features are undesired, one can subsequently apply a fairing procedure [5] to the surface mesh to smooth it (Fig. 10.8c).

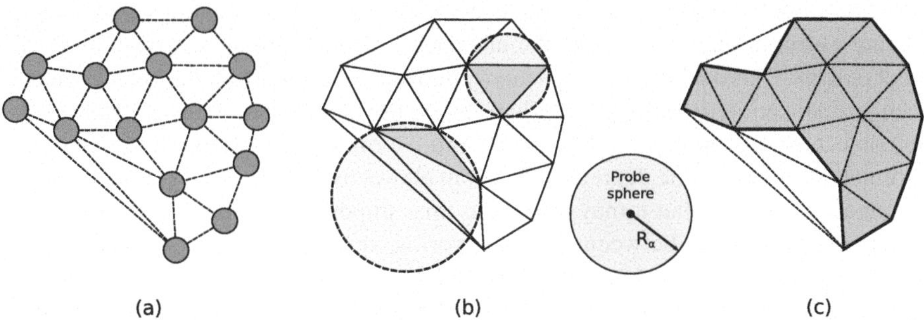

Fig. 10.7 Alpha-shape construction to determine the solid volume of a point set. (**a**) The Delaunay tessellation calculated from the input point set. (**b**) Two exemplary tessellation elements are highlighted, and their circumspheres are indicated. One element's circumsphere is larger than the reference probe sphere while the other is smaller. The corresponding elements are classified as *open* and *solid*, respectively. (**c**) The union of all solid Delaunay elements defines the geometric shape of the atomistic solid (*bold line*)

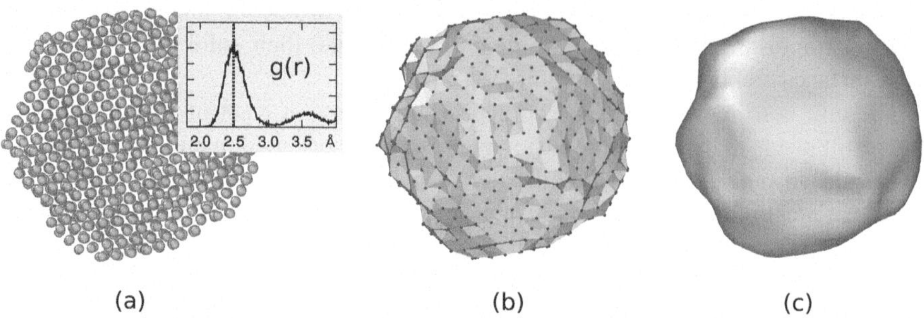

Fig. 10.8 (**a**) Atomistic model of a carbon nanoparticle with an inset showing the corresponding pair distribution function. The position of the first peak is used as probe sphere radius ($R_\alpha = 2.5$ Å) for the α-shape construction. (**b**) The resulting triangulated surface mesh. (**c**) Final surface model after six iterations of the smoothing algorithm were applied

It needs to be emphasized that the results of the described surface reconstruction method generally depend on the selected probe sphere radius parameter R_α. A rule of thumb is to choose R_α equal to the nearest neighbor atom separation in the material at hand as demonstrated in Fig. 10.8a. Definitely, the value of R_α should be reported in publications where the geometric surface reconstruction method is used to quantitatively measure surface areas, solid volumes, or porosities.

10.3.5 Dislocation Identification

Dislocation defects play an important role in the plasticity and the mechanical properties of most crystalline materials. Atomistic simulation methods such as MD represent ideal research tools to study these defects in a computer, because they allow us to fully resolve the irregular core structure of dislocations and readily capture essential phenomena like the nucleation of dislocations and reactions with other defects. Classical dislocations theory and corresponding simulation techniques such as dislocation dynamics (see Chap. 2), on the other hand, view dislocations as line objects that behave according to constitutive rules and mobility laws. Here, the elimination of the atomic degrees of freedom allows us to study much larger samples at longer time scales, but it may also let us miss important atomistic effects.

To bridge the gap between the two worlds, researchers have worked on the development of computational algorithms that can automatically identify dislocation defects in atomistic models of crystals. In addition to the multiscale modeling problem mentioned above, there are more reasons why the capability to convert dislocations from the atomistic to the continuous line representation is very useful. One is the possibility to considerably reduce the amount of output data generated by all-atom simulations. Massively parallel supercomputers enable the simulation of large models that include many millions or even billions of atoms. Frequently dumping all atomic positions to disk can quickly become too expensive in terms of required storage. Even though atomic filtering techniques like those presented in Sect. 10.3.3 help to mitigate this problem, more sophisticated computational approaches are needed to directly transform the simulation output to a more concise high-level representation of the relevant structural features.

Furthermore, interpreting the behavior of dislocations in an MD model requires the scientist to identify the dislocation lines. In many cases information like the Burgers vector, the dislocation character, or the glide plane is needed to follow and understand the observed dislocation processes. In the past, this meant that scientists had to identify each dislocation line in a simulation by hand using a manual method like the Burgers circuit procedure [8]. In addition to being cumbersome, such a manual approach is error-prone and quickly becomes unfeasible in complex situations (e.g., reactions of curved dislocations with grain boundaries in three dimensions).

The following section will introduce the *Dislocation Extraction Algorithm* (DXA) [23, 26], a computational method that has been developed to automatically identify and extract dislocations lines in atomistic simulation data. It is the most commonly used approach among several dislocation identification methods proposed in recent years [1, 4, 9, 22, 28], and, starting with program version 2.5, the DXA is readily available as an analysis function in OVITO.

10.3.5.1 Dislocation Extraction Algorithm

The fundamental concept underlying the DXA is the *Burgers circuit* construction [8], which is the canonical method [2] already proposed in the 1950s to discriminate dislocations from other crystal defects and to determine their Burgers vectors. In the formulation employed here, a Burgers circuit C is a path in the dislocated crystal consisting of a sequence of atom-to-atom steps (line elements $\Delta\mathbf{x}$), as shown in Fig. 10.9a. The circuit is closed, thus $\sum_C \Delta\mathbf{x} = \mathbf{0}$.

We assume that there exists a mapping $\Delta\mathbf{x} \rightarrow \Delta\mathbf{x}'$ that translates each line element of the path to a corresponding image, $\Delta\mathbf{x}'$, in a perfect crystal lattice

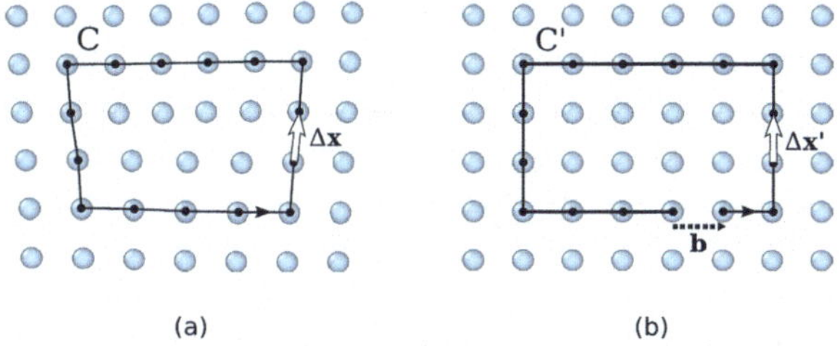

Fig. 10.9 Burgers circuit method to detect and identify a dislocation. A closed circuit around the dislocation is translated from (**a**) the dislocated crystal to (**b**) the perfect reference crystal. The closure failure is called the Burgers vector of the dislocation

(Fig. 10.9b). Summing these transformed line elements algebraically along the associated path, C', gives the true Burgers vector of the dislocation enclosed by C:

$$\mathbf{b} = -\sum_{C'} \Delta \mathbf{x}'. \tag{10.11}$$

The Burgers vector **b** is the closure failure of the path after transferring it to the perfect reference crystal. Notably, the resulting vector **b** stays the same if we change the original circuit C, as long as it still encloses the same dislocation. On the other hand, if $\mathbf{b} = \mathbf{0}$, we know that the Burgers circuit did not enclose any defect with dislocation character (deliberately ignoring the possibility that the circuit encloses multiple dislocations whose Burgers vectors cancel).

Typically the Burger circuit construction is performed by hand to analyze two-dimensional crystal images obtained from high-resolution microscopy or atomistic computer simulations. Human intuition and cognitive capabilities are required to spot irregularities in the crystal lattice which are potential dislocation defects and to map path steps in elastically distorted crystal regions to the ideal lattice. Automating these tasks poses a particular challenge when developing a dislocation identification algorithm. First of all, an efficient strategy is needed that guides the construction of Burgers circuits, given that there is no a priori knowledge of the dislocation positions, because it clearly is not feasible to enumerate all possible circuits in a crystal to find the contained dislocations.

Within the DXA framework, this problem is addressed by using the Delaunay tessellation of the dislocated input crystal (Fig. 10.10a). The edges of this tessellation define the set of elementary atom-to-atom steps from which Burgers circuits will be constructed. Before generating any circuits, the algorithm first tries to map each edge of the Delaunay tessellation to a corresponding vector in the perfect reference crystal (Fig. 10.10b). This is done with the help of the common neighbor analysis method discussed in Sect. 10.3.3.2, which finds atoms that form a perfect (but elastically strained) crystal lattice. Delaunay edges connecting a crystalline atom with one of its neighbors are mapped to the corresponding ideal lattice vectors by the algorithm.

Within the cores of dislocations, the atomic arrangement deviates considerably from a perfect crystal. Hence, the CNA will classify these core atoms as non-crystalline atoms. All tessellation edges adjacent to such atoms will be marked as "bad" by the algorithm, effectively excluding them from any Burgers circuits to be constructed. This corresponds to the original principle formulated by F. C. Frank [8], which states that a valid Burgers circuit must not pass through the so-called *bad crystal*. Good crystal regions, in contrast, are defined as those parts where the mapping to the perfect reference crystal is nonambiguous. In fact, the DXA also divides space into *good* and *bad* regions in this spirit as shown in Fig. 10.10b. Those Delaunay elements (triangles in 2D, tetrahedra in 3D systems) that are adjacent to one or more bad edges, which could not be mapped to an ideal lattice vector, are themselves marked as bad elements, while all others are considered good volume elements.

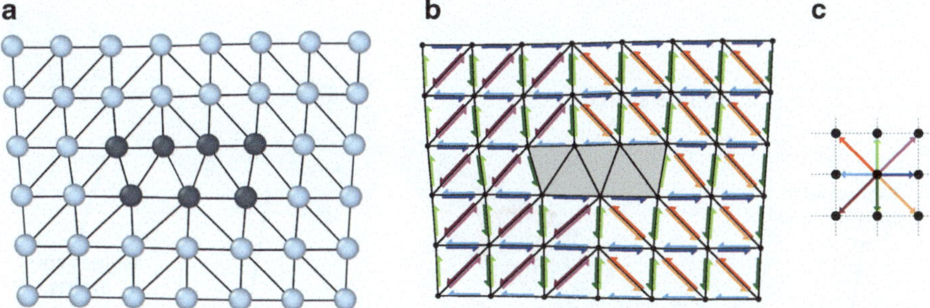

Fig. 10.10 (**a**) Delaunay tessellation of a dislocated crystal. Defect core atoms as identified by a structural characterization technique are shown in a darker color. (**b**) Colored arrows indicate the computed mapping of tessellation edges to corresponding ideal lattice vectors. Bad tessellation elements, for which the mapping to the perfect reference lattice cannot be determined, have been marked with a gray color. (**c**) Color legend for the eight different ideal lattice vectors appearing in (**b**)

Now it is time to think about how to efficiently construct trial Burgers circuits to find and classify the dislocations in the crystal. As mentioned above, the total number of possible circuits in a three-dimensional crystal is prohibitively large, and we need to find a way to considerably reduce the search space. The solution is provided by the aforementioned partitioning into good and bad regions, which defines a boundary surface separating the two regions. In three-dimensional systems this boundary is called the *interface mesh* and is constituted by those triangular Delaunay facets having a good tetrahedral element on one side and a bad element on the other.

The interface mesh, which is depicted in Fig. 10.11, is a two-dimensional manifold that encloses all defects in the crystal (including non-dislocation defects and even free surfaces of the crystal). Constructing trial Burgers circuits on this triangulated surface is sufficient to discover all dislocations. Moreover, this approach helps to ensure that the generated Burgers circuits enclose only single dislocation lines. Trial circuits generated by the DXA on the interface mesh are closed sequences of tessellation edges, and their Burgers vectors are computed from Eq. (10.11) by summing the respective ideal lattice vectors, which were determined in the first algorithm step. All possible trial circuits up to some prescribed maximum length (usually not more than ten edges) can be efficiently enumerated using a recursive search algorithm.

The algorithm enumerates all possible circuits on the interface mesh in order of increasing length until one with a non-zero Burgers vector is encountered. This seed circuit is subsequently used to discover the rest of the dislocation line. This happens by advancing the circuit on the interface mesh and sweeping along the dislocation line as indicated in Fig. 10.11. During this sweeping phase, a one-dimensional line representation of the dislocation is generated by computing the new center of mass of the circuit each time it advances along the boundary of the dislocation core. Here,

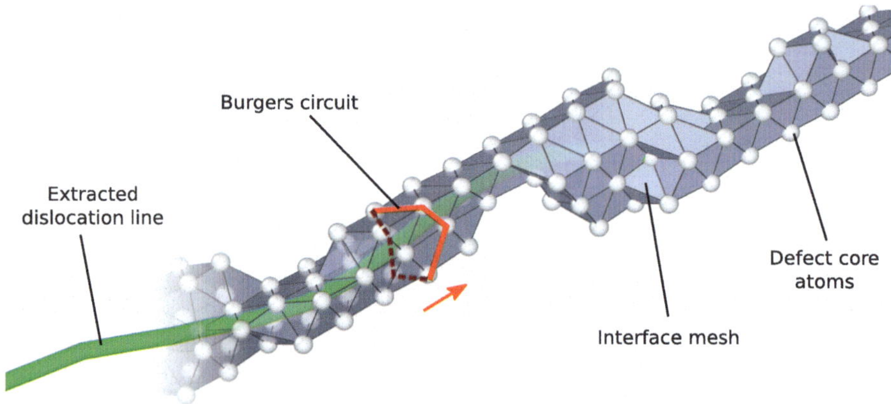

Fig. 10.11 Illustration of the line sweeping phase of the DXA. After constructing the interface mesh enclosing the defect core atoms, the algorithm uses a Burgers circuit on the interface mesh to sweep the dislocation line. While the Burgers circuit is being advanced in a step-wise fashion, triangle by triangle, a continuous line representation of the dislocation defect is produced

a circuit can be pictured as a rubber band tightly wrapped around the dislocation's core. As the circuit moves along the dislocation segment, it may need to locally expand to sweep over wider sections of the core, e.g., kinks or jogs. To prevent the circuit from sweeping past dislocation junctions or interfaces, a hard limit is imposed on the maximum circuit length.

10.3.5.2 Applications of the Dislocation Extraction Algorithm

The output of the DXA is a network of dislocation line segments, each of which is associated with its local Burgers vector and the crystallite (grain) it is embedded in. The method can be used to analyze polycrystalline structures, and, since the algorithm also determines the orientation of each crystallite, we can translate a dislocation's Burgers vector from the local lattice coordinate frame to the global simulation coordinate frame if needed. This, for example, makes it possible to determine the local character of dislocations and we can measure edge and screw dislocation fractions in a system.

Generally, the translation to a line representation of dislocation defects also enables the measurement of dislocation densities, simply by dividing the total line length by the volume of the crystal. Figure 10.12 shows an example from a simulation study of nanoporous gold [15]. Here, a gold monocrystal with an initial open porosity of 70 % is uniaxially compressed up to an engineering strain of 80 % to study its mechanical behavior. The severe deformation induces a large number of dislocations in the material, which can be visualized and quantified with the help of the DXA.

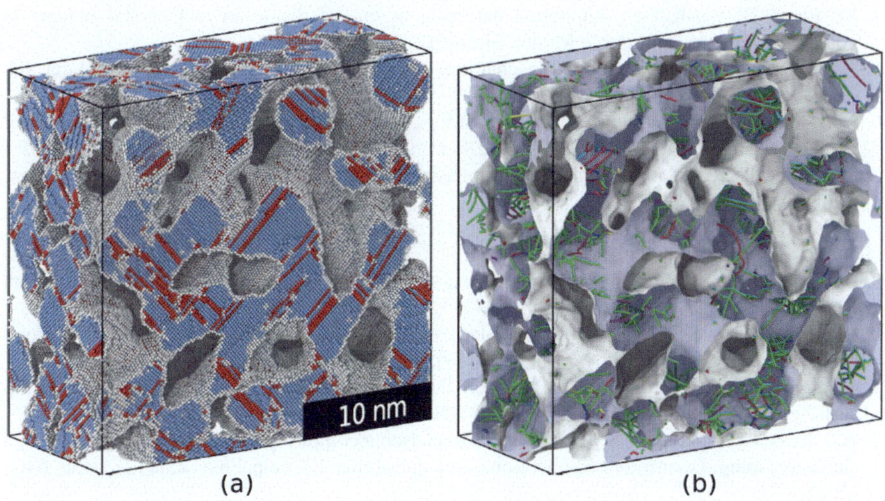

Fig. 10.12 MD simulation of a nanoporous gold sample [15] under uniaxial compression. (**a**) Conventional atomistic visualization with CNA-based coloring of atoms. Blue, red, and gray colors correspond to atoms in perfect, stacking fault, and other defect environments, respectively. (**b**) The simulation snapshot after being processed with the DXA and the surface reconstruction method described in Sect. 10.3.4. Line colors indicate different Burgers vector types and the reconstructed surface geometry enables measuring the porosity and specific surface area of the material during deformation

While the development of this method has been a recent step forward in simplifying the analysis of complex dislocation processes, there are still some limitations that will have to be addressed by future research. For instance, the Burgers vector alone, which is determined by the DXA for a dislocation segment, is generally not sufficient to determine the glide plane of the segment. Thus, one of the challenges is to correctly assign the extracted dislocations to glide systems of a crystal. Another problem lies in the fact that the DXA is a static analysis method, which takes into account only an instantaneous configuration of an atomistic crystal. This makes it difficult to correlate successive snapshots and measure dislocation velocities, for example, especially if dislocations undergo reactions.

References

1. C. Begau, J. Hua, A. Hartmaier, A novel approach to study dislocation density tensors and lattice rotation patterns in atomistic simulations. J. Mech. Phys. Solids **60**(4), 711–722 (2012)
2. V.V. Bulatov, W. Cai, *Computer Simulations of Dislocations* (Oxford University Press, Oxford, 2006)
3. H. Edelsbrunner, E.P. Mücke, Three-dimensional alpha shapes. ACM T. Graphic **13**(1), 43–72 (1994)

4. M. Elsey, B. Wirth, Fast automated detection of crystal distortion and crystal defects in polycrystal images. Multiscale Model. Simul. **12**, 1–24 (2014)
5. J. Erlebacher, I. McCue, Geometric characterization of nanoporous metals. Acta Mater. **60**, 6146–6174 (2012)
6. D. Faken, H. Jonsson, Systematic analysis of local atomic structure combined with 3d computer graphics. Comput. Mater. Sci. **2**(2), 279–286 (1994)
7. M.L. Falk, J.S. Langer, Dynamics of viscoplastic deformation in amorphous solids. Phys. Rev. E **57**, 7192–7205 (1998)
8. F.C. Frank, LXXXIII. Crystal dislocations – Elementary concepts and definitions. Philos. Mag. Ser. 7 **42**(331), 809–819 (1951)
9. C.S. Hartley, Y. Mishin, Characterization and visualization of the lattice misfit associated with dislocation cores. Acta Mater. **53**(5), 1313–1321 (2005)
10. J.D. Honeycutt, H.C. Andersen, Molecular dynamics study of melting and freezing of small Lennard-Jones clusters. J. Phys. Chem. **91**(19), 4950–4963 (1987)
11. C.L. Kelchner, S.J. Plimpton, J.C. Hamilton, Dislocation nucleation and defect structure during surface indentation. Phys. Rev. B **58**(17), 11085 (1998)
12. A.S. Keys, C.R. Iacovella, S.C. Glotzer, Characterizing complex particle morphologies through shape matching: Descriptors, applications, and algorithms. J. Comp. Phys. **230**(17), 6438–6463 (2011)
13. P.S. Landweber, E.A. Lazar, N. Patel, On fiber diameters of continuous maps. Am. Math. Mon. **123**, 392–397 (2016). doi: 10.4169/amer.math.monthly.123.4.392
14. E.A. Lazar, J. Han, D.J. Srolovitz, Topological framework for local structure analysis in condensed matter. PNAS **112**, E5769–E5776 (2015)
15. B.-N.D. Ngô, A. Stukowski, N. Mameka, J. Markmann, K. Albe, J. Weissmüller, Anomalous compliance and early yielding of nanoporous gold. Acta Mater. **93**, 144–155 (2015)
16. J. Schäfer, A. Stukowski, K. Albe, Plastic deformation of nanocrystalline {Pd-Au} alloys: {On} the interplay of grain boundary solute segregation, fault energies and grain size. J. Appl. Phys. **114**(14), 143501 (2013)
17. F. Shimizu, S. Ogata, J. Li, Theory of shear banding in metallic glasses and molecular dynamics calculations. Mater. Trans. **48**(11), 2923–2927 (2007)
18. A. Shrake, J.A. Rupley, Environment and exposure to solvent of protein atoms. Lysozyme and insulin. J. Mol. Biol. **79**(2), 351–371 (1973)
19. A. Stukowski, Visualization and analysis of atomistic simulation data with OVITO–the open visualization tool. Model. Simul. Mater. Sci. Eng. **18**(1), 015012 (2010)
20. A. Stukowski, Structure identification methods for atomistic simulations of crystalline materials. Model. Simul. Mater. Sci. Eng. **20**(4), 045021 (2012)
21. A. Stukowski, Computational analysis methods in atomistic modeling of crystals. JOM **66**(3), 399–407 (2014)
22. A. Stukowski, A triangulation-based method to identify dislocations in atomistic models. J. Mech. Phys. Solids **70**, 314–319 (2014)
23. A. Stukowski, K. Albe, Extracting dislocations and non-dislocation crystal defects from atomistic simulation data. Model. Simul. Mater. Sci. Eng. **18**(8), 085001 (2010)
24. A. Stukowski, K. Albe, D. Farkas, Nanotwinned fcc metals: Strengthening versus softening mechanisms. Phys. Rev. B **82**(22), 224103 (2010)
25. A. Stukowski, A. Arsenlis, On the elastic–plastic decomposition of crystal deformation at the atomic scale. Model. Simul. Mater. Sci. Eng. **20**(3), 035012 (2012)
26. A. Stukowski, V.V. Bulatov, A. Arsenlis, Automated identification and indexing of dislocations in crystal interfaces. Model. Simul. Mater. Sc. **20**, 085007 (2012)
27. E. Maras, O. Trushin, A. Stukowski, T. Ala-Nissila, H. Jónsson, Global transition path search for dislocation formation in Ge on Si(001). Comput. Phys. Commun. **205**, 13–21 (2016)
28. S. Wang, G. Lu, G. Zhang, A Frank scheme of determining the Burgers vectors of dislocations in a FCC crystal. Comput. Mater. Sci. **68**, 396–401 (2013)

Chapter 11
Advances in Discrete Dislocation Dynamics Modeling of Size-Affected Plasticity

Jaafar A. El-Awady, Haidong Fan, and Ahmed M. Hussein

11.1 Introduction

With increasing interests in nano- and micro-technologies, and the ever growing search for stronger materials, it has been apparent that the mechanical [1], thermal [2], magnetic [3], and electronic [4] properties of crystalline materials strongly depend on the physical size of the crystal (i.e., external geometry) or its microstructural features (e.g., grain size, precipitate spacing, twin boundary spacing, spacing between dislocation arrangements, etc.).

Over the past decade, numerous nano- and micro-scale experimental techniques have been utilized to study the size-dependent behavior of single and nanocrystalline materials under various loading conditions. These include, but not limited to, microcompression (e.g., [1, 5]), micro-tension (e.g., [6, 7]), and microbending (e.g., [8, 9]) of microcrystals, nanoindentation [10], and torsion of microwires [11]. In addition, many experimental techniques have also focused on the grain/twin-size effects on the strength of bulk nanocrystalline and ultrafine-grained bulk polycrystalline specimens (e.g., [12]). It has been generally argued that smaller leads to a stronger response, although this has been shown to depend strongly on the dislocation density [13]. Nevertheless, it can be shown that almost all materials display a universal size effect in the form

$$\sigma \propto \lambda^{-n} \tag{11.1}$$

J.A. El-Awady (✉) • A.M. Hussein
Department of Mechanical Engineering, Johns Hopkins University, Baltimore, MD 21218, USA
e-mail: jelawady@jhu.edu; ahussei4@jhu.edu

H. Fan
Department of Mechanics, Sichuan University, Chengdu, Sichuan, 610065, China
e-mail: haidongfan8@foxmail.com

C.R. Weinberger, G.J. Tucker (eds.), *Multiscale Materials Modeling for Nanomechanics*, Springer Series in Materials Science 245,
DOI 10.1007/978-3-319-33480-6_11

where σ is the crystal strength, λ is some characteristic size (e.g., sample size, grain size), and $0 \leq n \leq 1$. These phenomenological observations have resulted in many questions regarding the origins of such size-dependent response of materials.

A better understanding of the size-affected plastic response of metallic materials can greatly accelerate the future development of stronger materials. However, attaining such an understanding is difficult through only experimental investigations and/or simple theoretical models. On the other hand, experimentally validated physics-based numerical simulations can provide an avenue to bridge the gaps in our understanding of material deformation.

Molecular dynamics (MD) simulations is one common method for studying plasticity, deformation, and defect interactions in different material systems at the small scale. However, MD faces some major challenges, including the ability to capture reasonably large enough representative statistics of microstructural variations without worry about imposed boundary conditions; as well as the severe-limitations it suffers in the temporal domain. Thus, it is typically difficult to compare and upscale results from traditional MD simulations to predict experimental measurements at relevant length- and time-scales without sometimes extreme assumptions. As discussed in Chap. 6, a number of modeling methods have been developed over the years to overcome the time-scale limitations in MD such as the parallel-replica dynamics method [14], the hyperdynamics method [15], and the temperature accelerated dynamics method [16]. Nevertheless, such methods are still unable to predict the collective behavior of large number of defects and their contribution to damage accumulation and subsequent failure.

Length-scale limitations of MD have been typically overcome through coarse-graining methods, which are strictly material and mechanisms-based methods. In dislocation-mediated plasticity of crystalline materials, the discrete dislocation dynamics (DDD) method is one such coarse-grained approach that greatly mitigates both the time-scale and length-scale limitations of MD. The effectiveness of DDD is realized by direct numerical simulations of the collective motion of dislocation ensembles through the direct numerical solution of their equations of motion. Dislocations are represented as discrete line defects, from which their interactions among themselves as well as with other defects can be computed, and consequently their motion can be fully integrated over time [17].

A number of two-dimensional (2D) and three-dimensional (3D) computational DDD frameworks have been developed over the past two and a half decades. In 2D DDD, dislocations are modeled as infinite long straight dislocations, and therefore are regarded as point defects on a 2D plane with random obstacles dispersed on their slip planes (e.g., [18, 19]). In a slightly different approach, in the 2.5D DDD method some 3D dislocation mechanisms (e.g., dislocation junction formation, dynamic dislocation obstacles, dynamic dislocation sources, and line tension model for loop expansion) are incorporated into the 2D framework [20]. While 2D and 2.5D DDD simulations provide a quick approximate approach to model dislocation-mediated plasticity, they do not capture the true 3D nature of dislocations in materials. To address this, full 3D DDD simulations were developed to account for the dynamics of dislocation loops in 3D space. Each loop is typically discretized as connected straight segments having pure edge or pure screw segments [21], or mixed edge and screw segments [22]. Other 3D approaches represent dislocation loops as a series of connected curved splines [23]. Based on these methods a few widely used 3D DDD simulation codes have been developed over the past two decades [24–30]. While these codes vary in the approximations made, mechanisms accounted for, formulation, scalability, performance, and complexity, they predict similar material behavior to the extent captured by the deformation mechanisms accounted for in each code.

The theoretical development of 3D DDD is discussed in details in Chap. 2, while those for 2D and 2.5 DDD are available elsewhere [18–20]. In this chapter, the focus will be on practical applications of DDD in the field of dislocation-mediated plasticity. Over the past decade these DDD methods have been widely used to predict plastic deformation in a number of technologically important problems. These include simulations of different materials (metals, semi-conductors, and reactive materials), crystallographic systems (face centered cubic (FCC), body centered cubic (BCC), and hexagonal closed pack (HCP) crystals), loading conditions (thermal, uniaxial, multiaxial, torsion, bending, cyclic loading, etc.), and applications (bulk response, thin films, microcrystals, precipitate hardening, irradiation effects, etc.).

One of the greatest success stories of DDD simulations in the past decade is the major insights they have provided in identifying physical mechanisms associated with size-scale effects in metals. Unlike continuum mechanics models (e.g., crystal plasticity), DDD allows for modeling micro-scale plasticity in a physics-based framework, which is the root cause for the prevalence of DDD in predicting size-scale effects with minimal ad hoc assumptions if any. Some aspects of these success stories in studying size effects on the mechanical properties of crystals will be addressed in the following sections.

11.2 Size-Dependent Plasticity in Single Crystals

Along with the experimental studies on size effects, many multiscale simulations studies have been conducted in an attempt to identify the underlying mechanisms associated with size effects. These range from MD simulations at the nano-scale, DDD at the micro-scale, and crystal plasticity finite element (CPFE) at the meso-scale. In the following sections we will focus mostly on how DDD simulations have led to significant advances in our physics-based understanding of the mechanisms controlling size effects in metals under various loading conditions.

11.2.1 Microcrystal Response under Uniaxial Tension and Compression

A decade ago, Uchic et al. [5] developed an experimental methodology based on the combination of focused ion beam sample fabrication and subsequent testing using a simple extension of the nanoindentation technique. Following this methodology a vast number of experimental studies have been conducted to quantify size effects in micropillar like specimens fabricated from different bulk metallic crystals having diameters, D, in the range of tens of nanometers to tens of microns [1, 31]. These micropillars were tested in compression using a flat nanoindenter tip [32], or in tension using micron sized grips [33]. These experiments revealed some outstanding observations, in particular the strong dependence of the flow strength on the diameter following a power-law relationship similar to Eq. (11.1) with $\lambda = D$, as well as the dependency of the power low exponent, n, on the initial dislocation density [34, 35].

11.2.1.1 Size Effects on Strength in Metals

Some of the earliest modeling efforts, that attempted to enrich our understanding of the origins of size effects observed experimentally, were conducted using 2D DDD [36–40]. Due to the geometric constrains associated with these 2D simulations, they mimic a thin film specimen under uniaxial tension or compression. In these simulations, the dislocation image fields resulting from the enforcement of the boundary conditions (i.e., the finite sample geometry) are evaluated by coupling the 2D DDD framework with a finite element method (FEM) solver as discussed in great details elsewhere [19]. Figure 11.1 shows distributions of stresses, dislocation arrangements, and slip traces from 2D DDD simulations of two typical geometries of free-standing thin films. In general, these simulations do predict qualitatively that the strength increases with decreasing film thicknesses. However, these results were shown to be strongly influenced by many simulation variables (e.g., boundary conditions, loading conditions, dislocation source densities, and dislocation obstacle densities). As an example, it was shown that simulations of free-standing FCC thin films having dislocation source and obstacle densities both equal to 5.6×10^{13} m^{-2} exhibit a film thickness size effect on the strength with a power-law exponent of $n = 0.49$, while simulations of obstacle-free films exhibit a negligible size effect with an exponent of $n = 0.07$ [37]. However, these observations are in contrast with experimental observations of micropillar like specimens orientated for single slip in which the effect of obstacles is typically absent and the power-law exponent is on the order of $n = 0.7$ for FCC crystals [32]. In addition, in simulations with the tensile axis rotation being constrained, to mimic experiments in which tension grips prevent the rotation of the specimen, a weak size effect with exponent $n = 0.08$ was predicted. This again is in contrast with the experimental observations [1]. Finally, these simulations also predict a stronger size effect (i.e., larger exponent n) when

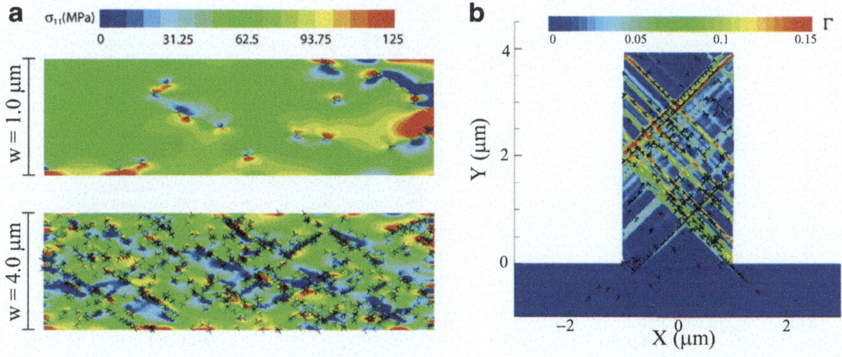

Fig. 11.1 Two-dimensional DDD simulations of single crystal thin films under tensile and compressive loadings: (**a**) The distribution of the tensile stress and corresponding dislocation structure in 1.0 and 4.0 μm thin films. Reprinted from [36] with permission. (**b**) Distribution of slip traces and dislocation pattern during compression of a thin film having height 4.0 μm, and width 2.0 μm, attached to a large bulk base. Reprinted from [40] with permission

increasing the initial dislocation density to 2.0×10^{14} m^{-2} [36, 37], which again is in contrast to experimental results showing a decrease in the power-law exponent with increasing dislocation density [34, 35].

It should be noted that simulations of thin films on an elastic substrate with the film/substrate interface modeled as a rigid interface show an enhanced size effect with an exponent of 0.7 [39]. This was attributed to the development of a hard layer of piled-up dislocations at the film/substrate interface. However, these boundary conditions are quantitatively different than those for micropillar compression and tension experiments.

On the other hand, simulations of similar thin film geometries and boundary conditions but using the 2.5D DDD framework seem to predict some of the features observed experimentally [41–44]. In addition to the typical size effects studies on flow strength, these simulations were also extended to large strains to investigate the effect of crystal size on the work-hardening rate. As shown in Fig. 11.2, these simulations predict a significant size effect for both the flow stress and the work-hardening rate [43]. The stage-II hardening rate was observed to increase with decreasing thin film thickness by an order of magnitude for simulations of film thicknesses in the range of 0.2–12.8 μm. Finally, it should be noted that these simulations have also indicated a decrease in the power-law exponent with increasing initial dislocation source density [41]. These results are in qualitative agreement with experimental observations [34, 35, 44].

Along with 2D and 2.5D DDD studies, 3D DDD simulations have also been extensively used to study the effects of crystal size on the deformation of cylindrical and rectangular micropillars. The results of these simulations arguably provide the best quantitative match to experiments. The 3D DDD simulations can be grouped into two sets. In one set, the dislocation image fields due to the finite geometry of

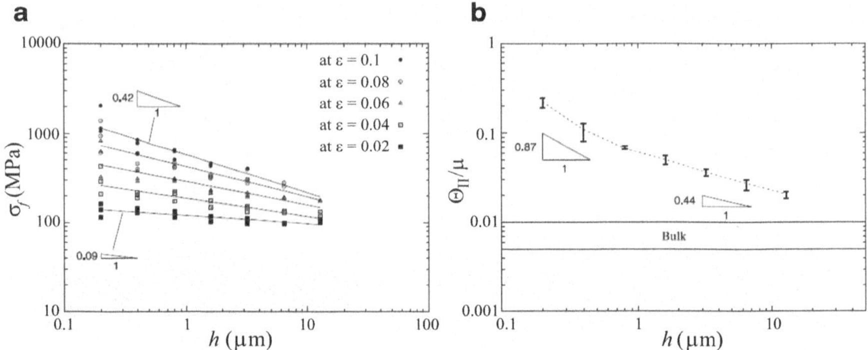

Fig. 11.2 (**a**) Flow strength, σ_f, at different strain levels; and (**b**) stage-II work-hardening rate, Θ_{II}, in units of the shear modulus μ, versus crystal thickness, H. Reprinted from [43] with permission

the crystals were ignored. In the second set, those image fields are computed by solving a boundary value problem through coupling the 3D DDD framework with the boundary element method (BEM) [45] or FEM [46]. It was shown that the image field can significantly alter the character of dislocations very close to the surface; however, its effect is insignificant in regards to the predicted size effect response for microcrystals having diameters $D \geq 1.0\,\mu$m [45].

In 3D DDD simulations, the initial dislocation network is typically introduced randomly in the simulation cell. In earlier size effects studies these networks were introduced in the form of random Frank–Read sources [45, 47–49], and later in the form of relaxed unpinned dislocation loops [46, 50]. It has been shown that both cases give quantitatively similar response in terms of size-scale effects when the starting dislocation densities in the pinned dislocation network simulations are the same as the relaxed dislocation densities in the unpinned simulations [13, 46].

The predicted power-law exponent for the strength versus crystal diameter relationship from 3D DDD simulations was shown to be in excellent agreement with those predicted experimentally for FCC (e.g., [47, 48]), BCC (e.g., [51]), and HCP (e.g., [52]) crystal structures. In addition 3D DDD simulations were also the first to show that the power-law exponent decreases with increasing initial dislocation density [48]. Furthermore, 3D DDD simulations [47, 48] also provided the first indication of the plausibility of the single-arm dislocation source model [53] as one of the controlling mechanisms of size effects a few years before it was affirmed experimentally [54]. Other controlling mechanisms were also identified from 3D DDD simulations depending on the crystal size and initial dislocation density including dislocation starvation [13], weakest-link mechanism [47], source-truncation hardening [48], dislocation source shutdown [50], exhaustion hardening [48, 49], and forest hardening [13].

In addition, El-Awady [13] conducted a set of comprehensive 3D-DDD simulations that spanned two orders of magnitude in crystal size and five orders of magnitude in initial dislocation densities. These simulations have revealed the

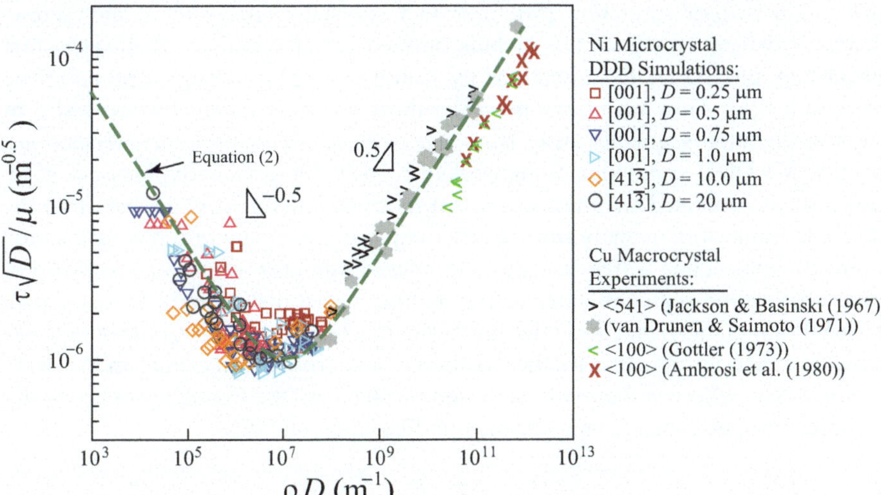

Fig. 11.3 Generalized size-dependent crystal strength from DDD simulations of microcrystals in the range $0.25 \leq D \leq 20\,\mu$m. Reprinted from [13] with permission

existence of a well-defined relationship between strength, dislocation density, and crystal size at all length scales, as shown in Fig. 11.3. The results predicted a transition from a dislocation source strengthening to forest-dominated strengthening at a size-dependent critical dislocation density. Based on these results a generalized size-dependent dislocation-based model was deduced as follows [13]:

$$\frac{\tau}{\mu} = \frac{\beta}{D\sqrt{\rho}} + \alpha b\sqrt{\rho} \tag{11.2}$$

where τ is the critical resolved shear stress of the crystal, b is the Burgers vector, and β and α are dimensionless material constants. This law was also shown to be in excellent agreement with experiments [13]. The first term on the right-hand side of Eq. (11.2) represents the strength of the weakest dislocation source in the crystal, which dominates below a size-dependent critical dislocation density on the order of $\rho D = 10^7\ \mathrm{m}^{-1}$. On the other hand, the second term in Eq. (11.2) accounts for forest strengthening, which dominates above the critical dislocation density [13].

11.2.1.2 Effects of Cross-Slip Activation

Cross-slip is a thermally activated deformation mechanism by which a screw dislocation can conservatively move from one glide plane to another containing its Burgers vector [55]. It is generally accepted that cross-slip influences many aspects of plasticity, including dislocation multiplication, work-hardening, and dislocation pattern formation [56]. For FCC crystals, cross-slip of any screw-character dislocation segment in the bulk (termed hereafter bulk cross-slip) was first introduced into 3D-DDD simulations by Kubin et al. [24] through a probabilistic procedure [24]. This model has been the most common model for cross-slip of screw dislocation segments away from free surfaces or dislocation intersections, in most 3D DDD simulation frameworks [27, 47]. More recently, Hussein et al. [56] also incorporated two new cross-slip mechanisms for FCC crystals into the 3D DDD framework, namely surface cross-slip (i.e., cross-slip of screw dislocation segments intersecting a free surface) and intersection cross-slip (i.e., cross-slip of screw dislocation segments intersecting another forest dislocation). It was shown from MD simulations that these two cross-slip mechanisms have significantly lower activation energies than that for the traditional bulk cross-slip mechanism [57–61].

Since cross-slip is a thermally activated mechanism, the frequency of cross-slip events can be computed from an Arrhenius-like equation [56]:

$$f = \omega_a \frac{L}{L_o} \exp\left(-\frac{E_a^c - V\Delta\tau_E}{k_B T}\right) \tag{11.3}$$

where E_a^c is the energy barrier required to form a constriction point on the screw dislocation, $\Delta\tau_E$ is the difference between the Escaig stress on the glide plane and that on the cross-slip plane, ω_a is the cross-slip attempt frequency, L is the length of the screw segment, L_o is a reference length of 1 μm, V is the activation volume, k_B is the Boltzmann constant, and T is the temperature. It should be noted that E_a and V depend on the types of cross-slip and were reported for Ni in [56].

Interestingly, 3D DDD simulations that account for these new cross-slip types predict for the first time dislocation microstructure evolution and surface slip localization. These simulations showed that bulk cross-slip is the least frequent in microcrystals and does not contribute much to the microstructure evolution for crystal sizes $D \leq 10\,\mu$m simulated to strain levels less than 2 % [56]. On the other hand, surface and intersection cross-slip were observed to contribute significantly to the microstructure evolution, with surface cross-slip being the most frequent, and contributes to surface slip localization, as shown in Figs. 11.4a and 11.2b. Furthermore, for crystals having diameters $D > 2.0\,\mu$m, intersection cross-slip plays a significant role in microstructure cell structure formation (see Fig. 11.4c). In contrast, simulations of crystals having diameters $D < 2.0$ showed no significant cross-slip events, dislocation buildup, or cell structure formations. These simulation results are in good agreement with experimental observations [62].

11.2.1.3 Surface Dislocation Nucleation

Dislocation nucleation from the free surfaces is typically observed during in situ transmission electron microscopy (TEM) experiments of microcrystals having diameters less than a few hundred nanometers [63]. However, in traditional 3D DDD simulations only dislocation activation/multiplication from pre-existing dislocations

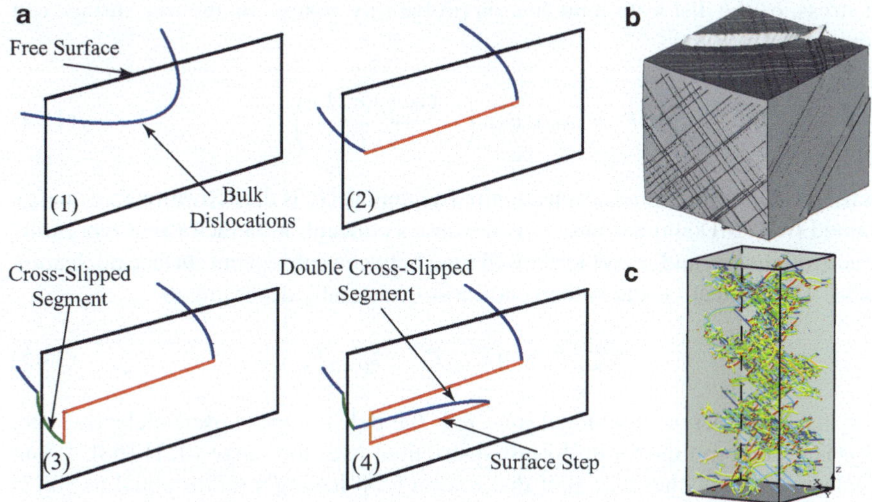

Fig. 11.4 (**a**) A schematic showing the sequence of a dislocation intersecting a free surface leading to surface slip localization. The *blue and green* dislocations lie inside the crystal on a glide and the corresponding cross-slip plane, respectively. The *red* surface segments represent dislocation segments that exited the crystal (i.e., surface steps). (**b**) 3D surface slip traces at 0.5 % strain in a cubical simulation cell with edge length 20 μm. (**c**) Dislocation microstructure at 0.5 % strain in a rectangular simulation cell having edge length 5.0 μm and height 15 μm and loaded in compression axially. Reprinted from [56] with permission

is typically accounted for. Thus, MD simulations have been favored to study dislocation nucleation in simulation cells mimicking nanowires or nanopillars having dimensions less than 100 nm [64–66]. In these simulations a perfect nanocrystal is deformed until the applied stress is high enough for dislocations to nucleate from the corners of the simulation cell. As the dislocation density increases the stress drops significantly to lower values. More recently, a network of pre-existing dislocations were introduced in Fe single crystal nanopillar like MD simulation cells by computing their anisotropic displacement fields then relaxing the simulation cell [67]. These MD simulations showed that for crystals having $D \leq 50$ nm dislocation nucleation dominates, while for crystals having $D = 80$ nm there competition between surface dislocation nucleation and slip/multiplication of pre-existing dislocations [67]. While such MD simulations provide insightful information on the nucleation and evolution of pre-existing dislocations in nanocrystals, they are still limited in the sense that the applied strain rates are typically orders of magnitude higher than their experimental counterparts.

To address some of these limitations, Ryu et al. [68] developed an algorithm to implement surface nucleation in the 3D DDD framework [68]. First, a number, N, of possible nucleation sites are randomly distributed on the surface. At a given time and stress, σ, the dislocation nucleation probability at site i on the free surface can be computed as [68]:

$$P_i = \nu_0 \Delta t \exp\left(-\frac{G_N(\chi_i\sigma, T)}{k_B T}\right) \tag{11.4}$$

where ν_0 is the surface nucleation attempt frequency, G_N is the activation free energy obtained from MD simulations, χ_i is the stress concentration factor that represents surface roughness, and Δt is the elapsed time between subsequent surface nucleation checks. The nucleation site is then chosen stochastically according to

$$\Sigma_{i=1}^{m-1} P_i \leq \eta \Sigma_{j=1}^{N} P_j < \Sigma_{k=1}^{m} P_k \tag{11.5}$$

where i, j, and k denote individual sites, m is the index of the chosen nucleation site, and η is a random number uniformly distributed over the range (0, 1) [68]. When a nucleation site is chosen, a half dislocation loop having a radius equal to $50b$ is introduced on the slip plane having the highest Peach–Koehler force. With such an algorithm in place, it is possible to quantify the potential competition between the activation/multiplication of pre-existing dislocations and surface dislocation nucleation in microcrystals having diameters in the submicron range, or when the initial dislocation density is low and dislocation starvation is expected [13].

Two simulation results using this method are shown in Fig. 11.5. In simulations of cylindrical microcrystals having diameters $D < 0.2\,\mu\text{m}$ and a relatively low initial dislocation density, all dislocations escape the crystal during the early stages of loading [68], thus, reaching a state of dislocation starvation. When the stress reaches the critical dislocation nucleation stress, half loop dislocations are introduced into the microcrystal from the surface. If the applied load is not allowed to drop, as is the case in these simulations, then these new dislocations will traverse the micropillar and subsequently escape. This is indicated by the regular dislocation density drops observed in Fig. 11.5a. The nucleation and escape process is then repeated with the flow stress reaching a plateau without any significant hardening. These results are qualitatively in agreement with experimental studies on micropillars of similar sizes with sparse initial dislocations [69]. On the other hand, as shown in Fig. 11.5b, for microcrystals having diameter $D \geq 1.0\,\mu\text{m}$ dislocation nucleation conditions are rarely reached and plasticity is mediated by the activation/multiplication of pre-existing sources. For microcrystals having $0.2 < D < 1.0\,\mu\text{m}$ the interplay between both mechanisms dominates in a stochastic manner [68]. For a more detailed multiscale model of the nucleation process, the reader is referred to Sect. 12 of this book.

11.2.1.4 Coated Solid and Annular Microcrystals

Simulations using 3D DDD were also performed to study size effects on the strength of annular pillars [70, 71]. These simulations showed that the mechanical response is solely controlled by the wall thickness (i.e., effective diameter), as shown in

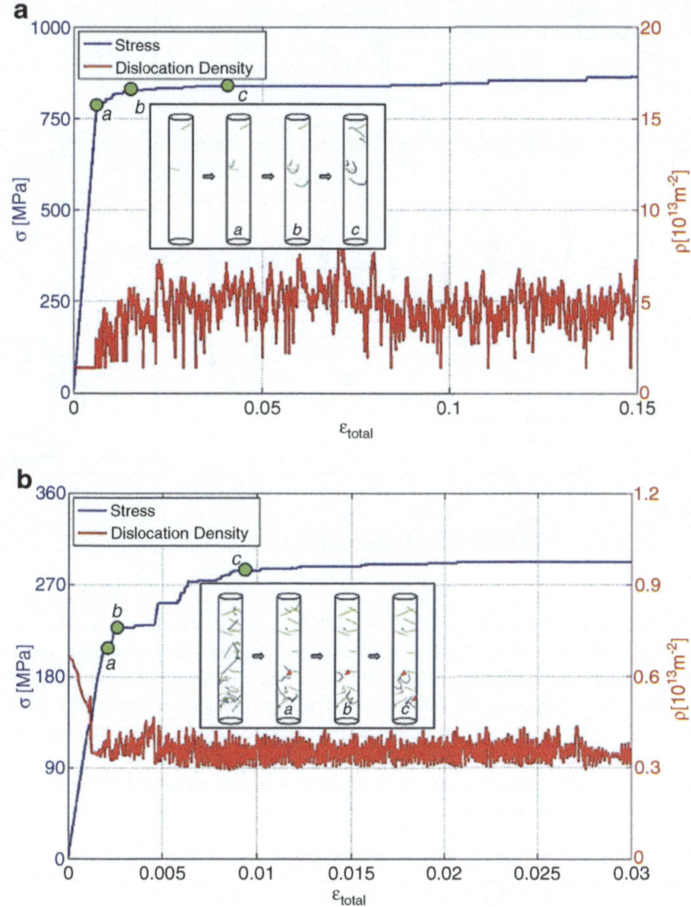

Fig. 11.5 Stress and dislocation density evolution versus strain for: (**a**) $D = 150\,\mathrm{nm}$; and (**b**) $D = 1000\,\mathrm{nm}$; cylindrical micropillars. Insets show the dislocation microstructures at different points along the stress–strain curve. Reprinted from [68] with permission

Fig. 11.6. These simulations were further extended to study the effect of a coatings on the response of Ni solid and annular micropillars [71]. In these simulations the coatings were modeled as surface interfaces acting as strong barriers to dislocation motion. Only when the local stress on a leading dislocation trapped at the interface reaches a large critical stress (i.e., interface strength) the coating is assumed to fracture and dislocations can escape the crystal locally. These simulations showed that the coatings lead to a dramatic increase in the crystal strength and dislocation density during deformation. The hardening rate was observed to increase with increasing interface strength, and weak-spots form on the interface due to dislocation breaking through the coating as the applied stress increases. These observations were in good agreement with experimental results [72].

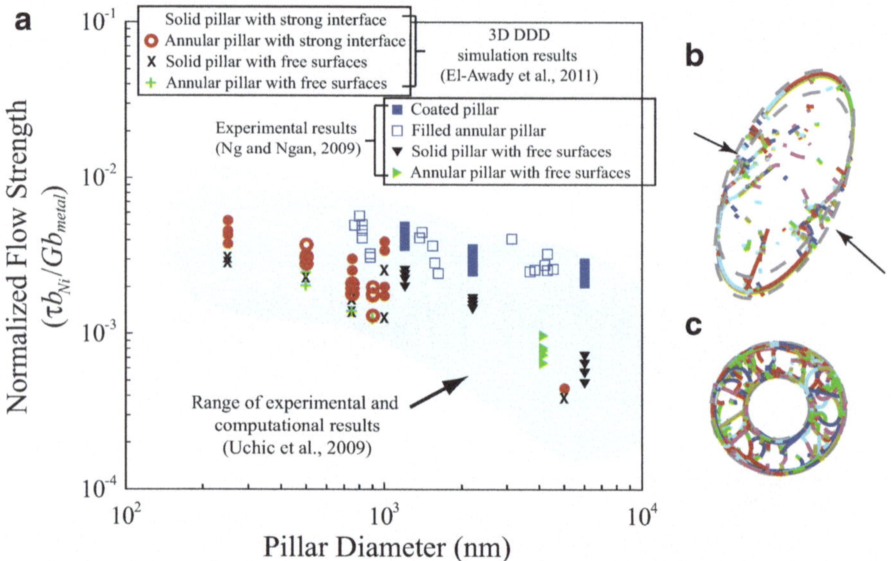

Fig. 11.6 (**a**) Flow strength versus micropillar size from 3D DDD simulations of coated and uncoated solid and annular Ni micropillars, as well as from experimental results of coated and center-filled Al pillars. The flow strength is normalized by the shear modulus and the Burgers vector of each material. For annular pillars, the effective diameter, $D_{\text{eff}} = (D - D_i)$, is used as the size of the pillar, while for all other cases, the external diameter is used. (**b**) Dislocation microstructure at 0.1 % strain in a thin foil of thickness 0.4 μm, extracted from a full 3D simulation of a $D = 5.0$ μm solid coated micropillar. The *arrows* show week spots on the surface. (**c**) Top view of the dislocation microstructure at 0.4 % strain in a coated annular pillar having internal and external diameters of 0.5 and 1.0 μm, respectively. Reprinted from [71] with permission

11.2.2 Microcrystal Response During Torsion

Given the significant interest in size effects on the response of single crystal microcrystals, it is surprising that far less attention has been given to understanding the plastic response of microwires in torsion either through modeling or experiments. At the nano-scale, a few MD simulations of twisted cylindrical nanowires have been recently reported. In FCC nanowires, the deformation was shown to be dependent on the crystal orientation along the twisting axis [73, 74]. For ⟨110⟩ oriented nanowires, dislocations parallel to the nanowire axis nucleate homogeneously from the free surfaces. On the other hand, for nanowires oriented along the ⟨111⟩ or ⟨100⟩ directions, a heterogeneous deformation was observed due to the formation of twist boundaries.

At the micro-scale, Senger et al. [75, 76] performed 3D DDD to investigate the response of twisted single crystal microwires having square cross-sections. The torsion loading is incorporated into these DDD simulations by coupling with FEM. A displacement filed on the top and bottom surfaces of the simulation cell are calculated every time step based on a prescribed twist angle rate, $\dot{\phi}$, while the microwire sidewalls were traction free. The twist can be imposed on the top surface with the bottom surface fully constrained, or anti-symmetrically imposed on both top and bottom surfaces.

From these simulations it was shown that the torsion moment required for plastic flow increases with decreasing microwire diameter [75, 76]. Furthermore, an orientation dependence was observed for the evolving dislocation microstructure. For ⟨234⟩ oriented microwires, dense coaxial dislocations pile-up at the sample center on the slip plane almost containing the sample axis, as shown in Fig. 11.7a. On the other hand, for ⟨100⟩ orientation microwires, dislocations are observed to pile-up almost perpendicular to the twist axis, leading to the formation of low angle grain boundaries, as shown in Fig. 11.7b. Both these results are in good agreement with MD simulation predictions [73].

Cross-slip was also observed to significantly influence the creation of an irreversible dislocation microstructure, with the size of dislocation pile-ups as well as the number of active slip systems in both orientations increasing when cross-slip is accounted for. Furthermore, the density of irreversible dislocations after untwisting increases with increasing twist angle [76].

The strong plastic anisotropy and the power-law relationship between stress and strain rate during torsion creep in HCP ice single crystals were also investigated using 3D DDD simulations [77]. While it was not reported how the torsion loading

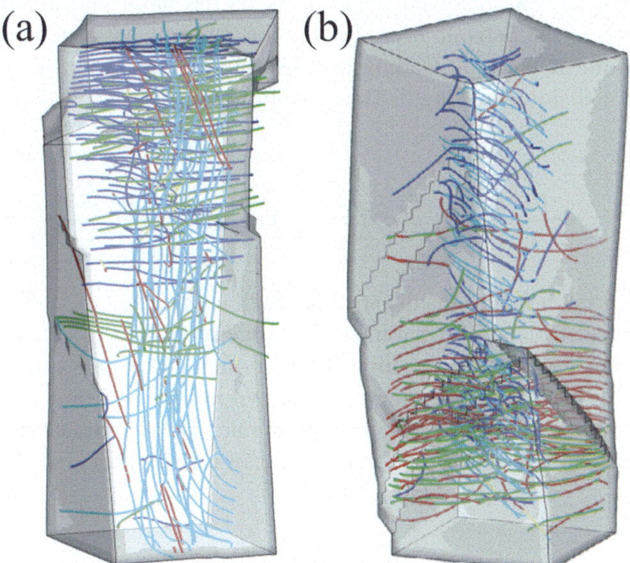

Fig. 11.7 The dislocation microstructure and the twisted geometry (magnified by a factor of 5) in $0.5 \times 1.5 \times 0.5\,\mu m^3$ samples having: (**a**) ⟨234⟩; and (**b**) ⟨001⟩; orientations at a plastic torsion angle 3°. Cross-slip is activated in both simulations. Reprinted from [76] with permission

was imposed in this study, in the absence of coupling with FEM one approximate approach is to impose the torsion by assuming a uniform shear stress, $\tau = Qr/J$ in the circular cross-section plane. Here, Q is the desired torque, r is the distance from the cylinder axis, and J is the polar moment of inertia. A double cross-slip mechanism for HCP crystals was implemented in these simulations [77]. It was assumed that a double kink nucleates from a constriction on a screw dislocation line, resulting in moving the dislocation from one basal plane to another. Furthermore, the motion of the edge kinks on the prismatic plane results in the extension of the dislocation on the new plane. In the DDD framework, the cross-slip probability of an isolated screw segment is based on the ratio of motion frequency on the basal, f_b, and prismatic, f_p, planes such that [77]:

$$P = \frac{1}{1 + f_b/f_p} \tag{11.6}$$

and

$$\begin{aligned} f_b &= \zeta\left(T, L\right) \tau_a \exp\left(-\frac{E_a^g}{k_B T}\right) \\ f_p &= \omega_a \frac{L}{x^*} \exp\left(-\frac{\Delta G^*\left(\tau_p\right)}{k_B T}\right) \end{aligned} \tag{11.7}$$

where τ_a and τ_p are the resolved shear stress on the basal and prismatic planes, respectively, E_a^g is the activation energy for the glide of basal dislocations, x^* is the critical width of the kink leading to maximum energy, and ζ is a function of temperature and dislocation length. The activation energy for cross-slip is expressed as [77]:

$$\Delta G^* = \frac{\mu b^2 h_c}{2\pi\left(1 - \nu\right)} + \Gamma d_c^2 \left(1 - \frac{\pi}{4}\right) - 2bh_c \sqrt{\frac{\mu b h_c \left(1 + \nu\right)}{8\pi\left(1 - \nu\right)} \tau_p} \tag{11.8}$$

Here, h_c is the kink height, Γ is the stacking fault energy, and d_c is the dissociation width on the basal plane. All parameters for ice are listed in ref. [77]. The 3D DDD simulation results incorporating the aforementioned cross-slip mechanism, in addition to those from experiments, are shown in Fig. 11.8 on a log–log plot of the maximum applied stress versus shear strain rate. These simulations suggest that cross-slip is an important factor in torsion creep of ice single crystals [77].

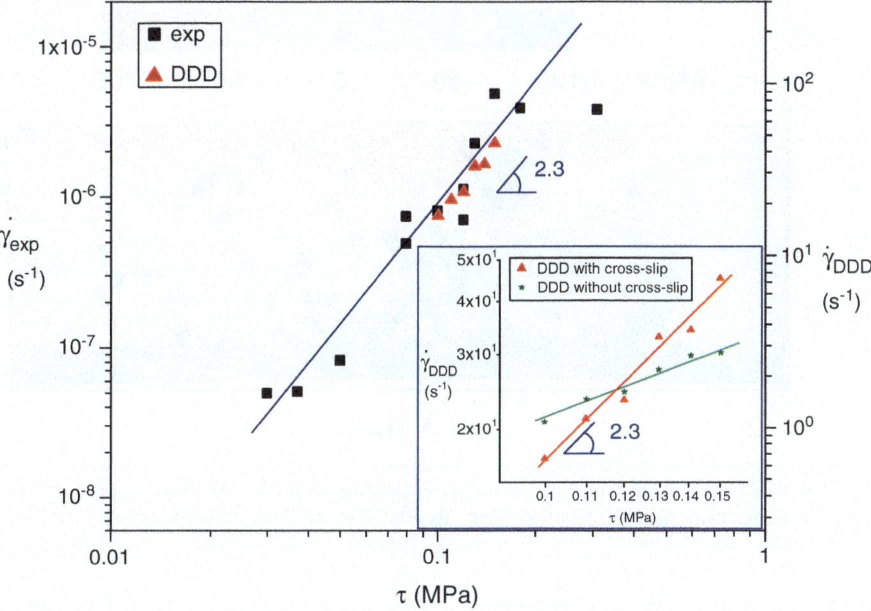

Fig. 11.8 Log–log plot of the maximum applied stress versus shear strain rate for torsion creep from 3D DDD simulations and experiments. The insert shows the simulation results with and without the double cross-slip mechanism. Reprinted from [77] with permission

11.2.3 Microcrystal Response Under Bending Loading

Size effects on the bending response of single crystal cantilever-like micro-specimens having square cross-sections were frequently studied through both experiments and simulations. Experimentally, these specimens are fabricated using FIB milling and the bending moment is imposed through a load applied on the free-end of the micro-cantilever beam using a nanoindenter tip [78]. Both 2D and 3D DDD simulations were performed by multiple researchers in an attempt to correlate dislocation mechanisms with experimental observations. In these DDD simulations the bending moment is imposed by coupling with an FEM, in which a displacement field is imposed on both ends of the simulation cell in one of two ways. In one case the displacement on one end is fixed, while a constant vertical displacement rate is imposed on the other end. This mimics a linear bending moment in a cantilever beam in a similar way to the experimental setup. In the other case two opposite and equal axial constant displacement rates that vary linearly through the beam thickness are imposed on both ends. This mimics the case of a beam under a pure bending moment.

As shown in Fig. 11.9, from 2D DDD simulations of thin films under uniform bending conditions, the deformation is accommodated by many dislocations that pile-up at the thin film neutral plane [38, 79, 80]. Furthermore, these 2D simulations

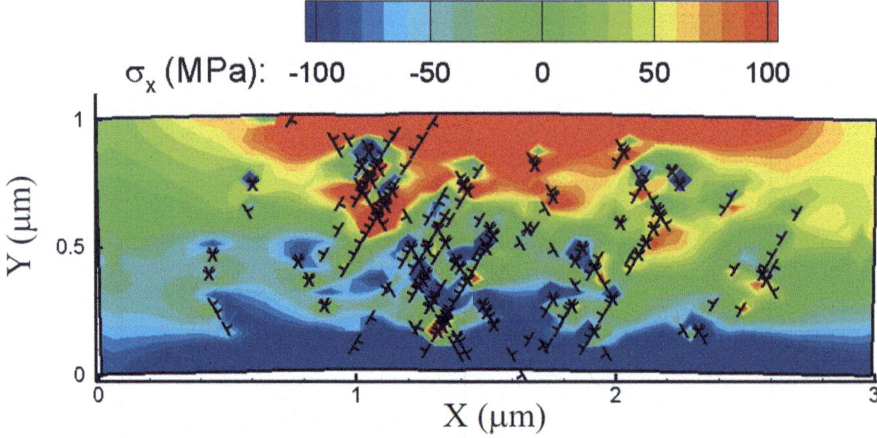

Fig. 11.9 Dislocation distribution and contours of bending stress at a bending angle of 0.03 in single crystalline films having sizes $1 \times 3\,\mu\text{m}^2$. Reprinted from [79] with permission

also predict a strong Bauschinger effect during cyclic bending [38, 79]. This agrees well with observations from micro-beam bending experiments [81].

Simulations from 3D DDD also show a power-law dependence between the flow strength and the beam thickness, similar to Eq. (11.1), with an exponent equal to $n = 1.0$ [82]. This indicates that the size-dependence of strength in bending is stronger than that during uniaxial tension or compression. From these simulations the strong bending size effects were attributed to pronounced dislocation pile-ups at the neutral plane of the beam [82]. In addition, the yield points predicted from these simulations agree well with the critical thickness theory [83]. These results were also in good agreement with experimental observations [82].

Cyclic heterogeneous stress fields and their effect on the dislocation activity of single crystal Cu micro-beams were also modeled using 3D DDD in a fully reversed bending simulation [81]. Neither cyclic hardening nor dislocation substructure formation was observed in these simulations. It should be noted that the dimensions of the simulated beams were smaller than the characteristic size of dislocation cell structures. Furthermore, the simulations were conducted for one loading cycle only, which might also explain the lack of dislocation cell-like structures.

The dislocation microstructure evolution in micro-cantilever beams was also studied using a highly scalable FEM framework coupled with the 3D DDD ParaDiS code through a distributed shared memory implementation [84]. This framework allows for the calculation of large numbers of dislocation segments interacting with an independently large number of surface finite elements [84]. In these simulations a rectangular computational cell having dimensions $23 \times 8.65 \times 8.65\,\mu\text{m}^3$ having an initial dislocation density of 10^{11} m^{-2}. For the FEM computations a fine mesh with more than five million quadratic tetrahedral elements was used. The boundary conditions were imposed such that one end is fixed while a constant distributed shear

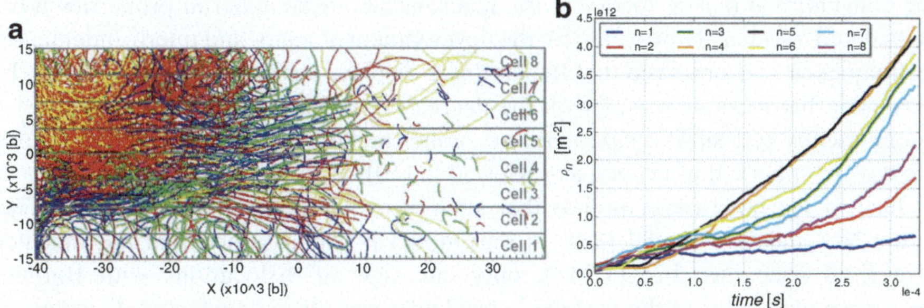

Fig. 11.10 (**a**) Side view of the dislocation microstructure in a micro-cantilever beam after 3.25×10^{-4} s. The locations of the eight cells used to compute the local dislocation densities are shown. (**b**) Time history of the dislocation density in each cell. The colors in (**a**) and (**b**) represent dislocations with common slip plane normal. Reprinted from [84] with permission

force equal to 10 MPa was imposed on the opposite end. The predicted dislocation microstructure after 3.25×10^{-4} s is shown in Fig. 11.10. These simulations show that the dislocation multiplication evolves in three stages, as shown in Fig. 11.10b. In the initial stage, the dislocation density increases significantly near the highly stressed top and bottom surfaces. In the second stage, dislocation pile-ups along the neutral plane lead to a local increase in the dislocation density at the neutral plane. Finally, the dislocation density increases with faster rates in the regions neighboring the neutral plane due to the spreading of the dislocation pile-ups to regions nearest the neutral plane [84].

In a different study, 3D DDD simulations of micro-beam bending were compared to those from a gradient crystal plasticity (CP) model that incorporates the density of geometrically necessary dislocations (GNDs) in the expression of mean glide path length to account for scale dependence. These comparisons showed the need for a size-dependent term in the expression for the strength of slip systems [85]. Finally, the bending deformation and the evolution of the dislocation microstructure were also analyzed for different orientations and specimen sizes using 2D DDD [80]. However, the effect of orientation on the bending size effects was only studied through higher order non-local CP simulations [86], which showed that crystals in the $\langle 110 \rangle$ orientation (i.e., double-slip) show a weaker size effect than crystals oriented for multi-slip (i.e., the $\langle 100 \rangle$ orientation).

11.2.4 Nano- and Micro-Indentation

As one of the most common experimental techniques extensively used for mechanical testing of materials, indentation has been the focus of many studies for many decades. Indentation testing has been one of the favored techniques to quickly extract metal bulk properties from hardness measurements [87]. From the conventional plastic theory at the macro-scale, these material properties were considered intrinsic. However, with the development of nano- and micro-indentation techniques, it was observed that the hardness displays a strong size-dependence. In particular, hardness increases with the decreasing indentation depth in the case of geometrically self-similar indenters (i.e., cones and pyramids) and indenter size in the case of spherical indenters, as reviewed in [10].

One of the most cited models that attempt to interpret indentation size effect is the Nix and Gao model [88]. In that model it was assumed that the indenter is a rigid cone (see Fig. 11.11a); only one type of dislocations with Burgers vector perpendicular to the surface is available; and all the nucleated dislocations reside within a hemispherical volume defined by the contact radius. With these

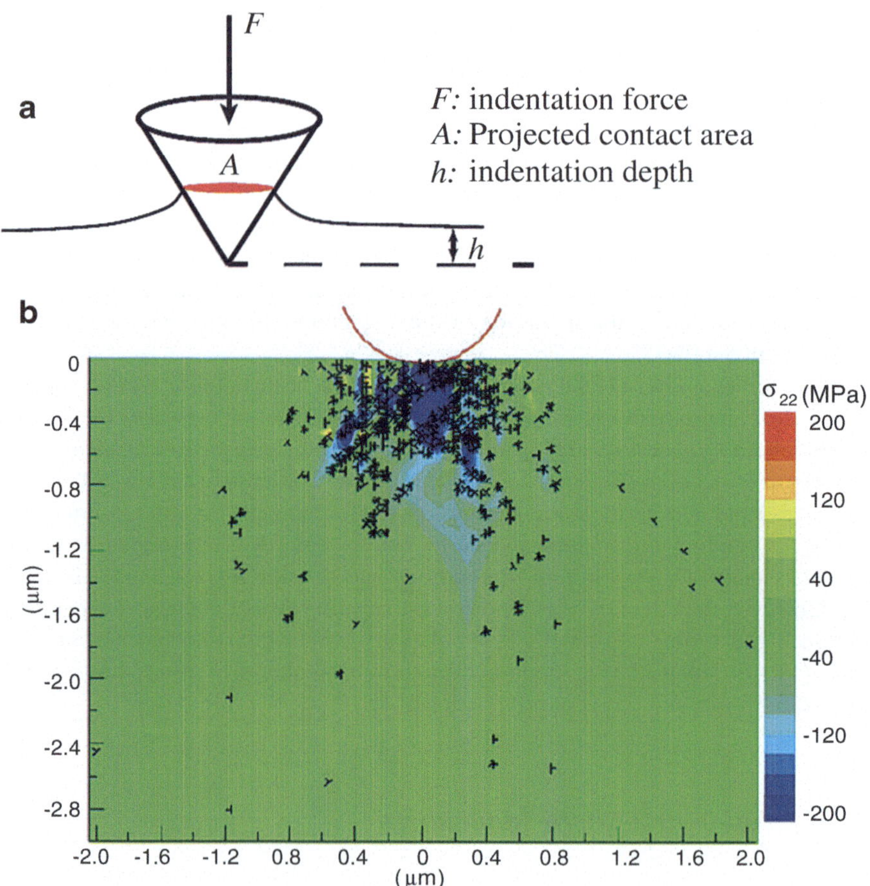

Fig. 11.11 (**a**) Schematic of the indentation test using a conical tip. (**b**) Distribution of stress and dislocation microstructure for a grain having three active slip systems at an indentation depth 30 nm as predicted from DDD simulations. Reprinted from [101] with permission

assumptions it was shown that the GND density is inversely proportional to the indentation depth, h, resulting in a higher hardness for a smaller depth. Based on this model the indentation hardness can be expressed as [88]:

$$\frac{H_v}{H_0} = \sqrt{1 + \frac{h^*}{h}} \tag{11.9}$$

where H_0 is the macroscopic hardness corresponding to a large enough indentation depth h, and h^* is a characteristic length that depends on the shape of the indenter. This model was shown to be in agreement with indentation measurements on Cu and Ag single crystals for depths in the range of hundreds of nanometers to a few microns. However, it was also shown that the Nix-Gao model breaks down for indentation depth in the range of tens of nanometers [89], potentially due to the spreading of GNDs in a larger volume than that assumed by the Nix-Gao model, or due to the scarcity of dislocations at such small scales.

Nanoindentation was commonly studied through MD simulations. For spherical nanoindenter simulations the simulation cell has been typically a thin film with a fixed bottom surface, representing a hard film-substrate interface. It was shown that for film thicknesses less than 12 nm both the elastic and plastic stages of the mean contact pressure versus depth curves are dependent on the film thickness [90]. For films thicker than 12 nm, the indentation force versus displacement curves follow the Hertzian theory, which is a prediction for a semi-infinite crystal, indicating the indentation is not influenced by the bottom surface and no thickness size effect is excepted [91]. In addition to film thickness effects, it was shown that for different radii of spherical indenters, the mean contact pressure is higher for smaller indenters [92], in good agreement with experiments, whereas the nucleation contact pressure is unaffected [93]. A more recent large scale MD study on Ni thin films showed that the size of the plastic zone below the nanoindenter tip divided by the contact radius is not a constant factor, but varies as the indentation depth changes [94]. These simulations also showed that for flat indenters the hardness increases with increasing depth and is in excellent agreement with the Swadner et al. [89] model. Finally for a conical indenter the results agree with the Nix-Gao model.

While such MD studies provide significant insights into indentation size effects, a complete mechanistic understanding is still lacking since indentation size effects are typically observed for indentation depth up to 2.0 μm. Therefore, MD simulations alone are not able to capture the full view. To address this, a number of 2D DDD simulations have been performed as a realization of plane strain indentation by Widjaja et al. [95–98], Kreuzer et al. [99, 100], and Ouyang et al. [101–103]. In such studies all the dislocations nucleate form the randomly pre-distributed dislocation sources inside the crystals. Figure 11.11b shows the predicted stress and dislocation microstructure distributions for a $4 \times 4\,\mu m^2$ grain with three active slip systems at indentation depth of 30 nm under a spherical micro-indenter having diameter 0.5 μm [101]. These studies have shown that the indenter diameter, grain size, indentation depth, and distance between the indenter and the grain boundary all have significant effects on the predicted hardness.

On the other hand, it is interesting to note that while fewer 3D DDD studies have focused on indentation, nevertheless, one of the earliest applications of 3D DDD simulations was to study the plastic zone in a 2.0 μm Cu single crystal cube for nanoindentation at a depth of 50 nm using a spherical tip of radius 420 nm [104]. Here the indenter induced heterogeneous stress field was computed based on the FEM elastic solution enforcing the boundary conditions then superimposed onto the DDD elastic solution [105]. To mimic the nucleation of dislocation during indentation, prismatic loops were introduced in the simulations with segments lying on the active glide systems and the loop center coinciding with the location of the maximum shear stress [104]. Loop nucleation was assessed based on comparisons with the experimentally macroscopic applied load on the indenter to generate a certain displacement. Thus, nucleation was stopped whenever the load was lower than the experimentally applied one. It was suggested that the predicted dislocation microstructure and the corresponding plastic volume from these simulations were qualitatively in agreement with experimental observations [104]. The dense dislocations were observed in the vicinity of the indentation, which serve as GNDs to accommodate the indented deformation field, providing substantial evidences for strain gradient models.

More recently, 3D DDD simulations were also conducted to model indentation on a 5 μm Cu single crystal using a 10 μm radius spherical indenter as well as a conical indenter with half cone angle of 85 ° [106]. This study focused on correlating the hardness with the evolving dislocation microstructure. The crystal was assumed to be initially dislocation free except for a distribution of random sized Frank–Read sources placed underneath the indented surface to mimic dislocation nucleation. The inhomogeneous displacement and stress fields induced by the penetrating indenter moving with a speed of 1.0 m/s were computed every time step by imposing contact boundary conditions in FEM [106]. After the FEM solution was obtained, the dislocations were allowed to evolve under the stress induced by the indenter for a time Δt, over which the FEM solution was fixed. This time step was computed based on the prescribed indenter speed and depth increment. The FEM solution was also updated at any time if the evolving dislocation microstructure led to significant changes to the infinite medium traction and displacement fields. The predicted hardness as computed from the indentation load divided by the actual contact area versus the indentation depth for both spherical and conical indenters is shown in Fig. 11.12a. For the spherical indenter it was shown that the hardness does not depend on the indent depth but only on the radius of the indenter [106]. On the other hand, the conical indenter is geometrically self-similar, which clearly leads to the hardness decrease with increasing indent depth, in qualitative agreement with many experimental observations. Moreover, it is interesting to note that the results from the conical indent simulations underestimate significantly the hardness values predicted based on the Nix-Gao model. By careful examination of the evolving dislocation microstructure it is clear that plastic zone volume is much larger than the hemispherical region assumed in the Nix-Gao model.

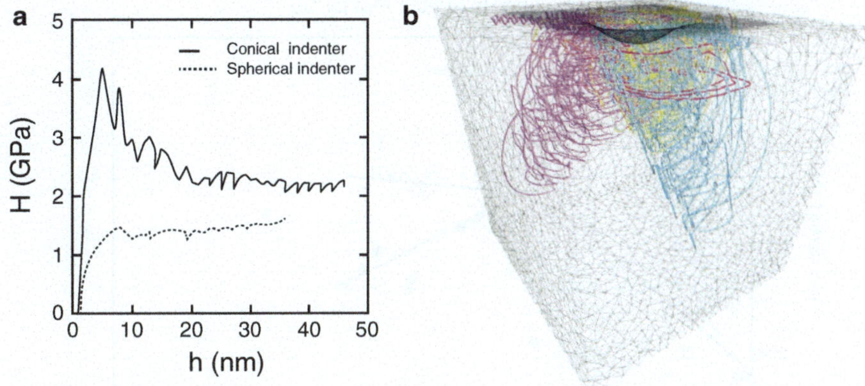

Fig. 11.12 Three-dimensional DDD simulation results of nanoindentation into Cu single crystals showing: (**a**) hardness versus indentation depth for spherical and conical indenters; and (**b**) predicted dislocation microstructure from the simulations with a spherical indenter. The mesh displacement in (**b**) is magnified to reveal the contact area. Reprinted from [106] with permission

11.3 Role of Interfaces on Dislocation-Mediated Plasticity

Interfaces, sometimes referred to as planner defects (e.g., grain boundaries (GBs) and twin boundaries (TBs)), have a considerable effect on the mechanical properties of materials [107]. These interfaces not only affect dislocation long range interactions (e.g., image fields), but also influence dislocation microstructure evolution through short range interactions (e.g., dislocation obstacles or dislocation sinks). Generally, interfaces are observed to impede dislocation movement in ultrafine-grained ($250 \leq d \leq 1000$ nm) and coarse-grained ($d \geq 1000$ nm) polycrystals, thus, leading to significant strengthening. This is directly displayed by the Hall–Petch relationship describing the dependence of the grain size d on the material strength σ_y as follows:

$$\sigma_y = \sigma_0 + kd^{-0.5} \tag{11.10}$$

Thus, an increasing yield strength is induced by grain refinement.

For nanocrystalline materials (i.e., $d < 25$ nm), the Hall–Petch relationship breaks down, and as summarized in [12], GBs become dislocation sources and sinks instead. In addition, GB sliding, migration, and grain rotation play a major role in the deformation. As a result, an inverse Hall–Petch relationship is expected for grain sizes less than 25 nm. Figure 11.13 shows a composite plot of the yield stress versus grain size for Cu polycrystals from various sources spanning several orders of magnitude of grain sizes, showing both the regimes of applications of the Hall–Petch relationship.

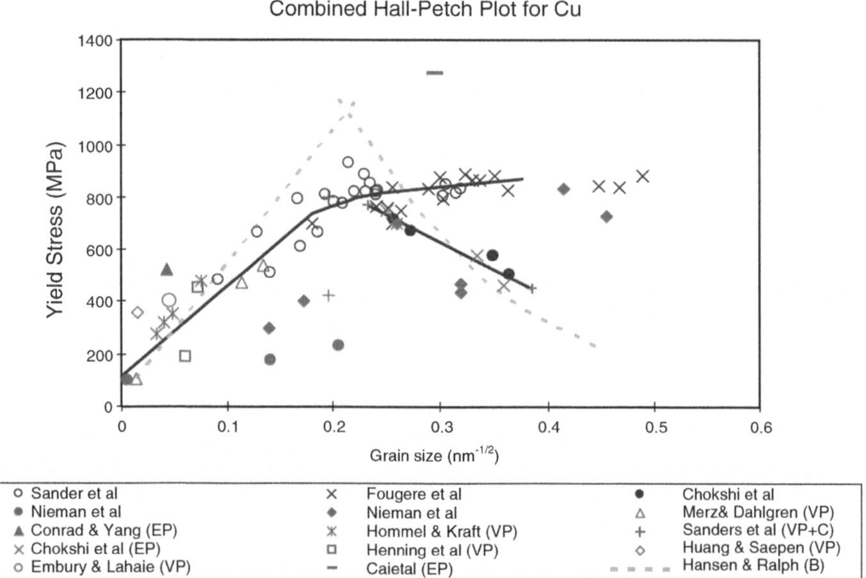

Fig. 11.13 Composite plot of the yield stress versus grain size for Cu from various sources (see [12] for full citations). Reprinted from [12] with permission

Unfortunately, to date, a fully realistic dislocation-interface interaction model based on substantial physical mechanisms is still lacking in DDD simulations, especially for twist GBs. In spite of this, many simplified models have been introduced into DDD simulations to study polycrystal intrinsic size effects (i.e., grain-size and twin-size effects) as well as polycrystal extrinsic size effects (i.e., effect of the sample external dimensions). These efforts will be discussed separately in the following two sections.

11.3.1 Grain-Size Effects

In ultrafine-grained and coarse-grained metals, deformation is predominantly mediated by dislocation slip. Therefore, DDD simulations can be an effective tool to explore the mechanisms involved in controlling grain-size and twin-size effects. It should be noted that the assumptions with regards to the implementation of GBs in DDD vary from study to study. One of the simplest GB models is to consider the GBs as rigid impenetrable interfaces to dislocations. This has been a typically assumption for simulations of high angle grain boundaries. Some of the earlier 2D DDD studies of size effects in polycrystals utilized this assumption to model columnar-like grain structures arranged in a series of hexagonal grains

[108, 109], or squared grains [110, 111], in a checkerboard-like arrangement. From these simulations it was shown that the flow stress versus grain size follows a power-law relationship with an exponent equal to 0.5 when only single slip is considered in each grain [108, 109]. Further detailed simulations also revealed that the exponent can vary between 0.3 and 1.6 depending on the number of slip systems, the grain arrangement, the dislocation source density, and the range of grain sizes to which power-law expression is fit [110, 111].

To alleviate some of the constrains introduced by an impenetrable GB mode, another phenomenological GB model was developed, in which dislocations can transmit across the GB from one grain into another if the local resolved shear stress on the leading dislocation at the GB is higher than a critical GB strength [112–114]. This GB strength is typically defined such that the yield strength predicted from the simulations matches those from experiments. This model was implemented in 3D DDD studies of Cu polycrystalline columnar-grained thin films [114], and in the study of orientation influence on the grain-size effects in HCP Mg polycrystals [112]. In the latter, the simulations were performed on *c*-axis oriented cubic simulation cell having periodic boundary conditions along the three directions. All six surfaces of the simulations cell were considered as nominal GBs. A dislocation segment was allowed to transmit a GB only if the local shear stress on the dislocation at the GB exceeded the GB strength. In this simplified model, the orientations of all the grains in the bulk materials were identical as a result of periodic boundary conditions, with the misorientation between the a-axes in neighboring grains not accounted for. This model mimics polycrystalline Mg having a strong basal texture. Since all grains have identical orientations, a transmitted dislocation does not require the formation of a residual dislocation on the GB plane.

The appropriate GB strength is determined by varying the GB strength and comparing the predicted yield strength with those measured experimentally for a textured Mg alloy under similar loading conditions and having similar grain sizes [112]. Some experimental observations suggest the absence of twinning for the range of grain sizes simulated here (i.e., $d < 4\,\mu\text{m}$). Thus, the contribution of twinning mediated plasticity is neglected in these simulations. Figure 11.14a–c shows the contours of effective plastic strain for three different loading orientations in a grain having edge length $d = 1.0\,\mu\text{m}$. These simulations predict that, for loading orientations leading to predominant basal slip, the grain-size effect is stronger with a power-law exponent $n = 1.16$. On the other hand, for loadings leading to either predominant prismatic or pyramidal slip the grain-size effect is weak with a exponent $n = 0.2$.

To explain these simulation results, a new model that accounts for both dislocation pile-up at GBs and transmission across the GB was developed to predict the critical resolved shear strength by taking the GB strength, τ_{GB}, and dislocation source strength, τ_{src}, into account such that [112]:

$$\tau_{\text{CRSS}} = \frac{1}{2}\left(\tau_{\text{src}} + \sqrt{\tau_{\text{src}}^2 + 16D_m\tau_{\text{GB}}/d}\right) \tag{11.11}$$

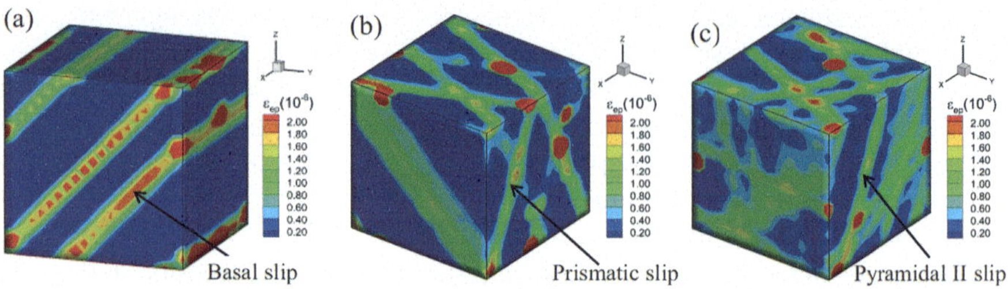

Fig. 11.14 Contours of effective plastic strain for: polycrystals with loading along the (**a**) *y* direction; (**b**) direction on the *yz* plane making a 45 ° with the *y* axis, (**c**) direction on the *yz* plane making a 135 ° angle with the *y* axis. Reprinted from [113] with permission

where $D_m = \mu b / 2\pi(1-\nu)$ is a material constant. It was also shown that this model agrees well with experimental results as well as predictions from 3D-DDD simulations [112]. Equation (11.11) indicates that in general the polycrystal strength does depend on orientation, which is accounted for explicitly in the definition of τ_{src} [112], and that the polycrystal strength does not always follow a simple power-law relationship with grain size and the loading orientation. Only when τ_{src} is low (e.g., large grains) or τ_{GB} is high (e.g., high angle GBs) does Eq. (11.11) recover the Hall–Petch relationship:

$$\tau_{CRSS} = \tau_{src} + 2\sqrt{D_m \tau_{GB}} d^{-0.5} \tag{11.12}$$

Furthermore, GB models that deal with the transmission of dislocations between neighboring grains with an actual misorientation have also been developed. The three main conditions that govern dislocation transmission are [115, 116]: (1) the angle between the incident and outgoing slip planes should be the smallest; (2) the magnitude of the Burgers vector of the residual dislocation on the grain boundary should be minimum; and (3) the resolved shear stress on the outgoing slip planes should be maximum. Again these rules were first implemented in 2D DDD [79, 115–117] and later in 3D DDD simulations [118].

More sophisticated GB models that accounted for GB sliding and the absorption, emission, and transmission of lattice dislocations at GBs were also incorporated in 2D DDD [119]. In this model dislocation sources are introduced into the GB, and the model was shown to describe many of the roles that GBs play in the deformation of nanocrystalline materials [119].

11.3.2 Twin-Size Effects

Fan et al. recently developed a systematic interaction model between dislocations and tension TBs in HCP Mg for 3D DDD simulations [113]. In essence, this TB model encompasses similar assumptions to those made in GB models with

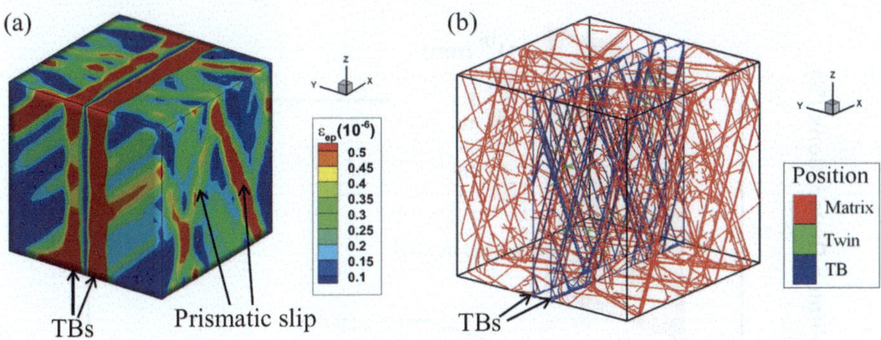

Fig. 11.15 (**a**) Contours of the effective plastic strain; and (**b**) dislocation configuration; for a $d = 3.0\,\mu m$ twinned grain at 0.2 % plastic strain. Reprinted from [121] with permission

two differences: (1) unlike GBs, TBs do not exhibit an explicit strength that a dislocation needs to overcome to cross; (2) based on MD simulations and geometric considerations of the HCP crystal structure the transmission of a dislocation across a TB will lead to the formation of both a residual dislocation and twinning dislocations [113]. This model has been utilized to study the hardening response of twinned polycrystalline Mg [113], twin-size effects [120], and the competition between dislocation slip and twinning as a function of grain size [121]. In the latter it was shown that twinning deformation exhibits a stronger grain-size effect than that for dislocation-mediated slip [121]. This leads to a critical grain size at $2.7\,\mu m$, above which twinning dominates, and below which dislocation slip dominates, indicating an excellent agreement with experimental observations. Figure 11.15 shows the contours of the effective plastic strain and the corresponding dislocation configuration in a $3.0\,\mu m$ twinned grain at 0.2 % plastic strain as predicted from these 3D DDD simulations. In addition, the plastic strain induced by twinning dislocations versus the total plastic strain as a function of grain size is shown in Fig. 11.16. Also, the insert in Fig. 11.16 shows the displacement contours for a TB in a $d = 3.0\,\mu m$ simulation cell, which results from the glide of twinning dislocation on the TB. These simulations showed for the first time the applicability of extending DDD simulations to study the competition between twinning plasticity and dislocation plasticity in a single framework.

11.3.3 Extrinsic (Sample) Size Effects

For polycrystalline materials, two characteristic dimensions can be defined: (1) grain size, d and (2) sample size, t. For polycrystals containing many grains across its smallest dimension, the grain size predominantly controls the strength of the material. On the other hand, when the number of grains across the polycrystal smallest dimension is less than some critical value this dimension predominantly

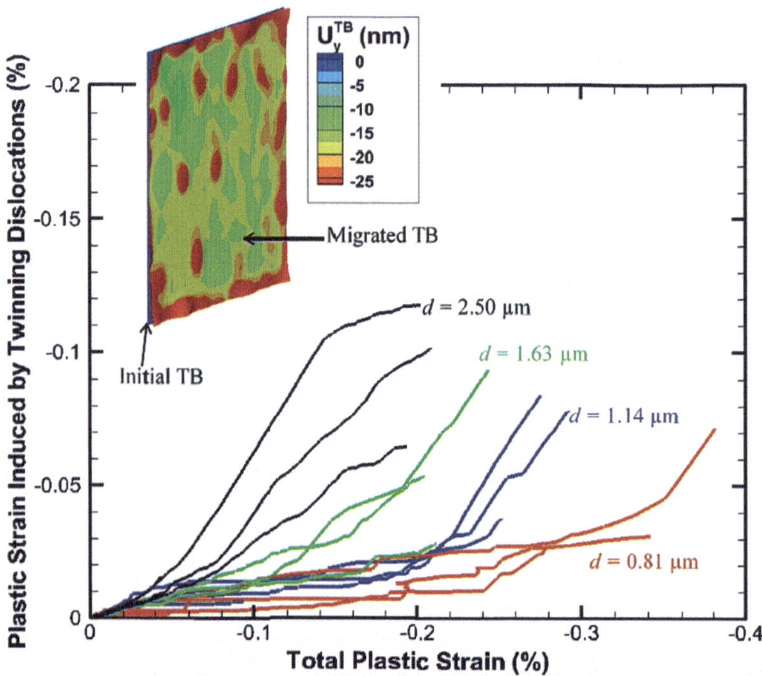

Fig. 11.16 Plastic strain induced by twinning dislocations versus the total plastic strain. The insert shows the initial and migrated TB for a $d = 3.0\,\mu$m simulation cell. The displacement contours, U_y^{TB}, are overlaid on the migrated TB. Reprinted from [121] with permission

controls the strength of the material for a constant grain size. This later effect has been studied using 2D DDD simulations in polycrystalline thin films [115, 122]. The dislocation distribution and contours of the tensile stress at 1.0 % strain as predicted from 2D DDD simulations are shown in Fig. 11.17. The GBs in these simulations are modeled as penetrable boundaries [115]. These simulations showed that for the same grain size, the strength increases with increasing film thickness. The underlying mechanism of this size effect is easy escape of dislocations in the surface grains versus those in interior grains. This leads to a lower dislocation density in the surface grains, and subsequently lower local stress levels as shown in Fig. 11.17. In the interior grains, dislocations are strongly impeded by the GBs, thus, high local stress levels are required to activate new dislocation sources in these internal grains. With increasing film thickness from $t = d$ to $t = 4d$, the ratio of surface grains to all grains decreases and accordingly the overall strength increases, leading to "thicker is stronger." As the film thickness increases further above $4d$, the effect of surface grains declines, and eventually no sample size effect is observed when the thickness is large enough. It should be noted that these predictions from 2D DDD simulations agree well with both experimental observations [123, 124] and

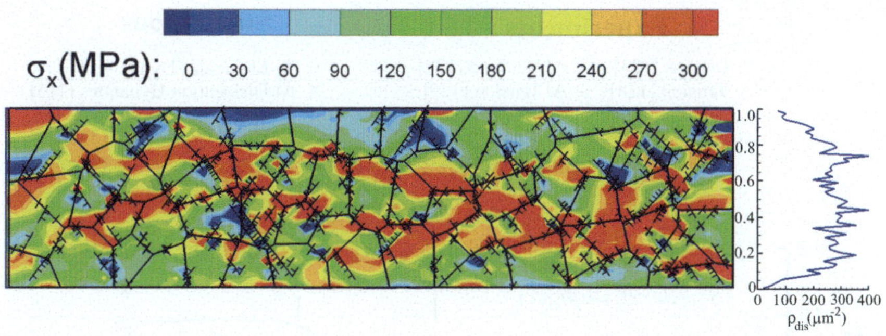

Fig. 11.17 Dislocation distributions, contour of tensile stress, and dislocation density variation through the thickness at strain 1.0 % in a film having thickness 1 μm and length 4.5 μm. Reprinted from [115] with permission

MD simulations [125]. Furthermore, when the thin film free surfaces are passivated or surface grains are refined to nano grains, a high stress layer develops near the passivated or refined surfaces, and opposite thickness effects would be predicted (i.e., smaller is stronger) [115].

Furthermore, 2D and 3D DDD simulations show that a change in the deformation mechanism is also expected when the film thickness $t \leq d$. From 2D DDD simulations it was shown that the decrease in the thickness leads to the decrease in the dislocation slip distance, and increase in the overall stress [126]. On the other hand, 3D DDD simulations showed that the size effect can be predicted by the single-arm source model [114], in a similar manner to size effects in single crystals [53]. In addition, dislocations can easily escape the crystal and the GB's effect is very minimal. Thus, in this case the film strength increases with decreasing film thickness. These results are also in agreement with experimental results [127–129].

Thus, as shown in Fig. 11.18 for the normalized stress as a function of the ratio of sample size (i.e., thickness) and grain size as predicted by MD simulations [125], DDD simulations [115, 130], strain gradient simulations [131], and experiments [123, 124, 128, 129, 132], the response can be classified into three stages:

1. quasi-single crystals response with smaller being stronger for $t < d$
2. multicrystals response with smaller being weaker for $d < t < 4d$
3. bulk polycrystalline response exhibiting no size effects for $t > 4d$

11.4 Closing Remarks

In summary, DDD is a coarse-grained method that is strictly material and mechanisms-based method. In this method dislocations are modeled as point defects in 2D or line defects in 3D, and plasticity is numerically computed from simulating the collective motion of dislocation ensembles through the direct numerical solution

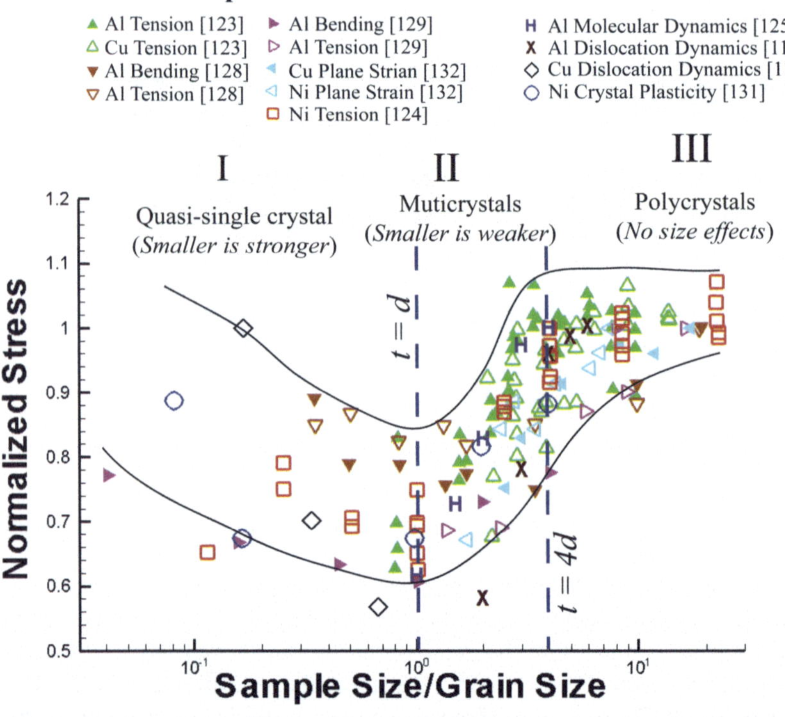

Fig. 11.18 Composite plot for the normalized stress as a function of the ratio of the sample size (i.e., thickness) and grain size. The normalized stress is defined as the flow or yield stress divided by that of the largest or smallest specimen

of their equations of motion. These simulations can provide insightful understanding and characterization of plasticity in a variety of crystals and for a variety of applications. The potential of DDD simulations has been mostly realized in recent years, especially with a wealth of predictions these simulations have led to regarding open questions in size-dependent dislocation-mediated plasticity.

There are still a number of challenges that are withholding DDD simulations from attaining their full potential. One of the limitations in DDD is reaching strains larger than 5 % for simulation volumes that are on the order of tens of microns. This is mostly due to the fact that dislocations profusely multiply with increasing strain, which subsequently translates to a significant increase in the number of degrees of freedom. This issue might be mitigated by running the simulations in parallel on large high performance computing (HPC) systems for extended periods of time. However, there are already limited computing resources given the significant demands on such HPC systems from different fields of research. Thus, the time might be ripe now to revisit the decade old architecture of DDD frameworks, and take advantage of accelerated computing through graphic processing units (GPUs), and hybrid GPU systems to help resolve such issues.

Another major challenge for DDD is the lack of appropriate experimental validations at the appropriate length scales. The micro-scale testing techniques, as discussed in Sect. 11.2, have been instrumental for validating DDD simulations. However, there is still a significant lack in direct experimental observations of mechanisms such as dislocation-GB interactions, dislocation-TB interactions, cross-slip, and climb at the appropriate strain levels and length scales of DDD simulations.

Dealing with dislocation climb mechanistically within the framework of DDD is also challenging due to the time-scale differences between dislocation glide and climb. Some DDD simulations do allow for a dislocation-climb mobility but that is mostly a fictitious model that does not account for the nonlinear vacancy-dislocation interactions inherent to climb [133]. This has been addressed recently in 2D DDD simulations that fully account for matter transport due to vacancy diffusion and its coupling with dislocation motion [133]. While more challenging, such formulation if incorporated into large scale 3D DDD simulations could potentially open up new avenues for DDD in predicting material properties at elevated temperatures, or for materials in environments where diffusion of species plays a major role in the deformation process (e.g., hydrogen embrittlement and stress assisted corrosion).

Acknowledgements The authors acknowledge the support by DARPA contract #N66001-12-1-4229, the Army Research Laboratory contract #W911NF-12-2-0022, and the National Science Foundation CAREER Award #CMMI-1454072. Author H.F. also acknowledges the financial support of the National Science Foundation of China (11302140) and Program for Innovative Research Team (IRT14R37).

References

1. M.D. Uchic, P.A. Shade, D.M. Dimiduk, Plasticity of micrometer-scale single crystals in compression: a critical review. Ann. Rev. Mater. Res. **39**(1), 361–386 (2009)
2. G.G. Chen, Size and interface effects on thermal conductivity of superlattices and periodic thin-film structures. ASME. J. Heat Tran. **119**(2), 220–229 (1997)
3. D.L. Leslie-Pelecky, R.D. Rieke, Magnetic properties of nanostructured materials. Chem. Mater. **8**(8), 1770–1783 (1996)
4. T.M. Shaw, S. Trolier-McKinstry, P.C. McIntyre, The properties of ferroelectric films at small dimensions. Ann. Rev. Mater. Sci. **30**, 263–298 (2000)
5. M.D. Uchic, D.M. Dimiduk, J.N. Florando, W.D. Nix, Sample dimensions influence strength and crystal plasticity. Science **305**, 986–989 (2004)
6. D. Kiener, A.M. Minor, Source truncation and exhaustion: Insights from quantitative in situ TEM tensile testing. Nano Lett. **11**, 3816–3820 (2011)
7. O. Kraft, P.A. Gruber, R. Monig, D. Weygand, Plasticity in confined dimensions. Ann. Rev. Mater. Res. **40**, 293–317 (2010)
8. J.S. Stolken, A.G. Evans, A microbend test method for measuring the plasticity length scale. Acta Mater. **46**, 5109–5115 (1998)
9. C. Motz, T. Schoberl, R. Pippan, Mechanical properties of micro-sized copper bending beams machined by the focused ion beam technique. Acta Mater. **53**, 4269–4279 (2005)
10. G.M. Pharr, E.G. Herbert, Y. Gao, The indentation size effect: a critical examination of experimental observations and mechanistic interpretations. Ann. Rev. Mater. Res. **40**, 271–292 (2010)

11. N.A. Fleck, G.M. Muller, M.F. Ashby, J.W. Hutchinson, Strain gradient plasticity: theory and experiment. Acta Metall. Mater. **42**, 475–487 (1994)
12. M.A. Meyers, A. Mishra, D.J. Benson, Mechanical properties of nanocrystalline materials. Prog. Mater. Sci. **51**(4), 427–556 (2006)
13. J.A. El-Awady, Unravelling the physics of size-dependent dislocation-mediated plasticity. Nat. Commun. **6**, 5926 (2015)
14. A.F. Voter, Parallel replica method for dynamics of infrequent events. Phys. Rev. B **57**, R13985 (1998)
15. A.F. Voter, Hyperdynamics: accelerated molecular dynamics of infrequent events. Phys. Rev. Lett. **78**, 3908–3911 (1997)
16. F. Montalenti, M.R. Sorensen, A.R. Voter, Closing the gap between experiment and theory: crystal growth by temperature accelerated dynamics. Phys. Rev. Lett. **87**, 126101 (2001)
17. R. LeSar, J.M. Rickman, *Coarse Graining of Dislocation Structure and Dynamics* (Wiley-VCH Verlag GmbH and Co. KGaA, Weinheim, 2005), pp. 429–444
18. J. Lepinoux, L.P. Kubin, The dynamic organization of dislocation structures: a simulation. Scripta Metall. **21**(6), 833 – 838 (1987)
19. E. Van der Giessen, A. Needleman, Discrete dislocation plasticity: a simple planar model. Model. Simul. Mater. Sci. Eng. **3**(5), 689–735 (1995)
20. A.A. Benzerga, Y. Brechet, A. Needleman, E. Van der Giessen, Incorporating three-dimensional mechanisms into two-dimensional dislocation dynamics. Model. Simul. Mater. Sci. Eng. **12**(1), 159–196 (2004)
21. B. Devincre, L.P. Kubin, Simulations of forest interactions and strain hardening in FCC crystals. Model. Simul. Mater. Sci. Eng. **2**, 559 (1994)
22. H.M. Zbib, M. Rhee, J.P. Hirth, On plastic deformation and the dynamics of 3-D dislocations. J. Mech. Sci. **40**(2–3), 113–127 (1998)
23. N.M. Ghoniem, L.Z. Sun, Fast sum method for the elastic field of 3-D dislocation ensembles. Phys. Rev. B **60**(1), 128–140 (1999)
24. L.P. Kubin, G. Canova, M. Condat, B. Devincre, V. Pontikis, Y. Bréechet, Dislocation microstructures and plastic flow: a 3-D simulation. Solid State Phenom. **23–24**, 455–472 (1992)
25. M. Verdier, M. Fivel, I. Groma, Mesoscopic scale simulation of dislocation dynamics in FCC metals: principles and applications. Model. Simul. Mater. Sci. Eng. **6**(6), 755–770 (1998)
26. K.W. Schwarz, Simulation of dislocations on the mesoscopic scale. J. Appl. Phys. **85**(1), 108–129 (1999)
27. D. Weygand, L.H. Friedman, E. van der Giessen, A. Needleman, Discrete dislocation modeling in three-dimensional confined volumes. Mater. Sci. Eng. A **309–310**, 420–424 (2001)
28. H.M. Zbib, M. Rhee, J.P. Hirth, A multiscale model of plasticity. Int. J. Plast. **18**, 1133–1163 (2002)
29. Z. Wang, N.M. Ghoniem, S. Swaminarayan, R. LeSar, A parallel algorithm for 3d dislocation dynamics. J. Comp. Phys. **219**(2), 608–621 (2006)
30. A. Arsenlis, W. Cai, M. Tang, M. Rhee, T. Oppelstrup, G. Hommes, T.G. Pierce, V.V. Bulatov, Enabling strain hardening simulations with dislocation dynamics. Model. Simul. Mater. Sci. Eng. **15**(6), 553–595 (2007)
31. J.R. Greer, J.Th.M. De Hosson, Plasticity in small-sized metallic systems: intrinsic versus extrinsic size effect. Prog. Mater. Sci. **56**(6), 654–724 (2011)
32. D.M. Dimiduk, M.D. Uchic, T.A. Parthasarathy, Size-affected single-slip behavior of pure nickel microcrystals. Acta Mater. **53**(15), 4065–4077 (2005)
33. D. Kiener, W. Grosinger, G. Dehm, R. Pippan, A further step towards an understanding of size-dependent crystal plasticity: in situ tension experiments of miniaturized single-crystal copper samples. Acta Mater. **56**, 580–592 (2008)
34. J.A. El-Awady, M.D. Uchic, P. Shade, S.-L Kim, S.I. Rao, D.M. Dimiduk, C. Woodward, Pre-straining effects on the power-law scaling of size-dependent strengthening in ni single crystals. Scrtipta Mater. **68**, 207–210 (2013)

35. A.S. Schneider, D. Kiener, C.M. Yakacki, H.J. Maier, P.A. Gruber, N. Tamura, M. Kunz, A.M. Minor, C.P. Frick, Influence of bulk pre-straining on the size effect in nickel compression pillars. Mater. Sci. Eng. A **559**(0), 147–158 (2013)
36. D.S. Balint, V.S. Deshpande, A. Needleman, E. Van der Giessen, Size effects in uniaxial deformation of single and polycrystals: a discrete dislocation plasticity analysis. Model. Simul. Mater. Sci. Eng. **14**, 409–422 (2006)
37. V.S. Deshpande, A. Needleman, E. Van der Giessen, Plasticity size effects in tension and compression of single crystals. J. Mech. Phys. Solids **53**(12), 2661–2691 (2005)
38. C. Hou, Z. Li, M. Huang, C. Ouyang, Discrete dislocation plasticity analysis of single crystalline thin beam under combined cyclic tension and bending. Acta Mater. **56**, 1435–1446 (2008)
39. L. Nicola, E. Van der Giessen, A. Needleman, Discrete dislocation analysis of size effects in thin films. J. Appl. Phys. **93**, 5920–5928 (2003)
40. C. Ouyang, Z. Li, M. Huang, L. Hu, C. Hou, Combined influences of micro-pillar geometry and substrate constraint on microplastic behavior of compressed single-crystal micro-pillar: two-dimensional discrete dislocation dynamics modeling. Mater. Sci. Eng. A **526**, 235–243 (2009)
41. A.A. Benzerga, Micro-pillar plasticity: 2.5D mesoscopic simulations. J. Mech. Phys. Solids **57**(9), 1459–1469 (2009)
42. A.A. Benzerga, N.F. Shaver, Scale dependence of mechanical properties of single crystals under uniform deformation. Scripta Mater. **54**(11), 1937–1941 (2006)
43. P.J. Guruprasad, A.A. Benzerga, Size effects under homogeneous deformation of single crystals: a discrete dislocation analysis. J. Mech. Phys. Solids **56**(1), 132–156 (2008)
44. D. Kiener, P.J. Guruprasad, S.M. Keralavarma, G. Dehm, A.A. Benzerga, Work hardening in micropillar compression: in situ experiments and modeling. Acta Mater. **59**(10), 3825–3840 (2011)
45. J.A. El-Awady, S.B. Biner, N.M. Ghoniem, A self-consistent boundary element, parametric dislocation dynamics formulation of plastic flow in finite volumes. J. Mech. Phys. Solids **56**(5), 2019–2035 (2008)
46. C. Motz, D. Weygand, J. Senger, P. Gumbsch, Initial dislocation structures in 3-D discrete dislocation dynamics and their influence on microscale plasticity. Acta Mater. **57**(6), 1744–1754 (2009)
47. J.A. El-Awady, M. Wen, N.M. Ghoniem, The role of the weakest-link mechanism in controlling the plasticity of micropillars. J. Mech. Phys. Solids **57**(1), 32–50 (2009)
48. S.I. Rao, D.M. Dimiduk, T.A. Parthasarathy, M.D. Uchic, M. Tang, C. Woodward, Athermal mechanisms of size-dependent crystal flow gleaned from three-dimensional discrete dislocation simulations. Acta Mater. **56**(13), 3245–3259 (2008)
49. C. Zhou, S.B. Biner, R. LeSar, Discrete dislocation dynamics simulations of plasticity at small scales. Acta. Mater. **58**(5), 1565–1577 (2010)
50. H. Tang, K.W. Schwarz, H.D. Espinosa, Dislocation-source shutdown and the plastic behavior of single-crystal micropillars. Phys. Rev. Lett. **100**, 185503 (2008)
51. I. Ryu, W.D. Nix, W. Cai, Plasticity of BCC micropillars controlled by competition between dislocation multiplication and depletion. Acta Mater. **61**(9), 3233–3241 (2013)
52. Z.H. Aitken, H. Fan, , J.A. El-Awady, J.R. Greer, The effect of size, orientation and alloying on the deformation of AZ31 nanopillars. J. Mech. Phys. Solids **76**, 208–223 (2015)
53. T.A. Parthasarathy, S.I. Rao, D.M. Dimiduk, M.D. Uchic, D.R. Trinkle, Contribution to size effect of yield strength from the stochastics of dislocation source lengths in finite samples. Scripta Mater. **56**(4), 313–316 (2007)
54. F. Mompiou, M. Legros, A. Sedlmayr, D.S. Gianola, D. Caillard, O. Kraft, Source-based strengthening of sub-micrometer al fibers. Acta Mater. **60**(3), 977–983 (2012)
55. D. Hull, D.J. Bacon, *Introduction to Dislocations*, (Butterworth-Heinemann, Oxford, 2011)
56. A.M. Hussein, S.I. Rao, M.D. Uchic, D.M. Dimiduk, J.A. El-Awady, Microstructurally based cross-slip mechanisms and their effects on dislocation microstructure evolution in FCC crystals. Acta Mater. **85**, 180–190 (2015)

57. S.I. Rao, D.M. Dimiduk, J.A. El-Awady, T.A. Parthasarathy, M.D. Uchic, C. Woodward, Atomistic simulations of athermal cross-slip nucleation at screw dislocation intersections in face-centered cubic nickel. Phil. Mag. **89**(34–36), 3351–3369 (2009)
58. S.I. Rao, D.M. Dimiduk, J.A. El-Awady, T.A. Parthasarathy, M.D. Uchic, C. Woodward, Activated states for cross-slip at screw dislocation intersections in face-centered cubic nickel and copper via atomistic simulation. Acta Mater. **58**, 5547–5557 (2010)
59. S.I. Rao, D.M. Dimiduk, T.A. Parthasarathy, J. El-Awady, Woodward C., M.D. Uchic, Calculations of intersection cross-slip activation energies in FCC metals using nudged elastic band method. Acta Mater. **59**(19), 7135–7144 (2011)
60. S.I. Rao, D.M. Dimiduk, J.A. El-Awady, T.A. Parthasarathy, M.D. Uchic, C. Woodward, Spontaneous athermal cross-slip nucleation at screw dislocation intersections in FCC metals and L12 intermetallics investigated via atomistic simulations. Phil. Mag. **93**(22), 3012–3028 (2013)
61. S.I. Rao, D.M. Dimiduk, T.A. Parthasarathy, M.D. Uchic, C. Woodward, Atomistic simulations of surface cross-slip nucleation in face-centered cubic nickel and copper. Acta Mater. **61**(7), 2500–2508 (2013)
62. Q. Yu, R.K. Mishra, J.W. Morris, A.M. Minor, The effect of size on dislocation cell formation and strain hardening in aluminum. Phil. Mag. **94**, 2062–2071 (2014)
63. L.Y. Chen, M.R. He, J. Shin, G. Richter, D.S. Gianola, Measuring surface dislocation nucleation in defect-scarce nanostructures. Nat. Mater. **14**, 707–713 (2015)
64. H. Zheng, A. Cao, C.R. Weinberger, J.Y. Huang, K. Du, J. Wang, Discrete plasticity in sub-10-nm-sized gold crystals. Nat. Commun. **1**, 144 (2010)
65. C.R. Weinberger, A.T. Jennings, K. Kang, J.R. Greer, Atomistic simulations and continuum modeling of dislocation nucleation and strength in gold nanowires. J. Mech. Phys. Solids **60**(1), 84–103 (2012)
66. Y. Tang, J.A. El-Awady, Formation and slip of pyramidal dislocations in hexagonal close-packed magnesium single crystals. Acta Mater. **71**, 319–332 (2014)
67. M. Wagih, Y. Tang, T. Hatem, J.A. El-Awady, Discerning enhanced dislocation plasticity in hydrogen-charged α-iron nano-crystals. Mater. Res. Lett. **3**, 184–189 (2015)
68. I Ryu, W. Cai, W.D. Nix, H. Gao, Stochastic behaviors in plastic deformation of face-centered cubic micropillars governed by surface nucleation and truncated source operation. Acta Mater. **95**, 176–183 (2015)
69. Z.W. Shan, R.K. Mishra, S.A.S. Asif, O.L. Warren, A.M. Minor, Mechanical annealing and source-limited deformation in submicrometer-diameter Ni crystals. Nat. Mater. **7**, 115–119 (2008)
70. H. Fan, Z. Li, M. Huang, Size effect on the compressive strength of hollow micropillars governed by wall thickness. Scripta Mater. **67**, 225–228 (2012)
71. J.A. El-Awady, S.I. Rao, C. Woodward, D.M. Dimiduk, M.D. Uchic, Trapping and escape of dislocations in micro-crystals with external and internal barriers. Int. J. Plast. **27**(3), 372–387 (2011)
72. K.S. Ng, A.H.W. Ngan, Effects of trapping dislocations within small crystals on their deformation behavior. Acta Mater. **57**(16), 4902–4910 (2009)
73. C.R. Weinberger, W. Cai, Plasticity of metal wires in torsion: molecular dynamics and dislocation dynamics simulations. J. Mech. Phys. Solids **58**, 1011–1025 (2010)
74. C.R. Weinberger, The structure and energetics of, the plasticity caused by Eshelby dislocations. Int. J. Plast. **27**, 1391–1408 (2011)
75. J. Senger, D. Weygand, C. Motz, P. Gumbsch, O. Kraft, Evolution of mechanical response and dislocation microstructures in small-scale specimens under slightly different loading conditions. Philos. Mag. **90**, 617–628 (2010)
76. J. Senger, D. Weygand, O. Kraft, P. Gumbsch, Dislocation microstructure evolution in cyclically twisted microsamples: a discrete dislocation dynamics simulation. Model. Simul. Mater. Sci. Eng. **19**, 074004 (2011)
77. J. Chevy, F. Louchet, P. Duval, M. Fivel, Creep behaviour of ice single crystals loaded in torsion explained by dislocation cross-slip. Phil. Mag. Lett. **92**, 262–269 (2012)

78. G. Dehm, C. Motz, C. Scheu, H. Clemens, P.H. Mayrhofer, C. Mitterer, Mechanical size-effects in miniaturized and bulk materials. Adv. Eng. Mater. **8**, 1033–1045 (2006)
79. H. Fan, Q. Wang, M.K. Khan, Cyclic bending response of single- and polycrystalline thin films: two dimensional discrete dislocation dynamics. Appl. Mech. Mater. **275–277**, 132–137 (2013)
80. S. Yefimov, E. van der Giessen, I. Groma, Bending of a single crystal: discrete dislocation and nonlocal crystal plasticity simulations. Model. Simul. Mater. Sci. Eng. **12**, 1069 (2004)
81. D. Kiener, C. Motz, W. Grosinger, D. Weygand, R. Pippan, Cyclic response of copper single crystal micro-beams. Scripta Mater. **63**, 500–503 (2010)
82. C. Motz, D. Weygand, J. Senger, P. Gumbsch, Micro-bending tests: a comparison between three-dimensional discrete dislocation dynamics simulations and experiments. Acta Mater. **56**(9), 1942–1955 (2008)
83. C. Motz, D.J. Dunstan, Observation of the critical thickness phenomenon in dislocation dynamics simulation of microbeam bending. Acta Mater. **60**, 1603–1609 (2012)
84. J.C. Crone, P.W. Chung, K.W. Leiter, J. Knap, S. Aubry, G. Hommes, A. Arsenlis, A multiply parallel implementation of finite element-based discrete dislocation dynamics for arbitrary geometries. Model. Simul. Mater. Sci. Eng. **22**(3), 035014 (2014)
85. F. Akasheh, H.M. Zbib, T. Ohashi, Multiscale modelling of size effect in FCC crystals: discrete dislocation dynamics and dislocation-based gradient plasticity. Philos. Mag. **87**, 1307–1326 (2007)
86. S. Gupta, A. Ma, A. Hartmaier, Investigating the influence of crystal orientation on bending size effect of single crystal beams. Comp. Mater. Sci. **101**, 201–210 (2015)
87. A. Gouldstone, N. Chollacoop, M. Dao, J. Li, A.M. Minor, Y.L. Shen, Indentation across size scales and disciplines: recent developments in experimentation and modeling. Acta Mater. **55**(12), 4015–4039 (2007)
88. W.D. Nix, H. Gao, Indentation size effects in crystalline materials: a law for strain gradient plasticity. J. Mech. Phys. Solids **46**(3), 411–425 (1998)
89. J.G. Swadener, E.P. George, G.M. Pharr, The correlation of the indentation size effect measured with indenters of various shapes. J. Mech. Phys. Solids **50**, 681–694 (2002)
90. A.K. Nair, E. Parker, P. Gaudreau, D. Farkas, Kriz R.D, Size effects in indentation response of thin films at the nanoscale: A molecular dynamics study. Int. J. Plast. **24**, 2016–2031 (2008)
91. V. Dupont, F. Sansoz, Molecular dynamics study of crystal plasticity during nanoindentation in ni nanowires. J. Mater. Res. **24**, 948–956 (2009)
92. C.Y. Chan, Y.Y. Chen, S.W. Chang, C.S. Chen, Atomistic studies of nanohardness size effects. Int. J. Theor. Appl. Multiscale Mech. **2**, 62–71 (2011)
93. T. Tomohito, S. Yoji, Atomistic simulations of elastic deformation and dislocation nucleation in al under indentation-induced stress distribution. Model. Simul. Mater. Sci. Eng. **14**, S55 (2006)
94. G.Z. Voyiadjis, M. Yaghoobi, Large scale atomistic simulation of size effects during nanoindentation: dislocation length and hardness. Mater. Sci. Eng. A. **634**, 20–31 (2015)
95. A. Widjaja, E. Van der Giessen, A. Needleman, Discrete dislocation modelling of submicron indentation. Mater. Sci. Eng. A. **400–401**, 456–459 (2005)
96. A. Widjaja, E. Van der Giessen, A. Needleman, Discrete dislocation analysis of the wedge indentation of polycrystals. Acta Mater. **55**, 6408–6415 (2007)
97. A. Widjaja, A. Needleman, E. Van der Giessen, The effect of indenter shape on sub-micron indentation according to discrete dislocation plasticity. Model. Simul. Mater. Sci. Eng. **15**, S121 (2007)
98. D.S. Balint, V.S. Deshpande, A. Needleman, E. Van der Giessen, Discrete dislocation plasticity analysis of the wedge indentation of films. J. Mech Phys. Solids **54**(11), 2281–2303 (2006)
99. H.G.M. Kreuzer, R. Pippan, Discrete dislocation simulation of nanoindentation: the effect of moving conditions and indenter shape. Mater. Sci. Eng. A. **387–389**, 254–256 (2004)
100. H.G.M. Kreuzer, R. Pippan, Discrete dislocation simulation of nanoindentation: indentation size effect and the influence of slip band orientation. Acta Mater. **55**, 3229–3235 (2007)

101. C. Ouyang, Z. Li, M. Huang, C. Hou, Discrete dislocation analyses of circular nanoindentation and its size dependence in polycrystals. Acta Mater. **56**, 2706–2717 (2008)
102. C. Ouyang, M. Huang, Z. Li, L. Hu, Circular nano-indentation in particle-reinforced metal matrix composites: simply uniformly distributed particles lead to complex nano-indentation response. Comp. Mater. Sci. **47**, 940–950 (2010)
103. C. Ouyang, Z. Li, M. Huang, H. Fan, Cylindrical nano-indentation on metal film/elastic substrate system with discrete dislocation plasticity analysis: a simple model for nano-indentation size effect. Int. J. Solids Struct. **47**, 3103–3114 (2010)
104. M.C. Fivel, C.F. Robertson, G.R. Canova, L. Boulanger. Three-dimensional modeling of indent-induced plastic zone at a mesoscale. Acta Mater. **46**, 6183–6194 (1998)
105. M. Fivel, M. Verdier, G. Canova, 3d simulation of a nanoindentation test at a mesoscopic scale. Mater. Sci. Eng. A **234**, 923–926 (1997)
106. G. Po, M.S. Mohamed, T. Crosby, C. Erel, A. El-Azab, N. Ghoniem, Recent progress in discrete dislocation dynamics and its applications to micro plasticity. J. Mater. **66**, 2108–2120 (2014)
107. J.P. Hirth, The influence of grain boundaries on mechanical properties. Metall. Trans. **3**, 3047–3067 (1972)
108. S.B. Biner, J.R. Morris, A two-dimensional discrete dislocation simulation of the effect of grain size on strengthening behaviour. Model. Sim. Mater. Sci. Eng. **10**, 617 (2002)
109. S.B. Biner, J.R. Morris, The effects of grain size and dislocation source density on the strengthening behaviour of polycrystals: a two-dimensional discrete dislocation simulation. Phil. Mag. **83**, 3677–3690 (2003)
110. D.S. Balint, V.S Deshpande, A. Needleman, E. Van der Giessen, A discrete dislocation plasticity analysis of grain-size strengthening. Mater. Sci. Eng. A **400–401**, 186–190 (2005)
111. D.S. Balint, V.S. Deshpande, A. Needleman, E. Van der Giessen, Discrete dislocation plasticity analysis of the grain size dependence of the flow strength of polycrystals. Int. J. Plast. **24**, 2149–2172 (2008)
112. H. Fan, S. Aubry, A. Arsenlis, J.A. El-Awady, Orientation influence on grain size effects in ultrafine-grained magnesium. Scripta Mater. **97**, 25–28 (2015)
113. H. Fan, S. Aubry, A. Arsenlis, J.A. El-Awady, The role of twinning deformation on the hardening response of polycrystalline magnesium from discrete dislocation dynamics simulations. Acta Mater. **92**, 126–139 (2015)
114. C. Zhou, R. Lesar, Dislocation dynamics simulations of plasticity in polycrystalline thin films. Int. J. Plast. **30–31**, 185–201 (2012)
115. H. Fan, Z. Li, M. Huang, X. Zhang, Thickness effects in polycrystalline thin films: surface constraint versus interior constraint. Int. J. Solids. Struct. **48**, 1754–1766 (2011)
116. Z. Li, C. Hou, M. Huang, C. Ouyang, Strengthening mechanism in micro-polycrystals with penetrable grain boundaries by discrete dislocation dynamics simulation and Hall–Petch effect. Comp. Mater. Sci. **46**(4), 1124–1134 (2009)
117. C. Hou, Z. Li, M. Huang, C. Ouyang, Cyclic hardening behavior of polycrystals with penetrable grain boundaries: two-dimensional discrete dislocation dynamics simulation. Acta Mater. Solida Sin. **22**, 295–306 (2009)
118. H. Fan, Z. Li, Toward a further understanding of intermittent plastic responses in the compressed single/bi-crystalline micro-pillars. Scripta Mater. **66**, 813–816 (2012)
119. S.S. Quek, Z. Wu, Y.W. Zhang, D.J. Srolovitz, Polycrystal deformation in a discrete dislocation dynamics framework. Acta Mater. **75**, 92–105 (2014)
120. H. Fan, S. Aubry, A. Arsenlis, J.A. El-Awady, Discrete dislocation dynamics simulations of twin size-effects in magnesium, in *MRS Proceedings*, vol. 1741 (2015)
121. H. Fan, S. Aubry, A. Arsenlis, J.A. El-Awady, Grain size effects on dislocation and twinning mediated plasticity in magnesium. Scripta Mater. **112**, 50–53 (2015)
122. R. Kumar, F. Szekely, E. Van der Giessen, Modelling dislocation transmission across tilt grain boundaries in 2D. Comput. Mater. Sci. **49**, 46–54 (2010)
123. S. Miyazaki, K. Shibata, H. Fujita, Effect of specimen thickness on mechanical properties of polycrystalline aggregates with various grain sizes. Acta Metall. **27**, 855–862 (1979)

124. C. Keller, E. Hug, X. Feaugas, Microstructural size effects on mechanical properties of high purity nickel. Int. J. Plast. **27**(4), 635–654 (2011)
125. Y. Zhu, Z. Li, M. Huang, Coupled effect of sample size and grain size in polycrystalline al nanowires. Scripta Mater. **68**, 663–666 (2013)
126. R. Kumar, L. Nicola, E. Van der Giessen, Density of grain boundaries and plasticity size effects: a discrete dislocation dynamics study. Mater. Sci. Eng. A **527**(1–2), 7–15 (2009)
127. C. Keller, E. Hug, R. Retoux, X. Feaugas, TEM study of dislocation patterns in near-surface and core regions of deformed nickel polycrystals with few grains across the cross section. Mech. Mater. **42**(1), 44–54 (2010)
128. J.T. Gau, C. Principe, J. Wang, An experimental study on size effects on flow stress and formability of aluminum and brass for microforming. J. Mater. Proc. Tech. **184**, 42–46 (2007)
129. L.V. Raulea, A.M. Goijaerts, L.E. Govaert, F.P.T. Baaijens, Size effects in the processing of thin metal sheets. J. Mater. Process. Technol. **115**, 44–48 (2001)
130. C. Zhou, I.J. Beyerlein, R. LeSar, Plastic deformation mechanisms of FCC single crystals at small scales. Acta Mater. **59**(20), 7673–7682 (2011)
131. C. Keller, E. Hug, A.M. Habraken, L. Duchene, Finite element analysis of the free surface effects on the mechanical behavior of thin nickel polycrystals. Int. J. Plast. **29**, 155–172 (2012)
132. P.A. Dubos, E. Hugm, S. Thibault, M. Ben Bettaieb, C. Keller, Size effects in thin face-centered cubic metals for different complex forming loadings. Metall. Mater. Trans. A **44**, 5478–5487 (2013)
133. S.M. Keralavarma, T. Cagin, A. Arsenlis, A.A. Benzerga, Power-law creep from discrete dislocation dynamics. Phys. Rev. Lett. **109**, 265504 (2012)

Chapter 12
Modeling Dislocation Nucleation in Nanocrystals

Matthew Guziewski, Hang Yu, and Christopher R. Weinberger

12.1 Introduction

In metallic materials, dislocations play a key role in the strength and plastic deformation of the bulk material. This remains true as the material's size decreases down to the nanoscale, however the role of pre-existing dislocations versus those that are nucleated changes. It has been well established that the strength of nanoscale metallic materials increases when the size of the crystal is decreased to the micron range and sub-micron range [11, 33, 59]. This sparked a surge in interest in small scale plasticity and a renewed interest in the role dislocation nucleation plays in plasticity since it should provide an upper bound to the strength of these materials [72].

Molecular simulations were initially used to explore the role free surfaces play in dislocation nucleation in nanowires [4, 5, 10, 16, 25, 34, 36, 47, 50, 60, 69, 70]. These results showed a plethora of interesting aspects of plasticity. First, in many cases plasticity can occur through the nucleation of both perfect dislocations and partial dislocations [8, 16]. Second, twinning became a dominant deformation mechanism as well if the nucleation of leading partial dislocations was favorable and this occurred sequentially on the same slip planes [16, 37]. The competition between these mechanisms is controlled at least in part through the orientation of the nanowire and its slip planes through the Schmid factor [7]. The strength of these nanowires was indeed high, approaching the ideal strength of the material [7]. However, the nucleation strength of the nanowires also varied with diameter, a consequence of the surface stress [8]. This resulted in nanowires being stronger in tension and weaker in compression, as the surface stress adds a pre-stress to the

M. Guziewski • H. Yu • C.R. Weinberger (✉)
Drexel University 3141 Chestnut St., Philadelphia PA, USA
e-mail: cweinberger@coe.drexel.edu

C.R. Weinberger, G.J. Tucker (eds.), *Multiscale Materials Modeling for Nanomechanics*, Springer Series in Materials Science 245,
DOI 10.1007/978-3-319-33480-6_12

nanowire, and if the wire were thin enough and in the right orientation, dislocations could spontaneously nucleate from the free surface.

Experimental results on the deformation of nanowires and nanocrystals with sizes below 100 nm (where the true nanoscale has been defined) were slow to emerge in part due to the difficulty of both fabricating the samples and designing devices that could reliably test the low dimensional materials. However, the number of experiments on the deformation of these true nanowires and nanocrystals has exploded over the last 5 years [38, 39, 51, 55–57, 61, 71, 73]. These experiments have been instrumental in exploring dislocation nucleation that previously was difficult to observe and quantify. These experiments have demonstrated that the strength of these small scale materials is indeed high and does approach the theoretical strength of the material [51, 56]. In addition, experiments have confirmed that small nanowires and nanocrystals can slip via perfect and partial dislocations as well as undergo deformation twinning [71]. Of particular interest has been in the behavior of $\langle 110 \rangle$ nanowires which exhibit complete deformation twinning which, upon completion, reorients the wire to the $\langle 100 \rangle$ orientation [55–57]. Many of these observations have helped validate the use of molecular simulations in describing the deformation behavior of nanowires and nanocrystals, despite the large difference in strain rates between the simulations ($\sim 10^8$ 1/s) and experiments ($\sim 10^{-3}$ 1/s).

As this book is focused on multiscale modeling applied to nanomechanics, this chapter will cover multiscale methods applied to dislocation nucleation. Specifically, we will present a method to describe the energetics of dislocation nucleation, both homogeneously and heterogeneously from the free surfaces. The energetics of dislocation nucleation will be combined with a statistical model for dislocation nucleation that depends on strain rate to predict the nucleation strength of the crystal. The energetics of dislocation nucleation will first be directly simulated in atomistics using methods of finding energy barriers, such as the nudged-elastic-band method (NEB) and string method. Then, we will develop analytical models of dislocation nucleation based on continuum theory and fit these models to atomistic data. The development of atomistically informed continuum models provides a method to not only understand the factors that control nucleation strength, but that allow for the development of dislocation nucleation models for different materials, different crystal orientations, and different nanocrystal geometries without undertaking the arduous molecular simulations. These continuum models then can also be used as inputs to mesoscale models, such as dislocation dynamics, which are able to track the evolution of the dislocations once they are nucleated and thus predict the evolution of stress with continued applied strain.

12.2 Factors That Influence Nucleation

Dislocation nucleation in nanostructures, in particular nanowires and nanopillars, was studied initially using atomistic methods. Thus, much of the initial insights regarding dislocation nucleation came directly from these simulations with experimental verification following. In this section, the literature that specifically studied the numerous factors that affect dislocation is reviewed as these factors are important to consider in creating a model of dislocation nucleation.

One of the most important factors that affects dislocation nucleation in the context of axial loading of nanostructures is the Schmid factor [54]. The Schmid factor is a geometrical relationship between the direction of the applied axial stress and the slip plane and slip direction of a dislocation. The shear stress acting on the dislocation can be written as:

$$\tau = m\sigma \tag{12.1}$$

where $m = \sin(\phi)\sin(\psi)$ and ϕ is the angle between the direction of axial loading and the slip plane while ψ is the angle between the axial loading direction and the slip direction. Thus, slip planes with higher Schmid factors are more likely to nucleate dislocations than slip planes with lower Schmid factors. In uniaxial loading of bulk FCC crystals, the Schmid factor associated with perfect dislocations controls the variation of the yield and flow stress of single crystals loaded in uniaxial loading. Schmid factors have been shown to be critical in understanding dislocation nucleation in nanocrystals as well [7, 8, 10, 16]. However, in these cases, the nucleating dislocation is often a partial dislocation and so the relevant Schmid factor for partial dislocation must consider the fact that the Burgers vector is different than perfect dislocations, i.e., along the $\langle 112 \rangle$. It is also important to take into consideration the fact that only leading partial dislocations initially nucleate. The Schmid factors for perfect dislocations in FCC crystals are plotted in Fig. 12.1 for perfect dislocations as well as partial dislocations.

One of the key observations made in molecular simulations was that the strength of FCC nanowires was considerably different in tension and compression [7, 8], which is not expected if perfect dislocations dominate deformation. This is because the Schmid factor in tension and compression for perfect dislocations is the same. However, since partial dislocations were observe to emit from the surface, the

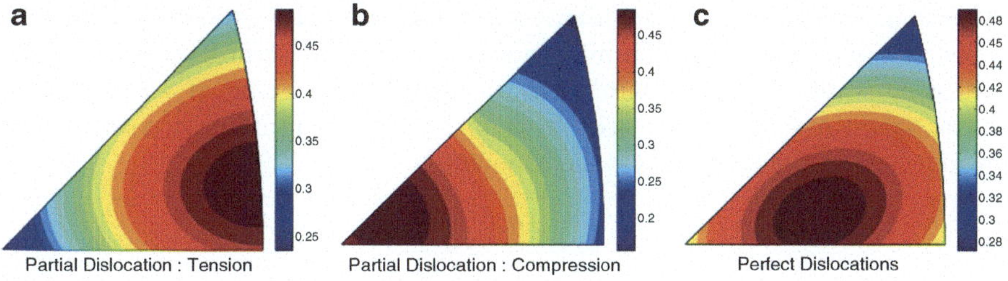

Fig. 12.1 The Schmid factors for (**a**) partial dislocations in tension, (**b**) partial dislocations in compression, and (**c**) perfect dislocation for typical FCC metals. Reproduced from [22] with permission

relevant Schmid factor that should govern strength is the Schmid factor of the leading partial dislocation, which can differ by a factor of ~2 for FCC crystals.

The Schmid factor was also useful in determining if the nanowires that deformed did so via twinning or slip [6, 16, 37, 48]. If the Schmid factor of the leading dislocation is smaller than the trailing dislocation, then leading partial dislocations would be favored which will preferentially lead to twinning, which is the repeated nucleation of leading partial dislocations on adjacent slip planes. However, if the Schmid factor of the trailing partial dislocation is smaller, then trailing partial dislocation would immediately nucleate after the leading partial, resulting in deformation via slip. This type of analysis agreed favorably with the observations of twinning in ⟨110⟩, nanowires in tension and ⟨100⟩ nanowires in compression. Molecular dynamics simulations have shown this to be true in some FCC metals such as gold and copper, though the simple Schmid factor argument must be augmented with the stacking fault energy to correctly account for observations of slip in aluminum. The twinning of FCC nanowires, notably gold and palladium, has been reproduced experimentally in ⟨110⟩ nanowires [57]. However, there also been a notable observation of a competition between twinning and slip, confirming that the Schmid factor is not the sole factor that governs the competition between twinning and slip [23, 24, 26, 64]. Nonetheless, it is clear from these simulations and experiments that the Schmid factor is a key component of describing the nucleation of dislocations as well as the ensuing deformation and that the Schmid factors of partial dislocations must be considered.

Another observation made in molecular simulations was the size dependent nucleation stress of nanowires and nanopillars [10, 40, 47, 50, 69], and for some very thin nanowires, spontaneous dislocation nucleation. Numerous studies have shown that the strength computed in atomistic simulations, both dynamic and static, shows a strength that increases in tension and decreases in compression with decreasing nanowire diameter. Sufficiently small nanowires, therefore, have vanishing nucleation strengths. The origin of this size dependence is through the surface stress [13, 32, 41]. The surface stress originates from the fact that the surface atoms are under-coordinated relative to atoms in the bulk and therefore thus, when cut out of a bulk crystal as is done in atomistic simulations, they are in a state of tension. When the atoms are allowed to relax, they reach equilibrium with the atoms that make up the core of the nanowire, leaving the surface atoms in a state of tension while the bulk atoms are in a state of compression. If we consider a square nanowire with side surfaces of width L that have a surface stress of f, then the induced compressive stress in the nanowire is

$$\sigma = -\frac{4fL}{A} = -\frac{4f}{L} \tag{12.2}$$

Thus, as the side length L is reduced, the induced compressive stress is increased. If we take some nominal values of $f = 1\,\mathrm{J/m^2}$, $L = 1\,\mathrm{nm}$, then the induced compressive stress is $\sigma = -4\,\mathrm{GPa}$. Taking this to its logical conclusion, if the nanowire is reduced small enough, dislocation nucleation will be spontaneous, a phenomenon that is observed in molecular dynamics simulations [6, 40, 48]. However, it is important to note that the surface stress is only important when the size of the nanostructure is small, a lateral dimension of $L = 40\,\text{nm}$ is no longer significantly affected by surface stress.

The size dependent strength of FCC metal nanowires in pristine or near pristine conditions has been measured [57]. However, the trends from these studied do not appear to show the L^{-1} dependence suggested by the surface stress effects and the size dependent strength appears to extend beyond the small length scales suggested by the role of surface stress. The strength of these materials is found to depend on diameter with power law form with an exponent of approximately −0.6. Most of the size dependent strength data found to date is for $\langle 110 \rangle$ nanowires in tension made from copper, gold, and platinum, and thus differences in tension compression (or changes in orientation) have not been explored to help rule out the role of surface stress.

The effects of temperature have been studied by a number of authors through atomistic simulations which is shown to have a moderate effect on the strength of the materials [3, 50]. Two different temperature trends can be found in atomistics for the temperature dependence of nucleation (or yield): a linear decrease and a variation with respect to the square root of the temperature. In other words, the nucleation strengths of the two forms:

$$\sigma = \sigma_0 - AT$$

$$\sigma = \sigma_0 - A\sqrt{T}$$

The temperature dependence is not surprising as plastic deformation is temperature dependent in the bulk; however, its characterization is quite important.

It is also worth pointing out that deformation and strength can depend on the shape of the nanowire even for the same orientations [3, 23, 24]. It is well known that square shaped nanowires and circular nanowires for the same material, same orientation, size, etc., will have different nucleation strengths. Perhaps even more striking is that the deformation mechanism, i.e., twinning or slip, can change with the geometry of the nanowire. While the change from twinning to slip deformation is very interesting and important, it is beyond the scope of this book chapter to discuss this phenomenon in detail. The interested reader is referred to a number of papers and review articles [23, 24, 26, 37, 64] for background information on this topic.

12.3 Nucleation Rates and Nanocrystal Strength

In nucleation theory, the nucleation of defects is controlled by the activation energy barrier of the nucleating defect. In terms of equilibrium thermodynamics, this corresponds to the activation Gibbs free energy under constant stress conditions, i.e.,

$\Delta G^*(\sigma, T)$, or the activation Helmholtz free energy under constant strain conditions $\Delta A^*(\boldsymbol{\epsilon}, T)$. From here on out, we typically will refer to the Gibbs free energy for convenience, though equivalent arguments can be made under isostrain conditions. The activation free energy is the maximum free energy difference between the perfect and defected state, i.e.,

$$\Delta G^*(\sigma, T) = \max_R \Delta G(\boldsymbol{\sigma}, T, R) \tag{12.3}$$

where R denotes the size of the defect. The nucleation rate can then be described as:

$$\nu = N\nu_0 \exp\left(-\frac{\Delta G^*(\boldsymbol{\sigma}, T)}{k_B T}\right) \tag{12.4}$$

where k_B is Boltzmann's constant, T is the absolute temperature, N is the number of equivalent nucleation sites, and ν_0 is the frequency prefactor. The functional form of the frequency prefactor depends on the exact theory used, either transition state theory (TST) [14, 49] or Becker–Doring (BD) theory and both approaches have been used to evaluate the absolute rate of dislocation nucleation [45, 53].

Since the nucleation rate is exponentially dependent on the activation free energy, this is the key factor to consider in accurately estimating the dislocation nucleation rate. Ryu et al. [53] have demonstrated that anharmonic effects in the activation entropy, in the form of lattice thermal expansion and the reduction of elastic constants, is the most important feature to capture in computing the absolute nucleation rate. Estimates for the prefactor need to be made and the full use of Becker–Doring theory or exact TST is intractable for any but the smallest nanocrystals. However, since the attempt frequency is high, of order the highest frequencies in the crystal, the Debye frequency can be taken as an initial estimate. Ryu et al. in fact computed the prefactor of a partial dislocation in copper and found the value to be $\sim 2.5 \times 10^{13}\,\text{s}^{-1}$ for homogeneous nucleation and $\sim 0.5 \times 10^{13}\,\text{s}^{-1}$ for heterogeneous. This compares to the Debye frequency of copper of $\sim 8 \times 10^{13}\,\text{s}^{-1}$. Thus, it seems reasonable to take the frequency prefactor as a constant as was done in [52] and estimate this value as a fraction of the Debye frequency, scaling it by a factor between $\frac{1}{3}$ and $\frac{1}{16}$. It is important to emphasize again that the frequency prefactor only needs to be estimated within an order of magnitude, a point demonstrated in Fig. 12.2. This figure shows the variation of the predicted strength of a gold nanowire plotted as a function of the prefect or ν_0, which shows a variation of the strength of about 3 % over two orders of magnitude variation in υ_0.

The second factor to consider in dislocation nucleation is the number of equivalent nucleation sites N. For homogeneous dislocation nucleation, the dislocations can form anywhere within the crystal (Fig. 12.3a) and thus a reasonable estimate of N would be the number of atoms in the crystal, n_a. This argument then creates a size scaling law for dislocation nucleation based solely on the number of equivalent nucleation sizes. For example, a cubic crystal with sides L, N will scale with L^3 as the cube is expanded self-similarly. Heterogeneous dislocation nucleation occurs

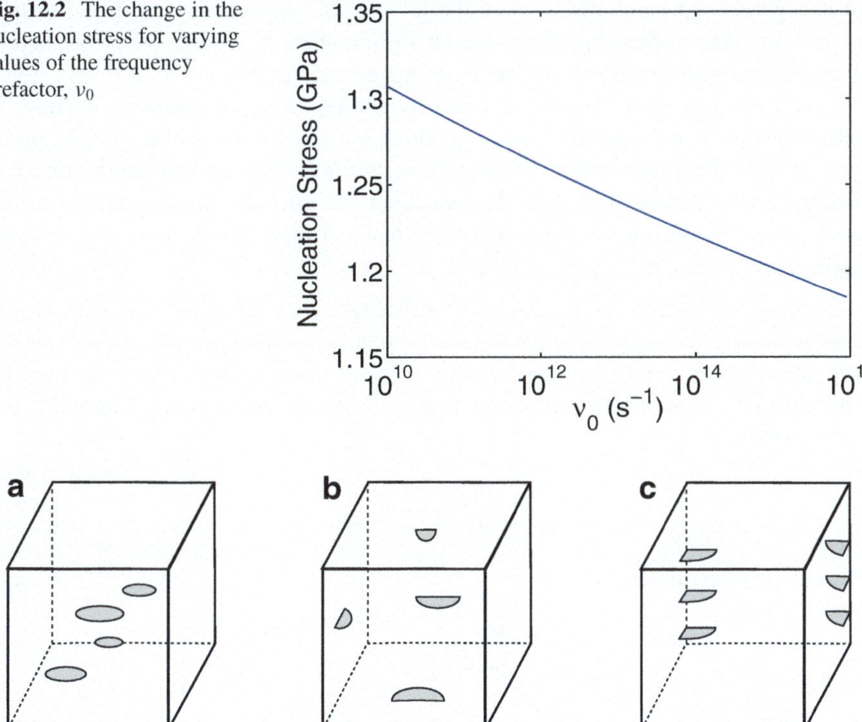

Fig. 12.2 The change in the nucleation stress for varying values of the frequency prefactor, ν_0

Fig. 12.3 (**a**) Homogeneous dislocation nucleation can occur anywhere inside the crystal and thus the number of sites scales with the volume. (**b**) Heterogeneous nucleation from free surfaces assuming the nucleation sites can be anywhere on the free surface and thus the number of sites scales with the surface area. (**c**) Site specific heterogeneous nucleation where the dislocations are only able to nucleate at the corners and thus the number of sites scales with the length of potential sites

on the surface of the crystal (Fig. 12.3b) , so an initial guess would be $N = n_a \frac{A}{V}$, which in the case of a cube to be $N = n_a \frac{6}{L}$. In this case, if we assume the cube changes size homogeneously, N scales on the order of L^2 since n_a scales with L^3. However if the dislocation is formed at specific surface feature, such as the corners of a cube (Fig. 12.3c), the number of possible sites can be approximated as $N = n_a \frac{l}{L}$ where l is the length of the surface feature that the dislocation nucleate from which gives a scaling relationship of order L. While both homogeneous and heterogeneous dislocation nucleation have the potential to occur within a crystal, heterogeneous tends to dominate due to its lower activation energy. Thus, it is best to approximate N based on the number of available heterogeneous nucleation sites. While the absolute nucleation rate scales with N, the dependence of strength on the number of sites is weak, similar to that of ν_0. This is because both ν scales linearly with ν_0 and N, but it depends exponentially on both the inverse of temperature and the activation free energy. Therefore an approximation is sufficient for N as well, with a prediction based on the order L and the number of atoms are acceptable.

If we wish to predict the strength of a nanocrystal such as a nanowire, we have to predict the stress or ranges of stresses at which the dislocation is able to nucleate. A semi-analytical expression for the nucleation strength of a nanowire under uniaxial loading can be derived [43, 65, 72], assuming the nucleation rate under constant strain rate or stress rate loading conditions is equivalent to that under equilibrium conditions, i.e., $\nu = N\nu_0 \exp\left(-\frac{\Delta G^*}{k_B T}\right)$.

Consider an ensemble of nanowires all being subjected to the same strain rate conditions and define f as the fraction of nanowires that have not nucleated a dislocation. Over any given time, if the dislocation nucleation rate is ν, then νf nanowires nucleated a dislocation and the rate of change of surviving nanowires can be described as:

$$\frac{\mathrm{d}f}{\mathrm{d}t} = -\nu f \tag{12.5}$$

This can be rewritten as:

$$\frac{\mathrm{d}f}{\mathrm{d}\sigma}\dot{\epsilon}E_T = -\nu f \tag{12.6}$$

where $\dot{\epsilon}E_T = \frac{\mathrm{d}\sigma}{\mathrm{d}t}$, $\dot{\epsilon}$ is the strain rate, and $E_T = \frac{\mathrm{d}\sigma}{\mathrm{d}\epsilon}$, which is the tangent modulus. Taking the derivative with respect to stress:

$$\frac{\mathrm{d}^2 f}{\mathrm{d}\sigma^2}\dot{\sigma} = -\nu\frac{\mathrm{d}f}{\mathrm{d}\sigma} - f\frac{\mathrm{d}\nu}{\mathrm{d}\sigma} \tag{12.7}$$

The most probable nucleation rate can be defined as:

$$\left.\frac{\mathrm{d}^2 f}{\mathrm{d}\sigma^2}\right|_{\sigma_{\mathrm{crit}}} = 0 \tag{12.8}$$

Combining Eqs. (12.7) and (12.8) yields the following expression:

$$\left.\frac{\mathrm{d}f}{\mathrm{d}\sigma}\right|_{\sigma_{\mathrm{crit}}} = -\left.\frac{\mathrm{d}\nu}{\mathrm{d}\sigma}\frac{f}{\nu}\right|_{\sigma_{\mathrm{crit}}} \tag{12.9}$$

This can be combined with Eq. (12.6) to remove all the f terms.

$$\left.\frac{\mathrm{d}\nu}{\mathrm{d}\sigma}\right|_{\sigma_{\mathrm{crit}}} = -\left.\frac{\nu^2}{\dot{\sigma}}\right|_{\sigma_{\mathrm{crit}}} \tag{12.10}$$

Taking the derivative of Eq. (12.4) with respect to stress:

$$\frac{d\nu}{d\sigma} = \frac{N\nu_0}{k_B T}\Omega \exp\left(-\Delta G^*(\sigma)/k_B T\right) = \frac{\nu\Omega}{k_B T} \tag{12.11}$$

where Ω is the activation volume for nucleation.

$$\Omega \equiv -\frac{\partial \Delta G^*(\sigma)}{\partial \sigma} \tag{12.12}$$

Combining Eqs. (12.10) and (12.11) and assuming a constant strain rate loading such that the stress is $\dot{\sigma} = E_T\dot{\epsilon}$ results in the following equation for the critical nucleation strength:

$$\frac{\Delta G^*}{k_B T} = \ln\left(\frac{k_B T N \nu_0}{E_T \dot{\epsilon} \Omega}\right) \tag{12.13}$$

This equation provides a direct method, albeit an implicit one, to calculate the critical stress from the known activation free energy as a function of stress, $\Delta G^*(\sigma, T)$. We note that if we approach the same problem under the assumption of constant stress rate rather than constant strain rate, then $E_T\dot{\epsilon}$ is simply replaced by the stress rate itself with no essential modifications of the theory. It is worth noting that in the limit of 0 K, the nucleation of defects can only occur when $\Delta G^* = 0$, which creates and athermal strength of the crystal, $\boldsymbol{\sigma}_{\text{ath}}$.

An explicit analytical model can be derived if we assume that the activation free energy is linearly dependent on the stress. This approach can be alternatively rationalized by taking a series expansion of the activation free energy about a given point. In either case, we assume that $\Delta G^* = \Delta G_0 - \sigma\Omega_0$, and the critical nucleation stress is then:

$$\sigma_{\text{crit}} = \frac{\Delta G_0}{\Omega_0} - k_B T \Omega_0 \ln\left(\frac{k_B T N \nu_0}{E_T \dot{\epsilon} \Omega_0}\right) \tag{12.14}$$

While in practice the activation energy does not vary linearly with stress, this does allow for the establishment of rough relations for the critical stress. It will decrease logarithmically with increases in the frequency prefactor, Fig. 12.2, the number of potential nucleation sites and the inverse of the strain rate. In addition there is a strong temperature dependence.

This framework provides a convenient way to model the nucleation of defects both at relevant time scales and a way to include dislocation nucleation in larger continuum-based models. However, this requires predictions of the activation Gibbs free energy, $\Delta G^*(\boldsymbol{\sigma}, T)$, the nucleation sites, and the frequency prefactor. Since the activation Gibbs free energy is exponential in the nucleation rate, most of the emphasis of this chapter will be on appropriately addressing the activation free energy.

12.4 Free Energy Estimates from Atomistic Simulations

Atomistic simulations continue to be a popular method to examine the nucleation of dislocations and incipient plasticity in small scale structures. However, as mentioned previously, direct atomistic simulations, i.e., those that directly integrate the equations of motion, can be limited to high strain rates. In addition, the results are often limited to the conditions simulated and do not necessarily provide insight into how factors such as temperature and strain rate effect dislocation nucleation rates and the strength of these nanostructures. One approach to overcome these limitations but keep the fully atomistic description of the nucleation process is to determine the energetics of the nucleation process from atomistic simulations. This approach was been used by a number of different groups to model dislocation nucleation in nanostructures [1, 2, 15, 19, 53, 65, 72].

To determine the activation energy of the dislocation nucleation process, the saddle point between the nanocrystal without a dislocation (the pristine crystal) and dislocated nanocrystal must be found. This saddle point can be determined in atomistic simulations using a number of methods that search the dislocation nucleation energy landscape using a set of replicas of the system (or chain-of-states). One candidate method to determine the saddle points is the NEB and its many modifications [17, 18], including the climbing image nudged elastic band (CINEB). Alternatively, many research groups have used the string method [62] as an alternative to the classic NEB method. Both of these methods attempt to search for the saddle point by evolving a set of replicas to minimize the total energy of the system and the maximum between the two ends is presumably the energy barrier between the two end states.

This class of methods will typically find the energy barrier between two ends states, which are either prescribed as local minima or, if not prescribed, will evolve to local minima. If the energy between the two end states is large compared to the energy barrier relative to the highest end state, these methods may have convergence issues and will in the very least provide a very poor estimate of the energy barrier. This situation is very common in dislocation nucleation under applied stresses since the energy difference between the critical dislocation nucleus and the undislocated state is very small compared to the energy difference between the unslipped and completely slipped nanocrystal. To remedy this issue, several groups have introduced alterations to the NEB and string methods to create a "free-end" that is able to stay on the energy landscape and evolve during the energy minimization but does not fall into a local energy minima [27].

An example of an energy barrier computed in this manner is shown in Fig. 12.4. The nanowire was constructed by cutting a 5 nm cylinder out of a bulk gold crystal such that the axis lies along the $\langle 100 \rangle$ and relaxed to an equilibrium position. To create a driving force for dislocation nucleation the nanowire was compressed to an axial strain of 5 % and relaxed. For the end state, a dislocation was introduced into the nanowire using the continuum displacement field to displace the atoms and the configuration was partially relaxed. The minimization must be truncated so that the

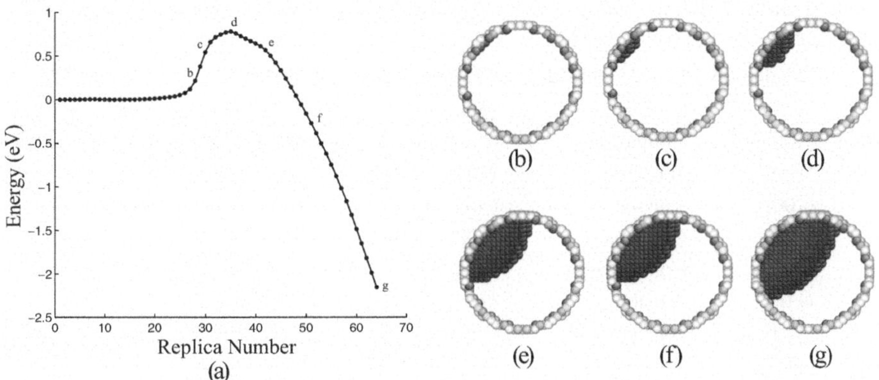

Fig. 12.4 (a) The energy barrier computed for a 5 nm diameter ⟨100⟩ gold nanowire under compression. Sixty-four replicas were used to describe the transition between the pristine nanowire and a nanowire with a single dislocation in it. (b)-(g) State of the nanowire at the corresponding points in (a). The atoms are colored according to a local centrosymmetry parameter [29] which identifies the free surfaces as well as the dislocation and its trailing stacking fault

dislocation does not exit the wire and so that the end state does not fall too far below the initial state. A set of 64 replicas were created by linearly interpolating between the initial and final configurations and the ensemble of replicas was relaxed using the free-end string method to determine the saddle point. In this particular configuration, the energy barrier is found to be around 0.8 eV and the corresponding axial stress is 1 GPa.

This provides a direct method to measure the activation of energy as a function of axial stress or axial strain, but to predict the nucleation strength according to Eq. (12.13) we need a continuous function for the activation energy. This can be done by evaluating the energy barrier as a function of stress (strain) and fitting a closed form function to the data. The most popular functional form is that proposed by Kocks et al. [31]:

$$\Delta E^* = \Delta E_0 \left[1 - \left(\frac{\sigma}{\sigma_0} \right)^p \right]^q \tag{12.15}$$

however, we have also used the form of

$$\Delta E^* = \Delta E_0 \left[1 - \exp\left(\alpha \left(1 - \frac{\sigma}{\sigma_0} \right) \right) \right] \tag{12.16}$$

and found similar standard errors of the fit with the form suggested by Kocks. It is worth noting that one of the advantages of closed form solutions over very general numerical methods, e.g., cubic splines, is that the activation volume can be derived in a simple closed form by taking the derivative of the activation energy model with respect to the applied stress. Table 12.1 shows the fitting parameters determined by

Table 12.1 Fitting parameters for energy barrier and activation volume of ∼5 nm gold nanowires using Eq. (12.16)

Orientation	Loading direction	Cross-section	ΔE_0(eV)	α	σ_0(GPa)
⟨100⟩	Compression	⟨100⟩ × ⟨100⟩	−0.0057	6.11	2.28
⟨100⟩	Compression	Circular	−0.452	2.19	1.79
⟨100⟩	Tension	⟨100⟩ × ⟨100⟩	−0.0082	9.68	4.41
⟨100⟩	Tension	Circular	−0.3670	3.64	4.34
⟨110⟩	Compression	⟨100⟩ × ⟨100⟩	−0.0019	7.37	18.4
⟨110⟩	Compression	Circular	−0.0095	5.88	21.8
⟨110⟩	Tension	⟨100⟩ × ⟨100⟩	−0.0487	7.57	3.46
⟨110⟩	Tension	Circular	−0.377	3.86	3.45
⟨110⟩	Tension	⟨111⟩ × ⟨111⟩	−0.0926	7.06	3.67

Weinberger et al. [65] to be used in Eq. (12.16) for various orientations and cross-sections of 5 nm diameter gold nanowires.

From fitting these equations, it is also possible to extrapolate the energy barrier at zero stress, which comes from ΔE_0, and the thermal strength from σ_0. However, it is probably advisable to only use these equations within the stress range to which they were fit and regard both ΔE_0 and σ_0 as simple fitting parameters. It is also important to keep in mind when using these equations that the stress in all these cases is the axial stress and therefore the activation volume, which is proportional to the area swept out by the dislocation times its Burgers vector, is also convoluted with the Schmid factor associated with the applied load.

Figure 12.5 shows the computed activation energy of dislocation in gold nanowires as a function of the applied axial stress [65]. The nanowires were modeled using the Foiles potential for gold [47] and have roughly 5 nm diameters. The nanowires are oriented such that the axial loading is applied along the ⟨110⟩ direction and three cross-sections were modeled. One is circular with a diameter of 5 nm, one is square with {110} and {001} side surfaces of lengths approximately 5 nm, and the third is a rhombic nanowire with {111} side surfaces that have 5 nm lengths. Similar activation energies for ⟨100⟩ 5 nm gold nanowires were computed using the same method and the activation energies for small square copper nanowires were computed by Zhu et al. [72]. These activation energies can be fitted using Eqs. (12.15) or (12.16) and then used to estimate the strength of the nanowires through Eq. (12.13).

12.4.1 Entropic Contributions to Dislocation Nucleation

In the previous section, the activation energy at 0 K was evaluated using atomistic simulations at 0 K. Unfortunately, these values are not the Gibbs free energy, and therefore are not formally correct for computing the dislocation nucleation rate of

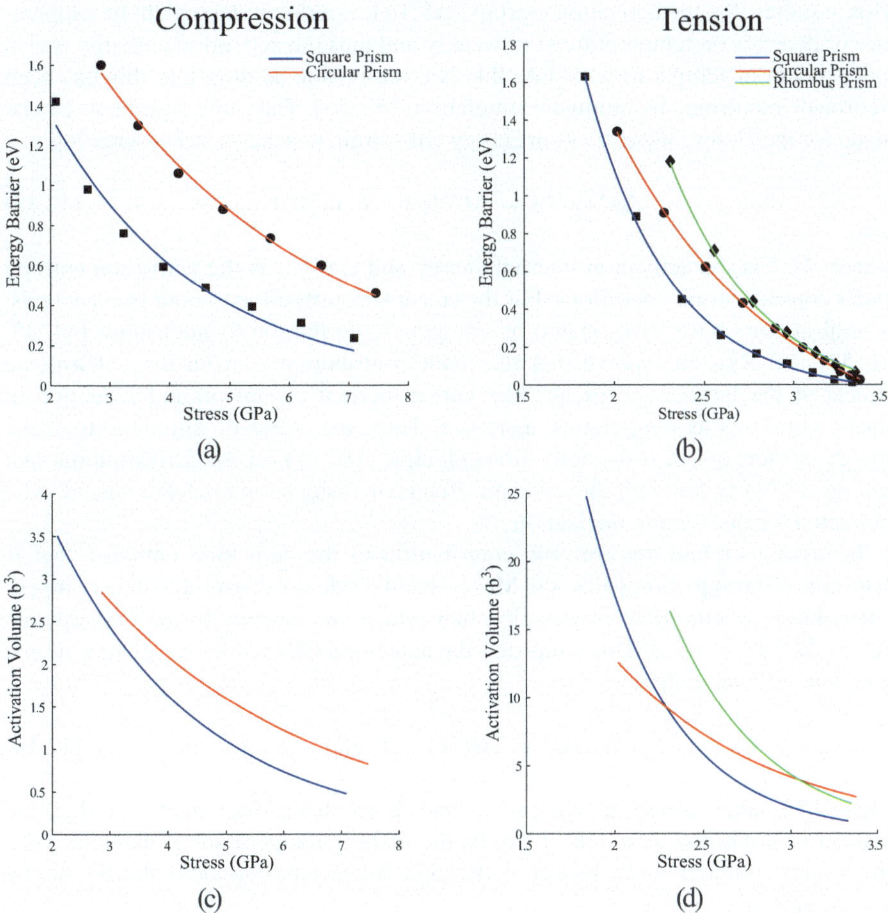

Fig. 12.5 The energy barriers and computed activation energies for ⟨110⟩ gold nanowires that are roughly 5 nm in size loaded uniaxially. Three different cross-sections were used, one that is cylindrical, one that is square, and the third is a rhombic cross-section whose side surfaces are comprised of only {111} surfaces. The activation energy for dislocation nucleation in (**a**) compression and (**b**) tension where the data points are from atomistic simulations and the curves are fits of Eq. (12.16). The activation volumes in (**c**) compression and (**d**) tension were computed by taking the derivative of Eq. (12.16). Reproduced from [65] with permission

the strength of the nanowire and non-zero temperatures. Formally, the energy barrier computed above is the activation enthalpy ($\Delta H^*(\sigma, T = 0)$), and this must be corrected for temperature effects to estimate the activation Gibbs free energy. To see how this works, the Gibbs free energy for dislocation nucleation can be written as:

$$\Delta G^*(\sigma, T) = \Delta H^*(\sigma) - T\Delta S^*_\sigma(\sigma) \tag{12.17}$$

This assumes that the activation entropy, $\Delta S^*_\sigma(\sigma)$, is either independent of temperature or depends on temperature very weakly and thus the activation enthalpy is also independent of temperature. While this is a reasonable assumption, this has been rigorously confirmed by atomistic simulations [52, 53]. The same arguments can be made for the Helmholtz activation energy with strain instead of stress, yielding:

$$\Delta A^*(\gamma, T) = \Delta U^*(\gamma) - T\Delta S^*_\gamma(\gamma) \tag{12.18}$$

where ΔU^* is the activation internal energy and $\Delta S^*_\gamma(\gamma)$ is the activation entropy under constant strain conditions. For the entropic contribution to both free energies, several authors have investigated its energetic contribution to nucleation [44, 45, 52, 53] and Ryu has shown that the main contribution is from the anharmonic effects of the lattice, specifically the lattice thermal expansion and reduction in elastic constants as temperature increases. Thus, the atomistic simulations of the energy barriers at 0 K is the activation enthalpy, $\Delta H^*(\sigma)$ (or the activation internal energy $\Delta U^*(\gamma)$), and only the entropic effects on dislocation nucleation need to be estimated for dislocation nucleation.

In order to relate the entropic contribution to the activation enthalpy that is determined through atomistics, the Meyer–Neldel rule (or thermodynamic compensation law), an empirical observation that relates the entropy to the enthalpy as: $\Delta S = \Delta H/T^\sigma_0$, is used. This simplifies the activation Gibbs free energy to a simple function of the enthalpy:

$$\Delta G^* = \Delta H^*(1 - T/T^\sigma_0) \tag{12.19}$$

where T^σ_0 is taken as a fitting parameter. One choice that is often made to reduce the number of unknowns is to take T^*_σ to be the melting temperature of the solid [72]. However, as pointed out by Ryu et al. the same rule can be applied to the Helmholtz free energy, i.e.,

$$\Delta A^* = \Delta U^*(1 - T/T^\gamma_0) \tag{12.20}$$

where ΔU^* is the activation internal energy. The new fitting constant, T^γ_0, is not the same as T^σ_0. This can be deduced from the fact that while the activation Gibbs free energy must be equal to the activation Helmholtz free energy (for the nucleation rates to be the same and hence for strength not to depend on the nature of the method of application of the loads), the activation enthalpy (ΔH^*) and activation internal energy (ΔU^*) are not equal. This implies the activation entropies ΔS^*_σ and ΔS^*_γ are not equal. The relationship between these entropies can be derived using the equality of $\Delta G^*(\sigma, T)$ and $\Delta A^*(\gamma, T)$.

$$
\begin{aligned}
\Delta S^*_\gamma &= -\frac{\partial \Delta A^*(\gamma,T)}{\partial T}\bigg|_\gamma = -\frac{\partial \Delta G^*(\sigma,T)}{\partial T}\bigg|_\gamma \\
&= -\frac{\partial \Delta G^*(\sigma,T)}{\partial T}\bigg|_\sigma - \frac{\partial \Delta G^*(\sigma,T)}{\partial \sigma}\bigg|_T \frac{\partial \sigma}{\partial T}\bigg|_\gamma \\
&= \Delta S^*_\sigma - \frac{\partial \Delta G^*(\sigma,T)}{\partial \sigma}\bigg|_T \frac{\partial \sigma}{\partial T}\bigg|_\gamma
\end{aligned}
\tag{12.21}
$$

If the difference between the entropies is defined as $\Delta(\Delta S^*) = \Delta S^*_\sigma - \Delta S^*_\gamma$ and the definition of activation volume is considered (Eq. 12.12), the change in activation entropy can be shown to be a product of the activation volume and the softening of elastic constants with temperature:

$$
\Delta(\Delta S^*) = -\Omega \frac{\partial \sigma}{\partial T}\bigg|_\gamma \tag{12.22}
$$

From a theoretical point of view, it would be beneficial if a rule were established to estimate T_0 across materials such that predictions can be made without the intensive atomistic calculations. Ryu et al. pointed out that this can be done for the activation Helmholtz free energy since they observed the Helmholtz free energy scales with the change of the shear modulus with temperature, i.e.,

$$
\Delta A^* \approx \frac{\mu(T)}{\mu(0)} \Delta U^* \tag{12.23}
$$

when this is combined with Eq. (12.20) and if $\mu(T)$ is assumed to be linear, such that $\mu(T) = \mu_0 - \mu' T$, it allows for a T_0^γ estimate of:

$$
T_0^\gamma \approx \frac{\mu_0}{\mu'} \tag{12.24}
$$

An estimation for T_0^σ can be found by relating the values for ΔH^* and ΔU^*. Once again assuming the activation energy for both the Helmholtz and Gibbs free energy are equal:

$$
\begin{aligned}
\Delta H^* - T\Delta S^*_\sigma &= \Delta U^* - T\Delta S^*_\gamma \\
\Delta U^* &= \Delta H^* - T(\Delta S^*_\sigma - \Delta S^*_\gamma) = \Delta H^* - T\Delta(\Delta S^*)
\end{aligned}
$$

Using the relation for Eq. (12.22) gives

$$
\Delta U^* = \Delta H^* + T\Omega \frac{\partial \sigma}{\partial T}\bigg|_\gamma \tag{12.25}
$$

If the crystal is assumed to be linearly elastic, such that $\sigma = \mu\gamma$:

$$\Delta U^* = \Delta H^* + T\Omega \frac{\sigma}{\mu} \frac{\partial \mu}{\partial T}$$

If the entropic contribution is considered to be independent of stress in the Gibbs free energy (Eq. 12.17), then $\Omega = -\partial \Delta H^*/\partial\sigma$. If the line tension model is used, which will be discussed in detail later in the chapter, it can be assumed that $\Delta H^* \propto \sigma^{-1}$. It then follows that $\Omega\sigma \approx \Delta H^*$. This shows that ΔU^* and ΔH^* are related by factor that is function of temperature and the shear modulus:

$$\Delta U^* = \Delta H^* \left(1 + T\frac{1}{\mu}\frac{\partial \mu}{\partial T}\right)$$

Substituting this relation into Eq. (12.23) and assuming that shear stress is linear with temperature again, $\mu(T) = \mu_0 - \mu' T$, the activation energy can be given as:

$$\Delta A^* = \Delta H^* \left(1 - \frac{2T\mu'}{\mu_0}\right)$$

Since the activation Helmholtz and Gibbs free energies are equal, it is possible to use the Meyer–Neldel rule (Eq. 12.19) to estimate the fitting parameter for a stress controlled system:

$$T_0^\sigma = \frac{\mu_0}{2\mu'} \tag{12.26}$$

This gives a relatively easy method to compute T_0^γ and T_0^σ if one knows the rate of decrease of the shear modulus with temperature for the material. These estimates can be made more concretely using a simple model of dislocation nucleation, which will be done later in this chapter.

It is interesting to note that the failure to account for thermal effects in dislocation nucleation results in large inaccuracies in the absolute nucleation rate, even orders of magnitude differences [45, 52], it appears that it is a marginal effect on the nucleation stress and hence strength of the material [53]. This weak relationship between the two was previously highlighted in the discussion of the frequency prefactor. To illustrate this point, Ryu et al. plotted the nucleation strengths predicted by Eq. (12.13) for a square copper nanowire with 5 nm side lengths as a function of temperature (Fig. 12.6a). Two methods were used to predict the strength of the nanowire, one directly from the activation free energies computed using their umbrella sampling and one from the activation energy at 0 K without modification for entropy. The difference between the two curves is then essentially the entropic contribution. Figure 12.6b shows the implementation of the Meyer–Neldel rule and finds that the T_0^γ that best fits the entropy to be 2450 K. This value can then be used in conjunction with the 0 K data to provide a fairly good approximation of the free energies found through umbrella sampling.

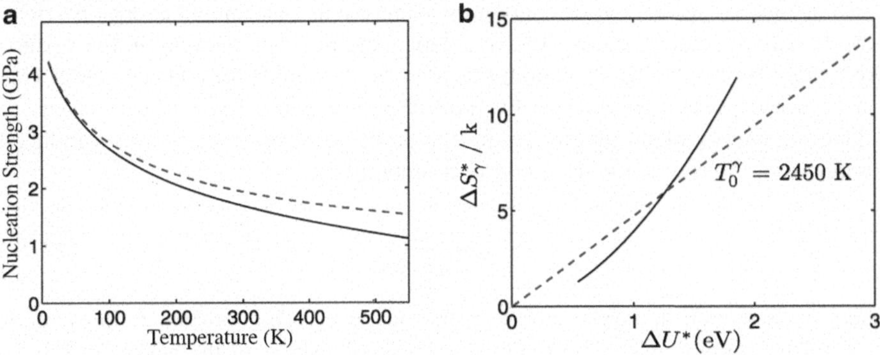

Fig. 12.6 (**a**) Compressive nucleation strength predictions in a square copper nanorod of 5 nm sides. Dashed line represents approximation based on 0 K atomistics, solid line represents data found from umbrella sampling. (**b**) T_0^γ fitting parameter used to fit the 0 K data to that determined from sampling. Reproduced from [53] with permission

While in this particular case the free energies with entropy taken into account are known, and an accurate T_0 term could be calculated, this is generally not the case. As a result, a poorly chosen fitting parameter could grossly underestimate the strength of the material. Even with experimental data, due to the scatter often observed, it can be difficult to accurately determine T_0. Therefore the use of only the 0 K data, without accounting for entropy can be useful in that it provides an upper bound to the strength of the material. In addition, given that many interatomic potentials fit experimental lattice constants and elastic constants (usually measured at 300 K) to the potential at 0 K, it is possible for some potentials, data at 0 K may be more representative of the material at 300 K.

12.4.2 Surface Stress Effects

The second important factor that can significantly alter the strength of nanowires is the surface stress. As noted earlier, the tensile surface stress creates a compressive stress in the core of the nanowire, effectively creating a prestress on the nanowire that scales inversely with the nanowire diameter. For nanowires with lateral dimensions of 40 nm or larger, this effect is minimal and can be neglected. However, for nanowires with diameters around 5 nm, this value is between 500 MPa and 1 GPa in gold and copper nanowires, and cannot be neglected. This is of particular importance because all of the activation energies computed in the literature [52, 53, 65, 72] were computed in nanowires that are roughly 5 nm to reduce computational costs. Thus, all of the atomistically determined activation energies, while quite accurate for the as described nanowire geometries, cannot be directly used to estimate strengths of large nanowires without correcting for the surface stress.

To accurately model larger nanowires using the as-determined energy barriers, as well as to accurately predict the size dependent nucleation strength, the surface stress that is present in these simulations must be accounted for and corrected. This can be accomplished in practice by modifying the applied stress, which acts to reduce the energy of the nanowire, by the addition of the surface stress induced compression. Thus, the applied stress σ is replaced by

$$\sigma = \sigma_a - \frac{C}{L} \tag{12.27}$$

where σ_a is the applied axial stress on the nanowire. The constant C must be fit to atomistic simulations of the induced compressive stress of the nanowire for the exact orientation and surface facets and L must selected as a specific length scale of the cross-section (e.g., side length of a square cross-section or diameter of a circular one). These parameters can be easily obtained from atomistic simulations by determining the stress in the nanowire as it is cut from the bulk crystal as a function of size. Figure 12.7 does just this for ⟨100⟩ gold nanowires, with the associated C values shown as well.

Using this surface stress allows us to correct the atomistic-based energy barrier calculations, like those shown in Fig. 12.5. For example, consider the 5 nm ⟨110⟩ gold nanowire in which the surface stress creates an approximate 740 MPa compressive stress. To correct the results of Fig. 12.5, the 740 MPa stress can be added to the effective applied stress, shifting the tensile stresses to the left and the compressive stresses to the right by this amount. The rigid shift of the curves is not, however, constant between different geometries since these nanowires have different surface stresses due to their different geometries. This approach is probably the easiest, and it is readily available to use in continuum models. However, it does neglect small higher order effects caused by the stress dependent surface elastic constants.

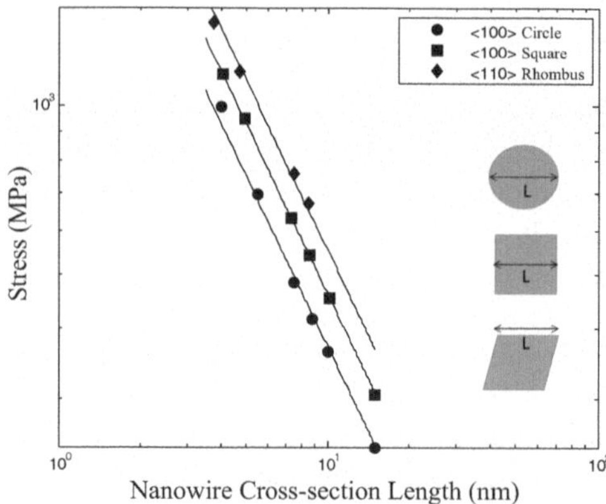

Cross-Section	C (MPa·nm)
Circular	$3.7\text{x}10^3$
Square	$4.6\text{x}10^3$
Rhombic	$5.5\text{x}10^3$

Fig. 12.7 Surface stress contribution for various cross-sections of ⟨100⟩ gold nanowires and associated C values

To account for these nonlinearities, the entire stress–strain curve of the nanowires can be used, like those in [65], which accounts for both the surface stress and surface elastic constants and the nonlinearity of the stress–strain curves.

This second approach has been applied to the data in Fig. 12.5 to compute the activation energies of very large nanowires using the potential predicted bulk stress–strain curves. These new values are show in Fig. 12.8 and the curve fits of the activation energies are listed in Table 12.2.

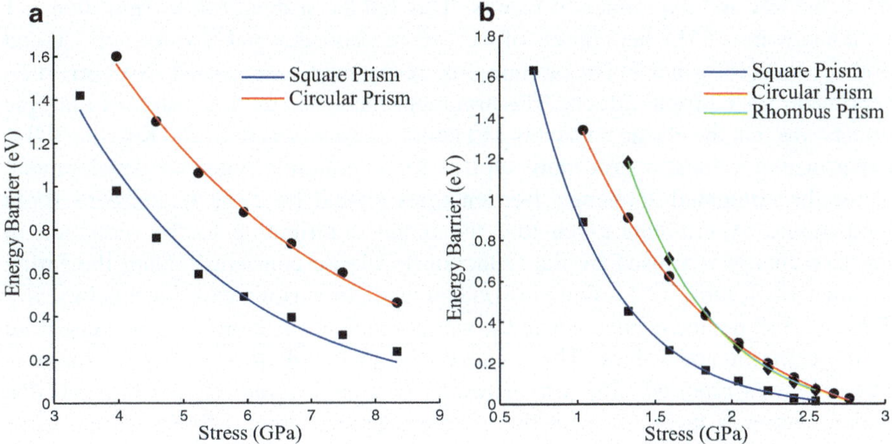

Fig. 12.8 The activation energies of the ⟨110⟩ gold nanowires correcting for the effects of the surface stress in (a) compression (b) tension. These values are representative of large diameter nanowires

Table 12.2 Fitting parameters for energy barrier and activation volume of ∼100 nm gold nanowires using Eq. (12.16)

Orientation	Loading direction	Cross-section	ΔE_0(eV)	α	σ_0(GPa)
⟨100⟩	Compression	⟨100⟩ × ⟨100⟩	−0.0289	7.10	2.47
⟨100⟩	Compression	Circular	−1.05	2.02	2.25
⟨100⟩	Tension	⟨100⟩ × ⟨100⟩	−0.0055	7.93	3.85
⟨100⟩	Tension	Circular	−0.275	3.31	4.28
⟨110⟩	Compression	⟨100⟩ × ⟨100⟩	−0.0015	8.17	20.42
⟨110⟩	Compression	Circular	−0.0124	5.96	21.25
⟨110⟩	Tension	⟨100⟩ × ⟨100⟩	−0.286	5.47	2.79
⟨110⟩	Tension	Circular	−0.198	3.22	2.84
⟨110⟩	Tension	⟨111⟩ × ⟨111⟩	−0.0854	5.02	2.87

12.4.3 Prediction of Strengths Using Atomistic Simulations

In the previous sections we outlined a method for computing the nucleation strength of nanowires using the activation energies computed directly from atomistic simulations. The advantage of using this method over direct molecular dynamics methods is that the strain rate can be controlled to approach those of experiments, $\sim 10^{-3}\text{s}^{-1}$ in addition to typical molecular dynamic strain rates of $\sim 10^{8}\text{s}^{-1}$. This gives the potential to make direct comparisons of nucleation strengths between these models and experimental results. This has been done for 100 nm diameter gold nanowires [65] where inputs of the Debye frequency and N values of 320 and 200 for the $\langle 100 \rangle$ and $\langle 110 \rangle$ orientations, respectively, were used. No correction was made for entropic effects. The argument made for the exclusion of entropic effects was that the elastic constants and lattice constants were fit at 0 K to the 300 K experimental values, a procedure typical for interatomic potential development. Thus, the computed activation free energies should be close to those of room temperature experiments given that the major contribution to the reduction in the free energy is caused by the reduction in elastic constants. Using the fitting parameters of Table 12.1 a comparison can made to existing experiments [55, 56]. Table 12.3 shows the results, where the values predicted by atomistics are very close to the experimental values. The nucleation strength of one $\langle 110 \rangle$ gold rhombic nanowire was reported [56], which results in a nucleation strength of 1.54 GPa, which compares favorably with the atomistic prediction of 1.5 GPa. In the same paper, the authors pulled a $\langle 100 \rangle$ nanowire with square cross-section to yield, obtaining a value of 1.2 GPa, close to the predicted strength of 1.1 GPa. It is worth pointing out that this initial work on gold nanowires was not available when the predictions were made. Also, the prediction of the strength of gold nanowires with hexagonal cross-sections, which are an energetically favorable truncated version of the rhombic nanowires tested here, are found to lie in the range of 0.8–1.6 GPa. While the simulations reported here were not conducted on these exact type of nanowires, the nucleation sites should be comparable since in both cases it is energetically favorable for the dislocations to nucleate in the same corner of the nanowire. Thus, our model predicts a value of 1.5 GPa, very close to the upper strength reported of 1.6 GPa with lower values from the experiments likely caused by as-fabricated surface defects.

The values predicted by atomistics, however, are for pristine nanowires only and cannot be taken as the strength observed in the micro-compression or micro-tension of sub-micron pillars [11, 33] often reported in the literature, as these very likely have pre-existing dislocations. In these cases, the atomistic results serve as an upper limit to the strength of these sub-micron and nanopillars [21, 72]. An example of this is shown in Fig. 12.9, where the strength of $\langle 100 \rangle$ gold nanowires in both tension and compression was calculated using atomistics and compared to experimental observations [12, 30]. While the experimental data shows the well-known smaller is stronger trend, the atomistic simulations also show a size effect. However, this trend, generally speaking, is stronger in tension and weaker in compression. The main

Table 12.3 Comparison of atomistic and experimental predicted nucleation strengths for pristine gold nanowires under tensile stress [63]

Orientation	Cross-section	Atomistic strength	Experimental strength
⟨100⟩	{100} × {100}	1.2 GPa	1.1 GPa
⟨110⟩	{111} × {111}	1.5 GPa	1.54 GPa
⟨110⟩	Hexahedral[a]	1.5 GPa	0.8–1.6 GPa

The atomistic results are for ∼100 nm diameter nanowires where surface stress contributions are negligible

[a]Hexahedral nucleation strength from atomistics taken to be the same as rhombic as nucleation occurs at equivalent locations

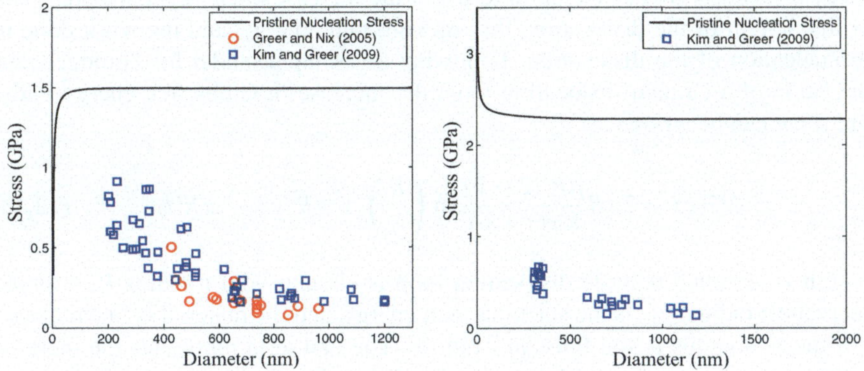

Fig. 12.9 Predicted nucleation strength of pristine gold nanowires using atomistic simulations is shown for ⟨100⟩ (**a**) compression and (**b**) tension; the results are a *solid black line*. These values are compared against experimental results for the micro-compression of gold sub-micron pillars. These pillars likely deform through the activation of already present dislocation sources rather than the nucleation of new sources and thus the simulations represent an upper bound on strength

factor contributing to the size effect is from surface stress, which makes nucleation more difficult in tension as the diameter decreases, and the opposite in compression. The number of nucleation sites also plays a role, as diameter increases so does the nucleation sites, causing an increase in the nucleation rate with size and thus making the nanowires weaker. This will always make larger wires comparatively weaker to smaller ones under the same conditions, but this effect is small compared to that of the surface stress.

12.5 Continuum Model with Atomistic Inputs

An alternative approach to using a fully atomistic simulation to determine the activation energy is the use of a continuum model. This type of analytical model attempts describes the activation Gibbs free energy in terms of the dislocation loop size and Burgers vector, dramatically reducing the number of degrees of freedom in the determination of the saddle point. However, a few of the terms in the continuum model must be informed from atomistic simulations. Some of these terms, like the dislocation core radius, can be treated as a material property and therefore are not orientation or loading dependent. Additionally, the calculations for those atomistic inputs that are orientation dependent are less computationally expensive than a full atomistic simulation. This allows for the broad analysis of trends across materials and orientations that would be prohibitively time consuming otherwise. These models can also be used as inputs to dislocation dynamic models, increasing the length scales over which the effects of dislocation nucleation can be studied.

Hirth and Lothe [20] introduced a simple model of the activation Gibbs free energy of dislocation nucleation in terms of the increase in energy associated with the line length of the dislocation, the stacking fault energy, and the work done in the nucleation of the dislocation. Following this simple model for homogeneous nucleation of a circular dislocation loop, the increase in Gibbs free energy under pure shear can be written as:

$$\Delta G(\tau) = 2\pi R \frac{\mu b^2(2-\nu)}{8\pi(1-\nu)} \ln\left(\frac{R}{r_c}\right) + \pi R^2 \gamma_{SF} - \pi R^2 b\tau \tag{12.28}$$

Here, it is assumed that the dislocation loop is circular with a radius R, r_c is the inner cutoff radius, γ_{SF} is the stacking fault energy, and the material is linear elastic with shear modulus μ and Poisson's ratio ν. The first term represents the increase in line energy associated with the circular dislocation loop from classical elastic dislocation theory, the second term is the energy of the stacking fault and the third term is the work done by the applied shear stress τ as the dislocation slips through the Burgers vector b. This model has been used by Hirth and Lothe to estimate dislocation nucleation strengths, but the activation energetics described by this model can significantly overestimate the stress for homogeneous nucleation and the results can be improved with atomistic input.

Recently, Aubry et al. [1] pointed out that this model could be dramatically improved by two simple changes: (1) allow the magnitude of the Burgers vector to change during nucleation (a well-known modification) which requires the use of the generalized stacking fault (GSF) energy curve and (2) accounting for the fact that the GSF energy curve is stress dependent. The expression for the activation Gibbs free energy in this new model is now:

$$\begin{aligned}\Delta G(\tau) = {} & 2\pi R \frac{\mu b_f^2(2-\nu)}{8\pi(1-\nu)} \ln\left(\frac{R}{r_c}\right) \\ & + \pi R^2(\gamma_{\text{GSF}}(\tau, u_0 + b_f) - \gamma_{\text{GSF}}(\tau, u_0)) - \pi R^2 b_f \tau \end{aligned} \tag{12.29}$$

where b_f is the fraction Burgers vector and γ_{GSF} is the GSF energy. With the exception of the GSF energy and r_c, all these values are material constants or physical properties that are easily determined. It becomes useful then to use

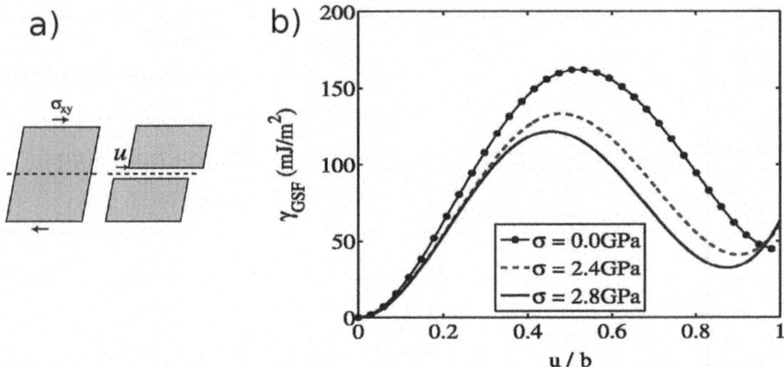

Fig. 12.10 (**a**) The generalized stacking fault (GSF) energy curve can be created by computing the excess energy between two slabs that are rigidly sheared relative to one another and can be done under applied stress, as shown here. (**b**) The resulting generalized stacking fault energy curves are now stress dependent and the local minima do not occur at the same shear. Reproduced from [1] with permission

atomistics as a means to generate the stacking fault energy curve as well as the inner cutoff radius r_c.

The GSF energy curve can be readily calculated in atomistic simulations, and if greater accuracy or transferability is needed, it can also be computed using electronic structure density functional theory. The GSF curve is the excess energy that is incurred when one rigid block of atoms is slid past another block of atoms (Fig. 12.10a), which is characterized by the plane normal (slip plane) and slip direction (Burgers vector). After each shift the energy is minimized and this value is used to create the GSF curve for the given stress (Fig. 12.10b). [1, 28]. Since the applied stress will result in an initial displacement, it is important to account for this value, u_0, before considering the contributions of the partial Burgers vector to the GSF energy. This value can be determined with the following:

$$\left.\frac{d\gamma_{\mathrm{GSF}}}{du}\right|_{u_0} - \tau = 0 \tag{12.30}$$

The activation free energy for this continuum model is given by the maximum to Eq. (12.29) with respect to R and b_f. Since the computation of the activation free energy at 0 K of a dislocation in pure shear is relatively straightforward atomistic computation, this model provides a very convenient way to extract the unknown parameter r_c. Figure 12.11 shows the activation free energies computed for the nucleation of a partial dislocation in five face-centered-cubic metals as well as the results of the continuum model proposed by Aubry. The continuum models were fit to the atomistic data solely by varying the inner cutoff radius r_c. With the determination of these values, it is reasonable to take r_c as a constant value in all further calculations.

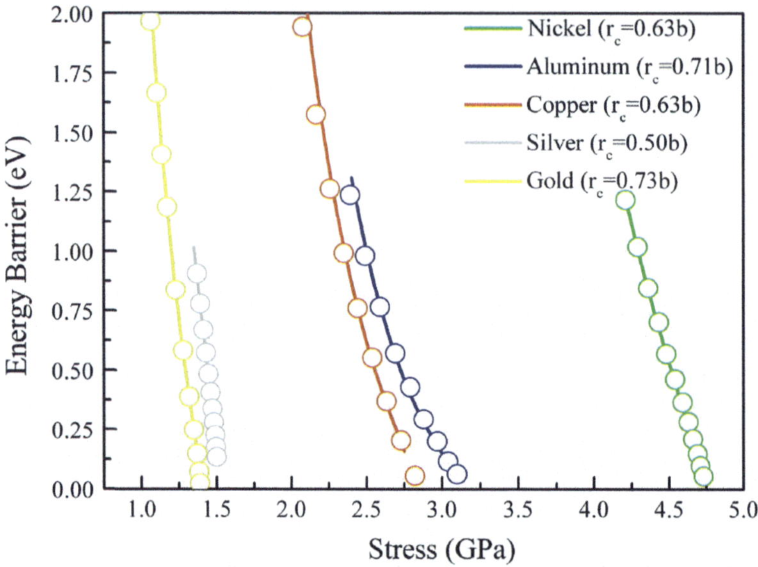

Fig. 12.11 Energy barriers for homogeneous nucleation. Points represent atomistic calculations. Reproduced from [22] with permission

While dislocations are unlikely to homogeneously nucleate, this results of this model can still be used to estimate the stress at which they do. This can be done by using the continuum model, complete with a stress dependent generalized stacking fault energy curve, to compute the activation free energies. The activation free energies can be fit to a functional form [i.e., Eqs. (12.15) or (12.16)] and the strengths predicted by Eq. (12.13). As discussed previously, it is still important to consider the contribution of entropy to the free energy and, similar to the fully atomistic approach, the Meyer–Neldel rule can be used. A subsequent section will evaluate this relation from a continuum model prospective.

The model described above was developed with pure shear in mind; however, it can be readily adapted to handle homogeneous nucleation under uniaxial load (or any homogeneous applied stress state). The easiest way to treat the model is to consider the dislocation nucleating under in the same orientation as before, but a state of non-pure shear is applied to the crystal. The term in the free energy that accounts for the work done through slip must then be modified using the Schmid factor for the nucleating dislocation, viz. $\tau = S\sigma$. It is important to use the correct Schmid factor for the type of dislocation being nucleated, either a leading partial dislocation or a perfect dislocation. The nucleation of leading partial dislocations results in an asymmetry in the nucleation because the Schmid factors in tension and compression are not the same since a leading partial dislocation in tension becomes a trailing partial dislocation in compression. The Schmid factors for leading partial dislocations in tension and compression in FCC metals as well as perfect dislocations are given in Fig. 12.1. It is important to note, however, that the uniaxial stress result in stresses that do not resolve onto the slip plane and direction. While these stresses do not affect the work done when the dislocation slips, they may affect the GSF energy. Therefore it is important that the GSF curve is created under the same uniaxial loading state that is being considered, accounting for the entire stress state. This additional stress component has been shown to have a large influence on the final curve.

Considering these modifications, the equation for homogeneous nucleation under uniaxial stress can then be given as:

$$\Delta G_{\text{uniaxial}}(\sigma) = 2\pi R \frac{\mu b_f^2 (2-\nu)}{8\pi(1-\nu)} \ln\left(\frac{R}{r_c}\right) + \pi R^2 (\gamma_{\text{GSF}}(\sigma, u_0 + b_f) - \gamma_{\text{GSF}}(\sigma, u_0)) - \pi R^2 b_f S \sigma \quad (12.31)$$

where, similar to Eq. (12.30), the initial displacement is given by $\frac{d\gamma_{\text{GSF}}}{du}\Big|_{u_0} - \sigma = 0$. To determine nucleation strength, Eq. (12.13) is again used. As a final note on homogeneous dislocation nucleation, while Eqs. (12.29) and (12.31) are defined in terms of a partial dislocation, the energy of a perfect dislocation can be determined by replacing the direction of b_f with the full Burgers vector and generating the appropriate γ_{GSF} curve.

Figure 12.12 shows the results of the continuum model of the nucleation strength of homogenous dislocation nucleation in nanopillars for multiple materials, orientations, and sizes. These results were obtained using the GSF and r_c values determined from atomistic simulations and the effect of surface stresses (a nominal value is taken as $C = 4 \times 10^3$ MPa nm, independent of the material) and the number of nucleation sites were also included. Several trends are worth noting, the first being that gold and silver tend to be the weakest, followed by copper and aluminum and

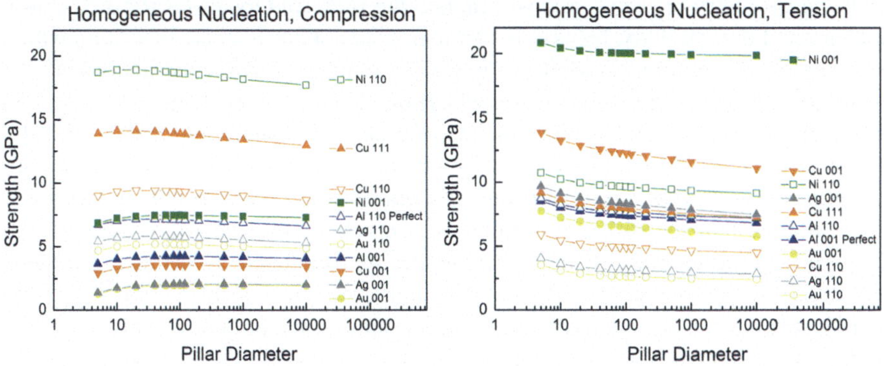

Fig. 12.12 Continuum model predictions of the homogeneous nucleation strength of nanopillars of varying materials, orientations, and sizes. Reproduced from [22] with permission

nickel are the strongest. In tension, the strength increases with decreasing pillar size while in compression, the trend is typically the opposite. These general trends are caused by the dominant surface stress. The number of nucleation sites always makes smaller pillars stronger, but this is secondary to the surface stress as mentioned previously. It is also worth noting that the Schmid factor plays a large role in the nucleation strength. The Schmid factor is the same for pillars in $\langle 110 \rangle$ tension as those in $\langle 100 \rangle$ compression and these strengths are similar. It will be shown that these homogeneous strengths are noticeably higher than those of heterogeneous nucleation and thus the need to accurately model heterogeneous nucleation.

12.5.1 Heterogeneous Nucleation

As mentioned previously, heterogeneous nucleation will generally be the preferred nucleation method of dislocations in nanostructures due to their lower activation free energies. Continuum modeling of this process becomes significantly more complicated than homogeneous nucleation because the dislocation will nucleate from a free surface and thus the energetics of its interaction with the free surface must be accounted for. In this section, we will develop a simple model, similar to the one described previously, that is able to handle heterogeneous nucleation with as few parameters as possible. To this end, we will make the assumption in all model developments that the dislocation nucleus will be a circular arc whose center is fixed on the free surface as shown in Fig. 12.13. This is an approximation at best as atomistic simulations have shown that nucleating dislocations are not circular arcs [65]. More advanced continuum models, for example, those based on the Peierls–Nabarro model [35, 66–68], can handle the relaxation of the line shape as well but at considerable numerical cost which may prevent its use in larger scale models.

In modeling heterogeneous dislocation nucleation, it is very important to consider the geometry of slip as this will dictate the preferred locations for dislocation nucleation. For simplicity, let us consider nanowires of constant cross-section which have the most common, as well as simplest, geometries of nanostructures. If the nanowire is circular, the slip plane on which the dislocation nucleates on will be an ellipse whose eccentricity is controlled by the angle the slip plane makes with the nanowire axis. Constructing the exact geometry of the nucleating dislocation with the elliptical slip plane can be a needlessly complex geometry problem. However, if we assume that the dislocation nucleus is much smaller than the radius of the nanowire, the nucleating dislocation can be approximated as occurring from a half space. If the nanowire is square, like that shown in Fig. 12.13, the slip plane will be some form of a quadrilateral. For example, the commonly analyzed $\langle 100 \rangle$ square nanowire with $\{100\}$ side surfaces has four potential slip planes (Fig. 12.13b) that are symmetric about the axis. Isolating one of these slip planes, Fig. 12.13c, shows that the slip plane is a rhombus and that there are several different types of heterogeneous nucleation sites. The first is at the corner where the included

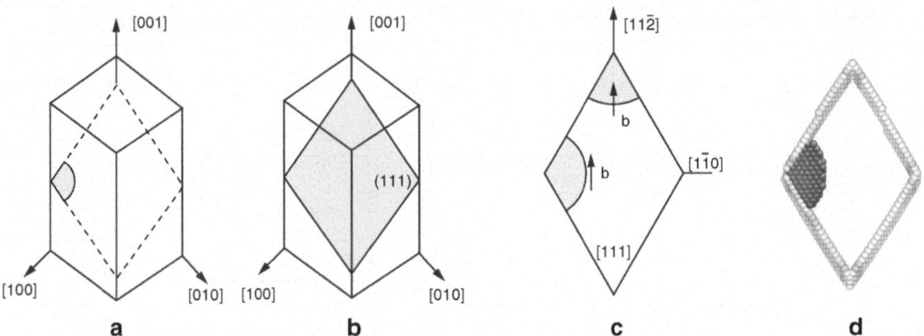

Fig. 12.13 An illustration of the nucleation of partial dislocations in a square FCC nanowire illustrating the importance of considering the geometry of the nucleating defect. (**a**) A schematic of the nucleation of a partial dislocation in a $\langle 100 \rangle$ FCC nanowire and (**b**) the isolation of the $\{111\}$ slip plane it glides on and its intersection with the free surfaces. (**c**) The glide plane of the dislocation rotated normal to the viewing direction with two possible nucleation sites highlighted with Burgers vectors consistent with leading partial dislocations in compression. (**d**) The observation of a nucleating dislocation in atomistics in a gold nanowire in $\langle 100 \rangle$ compression. Some figures are reproduced from [65] with permission

angle is obtuse, called the obtuse corner. The second is the other angle, which is termed the acute corner. These two sites are distinct as the included angle of the nucleating dislocation is different as is the character of the dislocation as illustrated in Fig. 12.13c. Finally, it is also possible (though improbable) that the dislocation will nucleate from a surface instead of the corner. Thus, in modeling dislocation nucleation it is very important to model the correct nucleation site, which is the one with the lowest activation energy. In certain cases, due to differences in character dependent line energy, this will not always be the dislocation that subtends the smallest angle.

To examine the energetics of the dislocation formation on a surface, we first considered nucleation on a flat surface of a nanocrystal, i.e., the half space solution, which we can take as an approximation for the nucleation in a circular nanowire. As discussed previously, we will model this dislocation as a semicircle with a Burgers vector parallel to the surface. While the first instinct would be to simply divide the line length in half to account for the reduction in line energy, it is also important to adjust the contributions of the logarithmic term to correctly account for the dislocations interaction with the free surface. This leads to the formulation first proposed by Beltz and Freund of:

$$U_{\text{line}}^{\text{half}} = \pi R \frac{\mu b_f^2 (2 - \nu)}{8\pi(1 - \nu)} \ln \left(m_{\text{BF}} \frac{R}{r_c} \right) \tag{12.32}$$

where m_{BF} is a correction factor to the logarithmic term due to the presence of the free surface. This value has been shown to be a function of the Poisson's ratio and is given by:

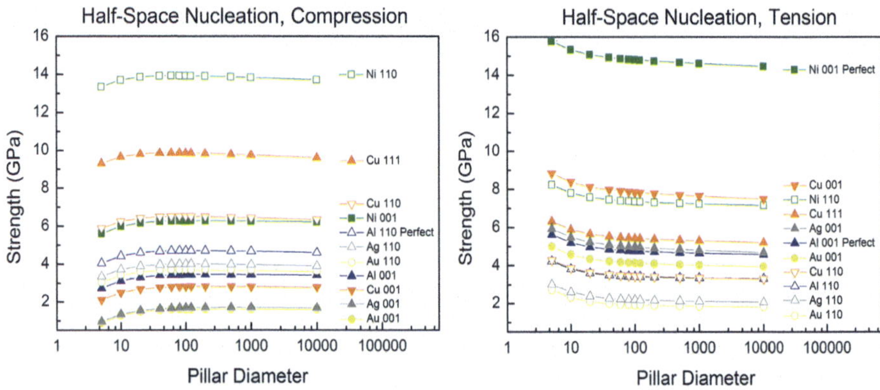

Fig. 12.14 Continuum model solutions for heterogeneous nucleation in the half space for nanopillars of varying materials, orientations, and sizes. Reproduced from [22] with permission

$$m_{\mathrm{BF}} = \exp\left(\frac{0.7734 - 1.059(2-\nu)}{2-\nu}\right) \tag{12.33}$$

For most metals, where $\nu \approx 0.3$, m_{BF} can be approximated by 0.55. Figure 12.14 shows the prediction of nucleation strengths using the continuum model for the half space of varying materials, orientations and sizes. This solution includes surface stresses, stress dependent GSF curves, and the number of nucleation sites, the same fashion as the calculations for the bulk materials with the only change being the line energy of the defect and the associated change in area associated with the stacking faults and work terms. Similar to the homogeneous solution, the pillars are stronger in tension and within the same loading similar behavior is exhibited as the diameter changes. The strength is different, however, highlighting the effect of the change in the dislocation line length, however, the reduction in strength is not a factor of two.

Eq. (12.32) can be further generalized for an arc whose center is at the corner of the nucleation site (Fig. 12.15), the line energy is:

$$U_{\mathrm{line}} = \int_{-\alpha}^{\alpha} \frac{\mu b_f^2}{4\pi(1-\nu)}(1-\nu\cos^2(\theta))R\ln\left(m\frac{R}{r_c}\right)\mathrm{d}\psi \tag{12.34}$$

where 2α is the angle of the dislocation arc and θ is the angle between the dislocation line tangent and the Burgers vector. The change in limits from the half space solution to an arc will reduce the line energy in a manner that is no longer linear since the line energy is dislocation character dependent. The value for m will also change, as it did between homogeneous and half space and formally should be included in the correct of the line energy. Jennings et al. [22] derived a solution to a 2-D problem of a screw dislocation in a wedge, which is similar to the current problem, and found the value of m to vary linearly with α, and to be independent of material properties. Therefore a linear approximation of m from the

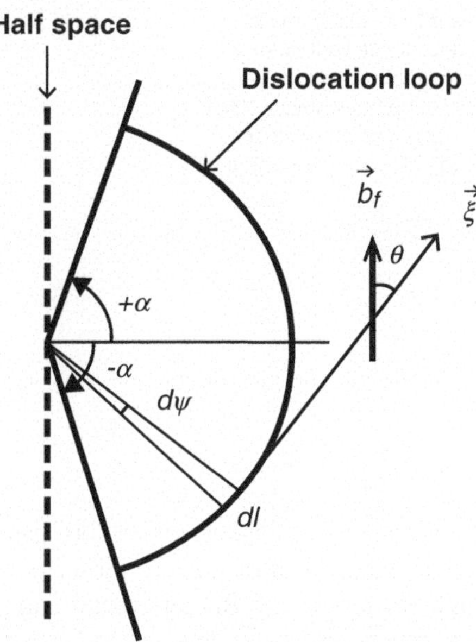

Fig. 12.15 A schematic of a dislocation heterogeneously nucleating from a corner with a circular profile. The angles α and θ are defined here for use in Eq. (12.34). Reproduced from [22] with permission

half space solution seems appropriate. This trend is already roughly observed in the change from the homogeneous nucleation problem to the heterogeneous one as m changes from 1.0 to 0.55 and, due to the logarithmic nature of the m term, a slight variation from the actual value will have only a minimal effect on the energy barrier calculation. Normalizing the angle to the half space solution where $\alpha = \pi/2$ gives

$$m = m_{\mathrm{BF}} \frac{2\alpha}{\pi} \tag{12.35}$$

As $\alpha \to \pi$ m does not approach the homogeneous solution of $m = 1.0$ but rather $m = 1.1$ an error created by the interpolation introduced above. This error, first is small since it appears in the logarithm and, second, is generally inconsequential since heterogeneous dislocations will generally nucleate in corners with $\alpha \leq \frac{\pi}{2}$. The error can be removed in homogeneous nucleation by treating the energy exactly within the formulation described previously for homogeneous nucleation.

The other energetic factor that needs to be considered in computing the activation free energy for heterogeneous nucleation of a dislocation is the energy associated with the formation of a surface ledge. When a dislocation intersects a surface, it forms a surface ledge in its wake, which adds to the energy of the nucleating defect. A first order approximation of this energy is a product of the ledge area times the energy of the surface ledge per-unit-area, $\Delta E = \gamma_{\mathrm{ledge}} A_{\mathrm{ledge}}$. The area of the ledge can be found from the magnitude of the fractional Burgers vector and its direction relative to the surface. Therefore, the ledge area associated with a nucleating circular dislocation loop is a product of the fractional Burgers vector, the length of the ledge, and the sine of the angle between them: $A_{\mathrm{ledge}} = R b_f \sin\theta$. Figure 12.16 illustrates

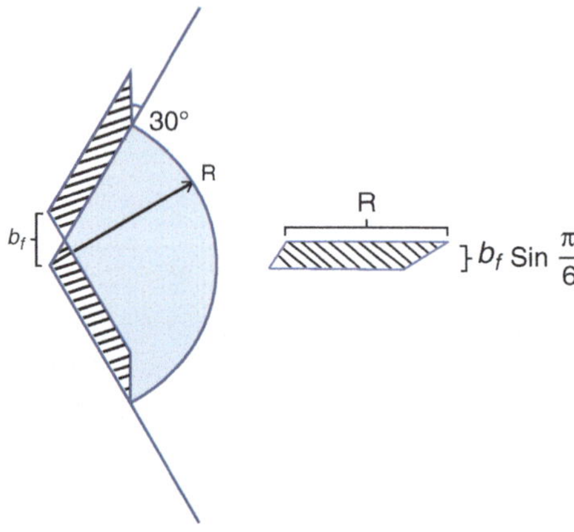

Fig. 12.16 Diagram of surface ledge formed by a dislocation nucleating on a ⟨111⟩ slip plane under ⟨001⟩ compression. Reproduced from [22] with permission

this for the specific case of the ⟨100⟩ nanowire deformed in compression discussed above, the dislocation nucleus forms on the obtuse corner and creates a ledge at the angle formed by the Burgers vector and the free surface: 30°. The total area of the ledges is simply twice this value. The formation of this ledge causes an increase in the free energy of the system that can be defined as:

$$\Delta E_{\text{ledge}} = 2\gamma_{\text{ledge}} R b_f \sin\theta \tag{12.36}$$

It is certainly possible, within this approximation, to consider a step energy that depends on orientation. Such a model, however, would be quite complex and would again defeat the utility of the simple continuum models developed here. So, for this simple model, we will consider the ledge energy per-unit-area to be a constant. The method for calculating this value is similar to that of the stacking fault energy. Two boxes are welded together and moved relative to each other. In this case, however, the box has a free surface along the Burgers vector direction instead periodic. The resulting increase in energy is then a function of the creation of the ledge as well as the stacking fault energy. Subtracting the stacking fault energy, a value also necessary for the continuum approximation, then gives the ledge energy per unit area.

When all of these contributions are considered, they yield a continuum model as follows:

$$\begin{aligned}\Delta G^* = &\int_{-\alpha}^{\alpha} \frac{\mu b_f^2}{4\pi(1-\nu)}(1-\nu\cos^2(\theta))R\ln\left(m\frac{R}{r_c}\right)\mathrm{d}\psi \\ &+ A(\gamma_{\text{GSF}}(\tau, u_0 + b_f) - \gamma_{\text{GSF}}(\tau, u_0)) - A b_f \tau + 2\gamma_{\text{ledge}} R b_f \sin\theta\end{aligned} \tag{12.37}$$

Figure 12.17 shows the results of the continuum model both with and without the ledge energy approximation as well as the atomistic prediction. In addition,

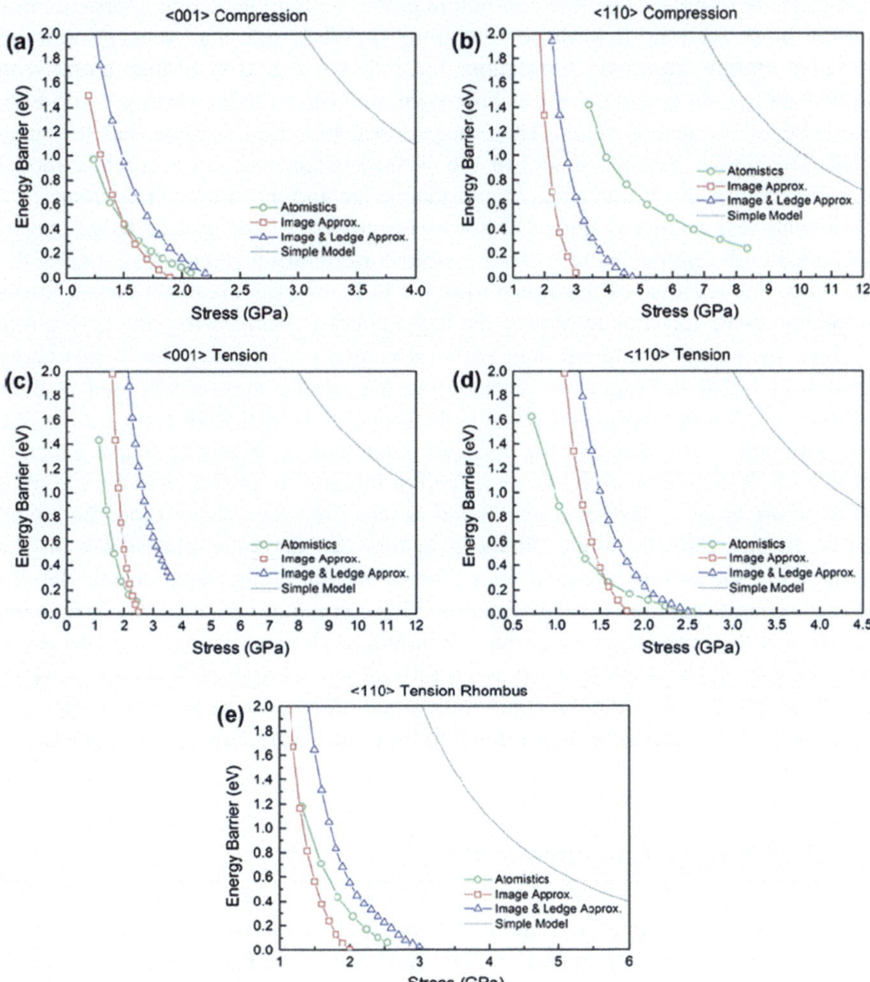

Fig. 12.17 Comparison of atomistic energy barrier calculations with various continuum model approximations for (**a**)–(**d**) square nanowires oriented along the ⟨110⟩ and ⟨100⟩ orientations in both tension and compression. (**e**) A comparison of energy barriers for a ⟨110⟩ rhombic nanowire in tension using both atomistics and continuum models. Reproduced from [22] with permission

it also compares the results of the original simple model introduced by Hirth and Lothe [20] where $r_c = b$, the Burgers vector is fixed, the stacking fault energy is constant, and the dislocation is treated as a complete circle. It can easily be seen that while the continuum model is generally close to the values predicted by atomistics, there are differences. This suggests that the continuum model is not completely accurate in representation of some of the energetics of the nucleation process, in contrast to the homogenous case where exceptionally close matches were made. One characteristic of note is that the continuum model without the ledge approximation is often more accurate than those containing it. While one may suggest omitting the ledge energy, atomistic simulations have shown that it is formed along with the dislocation suggesting that our approximation of the ledge energy is too high compared to the actual value. There are several potential reasons that this may occur. The first is that the model of the surface ledge may not accurately model the relaxation of the atoms about the surface ledge and the simple simulation used to calculate this energy is therefore not representative of the surface ledge energy associated with real nucleation. One possible consideration is the stress state of the nanowire. The removal of the atoms from the bulk material creates a surface stress on the nanowire and thus may alter the ledge energy. Additionally, the GSF curve has been shown to be a function of stress, it is also possible that the ledge energy curve would also be, especially considering the similar approaches used in their calculations. Another issue that arises is the shape of the dislocation nucleation. We have assumed in all cases that the dislocation nucleus is circular in shape, which is not true of all the dislocations as some tend to look more straight than curved [65]. All these are areas of further research and as they are better understood, the model can be further refined. While all these factors can certainly play a role in the energetics of the system, these additional terms would require separate calculations for each separate geometry and orientation. This computational price would in many ways defeat the purpose of using the continuum model and, if such considerations are important to the work being done, a fully atomistic approach would likely be more appropriate. So while the continuum model in its current form is not perfect, it does provide a reasonable approximation for inclusion in larger scale models.

12.5.2 Entropic Considerations

As was discussed in the case of atomistic simulations, the role of entropy must be accounted for in these continuum models. However, if we follow the arguments of Ryu et al. [53], the major entropic contributions come from the thermal expansion and subsequent elastic constant reduction. Thus, using experimental values of the elastic constants and lattice constants at 300 K should, by and large, accurately account for entropic contributions and the continuum model does not need to be modified. If the inputs into the continuum model, e.g., the shear modulus, are at 0 K, then the continuum model would not include the effects of entropy on nucleation just as in the atomistic models. Similarly, if we wish to use the continuum models to predict strength accurately as a function of temperature, we also need to account for entropy in an appropriate manner. This can be done, under most circumstances, using the Meyer–Neldel rule and the insight into this can be obtained using a simple continuum model.

To understand the connection between the continuum model and the Meyer–Neldel rule, consider a simplified version of Eq. (12.29) that assumes a line orientation independent energy:

$$\Delta G = 2\pi R k \mu(T) - \pi R^2 \tau b$$

where $k = \frac{b^2(2-\nu)}{8\pi(1-\nu)}$ and the logarithmic term has been omitted for simplicity. This allows for an easy approximation of the activation energy:

$$\frac{\partial \Delta G}{\partial R} = 2\pi k \mu(T) - 2\pi R_c \tau b = 0$$

$$R^* = \frac{\mu(T)k}{\tau b}$$

$$\Delta G^* = \frac{\pi}{\tau b}(\mu(T)k)^2$$

Since $\frac{\partial \Delta G^*}{\partial \tau} = -\Omega$, it can be shown that $\Omega\tau = \Delta G^*$. This relation can be used to justify the derivation of Eq. (12.26). Additionally, if as previously discussed, it is assumed that ΔH^* is a function of stress only, the entropic contribution can be calculated by taking the derivative with respect to temperature:

$$\Delta S_\tau^* = -\frac{\partial \Delta G_c}{\partial T} = -\frac{2\pi\mu(T)k^2}{\tau b}\frac{\partial \mu(T)}{\partial T} \tag{12.38}$$

A similar approach with the Helmholtz free energy yields

$$\Delta S_\gamma^* = -\frac{\pi k^2}{\gamma b}\frac{\partial \mu(T)}{\partial T} \tag{12.39}$$

If linear elasticity is assumed, $\tau = \mu\gamma$, the relationship between the entropic contribution in stress and strain controlled simulations is

$$\Delta S_\tau^* = 2\Delta S_\gamma^* \tag{12.40}$$

If these relations are used in conjunction with the Meyer–Neldel rule [Eqs. (12.19) and (12.20)], the fitting parameters are given by:

$$T_0^\gamma = -\mu(T)\frac{\partial T}{\partial \mu(T)} + T \tag{12.41}$$

$$T_0^\tau = -\frac{\mu(T)}{2}\frac{\partial T}{\partial \mu(T)} + T \tag{12.42}$$

This can be further simplified if it is assumed $\mu(T)$ changes linearly with T, $\mu(T) = \mu_0 - \mu_1 T$:

$$T_0^{\gamma} = \frac{\mu_0}{\mu_1} \tag{12.43}$$

$$T_0^{\tau} = \frac{\mu_0}{2\mu_1} + \frac{T}{2} \tag{12.44}$$

These values are very similar to those of Eqs. (12.24) and (12.26), which were determined by Ryu et al. [53]. The term that is linear in T for T_0^{τ} is small and can be neglected in many calculations, resulting in a constant value for T_0^{τ} that is half the value of T_0^{γ}. This yields a straightforward way in which to approximate the entropic contribution when the variation of the shear modulus with temperature is known. Figure 12.18 does just this, using existing values from literature [9, 42, 46, 58] and fitting the data with a line. Using the intercept and slope of these fitting lines and combining them with Eq. (12.41) allows for the estimation of the T_0^{γ} of several FCC metals. It can be seen that these lines tend to overestimate the shear modulus as the temperature approaches 0 K, presumably a quantum effect. However, at these temperatures the effect of entropy is minimal so any error caused will be negligible. Table 12.4 lists these fitting parameters for constant strain conditions as well as shows its ratio to the melting temperature of the individual materials:

It can be seen that the ratios of T_0^{γ}/T_m are in the range of approximately 2 to 3. This allows for a rough but simple approximation for the fitting parameter based on

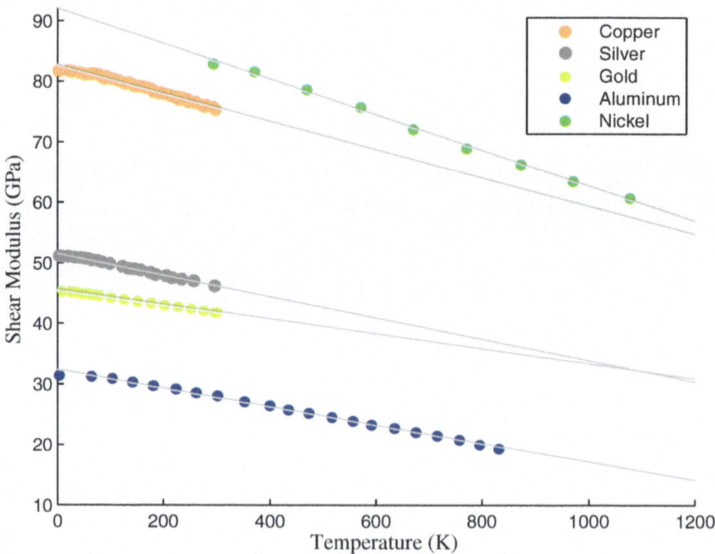

Fig. 12.18 Experimental data for shear modulus of FCC metals at various temperatures with an associated linear fit

Table 12.4 Approximate T_0^{γ} values for FCC metals obtained from fitting experimental data for the temperature dependent shear modulus, Fig. 12.18

Material	T_0^{γ} (K)	T_0^{γ}/T_m
Nickel	3140.8	1.82
Aluminum	2125.9	2.28
Copper	3551.1	2.62
Silver	2927.2	2.37
Gold	3709.4	2.77

the melting temperature. Since the value for T_0^{σ} is approximately half of T_0^{γ}, this provides evidence for the use of the melting temperature as the fitting parameter for T_0^{τ} in the correction for entropy in the Gibbs free energy.

If we revisit the work of Ryu et al. [53], the temperature dependent shear modulus of their atomistic simulations would yield a value for T_0^{γ} of 2800 K when Eq. (12.43) is used. This matches well with their results when comparing the use of the Meyer–Neldel rule with results obtained from umbrella sampling. However, experimental results yield a fitting temperature of T_0^{γ} =3500 K. This suggests that the potential used in their simulations softened too much with temperature and highlights a potential foil of atomistic simulations, it is very dependent on the potential used. Thus it will only predict the correct entropic contribution if it models the softening of the shear modulus correctly. This is why the Meyer–Neldel rule is so useful in the calculation of the entropic contribution, it allows the use of well established, simple experimental results in conjunction with 0 K atomistics, potentially bypassing many of the issues that can occur in higher temperature simulations. These results also suggest that entropy can be included in models of nucleation through the appropriate use of the temperature dependent shear modulus. Since the shear modulus generally decreases linearly with temperature, the change in the free energy with temperature can be easily included to improve the strength predictions.

12.6 Summary

This chapter presents case study on multiscale molding of dislocation nucleation, a key process that controls the strength of nanoscale materials. The multiscale nature is hierarchical in that information found in the atomistic simulations are passed up to create a simple continuum model of the nucleation process. This is accomplished by performing atomistic simulations to determine the energy barriers of dislocation motion and attempting to match the continuum model to the atomistic data. The simplicity of the model provides the potential to include nucleation rates directly into dislocation dynamics models (See Chap. 2) that can subsequently and accurately treat the dynamics of the nucleating defects, which are not treated in any way in this chapter.

Specifically, the atomistics provide a much more robust estimate for the inner core radius by matching continuum models with the atomistic results for the homogeneous nucleation of dislocations and demonstrate in need for the inclusion of stress dependent GSF energy curves. The stress dependent stacking fault energies have been shown to be necessary to match atomistic and continuum models [1, 22]. Estimates of stress dependent stacking fault energies, now that they have been determined as significant, can be informed by more robust methods such as DFT, if need be. Classical atomistic methods have also demonstrated that the surface stress can dramatically alter the stress in the nanowires and therefore can have a huge influence on their strength below about 40 nm; atomistics can provide a robust estimate for the appropriate scaling law for these induced compressive stresses. Atomistic methods have also identified that the activation free energy is mainly affected by the reduction of elastic constants with temperature, and that the effects of temperature can be estimated by using experimentally determined temperature dependent elastic constants. In addition, comparisons between the atomistic simulations and continuum models demonstrate that little accuracy is gained by including surface ledge energetics. Finally, using atomistic methods or atomistic informed continuum models, the nucleation strength of gold nanowires appears to agree well experimentally reported nucleation strengths.

References

1. S. Aubry, K. Kang, S. Ryu, W. Cai, Energy barrier for homogeneous nucleation: Comparing atomistic and continuum models. Scr. Mater. **64**, 1043 (2011)
2. S. Brochard, P. Hirel, L. Pizzagalli, J. Godet, Elastic limit for surface step dislocation nucleation in face-centered cubic metals: temperature and step height dependence. Acta Mater. **58**, 4182–4190 (2010)
3. A. Cao, E. Ma, Sample shape and temperature strongly influence the yield strength of metallic nanopillars. Acta Mater. **56**(17), 4816–4828 (2008)
4. A. Cao, Y. Wei, Atomistic simulations of the mechanical behavior of fivefold twinned nanowires. Phys. Rev. B **74**, 214108 (2006)
5. J. Diao, K. Gall, M. Dunn, Surface-stress-induced phase transformation in metal nanowires. Nat. Mater. **2**, 656–660 (2003)
6. J. Diao, K. Gall, M. Dunn, Surface stress driven reorientation of gold nanowires. Phys. Rev. B **70**, 075413 (2004)
7. J. Diao, K. Gall, M. Dunn, Yield strength asymmetry in metal nanowires. Nano Lett. **4**(10), 1863–1867 (2004)
8. J. Diao, K. Gall, M. Dunn, J. Zimmerman, Atomistic simulations of the yielding of gold nanowires. Acta Mater. **54**, 643–653 (2006)
9. R. Farraro, R.B. Mclellan, Temperature dependence of the young's modulus and shear modulus of pure nickel, platinum, and molybdenum. Metall. Trans. A **8**(10), 1563–1565 (1977)
10. K. Gall, J. Diao, M. Dunn, The strength of gold nanowires. Nano Lett. **4**(12), 2431–2436 (2004)
11. J. Greer, J.D. Hosson, Plasticity in small-sized metallic systems: Intrinsic versus extrinsic size effect. Prog. Mater. Sci. (2011). doi:10.1016/j.pmatsci.2011.01.005
12. J.R. Greer, W.C. Oliver, W.D. Nix, Size dependence of mechanical properties of gold at the micron scale in the absence of strain gradients. Acta Mater. **52**, 1821–1830 (2005)
13. M.E. Gurtin, A.I. Murdoch, Surface stress in solids. Int. J. Solids Struct. **14**(6), 431–440 (1978)

14. P. Hänggi, P. Talkner, M. Borkovec, Reaction-rate theory: fifty years after Kramers. Rev. Mod. Phys. **62**(2), 251 (1990)
15. S. Hara, S. Izumi, S. Sakai, Reaction pathway analysis for dislocation nucleation from a Ni surface step. J. Appl. Phys. **106**, 093507 (2009)
16. K.G. Harold Park, J. Zimmerman, Deformation of FCC nanowires by twinning and slip. J. Mech. Phys. Solids **54**, 1862–1881 (2006)
17. G. Henkelman, H. Jónsson, Improved tangent estimate in the nudged elastic band method for finding minimum energy paths and saddle points. J. Chem. Phys. **113**, 9978–9985 (2000)
18. G. Henkelman, G. Johannesson, H. Jónsson, Methods for finding saddle points and minimum energy paths, in *Progress on Theoretical Chemistry and Physics*, ed. by S.D. Schwartz (Kluwer Academics, Dordrecht, 2000), pp. 269–300
19. P. Hirel, J. Godet, S. Brochard, L. Pizzagalli, P. Beauchamp, Determination of activation parameters for dislocation formation from a surface in fcc metals by atomistic simulations. Phys. Rev. B **78**, 064109 (2008)
20. J. Hirth, J. Lothe, *Theory of Dislocations* (Krieger, Malabar, FL, 1982)
21. A.T. Jennings, J. Li, J.R. Greer, Emergence of strain-rate sensitivity in Cu nanopillars: transition from dislocation multiplication to dislocation nucleation. Acta Materialia **59**(14), 5627–5637 (2011)
22. A.T. Jennings, C.R. Weinberger, S.W. Lee, Z.H. Aitken, L. Meza, J.R. Greer, Modeling dislocation nucleation strengths in pristine metallic nanowires under experimental conditions. Acta Mater. **61**(6), 2244–2259 (2013)
23. C. Ji, H.S. Park, Geometric effects on the inelastic deformation of metal nanowires. Appl. Phys. Lett. **89**, 181916 (2006)
24. C. Ji, H.S. Park, The coupled effects of geometry and surface orientation on the mechanical properties of metal nanowires. Nanotechnology **18**(30), 305704 (2007)
25. C. Ji, H. Park, The effect of defects on the mechanical behavior of silver shape memory nanowires. J. Comput. Theor. Nanosci. **4**(3), 578 (2007)
26. J.W. Jiang, A.M. Leach, K. Gall, H.S. Park, T. Rabczuk, A surface stacking fault energy approach to predicting defect nucleation in surface-dominated nanostructures. J. Mech. Phys. Solids **61**(9), 1915–1934 (2013)
27. K. Kang, Atomistic modeling of fracture mechanisms in semiconductor nanowires under tension. Ph.D. Thesis, Stanford University, 2010
28. K. Kang, W. Cai, Brittle and ductile fracture of semiconductor nanowires–molecular dynamics simulations. Philos. Mag. **87**(14–15), 2169–2189 (2007)
29. C.L. Kelchner, S.J. Plimpton, J.C. Hamilton, Dislocation nucleation and defect structure during surface indentation. Phys. Rev. B **58**, 11085–11088 (1998)
30. J.Y. Kim, J.R. Greer, Tensile and compressive behavior of gold and molybdenum single crystals at the nano-scale. Acta Mater. **57**(17), 5245–5253 (2009)
31. U.F. Kocks, A.S. Argon, M.F. Ashby, Thermodynamics and kinetics of slip. Prog. Mater. Sci. **19**, 1–289 (1975)
32. J. Kollar, L. Vitos, J.M. Osorio-Guillen, R. Ahuja, Calculation of surface stress for fcc transition metals. Phys. Rev. B **68**, 245417 (2003)
33. O. Kraft, P.A. Gruber, R. Monig, D. Weygand, Plasticity in confined dimensions. Annu. Rev. Mater. Res. **40**, 293–317 (2010)
34. A. Leach, M. McDowell, K. Gall, Deformation of top-down and bottom-up silver nanowires. Adv. Funct. Mater. **17**, 43–53 (2007)
35. C. Li, G. Xu, Critical conditions for dislocation nucleation at surface steps. Philos. Mag. **86**, 2957–2970 (2007)
36. W. Liang, M. Zhou, Response of copper nanowires in dynamic tensile deformation. Proc. Inst. Mech. Eng. C J. Mech. Eng. Sci. **218**(6), 599–606 (2004)
37. W. Liang, M. Zhou, Atomistic simulations reveal shape memory of fcc metal nanowires. Phys. Rev. B **73**, 115409 (2006)
38. Y. Lu, C. Peng, Y. Ganesan, J.Y. Huang, J. Lou, Quantitative in situ TEM tensile testing of an individual nickel nanowire. Nanotechnology **22**(35), 355702 (2011)

39. Y. Lu, J. Song, J.Y. Huang, J. Lou, Fracture of sub-20 nm ultrathin gold nanowires. Adv. Funct. Mater. **21**, 3982–3989 (2011)
40. J. Marian, J. Knap, Breakdown of self-similar hardening behavior in au nanopillar microplasticity. Int. J. Multiscale Comput. Eng. **5**, 287–294 (2007)
41. R.J. Needs, M.J. Godfrey, M. Mansfield, Theory of surface stress and surface reconstruction. Surf. Sci. **242**, 215–221 (1991)
42. J. Neighbours, G. Alers, Elastic constants of silver and gold. Phys. Rev. **111**(3), 707 (1958)
43. A. Ngan, L. Zhuo, P. Wo, Size dependence and stochastic nature of yield strength of micron-sized crystals: a case study on Ni3Al. Proc. R. Soc. **462**, 1661 (2006)
44. L. Nguyen, D. Warner, Improbability of void growth in aluminum via dislocation nucleation under typical laboratory conditions. Phys. Rev. Lett. **108**(3), 035501 (2012)
45. L. Nguyen, K. Baker, D. Warner, Atomistic predictions of dislocation nucleation with transition state theory. Phys. Rev. B **84**(2), 024118 (2011)
46. W. Overton Jr., J. Gaffney, Temperature variation of the elastic constants of cubic elements. i. Copper. Phys. Rev. **98**(4), 969 (1955)
47. H.S. Park, J.A. Zimmerman, Modeling inelasticity and failure in gold nanowires. Phys. Rev. B **72**, 054106 (2005)
48. H. Park, K. Gall, J. Zimmerman, Shape memory and pseudoelasticity in metal nanowires. Phys. Rev. Lett. **95**(25), 255504 (2005)
49. P. Pechukas, Transition state theory. Annu. Rev. Phys. Chem. **32**(1), 159–177 (1981)
50. E. Rabkin, H. Nam, D. Srolovitz, Atomistic simulation of the deformation of gold nanopillars. Acta Mater. **55**(6), 2085–2099 (2007)
51. G. Richter, K. Hillerich, D. Gianola, R. Monig, O. Kraft, C. Volkert, Ultrahigh strength single crystalline nanowhiskers grown by physical vapor deposition. Nano Lett. **9**, 3048–3052 (2009)
52. S. Ryu, K. Kang, W. Cai, Entropic effect on the rate of dislocation nucleation. Proc. Natl. Acad. Sci. USA **108**, 5174 (2011)
53. S. Ryu, K. Kang, W. Cai, Predicting the dislocation nucleation rate as a function of temperature and stress. J. Mater. Res. **26**(18), 2335–2354 (2011)
54. E. Schmid, W. Boas, *Plasticity of Crystals* (F. A. Hughes Co., London, 1950)
55. A. Sedlmayr, E. Bitzek, D.S. Gianola, G. Richter, R. Mönig, O. Kraft, Existence of two twinning-mediated plastic deformation modes in au nanowhiskers. Acta Mater. **60**(9), 3985–3993 (2012)
56. J.H. Seo, Y. Yoo, N.Y. Park, S.W. Yoon, H. Lee, S. Han, S.W. Lee, T.Y. Seong, S.C. Lee, K.B. Lee et al., Superplastic deformation of defect-free au nanowires via coherent twin propagation. Nano Lett. **11**(8), 3499–3502 (2011)
57. J.H. Seo, H.S. Park, Y. Yoo, T.Y. Seong, J. Li, J.P. Ahn, B. Kim, I.S. Choi, Origin of size dependency in coherent-twin-propagation-mediated tensile deformation of noble metal nanowires. Nano Lett. **13**, 5112–5116 (2013)
58. J. Tallon, A. Wolfenden, Temperature dependence of the elastic constants of aluminum. J. Phys. Chem. Solids **40**(11), 831–837 (1979)
59. M.D. Uchic, P.A. Shade, D. Dimiduk, Plasticity of micrometer-scale single crystals in compression. Annu. Rev. Mater. Res. **39**, 361–386 (2009)
60. J. Wang, H. Huang, Novel deformation mechanism of twinned nanowires. Appl. Phys. Lett. **88**, 203112 (2006)
61. J. Wang, F. Sansoz, J. Huang, Y. Liu, S. Sun, Z. Zhang, S.X. Mao, Near-ideal theoretical strength in gold nanowires containing angstrom scale twins. Nat. Commun. **4**, 1742 (2013)
62. E. Weinan, W. Ren, E. Vanden-Eijnden, Simplified and improved string method for computing the minimum energy paths in barrier-crossing events. J. Chem. Phys. **126**, 164103 (2007)
63. C.R. Weinberger, The fundamentals of plastic deformation: several case studies of plasticity in confined volumes. Sandia Report **7736** (2012)
64. C.R. Weinberger, W. Cai, Plasticity in metallic nanowires. J. Mater. Chem. **22**, 3277 (2012)
65. C.R. Weinberger, A.T. Jennings, K. Kang, J.R. Greer, Atomistic simulations and continuum modeling of dislocation nucleation and strength in gold nanowires. J. Mech. Phys. Solids **60**(1), 84–103 (2012)

66. Y. Xiang, H. Wei, P. Ming, W. E, : A generalized Peierls-Nabarro model for curved dislocations and cores structures of dislocation loops in Al and Cu. Acta Mater. **56**, 1447–1460 (2008)
67. G. Xu, A.S. Argon, Homogeneous nucleation of dislocation loops under stress in perfect crystals. Philos. Mag. Lett. **80**, 605 (2000)
68. G. Xu, C. Zhang, Analysis of dislocation nucleation from a crystal surface based on the Peierls-Nabarro dislocation model. J. Mech. Phys. Solids **51**, 1371–1394 (2003)
69. L. Zepeda-Ruiz, B. Sadigh, J. Biener, A. Hodge, A. Hamza, Mechanical response of freestanding au nanopillars under compression. Appl. Phys. Lett. **91**, 101907 (2007)
70. Y. Zhang, H. Huang, Do twin boundaries always strengthen metal nanowires? Nanoscale Res. Lett. **4**, 34–38 (2009)
71. H. Zheng, A. Cao, C.R. Weinberger, J.Y. Huang, K. Du, J. Wang, Y. Ma, Y. Xia, S.X. Mao, Discrete plasticity in sub-10 nm au crystals. Nat. Commun. **1**, 144 (2010)
72. T. Zhu, J. Li, A. Samanta, A. Leach, K. Gall, Temperature and strain-rate dependence of surface dislocation nucleation. Phys. Rev. Lett. **100**, 025502 (2008)
73. Y. Zhu, Q. Qin, F. Xu, F. Fan, Y. Ding, T. Zhang, B.J. Wiley, Z.L. Wang, Size effects on elasticity, yielding, and fracture of silver nanowires. Phys. Rev. B **85**, 045443 (2012)

Chapter 13
Quantized Crystal Plasticity Modeling of Nanocrystalline Metals

Lin Li and Peter M. Anderson

13.1 Introduction

Nanocrystalline (NC) metals are often defined as polycrystalline metals with an average grain size less than 100 nm. Their appealing mechanical properties make them promising structural materials. Compared to coarse-grained (CG) counterparts, NC metals exhibit: (1) ultrahigh yield and fracture strength; (2) an extended elastic–plastic transition regime; (3) large recoverable plastic deformation; (4) high strain-rate sensitivity; and (5) limited ductility due to strain localization [1–3]. These unique features pertain to the physics of deformation that emerge as grain size decreases below $\sim$100 nm [4–6]. First, the plastic flow shifts from multiple to single dislocation slip. In stark contrast to CG metals, intragranular dislocation activities—including the operation of dislocation sources (e.g., Frank-Read) as well as dislocation multiplication and intersection—are highly prohibited. Instead, dislocations in NC metals tend to nucleate at grain boundaries (GBs) and traverse entire grains without impediment until they are absorbed into GBs. GB sites therefore act as sources and sinks for dislocations. If the grain size is refined below 10–20 nm, GB accommodation mechanisms (e.g., GB sliding and rotation) become more important and begin to control plastic flow [7, 8].

This chapter presents a quantized crystal plasticity (QCP) model that connects emerging deformation physics with the unique mechanical properties of NC metal

L. Li (✉)
Department of Metallurgical and Materials Engineering, The University of Alabama, Tuscaloosa, AL 35487, USA
e-mail: lin.li@eng.ua.edu

P.M. Anderson
Department of Materials Science and Engineering, The Ohio State University, Columbus, OH 43210, USA

C.R. Weinberger, G.J. Tucker (eds.), *Multiscale Materials Modeling for Nanomechanics*, Springer Series in Materials Science 245,
DOI 10.1007/978-3-319-33480-6_13

in the grain size regime between 20 and 100 nm, where single dislocation slip controls the plastic flow [9–12]. Notably, the QCP model adopts a discrete/quantized constitutive flow rule associated with a single dislocation slip event in a nano grain. The flow rule is implemented for a NC assembly within the framework of crystal plasticity. By calibrating the model to measurements of aggregate flow strength and internal stress within subpopulations of grains, we infer that the critical resolved shear stress (CRSS) for quantized slip is heterogeneous from one grain to another. The CRSS distribution is skewed (asymmetric) so that a larger fraction of low-strength grains is balanced by a smaller fraction of high-strength grains. Further, the CRSS of a grain depends on the grain size but not grain orientation. In Sect. 13.2 to follow, the QCP model is developed. In Sect. 13.3, the model is calibrated and validated under monotonic and cyclic mechanical loadings. In Sect. 13.4, the evolution of lattice strain with deformation in NC metals is explored by capturing the deformation footprints in in-situ X-ray diffraction tests. Section 13.5 summarizes the key aspects and motivates areas of future development.

13.2 Model Development

13.2.1 Dislocation Depinning from Grain Boundaries

Atomic simulations have provided considerable insight into the deformation mechanisms in NC metals [5, 13–15]. One of the emerging deformation processes is the nucleation and propagation of dislocations from grain boundaries [5]. In this process, dislocations often become pinned at GB ledges as they propagate across the grain. Thus, the rate-limiting process is the depinning of the dislocations and the characteristic length scale is the spacing between boundary pinning points. A molecular dynamics (MD) simulation of three-dimensional NC Al with an averaged grain size $\sim$12 nm was performed at a constant strain rate 1×10^8 s^{-1} and at a constant temperature 300 K. The simulated sample consisted of 100 grains, and periodic boundary conditions were applied to all the three directions. The embedded atom model potential for Al of Mishin et al. [16] was used, it has an unstable-to-stable stacking fault energy ratio close to unity, so that the nucleation of trailing partial dislocations after the leading partial dislocation can be observed within the time frame of the MD simulation. The simulated results were analyzed in terms of grain-averaged shear produced by dislocation slip and the grain-averaged resolved shear stress during deformation [17]. Figure 13.1a–c illustrates three crucial snapshots from the MD simulations and Fig. 13.1d shows the corresponding evolution of resolved shear stress and strain within a grain, measured on a grain-averaged basis. A full dislocation loop is observed to persist in a pinned, embryonic state for 25 ps (Fig. 13.1a) until the grain-averaged stress τ reaches a critical value (Fig. 13.1b). The full dislocation then depins from the GB obstacles and sweeps across the grain (Figs. 13.1b, c). The grain-averaged shear γ^* then increments by a

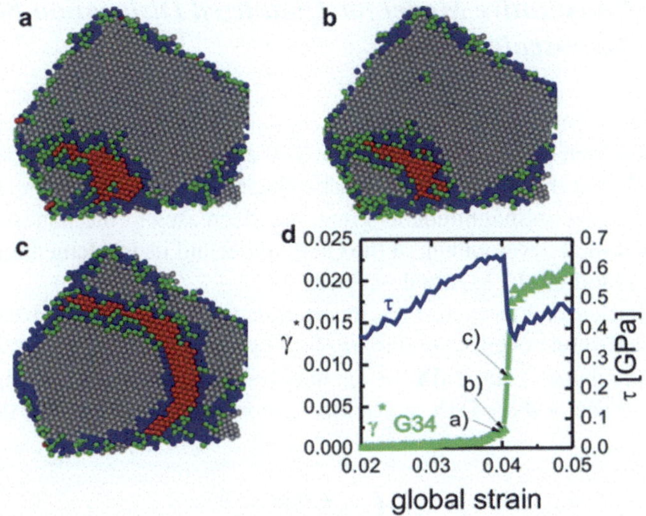

Fig. 13.1 Molecular dynamics simulations of a cross section of an FCC Al crystal. With increasing applied deformation, a perfect dislocation consisting of two partials depins from a GB ledge [17]. (**a–c**) Shows a cut through a selected grain with the atoms colored according to their local crystallographic order. Atoms with an FCC environment are colored in *grey*, those with HCP are *red*, others with coordination = 12 are *green*, and those with coordination ≠ 12 are *blue*. (**d**) Shows the corresponding evolution of grain-averaged shear γ^* and resolved shear stress τ. Reprinted figure from [17] with permission from Elsevier

discrete amount and the resolved shear stress τ simultaneously drops abruptly due to load shedding (Fig. 13.1d).

The grain-averaged analysis obtained from MD simulations highlights the quantized/discrete jumps in grain-averaged resolved shear stress and strain in nanoscale grains. Furthermore, the distribution of the depinning stress (CRSS) controls the onset of avalanches consisting of numerous quantized slip events. A statistical analysis of dislocation slip processes reveals a broad CRSS distribution that peaks between 600 and 700 MPa. Slip processes are seldom observed in a grain if the average resolved shear stresses in the grain is <500 MPa [17]. These MD observations of dislocation slip in nano grains motivate our meso-scale, QCP model. It is noteworthy that the MD simulations have the limits of grain sizes and the deformation strain rates that would be comparable to experimentally relevant values. The QCP model, therefore, adopts the quantized and stochastic single dislocation slip features, performs simulations at meso-scale level, and calibrates the characteristics of quantized slip with experimental measurements.

13.2.2 A Constitutive Model for Quantized Dislocation Slip in Nanoscale Grains

The QCP model employs a conventional crystal plasticity formulization [18, 19], the details of which can be found in Chap. 3. We augment the conventional crystal plasticity such that the increments in plastic shear strain take on discrete/quantized values and increase in magnitude as grain size decreases to the nanoscale. In this section, the detailed development of the QCP model and its implementation into the finite element method (FEM) are described.

Large-deformation kinematics are incorporated into the constitutive framework because these discrete events can impart plastic strain increments ~1 %. Following [20], an infinitesimal vector $\mathbf{dX}$ in an undeformed (reference) configuration is distorted into a vector $\mathbf{dx} = \mathbf{FdX}$ in a deformed configuration. Here, the deformation gradient

$$\mathbf{F} = \mathbf{F}^*\mathbf{F}^p \tag{13.1}$$

is multiplicatively decomposed into an elastic part $\mathbf{F}^*$ and a plastic part $\mathbf{F}^p$. The elastic strain is defined as

$$\mathbf{E}^* = \frac{1}{2}\left\{\mathbf{F}^{*^T}\mathbf{F}^* - \mathbf{I}\right\} \tag{13.2}$$

The stress at a material point is given by

$$\mathbf{T}^* = \mathbf{CE}^* \tag{13.3}$$

Here, $\mathbf{C}$ is the fourth-order elasticity tensor for a grain. For FCC materials, there are only three independent constants and they are traditionally denoted by C_{11}, C_{12}, and C_{44} within the cubic crystal basis. $\mathbf{T}^*$ is the symmetric Piola-Kirchhoff stress, which is the work conjugate to $\mathbf{E}^*$. It is related to the Cauchy stress $\mathbf{T}$ by

$$\mathbf{T}^* = \mathbf{det}\left(\mathbf{F}^*\right)\mathbf{F}^{*^{-1}}\mathbf{T}\mathbf{F}^{*^{-T}} \tag{13.4}$$

The time evolution equation for $\mathbf{F}^p$ is given by flow rule

$$\dot{\mathbf{F}}^p = \mathbf{L}^p\mathbf{F}^p \tag{13.5}$$

with

$$\mathbf{L}^p = \sum_\alpha \dot{\gamma}^\alpha \mathbf{S}_0^\alpha, \quad \mathbf{S}_0^\alpha = \mathbf{s}_0^\alpha \otimes \mathbf{m}_0^\alpha \tag{13.6}$$

The orthonormal vector pairs $(\mathbf{s}_0^\alpha, \mathbf{m}_0^\alpha)$ define the respective slip direction and slip plane normal for slip system α. For FCC materials, $\mathbf{S}_0^\alpha$ is chosen to be 1 of the 12 $\langle 110\rangle/\{111\}$ slip systems. The shear rate $\dot{\gamma}^\alpha$ is determined by a constitutive relation of the form $\dot{\gamma}^\alpha = \widehat{\dot{\gamma}}^\alpha\left(\tau^\alpha, \tau_c^\alpha\right)$, where the resolved shear stress τ^α is determined by

$$\tau^\alpha = \mathbf{s}_0^\alpha \cdot \left(\mathbf{F}^{*^{\mathrm{T}}}\mathbf{F}^*\mathbf{T}^*\right)\mathbf{m}_0^\alpha \tag{13.7}$$

and τ_c^α is the CRSS for a slip event. An initial assumption is that τ_c^α at a material point is the same for all 12 slip systems and remains constant with deformation (denoted by τ_c for brevity).

The discrete jumps in shear strain as a result of dislocation depinning from GBs observed in the MD simulations (ref. to Fig. 13.2a) can be rationalized in terms of the grain-averaged increment in plastic strain. In particular, the discrete/quantized shear increment γ_{target} associated with a single dislocation gliding on slip system α across a grain with grain size d can be expressed as

$$\Delta\gamma^\alpha = \gamma_{\text{target}} = \frac{A_s b}{V_g} = g\frac{b}{d} \tag{13.8}$$

where b is the Burger vector magnitude of the dislocation, A_s is the cross sectional area of the slip plane, V_g is volume of the grain, and g is a geometric factor. In principle, γ_{target} depends on the specific glide plane and the grain shape. These factors are incorporated into g. For example, $g = 1.5$ for a center-cut glide plane within a spherical grain and $g = 1.2$ for the maximum cross sectional area of a cubic grain with edge length d. Furthermore, when multiple slip events occur on slip system α, the accumulated shear strain takes on a quantized value of γ_{target}, given as

$$\gamma^\alpha = q\gamma_{\text{target}} \tag{13.9}$$

The coefficient $q = 1$, 2, etc., is an integer that characterizes the quantized plastic state.

The discrete or quantized nature of slip in Eq. (13.8) is implemented via a modification to conventional kinetic flow law proposed by Peirce et al. [21]. In particular, the rate of plastic shear strain on the 12 FCC $\langle 110\rangle/\{111\}$ slip systems is specified by

$$\dot{\gamma}^\alpha = \begin{cases} \dot{\gamma}_0\left|\dfrac{\tau^\alpha}{R\tau_c}\right|^{1/m}\operatorname{sign}(\tau^\alpha) & \text{inactive slip} \\ \dot{\gamma}_0\operatorname{sign}(\tau^\alpha) & \text{active slip} \end{cases} \tag{13.10}$$

where $\dot{\gamma}_0$ is the local reference shear rate of slip. The *inactive slip condition* on slip system α applies when $|\tau^\alpha| < \tau_c$. Equation (13.10) differs from a standard flow law formalism in that a coefficient R has been inserted to reinforce a numerically *inactive* condition and also τ_c remains constant with deformation. The coefficient R is determined such that if an inactive slip condition applies, then $\dot{\gamma}^{\alpha} < 10^{-20}\dot{\gamma}_0$ so that grains deform primarily by anisotropic elasticity. For instance, if the power-law exponent $m = 0.1$ is chosen, then the numerically inactive condition is satisfied by selecting $R \sim 100$, which approximates the value of τ_c expressed in MPa, i.e., $R = \tau_c$/MPa. The use of the modified power function with R and m can avoid a numerical issue when $\dot{\gamma}^{\alpha}$ transitions upon activation of a quantized slip event. The *active slip condition* for slip system α applies if the condition $|\tau^{\alpha}| \geq \tau_c$ is met. It prevails until the grain-averaged resolved shear strain γ^{α} achieves the quantized amount, $\gamma_{\text{target}} \cdot \text{sign}(\tau^{\alpha})$, even if load shedding from the deforming grain causes $|\tau^{\alpha}| < \tau_c$ to occur before γ^{α} reaches the target value. A propagation condition is imposed to ensure that τ^{α} on the active slip system does not change sign during a slip event. During the active condition, $\left|\dot{\gamma}^{\alpha}\right| = \dot{\gamma}_0$, where $\dot{\gamma}_0 > 10\dot{\varepsilon}_{\text{global}}$ is required to ensure that the increment in macroscopic strain on the polycrystal is negligible during the course of a local slip event.

13.2.3 Statistical Distribution of Critical Stress of Quantized Slips

In addition to quantized plasticity, the QCP model adopts a grain-to-grain variation in the CRSS τ_c at which a quantized plastic event is triggered. Such a variation is suggested by the MD simulation results [17], in which a very broad distribution of grain-level critical stress is observed in the simulated nanocrystals. In principle, τ_c depends on numerous factors including the detailed structure of the bounding GBs, the orientation of slip planes to the GBs, elastic anisotropy, and the accommodation of incoming or outgoing dislocation content within GBs [5, 14, 17, 22]. In the absence of a deterministic relationship between τ_c and GB structure, two types of probability distribution functions for τ_c are explored. **Type S** is a normal distribution with probability density

$$\rho\left(\tau_c; \mu, \sigma\right) = \frac{1}{\sigma\sqrt{2\pi}} \exp\left(-\frac{\left(\tau_c - \mu\right)^2}{2\sigma^2}\right) \tag{13.11}$$

where μ is the mean value and σ^2 is the variance. **Type A** is a gamma distribution with probability density

$$\rho\left(\tau_c; k, \theta, \tau_{\min}\right) = \left(\tau_c - \tau_{\min}\right)^{k-1} \frac{\exp\left(-\left(\tau_c - \tau_{\min}\right)/\theta\right)}{\Gamma(k)\theta^k} \tag{13.12}$$

where k is a shape parameter, θ is a scale parameter, and $\tau_{\min}$ is a reference value below which $\rho(\tau_c) = 0$. The corresponding mean and variance are $\tau_{\min} + k\theta$ and $k\theta^2$, respectively. These two types differ in skewness. The Type S normal distribution is symmetric about the mean value with skewness = 0 whereas the Type A gamma distribution is asymmetric and positively skewed (i.e., the tail at larger τ_c is longer than at smaller τ_c). The skewness of the gamma distribution is $2/k$ and therefore the gamma distribution approaches a normal distribution as k increases.

The stochastic nature of τ_c and its effect on the mechanical properties (e.g., strength) of NC metals have been reported experimentally [23–26]. However, the correlation with GB structure or GB network structure and the evolution with deformation are open issues that require future investigation. In Sect. 13.4, the relation between τ_c and grain orientation is investigated by calibrating $\rho(\tau_c)$ in the QCP model to capture in-situ X-ray diffraction measurements in NC Ni.

13.2.4 Propagation Condition

A *propagation condition* for a slip event imposes a lower bound on τ_c or equivalently an upper bound on γ_{target}. It is based on the physical requirement that the resolved shear stress in a grain does not change sign during a slip event. The purpose is to ensure sufficient stress to complete the expansion of the loop. An Eshelby-inclusion analysis [27, 28] indicates that a transformation strain γ_{target} in an ellipsoidal region will induce a stress drop

$$\Delta\tau = c\Delta\tau^{\text{Eshelby}} = -cM\gamma_{\text{target}}; \quad M = \mu \cdot \frac{7-5\nu}{15\cdot(1-\nu)} \tag{13.13}$$

where μ and ν are the elastic shear modulus and Poisson's ratio of a homogeneous, isotropic-elastic medium, respectively. A parameter c is introduced to calibrate the deviation of non-ellipsoidal region in an elastic anisotropic medium to the Eshelby solution.

The propagation condition that the resolved shear stress of an active slip event does not change sign can be applied to Eq. (13.13), given

$$\tau_c \geq cM\gamma_{\text{target}} \text{ or equivalently, } \gamma_{\text{target}} \leq \tau_c/cM \tag{13.14}$$

Upon implementation, the geometric prescription for γ_{target} (Eq. (13.8)) is generally applied for a given grain and slip plane geometry. Experimentally measured stress–strain responses are used to parameterize τ_c distribution. Sometimes for small grain sizes (e.g., $d < 50$ nm) in particular, the fitted τ_c distribution has a small population (<20 %) that does not satisfy the propagation condition given in Eq. (13.14). For such small populations, the value of γ_{target} is adjusted to satisfy Eq. (13.14) using the prescribed value of τ_c. In practice, γ_{target} for a given grain is specified by the minimum,

$$\gamma_{\text{target}} = \min\left(\frac{\tau_c}{cM}, g\frac{b}{d}\right) \tag{13.15}$$

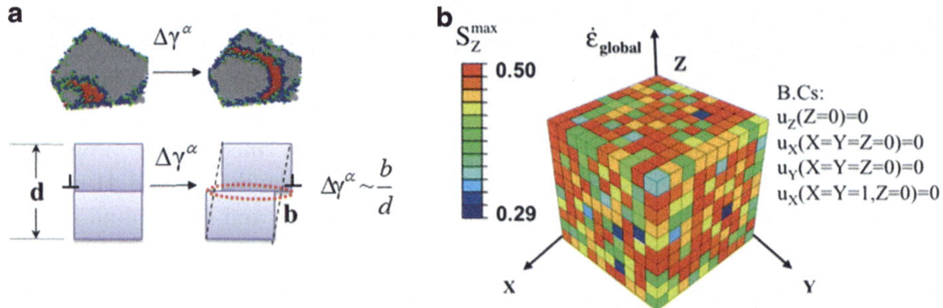

Fig. 13.2 (**a**) A schematic representation of coarse-graining dislocation depinning in the QCP model. (**b**) A finite element model of a polycrystal with 10 × 10 × 10 grains, each represented by an 8-node brick element. Uniaxial tension is applied along the *z*-direction. Each grain is colored according to S_z^{max} the maximum Schmid factor among the 12 FCC slip systems in that grain. Numerical values are as shown in color bar

where τ_c is chosen from a statistical distribution following Eqs. (13.11) or (13.12).

13.2.5 QCP/FE Simulations

The QCP model is implemented in the commercial finite element (FE) packet ABAQUS via a user defined material subroutine (UMAT) [29]. A three-dimensional (3D) polycrystal is defined with $N \times N \times N$ grains, each represented by an 8-node, 3D cubic element (type = C3D8). A crystallographic orientation, CRSS τ_c and quantized plastic strain jump γ_{target} are assigned to each grain. Figure 13.2b shows a FE sample for $N = 10$ in which each grain (element) is colored according to the maximum Schmid factor with respect to the loading axis among the 12 ½⟨110⟩/{111} face-centered cubic (FCC) slip systems. A uniaxial strain rate $\dot{\varepsilon}_{\text{global}}$ is imposed via a constant displacement rate $\dot{u}_Z$ applied normal to the $+Z$ (top) surface while $u_Z = 0$ is imposed on the $-Z$ (bottom) surface. The tractions $t_X = t_Y = 0$ are imposed on these surfaces and the remaining $\pm X$ and $\pm Y$ surfaces are traction-free, as shown in Fig. 13.2b.

13.3 Model Calibration

13.3.1 Quantized Jumps in Shear Strain and Stress

The QCP simulation results capture the quantized jumps in shear strain and stress at grain level. Figure 13.3 shows the resolved shear strain γ^α and resolved shear stress τ^α vs. global plastic strain $\varepsilon^p{}_{\text{global}}$, for a slip system with a moderate Schmid factor $S_Z{}^{\alpha} = 0.465$ at a material point in an interior grain. The grain takes different quantized states of γ^{α}, and jumps from one state to another when the resolved shear stress reaches a critical value. The black stepped (lower) trace shows that γ^{α} jumps successively by γ_{target} (=4×10^{-3}). The red (upper) trace confirms that γ^{α} jumps when $\tau^{\alpha} = \tau_c$ (=500 MPa). During a jump by γ_{target}, τ^{α} monotonically decreases, reaching a minimum at the completion of the jump. Subsequently, τ^{α} monotonically increases via elastic deformation as the applied macroscopic strain increases. Occasionally, τ^{α} drops abruptly (arrow A) or deviates from linearity (arrow A′), even though slip system α is not active. This is a consequence of slip events in adjoining grains or sometimes in the same grain. Quantized slip is therefore predicted to generate violent redistributions of stress at the grain scale. This captures the key features of dislocation slip and depinning processes as observed in MD simulations.

13.3.2 Grain Size Effect on Strength

Plastic strengthening with decreasing grain size emerges as a consequence of quantized slip and the associated propagation condition. Figure 13.4a displays the evolution of resolved shear strain γ^{α} for discrete values of γ_{target}. The corresponding traces of τ^{α} vs. $\varepsilon^{p}{}_{\text{global}}$ for $\gamma_{\text{target}} = 1 \times 10^{-4}$ and 6×10^{-3} are shown in Fig. 13.4b, c, respectively. The stress redistribution in τ^{α} is more violent for the larger γ_{target} case. In particular, the stress drop $\Delta\tau$ due to a quantized slip event increases as γ_{target} increases. Figure 13.4d shows that $\Delta\tau$ increases linearly with γ_{target} with a slope of 65 GPa. This is consistent with the Eshelby solution (Eq. (13.13)) with $c = 1.6$, $\mu = 76$ GPa, and $\upsilon = 0.31$ for Ni. Since γ_{target} is linearly dependent on 1/grain size (Eq. (13.15)), $\Delta\tau$ therefore scales as 1/grain size. In the limit of large grain size, both γ_{target} and $\Delta\tau$ diminish and the QCP model reverts to a conventional, elastic-perfectly plastic crystal plasticity model. In the limit of small grain size, the *propagation condition* requires that $\tau_c \geq \Delta\tau$. Combining Eqs. (13.8) and (13.13),

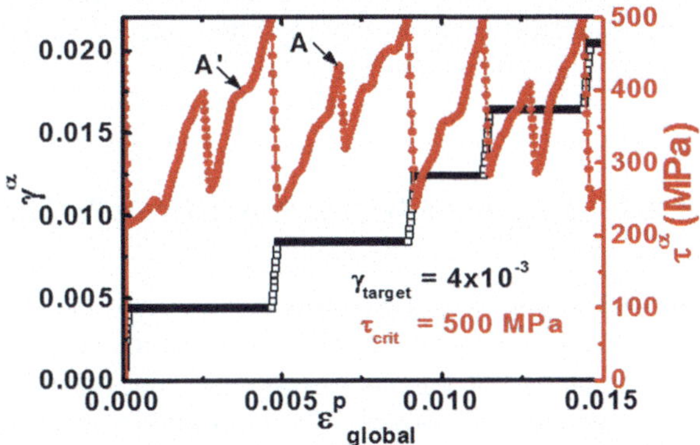

Fig. 13.3 The "quantized" evolution of local shear strain and the associated violent stress redistribution in shear stress at a material point of an interior grain with $\gamma_{\text{target}} = 4 \times 10^{-3}$ and $\tau_c = 500$ MPa. The elastic constants of Ni are used. Figure reproduced from [9]

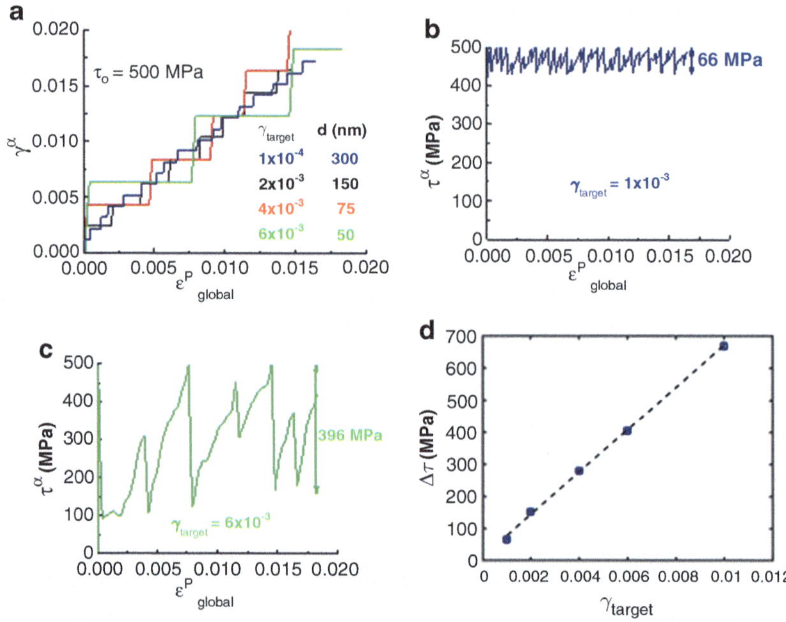

Fig. 13.4 (**a**) Evolution of plastic shear strain γ^{α} at a material point in an interior grain with various values of γ_{target} and $\tau_c = 500$ MPa. The corresponding evolution of resolved shear stress τ^{α} for $\gamma_{\text{target}} = 1 \times 10^{-4}$ and 6×10^{-3} are shown in (**b, c**), respectively. (**d**) The stress drop after a quantized slip $\Delta\tau$ vs. γ_{target}. The resulting linear relationship follows the prediction of Eq. (13.13), with $c = 1.6$, $\mu = 76$ GPa, and $\upsilon = 0.31$ for Ni. Figure (**a–c**) reproduced from [9]

we obtain

$$\tau_{c(\min)} = cM\gamma_{\text{target}} = c'\mu \cdot \frac{7 - 5\nu}{15 \cdot (1 - \nu)} \frac{b}{d} \tag{13.16}$$

where $c = 1.6$, $M = 40$ GPa, and $c' = 1.92$ for Ni. This is the origin of source strengthening with decreasing grain size in the QCP model. The resultant $1/d$ relationship differs from the $1/d^{1/2}$ dependence in the Hall–Petch relationship for conventional polycrystals [30, 31].

13.3.3 Extended Micro-Plasticity in NC Metals

The QCP model requires a valid distribution $\rho(\tau_c)$, and this can be achieved by fitting the experimental tensile stress–strain response for various average grain sizes. A particular goal is to reproduce the extended experimental micro-plastic regime, which is hypothesized to represent the percolation of quantized slip events throughout the polycrystal, from soft to hard grains. The extended regime is quantified by $\varepsilon_{p(\text{trans})}$, the elastic–plastic transition strain over which the flow stress increases from an initial value σ_0 to approximately 95 % of the plateau stress σ_{plateau}. $\varepsilon_{p(\text{trans})} \sim 0.2\,\%$ for polycrystals with a conventional average grain size but it can approach 1 % for NC metals and exhibit apparent strain hardening under tension in this regime [32–34].

Figure 13.5a displays the fitted QCP simulation results for grain size $d = 50$, 150, 300 nm along with the experimental data for electro-deposited Ni from [35]. Two candidate distributions are considered for $\rho(\tau_c)$: a normal distribution (Fig. 13.5b, Eq. (13.11)) and a gamma distribution with the shape factor $k = 1$ (Fig. 13.5c, Eq. (13.12)) and $k = 2$ (Eq. (13.12)). The quantities to be calibrated (e.g., $\tau_{c(\text{min})}$, $\tau_{c(\text{mean})}$) are estimated from the experimental values of initial yield stress σ_0 and saturated flow stress σ_{plateau}. Iterative QCP simulations are performed to determine the values of $\tau_{c(\text{min})}$ and $\tau_{c(\text{mean})}$ that produce the best match to the data in Fig. 13.5a. The gamma distribution with $k = 1$ captures the gradual elastic-to-plastic transition, with $\varepsilon_{p(\text{trans})} = 0.7\,\%$ for $d = 50$ nm. These distributions are asymmetric and are denoted by "Type A" in Fig. 13.5a. In contrast, the normal distributions are symmetric and denoted by "Type S." They predict too small a transition, with $\varepsilon_{p(\text{trans})} = 0.08\,\%$ for $d = 50$ nm. The fitting process is detailed in [9] and the material and computational parameters are summarized in Table 13.1.

Figure 13.5c–e shows the gamma, $k = 1$ (Type A) $\rho(\tau_c)$ distributions that best fit the experimental data for various grain sizes. Both the mean value and standard deviation in τ_c increase with decreasing grain size. This reflects that yield in NC metals is very heterogeneous on a grain-to-grain scale. Moreover, the τ_c distribution becomes more asymmetric/skewed with decreasing grain size. In the limit of larger grain size, on the other hand, a gamma distribution with either $k = 1$ and $k = 2$ captures the experimental stress–strain data for $d = 300$ nm.

The QCP simulations predict that deformation is more localized as grain size decreases. Figure 13.6 shows the fraction (f_{slipped}) of grains with at least one slip event as a function of applied macro strain, as predicted by the calibrated QCP simulations. At smaller grain size ($d = 50$ nm), f_{slipped} increases gradually, indicating that plastic flow percolates gradually through the polycrystal. This produces the extended micro-plastic regime in Fig. 13.5a. At larger grain size ($d = 300$ nm), f_{slipped} is predicted to increase more rapidly with applied strain, consistent with more conventional crystal plasticity models. Compared to metals with a conventional grain size, NC metals accommodate global strain with larger local plastic strain events that are distributed among fewer grains; in other words, plastic deformation in NC metals tends to localize in fewer grains. For example, f_{slipped} at 0.2 % strain is ~40 % for $d = 50$ nm but ~80 % for $d = 300$ nm. Thus, a conventional 0.2 % macro plastic strain can capture fully plastic flow at larger grain size ($d = 300$ nm) but not at smaller grain size ($d = 50$ nm) [32–34].

Table 13.1 Summary of material and computational parameters employed by QCP simulations

Ni elastic moduli [55] (GPa)	C_{11}: 246.5	C_{12}: 147.3	C_{44}: 124.7
Crystal texture	Random with average maximum Schmid factor = 0.45		
Ave. grain size d (nm)	50	150	300
Crit. stress distribution (MPa): $\rho\,(\tau_c; k, \theta, \tau_{min}) = (\tau_c - \tau_{min})^{k-1} \frac{\exp(-(\tau_c-\tau_{min})/\theta)}{\Gamma(k)\theta^k}$	$k = 1$ $\theta = 420$ $\tau_{c(min)} = 280$	$k = 1$ $\theta = 285$ $\tau_{c(min)} = 160$	$k = 1$ $\theta = 110$ $\tau_{c(min)} = 80$
Quantized plastic strain: $\gamma_{target} = \min(\tau_c/60, 1.2\ b/d)$	$\gamma_{target(min)} = 4.7 \times 10^{-3}$ $\gamma_{target(max)} = 6.0 \times 10^{-3}$	2.0×10^{-3}	1.0×10^{-3}
Computational parameters	$\dot{\gamma}_0\ (s^{-1})$: 2×10^{-2}	m: 0.1	Δt: $\gamma_{target,\ max}/\dot{\gamma}_0$

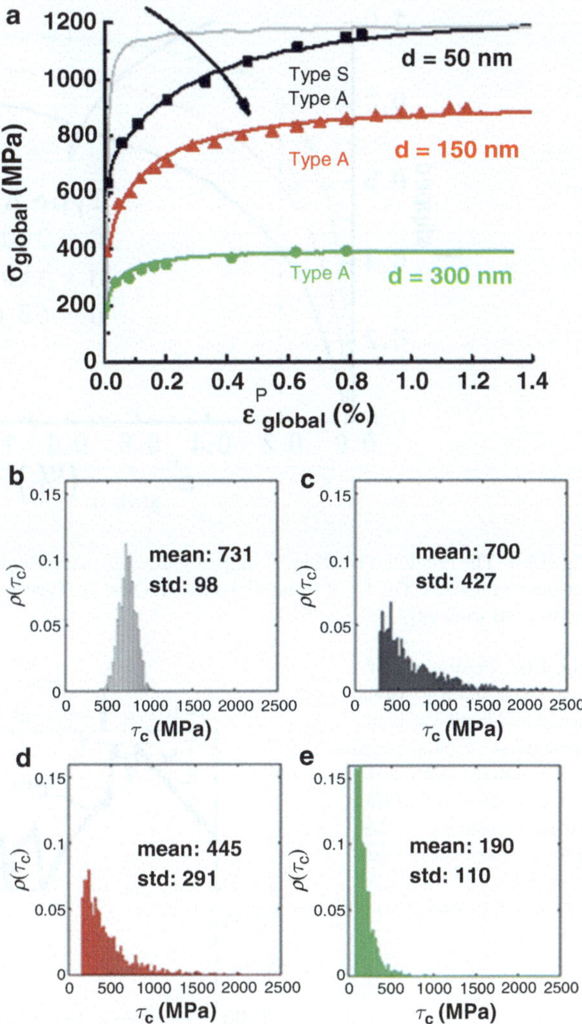

Fig. 13.5 (**a**) The simulated stress–strain responses for various grain sizes, along with the experimental data adopted from [35]. (**b**) A Type S (symmetric normal) distribution for $d = 50$ nm. (**c–e**) The best-fit Type A (asymmetric gamma, with $k = 1$) distributions for $d = 50$ nm, 150 nm and 300 nm, respectively. Figure reproduced from [9]

13.3.4 Enhanced Plastic Recovery

An asymmetric $\rho(\tau_c)$ distribution for NC metals is also supported by cyclic stress–strain data, in which an enhanced plastic recovery is usually observed [36–39]. The QCP model predicts reversible plastic strain to occur when NC metals are unloaded [10]. This arises when backward (or reverse) quantized slips occur in grains in which the resolved shear stress reverses sign and reaches a critical value $\tau_{c,b}$ for backward slip. Figure 13.7 shows the evolution of γ^α and τ^α on slip system α at a material point within a grain during macroscopic cyclic loading of a polycrystal. Here, $\gamma_{\text{target}} = 6 \times 10^{-3}$, $\tau_c = 428.4$ MPa, and $\tau_{c,b} = 228.4$ MPa. The

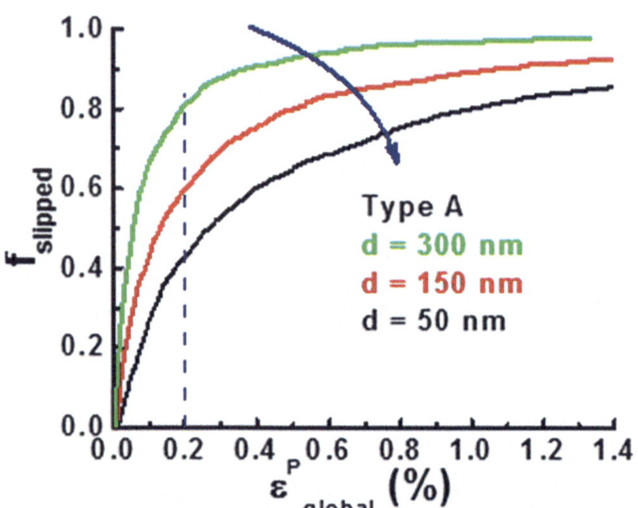

Fig. 13.6 The predicted fraction of slipped grains vs. macroscopic strain during a tension test, for various grain sizes. The QCP simulations use the best-fit Type A distributions in Fig. 13.5. Figure reproduced from [9]

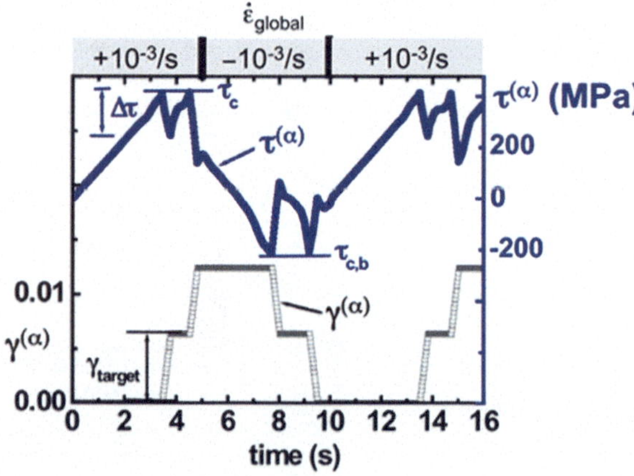

Fig. 13.7 Predictions of local shear strain γ^{α} (lower curve) and local resolved shear stress τ^{α} on a specific slip system α in an interior grain, as a function of time during a tension test. The applied global strain rate reverses sign at 5 and 10 s. Figure reproduced from [10]

smaller magnitude of $\tau_{c,b}$ signifies the case in which the pinning strength to reverse a quantized slip event is smaller than that for the initial forward event. For instance, deposition of the dislocation loop into a grain boundary could be energetically unfavorable, so that there is a driving force to reverse the process. During the first loading period, τ_c is reached twice so that the quantized slip number q increases from 0 to 1 and then 1 to 2. During the first unloading period, $-\tau_{c,b}$ is reached twice so that q decreases from 2 to 1 and then 1 to 0. During the second loading period, two forward events also occur and they produce $q = 2$.

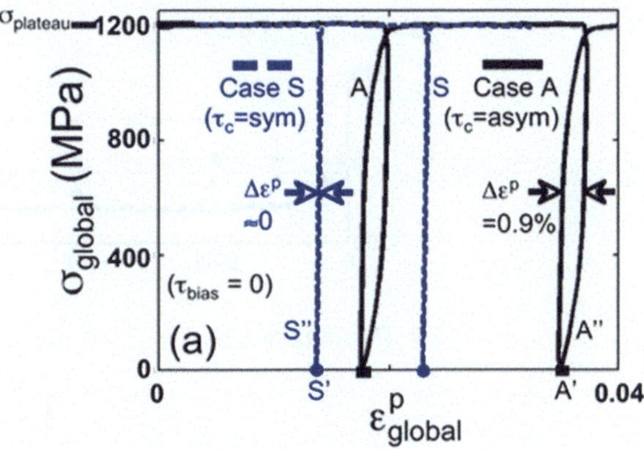

Fig. 13.8 (**a**) Predicted cyclic stress-plastic strain response of NC metals with $d = 50$ nm at large strain, using the Type A (asymmetric gamma, $k = 1$) and Type S (symmetric normal) distributions $\rho(\tau_c)$ shown in Fig. 13.5. Figure reproduced from [10]

Figure 13.8 shows that the calibrated asymmetric (Type A) distribution for $d = 50$ nm can predict comparable values of hysteretic strain to experiments, where $\varepsilon_{p(\mathrm{hyst})} = 0.9\,\%$ results when the macro stress is cycled from 0 to $\sigma = 1200$ MPa. A proviso is that the cyclic stress is applied after straining the NC metal into the fully plastic regime (>2.5 %). In contrast, the calibrated symmetric (Type S) distribution predicts negligible width ($\varepsilon_{p(\mathrm{hyst})} \sim 0$). A physical explanation is that the asymmetric distribution produces a relatively large fraction of weaker (smaller τ_c) grains and a relatively small fraction of stronger grains. The weaker grains produce a large number of forward events during loading and redistribute internal stress to the stronger grains. During unloading, the stronger grains unload elastically and attempt to restore the polycrystal to the original shape. This drives reverse slip events in the weaker grains.

13.4 Model Application to Lattice Strain Evolution in NC Ni

In this section, we apply the QCP model to investigate the evolution of lattice strains measured by in-situ X-ray diffraction and understand its dependence on grain size. In particular, in-situ X-ray diffraction experiments show that the evolution of residual lattice strain during interrupted uniaxial tension is different for NC vs. coarse-grained (CG) metals [11, 40–43]. The QCP model is applied to identify and understand the conditions to reproduce the experimental diffraction data, including the shift in the position and width of various $\langle hkl \rangle$ diffraction peaks after tensile elongation to various macroscopic strains. The QCP simulations are able to link these X-ray diffraction "footprints" to the violent redistribution of stress at the intergranular scale, the dependence on grain size, the critical stress distribution $\rho(\tau_c)$, and the prior deformation history that precedes the tensile test. The capacity to capture the lattice strain evolution of NC metals provides a robust test of the QCP model.

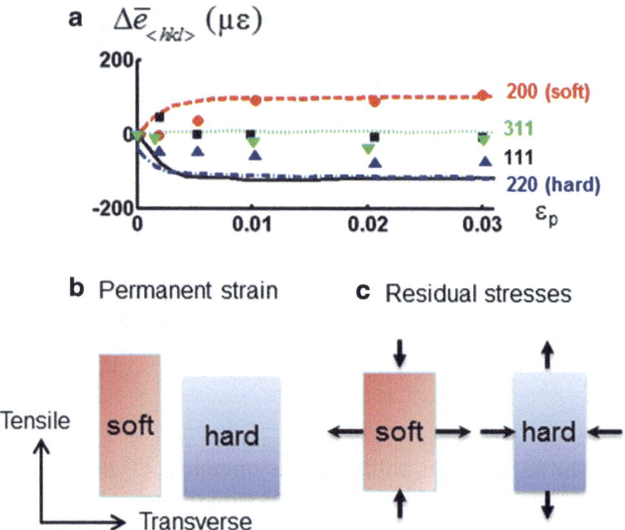

Fig. 13.9 (**a**) The change in residual lattice strain in CG Cu along the transverse direction, as a function of plastic strain (adapted from [44]). These trends can be rationalized from (**b**) the accumulation of a larger axial plastic strain in soft grains under axial tension and (**c**) the development of axial compression and transverse tension in soft grains after unloading. Figure reproduced from [11]

13.4.1 Lattice Strain Evolution in CG vs. NC Ni

For many coarse-grained, elastically anisotropic FCC metals, tensile elongation will shift diffraction peaks due to the deformation anisotropy. In particular, the residual lattice strain along ⟨200⟩ transverse directions increases while that along ⟨220⟩ transverse directions decreases as the imposed tensile elongation is increased. Figure 13.9 confirms these trends for CG Cu [44]. An explanation is that the "family" of FCC grains with a ⟨200⟩ transverse direction has a smaller yield strength in uniaxial tension, on average, while the corresponding ⟨220⟩ family has a larger yield strength on average. The transverse ⟨200⟩ family of grains is therefore soft and accumulates a larger-than-average axial plastic strain during tensile loading (Fig. 13.9b). Upon unloading, the ⟨200⟩ family is overextended plastically and thus tends to have a compressive axial stress and tensile transverse stress. Therefore, the transverse strain for the ⟨200⟩ family tends to increase. The reverse is true for the hard ⟨220⟩ family (Fig. 13.9c).

NC FCC metals do not follow this CG trend. Instead, the measured ⟨*hkl*⟩ families have been reported to exhibit all tensile shifts [42], all compressive shifts [41], or remain unchanged [40]. The varied results were interpreted in terms of a crossover from intragranular slip for larger grains to GB-mediated mechanisms for NC metals.

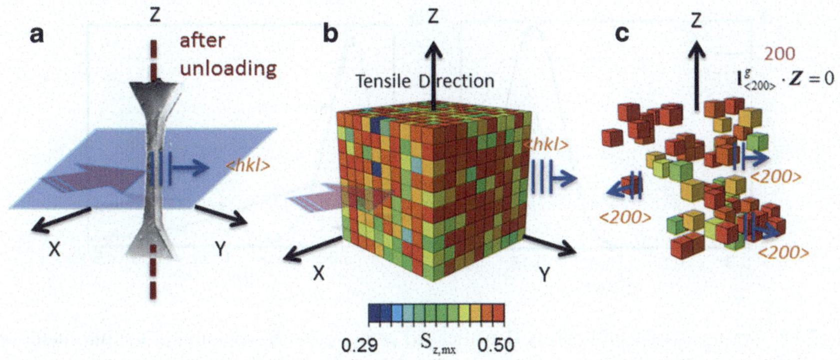

Fig. 13.10 (**a**) A schematic of the in situ X-ray tensile setup. (**b**) A three-dimensional finite element model of a polycrystal with 10 × 10 × 10 grains, each represented by an 8-node brick element. Uniaxial loading/unloading is applied along the *Z*-direction. Each grain is colored according to the maximum Schmid factor ($S_{z,mx}$) among the 12 FCC slip systems with respect to the loading direction. Numerical values are shown in the color legend. (**c**) An example of a ⟨200⟩ diffraction group, for which the diffraction vector satisfies $\mathbf{l}^{\mathbf{g}}_{\langle 200 \rangle} \cdot \mathbf{Z} = 0$. Figure reproduced from [11]

An alternative view is supported by the application of the QCP model to capture the lattice strain evolution for NC and ultrafine grained (UFG) Ni.

13.4.2 QCP/FE Simulations of In Situ X-ray Diffraction

A combined experimental and modeling effort is made to unveil the deformation mechanisms that lead to the unique NC lattice strain evolution. The experimental measurements of NC lattice strain were obtained by in situ X-ray diffraction performed at the Swiss Light Source (Paul Scherrer Institute (PSI), Switzerland) [45]. Figure 13.10a illustrates the experimental diffraction geometry. Three millimeter dogbone-shaped specimens with a cross section of 200 × 200 μm^2 were loaded in tension along the axial (*Z*) direction at various strain rates $1 \times 10^{-4} \sim 1 \times 10^{-3}$ s^{-1} [43]. During testing, an incident X-ray beam along the X direction diffracted and the corresponding diffracted intensities were measured in various directions in the *X*–*Y* (transverse) plane, giving the diffraction vector and corresponding lattice strain along transverse ⟨*hkl*⟩ directions. Residual lattice strains were measured after unloading at several points along a stress–strain curve (Fig. 13.11a). Average values were derived from the change in peak position before straining. Electro-deposited NC Ni with a number-averaged grain size $d \sim 30$ nm and a minor out-of-plane 200 texture was investigated.

The QCP model is used to mimic the experimental lattice strains by calculating the elastic strains of subgroup grains that satisfy the diffraction condition. Figure 13.10b shows a QCP/FE model setup with a 10 × 10 × 10 array of 3D cubic finite

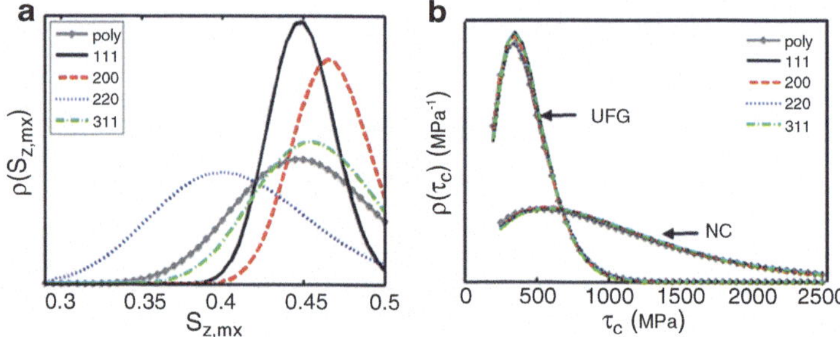

Fig. 13.11 Grain-to-grain probability densities (**a**) $\rho(S_{z,mx})$ for the maximum Schmid factor and (**b**) $\rho(\tau_c)$ for critical resolved shear stress to activate a slip event. Both are shown for the overall polycrystal and for various $\langle hkl \rangle$ families. $\rho(\tau_c)$ is shown for the ultrafine grain (UFG) and nanocrystalline (NC) cases (see Table 13.2 for details). Figure reproduced from [11]

elements (type = C3D8), where each element represents one grain with a particular crystallographic orientation, CRSS τ_c, and a quantized plastic strain jump γ_{target} that can occur on any of the 12 $\langle 110 \rangle / \{111\}$ FCC slip systems. Uniaxial loading and unloading is imposed along the Z direction at a nominal strain rate of 10^{-3}s^{-1}. A *transverse* $\langle hkl \rangle$ diffraction family consists of grains with plane normal $\mathbf{l}_{\langle hkl \rangle}$ perpendicular to the loading axis, similar to the experimental diffraction geometry. In practice, a 2° tolerance in $\mathbf{l}_{\langle hkl \rangle}$ is allowed to reflect the experimental conditions. Figure 13.10c displays a transverse $\langle 200 \rangle$ diffraction family of grains that have a $\langle 200 \rangle$ plane normal satisfying the diffraction condition (i.e., $\mathbf{l}^{\mathbf{g}}_{<200>} \cdot \mathbf{Z} = 0.$) in the QCP/FE simulations. In the simulations, $\mathbf{l}_{\langle 200 \rangle}$ can lie in any direction in the X–Y plane by assuming that the virtual incident beam can come from any in-plane direction. This increases the grain statistics in the relatively small (1000 grain) simulations.

Two grain sizes are investigated: $d = 30$ nm (NC Ni) and $d = 100$ nm (UFG Ni). Grain orientations are randomly assigned among the 1000 grains. Figure 13.11a shows the resulting grain-to-grain probability distributions $\rho(S_{z,mx})$ for the maximum Schmid factor among the 12 slip systems, assuming a polycrystal with random texture. The average value for all grains is $\overline{S}_{z,mx,\text{poly}} = 0.45$, but a larger value $\overline{S}_{z,mx,\langle 200 \rangle} = 0.47$ is obtained for transverse $\langle 200 \rangle$ grains and a smaller value $\overline{S}_{z,mx,\langle 220 \rangle} = 0.41$ is obtained for *transverse* $\langle 220 \rangle$ grains. The remaining $\langle 311 \rangle$ and $\langle 111 \rangle$ families have intermediate values with $\overline{S}_{z,mx,\langle 311 \rangle} \approx \overline{S}_{z,mx,\text{poly}}$ and $\overline{S}_{z,mx,\langle 111 \rangle} \approx \overline{S}_{z,mx,\text{poly}}$.

Figure 13.11b compares two candidate distributions $\rho(\tau_c)$ for the critical shear stress: a *relatively narrow* one that will be identified with UFG material and a *relatively wide, asymmetric* one to be identified with NC material. As noted, the asymmetric $\rho(\tau_c)$ distribution for NC material signifies a large fraction of soft grains that are balanced by a small fraction of hard grains. The $\rho(\tau_c)$ distributions are applied randomly, independent of grain orientation. This is consistent with the concept that the pinning strength is independent of grain orientation. The simulations may impose a plastic prestrain $\varepsilon_{p(\text{pre})}$ prior to the uniaxial tensile test. The prestrain imposes a pre-existing residual lattice strain, which is known to be high in NC metals. These simulation parameters are summarized in Table 13.2.

Key output quantities are the macroscopic uniaxial stress-plastic strain response $(\sigma - \varepsilon_p)$ for loading, unloading, and reloading, as well as the lattice strain $e_{\langle hkl\rangle,i}$ for each grain i in an $\langle hkl\rangle$ family. The latter is obtained by first computing the components $e_{ij(c)}$ of the elastic strain in terms of the Cauchy stress components $\sigma_{ij(c)}$ and the anisotropic elastic moduli $C_{ijkl(c)}$

$$e_{ij(c)} = C^{-1}_{ijkl}\sigma_{kl(c)},\ \sigma_{kl(c)} = Q_{km}\sigma_{mn(g)}Q_{ln} \tag{13.17}$$

The first equation expresses all quantities in the crystal basis. The second equation determines $\sigma_{ij(c)}$ in terms of the stress components $\sigma_{kl(g)}$ in the fixed global basis and the transformation matrix Q_{km} between the crystal and global bases. Values of $C_{ijkl(c)}$ for Ni are provided in Table 13.2. The lattice strain along the unit normal $\mathbf{l}_{\langle hkl\rangle(c)}$ to $\{hkl\}$ planes is then

$$e_{\langle hkl\rangle} = \mathbf{l}_{\langle hkl\rangle(c)} \cdot \mathbf{e}_{(c)} \cdot \mathbf{l}_{\langle hkl\rangle(c)} \tag{13.18}$$

Finally, the average and deviation in transverse lattice strain among the number $n_{\langle hkl\rangle}$ of grains with a transverse $\langle hkl\rangle$ direction are given by

$$\begin{aligned} \bar{e}_{\langle hkl\rangle} &= \frac{1}{n_{hkl}}\sum_{i=1}^{n_{hkl}} e_{\langle hkl\rangle i} \\ s_{\langle hkl\rangle} &= \left(\frac{1}{n_{hkl}-1}\sum_{i=1}^{n_{hkl}}\left(e_{\langle hkl\rangle i} - \bar{e}_{\langle hkl\rangle}\right)^2\right)^{1/2} \end{aligned} \tag{13.19}$$

The small sample statistics inherent in 1000 grain simulations are addressed by averaging the results for $\bar{e}_{<hkl>}$ and $s_{\langle hkl\rangle}$ over five instantiations with the same $\rho(\tau_c)$ and $\rho(S_{z,mx,\langle hkl\rangle})$.

13.4.3 Lattice Strain Evolution of NC Ni

Six features are reported from the in situ X-ray diffraction-tensile tests for NC Ni ($d = 30$ nm). Figure 13.12a shows the experimental stress vs. plastic strain response. It is characterized by (I) a large elastic–plastic transition strain ($\varepsilon_{p(\text{trans})}$ $\sim$1 %) and (II) a large hysteretic strain ($\varepsilon_{p(\text{hyst})}$ $\sim$0.2 %) from unloading–reloading. Figure 13.12b shows the change in average transverse residual lattice strain ($\Delta\bar{e}_{<hkl>}$) relative to the beginning of the tensile test. The error bars indicate the standard deviation, based on 100 spectra at each unloaded state. The characteristic trends include: (III) nearly constant average residual lattice strain at smaller plastic

Table 13.2 QCP simulation parameters for NC and UFG Ni

Elast. const. (GPa) [55]	C_{11}: 246.5; C_{12}: 147.3; C_{44}: 124.7
Burgers vector (nm)	b: 0.25
Grain size (nm)	d: 30 (NC); 100 (UFG)
Texture	Random
Schmid factors(avg. of max. in ea. grain)	S_{poly}: 0.45 $S_{\langle 200\rangle}$: 0.47; $S_{\langle 311\rangle}$: 0.46 $S_{\langle 111\rangle}$: 0.45; $S_{\langle 220\rangle}$: 0.41
Quantized pl. strain	$\gamma_{\text{target}} = \min(1.2\, b/d,\ \tau_c/60\ \text{GPa})$ NC: 1.2 b/d ~1 % UFG: 1.2 b/d ~0.3 %
Crit. stress distrib. (MPa)	$\rho(\tau_c) = \begin{cases} 0 & \tau_c \geq \tau' \\ \dfrac{\tau_c^{k-1} \exp(-\tau_c/\theta)}{\Gamma(k)\theta^k} & \tau_c < \tau_c' \end{cases}$ NC: $\tau_c' = 210, k = 1, \theta = 850$ UFG: $\tau_c' = 200, k = 1, \theta = 220$
Axial plastic prestrain	$\varepsilon_{p(\text{pre})} = -0.5\,\%$

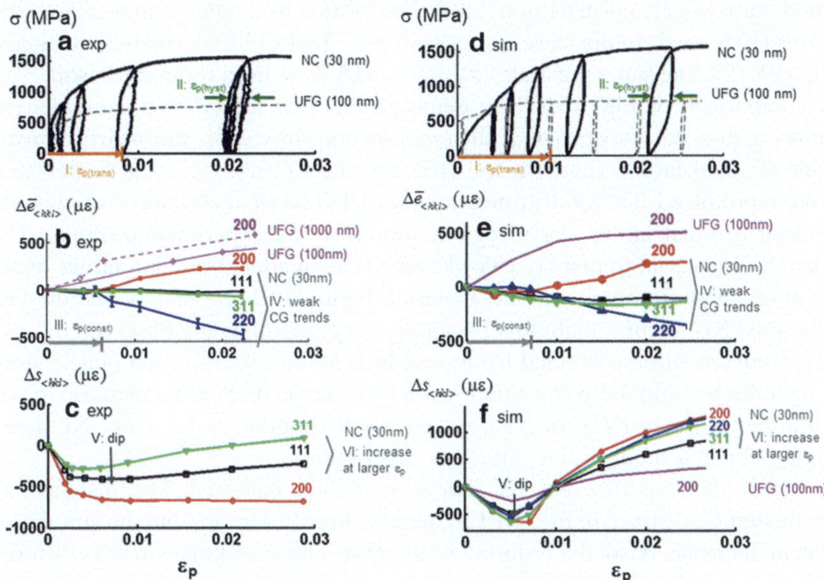

Fig. 13.12 Comparison between in situ X-ray diffraction measurements (**a–c**) vs. QCP simulation results (**d–f**) for NC and UFG Ni. (**a**, **d**) The stress-plastic strain ($\sigma - \varepsilon_p$) responses. (**b**, **e**) Change in transverse residual lattice strain $\Delta\bar{e}_{\langle hkl\rangle}$ vs. axial plastic strain ε_p. (**c**, **f**) Change in standard deviation in residual transverse lattice strain $\Delta s_{\langle hkl\rangle}$ vs. axial plastic strain ε_p. The experimental UFG data are adopted from [40]. The simulation parameters are shown in Table 13.2. Figure reproduced from [11]

strain (up to $\varepsilon_p \sim 0.6\,\%$); and (IV) CG-like trends at larger strain that are similar to but weaker than for UFG and CG counterparts (i.e., $\Delta\bar{e}_{\langle 200\rangle} > 0$, $\Delta\bar{e}_{\langle 220\rangle} < 0$, and $\Delta\bar{e}_{\langle 111\rangle} \sim \Delta\bar{e}_{\langle 311\rangle} \sim 0$) [44]. In contrast, the UFG response in Fig. 13.12b (e.g., $d = 100$ nm) has no constant region; rather, the residual lattice strains change at the onset of plastic deformation [40]. Figure 13.12c shows the corresponding evolution in lattice strain deviation $s_{\langle hkl\rangle}$ for different diffraction families. Features include: (V) an initial dip at small strain and (VI) a monotonic increase in peak width at larger strain (at least for $\langle 311\rangle$ and $\langle 111\rangle$). In contrast, the experimental peak widths for UFG Ni [46] and CG Cu [47] show no perceptible dip; rather, they gradually increase to a saturation value.

Figure 13.12d–f shows the corresponding QCP simulation results. A comparison of Fig. 13.12a, d shows that the QCP simulations for NC Ni ($d = 30$ nm) can capture the experimental $\sigma - \varepsilon_p$ features (I, II) provided the *wide*, *asymmetric* $\rho(\tau_c)$ distribution (Case NC, Fig. 13.11b) is used. Furthermore, the same distribution is used for all grain orientations and diffraction groups. An axial plastic prestrain ($\varepsilon_{p(\text{pre})} = -0.5\,\%$) is imposed prior to the tensile test. This plastic prestrain is needed to capture the experimental trends in lattice strain in Fig. 13.12b, c. The simulations can also capture the experimental $\sigma - \varepsilon_p$ response for UFG Ni by Cheng et al. [40] if the more narrow $\rho(\tau_c)$ distribution that is also shifted to smaller values (Case UFG, Fig. 13.11b) is used. In this case, γ_{target} (=0.3 %, Table 13.2) is smaller, as dictated by Eq. (13.15). The same prestrain ($\varepsilon_{p(\text{pre})} = -0.5$ %) is used for comparison.

A comparison of Fig. 13.12b, e demonstrates that the QCP simulations (NC 30 nm) capture the experimental diffraction footprints well: the nearly constant average residual lattice strain region (III) and the CG-like trends at larger strain (IV) are reproduced. The QCP simulations for UFG Ni ($d = 100$ nm) do not predict a constant residual lattice strain region, similar to the experimental trends. They predict UFG material to display CG-like trends at smaller strain and larger overall shifts at larger strain relative to NC material. Figure 13.12c, f shows that the QCP simulations (NC 30 nm) qualitatively capture the experimentally observed trends in lattice strain deviation (obtained from peak broadening) vs. imposed plastic strain. This includes the initial dip at a small strain (V) and the monotonic increase in peak width at larger strain (VI). In contrast, the QCP simulations for UFG Ni show a negligible dip at a small strain, similar to experiments.

Some of the experimental NC features are not captured. For instance, the experimental ⟨200⟩ trace in Fig. 13.12b remains flat after the dip but the simulations predict an increase. Also, the ordering of the ⟨*hkl*⟩ curves at large strain is different. We have shown that this discrepancy can be reduced when the same magnitude of axial plastic prestrain (−0.5 %) is imposed by applying biaxial transverse tension rather than the axial compression in the present examples [11]. This highlights the sensitivity of the mechanical and diffraction footprints to plastic prestrain. It also offers an explanation for the variation in experimental lattice strains in NC metals as discussed earlier and as reported in [40–42]. Overall, the comparison of experimental (Fig. 13.12c) vs. simulated (Fig. 13.12f) values of lattice strain deviation is essentially qualitative in nature. In particular, the simulations account primarily for the distribution of internal stress arising from grain-to-grain variations in τ_c and orientation whereas the experimental measurements capture additional contributions to stress from dislocations and other defects. This will have an effect on both the magnitude of the dip and the ordering of the ⟨*hkl*⟩ curves. Unfortunately, current theoretical models are not able to deconvolute all of these contributions.

13.4.4 A Physical View Based on Stress Redistribution

The QCP simulations provide a physical view of NC deformation in terms of the stress redistribution and criticality associated with intragranular slip. Figure 13.13a shows the evolution in residual lattice strain as averaged over the family of all transverse ⟨200⟩ grains, the plastically soft ($\tau_c < \overline{\tau}_c$) subgroup in this family, and the plastically hard ($\tau_c > \overline{\tau}_c$) subgroup ($\overline{\tau}_c$ = the polycrystalline average). Prior to tensile testing, at $\varepsilon_p = 0$, these subgroups have oppositely signed lattice strains. This arises from the axial pre-compression (−0.5 %), which creates residual compressive slip events in soft grains, leaving the soft subgroup in transverse compression and the hard subgroup in transverse tension (Fig. 13.13b). These residual slip events are

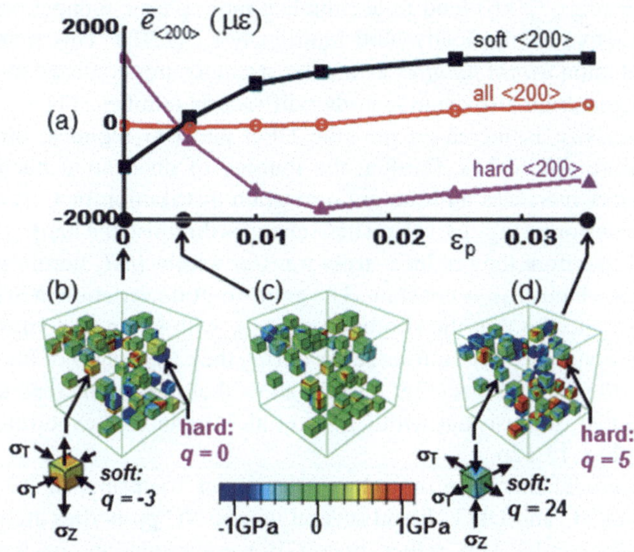

Fig. 13.13 (**a**) QCP simulations of residual transverse lattice strain $e_{\langle 200 \rangle}$ vs. macroscopic plastic strain ε_p for the entire transverse ⟨200⟩ family of grains as well as the plastically soft vs. hard subsets in this family. (**b–d**) Colored insets of the residual axial stress σ_Z in the transverse ⟨200⟩ family at $\varepsilon_p = 0$, 0.5 %, and 3.5 %, respectively. $q = \#$ of residual slip events in a grain. See Table 13.2 for simulation parameters. Figure reproduced from [11]

removed during the initial stages of the tension test, creating a tensile shift in the soft subgroup and a compressive shift in the hard one. At $\varepsilon_p \sim 0.5\,\%$, the average transverse residual lattice strains $\bar{e}_{\langle 200 \rangle}$ for the subgroups converge (Fig. 13.13a) and the corresponding average over the entire ⟨200⟩ family reaches a shallow minimum (Fig. 13.13c). This corresponds to a minimum in the deviation in residual transverse lattice strain (Fig. 13.13b).

During this initial straining from $\varepsilon_p = 0$ to $\sim 0.5\,\%$, the average $\bar{e}_{\langle 200 \rangle}$ for the entire ⟨200⟩ family is relatively constant (Fig. 13.13a). This signifies that during initial straining, there is a net transfer of stress between the soft and hard subgroups within the ⟨200⟩ family but not a net transfer of stress between the ⟨200⟩ and other families. Beyond $\varepsilon_p \sim 0.5\,\%$, $\bar{e}_{\langle 200 \rangle}$ for the soft and hard subgroups diverge (Fig. 13.13a). This arises primarily from tensile slip events in the soft ⟨200⟩ subgroup. The deviation in residual transverse lattice strain therefore increases (Fig. 13.12f), yet there is little shift in the ⟨200⟩ peak position (Fig. 13.12e). At $\varepsilon_p > \sim 1.5\,\%$, $\bar{e}_{\langle 200 \rangle}$ for the ⟨200⟩ family begins to increase, consistent with a weak CG trend. Here, the soft ⟨200⟩ subgroup approaches a maximum tensile shift as the grains within it achieve fully plastic flow. However, $\bar{e}_{\langle 200 \rangle}$ for the hard ⟨200⟩ subgroup reverses the trend and begins to increase (Fig. 13.13d). This signifies the onset of a net redistribution in stress between the ⟨200⟩ and other families at larger strain. In particular, plastically soft families with a larger average

Schmid factor (e.g., ⟨200⟩) tend to accumulate more plastic elongation during the tensile test relative to plastically hard families (e.g., ⟨220⟩). This transition from stress redistribution *within families* at smaller strain to stress redistribution *between families* at larger strain also occurs in other diffraction families [11].

If the grain size is increased into the UFG and CG regimes, the quantized plasticity feature diminishes. Further, the number of dislocation nucleation sites in grain interiors increases and the grain-to-grain distribution in τ_c is expected to become more uniform (Fig. 13.11b). These changes diminish the hard vs. soft nature of grains and therefore they reduce stress transfer within ⟨*hkl*⟩ families and shrink the region over which $\bar{e}_{\langle hkl \rangle}$ is constant. The grain-to-grain variation in Schmid factor gains in importance so that the evolution of $\bar{e}_{\langle hkl \rangle}$ with strain is dominated by the stress transfer *between* ⟨*hkl*⟩ families, even during the early stages of the tensile test. For example, the simulations (Fig. 13.12e) show that $\Delta\bar{e}_{\langle 200 \rangle}$ increases with grain size—a trend that is consistent with Cheng et al. [40] and the experimental results for $\Delta\bar{e}_{\langle 200 \rangle}$ in Fig. 13.12b.

The QCP model therefore reproduces the major footprints of the diffraction experiments on NC and UFG Ni and several unique NC properties are rationalized based on intragranular slip rather than GB-accommodation mechanisms. The successful predictions hinge on the assumptions that plastic flow at the grain scale is quantized and that the critical stress for slip takes on a spatially nonuniform, asymmetric grain-to-grain distribution. The quantized slip results in violent stress redistribution. The nonuniform critical stress distribution creates soft and hard grain subgroups, so that a net stress redistribution occurs *within* each diffraction family at smaller strain, and *between* diffraction families at larger strain.

13.5 Conclusions and Future Advances in QCP Modeling

The QCP model illustrates an extension of a conventional crystal plasticity model to successfully capture several unique properties of NC metals. In particular, a new constitutive flow rule is developed based on key observations from atomistic simulations of NC metals. As grain size decreases to the nanometer scale ($\sim$100 nm), the QCP model predicts a transition from continuous and relatively homogenous multiple-dislocation slip to quantized and highly heterogeneous single dislocation slip within grains. This chapter concludes with some key observations about the applications and further development of the model:

- A fit of the QCP model predictions to experimental data suggests that $\rho(\tau_c)$, the grain-to-grain distribution of CRSS for a slip event, is highly positive and skewed, with a larger fraction of relatively soft grains that is offset by an extended tail of relatively hard grains. The present simulations assume that this distribution does not change with plastic deformation and that specific grains retain the same value of τ_c on all 12 FCC slip systems during deformation. Thus, advances in the model might incorporate the evolution of $\rho(\tau_c)$ with strain, including the possibility that a slip event on a parent slip system in a specific grain changes not only τ_c for a subsequent event on the same system, but also τ_c on other slip systems in the same grain and in neighboring grains. The relative orientation of neighboring grains could be incorporated. In principle, this information could be mined from more fundamental dislocation dynamics studies of single and polycrystalline samples [48, 49].

- The present QCP simulations assume that slip events are driven purely by mechanical stress and that slip occurs instantaneously when the grain-averaged resolved shear stress reaches a critical value. At present, the simulations do not depend on the macroscopic strain rate because τ_c does not change with time and the local strain rate associated with a slip event is assumed to be much larger than the macroscopic rate. However, MD simulations suggest that the *rate-limiting* process for slip events is controlled by the depinning of dislocations from GB ledges [5, 50]. Advances in the model could include a thermally activated feature and utilize transition rate theory with a stress-dependent activation barrier. The strength of GB pinning sites could therefore incorporate a temporal component.
- As the grain size is reduced to <10–20 nm, GB-accommodation mechanisms will be highly interwoven as to affect both the GB structure and stress distribution within grains [8, 51]. Several crystalline-amorphous nanostructures/architectures have been reported to achieve an exceptional combination of strength and ductility by balancing the contributions from dislocation slip and GB accommodation [52, 53]. The QCP model can be modified in principle to capture the effects of both processes by discretizing grains into more elements and incorporating GB sliding [38, 54]. Thermodynamic state variables such as the degree of boundary relaxation could be introduced to qualitatively capture the dependence of GB pinning sites on GB-mediated processes.

References

1. K.S. Kumar, H. Van Swygenhoven, S. Suresh, Mechanical behavior of nanocrystalline metals and alloys. Acta Mater. **51**(19), 5743–5774 (2003)
2. M.A. Meyers, A. Mishra, D.J. Benson, Mechanical properties of nanocrystalline materials. Prog. Mater. Sci. **51**(4), 427–556 (2006)
3. M. Dao, L. Lu, R.J. Asaro, J.T.M. De Hosson, E. Ma, Toward a quantitative understanding of mechanical behavior of nanocrystalline metals. Acta Mater. **55**(12), 4041–4065 (2007)
4. V. Yamakov, D. Wolf, S.R. Phillpot, A.K. Mukherjee, H. Gleiter, Deformation-mechanism map for nanocrystalline metals by molecular-dynamics simulation. Nat. Mater. **3**(1), 43–47 (2004)
5. H. Van Swygenhoven, P.M. Derlet, A.G. Froseth, Nucleation and propagation of dislocations in nanocrystalline fcc metals. Acta Mater. **54**(7), 1975–1983 (2006)
6. H. Van Swygenhoven, J.R. Weertman, Deformation in nanocrystalline metals. Mater. Today **9**(5), 24–31 (2006)
7. J. Schiotz, F.D. Di Tolla, K.W. Jacobsen, Softening of nanocrystalline metals at very small grain sizes. Nature **391**(6667), 561–563 (1998)
8. H. Van Swygenhoven, P.M. Derlet, Grain-boundary sliding in nanocrystalline fcc metals. Phys. Rev. B **64**(22), 224105 (2001)

9. L. Li, P.M. Anderson, M.G. Lee, E. Bitzek, P. Derlet, H. Van Swygenhoven, The stress-strain response of nanocrystalline metals: a quantized crystal plasticity approach. Acta Mater. **57**(3), 812–822 (2009)
10. L. Li, M.-G. Lee, P. Anderson, Critical strengths for slip events in nanocrystalline metals: predictions of quantized crystal plasticity simulations. Metall. Mater. Trans. A **42**(13), 3875–3882 (2011)
11. L. Li, S. Van Petegem, H. Van Swygenhoven, P.M. Anderson, Slip-induced intergranular stress redistribution in nanocrystalline Ni. Acta Mater. **60**(20), 7001–7010 (2012)
12. L. Li, M.-G. Lee, P.M. Anderson, Probing the relation between dislocation substructure and indentation characteristics using quantized crystal plasticity. J. Appl. Mech. **79**(3), 031009 (2012)
13. D. Wolf, V. Yamakov, S.R. Phillpot, A. Mukherjee, H. Gleiter, Deformation of nanocrystalline materials by molecular-dynamics simulation: relationship to experiments? Acta Mater. **53**(1), 1–40 (2005)
14. A.G. Froseth, P.M. Derlet, H. Van Swygenhoven, Dislocations emitted from nanocrystalline grain boundaries: nucleation and splitting distance. Acta Mater. **52**(20), 5863–5870 (2004)
15. D. Farkas, A. Frøseth, H. Van Swygenhoven, Grain boundary migration during room temperature deformation of nanocrystalline Ni. Scr. Mater. **55**(8), 695–698 (2006)
16. Y. Mishin, D. Farkas, M.J. Mehl, D.A. Papaconstantopoulos, Interatomic potential for Al and Ni from experimental data and ab initio calculations. Mater. Res. Soc. Symp. Proc. **538**, 533 (1999)
17. E. Bitzek, P.M. Derlet, P. Anderson, H. Van Swygenhoven, The stress-strain response of nanocrystalline metals: a statistical analysis of atomistic simulations. Acta Mater. **56**(17), 4846–4857 (2008)
18. R.J. Asaro, A. Needleman, Overview no. 42 texture development and strain hardening in rate dependent polycrystals. Acta Metall. **33**(6), 923–953 (1985)
19. D. Peirce, R.J. Asaro, A. Needleman, An analysis of nonuniform and localized deformation in ductile single crystals. Acta Metall. **30**(6), 1087–1119 (1982)
20. S.R. Kalidindi, C.A. Bronkhorst, L. Anand, Crystallographic texture evolution in bulk deformation processing of FCC metals. J. Mech. Phys. Solids **40**(3), 537–569 (1992)
21. D. Peirce, R.J. Asaro, A. Needleman, Material rate dependence and localized deformation in crystalline solids. Acta Metall. **31**(12), 1951–1976 (1983)
22. Y. Shen, P.M. Anderson, Transmission of a screw dislocation across a coherent, non-slipping interface. J. Mech. Phys. Solids **55**(5), 956–979 (2007)
23. C.E. Packard, O. Franke, E.R. Homer, C.A. Schuh, Nanoscale strength distribution in amorphous versus crystalline metals. J. Mater. Res. **25**(12), 2251–2263 (2010)
24. H. Askari, M.R. Maughan, N. Abdolrahim, D. Sagapuram, D.F. Bahr, H.M. Zbib, A stochastic crystal plasticity framework for deformation of micro-scale polycrystalline materials. Int. J. Plast. **68**, 21–33 (2015)
25. M.D. Uchic, D.M. Dimiduk, J.N. Florando, W.D. Nix, Sample dimensions influence strength and crystal plasticity. Science **305**(5686), 986–989 (2004)
26. J.R. Greer, J.T.M. De Hosson, Plasticity in small-sized metallic systems: intrinsic versus extrinsic size effect. Prog. Mater. Sci. **56**(6), 654–724 (2011)
27. J.D. Eshelby, The determination of the elastic field of an ellipsoidal inclusion, and related problems. Proc. R. Soc. Lond. A Math. Phys. Sci. **241**(1226), 376–396 (1957)
28. T. Mura, *Micromechanics of Defects in Solids* (Martinus Nijhoff, The Hague, 1982)
29. Hibbitt, Karlsson & Sorensen Inc., *ABAQUS Reference Manuals* (Hibbitt, Karlsson & Sorensen Inc., Pawtucket, RI, 2005)
30. J.R. Trelewicz, C.A. Schuh, The Hall–Petch breakdown in nanocrystalline metals: a crossover to glass-like deformation. Acta Mater. **55**(17), 5948–5958 (2007)
31. G. Saada, Hall–Petch revisited. Mater. Sci. Eng. A Struct. Mater. **400**, 146–149 (2005)
32. S. Brandstetter, H. Van Swygenhoven, S. Van Petegem, B. Schmitt, R. Maass, P.M. Derlet, From micro- to macroplasticity. Adv. Mater. **18**(12), 1545–1548 (2006)

33. Y.M. Wang, R.T. Ott, T. van Buuren, T.M. Willey, M.M. Biener, A.V. Hamza, Controlling factors in tensile deformation of nanocrystalline cobalt and nickel. Phys. Rev. B **85**(1), 014101 (2012)
34. L. Thilly, S.V. Petegem, P.-O. Renault, F. Lecouturier, V. Vidal, B. Schmitt, H.V. Swygenhoven, A new criterion for elasto-plastic transition in nanomaterials: application to size and composite effects on Cu–Nb nanocomposite wires. Acta Mater. **57**(11), 3157–3169 (2009)
35. F. Ebrahimi, G.R. Bourne, M.S. Kelly, T.E. Matthews, Mechanical properties of nanocrystalline nickel produced by electrodeposition. Nanostruct. Mater. **11**(3), 343–350 (1999)
36. J. Rajagopalan, J.H. Han, M.T.A. Saif, Bauschinger effect in unpassivated freestanding nanoscale metal films. Scr. Mater. **59**(7), 734–737 (2008)
37. J. Rajagopalan, J.H. Han, M.T.A. Saif, Plastic deformation recovery in freestanding nanocrystalline aluminum and gold thin films. Science **315**(5820), 1831–1834 (2007)
38. Y.J. Wei, A.F. Bower, H.J. Gao, Recoverable creep deformation and transient local stress concentration due to heterogeneous grain-boundary diffusion and sliding in polycrystalline solids. J. Mech. Phys. Solids **56**(4), 1460–1483 (2008)
39. X.Y. Li, Y.J. Wei, W. Yang, H.J. Gao, Competing grain-boundary- and dislocation-mediated mechanisms in plastic strain recovery in nanocrystalline aluminum. Proc. Natl. Acad. Sci. U. S. A. **106**(38), 16108–16113 (2009)
40. S. Cheng, A.D. Stoica, X.L. Wang, Y. Ren, J. Almer, J.A. Horton, C.T. Liu, B. Clausen, D.W. Brown, P.K. Liaw, L. Zuo, Deformation crossover: from nano- to mesoscale. Phys. Rev. Lett. **103**(3), 4 (2009)
41. Y.M. Wang, R.T. Ott, A.V. Hamza, M.F. Besser, J. Almer, M.J. Kramer, Achieving large uniform tensile ductility in nanocrystalline metals. Phys. Rev. Lett. **105**(21), 215502 (2010)
42. H.Q. Li, H. Choo, Y. Ren, T.A. Saleh, U. Lienert, P.K. Liaw, F. Ebrahimi, Strain-dependent deformation behavior in nanocrystalline metals. Phys. Rev. Lett. **101**(1), 4 (2008)
43. S. Van Petegem, L. Li, P.M. Anderson, H. Van Swygenhoven, Deformation mechanisms in nanocrystalline metals: insights from in-situ diffraction and crystal plasticity modelling. Thin Solid Films **530**, 20–24 (2013)
44. B. Clausen, T. Lorentzen, T. Leffers, Self-consistent modelling of the plastic deformation of FCC polycrystals and its implications for diffraction measurements of internal stresses. Acta Mater. **46**(9), 3087–3098 (1998)
45. H. Van Swygenhoven, B. Schmitt, P.M. Derlet, S. Van Petegem, A. Cervellino, Z. Budrovic, S. Brandstetter, A. Bollhalder, M. Schild, Following peak profiles during elastic and plastic deformation: a synchrotron-based technique. Rev. Sci. Instrum. **77**(1), 10 (2006)
46. Z. Budrovic, S. Van Petegem, P.M. Derlet, B. Schmitt, H. Van Swygenhoven, E. Schafler, M. Zehetbauer, Footprints of deformation mechanisms during in situ x-ray diffraction: nanocrystalline and ultrafine grained Ni. Appl. Phys. Lett. **86**(23), 3 (2005)
47. Z. Budrovic, H. Van Swygenhoven, P.M. Derlet, S. Van Petegem, B. Schmitt, Plastic deformation with reversible peak broadening in nanocrystalline nickel. Science **304**(5668), 273–276 (2004)
48. F.F. Csikor, C. Motz, D. Weygand, M. Zaiser, S. Zapperi, Dislocation avalanches, strain bursts, and the problem of plastic forming at the micrometer scale. Science **318**(5848), 251–254 (2007)
49. V.V. Bulatov, W. Cai, *Computer Simulations of Dislocations* (Oxford University Press, Oxford, 2006)
50. C. Brandl, P.M. Derlet, H.V. Swygenhoven, Dislocation mediated plasticity in nanocrystalline Al: the strongest size. Model. Simul. Mater. Sci. Eng. **19**(7), 074005 (2011)
51. T.J. Rupert, D.S. Gianola, Y. Gan, K.J. Hemker, Experimental observations of stress-driven grain boundary migration. Science **326**(5960), 1686–1690 (2009)
52. Y.M. Wang, E. Ma, Three strategies to achieve uniform tensile deformation in a nanostructured metal. Acta Mater. **52**(6), 1699–1709 (2004)
53. Y. Wang, J. Li, A.V. Hamza, T.W. Barbee, Ductile crystalline–amorphous nanolaminates. Proc. Natl. Acad. Sci. **104**(27), 11155–11160 (2007)

54. Y.J. Wei, L. Anand, Grain-boundary sliding and separation in polycrystalline metals: application to nanocrystalline fcc metals. J. Mech. Phys. Solids **52**(11), 2587–2616 (2004)
55. H.B. Huntington, The elastic constants of crystals. Solid State Phys. Adv. Res. Appl. **7**, 213–351 (1958)

Chapter 14
Kinetic Monte Carlo Modeling of Nanomechanics in Amorphous Systems

Eric R. Homer, Lin Li, and Christopher A. Schuh

14.1 Introduction

Modeling the nanomechanics of amorphous systems presents a unique challenge as a result of the disparate time and length scales and diverse modes of deformation exhibited by this class of materials [1, 2]. At high temperatures, amorphous systems exhibit homogeneous deformation that follows Newtonian and non-Newtonian flow laws. At low temperatures, deformation is localized into shear bands with nanometer scale thickness but which can extend over hundreds of micrometers. Interestingly, a single fundamental unit of deformation is hypothesized to underlie these diverse modes of deformation. This fundamental unit of deformation is known as the shear transformation zone or STZ, and is characterized by the transient motion of several dozen atoms that deform inelastically in response to an applied shear stress [3]. The STZ is illustrated in Fig. 14.1 where the final state has been sheared by an increment of shear strain γ_o. The two modes of deformation are then expected to be comprised of either the uncorrelated activation of STZs uniformly distributed throughout the sample—in the case of homogeneous deformation—or of highly correlated activation of STZs in a localized band—in the case of inhomogeneous deformation [4].

E.R. Homer (✉)
Department of Mechanical Engineering, Brigham Young University, Provo, UT 84602, USA
e-mail: eric.homer@byu.edu

L. Li
Department of Metallurgical and Materials Engineering, University of Alabama, Tuscaloosa, AL 35487, USA

C.A. Schuh
Department of Materials Science and Engineering, Massachusetts Institute of Technology, Cambridge, MA 02139, USA

C.R. Weinberger, G.J. Tucker (eds.), *Multiscale Materials Modeling for Nanomechanics*, Springer Series in Materials Science 245,
DOI 10.1007/978-3-319-33480-6_14

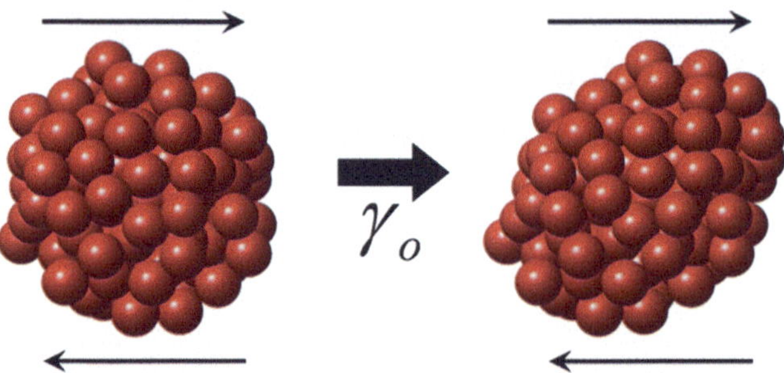

Fig. 14.1 Shear transformation zone (STZ), which represents the response of a collection of atoms to an applied shear stress. The final state of the atoms has been sheared from the initial state by a plastic strain of γ_o. Figure adapted with permission from [29]

Modeling this behavior is difficult because STZ activations in these cases can occur over several decades of strain rate. Furthermore, simultaneously capturing the length scale of nanometer shear band thickness and micrometer shear band slip is challenging. To truly resolve the nature of deformation within these materials, the mechanics must be modeled using a combination of modeling approaches [2]. Atomistic simulations are critical in resolving the mechanics associated with individual STZ activations as well the nature of STZ–STZ interactions [5, 6]. Unfortunately, the time and length scale limitations of atomistic simulations preclude simulating the behaviors of these materials at engineering time and length scales [2]. Continuum approaches, on the other hand, provide the ability to model deformation at engineering scales and provide ideal comparison to experiments [7–10]. However, the continuum approach is limited by its constitutive laws, which are often changed to accommodate the different modes of deformation [8, 9]. This can lead to the possibility of incorrectly capturing the deformation physics. As a result, modeling the behavior across the entire spectrum is best enabled by the addition of an intermediate or mesoscale technique [2]. Mesoscale techniques focus on capturing the physics that bridge the time and length scales between atomistic and continuum methods [11].

14.2 Mesoscale Modeling

The key to a mesoscale modeling technique that successfully bridges the time and length scales of interest is to adopt elements that accurately capture the physics associated with the fundamental events. In the mesoscale model adopted here, we employ two separate elements that individually bridge the time and length scales associated with deformation in amorphous systems. The kinetic Monte Carlo (kMC) algorithm is adopted to bridge the time scales, while a coarse-graining technique is utilized to bridge the length scales. It is noted, however, that there are a variety of mesoscale approaches that can be adopted [11, 12].

14.2.1 Coarse-Graining

The technique of coarse-graining involves the identification of features of a given process such as a cluster of atoms, an entire crystal of atoms, or region of material, and then treating the cluster, crystal, or region as a single unit. Atomic interactions, vibrations, defects, molecular interactions, and various other phenomena that do not play a governing role in the process of interest can either be ignored or their contributions can be incorporated into the constitutive law that governs the coarse-grained region. For example, polymer or protein modeling often treats the molecules or different clusters of atoms as different individual units [12]. This is efficient because the individual atomic interactions no longer need to be considered, and the net interaction between the clusters of atoms can be accommodated with a constitutive law. The result of coarse-graining is that one can model larger systems sizes more efficiently while ensuring that there is little loss in accurate physical interactions between the coarse-grained regions.

14.2.2 The kMC Algorithm

The kMC algorithm is an adaptation of the well-known Monte Carlo technique. The Monte Carlo technique uses probability distribution functions to sample the phase space of systems in equilibrium states. That is, in these algorithms, a system is examined for many different equilibrium configurations it can adopt and the system is moved into states that have higher probability of existence based on the equilibrium energy. The probability distribution most frequently used is the Maxwell–Boltzmann distribution

$$p\left(E_i\right) = \exp\left(-E_i/k_B T\right) / \Omega \tag{14.1}$$

where E_i is the energy of the system in configuration i, k_B is Boltzmann's constant, T is the temperature in Kelvin, and Ω is the partition function, which is defined as

$$\Omega = \sum_i \exp\left(-E_i/k_B T\right). \tag{14.2}$$

In this definition, the probabilities of all the states sum to one.

Consider the following system with five possible configurations, whose energetic landscape is illustrated in Fig. 14.2. If the system starts in state 3 at low temperature,

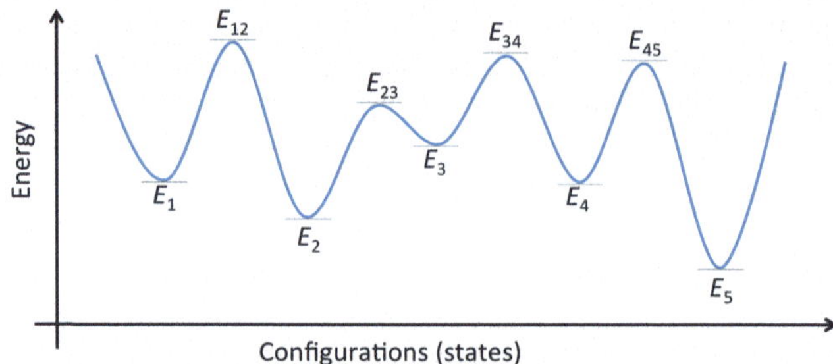

Fig. 14.2 Energetic landscape with various equilibrium states and transition states identified

it has the highest probability of visiting state 5, although it will visit all states according to the probabilities calculated by Eq. (14.1). Whether a different state can be accessed is dependent only on the change in energy between the equilibrium states, and the magnitude of the energy barrier separating the two states is irrelevant. MC algorithms can even transition to states that are not adjacent (state 1 directly to state 5) and time is not accounted for because the time to overcome the transition state barriers is not considered.

The kMC algorithm on the other hand is focused on the energy barriers and transition states between individual states and tracks time evolution during this process. The algorithm and the underlying transition state theory (TST) are explained in detail by Voter in [13]. According to TST, a system can only transition into adjacent states. The probability for transitioning between states is now based on the energetic difference between the current equilibrium state and transition state between the adjacent states. For a transition from state i to state j, this energetic barrier works out to

$$\Delta E_{i\to j} = \Delta E_{ij} = E_{ij} - E_i \tag{14.3}$$

The transition rate k from state i to j is then given by the Arrhenius equation:

$$k_{ij} = \nu \ \exp\left(-\Delta E_{ij}/k_B T\right) \tag{14.4}$$

where ν is the attempt frequency of the system to transition out of a given state. For most atomic processes, ν is related to the Debye frequency for the elements involved. It is assumed that once the system moves from the equilibrium state to the transition state, it will continue along its trajectory into the new equilibrium state.

As an example, consider again the system in Fig. 14.2, with the system starting in state 3. Since the system can only transition from state 3 into state 2 or 4, the two probabilities will be based on the energetic differences ΔE_{32} and ΔE_{34}. Since ΔE_{32} is smaller, the system is more likely to transition into state 2 than state 4. Going further, getting to state 5 is less likely than getting to state 2 because the two transitions required to get to state 5, ΔE_{34} and then ΔE_{45}, are so much larger than the single transition required to get to state 2, ΔE_{23}. This contrasts the MC results that would suggest state 5 as the most likely state to be occupied. However, with sufficient time and/or thermal energy, the kMC algorithm can and will visit state 5.

The kMC algorithm proceeds according to the following steps [14], which are repeated for every transition:

1. Make a list of transition states, enumerated by the index j, out of the current equilibrium state and calculate the rate for every transition k_j. (Note that the current equilibrium state is no longer listed as a subscript in this derivation.) Calculate the cumulative transition rate k_T and normalize each individual transition $\eta_j = k_j/k_T$, such that the sum over the normalized transitions, η_j, is equal to one.
2. Generate two random numbers, ξ_1 and ξ_2, uniformly on the interval (0,1].
3. Update the elapsed time of the simulation with the residence time in the current configuration, calculated according to

$$\Delta t = -\ln\ \xi_1/k_T \tag{14.5}$$

4. Select a single transition to implement by first calculating the partial cumulative rate according to

$$H_j = \sum_{n=1}^{j} \eta_n \tag{14.6}$$

and then use the random number, ξ_2, to find the single transition i to implement, which satisfies

$$H_{i-1} < \xi_2 \le H_i. \tag{14.7}$$

Note that when the transition events are listed in a successive fashion, ξ_2 falls in the subinterval η_i in the list of normalized rates.
5. Move the system into the new state by implementing the transition selected in the previous step.
6. Update any system calculations and return to step 1 to allow the system to evolve again.

The kMC algorithm repeats the steps above for any desired number of transitions, and allows the system to evolve by passing through transitions based on their probability for occurring and tracking the elapsed time between transitions. The kMC algorithm is powerful and has been used in numerous studies to investigate diverse phenomena [15–21].

One important aspect of the kMC algorithm and TST is that detailed balance must be obeyed. Essentially, detailed balance ensures that one is accurately sampling the phase space of possible equilibrium states and their transition pathways for microscopic reversibility. By definition, all processes at equilibrium must be balanced. Voter [13] explains detailed balance as:

> For every pair of connected states *i* and *j*, the number of transitions per unit time (on average) from *i* to *j* must equal the number of transitions per unit time from *j* to *i*. Because the number of escapes per time from *i* to *j* is proportional to the population of state *i* [based on MC probability of population] times the rate constant for escape from *i* to *j*, we have $p(E_i)\cdot k_{ij} = p(E_j)\cdot k_{ji}$ and the system is said to "obey detailed balance."

This can also be viewed as ensuring that the energetic landscape between states *i* and *j* be identical whether the system is transitioning from *i* to *j* or from *j* to *i* [4]. Additional discussion of detailed balance is available in referenced works [13, 22].

The kMC algorithm will provide an accurate model of a physical system so long as the transition states available to the algorithm accurately model the actual transition states of the physical system. A complete description of the kMC algorithm and many features and requirements of the technique are discussed in detail by Voter [13].

14.3 STZ Dynamics

To simulate the nanomechanics of amorphous systems, a mesoscale STZ dynamics modeling framework has been developed. This framework is inspired by the work of Bulatov and Argon [23–25], but extends the work in several ways. The method centers on the STZ introduced in Sect. 14.1. Since the cluster of atoms in an STZ consistently exhibit a transient shearing motion [5, 6, 26–28], it is possible to coarse-grain a theoretical sample into a system of potential STZs. Then each potential STZ has the ability to shear in the same manner as a cluster of atoms. This coarse-graining enables more efficient sampling of larger system sizes, thereby fulfilling one of the two improvements to modeling amorphous systems.

The second aspect of the STZ dynamics framework involves simulating longer system times more efficiently. This is accomplished by considering the transient STZ activation as a transition state between the initial and final equilibrium configurations. Thus, TST and the kMC algorithm can be employed as long as knowledge of the energetic landscape, including the transition states, is available.

14.3.1 STZ Coarse-Graining

In the present work, coarse-graining is accomplished by replacing the cluster of atoms that represent a potential STZ with features of a finite element mesh. In the development of the STZ dynamics framework, three criteria were identified as critical to proper representation of an STZ using a finite element mesh [14, 29]. First, the coarse-grained representation should approximate the shape of an STZ, which is generally believed to be roughly spherical. Second, since the STZ is a transient event, the definition of any potential STZs should enable them to overlap. The reason for this is that atoms would never be restricted to participate in only

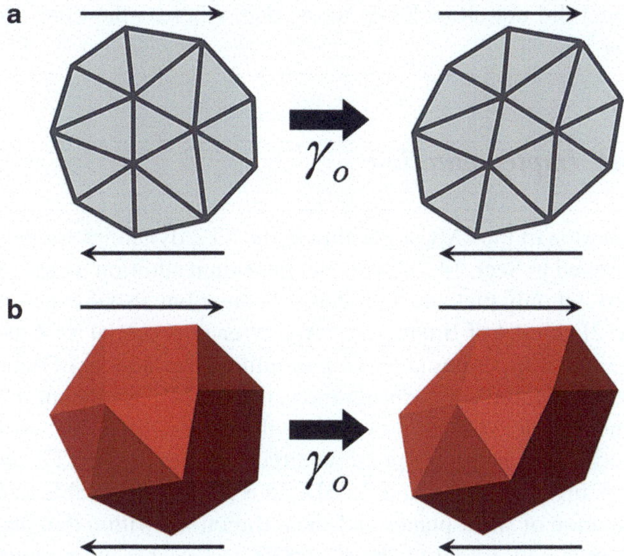

Fig. 14.3 Coarse-graining of an STZ using features of a finite element mesh in (**a**) 2D and (**b**) 3D. Figures adapted with permission from [14, 29]

one potential STZ. By defining them in a way that they overlap, any given element, which represents smaller groups of atoms, can participate in multiple potential STZ activations. Third, the coarse-grained STZ should accurately capture the expected behavior of an STZ. In his original proposal of the STZ, Argon modeled the STZ as an Eshelby inclusion [3]. As a result, analytical solutions can be utilized as a reference against which to compare a given coarse-graining technique and ascertain its accuracy [30, 31].

Following these criteria, techniques for coarse-graining STZs onto 2D and 3D finite element meshes were identified. In 2D, the technique involves the identification of a node and all surrounding elements or an element and all surrounding elements to represent a single STZ. This is illustrated in Fig. 14.3a for a 2D mesh that shows an element-centered, coarse-grained STZ. In 3D, a node and all surrounding elements are used to represent a single STZ, which is illustrated in Fig. 14.3b. These coarse-graining techniques have been shown to satisfy the three criteria stated above. Namely, the shapes are roughly equiaxed, potential STZs overlap, and as shown in published work, the error between the finite element solution and Eshelby inclusions solution is small [14, 29].

The use of finite elements builds on the previous work by Bulatov and Argon in an important way. In their original work, Bulatov and Argon employed a rigid lattice STZ model, which could track stress and strain but not displacement [23]. In the STZ dynamics modeling framework, the finite element mesh enables the framework to track displacement and deformation in addition to stress and strain. Additionally, the lattice model did not allow STZs to overlap, whereas the finite element mesh does allow this.

14.3.2 kMC Implementation

The implementation of the kMC algorithm in the STZ dynamics framework follows all the steps defined in Sect. 14.2.2 with one important addition. In selecting a single event, one must not only pick the STZ that will shear but also the direction of shear. In the original 2D model of Bulatov and Argon, each potential STZ had the ability to shear in one of six directions [23]. This simplified the number of transition events that had to be considered for each iteration of the kMC algorithm. However, just as atoms are not restricted to participate in only one possible STZ event, STZs are not restricted to shear in only six possible directions. In 2D, an STZ should be able to shear in any direction in a plane. In 3D, an STZ should be able to shear in any unique combination of shear planes and shear directions within that plane.

While it is relatively trivial to shear an STZ in any direction, enumerating and calculating the transition rates for all these directions is more challenging. We first begin with the calculation of the rate for an STZ to shear in one direction. The STZ activation rate is defined as

$$\dot{s} = \nu_o \exp\left(-\frac{\Delta G}{k_B T}\right) \tag{14.8}$$

where the prefactor ν_o is of the order of the Debye frequency, ΔG is the activation energy barrier, and $k_B T$ is the thermal energy of the system.

In order to calculate the activation energy barrier for a given transition, one must have knowledge of the transition itself. Methods such as the nudged-elastic band [32] or the activation–relaxation technique [33, 34] can be used in atomistic simulations to find the exact activation energy barrier from any given equilibrium state. One can then use Monte Carlo techniques to evolve the system as in [35]. However, these atomistic energy barrier search methods are computationally intensive and do not readily translate to mesoscale models. Since we desire to have a catalog with a large number of transitions, we favor another approach.

More traditional approaches to determining the energy in mesoscale systems will follow three steps: (1) calculate the energy, E_I, at the initial equilibrium state, (2) move the system to and calculate the energy, E_F, in the final state, and (3) add a predefined barrier height, ΔF, to the intermediate energy, which is the average energy of the initial and final equilibrium states. This is illustrated in Fig. 14.4a, b. Unfortunately, this approach is time consuming because one must move the system to all of the possible final equilibrium states in order to calculate the transition state.

In their model, Bulatov and Argon [23] exploited the fact that the Eshelby approach results in a quadratic solution to the strain energy for shearing an STZ, illustrated in Fig. 14.4 as a dashed line. This quadratic solution can be used to

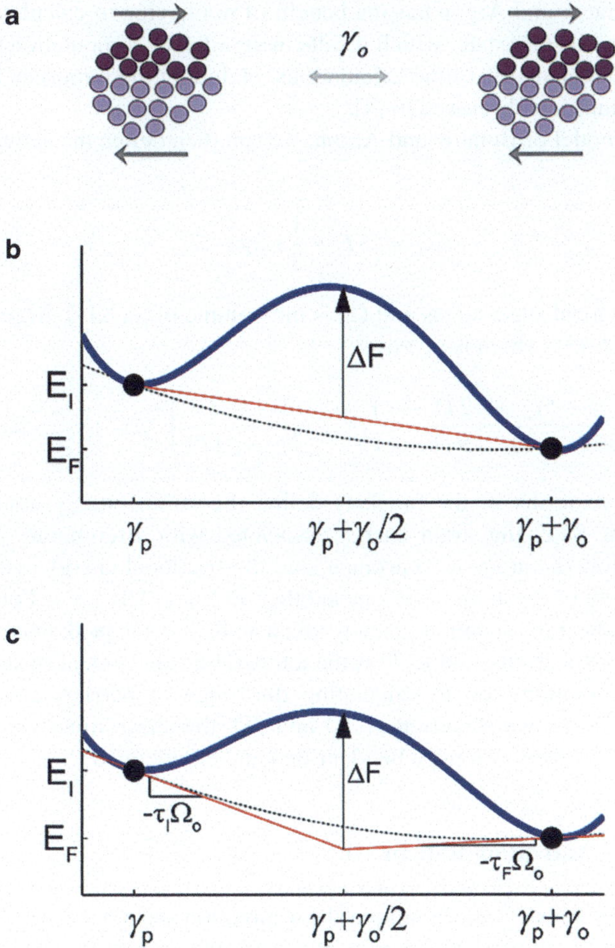

Fig. 14.4 (**a**) Representation of the two states before and after the activation of an STZ. Illustration of the energy landscape and the method of identifying the activated state of an STZ in a (**b**) traditional kMC model where the activation energy, ΔG, is obtained by adding the fixed energy barrier ΔF to the average of the initial and final states and (**c**) the energy landscape for the model proposed by Bulatov and Argon [23] where ΔG is obtained by adding ΔF to the projection of the slope (equal to the local shear stress) at the initial and final states. The variation in energy between the two states is given by *dashed line* and is the same in both (**b**) and (**c**). Figure reproduced with permission from [4]

predict the energy in the final state, but rather than calculate that final state energy, Bulatov and Argon simply extrapolate from the initial state, using the slope at that point, to the midpoint of the transition and then add the fixed barrier height, ΔF. Figure 14.4 accentuates the energy difference between the traditional approach and that of Bulatov and Argon for pedagogical purposes, but in actuality the difference in barrier height between the two methods is very small [4, 23]. Furthermore, the method of Bulatov and Argon has the benefit of not having to calculate the energy of the final equilibrium state, which results in significant computational savings. A more detailed explanation of the calculations, including the important requirement of detailed balance, is discussed in [4].

Using the model of Bulatov and Argon, we can then define the activation energy barrier as

$$\Delta G = \Delta F - \frac{1}{2}\tau \cdot \gamma_o \cdot \Omega_o \tag{14.9}$$

where τ is the local shear stress and Ω_o is the volume of an STZ. Argon developed a model for ΔF [3], which is given as

$$\Delta F = \left[\frac{7-5v}{30\,(1-v)} - \frac{2\,(1+v)}{9\,(1-v)}\beta^2 + \frac{1}{2\gamma_o} \cdot \frac{\hat{\tau}}{\mu(T)}\right] \cdot \mu(T) \cdot \gamma_o^2 \cdot \Omega_o \tag{14.10}$$

where the three terms in the brackets define the strain energy associated with shearing of the STZ, the strain energy associated with a temporary dilatation of the STZ to allow the atoms to rearrange, and the frictional energy associated with the free shearing of the atoms over one another. In Eq. (14.10), ν is Poisson's ratio, β is a ratio of shear to dilatation (usually taken as 1), $\hat{\tau}$ is the peak interatomic shear resistance between atoms, and $\mu(T)$ is the temperature-dependent shear modulus.

Since we are interested in calculating the range of barriers associated with shearing an STZ in any direction in 2D and 3D, the shear stress associated with each unique shear direction must be identified and enumerated.

14.3.2.1 STZ Activation Rate in 2D

In 2D, the shear stress for each unique shear direction around a circle can easily be evaluated using a Mohr's circle construct. Using this construct, the magnitude and sign of the shear stress along any given direction of the circle is equal to

$$\tau = \tau_{\max} \sin(\theta) \tag{14.11}$$

where θ is the angle to the stress state with stress τ, which is measured relative to the stress state with the highest principal stress. One can then integrate all shear directions by integrating θ over the interval [0°,360°). By combining Eqs. (14.8), (14.9), and (14.11), the integral STZ activation rate becomes

$$\dot{s} = \frac{v_o}{2\pi} \cdot \exp\left(-\frac{\Delta F}{k_B T}\right) \cdot \int_0^{2\pi} \exp\left(\frac{\tau_{\max} \cdot \sin(\theta) \cdot \gamma_o \cdot \Omega_o}{2k_B T}\right) d\theta \tag{14.12}$$

which evaluates to a modified Bessel function of the first kind, of order zero

$$\dot{s} = \frac{v_o}{2\pi} \cdot \exp\left(-\frac{\Delta F}{k_B T}\right) \cdot I_o\left(\frac{\tau_{\max} \cdot \gamma_o \cdot \Omega_o}{2k_B T}\right). \tag{14.13}$$

This particular form of the STZ activation rate is convenient because the analytical solution gives the rate for shearing an STZ in any direction in two dimensions with only one function evaluation. This integral rate is evaluated for each STZ and utilized in the steps listed in Sect. 14.2.2. Upon selection of a given STZ for activation, the angle of shear is selected for that particular STZ by numerically evaluating the integral equation up to the fraction of overlap by the random number on that state. This angle selection is explained in more detail in [14]. Fortunately, the method can provide the STZ for activation and the angle of shear with the single random number ξ_2.

14.3.2.2 STZ Activation Rate in 3D

The evaluation of the STZ activation rate in 3D is more complex than the 2D case due to the larger set of possible shear planes and shear directions, as well as the need to only evaluate unique combinations of shear planes and directions. In a generalized form the integral activation rate can be defined as

$$\dot{s} = v_o \cdot \exp\left(-\frac{\Delta F}{k_B T}\right) \cdot \iiint_{g \in G} \exp\left(\frac{\tau(\sigma, g) \cdot \gamma_o \cdot \Omega_o}{2k_B T}\right) \mathrm{d}g \tag{14.14}$$

where g is the orientation of any shear plane–shear direction combination belonging to the set G of all unique combinations of shear planes and shear directions. The integral is three-dimensional because the specific orientation of a shear plane and shear direction requires three parameters. The shear stress of that orientation g is defined as $\tau(\sigma, g)$ to denote the fact that the triaxial stress state that exists in a given STZ must be transformed by g to obtain the shear stress for that given shear plane and shear direction. Due to the complexity of this calculation, the details are not discussed here but are available in [29]. In short, these details describe the process for defining all unique combinations of shear planes and directions, and note that no analytical solution to the integral in Eq. (14.14) could be found. However, interesting trends indicating the equation's sole dependence on the deviatoric stress, as well as symmetries in similar stress states, lead to an efficient solution. The integral is numerically evaluated and tabulated for rapid recall during the modeling process while maintaining an error less than 0.01 %.

In a similar fashion to the 2D case, the integral STZ activation rates are used to run the kMC algorithm and select an STZ for activation. Once again, the shear plane and direction for the selected STZ are chosen by numerically integrating the rate equation until the magnitude of overlap of the random number ξ_2 on the interval of the selected STZ is matched.

Table 14.1 Material properties commonly employed by the STZ dynamics framework

Property/variable	Symbol and value
Temperature-dependent shear modulus	$\mu(T) = -0.004\left[\text{GPa K}^{-1}\right] * T + 37\,[\text{GPa}]$
Poisson's ratio	$\nu = 0.352$
Debye temperature	$\theta_D = 327$ K
Fixed activation energy barrier	$\Delta F(T) = 0.822 \times 10^{-29}\left[\text{J Pa}^{-1}\right] * \mu(T)$
STZ volume	$\Omega_o = 2.0\ \text{nm}^3$
STZ strain	$\gamma_o = 0.1$

14.3.3 Overall STZ Dynamics Framework

The STZ dynamics modeling approach incorporates the coarse-graining techniques and kMC algorithm evaluations of the STZ activation rates into an overall framework. The application of the framework to a given modeling problem requires that several steps be followed. First, a 2D or 3D finite element mesh is defined to match the geometry of the model material being simulated. Second, potential STZs are mapped onto the finite element mesh based on the technique determined according to the criteria discussed in Sect. 14.3.1. Third, the finite element mesh is assigned the set of material properties that will influence the material model and the kMC algorithm. In the present work, this limited set of properties is discussed in the following paragraph. Fourth, implement the kMC algorithm and repeat the following steps: (1) determine which STZ should be selected for activation, and which shearing angle should be applied, based on the current system state. (2) Apply plastic (Eigen) strains to the elements belonging to the STZ according to the selected shearing angle. (3) Use finite element analysis to determine the response of the system to the applied plastic strains. (4) Update the current system state, including stress, strain, and any functional material properties, to reflect the response to STZ activation. These last four steps involving the kMC algorithm are repeated many times in succession to determine the evolution of the system.

Due to the fact that the kMC algorithm determines and applies the plastic strains, the finite element analysis solver determines the elastic response of the system to the applied plastic strain, requiring only the use of a linear elastic solver.

The material properties and simulation variables used by many of the published STZ dynamics papers are listed in Table 14.1. This list is intentionally kept short to simplify the model and variables simply for the purpose of obtaining an intended response. The attempt frequency ν_o is taken as the Debye frequency, which can be calculated from the Debye temperature θ_D. The variables $\mu(T)$, ν, and θ_D are values for the commonly studied Vitreloy 1 with composition $Zr_{41.2}Ti_{13.8}Cu_{12.5}Ni_{10}Be_{22.5}$, and are obtained from [36], [36], and [37], respectively. Rather than using the complex form of the fixed barrier height in Eq. (14.10), we reduce ΔF to a simple functional form that is dependent upon the shear modulus. The STZ volume is in the range commonly reported in the literature [1, 38, 39] and the STZ strain is equal to the commonly accepted value [1].

14.3.4 Adaptations of the STZ Dynamics Framework

Since its original inception, the STZ dynamics framework has been adapted for different implementations and purposes.

First, the framework has been adapted for contact mechanics in finite element analysis to simulate nanoindentation [40]. In fact, the use of a finite element mesh and finite element analysis solver enables just about any set of boundary conditions regularly used in finite element analysis to be incorporated into the STZ dynamics framework.

Second, the kMC algorithm has been modified for specific scenarios to suppress the selection of a transition when that transition occurs on irrelevant time scales. This is crucial since standard kMC is designed to implement any transition selected by the algorithm no matter the elapsed time. Since stresses (and STZ activation rates) can vary by large magnitudes during nanoindentation or other testing conditions, this modified-kMC algorithm suppresses events that would occur on time scales greater than a predetermined maximum allowed time step. This ensures that stress and other conditions cannot change by a radical amount before being updated. In a given iteration, the modified-kMC algorithm determines the elapsed time before activating an STZ. If that elapsed time is larger than the maximum time, the STZ activation (plasticity) is suppressed and the system increments time by the maximum allowed time. If the elapsed time is smaller than the maximum time, the STZ activation (plasticity) is implemented and the system increments time by the amount suggested by the algorithm. In both cases, the elastic response of the system and all variables (stress, loading, displacement, etc.) are updated at the end of the step and the algorithm repeats with the updated values. In each iteration, an STZ may or may not activate, but the system can never evolve so rapidly that unphysical activations occur.

Third, state variables have been incorporated in some cases to account for the fact that the evolution of stress and strain cannot retain important information about the evolution of the structure. Free volume has been established as an important state variable for amorphous materials, so some implementations of the STZ dynamics framework track this state variable and adjust the material model to account for the current state of the structure at any given state.

Fourth, composite materials, such as metallic glass matrix (MGM) composites, have been simulated by partitioning the finite element mesh into two or more phases. Each element in the mesh is assigned a phase and the finite element solver uses the material model for that phase when that element is evaluated.

These various adaptations demonstrate the flexibility provided by the STZ dynamics framework, though other adaptations are certainly possible.

14.4 Applications of the STZ Dynamics Framework

Modeling amorphous materials using the kMC algorithm in the STZ dynamics framework provides an opportunity to study many different aspects of their nanomechanical behavior. In the following sections, we demonstrate: (1) how the framework can be used to study the various modes of deformation and understand the overall behavior of the model, (2) how thermomechanical processing can be investigated and how state variables are critical to capturing the evolution of the system, (3) how one can gain insight into the physics that control the nanomechanical behaviors of shear banding, (4) how mechanical contacts can be studied using the STZ dynamics modeling framework to gain insight into nanoindentation experiments on metallic glasses, and (5) how microstructural factors in MGM composites influence their mechanical properties.

14.4.1 General Behaviors

As noted in the introduction, amorphous materials exhibit homogeneous deformation at high temperatures and inhomogeneous deformation at low temperatures. To test the STZ dynamics framework, a model metallic glass with no preexisting distribution of stress and strain was subjected to two scenarios: high temperature (near Tg), intermediate stress and low temperature, high stress. Model responses for these two scenarios are shown in Fig. 14.5 for both 2D and 3D implementations. As can be seen, the first case exhibited the expected homogeneous deformation and STZs activated uniformly throughout the simulation cell. In the second case, the STZs localized the plastic strain into a shear band to give the expected inhomogeneous deformation.

Metallic glasses are expected to follow very specific rheological behaviors at higher temperatures. While the results are not shown here, both the 2D and 3D simulations closely follow the constitutive laws that are frequently used to fit experimental data.

To investigate the model responses over a range of conditions, we construct deformation maps for simulation cells subjected to a range of applied stresses at various temperatures. The deformation maps are illustrated for both 2D and 3D in Fig. 14.6. The steady-state strain rate is measured from each simulation and contours of constant strain rate are overlaid on the map for rates ranging form 10^{-10} to 1 s^{-1}. Simulations that had strain rates slower than 10^{-10} s^{-1} are deemed to be "elastic" and marked with an "×" as the deformation would be too slow to observe in experiments. Local values of the strain rate sensitivity are presented according to the shading inside each data point. As the stress is increased, the strain rate sensitivity decreases from unity (Newtonian flow), and trends toward zero (non-Newtonian flow). Finally, the regions of homogenous and inhomogeneous deformation are shaded. The 3D simulations are further broken down into shear banding versus

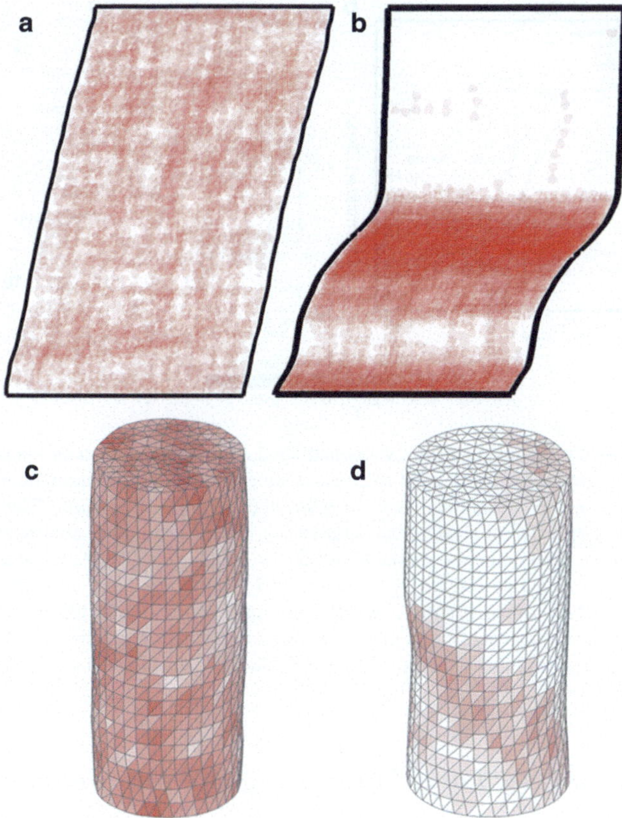

Fig. 14.5 High and low temperature representative responses of the STZ dynamics framework. The high temperature response is shown in (**a**) 2D and (**c**) 3D while the low temperature response is shown in (**b**) 2D and (**d**) 3D. Figures adapted with permission from [14, 29]

necking. This ability to distinguish between shear banding and necking is possible in the case of the 3D simulations because they were subjected to tension (3D) rather than pure shear (2D). These maps compare favorably with experimental deformation maps [1], in that they capture the basic features of amorphous deformation.

This match between experimental and simulated deformation maps is a testament to the strength of a mesoscale model that successfully coarse-grains a process and determines the transition states that control the evolution of the system.

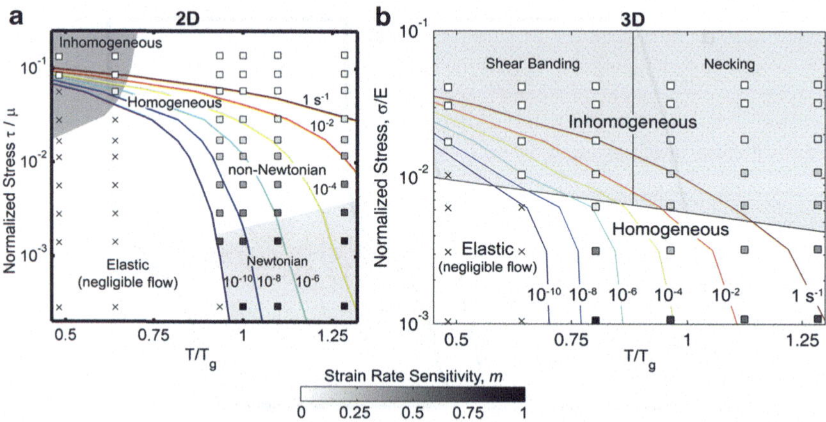

Fig. 14.6 Deformation maps for Vitreloy 1 constructed from data obtained by loading (**a**) 2D and (**b**) 3D model geometries over a range of loads and temperatures. The 2D simulations are subjected to pure shear while the 3D simulations are subjected to uniaxial tension. The *colored lines* represent contours of different steady-state strain rates, where strain rates slower than 10^{-10} s^{-1} are considered to be elastic and are marked with an "×." Other data points are shaded according to their respective strain rate sensitivity as indicated by the *color bar* below the map. Further regions marked as Newtonian (*lightly shaded*) and non-Newtonian are differentiated in the 2D map. The regions of homogeneous and inhomogeneous deformation are distinguished. In 3D this is further divided into samples which exhibit shear banding versus necking. Figures reproduced with permission from [14, 29]

14.4.2 Thermomechanical Processing and Free Volume

It is well known that the deformation behavior of an amorphous metal is sensitive to its processing history. Slowly quenched and well-annealed glasses show more serrated flow than quickly quenched glasses of the same composition [1, 2, 41]. As such, one desired capability of the STZ dynamics framework is to simulate thermomechanical processing and capture the effects of that processing history in subsequent deformation.

In the initial implementation of the STZ dynamics framework, processing history was captured only through the redistribution of stress and strain. As a result, when models with different processing history were tested, shear banding was not observed. The activation of STZs during thermal processing created a distribution of stress and strain whose magnitude was of the same order as the stress from an STZ activation attempting to form a shear band. As a result of this low signal-to-noise ratio, homogeneous deformation is observed even at low temperatures and high stresses in models that have been thermally processed [14].

To solve this problem, a state variable has been added to the STZ dynamics modeling framework [55]. The purpose of this state variable is to capture the evolution of the structure beyond the redistribution of stress and strain when an STZ is activated. One could choose from a range of state variables, such as atomic stress and strains, topological or chemical order, etc., but the state variable of "free volume" has been widely adopted in such a way that it incorporates these effects indirectly [42–44].

In fact, in Argon's original definition of the STZ, he includes free volume as a state variable to capture the structural evolution of the system. In the free volume adaptation of the STZ dynamics framework, excess free volume, f_v, is defined as a normalized quantity where $f_v = 0$ corresponds to no excess free volume above the average polyhedral volume V^* in a dense random hard sphere glass, while $f_v = 1$ is an upper bound corresponding to a state where an STZ can be activated without accumulating extra free volume [2].

Since excess free volume influences the energy barrier for activation, the fixed barrier is redefined as

$$\Delta F_{\mathrm{STZ}}(f_v) = \Delta F_{\mathrm{shear}} + \Delta F_{v0} \cdot g_{\mathrm{stz}}(f_v) \quad (14.15)$$

where $\Delta F_{\mathrm{shear}}$ captures the strain energy associated with shear (not dependent upon excess free volume) and ΔF_{v0} captures the strain energy associated with dilatation and friction of the atoms sliding over each other (dependent upon excess free volume). Equation (14.15) essentially alters Eq. (14.10) by making the first term in the bracket of Eq. (14.10) equal to $\Delta F_{\mathrm{shear}}$, which is not dependent on the magnitude of excess free volume, and makes the last two terms in the bracket of Eq. (14.10) equal to ΔF_{v0}, since these are dependent upon the magnitude of excess free volume. In fact, ΔF_{v0} is smaller when greater excess free volume exists since the STZ will need to dilate less and the friction will be lowered. This change in the energy is captured by the function g_{stz}, which lowers the activation energy barrier as the excess free volume is increased.

Following a given STZ activation, the excess free volume within the activated STZ is increased since it is believed that the atoms are not able immediately return to the original magnitude of excess free volume.

In this model however, a competing process to the activation of STZs is introduced. This competing process is the diffusive rearrangement (and destruction) of excess free volume. As atoms are constantly vibrating, their thermal energy enables diffusive rearrangements that would enable a given region of material to lower its excess free volume. The rate of diffusive rearrangement is given as

$$\dot{s}_D = (1 - f_v)\, \nu_D \exp\left(-\frac{\Delta G_D(f_v)}{k_B T}\right) \quad (14.16)$$

where $\Delta G_D(f_v)$ is the activation energy barrier for diffusive rearrangement, which is dependent upon the current magnitude of excess free volume. Higher excess free volume has a lower energy barrier than lower excess free volume because it is farther from its equilibrium state. The quantity $(1 - f_v)$ reflects a decrease of available atomic sites for free volume diffusion as f_v increases. The prefactor ν_D for the diffusive rearrangements is once again of the order of the Debye frequency.

In this free volume STZ dynamics framework, these diffusive rearrangement events are listed in the catalog of possible events. At any given kMC step, the system can activate an STZ or diffusely rearrange excess free volume. With this addition, the model retains information about its processing history beyond the simple redistribution of stress and strain.

The high temperature rheological behavior with the excess free volume state variable conforms to the Vogel–Fulcher–Tammann (VFT) relationship for viscosity. This is a nontrivial result because the model does not contain the input parameters to the VFT relationship; the behavior is an emergent result from the two competing processes.

At low temperatures, the excess free volume state variable enables the simulation to localize even in the presence of a preexisting stress and strain distribution. This can be seen in Fig. 14.7 where both the plastic strain and excess free volume is plotted for several snapshots. Here it can be seen that the excess free volume accumulates ahead of the formation of the shear band.

The excess free volume model has also been used to study the transient response of deformation, specifically the stress overshoot common in homogeneous deformation at high temperatures as well as the suppression of free volume annihilation at low temperatures [45]. The excess free volume provides an opportunity to observe the evolution of the structure during deformation and provides unique insight in this mesoscale model.

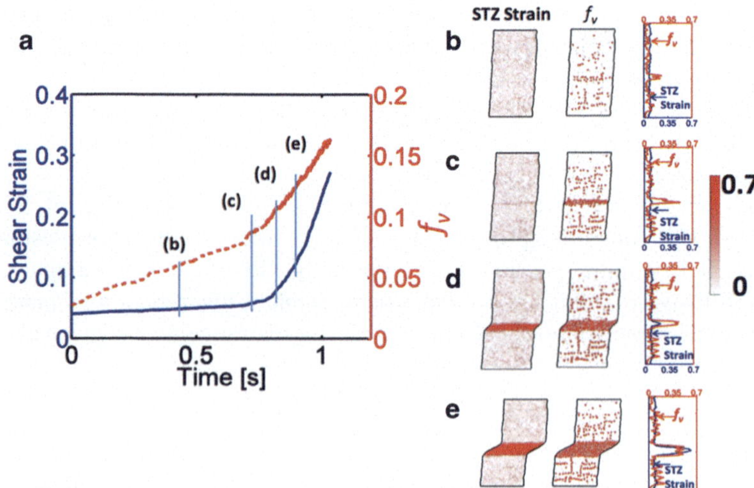

Fig. 14.7 (**a**) The shear strain and excess free volume f_v vs. time data of a cooled structure deformed at 300 K and 2 GPa shear load. (**b–e**) correspond to the snapshots at different times during the creep test. For each time, the physical deformation along with the magnitude of STZ strain and f_v are displayed; additionally, a plot with the 1D profile of STZ strain and f_v distributions along the vertical direction of the deformed sample is provided. Figure reproduced with permission from [55]

14.4.3 Correlations Between STZ Activations and Shear Banding

While a state variable such as free volume can capture the effects of processing or the evolution of the structure, we find that studying STZ behaviors in the absence of preexisting stress distributions also provides significant insight into the correlations between STZs.

Analysis of the 2D simulations discussed in Sect. 14.4.1 provides significant insight into correlations between STZ activations and the activation energy barrier distributions for the STZ activations [4]. Time-dependent radial distribution functions (TRDFs) can be constructed to show how STZ activations are correlated in both space and time. These TRDF functions indicate the likelihood of shearing an STZ at a nearby position and after a certain number of steps relative to a given STZ activation; magnitudes less than 1 are less likely to occur at a given position and time than if it occurred randomly throughout the simulation cell, and magnitudes greater than 1 are more likely to occur at a given position and time than if it occurred randomly throughout the simulation cell.

Representative TRDFs for low temperature, high stress and high temperature, high stress simulations are shown in Fig. 14.8. As can be seen in the figure, the low temperature, high stress conditions correspond to the preference for nearest-neighbor activations that is a maximum at the subsequent step and which decays over time. Despite the decay, however, there remains a preference to activate nearest-neighbor STZs in the region long after a given STZ is activated. This correlated behavior is the source for the shear localization that underlies the macroscopic shear bands that can be observed in experiments. Under conditions of high temperature, high stress it can be seen that there is no noticeable correlation between STZ activations. In other words, the additional thermal energy cancels the effect of stress concentrations that might otherwise cause shear localization. As a result, the STZ activations are uncorrelated, leading to uniform, random activation throughout the simulation cell causing homogeneous deformation. These distributions are representative of simulations that exhibited inhomogeneous and homogeneous deformation in Fig. 14.6a, across the range of conditions studied.

In addition to the ability to study correlations between STZ activations, one can study the formation of a shear band in detail [56]. Snapshots of a simulation cell subjected to a constant strain rate, uniaxial tension test are shown in Fig. 14.9. Analysis of the simulation indicates five different stages: (I) purely elastic, with no STZ activity, (II) STZ clustering, where correlated STZ activations lead to the formation of clusters, (III) growth following nucleation of a shear band, where all STZ activity transitions from being distributed throughout the simulation cell to being concentrated in the shear band, (IV) relaxation thickening, which is manifest by the continued thickening of the shear band while the stress is still dropping even after it has propagated across the simulation cell, and (V) flow thickening, which is indicated by the continued thickening of a single shear band at a constant flow stress. Most of the plastic strain is accumulated during prolonged flow thickening,

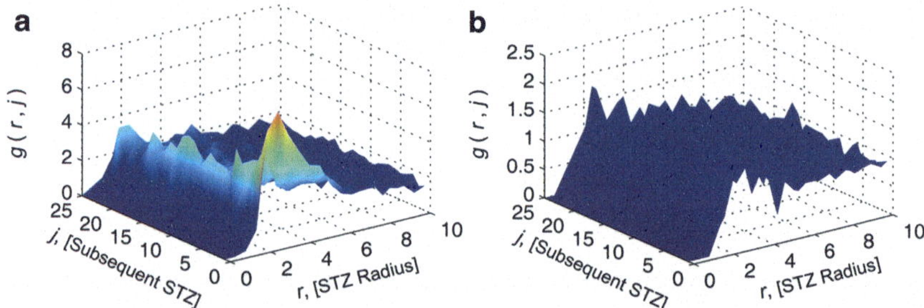

Fig. 14.8 TRDFs of STZ activation, where the behaviors and their corresponding conditions are (**a**) nearest-neighbor STZ activation: high stress and low temperature, and (**b**) independent STZ activation: high stress and high temperature. The shading of all the surfaces uses the same color scheme, permitting comparison of the magnitudes of the different trends. Figure reproduced with permission from [4]

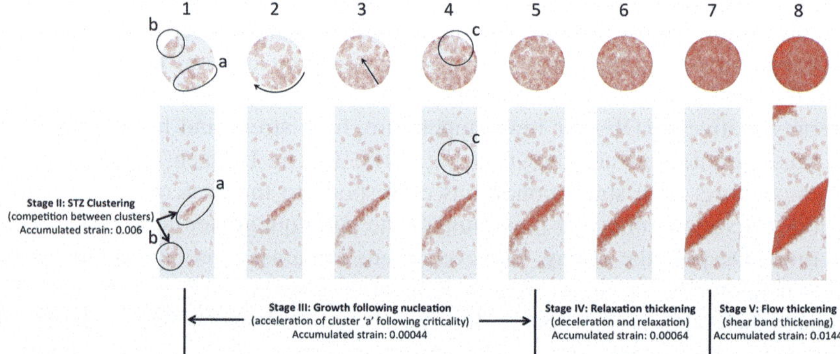

Fig. 14.9 Numbered snapshots of a model metallic glass at various times during a constant strain-rate, uniaxial tensile simulation. Each snapshot includes a semitransparent top and side view of the simulation cell. The evolution of the shear band is divided into four different stages following the initial elastic response. The shear band initiates at one of multiple potential nucleation sites formed during stage II. Once it achieves a critical state it exhibits a propagating shear front as it crosses the sample during stage III. The front propagation is followed by thickening of the shear band in stages IV and V, of which the majority of strain is accumulated during stage V. Figure reproduced with permission from [56]

indicating that nucleation and initial propagation of a shear band are very brief. The yield point matches a thermodynamic model for nucleation, and the work supports evidence that once a shear band nucleates, there is very little that can be done to stop it.

These types of studies demonstrate the strength of mesoscale models in elucidating the nanomechanics behind macroscopic processes. Some of these nanomechanical phenomena would be difficult to observe by other techniques.

14.4.4 Comparison with Experiments

Metallic glasses exhibit a broad range of interesting phenomena, one of which is that they can exhibit nanoscale strengthening when they are cyclically loaded in the elastic regime. This has been demonstrated in nanoindentation experiments where a metallic glass subjected to elastic cycling, which would otherwise leave behind no visible plastic deformation, shows a statistical increase in strength as a result of the cycling [59]. Interestingly, the cyclic strengthening can only occur if the cycling is of a sufficient magnitude, if the indenter is actually cycled (holding a constant load of equal magnitude and time does not lead to strengthening), and the strengthening saturates (one cannot cycle indefinitely to increase strength).

Since the cyclic loading is in the elastic regime, no structural changes can be detected during the cycling. Thus, modeling is an ideal method to elucidate the nanomechanics that cause the strengthening. The STZ dynamics framework is adapted to include contact mechanics in the finite element analysis solver and the modified-kMC algorithm is implemented due to the fact that the stress varies significantly as the indentation load is increased from zero to the point of plastic deformation [40].

Results from a simulation with monotonic loading are illustrated in Fig. 14.10. The deviation from the purely elastic solution can be observed, as well as the fact that a large region of the material exceeds the model yield point prior to the activation of any STZs, which is consistent with theory [46]. The STZs form slip lines early on but no distinct shear bands.

When subjected to cycling at different depths, the simulations indicate that significant plasticity can occur under the indented region without being evident in the load–displacement curve, the cyclic action leads to progressive STZ activity, although there is an indication for saturation of that activity. In other words, the simulations indicate that the energetics and time scales of STZ activity are plausible as a mechanism to cause structural evolution that would be consistent with nanoscale strengthening.

The nanoindentation simulation technique has also been used to detect the cause of the fluctuations in strength values for a metallic glass, which are reported as nanoscale strength distributions (NSDs) [57]. For example, in crystalline metals, the well-ordered structure leads to very specific energetic values to initiate plasticity and thermal excitations account for the fluctuations in the NSD. Amorphous metals on the other hand have larger fluctuations in strength and they do not have well-ordered structure, so any fluctuations could be due to structure or to thermal excitation. To help answer this question, simulated nanoindentation using the STZ dynamics framework is carried out on a model glass with a preexisting distribution of stress and strain but with no excess free volume state variable. The model is indented at the same position 10 times with different random number seeds to simulate thermal fluctuations. The same model is also indented at ten different locations (where the preexisting stress and strains are different) to simulate structural fluctuations. Figure 14.11 shows the NSDs as a function of the normalized load at the first

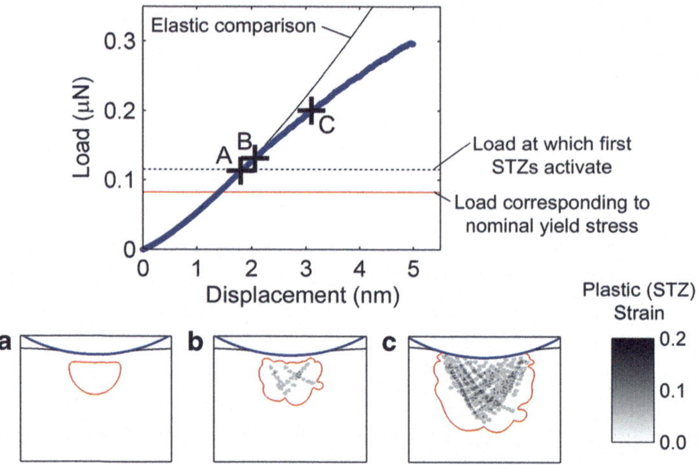

Fig. 14.10 Simulated nanoindentation results for monotonic loading. The graph shows the load–displacement curve for a single monotonic indentation test, with results for a purely elastic contact for comparison. Snapshots of the system during the simulation are provided below the graph as marked by "A," "B," and "C." The *red contour* on the snapshots shows the region of material that has exceeded the model yield stress, while the *gray* regions denote the operation of STZs. Figure reproduced with permission from [40]

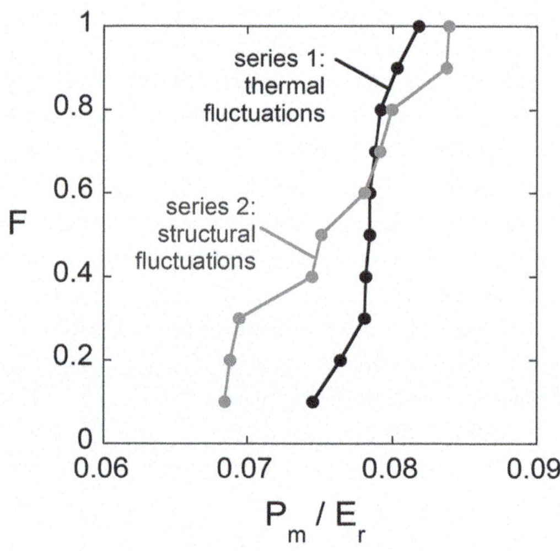

Fig. 14.11 Nanoscale strength distributions (NSDs) for the glass simulated by STZ dynamics. Here the first plasticity is measured as the point of the first STZ activation for two sets of simulations with different fluctuations imposed. For series 1, only thermal fluctuations are different among the tests in the series, while for series 2, the structure under the indenter was different in each of the ten cases. Figure reproduced with permission from [57]

STZ activation. Here it can be seen that the structural fluctuations cause more variation in load than the thermal fluctuations. In fact, the magnitude of the structural fluctuations matches well with the experimental variation in load, indicating that the structural fluctuations are the dominant source of variability in amorphous NSDs.

This application of a mesoscale technique to investigate the nanomechanics of experiments indicates the strong potential to elucidate phenomena that are difficult to measure by experimental techniques.

14.4.5 Composites

Due to the limited ductility in bulk metallic glasses, MGM composites have emerged as an attractive solution that retains most of the strength while providing significant ductility. These MGM composites can be created either by crystallizing regions of the material or by introducing a second phase during processing. The result is a metallic glass matrix that surrounds crystalline inclusions.

Research has demonstrated that the length scales, volume fraction, and ductility of the composite influence strength and ductility [47–54]. However, the exact relationships between the various microstructural factors and the material properties are not well understood.

The STZ dynamics framework can be adapted to allow two material models to be simulated [58], just as discussed in Sect. 14.3.4. This is accomplished by partitioning the mesh into the various phases and then directing the finite element analysis solver to use the appropriate material model for the phase of a given finite element. The STZ dynamics model is used for the amorphous matrix, while an isotropic hardening metal plasticity model is used for the crystalline inclusions. The kMC algorithm controls the time evolution of the system, though the metal plasticity model is not time dependent.

An n-factorial design of experiments is used to study three factors that are known to influence MGM composites: volume fraction of the crystalline phase, inclusion size of the crystalline phase, and yield point of the crystalline phase relative to the yield point of the amorphous matrix. High, low, and intermediate values of the three factors are studied, and simulation examples from the high and low values are provided in Fig. 14.12.

In Fig. 14.12 it can be seen that some combinations of the factors lead to improved plasticity while others lead to improved strength. Explicit, statistically significant functional forms for various dependent quantities, such as ductility, yield point, etc., can be extracted from the design of experiments. The functional forms of these dependent variables are in line with general expectations but demonstrate that increased volume fraction of a second phase alone is insufficient to improve ductility. The actual yield point of the second phase is critical to delocalizing the strain and improving ductility. These functional forms can even be used in the design and optimization of future MGM composites. The work also indicates different ways by which the crystalline inclusions delocalize the strain.

This simple adaptation of the STZ dynamics framework to more complex composites demonstrates the utility of a mesoscale framework to capture phenomena of multiphase materials while retaining the necessary view of the nanomechanics that control the behavior.

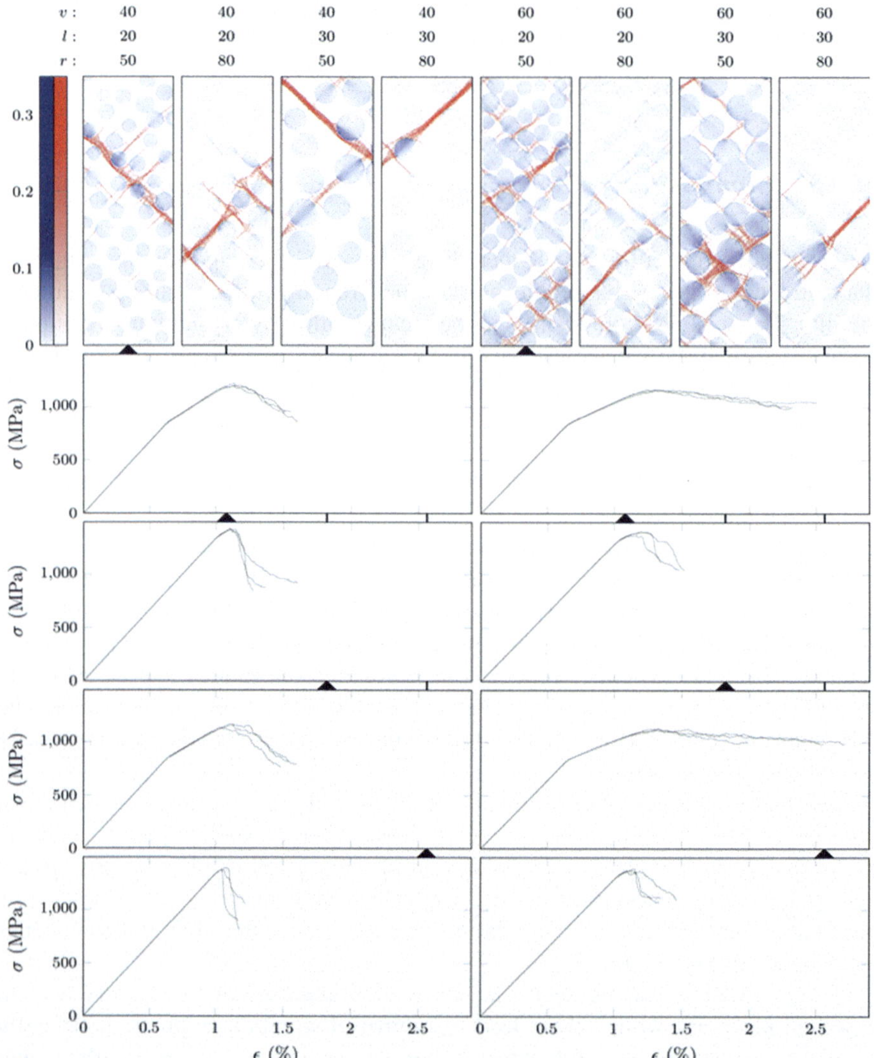

Fig. 14.12 Snapshots of the plastic strain in representative MGM composite simulations for different combinations of microstructural factors are shown. The microstructural factors and their values are v volume fraction of the crystalline phase at 40 and 60 %, l length scale of the crystalline inclusions at diameters of 20 and 30 nm, and r ratio of the crystalline yield point relative to the glass yield point at values of 50 and 80 %. The snapshots show crystalline inclusions in *gray*, glass plasticity in *red*, and crystalline plasticity in *blue*. Stress–strain curves are shown below each treatment to show the model response and the variability due to replicate simulations. Figure reproduced with permission from [58]

14.5 Conclusions

The combination of the kMC algorithm and a coarse-grained approach provides a useful framework with which to investigate the nanomechanical behavior of amorphous systems. The coarse-graining enables collections of atoms, or STZs, to be tracked rather than individual atoms and the kMC algorithm allows the shearing of these STZs to be controlled by stochastic processes based on the energetics of the system. The combination of these two features enables access to the relevant time and length scales while preserving a microscopic view of the processes that dominate the nanomechanical behavior.

As a result of the adaptability of the STZ dynamics framework, it has been used to investigate amorphous metal behavior in a variety of conditions. The modeling technique captures the overall deformation behaviors expected of a metallic glass. The technique also provides insight into the correlations between STZs and their formation into shear bands. The technique enables the study of the structural evolution and processing history to be readily accessed via a free volume state variable. Insight into nanoindentation experiments is possible through contact mechanics adaptations of the technique. MGM composites can be modeled through the combination of multiple materials models. It is expected that the STZ dynamics framework will continue to be useful in the investigation of amorphous metal behavior. In addition, the commonalities in deformation mechanics with other amorphous systems provide opportunities to study their behaviors as well.

Finally, the combination of the kMC algorithm with coarse-graining techniques has a wide range of possible applications outside amorphous systems or deformation nanomechanics. These mesoscale modeling techniques can be applied in myriad ways.

References

1. C.A. Schuh, T.C. Hufnagel, U. Ramamurty, Mechanical behavior of amorphous alloys. Acta Mater. **55**, 4067–4109 (2007)
2. D. Rodney, A. Tanguy, D. Vandembroucq, Modeling the mechanics of amorphous solids at different length scale and time scale. Model. Simul. Mater. Sci. Eng. **19**, 083001 (2011)
3. A.S. Argon, Plastic deformation in metallic glasses. Acta Metall. Mater. **27**, 47–58 (1979)
4. E.R. Homer, D. Rodney, C.A. Schuh, Kinetic Monte Carlo study of activated states and correlated shear-transformation-zone activity during the deformation of an amorphous metal. Phys. Rev. B. **81**, 064204 (2010)
5. A. Tanguy, F. Leonforte, J.-L. Barrat, Plastic response of a 2D Lennard-Jones amorphous solid: detailed analysis of the local rearrangements at very slow strain rate. Eur. Phys. J. E. **20**, 355–364 (2006)
6. C.E. Maloney, A. Lemaitre, Amorphous systems in athermal, quasistatic shear. Phys. Rev. E **74**, 016118 (2006)
7. F. Abdeljawad, M. Haataja, Continuum modeling of bulk metallic glasses and composites. Phys. Rev. Lett. **105**, 125503 (2010)

8. L. Anand, C. Su, A constitutive theory for metallic glasses at high homologous temperatures. Acta Mater. **55**, 3735–3747 (2007)
9. L. Anand, C. Su, A theory for amorphous viscoplastic materials undergoing finite deformations, with application to metallic glasses. J. Mech. Phys. Solids **53**, 1362–1396 (2005)
10. C. Su, L. Anand, Plane strain indentation of a Zr-based metallic glass: experiments and numerical simulation. Acta Mater. **54**, 179–189 (2006)
11. S. Yip, M.P. Short, Multiscale materials modelling at the mesoscale. Nat. Mater. **12**, 774–777 (2013)
12. S. Yip (ed.), *Handbook of Materials Modeling* (Springer, Dordrecht, 2005)
13. A.F. Voter, Introduction to the kinetic Monte Carlo method, in *Radiation Effects in Solids*, ed. by K.E. Sickafus, E.A. Kotomin, B.P. Uberuaga (Springer, Dordrecht, 2007), pp. 1–23
14. E.R. Homer, C.A. Schuh, Mesoscale modeling of amorphous metals by shear transformation zone dynamics. Acta Mater. **57**, 2823–2833 (2009)
15. C. Domain, C.S. Becquart, L. Malerba, Simulation of radiation damage in Fe alloys: an object kinetic Monte Carlo approach. J. Nucl. Mater. **335**, 121–145 (2004)
16. C.S. Deo, D.J. Srolovitz, W. Cai, V.V. Bulatov, Kinetic Monte Carlo method for dislocation migration in the presence of solute. Phys. Rev. B **71**, 014106 (2005)
17. F. Roters, P. Eisenlohr, L. Hantcherli, D.D. Tjahjanto, T.R. Bieler, D. Raabe, Overview of constitutive laws, kinematics, homogenization and multiscale methods in crystal plasticity finite-element modeling: theory, experiments, applications. Acta Mater. **58**, 1152–1211 (2010)
18. Y. Mishin, M. Asta, J. Li, Atomistic modeling of interfaces and their impact on microstructure and properties. Acta Mater. **58**, 1117–1151 (2010)
19. R. Devanathan, L. Van Brutzel, A. Chartier, C. Gueneau, A.E. Mattsson, V. Tikare, T. Bartel, T. Besmann, M. Stan, P. Van Uffelen, Modeling and simulation of nuclear fuel materials. Energy Environ. Sci. **3**, 1406–1426 (2010)
20. P. Zhao, J. Li, Y. Wang, Heterogeneously randomized STZ model of metallic glasses: softening and extreme value statistics during deformation. Int J Plast **40**, 1–22 (2013)
21. Y. Chen, C.A. Schuh, A coupled kinetic Monte Carlo–finite element mesoscale model for thermoelastic martensitic phase transformations in shape memory alloys. Acta Mater. **83**, 431–447 (2015)
22. D. Frenkel, B. Smit, *Understanding Molecular Simulation* (Academic, San Diego, 2002)
23. V.V. Bulatov, A.S. Argon, A stochastic model for continuum elasto-plastic behavior: I. Numerical approach and strain localization. Model. Simul. Mater. Sci. Eng. **2**, 167–184 (1994)
24. V.V. Bulatov, A.S. Argon, A stochastic model for continuum elasto-plastic behavior: II. A study of the glass-transition and structural relaxation. Model. Simul. Mater. Sci. Eng. **2**, 185–202 (1994)
25. V.V. Bulatov, A.S. Argon, A stochastic model for continuum elasto-plastic behavior: III. Plasticity in ordered versus disordered solids. Model. Simul. Mater. Sci. Eng. **2**, 203–222 (1994)
26. D.J. Srolovitz, V. Vitek, T. Egami, An atomistic study of deformation of amorphous metals. Acta Metall. Mater. **31**, 335–352 (1983)
27. C.E. Maloney, A. Lemaitre, Universal breakdown of elasticity at the onset of material failure. Phys. Rev. Lett. **93**, 195501 (2004)
28. D. Rodney, C.A. Schuh, Distribution of thermally activated plastic events in a flowing glass. Phys. Rev. Lett. **102**, 235503 (2009)
29. E.R. Homer, C.A. Schuh, Three-dimensional shear transformation zone dynamics model for amorphous metals. Model. Simul. Mater. Sci. Eng. **18**, 065009 (2010)
30. T. Mura, *Micromechanics of Defects in Solids* (Kluwer, Dordrecht, 1991)
31. T.W. Clyne, P.J. Withers, *An Introduction to Metal Matrix Composites* (Cambridge University Press, Cambridge, 1993)
32. E.B. Tadmor, R.E. Miller, *Modeling Materials* (Cambridge University Press, New York, 2011)
33. R. Malek, N. Mousseau, Dynamics of Lennard-Jones clusters: a characterization of the activation-relaxation technique. Phys. Rev. E **62**, 7723–7728 (2000)

34. E. Cancès, F. Legoll, M.C. Marinica, K. Minoukadeh, F. Willaime, Some improvements of the activation-relaxation technique method for finding transition pathways on potential energy surfaces. J. Chem. Phys. **130**, 114711 (2009)
35. D. Rodney, C.A. Schuh, Yield stress in metallic glasses: the jamming-unjamming transition studied through Monte Carlo simulations based on the activation-relaxation technique. Phys. Rev. B **80**, 184203 (2009)
36. W.L. Johnson, K. Samwer, A universal criterion for plastic yielding of metallic glasses with a $(T/Tg)^{2/3}$ temperature dependence. Phys. Rev. Lett. **95**, 195501 (2005)
37. W.H. Wang, P. Wen, D.Q. Zhao, M.X. Pan, R.J. Wang, Relationship between glass transition temperature and Debye temperature in bulk metallic glasses. J. Mater. Res. **18**, 2747–2751 (2003)
38. M. Zink, K. Samwer, W.L. Johnson, S.G. Mayr, Plastic deformation of metallic glasses: size of shear transformation zones from molecular dynamics simulations. Phys. Rev. B **73**, 172203 (2006)
39. X.L. Fu, Y. Li, C.A. Schuh, Homogeneous flow of bulk metallic glass composites with a high volume fraction of reinforcement. J. Mater. Res. **22**, 1564–1573 (2007)
40. C.E. Packard, E.R. Homer, N. Al-Aqeeli, C.A. Schuh, Cyclic hardening of metallic glasses under Hertzian contacts: experiments and STZ dynamics simulations. Philos. Mag. **90**, 1373–1390 (2010)
41. Y. Shi, M.L. Falk, Stress-induced structural transformation and shear banding during simulated nanoindentation of a metallic glass. Acta Mater. **55**, 4317–4324 (2007)
42. M.L. Falk, Molecular-dynamics study of ductile and brittle fracture in model noncrystalline solids. Phys. Rev. B **60**, 7062–7070 (1999)
43. F. Spaepen, Homogeneous flow of metallic glasses: a free volume perspective. Scr. Mater. **54**, 363–367 (2006)
44. T. Egami, Formation and deformation of metallic glasses: atomistic theory. Intermetallics **14**, 882–887 (2006)
45. L. Li, N. Wang, F. Yan, Transient response in metallic glass deformation: a study based on shear transformation zone dynamics simulations. Scr. Mater. **80**, 25–28 (2014)
46. C.E. Packard, C.A. Schuh, Initiation of shear bands near a stress concentration in metallic glass. Acta Mater. **55**, 5348–5358 (2007)
47. A.C. Lund, C.A. Schuh, Critical length scales for the deformation of amorphous metals containing nanocrystals. Philos. Mag. Lett. **87**, 603–611 (2007)
48. Y. Wang, J. Li, A.V. Hamza, T.W. Barbee, Ductile crystalline-amorphous nanolaminates. Proc. Natl. Acad. Sci. U. S. A. **104**, 11155–11160 (2007)
49. X.L. Fu, Y. Li, C.A. Schuh, Mechanical properties of metallic glass matrix composites: effects of reinforcement character and connectivity. Scr. Mater. **56**, 617–620 (2007)
50. T.G. Nieh, J. Wadsworth, Bypassing shear band nucleation and ductilization of an amorphous-crystalline nanolaminate in tension. Intermetallics **16**, 1156–1159 (2008)
51. D.C. Hofmann, J.-Y. Suh, A. Wiest, G. Duan, M.-L. Lind, M.D. Demetriou, W.L. Johnson, Designing metallic glass matrix composites with high toughness and tensile ductility. Nature **451**, 1085–1089 (2008)
52. S. Scudino, B. Jerliu, S. Pauly, K.B. Surreddi, U. Kühn, J. Eckert, Ductile bulk metallic glasses produced through designed heterogeneities. Scr. Mater. **65**, 815–818 (2011)
53. Y.S. Oh, C.P. Kim, S. Lee, N.J. Kim, Microstructure and tensile properties of high-strength high-ductility Ti-based amorphous matrix composites containing ductile dendrites. Acta Mater. **59**, 7277–7286 (2011)
54. R.T. Qu, J.X. Zhao, M. Stoica, J. Eckert, Z.F. Zhang, Macroscopic tensile plasticity of bulk metallic glass through designed artificial defects. Mater. Sci. Eng. A **534**, 365–373 (2012)
55. L. Li, E.R. Homer, C.A. Schuh, Shear transformation zone dynamics model for metallic glasses incorporating free volume as a state variable. Acta Mater. **61**, 3347–3359 (2013)
56. E.R. Homer, Examining the initial stages of shear localization in amorphous metals. Acta Mater. **63**, 44–53 (2014)

57. C.E. Packard, O. Franke, E.R. Homer, C.A. Schuh, Nanoscale strength distribution in amorphous versus crystalline metals. J. Mater. Res. **25**, 2251–2263 (2010)
58. T.J. Hardin, E.R. Homer, Microstructural factors of strain delocalization in model metallic glass matrix composites. Acta Mater. **83**, 203–215 (2015)
59. C.E. Packard, L.M. Witmer, C.A. Schuh, Hardening of a metallic glass during cyclic loading in the elastic range, Appl Phys Letters. **92**, (2008) 171911. doi:10.1063/1.2919722.

Chapter 15
Nanomechanics of Ferroelectric Thin Films and Heterostructures

Yulan Li, Shengyang Hu, and Long-Qing Chen

15.1 Ferroelectrics

Ferroelectricity was first observed by Valasek in 1920 in Rochelle salt [1]. Ferroelectricity refers to materials that possess a spontaneous electric polarization along a unique crystallographic direction with the additional property that the polarization can be reversed by the application of an external sufficiently strong electric field. Ferroelectrics are a group of materials that posses the property of ferroelectricity. The prefix *ferro*, meaning iron, was borrowed from ferromagnetism, as they both exhibit hysteresis loops, although most ferroelectric materials do not contain iron. Typically, materials only demonstrate ferroelectricity below a certain phase transition temperature, called the Curie temperature, T_c, and are paraelectric above this temperature. Ferroelectrics are very useful for devices and are used in many different ways today such as ferroelectric capacitors, random-access memory (RAM), and radio-frequency identification (RFID) cards.

Barium titanate ($BaTiO_3$) is the most widely used ferroelectric material and is a member of the perovskite family, which is based on the mineral $CaTiO_3$. $BaTiO_3$ has its titanium ion occupying the octahedrally coordinated site and the Ba ion in the 12-fold coordinated site in a high temperature Pm3m cubic symmetry. The ferroelectric phase at room temperature is tetragonal with oxygen and titanium ions shifting to produce a spontaneous polarization. As the temperature

Y. Li (✉) • S. Hu
Pacific Northwest National Laboratory, 902 Battelle Boulevard, Richland, WA 99352, USA
e-mail: yulan.li@pnnl.gov; shenyang.hu@pnnl.gov

L.-Q. Chen
Pennsylvania State University, N-321 Millennium Science Complex, University Park, PA 16802, USA
e-mail: lqc3@psu.edu

C.R. Weinberger, G.J. Tucker (eds.), *Multiscale Materials Modeling for Nanomechanics*, Springer Series in Materials Science 245,
DOI 10.1007/978-3-319-33480-6_15

is reduced, $BaTiO_3$ undergoes a series of ferroelectric–ferroelectric transitions from tetragonal (P4mm) to orthorhombic (Bmm2), around 5 °C, and a transition from orthorhombic to rhombohedral (R3m) at around −80 °C [2, 3]. Recent theoretical and experimental studies showed that the ferroelectric transition temperatures and polarization magnitudes of an epitaxial $BaTiO_3$ thin film can be dramatically altered by a substrate constraint [4–7]. For example, it was recently discovered that a 1.7 % biaxial compressive strain on a (001)-oriented epitaxial $BaTiO_3$ thin film could increase its ferroelectric transition temperature to over 600 °C [6]. The ferroelectric transition between orthorhombic and rhombohedral phases at lower temperatures can disappear under a sufficiently large biaxial tension strain [8].

When a ferroelectric transition occurs from its original cubic paraelectric state to a tetragonal ferroelectric state, the spontaneous polarization can be parallel to one of the $\langle 100\rangle$ directions of the cubic crystallographic directions resulting in a total of six tetragonal domains. Correspondingly, there are 12 possible domains in the orthorhombic phase with the spontaneous polarizations parallel to one of the $\langle 110\rangle$ directions, and eight possible domains in rhombohedral phase with the spontaneous polarization parallel to one of the $\langle 111\rangle$ directions, respectively. A domain refers to a region in the crystal that possesses the same polarization orientations. A typical feature of ferroelectrics is the formation of domain structures when a paraelectric phase is cooled through the ferroelectric transition temperature. The formation of domain structures in ferroelectric materials occurs to accommodate the boundary conditions imposed on the system thus to reduce the depolarization energy and the elastic energy of the material system. The interface between two different domains is called domain wall, which is a thin layer with a thickness usually of only several nanometers. For a given temperature, the stable phases are determined by the minimization of the bulk free energy. The bulk free energy density in the absence of any constraint can be expanded as a polynomial of the components $P_i\,(i = 1, 2, 3)$ of the polarization vector $\mathbf{P} = (P_1, P_2, P_3)$ known as the Landau–Devonshire description [9–11]:

$$\begin{aligned} f_{\text{bulk}}(P_i) = {} & \alpha_1\left(P_1^2+P_2^2+P_3^2\right)+\alpha_{11}\left(P_1^4+P_2^4+P_3^4\right)+\alpha_{12}\left(P_1^2P_2^2+P_2^2P_3^2+P_1^2P_3^2\right) \\ & +\alpha_{111}\left(P_1^6+P_2^6+P_3^6\right)+\alpha_{112}\left[P_1^4\left(P_2^2+P_3^2\right)+P_2^4\left(P_3^2+P_1^2\right)+P_3^4\left(P_1^2+P_2^2\right)\right]+\alpha_{123}P_1^2P_2^2P_3^2 \\ & +\alpha_{1111}\left(P_1^8+P_2^8+P_3^8\right)+\alpha_{1112}\left[P_1^6\left(P_2^2+P_3^2\right)+P_2^6\left(P_3^2+P_1^2\right)+P_3^6\left(P_1^2+P_2^2\right)\right] \\ & +\alpha_{1122}\left(P_1^4P_2^4+P_2^4P_3^4+P_1^4P_3^4\right)+\alpha_{1123}\left(P_1^4P_2^2P_3^2+P_2^4P_1^2P_3^2+P_3^4P_1^2P_2^2\right), \end{aligned} \tag{15.1}$$

where the coefficients α_{ij}, α_{ijk}, and α_{ijkl} are constants. α_1 is linearly dependent on temperature (T) and obeys the Curie–Weiss law $\alpha_1 = 1/(2\varepsilon_0\chi) = (T - T_0)/(2\varepsilon_0 C)$, where C is the Curie–Weiss constant, T_0 is the Curie–Weiss temperature, and ε_0 is the dielectric susceptibility of vacuum. For $BaTiO_3$, these coefficients can be found in Ref. [12] and are listed in Table 15.1. The energy density of Eq. (15.1) with the given coefficients in Table 15.1 yields the transition temperatures $T_{c(\text{Cubic}\leftrightarrow\text{Tetragonal})} = 125$ °C, $T_{c(\text{Tetragonal}\leftrightarrow\text{Orthorhombic})} = 8$ °C, and $T_{c(\text{Orthorhombic}\leftrightarrow\text{Rhombohedral})} = -71$ °C for stress-free $BaTiO_3$ single crystals. The polarizations as a function of temperature obtained from minimizing Eq. (15.1) are displayed in Fig. 15.1.

Table 15.1 Coefficients of Landau–Devonshire description of Eq. (15.1) for $BaTiO_3$ where T is temperature in °C

Coefficients	Values	Units
α_1	$4.124 \times 10^5\ (T - 115)$	$C^{-2}m^2N$
α_{11}	-2.097×10^8	$C^{-4}m^6N$
α_{12}	7.974×10^8	$C^{-4}m^6N$
α_{111}	1.294×10^9	$C^{-6}m^{10}N$
α_{112}	-1.950×10^9	$C^{-6}m^{10}N$
α_{123}	-2.500×10^9	$C^{-6}m^{10}N$
α_{1111}	3.863×10^{10}	$C^{-8}m^{14}N$
α_{1112}	2.529×10^{10}	$C^{-8}m^{14}N$
α_{1122}	1.637×10^{10}	$C^{-8}m^{14}N$
α_{1123}	1.367×10^{10}	$C^{-8}m^{14}N$

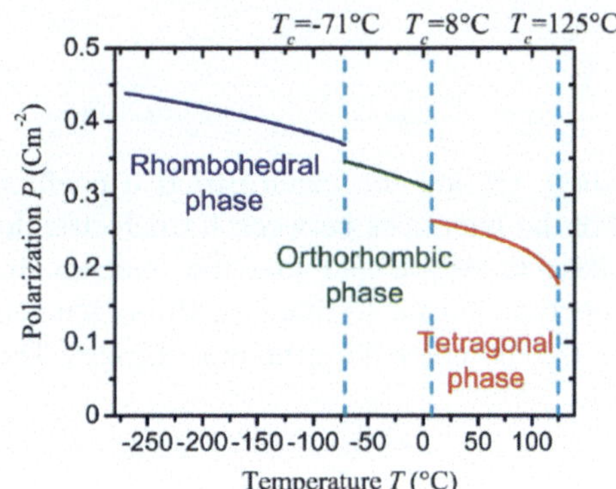

Fig. 15.1 Polarizations versus temperature in $BaTiO_3$ single crystal under stress-free condition, where $P = |\mathbf{P}|, \mathbf{P} = (0, 0, P_T)$ in tetragonal phase, $\mathbf{P} = (P_O, 0, P_O)$ in orthorhombic phase, $\mathbf{P} = (P_R, P_R, P_R)$ in rhombohedral phase [12]

15.2 Nanomechanics of Ferroelectric Thin Films

As all ferroelectric phase transitions are accompanied by a change in their prototypic crystal structure, the lattice parameters and phase stability in ferroelectrics could be drastically changed by external constraints. For example, when a ferroelectric thin film is constrained by a substrate (Fig. 15.2), the strains inside the film can significantly affect both the ferroelectric transition temperature and the domain configuration.

The structural changes associated with ferroelectric phase transition can be described by stress-free strains (also called eigenstrains or spontaneous strains) ε^0_{ij}. In rectangular coordinate $\mathbf{x} = (x_1, x_2, x_3)$, these strains are related to the polarization as:

$$\varepsilon^0_{ij} = q_{ijkl}P_kP_l, \tag{15.2}$$

where q_{ijkl} are the electrostrictive coefficients, which are usually measured experimentally. For a cubic crystal material, Eq. (15.2) degenerates to

$$\begin{aligned} &\varepsilon^0_{11} = Q_{11}P_1^2 + Q_{12}\left(P_2^2 + P_3^2\right), \quad \varepsilon^0_{22} = Q_{11}P_2^2 + Q_{12}\left(P_1^2 + P_3^2\right), \\ &\varepsilon^0_{33} = Q_{11}P_3^2 + Q_{12}\left(P_1^2 + P_2^2\right), \quad \varepsilon^0_{23} = Q_{44}P_2P_3, \varepsilon^0_{13} = Q_{44}P_1P_3, \varepsilon^0_{12} = Q_{44}P_1P_2, \end{aligned} \tag{15.3}$$

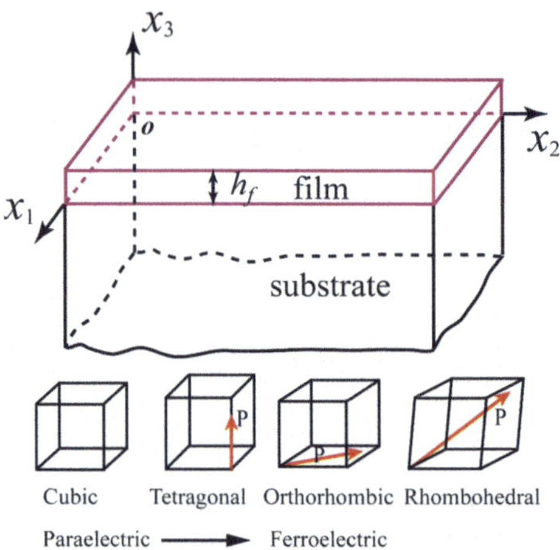

Fig. 15.2 Schematic illustrations of a thin film coherently constrained by a very thick substrate and transition from paraelectric to ferroelectric with changes of polarization directions and crystal structures. h_f refers to the film thickness

where Q_{ij} are the electrostrictive coefficients in Voigt notation. If we assume that the interfaces between ferroelectric domains developed during a ferroelectric phase transition and that the interface between the film and the substrate are coherent, elastic strains e_{ij} will be generated during the phase transition in order to accommodate the structural changes. They are given by

$$e_{ij} = \varepsilon_{ij} - \varepsilon_{ij}^{0}, \quad i, j = 1, 2, 3, \tag{15.4}$$

where ε_{ij} are the total strains. The corresponding elastic strain energy density can be expressed as

$$f_{\text{elastic}}\left(P_i, \varepsilon_{ij}\right) = \frac{1}{2} c_{ijkl} e_{ij} e_{kl} = \frac{1}{2} c_{ijkl}\left(\varepsilon_{ij} - \varepsilon_{ij}^{0}\right)\left(\varepsilon_{kl} - \varepsilon_{kl}^{0}\right), \tag{15.5}$$

where c_{ijkl} is the elastic stiffness tensor. Summation convention for repeated indices is employed and Latin letters i, j, k, l take on values of 1,2,3. For a cubic anisotropic material, elastic strain energy density can be written as

$$\begin{aligned} f_{\text{elastic}} = \tfrac{1}{2} C_{11}\left[\left(\varepsilon_{11} - \varepsilon_{11}^{0}\right)^2 + \left(\varepsilon_{22} - \varepsilon_{22}^{0}\right)^2 + \left(\varepsilon_{33} - \varepsilon_{33}^{0}\right)^2\right] \\ + C_{12}\left[\left(\varepsilon_{11} - \varepsilon_{11}^{0}\right)\left(\varepsilon_{22} - \varepsilon_{22}^{0}\right) + \left(\varepsilon_{22} - \varepsilon_{22}^{0}\right)\left(\varepsilon_{33} - \varepsilon_{33}^{0}\right) + \left(\varepsilon_{11} - \varepsilon_{11}^{0}\right)\left(\varepsilon_{33} - \varepsilon_{33}^{0}\right)\right] \\ + 2C_{44}\left[\left(\varepsilon_{12} - \varepsilon_{12}^{0}\right)^2 + \left(\varepsilon_{23} - \varepsilon_{23}^{0}\right)^2 + \left(\varepsilon_{13} - \varepsilon_{13}^{0}\right)^2\right], \end{aligned} \tag{15.6}$$

where C_{11}, C_{12}, and C_{44} are the elastic stiffness components in Voigt notation. By expanding Eq. (15.6), we have

$$f_{\text{elastic}}=\tfrac{1}{2}C_{11}\left(\varepsilon_{11}^2+\varepsilon_{22}^2+\varepsilon_{33}^2\right)+C_{12}\left(\varepsilon_{11}\varepsilon_{22}+\varepsilon_{22}\varepsilon_{33}+\varepsilon_{11}\varepsilon_{33}\right)+2C_{44}\left(\varepsilon_{12}^2+\varepsilon_{23}^2+\varepsilon_{13}^2\right)$$
$$+\beta_{11}\left(P_1^4+P_2^4+P_3^4\right)+\beta_{12}\left(P_1^2P_2^2+P_2^2P_3^2+P_1^2P_3^2\right)$$
$$-\left(\lambda_{11}\varepsilon_{11}+\lambda_{12}\varepsilon_{22}+\lambda_{12}\varepsilon_{33}\right)P_1^2-\left(\lambda_{12}\varepsilon_{11}+\lambda_{11}\varepsilon_{22}+\lambda_{12}\varepsilon_{33}\right)P_2^2-\left(\lambda_{12}\varepsilon_{11}+\lambda_{12}\varepsilon_{22}+\lambda_{11}\varepsilon_{33}\right)P_3^2$$
$$-2\lambda_{44}\left(\varepsilon_{12}P_1P_2+\varepsilon_{23}P_2P_3+\varepsilon_{13}P_1P_3\right), \tag{15.7}$$

where,

$$\beta_{11}=\tfrac{1}{2}C_{11}\left(Q_{11}^2+2Q_{12}^2\right)+C_{12}Q_{12}\left(2Q_{11}+Q_{12}\right),$$
$$\beta_{12}=C_{12}\left(Q_{11}^2+3Q_{12}^2+2Q_{11}Q_{12}\right)+C_{11}Q_{12}\left(2Q_{11}+Q_{12}\right)+2C_{44}Q_{44}^2,$$
$$\lambda_{11}=C_{11}Q_{11}+2C_{12}Q_{12},\quad \lambda_{12}=C_{11}Q_{12}+C_{12}\left(Q_{11}+Q_{12}\right),\quad \lambda_{44}=2C_{44}Q_{44}. \tag{15.8}$$

It can be seen that the presence of the strains and the elastic energy alters the coefficients of the polarization polynomial of Eq. (15.1), thus changes both the Curie temperature and polarization direction as well as the relative volume fractions of different domains when multiple domains coexist.

In order to calculate the strain field in a constrained film, small strains are assumed such that linear elasticity can be employed. The associated stresses σ_{ij} obey the Hooke's law,

$$\sigma_{ij}=c_{ijkl}e_{kl}=c_{ijkl}\left(\varepsilon_{kl}-\varepsilon_{kl}^0\right). \tag{15.9}$$

For a cubic anisotropic material, the stresses can be expanded as

$$\sigma_{11}=C_{11}\left(\varepsilon_{11}-\varepsilon_{11}^0\right)+C_{12}\left(\varepsilon_{22}-\varepsilon_{22}^0\right)+C_{12}\left(\varepsilon_{33}-\varepsilon_{33}^0\right),$$
$$\sigma_{22}=C_{12}\left(\varepsilon_{11}-\varepsilon_{11}^0\right)+C_{11}\left(\varepsilon_{22}-\varepsilon_{22}^0\right)+C_{12}\left(\varepsilon_{33}-\varepsilon_{33}^0\right),$$
$$\sigma_{33}=C_{12}\left(\varepsilon_{11}-\varepsilon_{11}^0\right)+C_{12}\left(\varepsilon_{22}-\varepsilon_{22}^0\right)+C_{11}\left(\varepsilon_{33}-\varepsilon_{33}^0\right),$$
$$\sigma_{23}=2C_{44}\left(\varepsilon_{23}-\varepsilon_{23}^0\right),\quad \sigma_{13}=2C_{44}\left(\varepsilon_{13}-\varepsilon_{13}^0\right),\quad \sigma_{12}=2C_{44}\left(\varepsilon_{12}-\varepsilon_{12}^0\right). \tag{15.10}$$

The stresses satisfy mechanical equilibrium which can be expressed as

$$\sigma_{ij,j}=\partial\sigma_{ij}/\partial x_j=0,\ (i=1,2,3), \tag{15.11}$$

in the absence of body forces. Moreover, these stresses are required to satisfy any appropriate traction boundary conditions imposed on the studied film. If we consider a film with a stress-free surface, this means

$$\sigma_{i3}\big|_{\text{film-surface: } x_3=h_f}=0\quad (i=1,2,3), \tag{15.12}$$

when the x_3-axis is perpendicular to the film plane, as shown in Fig. 15.2.

If the thin film is coherent with the substrate, then the total strain field in the thin film is controlled by the thick substrate. For example, for a cubic substrate of lattice parameter, a_s, and a thin film with a stress-free lattice parameter, a_f, the corresponding in-plane strains, $\bar{\varepsilon}_{\alpha\beta}$ $(\alpha, \beta = 1, 2)$, are given by

$$\bar{\varepsilon}_{11} = \bar{\varepsilon}_{22} = \frac{a_s - a_f}{a_s} = e_0, \quad \bar{\varepsilon}_{12} = 0. \tag{15.13}$$

In the following we will show how to solve the strains and stresses as well as the elastic energy for a film containing different domain structures.

(1) Single tetragonal c-domain

When a ferroelectric film contains only a single tetragonal c-domain (a domain whose polarization is normal to the film), the polarization in such a case has the form $\mathbf{P} = (0, 0, P_s)$, where P_s is the value of the spontaneous polarization. The stress-free strain caused by the polarization is given by $\varepsilon^0_{11} = Q_{12}P_s^2$, $\varepsilon^0_{22} = Q_{12}P_s^2$, $\varepsilon^0_{33} = Q_{11}P_s^2$, $\varepsilon^0_{23} = 0$, $\varepsilon^0_{13} = 0$, $\varepsilon^0_{12} = 0$.

Because it is a single domain film, the corresponding stresses and strains are constants over the whole film. Therefore, the mechanical equilibrium Eq. (15.11) is satisfied automatically. The boundary condition (15.12) at the film surface requires $\sigma_{13} = 2C_{44}\varepsilon_{13} = 0$, $\sigma_{23} = 2C_{44}\varepsilon_{23} = 0$, $\sigma_{33} = C_{12}\left(\varepsilon_{11} - Q_{12}P_s^2\right) + C_{12}\left(\varepsilon_{22} - Q_{12}P_s^2\right) + C_{11}\left(\varepsilon_{33} - Q_{11}P_s^2\right) = 0$, and the boundary condition (15.13) from the substrate makes $\varepsilon_{12} = \bar{\varepsilon}_{12} = 0$ and $\varepsilon_{11} = \varepsilon_{22} = \bar{\varepsilon}_{11} = \bar{\varepsilon}_{22} = e_0$. Therefore, we have $\varepsilon_{13} = 0$, $\varepsilon_{23} = 0$, and $\varepsilon_{33} = \frac{(C_{11}Q_{11} + 2C_{12}Q_{12})P_s^2 - 2C_{12}e_0}{C_{11}}$. Thus, the corresponding elastic energy density becomes

$$f_{\text{elastic}/c} = \frac{(C_{11} - C_{12})(C_{11} + 2C_{12})}{C_{11}}\left(e_0 - Q_{12}P_s^2\right)^2. \tag{15.14}$$

At given substrate constraint strain (e_0) and temperature (T), the value of the polarization (P_s) can be obtained by minimizing the total energy of the film with respect to the polarization. The total energy density is the sum of the elastic energy in Eq. (15.14) and the bulk free energy density in Eq. (15.1) with $P_1 = P_2 = 0$, $P_3 = P_s$, i.e.,

$$f_{c\text{-domain}}(P_s) = (\alpha_1 + a_3)P_s^2 + (\alpha_{11} + a_{33})P_s^4 + \alpha_{111}P_s^6 + \alpha_{1111}P_s^8 + a_0, \tag{15.15}$$

where,

$$a_3 = -\frac{2(C_{11} - C_{12})(C_{11} + 2C_{12})Q_{12}e_0}{C_{11}},$$
$$a_{33} = \frac{(C_{11} - C_{12})(C_{11} + 2C_{12})Q_{12}^2}{C_{11}}, \quad a_0 = \frac{(C_{11} - C_{12})(C_{11} + 2C_{12})e_0^2}{C_{11}}. \tag{15.16}$$

Therefore, the polarization, P_s, can be obtained by solving $\partial f_{c\text{-domain}}(P_s)/\partial P_s$.

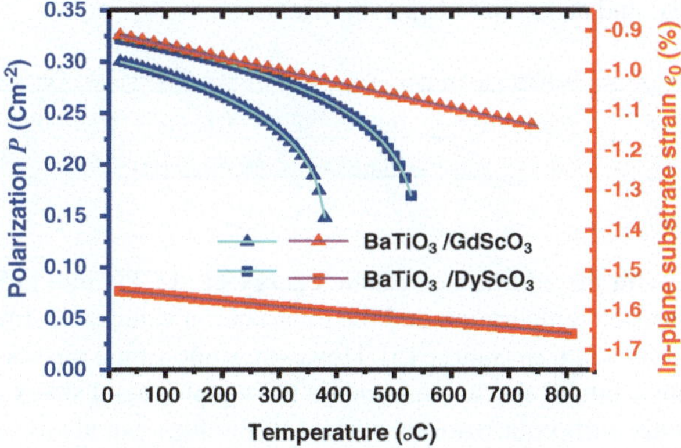

Fig. 15.3 In-plane strain (e_0) and polarization ($P = P_s$) versus temperature in the (001) $BaTiO_3$ films commensurately grown on (110) $GdScO_3$ and $DyScO_3$ substrates, respectively

Experimentally, single crystal $BaTiO_3$ thin films were successfully grown on (110) $GdScO_3$ and (110) $DyScO_3$ single-crystal substrates by both reactive molecular beam epitaxy (MBE) and pulsed-laser deposition (PLD) with in situ high-pressure reflection high-energy electron diffraction [6]. These films are epitaxial, purely *c*-axis oriented, and fully coherent with the substrates without any resolvable lattice relaxation. The in-plane and out-of-plane lattice parameters of the films and substrates as a function of temperature were measured with a variable-temperature four-circle X-ray diffractometer equipped with a two-dimensional area detector with an angular resolution of ~0.02°. From these measurements, the corresponding substrate constraint strain, e_0, is calculated and plotted in Fig. 15.3. For the constraint strain at the associated temperature, the corresponding spontaneous polarization can be calculated and is plotted in the same figure. In the calculations, the elastic constants and electrostrictive coefficients used were $C_{11} = 1.78 \times 10^{11}$, $C_{12} = 0.964 \times 10^{11}$, $C_{44} = 1.22 \times 10^{11}$ (Nm^{-2}), $Q_{11} = 0.10$, $Q_{12} = -0.034$, $Q_{44} = 0.029$ ($C^{-2}m^4$) [7, 13–15], and the coefficients of Landau–Devonshire description of Eq. (15.1) were from Table 15.1. Comparing with the tetragonal phase in Fig. 15.1, both the ferroelectric transition temperature, T_c, and spontaneous polarization, P_s, changed significantly.

(2) Arbitrary single domain

For a film with a single phase of an arbitrary polarization $\mathbf{P} = (P_1, P_2, P_3)$, the film boundary conditions of Eqs. (15.12) and (15.13) yield: $\varepsilon_{12}=0$, $\varepsilon_{11}=\varepsilon_{22}=e_0$, $\varepsilon_{23}=Q_{44}P_2P_3$, $\varepsilon_{13}=Q_{44}P_1P_3$, $\varepsilon_{33}=\frac{C_{11}Q_{12}+C_{12}(Q_{11}+Q_{12})}{C_{11}}\left(P_1^2+P_2^2\right)+\frac{(C_{11}Q_{11}+2C_{12}Q_{12})}{C_{11}}P_3^2-\frac{2C_{12}}{C_{11}}e_0$. The corresponding elastic energy density is formulated as

$$f_{\text{elastic}/s} = a_0 + a_1\left(P_1^2 + P_2^2\right) + a_3P_3^2 + a_{11}\left(P_1^4 + P_2^4\right) + a_{33}P_3^4 + a_{12}P_1^2P_2^2 + a_{13}\left(P_1^2P_3^2 + P_2^2P_3^2\right), \quad (15.17)$$

with a_3, a_{33}, and a_0 given in Eq. (15.16), and

$$a_1 = -\frac{(C_{11}-C_{12})(C_{11}+2C_{12})(Q_{11}+Q_{12})e_0}{C_{11}}, \quad a_{11} = \frac{C_{11}^2\left(Q_{11}^2+Q_{12}^2\right)-C_{12}^2(Q_{11}+Q_{12})^2+2C_{11}C_{12}Q_{11}Q_{12}}{C_{11}},$$
$$a_{12} = \frac{2C_{11}^2Q_{11}Q_{12}-C_{12}^2(Q_{11}+Q_{12})^2+C_{11}C_{12}\left(Q_{11}^2+Q_{12}^2\right)+2C_{11}C_{44}Q_{44}^2}{C_{11}},$$
$$a_{13} = \frac{(C_{11}-C_{12})(C_{11}+2C_{12})Q_{12}(Q_{11}+Q_{12})}{C_{11}}. \quad (15.18)$$

Obviously, with the addition of elastic energy of (15.17) into (15.1), the ferroelectric transition temperature and the polarization values and directions can be alerted notably when e_0 is large [5]. However, some single phases with arbitrary polarizations cannot fully accommodate the equal biaxial stress. Nevertheless, single crystals with polarizations normal to the film can always accommodate any equal biaxial strain. Thus, if ferroelectrics are assumed to always form single domains, the results could be inaccurate resulting in incorrect predictions of phase diagrams of the ferroelectric material as a function of temperature and constraint strain [5].

(3) Twinned tetragonal domains

The phase transition from paraelectric to ferroelectric can lead to the formation of a twin domain structure, i.e., a structure consisting of two kinds of domains. In order to form a twin domain structure, the polarizations in the twin domains must have a specific relationship. For example, the tetragonal phase $a_1:(P_s,0,0)/a_2:(0,P_s,0)$ can form a twin structure as shown in Fig. 15.4. This twin domain structure has domain walls orthogonal to the film/substrate interface and orientated along the $\{110\}$ planes of the prototypic cubic phase. The strain and stress distributions in the twin structure are calculated below.

In the twin structure shown in Fig. 15.4, the equilibrium volume fractions of $a_1:(P_s,0,0)$ and $a_2:(0,P_s,0)$ are equal to each other and the spontaneous polarization P_s has the same magnitude in the two domains. The corresponding elastic energy density is calculated as

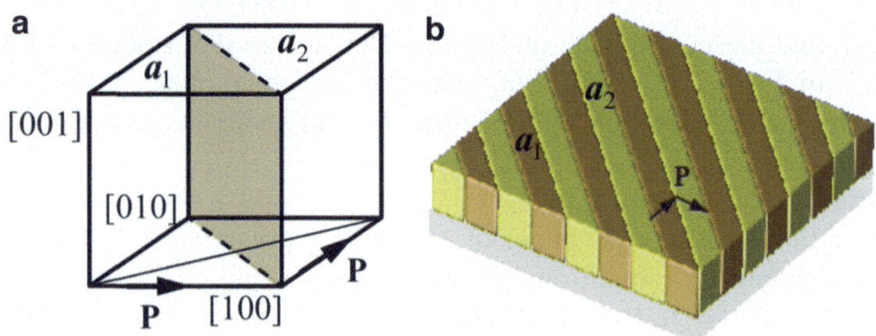

Fig. 15.4 (**a**) Schematic illustration of a twin tetragonal a_1/a_2 structure; (**b**) A domain structure of twinned a_1/a_2 tetragonal phases obtained in a $BaTiO_3$ film

$$f_{\text{elastic}/a_1a_2} = \frac{1}{2}f_{\text{elastic}/a_1} + \frac{1}{2}f_{\text{elastic}/a_2}, \tag{15.19}$$

where the factor ½ represents the volume fraction of one of the twin domains.

In domain $a_1 : (P_s, 0, 0)$, the corresponding strains and stresses are $\{\varepsilon_{11}^{a_1}, \varepsilon_{22}^{a_1}, \varepsilon_{33}^{a_1}, \varepsilon_{23}^{a_1}, \varepsilon_{13}^{a_1}, \varepsilon_{12}^{a_1}\}$ and $\{\sigma_{11}^{a_1}, \sigma_{22}^{a_1}, \sigma_{33}^{a_1}, \sigma_{23}^{a_1}, \sigma_{13}^{a_1}, \sigma_{12}^{a_1}\}$, respectively. In domain $a_2 : (0, P_s, 0)$, the corresponding strains and stresses are likewise $\{\varepsilon_{11}^{a_2}, \varepsilon_{22}^{a_2}, \varepsilon_{33}^{a_2}, \varepsilon_{23}^{a_2}, \varepsilon_{13}^{a_2}, \varepsilon_{12}^{a_2}\}$ and $\{\sigma_{11}^{a_2}, \sigma_{22}^{a_2}, \sigma_{33}^{a_2}, \sigma_{23}^{a_2}, \sigma_{13}^{a_2}, \sigma_{12}^{a_2}\}$, respectively. They are required to satisfy the film surface boundary condition and substrate constraint conditions together. These conditions in the twinned structure are converted to

$$\frac{1}{2}\sigma_{33}^{a_1} + \frac{1}{2}\sigma_{33}^{a_2} = 0, \quad \frac{1}{2}\sigma_{13}^{a_1} + \frac{1}{2}\sigma_{13}^{a_2} = 0, \quad \frac{1}{2}\sigma_{23}^{a_1} + \frac{1}{2}\sigma_{23}^{a_2} = 0, \tag{15.20}$$

$$\frac{1}{2}\varepsilon_{11}^{a_1} + \frac{1}{2}\varepsilon_{11}^{a_2} = e_0, \quad \frac{1}{2}\varepsilon_{22}^{a_1} + \frac{1}{2}\varepsilon_{22}^{a_2} = e_0, \quad \frac{1}{2}\varepsilon_{12}^{a_1} + \frac{1}{2}\varepsilon_{12}^{a_2} = 0. \tag{15.21}$$

At the interface between the two domains, i.e., at the plane of (110) shown in Fig. 15.4a, the associated stresses and strains are continuous across the interface, i.e.,

$$\begin{gathered}\sigma_{11}^{a_1}+\sigma_{22}^{a_1}+2\sigma_{12}^{a_1}=\sigma_{11}^{a_2}+\sigma_{22}^{a_2}+2\sigma_{12}^{a_2}, \quad \sigma_{11}^{a_1}-\sigma_{22}^{a_1}=\sigma_{11}^{a_2}-\sigma_{22}^{a_2}, \quad \sigma_{13}^{a_1}+\sigma_{23}^{a_1}=\sigma_{13}^{a_2}+\sigma_{23}^{a_2}, \\ \varepsilon_{33}^{a_1}=\varepsilon_{33}^{a_2}, \quad \varepsilon_{11}^{a_1}+\varepsilon_{22}^{a_1}-2\varepsilon_{12}^{a_1}=\varepsilon_{11}^{a_2}+\varepsilon_{22}^{a_2}-2\varepsilon_{12}^{a_2}, \quad \varepsilon_{23}^{a_1}-\varepsilon_{13}^{a_1}=\varepsilon_{23}^{a_2}-\varepsilon_{13}^{a_2}.\end{gathered} \tag{15.22}$$

The stresses $\sigma_{ij}^{a_1}$ and $\sigma_{ij}^{a_2}$ are related to strains $\varepsilon_{ij}^{a_1}$ and $\varepsilon_{ij}^{a_2}$ through Eq. (15.10) in domains a_1 and a_2, respectively, with $\varepsilon_{11}^{0a_1} = Q_{11}P_s^2$, $\varepsilon_{22}^{0a_1} = Q_{12}P_s^2$, $\varepsilon_{33}^{0a_1} = Q_{12}P_s^2$, $\varepsilon_{12}^{0a_1} = 0$, $\varepsilon_{23}^{0a_1} = 0$, $\varepsilon_{13}^{0a_1} = 0$, and $\varepsilon_{11}^{0a_2} = Q_{12}P_s^2$, $\varepsilon_{22}^{0a_2} = Q_{11}P_s^2$, $\varepsilon_{33}^{0a_2} = Q_{12}P_s^2$, $\varepsilon_{12}^{0a_2} = 0$, $\varepsilon_{23}^{0a_2} = 0$, $\varepsilon_{13}^{0a_2} = 0$. By solving Eqs. (15.20)–(15.22), we have $\varepsilon_{11}^{a_1} = e_0 + \frac{1}{2}P_s^2(Q_{11} - Q_{12})$, $\varepsilon_{22}^{a_1} = e_0 - \frac{1}{2}P_s^2(Q_{11} - Q_{12})$, $\varepsilon_{33}^{a_1} = \frac{-2C_{12}e_0+[C_{11}Q_{12}+C_{12}(Q_{11}+Q_{12})]P_s^2}{C_{11}}$, $\varepsilon_{12}^{a_1} = 0$, $\varepsilon_{13}^{a_1} = 0$, $\varepsilon_{23}^{a_1} = 0$, $\sigma_{11}^{a_1} = \sigma_{22}^{a_1} = \frac{(C_{11}-C_{12})(C_{11}+2C_{12})\left[2e_0-(Q_{11}+Q_{12})P_s^2\right]}{2C_{11}}$, $\sigma_{33}^{a_1} = \sigma_{12}^{a_1} = \sigma_{13}^{a_1} = \sigma_{23}^{a_1} = 0$, and $\varepsilon_{11}^{a_2} = e_0 - \frac{1}{2}P_s^2(Q_{11} - Q_{12})$, $\varepsilon_{22}^{a_2} = e_0 + \frac{1}{2}P_s^2(Q_{11} - Q_{12})$, $\varepsilon_{33}^{a_2} = \frac{-2C_{12}e_0+[C_{11}Q_{12}+C_{12}(Q_{11}+Q_{12})]P_s^2}{C_{11}}$, $\varepsilon_{12}^{a_2} = 0$, $\varepsilon_{13}^{a_2} = 0$, $\varepsilon_{23}^{a_2} = 0$, $\sigma_{11}^{a_2} = \sigma_{22}^{a_2} = \frac{(C_{11}-C_{12})(C_{11}+2C_{12})\left[2e_0-(Q_{11}+Q_{12})P_s^2\right]}{2C_{11}}$, $\sigma_{33}^{a_2} = \sigma_{12}^{a_2} = \sigma_{13}^{a_2} = \sigma_{23}^{a_2} = 0$. The corresponding elastic energy density is of the expression:

$$f_{\text{elastic}/a_1a_2} = \frac{(C_{11} - C_{12})(C_{11} + 2C_{12})\left[-2e_0 + P_s^2(Q_{11} + Q_{12})\right]^2}{4C_{11}}. \tag{15.23}$$

It will be seen later in Sect. 15.4 that such a tetragonal $a_1 : (P_s, 0, 0)/a_2 : (0, P_s, 0)$ twin structure can be stabilized under tensile strain e_0 at relative higher temperature.

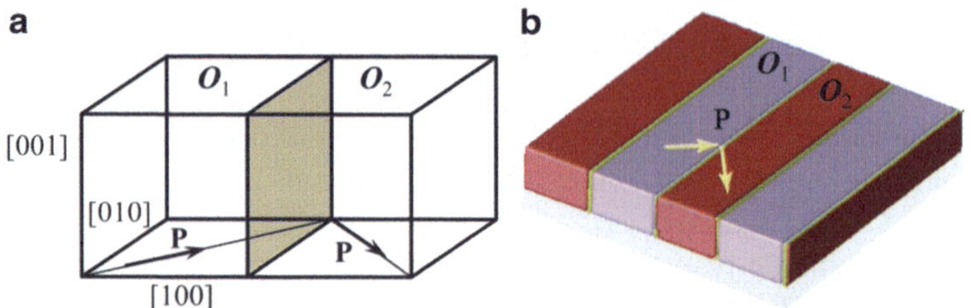

Fig. 15.5 (**a**) Schematic illustration of a twin orthorhombic O_1/O_2 structure; (**b**) A domain structure of twinned O_1/O_2 orthorhombic phases obtained in a $BaTiO_3$ film

(4) Twinned orthorhombic domains

The twin structure shown in Fig. 15.5 also exists for the orthorhombic phases $O_1 : \left(P_s/\sqrt{2}, P_s/\sqrt{2}, 0\right)$ and $O_2 : \left(P_s/\sqrt{2}, -P_s/\sqrt{2}, 0\right)$. This twinned structure has domain walls orthogonal to the film/substrate interface but orientated along the (100) or (010) planes of the prototypic cubic phase. In such a structure, the equilibrium volume fractions of O_1-domain and O_2-domain are also equal to each other and the spontaneous polarization P_s has the same magnitude in the two domains.

The corresponding elastic energy density can be evaluated by

$$f_{\text{elastic}/O_1O_2} = \frac{1}{2} f_{\text{elastic}/O_1} + \frac{1}{2} f_{\text{elastic}/O_2}. \tag{15.24}$$

In domain $O_1 : \left(P_s/\sqrt{2}, P_s/\sqrt{2}, 0\right)$, the corresponding strains and stresses are $\{\varepsilon_{11}^{O_1}, \varepsilon_{22}^{O_1}, \varepsilon_{33}^{O_1}, \varepsilon_{23}^{O_1}, \varepsilon_{13}^{O_1}, \varepsilon_{12}^{O_1}\}$ and $\{\sigma_{11}^{O_1}, \sigma_{22}^{O_1}, \sigma_{33}^{O_1}, \sigma_{23}^{O_1}, \sigma_{13}^{O_1}, \sigma_{12}^{O_1}\}$, respectively. Similarly, in domain $O_2 : \left(P_s/\sqrt{2}, -P_s/\sqrt{2}, 0\right)$, the corresponding strains and stresses are $\{\varepsilon_{11}^{O_2}, \varepsilon_{22}^{O_2}, \varepsilon_{33}^{O_2}, \varepsilon_{23}^{O_2}, \varepsilon_{13}^{O_2}, \varepsilon_{12}^{O_2}\}$ and $\{\sigma_{11}^{O_2}, \sigma_{22}^{O_2}, \sigma_{33}^{O_2}, \sigma_{23}^{O_2}, \sigma_{13}^{O_2}, \sigma_{12}^{O_2}\}$, respectively. They are required to satisfy the film surface stress-free boundary condition and substrate constraint condition:

$$\frac{1}{2}\sigma_{33}^{O_1} + \frac{1}{2}\sigma_{33}^{O_2} = 0, \quad \frac{1}{2}\sigma_{13}^{O_1} + \frac{1}{2}\sigma_{13}^{O_2} = 0, \quad \frac{1}{2}\sigma_{23}^{O_1} + \frac{1}{2}\sigma_{23}^{O_2} = 0, \tag{15.25}$$

$$\frac{1}{2}\varepsilon_{11}^{O_1} + \frac{1}{2}\varepsilon_{11}^{O_2} = e_0, \quad \frac{1}{2}\varepsilon_{22}^{O_1} + \frac{1}{2}\varepsilon_{22}^{O_2} = e_0, \quad \frac{1}{2}\varepsilon_{12}^{O_1} + \frac{1}{2}\varepsilon_{12}^{O_2} = 0. \tag{15.26}$$

At the interface between the two domains, the associated stresses and strains are required to be continuous, i.e.,

$$\sigma_{11}^{O_1} = \sigma_{11}^{O_2}, \quad \sigma_{12}^{O_1} = \sigma_{12}^{O_2}, \quad \sigma_{13}^{O_1} = \sigma_{13}^{O_2}, \quad \varepsilon_{22}^{O_1} = \varepsilon_{22}^{O_2}, \quad \varepsilon_{33}^{O_1} = \varepsilon_{33}^{O_2}, \quad \varepsilon_{23}^{O_1} = \varepsilon_{23}^{O_2}. \tag{15.27}$$

The solutions of Eqs. (15.25)–(15.27) are

$$\varepsilon_{11}^{O_1} = \varepsilon_{11}^{O_2} = e_0, \quad \varepsilon_{12}^{O_1} = -\varepsilon_{12}^{O_2} = \tfrac{1}{2}Q_{44}P_s^2 \quad \varepsilon_{13}^{O_1} = \varepsilon_{13}^{O_2} = 0,$$
$$\varepsilon_{23}^{O_1}=\varepsilon_{23}^{O_2}=0, \quad \varepsilon_{22}^{O_1}=\varepsilon_{22}^{O_2}=e_0, \quad \varepsilon_{33}^{O_1}=\varepsilon_{3}^{O_2}= -\tfrac{C_{12}}{C_{11}}\left[2e_0-(Q_{11}+Q_{12})\,P_s^2\right]+Q_{12}P_s^2, \tag{15.28}$$

since $\varepsilon_{11}^{0O_1}=\frac{1}{2}(Q_{11}+Q_{12})\,P_s^2$, $\varepsilon_{22}^{0O_1}=\frac{1}{2}(Q_{11}+Q_{12})\,P_s^2$, $\varepsilon_{33}^{0O_1}=Q_{12}P_s^2$, $\varepsilon_{23}^{0O_1}=0$, $\varepsilon_{13}^{0O_1}=0$, $\varepsilon_{12}^{0O_1}=\frac{1}{2}Q_{44}P_s^2$, and $\varepsilon_{11}^{0O_2} = \frac{1}{2}(Q_{11}+Q_{12})\,P_s^2$, $\varepsilon_{22}^{0O_2} = \frac{1}{2}(Q_{11}+Q_{12})$ P_s^2, $\varepsilon_{33}^{0O_2} = Q_{12}P_s^2$, $\varepsilon_{23}^{0O_2} = 0$, $\varepsilon_{13}^{0O_2} = 0$, $\varepsilon_{12}^{0O_2} = -\frac{1}{2}Q_{44}P_s^2$. Therefore, we have

$$f_{\text{elastic}/O_1O_2}=\frac{(C_{11}-C_{12})\,(C_{11}+2C_{12})}{4C_{11}}\left[(Q_{11}+Q_{12})^2P_s^4-4e_0\,(Q_{11}+Q_{12})\,P_s^2+4e_0^2\right] \tag{15.29}$$

We will see later in Sect. 15.4 that such an orthorhombic O_1: $\left(P_s/\sqrt{2}, P_s/\sqrt{2}, 0\right)$ and O_2 : $\left(P_s/\sqrt{2}, -P_s/\sqrt{2}, 0\right)$ twin structure can be stabilized under larger tensile strain e_0 at quite a range of temperature down to 0 K.

In addition to the two kinds of twin domain structures discussed above, Koukhar et al. have considered more twin domain structures in their work [16]. It should be borne in mind, however, that not all these considered twin structures can completely accommodate with the equal biaxial in-plane constraint and not all domain walls lie along the {100} or {110} planes. For example, the twin domain structures, either $c/a_1/$ or $c/a_2/$, cannot exist alone to accommodate the equal biaxial substrate constraint. Their combination, i.e., $c/a_1/a_2$ domain structures can be stabilized under certain constraints and temperatures. This will be seen from Sect. 15.4 of ferroelectric domain heterostructure.

(5) Heterostructure case

In the general case, the domain shapes and domain wall configurations in constrained films can be much more complicated than we have seen so far. For the general case, the internal strains and stresses cannot be solved analytically. They have to be solved numerically.

Consider a ferroelectric thin film grown on a substrate. Below the Curie temperature such that a ferroelectric transition occurs and a spontaneous polarization $\mathbf{P} = (P_1, P_2, P_3)$ exists. However, the polarization is not homogeneous over the film but spatially dependent, i.e., $\mathbf{P}(\mathbf{x})$. Since the proper ferroelectric phase transition involves structural change, a relation of Eqs. (15.2) or (15.3) between spontaneous or stress-free strains and spontaneous polarization is employed.

In order to consider the constraint strains on the ferroelectric film from its underlying substrate, the total strain of the film is separated into two parts. One part is a homogeneous strain $\overline{\varepsilon}_{ij}$ that is the same for each point of the film. The other part is a heterogeneous strain, $\eta_{ij}(\mathbf{x})$. Therefore,

$$\varepsilon_{ij}(\mathbf{x}) = \overline{\varepsilon}_{ij} + \eta_{ij}(\mathbf{x}). \tag{15.30}$$

We let $\overline{\varepsilon}_{\alpha\beta}\ (\alpha, \beta = 1, 2)$ represent the macroscopic shape deformation of the film in the film plane. This means that $\iiint\limits_V \eta_{\alpha\beta}(\mathbf{x})\, dV = 0$, $(\alpha, \beta = 1, 2)$. For a thin film grown coherently on a thick substrate, the macroscopic deformation of the film in the film plane is totally controlled by the sufficiently thick substrate. Equation (15.13) gives the corresponding $\overline{\varepsilon}_{\alpha\beta}\ (\alpha, \beta = 1, 2)$ for the case of a film with cubic crystal structure grown coherently on a substrate of cubic crystal structure. The macroscopic deformation of the film along the x_3 direction is determined by both the substrate constraint and the domain structure in the film. Since at film surface, the stress-free condition requires $\sigma_{i3}\left|_{\text{film-surface: } x_3=h_f}\right. = 0$, i.e., $c_{i3kl}\left(\overline{\varepsilon}_{kl} + \eta_{kl} - \varepsilon^0_{kl}\right)\left|_{\text{film-surface: } x_3=h_f}\right. = 0$, we choose the quantity $\overline{\varepsilon}_{i3}\ (i = 1, 2, 3)$ in such a way that makes $c_{i3kl}\overline{\varepsilon}_{kl} = 0$. Then the stress-free of the film top surface becomes

$$c_{i3kl}\left(\eta_{kl} - \varepsilon^0_{kl}\right)\left|_{\text{film-surface: } x_3=h_f}\right. = 0. \tag{15.31}$$

However, it needs to be pointed out that $\overline{\varepsilon}_{i3}\ (i = 1, 2, 3)$ employed here is only part of the total deformation of the film.

Assuming that the displacement $u_i(\mathbf{x})$ is associated with the heterogeneous strain, $\eta_{ij}(\mathbf{x})$, i.e.,

$$\eta_{ij}(\mathbf{x}) = \frac{1}{2}\left[u_{i,j}(\mathbf{x}) + u_{j,i}(\mathbf{x})\right], \tag{15.32}$$

the mechanical equilibrium equations can be rewritten as

$$c_{ijkl}u_{k,lj}(\mathbf{x}) = c_{ijkl}\varepsilon^0_{kl,j}(\mathbf{x}). \tag{15.33}$$

And the film surface stress-free condition becomes

$$c_{i3kl}\left(u_{k,l} - \varepsilon^0_{kl}\right)\left|_{\text{film-surface: } x_3=h_f}\right. = 0. \tag{15.34}$$

Because the elastic perturbation resulting from the heterogeneous strain disappears in the substrate far away from the film-substrate interface, one can use the following condition:

$$u_i\left|_{x_3=-h_s}\right. = 0 \tag{15.35}$$

to replace the constraint of the substrate. In the equation, h_s is the distance from the film-substrate interface into the substrate, beyond which the elastic deformation is ignored. Actually it was verified numerically that the domain shapes practically do not change when h_s exceeds about half of the film thickness [17].

After solving Eq. (15.33) under boundary conditions (15.34) and (15.35) with given polarization distribution, the strain field $\varepsilon_{ij}(\mathbf{x})$ can be calculated from Eqs. (15.30) and (15.32). For the sake of simplicity, the elastic properties of the film and the substrate are assumed to be the same although the elastic anisotropy can be arbitrary. An efficient and accurate numerical method was proposed to solve $u_i(\mathbf{x})$ from Eqs. (15.33)–(15.35) [17]. It converts a three-dimensional (3D) problem into an out-of-plane one-dimensional (1D) problem by using the fast Fourier transformation (FFT) and gives an analytical solution along the out-of-plane direction. Thus the obtained solution is a semi-analytical solution which satisfies the boundary condition (15.34) and (15.35) analytically.

Figure 15.6a illustrates the distribution of the displacement component $u_3(\mathbf{x})$ on the top surface of a film with a given domain structure of Fig. 15.6b. The domain structure of Fig. 15.6b is from a phase field simulation [17]. The displacements were the solution of Eqs. (15.33)–(15.35) with the given domain structure, i.e., the distribution of $P_i\,(\mathbf{x})$ $(i = 1, 2, 3)$. This shows that the wrinkly surface reflects the domain morphology of the film. Since ferroelectric domain sizes are on the order of nanometers, the displacements at the thin film free surface are also on the order of nanometers or less.

15.3 Phase Field Method

To investigate the stability of ferroelectric phases and the details of the associated ferroelectric domain structure in a thin film constrained by its underlying substrate, the phase field method (PFM) is widely used [17–36] and is reviewed in detail by Chen [37]. The PFM for ferroelectrics is based on the Landau–Devonshire theory of phase transition and searched for the minimum energy state of the system using the time-dependent Ginzburg–Landau (TDGL) evolution equation. The TDGL

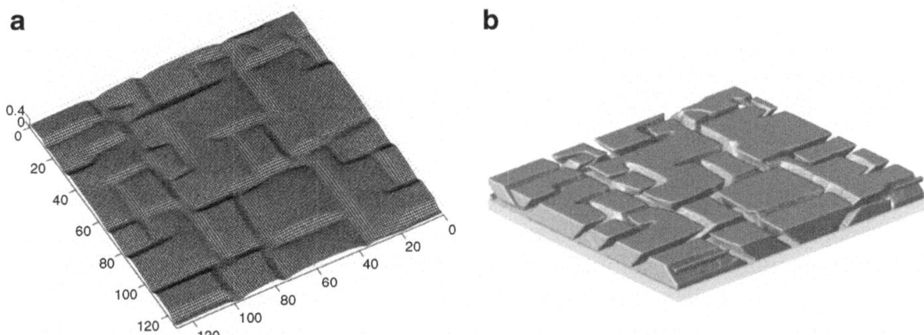

Fig. 15.6 (**a**) Displacement pattern of the x_3 component u_3 on the surface of the ferroelectric film with domain structure of (**b**). In (**b**), the solid areas are tetragonal c-domains and the open areas are tetragonal a_1/a_2-domains [17]

equations in terms of the polarization components $P_i(\mathbf{x})$ are

$$\frac{\partial P_i(\mathbf{x},t)}{\partial t} = -L\frac{\delta F}{\delta P_i(\mathbf{x},t)},\ (i = 1,2,3), \tag{15.36}$$

where L is the kinetic coefficient, and F is the total free energy of the system. $\delta F/\delta P_i(\mathbf{x},t)$ is the thermodynamic driving force for the spatial and temporal evolution of $P_i(\mathbf{x},t)$. The total free energy of the system includes bulk free energy, elastic energy, domain wall energy, and the depolarization energy. In order to distinguish the impact of the substrate mechanical constraint on the domain structures of the film, the depolarization energy considered in Ref. [19] is neglected here. Therefore, the total energy is

$$F = \iiint_V [f_{\text{bulk}} + f_{\text{elastic}} + f_{\text{wall}}]\,dV, \tag{15.37}$$

where the densities of bulk free energy and elastic energy are given in Eqs. (15.1) and (15.5), respectively.

The domain wall energy is evaluated through the gradients of the polarization field. For a general anisotropic system, the gradient energy density may be calculated by

$$f_{\text{wall}} = \frac{1}{2}g_{ijkl}P_{i,j}P_{k,l}, \tag{15.38}$$

where $P_{i,j} = \partial P_i/\partial x_j$ and g_{ijkl} are the gradient energy coefficients with the property of $g_{ijkl} = g_{klij}$. For a cubic system, this degenerates into

$$\begin{aligned} f_{\text{wall}} = {} & \tfrac{1}{2}G_{11}\left(P_{1,1}^2 + P_{2,2}^2 + P_{3,3}^2\right) + G_{12}\left(P_{1,1}P_{2,2} + P_{2,2}^2P_{3,3}^2 + P_{1,1}P_{3,3}\right) \\ & + \tfrac{1}{2}G_{44}\left[(P_{1,2} + P_{2,1})^2 + (P_{2,3} + P_{3,2})^2 + (P_{1,3} + P_{3,1})^2\right] \\ & + \tfrac{1}{2}G'_{44}\left[(P_{1,2} - P_{2,1})^2 + (P_{2,3} - P_{3,2})^2 + (P_{1,3} - P_{3,1})^2\right], \end{aligned} \tag{15.39}$$

where G_{ij} are also called as the gradient energy coefficients and are related to g_{ijkl} through Voigt notation, for instance, $G_{11} = g_{1111}$.

The temporal evolution of the polarization fields is obtained by numerically solving the TDGL Eq. (15.36). The semi-implicit Fourier spectral method for the time-stepping and spatial discretization proposed in Ref. [38] gives more efficient and accurate numerical solution for Eq. (15.36) so is recommended. Since the strains, $\varepsilon_{ij}(\mathbf{x})$, as well as the displacements, $u_i(\mathbf{x})$, are coupled with polarization components, $P_i(\mathbf{x})$, Eqs. (15.33)–(15.36) are solved iteratively. Once we have solutions for the polarization components, their spatial distribution gives the detail of a ferroelectric domain structure.

15.4 Ferroelectric Domain Heterostructure

The ferroelectric domain structure in the film can be obtained by solving the TDLG Eq. (15.36) and its coupling mechanical Eqs. (15.33)–(15.35) for a given temperature (T) and mismatch strain (e_0). Figure 15.7 illustrates some ferroelectric domain structures obtained through solving Eqs. (15.33)–(15.36) for $BaTiO_3$ films. All the data points shown in Fig. 15.7 were obtained by starting from an initial paraelectric state with small random perturbations. The data points simply represent the type of domain structures that exist at the end of a sufficiently long simulation for minimizing the total free energy. The various ferroelectric phases were determined by the non-zero components of local polarization.

In obtaining the domain structures of Fig. 15.7, a model size of $128\Delta x \times 128\Delta x \times 36\Delta x$ cubic grids was employed and periodic boundary conditions were applied along the x_1 and x_2 axes. The thickness of the film is $h_f = 20\Delta x$ so $\mathbf{P}\left(x_1, x_2, x_3 > h_f\right) = 0$ is assumed. The region of the substrate allowed to deform was set to be $h_s = 12\Delta x$ as it was pointed out previously that little change in results when h_s exceeds about half of the film thickness. Isotropic domain wall energy was assumed in Eq. (15.39) where $G_{44} = G'_{44} = G_{11}/2$, and $G_{12} = 0$. $\Delta x = \sqrt{G_{11}/\alpha_0}$ and $\Delta t = 1/(\alpha_0 L)$ with $\alpha_0 = |\alpha_1|_{T=25^\circ\mathrm{C}}$ were utilized for the normalization of Eq. (15.36).

It should be pointed out that the phase field simulations do not assume the domain wall orientations a priori. All the stable phases and domain structure that were determined under given temperature and substrate constraint strain were found

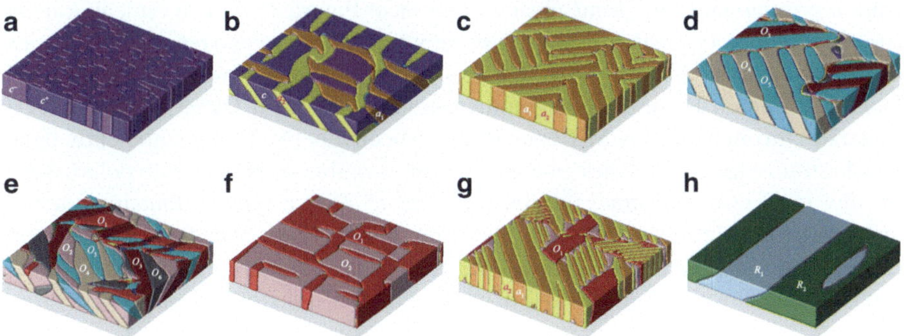

Fig. 15.7 Domain morphologies in $BaTiO_3$ films as a function of temperature (T) and substrate constraint strain (e_0). Domain definitions and the corresponding polarizations: a_1: $(P_1,0,0)$; a_2: $(0,P_1,0)$; c: $(0,0,P_3)$; O_1: $(P_1,P_1,0)$; O_2: $(P_1,-P_1,0)$; O_3:$(P_1,0,P_3)$; O_4:$(P_1,0,-P_3)$; O_5: $(0,P_1,P_3)$; O_6: $(0,P_1,-P_3)$; R_1: $(-P_1,-P_1,P_3)$; R_2: $(P_1,-P_1,P_3)$. (**a**) tetragonal c-phase at $T = 25$ °C and $e_0 = -1.0\,\%$; (**b**) tetragonal c-phase and a_1/a_2-phases at $T = 75$ °C and $e_0 = 0.0$; (**c**) tetragonal a_1/a_2-phases at $T = 75$ °C and $e_0 = 0.0$; (**d**) phases $O_3/O_4/O_5$ at $T = -25$ °C and $e_0 = -0.05\,\%$; (**e**) phases $O_1/O_2/O_3/O_4/O_5/O_6$ at $T = -25$ °C and $e_0 = 0.1\,\%$; (**f**) phases O_1/O_2 at $T = 25$ °C and $e_0 = 1.0\,\%$; (**g**) phases $a_1/a_2/O_1/O_2$ at $T = 25$ °C and $e_0 = 0.25\,\%$; (**h**) phases R_1/R_2 at $T = -100$ °C and $e_0 = 0.1\,\%$ [7]

from the minimization of total energy, i.e., the TDLG Eq. (15.36). These solutions were found by assigning small random numbers for $P_i(\mathbf{x})$ at beginning to represent an initial paraelectric state. It is seen that under larger tensile strains that the polarization of ferroelectric phases is parallel to the film/substrate interface, either along $\langle 100 \rangle$ or $\langle 110 \rangle$ direction depending on the temperature and the magnitude of the strain. The corresponding domain structures are similar to either the a_1/a_2 twins as shown in Fig. 15.7c, or the O_1/O_2 twins of Fig. 15.7f, or the mixture of a_1/a_2 twins and O_1/O_2 twins shown in Fig. 15.7g. Under relative smaller strains the polarization changes its orientation from $\langle 100 \rangle$ to $\langle 10\gamma \rangle$ then to $\langle 11\gamma \rangle$ as temperature decreases. This sequence is similar to that in bulk single crystals but γ is not always equal to 1 as in the bulk. The domain variants vary with the in-plane strain, e_0, which is clearly shown in Fig. 15.7a–c for the $\langle 100 \rangle$ orientated polarizations and in Fig. 15.7d–f for the $\langle 10\gamma \rangle$ orientated polarizations. It is interesting to note that the domain structures with $\langle 10\gamma \rangle$ orientated polarizations can be rather complicated: the domain walls between different variants are not only along the $\{100\}$ and $\{110\}$ planes, i.e., the crystallographically prominent walls with a fixed orientation, but also along the so-called S-walls whose orientations depend on the magnitude of the polarization, electrostrictive coefficients, and substrate constraint [39]. When the strain is small, its effect on the orientation of orthorhombic domain walls can be ignored. In such a case, the permissible domain walls are either $\{100\}$ or $\{110\}$ or $\{11\gamma\}$ planes with $\gamma = 2Q_{44}/\left(Q_{11} - Q_{12}\right) = 0.4403$.

Based on the simulation results, a phase diagram, i.e., a representation of stable ferroelectric phases and domain structures as a function of temperature and strain, was constructed and is shown in Fig. 15.8. Under sufficiently large compressive strains, the ferroelectric phase is of tetragonal symmetry with polarization orthogonal to the film/substrate interface. Figure 15.7a is a typical domain structure under large compressive strains, in which there are two types of c-domains separated by 180° domain walls. This result has been confirmed by experimental measurements on $BaTiO_3$ films commensurately grown on $DyScO_3$ and $GdScO_3$ substrates through reactive MBE and PLD, respectively [6]. Under sufficiently large tensile strains, the ferroelectric phase is of orthorhombic symmetry with polarization parallel to the film/substrate interface. Figure 15.7f is a typical domain structure under large tensile strains. It is seen that from the phase diagram that under certain ranges of strains, the ferroelectric phases with both symmetries P4mm and Bmm2 can be stable down all the way to 0 K without further transformation.

15.5 Summary

In summary, ferroelectrics and nanomechanics are coupled. Ferroelectric transitions involve material crystal structure change. The domain structure in ferroelectric films organizes or forms in a way to reduce the total free energy of the system and accommodate the constraint imposed on the film by its underlying substrate. On the other hand, the film ferroelectric properties are determined by its domain structure.

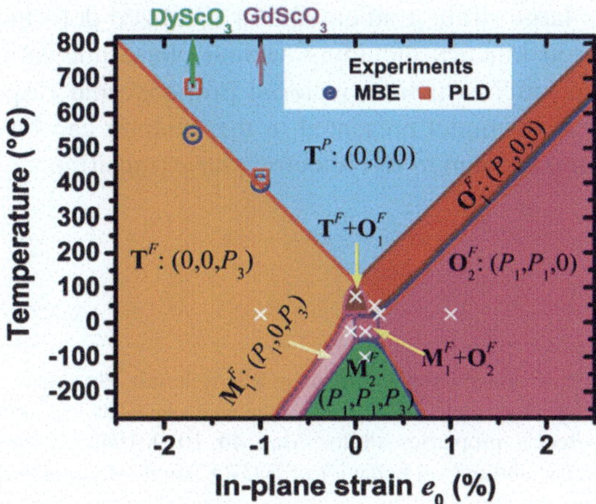

Fig. 15.8 Phase diagram of $BaTiO_3$ films as a function of temperature and substrate in-plane strain. The scattered circles and squares denote the ferroelectric transition temperatures measured from experiments on the $BaTiO_3$ films commensurately grown on $DyScO_3$ and $GdScO_3$ substrates, respectively [6]. The "×" indicates the locations of the domain structures shown in Fig. 15.7. Each single phase has equivalent variants with polarization vectors of $\mathbf{P} = (0, 0, 0)$ in T^P; $\mathbf{P} = (0, 0, \pm P_3)$ in T^F; $\mathbf{P} = (\pm P_1, 0, 0)/\mathbf{P} = (0, \pm P_1, 0)$ in O_1^F; $\mathbf{P} = (\pm P_1, \pm P_1, 0)/(\pm P_1, \mp P_1, 0)$ in O_2^F; $\mathbf{P} = (\pm P_1, 0, \pm P_3)/(\pm P_1, 0, \mp P_3)/(0, \pm P_1, \pm P_3)/(0, \pm P_1, \mp P_3)$ in M_1^F; $\mathbf{P} = (\pm P_1, \pm P_1, \pm P_3)/(\pm P_1, \mp P_1, \pm P_3)/(\pm P_1, \pm P_1, \mp P_3)/(\pm P_1, \mp P_1, \mp P_3)$ in M_2^F [7]

Therefore, this coupling can be used to engineer ferroelectric material properties. Such applications can be found in Refs. [6, 40–43].

This chapter presented methods for calculating strains and stresses in ferroelectric thin films containing nanoscale heterogeneous domain structures. The effect of strains on the ferroelectric transition temperature, stability of ferroelectric phases, and the domain structure were demonstrated for $BaTiO_3$ films. The PFM has been proven a powerful numerical method in predicting the effect of substrate constraint on the phase transitions and the details of domain structures in ferroelectrics without any priori assumptions on the possible domain structure. In addition to $BaTiO_3$ films, phase field simulations have been applied to predicting ferroelectric heterogeneous domain structures in $PbTiO_3$ [17–19, 32, 44], PZT [21, 45–47], $SrTiO_3$ [48], $BiFeO_3$ [49, 50], and $SrBi_2Nb_2O_9$ [51] thin films and $BaTiO_3/SrTiO_3$ [52, 53], PZT/PZT [36], $PbTiO_3/BaTiO_3$ [54, 55] superlattices. In calculating strain/stress distribution in thin films, homogeneous elastic properties in both the film and the substrate were assumed in order to use a semi-analytical solution for numerical accuracy and efficiency. This limitation, however, can be removed by utilizing an efficient iteration method proposed by Hu and Chen [56]. If one uses the concept of eigenstrains or stress-free strains, it is possible to consider the effect of any arbitrary distribution of dislocations and defects on microstructure evolution [57]. Very recently, efficient numerical methods for solving finite/large deformation have been developed [58–60]. Unlike small deformation, the large deformation causes large strain gradient and/or localized deformation, which may affect phase transition kinetics including second phase nucleation and transition sequence, hence, the microstructure, material property, and response. With these methods, the phase field model presented in this chapter can be extended to take large deformation into account in ferroelectric phase transition.

References

1. J. Valasek, Piezoelectric and allied phenomena in Rochelle salt. Phys. Rev. **15**, 537–538 (1920)
2. H.F. Kay, P. Vousden, Symmetry changes in barium titanate at low temperatures and their relation to its ferroelectric properties. Philos. Mag. **40**, 1019–1040 (1949)
3. W.J. Merz, The electric and optical behavior of $BaTiO_3$ single-domain crystals. Phys. Rev. **76**, 1221–1225 (1949)
4. Y. Yoneda, T. Okabe, K. Sakaue, H. Terauchi, H. Kasatani, K. Deguchi, Structural characterization of $BaTiO_3$ thin films grown by molecular beam epitaxy. J. Appl. Phys. **83**, 2458–2461 (1998)
5. N.A. Pertsev, A.G. Zembilgotov, A.K. Tagantsev, Effect of mechanical boundary conditions on phase diagrams of epitaxial ferroelectric thin films. Phys. Rev. Lett. **80**, 1988–1991 (1998)
6. K.J. Choi, M. Biegalski, Y.L. Li, A. Sharan, J. Schubert, R. Uecker, P. Reiche, Y.B. Chen, X.Q. Pan, V. Gopalan, L.Q. Chen, D.G. Schlom, C.B. Eom, Enhancement of ferroelectricity in strained $BaTiO_3$ thin films. Science **306**, 1005–1009 (2004)
7. Y.L. Li, L.Q. Chen, Temperature-strain phase diagram for $BaTiO_3$ thin films. Appl. Phys. Lett. **88**, 072905 (2006)
8. D.A. Tenne, X.X. Xi, Y.L. Li, L.Q. Chen, A. Soukiassian, M.H. Zhu, A.R. James, J. Lettieri, D.G. Schlom, W. Tian, X.Q. Pan, Absence of low-temperature phase transitions in epitaxial $BaTiO_3$ thin films. Phys. Rev. B **69**, 174101 (2004)
9. A.F. Devonshire, Theory of ferroelectrics. Adv. Phys. **3**, 85–130 (1954)
10. A.F. Devonshire, Theory of barium titanate 1. Philos. Mag. **40**, 1040–1063 (1949)
11. F. Jona, G. Shirane, *Ferroelectric Crystals* (Pergamon, Oxford, New York, 1962)
12. Y.L. Li, L.E. Cross, L.Q. Chen, A phenomenological thermodynamic potential for $BaTiO_3$ single crystals. J. Appl. Phys. **98**, 064101 (2005)
13. A.F. Devonshire, Theory of barium titanate 2. Philos. Mag. **42**, 1065–1079 (1951)
14. D. Berlincourt, H. Jaffe, Elastic and piezoelectric coefficients of single-crystal barium titanate. Phys. Rev. **111**, 143–148 (1958)
15. T. Yamada, Electromechanical properties of oxygen-octahedra ferroelectric crystals. J. Appl. Phys. **43**, 328–338 (1972)
16. V.G. Koukhar, N.A. Pertsev, R. Waser, Thermodynamic theory of epitaxial ferroelectric thin films with dense domain structures. Phys. Rev. B **64**, 214103 (2001)
17. Y.L. Li, S.Y. Hu, Z.K. Liu, L.Q. Chen, Effect of substrate constraint on the stability and evolution of ferroelectric domain structures in thin films. Acta Mater. **50**, 395–411 (2002)
18. Y.L. Li, S.Y. Hu, Z.K. Liu, L.Q. Chen, Phase-field model of domain structures in ferroelectric thin films. Appl. Phys. Lett. **78**, 3878–3880 (2001)
19. Y.L. Li, S.Y. Hu, Z.K. Liu, L.Q. Chen, Effect of electrical boundary conditions on ferroelectric domain structures in thin films. Appl. Phys. Lett. **81**, 427–429 (2002)
20. L.Q. Chen, Phase-field models for microstructure evolution. Annu. Rev. Mater. Res. **32**, 113–140 (2002)
21. Y.L. Li, S. Choudhury, Z.K. Liu, L.Q. Chen, Effect of external mechanical constraints on the phase diagram of epitaxial $PbZr_{1-x}Ti_xO_3$ thin films: thermodynamic calculations and phase-field simulations. Appl. Phys. Lett. **83**, 1608–1610 (2003)

22. A. Artemev, Phase field modeling of domain structures and *P–E* hysteresis in thin ferroelectric layers with deadlayers. Philos. Mag. **90**, 89–101 (2010)
23. A. Artemev, B. Geddes, J. Slutsker, A. Roytburd, Thermodynamic analysis and phase field modeling of domain structures in bilayer ferroelectric thin films. J. Appl. Phys. **103**, 074104 (2008)
24. W.J. Chen, Y. Zheng, B. Wang, Phase field simulations of stress controlling the vortex domain structures in ferroelectric nanosheets. Appl. Phys. Lett. **100**, 062901 (2012)
25. P. Chu, D.P. Chen, J.M. Liu, Multiferroic domain structure in orthorhombic multiferroics of cycloidal spin order: phase field simulations. Appl. Phys. Lett. **101**, 042908 (2012)
26. K. Dayal, K. Bhattacharya, A real-space non-local phase-field model of ferroelectric domain patterns in complex geometries. Acta Mater. **55**, 1907–1917 (2007)
27. W.D. Dong, D.M. Pisani, C.S. Lynch, A finite element based phase field model for ferroelectric domain evolution. Smart Mater. Struct. **21**, 094014 (2012)
28. C. Fang, Phase-field simulation study on size effect of the microstructure evolution of a single-domain barium titanate 2D lattice square. Phys. Status Solidi B **251**, 1619–1629 (2014)
29. T. Koyama, H. Onodera, Phase-field simulation of ferroelectric domain microstructure changes in $BatiO_3$. Mater. Trans. **50**, 970–976 (2009)
30. P.L. Liu, J. Wang, T.Y. Zhang, Y. Li, L.Q. Chen, X.Q. Ma, W.Y. Chu, L.J. Qiao, Effects of unequally biaxial misfit strains on polarization phase diagrams in embedded ferroelectric thin layers: phase field simulations. Appl. Phys. Lett. **93**, 132908 (2008)
31. D.C. Ma, Y. Zheng, C.H. Woo, Phase-field simulation of domain structure for $PbTiO_3/SrTiO_3$ superlattices. Acta Mater. **57**, 4736–4744 (2009)
32. G. Sheng, J.X. Zhang, Y.L. Li, S. Choudhury, Q.X. Jia, Z.K. Liu, L.Q. Chen, Domain stability of $PbTiO_3$ thin films under anisotropic misfit strains: phase-field simulations. J. Appl. Phys. **104**, 054105 (2008)
33. Y. Shindo, F. Narita, T. Kobayashi, Phase field simulation on the electromechanical response of poled barium titanate polycrystals with oxygen vacancies. J. Appl. Phys. **117**, 234103 (2015)
34. B. Winchester, P. Wu, L.Q. Chen, Phase-field simulation of domain structures in epitaxial $BiFeO_3$ films on vicinal substrates. Appl. Phys. Lett. **99**, 052903 (2011)
35. P.P. Wu, X.Q. Ma, J.X. Zhang, L.Q. Chen, Phase-field model of multiferroic composites: domain structures of ferroelectric particles embedded in a ferromagnetic matrix. Philos. Mag. **90**, 125–140 (2010)
36. F. Xue, J.J. Wang, G. Sheng, E. Huang, Y. Cao, H.H. Huang, P. Munroe, R. Mahjoub, Y.L. Li, V. Nagarajan, L.Q. Chen, Phase field simulations of ferroelectrics domain structures in $PbZr_xTi_{1-x}O_3$ bilayers. Acta Mater. **61**, 2909–2918 (2013)
37. L.-Q. Chen, Phase-field method of phase transitions/domain structures in ferroelectric thin films: a review. J. Am. Ceram. Soc. **91**, 1835–1844 (2008)
38. L.Q. Chen, J. Shen, Applications of semi-implicit Fourier-spectral method to phase field equations. Comput. Phys. Commun. **108**, 147–158 (1998)
39. J. Fousek, V. Janovec, Orientation of domain walls in twinned ferroelectric crystals. J. Appl. Phys. **40**, 135–142 (1969)
40. J.H. Haeni, P. Irvin, W. Chang, R. Uecker, P. Reiche, Y.L. Li, S. Choudhury, W. Tian, M.E. Hawley, B. Craigo, A.K. Tagantsev, X.Q. Pan, S.K. Streiffer, L.Q. Chen, S.W. Kirchoefer, J. Levy, D.G. Schlom, Room-temperature ferroelectricity in strained $SrTiO_3$. Nature **430**, 758–761 (2004)
41. D.A. Tenne, A. Bruchhausen, N.D. Lanzillotti-Kimura, A. Fainstein, R.S. Katiyar, A. Cantarero, A. Soukiassian, V. Vaithyanathan, J.H. Haeni, W. Tian, D.G. Schlom, K.J. Choi, D.M. Kim, C.B. Eom, H.P. Sun, X.Q. Pan, Y.L. Li, L.Q. Chen, Q.X. Jia, S.M. Nakhmanson, K.M. Rabe, X.X. Xi, Probing nanoscale ferroelectricity by ultraviolet Raman spectroscopy. Science **313**, 1614–1616 (2006)
42. D.G. Schlom, L.-Q. Chen, C.-B. Eom, K.M. Rabe, S.K. Streiffer, J.-M. Triscone, Strain tuning of ferroelectric thin films. Annu. Rev. Mater. Res. **37**, 589–626 (2007)
43. D.G. Schlom, L.-Q. Chen, C.J. Fennie, V. Gopalan, D.A. Muller, X. Pan, R. Ramesh, R. Uecker, Elastic strain engineering of ferroic oxides. MRS Bull. **39**, 118–130 (2014)

44. G. Sheng, J.M. Hu, J.X. Zhang, Y.L. Li, Z.K. Liu, L.Q. Chen, Phase-field simulations of thickness-dependent domain stability in $PbTiO_3$ thin films. Acta Mater. **60**, 3296–3301 (2012)
45. Y.L. Li, S.Y. Hu, L.Q. Chen, Ferroelectric domain morphologies of (001) $PbZr_{1-x}Ti_xO_3$ epitaxial thin films. J. Appl. Phys. **97**, 034112 (2005)
46. Y. Cao, G. Sheng, J.X. Zhang, S. Choudhury, Y.L. Li, C.A. Randall, L.Q. Chen, Piezoelectric response of single-crystal $PbZr_{1-x}Ti_xO_3$ near morphotropic phase boundary predicted by phase-field simulation. Appl. Phys. Lett. **97**, 252904 (2010)
47. S. Choudhury, Y.L. Li, L.Q. Chen, A phase diagram for epitaxial $PbZr_{1-x}Ti_xO_3$ thin films at the bulk morphotropic boundary composition. J. Am. Ceram. Soc. **88**, 1669–1672 (2005)
48. Y.L. Li, S. Choudhury, J.H. Haeni, M.D. Biegalski, A. Vasudevarao, A. Sharan, H.Z. Ma, J. Levy, V. Gopalan, S. Trolier-McKinstry, D.G. Schlom, Q.X. Jia, L.Q. Chen, Phase transitions and domain structures in strained pseudocubic (100) $SrTiO_3$ thin films. Phys. Rev. B **73**, 184112 (2006)
49. J.X. Zhang, Y.L. Li, S. Choudhury, L.Q. Chen, Y.H. Chu, F. Zavaliche, M.P. Cruz, R. Ramesh, Q.X. Jia, Computer simulation of ferroelectric domain structures in epitaxial $BiFeO_3$ thin films. J. Appl. Phys. **103**, 094111 (2008)
50. J.X. Zhang, Y.L. Li, Y. Wang, Z.K. Liu, L.Q. Chen, Y.H. Chu, F. Zavaliche, R. Ramesh, Effect of substrate-induced strains on the spontaneous polarization of epitaxial $BiFeO_3$ thin films. J. Appl. Phys. **101**, 114105 (2007)
51. Y.L. Li, L.Q. Chen, G. Asayama, D.G. Schlom, M.A. Zurbuchen, S.K. Streiffer, Ferroelectric domain structures in $SrBi_2Nb_2O_9$ epitaxial thin films: electron microscopy and phase-field simulations. J. Appl. Phys. **95**, 6332–6340 (2004)
52. Y.L. Li, S.Y. Hu, D. Tenne, A. Soukiassian, D.G. Schlom, L.Q. Chen, X.X. Xi, K.J. Choi, C.B. Eom, A. Saxena, T. Lookman, Q.X. Jia, Interfacial coherency and ferroelectricity of $BaTiO_3/SrTiO_3$ superlattice films. Appl. Phys. Lett. **91**, 252904 (2007)
53. Y.L. Li, S.Y. Hu, D. Tenne, A. Soukiassian, D.G. Schlom, X.X. Xi, K.J. Choi, C.B. Eom, A. Saxena, T. Lookman, Q.X. Jia, L.Q. Chen, Prediction of ferroelectricity in $BaTiO_3/SrTiO_3$ superlattices with domains. Appl. Phys. Lett. **91**, 112914 (2007)
54. L. Hong, Y.L. Li, P.P. Wu, L.Q. Chen, Minimum tetragonality in $PbTiO_3/BaTiO_3$ ferroelectric superlattices. J. Appl. Phys. **114**, 144103 (2013)
55. L. Hong, P. Wu, Y. Li, V. Gopalan, C.-B. Eom, D.G. Schlom, L.-Q. Chen, Piezoelectric enhancement of $(PbTiO_3)_m/(BaTiO_3)_n$ ferroelectric superlattices through domain engineering. Phys. Rev. B **90**, 174111 (2014)
56. S.Y. Hu, L.Q. Chen, A phase-field model for evolving microstructures with strong elastic inhomogeneity. Acta Mater. **49**, 1879–1890 (2001)
57. S.Y. Hu, Y.L. Li, L.Q. Chen, Effect of interfacial dislocations on ferroelectric phase stability and domain morphology in a thin film: a phase-field model. J. Appl. Phys. **94**, 2542–2547 (2003)
58. J.D. Clayton, J. Knap, A phase field model of deformation twinning: nonlinear theory and numerical simulations. Physica D **240**, 841–858 (2011)
59. W. Hong, X. Wang, A phase-field model for systems with coupled large deformation and mass transport. J. Mech. Phys. Solids **61**, 1281–1294 (2013)
60. L. Chen, F. Fan, L. Hong, J. Chen, Y.Z. Ji, S.L. Zhang, T. Zhu, L.Q. Chen, A phase-field model coupled with large elasto-plastic deformation: application to lithiated silicon electrodes. J. Electrochem. Soc. **161**, F3164–F3172 (2014)

Chapter 16
Modeling of Lithiation in Silicon Electrodes

Feifei Fan and Ting Zhu

16.1 Introduction

Lithium-ion batteries (LIBs) are critically important for portable electronics, electric vehicles, and grid-level energy storage [1]. Silicon (Si) is a promising anode material for next-generation LIBs, since it has the highest known theoretical capacity of lithium (Li) storage. However, such a high capacity is accompanied with a large volume change of ~280 % in Si anodes during insertion and extraction of Li. The volume expansion, when constrained, can cause fracture or even pulverization of the anodes, thereby leading to performance degradation [2–11]. Although considerable effort has been devoted to study the electrochemical behavior of LIBs, the mechanics and failure mechanisms in the high-capacity anode materials such as Si during charging and discharging remain largely unknown. The electro-chemo-mechanical response in LIBs is a multiscale phenomenon, involving multiple time and length scales (Fig. 16.1) [12]. For example, the electrochemical lithiation of Si anodes involves atomic-scale reactions at phase boundaries, microstructural evolution caused by species diffusion and phase separation, as well as fracture due to stress generation.

The multiscale nature of the charging/discharging processes in Si electrodes requires theoretical and computational studies at different time and length scales using the appropriate methods. In this chapter, we present examples of continuum

F. Fan (✉)
Department of Mechanical Engineering, University of Nevada, Reno, NV 89557, USA
e-mail: ffan@unr.edu

T. Zhu
Woodruff School of Mechanical Engineering, Georgia Institute of Technology, Atlanta, GA 30332, USA
e-mail: ting.zhu@me.gatech.edu

C.R. Weinberger, G.J. Tucker (eds.), *Multiscale Materials Modeling for Nanomechanics*, Springer Series in Materials Science 245,
DOI 10.1007/978-3-319-33480-6_16

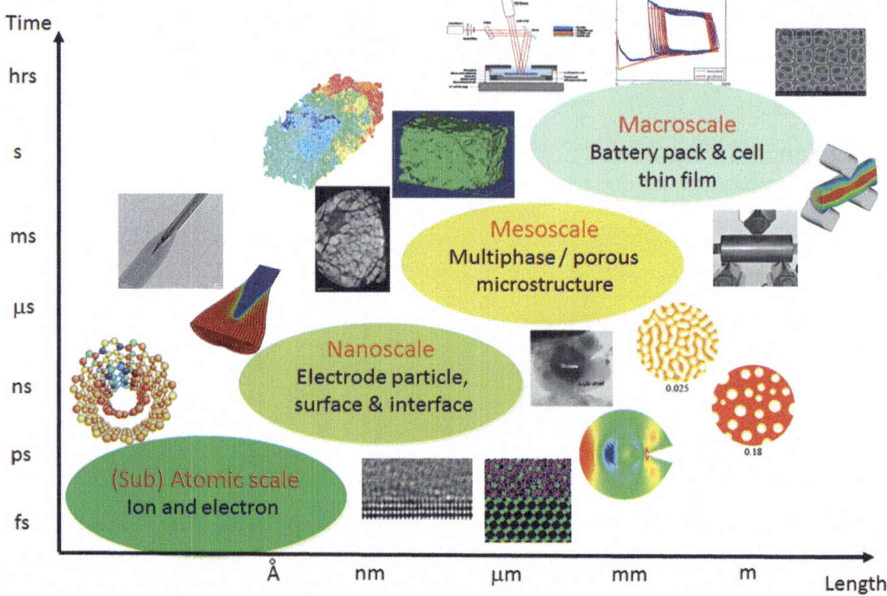

Fig. 16.1 An overview of the multiscale electro-chemo-mechanics in LIBs (reprinted from [12] with permission)

and atomistic modeling of the deformation and stress generation during electrochemical lithiation of Si electrodes [13, 14]. These models are developed by tightly coupling with in situ transmission electron microscopy (TEM) experiments. The results provide novel mechanistic insights into the electrochemically driven structural evolution, stress generation, and mechanical failures in Si electrodes.

16.2 Background

16.2.1 Lithiation Mechanisms in Si Nanostructures

Recent experiments indicate that plastic deformation can readily occur in the high-capacity electrode materials during lithiation [10]. Moreover, in situ TEM experiments reveal that lithiation of Si typically proceeds through movement of an atomically sharp phase boundary that separates the unlithiated and lithiated phases [7]. For example, the TEM images in Fig. 16.2a, b show the formation of a core–shell structure in a partially lithiated Si nanoparticle, involving a two-phase boundary that separates an inner core of crystalline Si (c-Si) with an outer shell of amorphous Li_xSi (a-Li_xSi) ($x \sim 3.75$) [13]. The observation of such a sharp phase boundary suggests that the Li-poor and Li-rich phases do not transform continuously

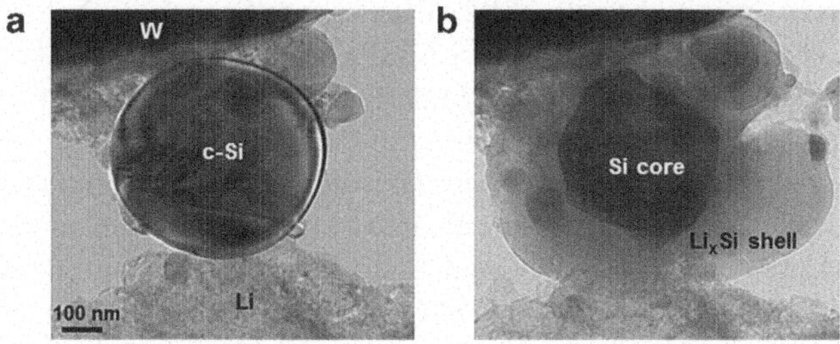

Fig. 16.2 Lithiation and cracking in an Si nanoparticle from in situ TEM experiments (reprinted from [13] with permission). (**a**) A crystalline Si (c-Si) nanoparticle is in contact with a tungsten (W) electrode and an Li metal counter electrode, whose surface is covered with Li_2O that acts as a solid electrolyte. (**b**) Under an applied voltage between the two electrodes, Li quickly covers the particle surface and then flows radially into the particle, forming the structure of a pristine inner Si core (*black*) and an amorphous Li_xSi ($x \sim 3.75$) alloy shell (*gray*) with a sharp interface in between. The lithiation-induced swelling causes crack initiation from the particle surface

into each other with changing composition. Namely, there is a large solubility gap between the two phases, manifested as an abrupt change of Li concentrations across the phase boundary. Moreover, Fig. 16.2b reveals the formation of a surface crack, indicating the development of hoop tension in the surface layer during Li insertion. While theoretical models have been developed to analyze the Li diffusion, reaction, and associated stress states in lithiated materials, the effects of plastic deformation and two-phase microstructures on the stress generation are not yet well understood.

Nanomaterials, such as nanoparticles and nanowires, are being intensively studied as the basic building blocks of electrodes for LIBs. This is motivated by the notion that the nanometer size scale can shorten the diffusion path and enable facile strain relaxation, thus enhancing both the rate capability and flaw tolerance of the electrodes composed of nanomaterials. Here we focus on the lithiation of individual nanoparticles and nanowires, and investigate stress generation associated with a curved two-phase boundary, i.e., core–shell interface [15]. It is important to note that lithiation at a moving core–shell interface affects not only the local stress states, but also the stress distributions behind the moving boundary. This may have a dramatic effect on the fracture behavior of lithiated nanoparticles, e.g., leading to cracking in the outer surface during Li insertion, as shown in Fig. 16.2b.

16.2.2 Mechanical Properties of Amorphous Li_xSi

Several recent experimental studies have investigated the mechanical properties of a-Li_xSi. Hertzberg et al. conducted nanoindentation testing on polycrystalline Si thin films at various stages of lithiation [16]. They reported a strong dependence of Young's modulus and hardness on the Li to Si ratio. Young's modulus was found to decrease from an initial value of 92 GPa for pure Si to 12 GPa for $Li_{3.75}Si$, and the hardness changes from 5 GPa to 1.5 GPa, correspondingly. Chon et al. used the curvature measurement technique to determine the yield stress in a lithiated Si thin film [17]. They reported a yield stress of 0.5 GPa for a-$Li_{3.5}Si$, and other thin-film lithiation experiments reported similar values [18]. Kushima et al. measured the fracture strength and plasticity of lithiated Si nanowires by in situ tensile testing inside a TEM [19]. The tensile strength decreased from the initial value of about 3.6 GPa for pristine Si nanowires to 0.72 GPa for a-Li_xSi ($x \sim 3.5$) nanowires. They also reported the large fracture strains that range from 8 to 16 % for a-Li_xSi ($x \sim 3.5$) nanowires, 70 % of which remained unrecoverable after fracture. Recently, Wang et al. performed fracture toughness measurements by nanoindentation on lithiated a-Si thin films [11]. Their results showed a rapid brittle-to-ductile transition of fracture as the Li:Si ratio was increased to above 1.5. They reported the fracture energy of a-Si as 2.85 ± 0.15 J/m^2 and $Li_{1.09}Si$ as 8.54 ± 0.72 J/m^2, respectively.

The mechanical properties of a-Li_xSi alloys have also been recently studied using quantum mechanical and interatomic potential calculations. Shenoy et al. performed the density functional theory (DFT) calculations of the elastic properties of a-Li_xSi alloys [20]. They showed the elastic softening with increasing Li concentration. Zhao et al. conducted DFT calculations for both the biaxial yield stress in a-Li_xSi thin films and the uniaxial yield stress in bulk a-Li_xSi under tension [21]. They found large differences between the characteristic yield stress levels of the two cases. Cui et al. developed a modified embedded atomic method (MEAM) interatomic potential for the Li_xSi alloys [22]. They performed molecular dynamics (MD) simulations of uniaxial tension for bulk a-Li_xSi at 300 K and showed that the yield strength decreased from about 2 GPa for a-LiSi to 0.5 GPa for a-$Li_{3.75}Si$. In addition, quantum mechanical calculations have been performed to study the atomic structures and energetics associated with Li diffusion and reaction in a-Li_xSi alloys [23–31].

The aforementioned modeling studies not only provide valuable insights into the mechanical properties of a-Li_xSi, but also highlight the challenges to understand the structure–property relationship in a-Li_xSi. Namely, the composition and atomic structure can change drastically during electrochemical cycling of Si electrodes, involving, for example, the transition from an Li-poor network glass to an Li-rich metallic glass with increasing Li concentration during lithiation.

16.3 Continuum Modeling of Lithiation in Si Particle

A continuum model has been developed to study the two-phase electrochemical lithiation and resulting stress generation in a sphere particle electrode [13]. A nonlinear diffusion model with Li concentration-dependent diffusivities is used to produce a sharp phase boundary between pristine Si and lithiated Si. The large-strain elastic–plastic deformation theory is adopted for capturing the lithiation-induced morphological evolution, stress generation, and crack formation. Effects of dilational vs. unidirectional lithiation strains are examined on stress generation. The modeling results highlight the important effects of plastic deformation, phase boundary-mediated lithiation mechanism, finite curvature of particle geometry, and large lithiation strain at the phase boundary.

To focus on the essential physical effects of lithiation on stress generation, we adopt a simple elastic–plastic model to evaluate the deformation and stress states during Li insertion. The total strain rate, $\dot{\varepsilon}_{ij}$, is taken to be the sum of three contributions,

$$\dot{\varepsilon}_{ij} = \dot{\varepsilon}_{ij}^{c} + \dot{\varepsilon}_{ij}^{e} + \dot{\varepsilon}_{ij}^{p} \tag{16.1}$$

where $\dot{\varepsilon}_{ij}^{c}$ is the chemical strain rate caused by lithiation and is proportional to the rate of the normalized Li concentration $\dot{c}$,

$$\dot{\varepsilon}_{ij}^{c} = \beta_{ij}\dot{c} \tag{16.2}$$

where β_{ij} is the lithiation expansion coefficient and c varies between 0 (pristine Si) and 1 (fully lithiated $Li_{3.75}Si$). In Eq. (16.1), $\dot{\varepsilon}_{ij}^{e}$ denotes the elastic strain rate and obeys Hooke's law,

$$\dot{\varepsilon}_{ij}^{e} = \frac{1}{E}\left[(1+\nu)\,\dot{\sigma}_{ij} - \nu\dot{\sigma}_{kk}\delta_{ij}\right] \tag{16.3}$$

where E is Young's modulus, ν is Poisson's ratio, $\delta_{ij} = 1$ when $i = j$ and $\delta_{ij} = 0$ otherwise; repeated indices mean summation. In Eq. (16.1), the plastic strain rate, $\dot{\varepsilon}_{ij}^{p}$,

$$\dot{\varepsilon}_{ij}^{p} = \dot{\bar{\varepsilon}}^{p}\frac{3\sigma_{ij}'}{2\sigma_e} \tag{16.4}$$

where $\sigma_e = \sqrt{3\sigma_{ij}'\sigma_{ij}'/2}$ is the von Mises effective stress, $\sigma_{ij}' = \sigma_{ij} - \sigma_{kk}\delta_{ij}/3$ is the deviatoric stress, and $\dot{\bar{\varepsilon}}^{p}$ is the effective plastic strain rate given by

$$\dot{\bar{\varepsilon}}^{p} = \dot{\bar{\varepsilon}}_0^{p}\left(\frac{\sigma_e}{s}\right)^{1/m} \tag{16.5}$$

here $\dot{\bar{\varepsilon}}_0^{p}$ is the effective strain rate constant, s is the plastic flow resistance, and m is the rate sensitivity exponent. In the limit of the rate-independent deformation, $m \to 0$ and $\sigma_e \to s$. In Eq. (16.5), s is often taken as a function of the accumulated plastic strain to model the strain hardening [32]. Here we assume a constant s equal to the yield stress σ_Y, corresponding to perfectly plastic deformation.

We study the evolution of a two-phase core–shell particle and associated stress generation by using finite element simulations. A nonlinear diffusion model is used to produce the two-phase core–shell structure. To capture the coexistence of Li-poor and Li-rich phases, we assume that diffusivity D is non-linearly dependent on the local Li concentration c, normalized to vary between zero and one. Note that our diffusion simulations mainly serve to generate a sequence of core–shell structures for stress analysis, rather than provide a precise description of the dynamic lithiation process, which would be difficult due to lack of experimental measurements for model calibration. To this end, we take a simple nonlinear function of $D = D_0\,[1/(1-c) - 2\Omega c]$, where D_0 is the diffusivity constant and Ω is tuned to control the concentration profile near the reaction front; this nonlinear function gives a low Li diffusivity when the normalized Li concentration c is less than one and a much higher diffusivity as c approaches one. As a result, in diffusion simulations, the c values behind the reaction front can quickly attain the high values (slightly below one), while those ahead of the front remain nearly zero. This produces a sharp reaction front that is consistent with experimental observation, thereby providing a basis of further stress analysis. It should be noted that a small gradient of Li concentration still exists behind the reaction front, so that Li can continuously diffuse through the lithiated shell, so as to advance the reaction front toward the particle center. Finally, we emphasize that the above non-linear diffusivity function is empirical and taken as a numerical convenience for generating a sharp phase boundary for stress analyses. In the future study, a mechanistically based model is needed to characterize both the phase boundary migration and Li diffusion in each phase.

The above diffusion and mechanics models have been implemented in the finite element package ABAQUS. The distributions of Li and stress–strain are solved with an implicit, coupled temperature–displacement procedure in ABAQUS/Standard. That is, the normalized concentration is surrogated by temperature and the lithiation expansion coefficient β_{ij} is equivalently treated as the thermal expansion coefficient. The user material subroutine for heat transfer (UMATHT) is programmed to interface with ABAQUS to update diffusivities based on the current Li concentration (i.e., temperature). The Li distribution and accordingly elastic–plastic deformation are updated incrementally. In finite element simulations, we take $\Omega = 1.95$ and assign a constant Li concentration $c = 1$ at the surface; the alternative flux boundary condition gives similar results of stress generation. For numerical stability, the maximum of D is capped at $10^4 D_0$. The axis-symmetric condition is used to reduce the computational cost in ABAQUS.

We choose the material properties that are representative of high-capacity electrode materials such as Si: yield stress $\sigma_Y = 0.05E$ and Poisson's ratio $v = 0.3$. It is important to emphasize that a physically based assignment of lithiation strains is not yet possible due to the lack of experimental measurement. Assuming that the lithiation strains are dilational, we take the lithiation expansion coefficients $\beta_r = \beta_\theta = \beta_\phi = 0.26$, yielding a volume increase of $\sim$100 %. Effects of lithiation strains on the stress generation will be further studied later by examining a limiting case of uniaxial lithiation strain in the radial direction. In addition, we take the rate sensitivity exponent $m = 0.01$ to approximate the rate-independent limit, and the effective strain rate constant $\dot{\varepsilon}_0^p = 0.001$.

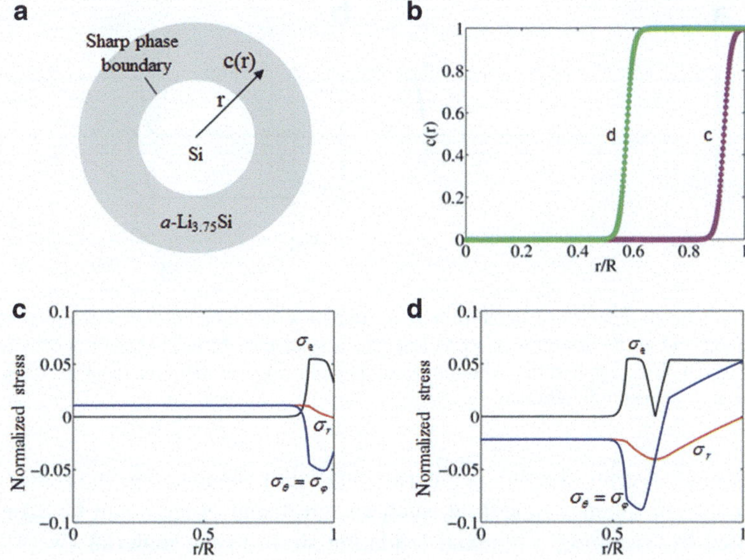

Fig. 16.3 Numerical results of a two-phase spherical particle with dilational lithiation strain (reprinted from [13] with permission). (**a**) Schematic of the two-phase model with a sharp phase boundary between the Si core and a-$Li_{3.75}Si$ shell. (**b**) Simulated radial distributions of Li concentration c, normalized by its maximum value at the fully lithiated state, and the radial distance r is normalized by the current radius R of a partially lithiated particle which increases as lithiation proceeds. (**c**) Radial distributions of the von Mises effective stress σ_e (as defined in the text), radial stress σ_r, and hoop stress $\sigma_\theta = \sigma_\phi$, which corresponds to the Li concentration profile of c in (**b**). (**d**) Same as (**c**) except corresponding to the Li concentration profile of c in (**b**). All the stresses in (**c**) and (**d**) are normalized by Young's modulus E

Figure 16.3b, c shows the simulated results of radial stress distributions that correspond to Li profiles in Fig. 16.3a. The most salient feature is the development of tensile hoop stresses in the surface layer in the late stage of lithiation, Fig. 16.3c, and such hoop tension reverses the initial compression, Fig. 16.3b. Correspondingly, the core is first subjected to hydrostatic tension and later changed to hydrostatic compression. This change results from the buildup of hoop tension in the surface layer, as well as the requirement that the integrated hoop stresses over any cross section must be zero in a free-standing particle.

Previous studies of oxidation-induced stresses have indicated that a correct assignment of the oxidation strain is important to understanding the stress generation. It has been noted that the assumed dilational oxidation strain would cause a large strain mismatch between the oxidized and un-oxidized materials at a sharp phase boundary, where the oxygen concentration changes abruptly. To reduce the mismatch strain and associated strain energy, the reactive layer model has been proposed [33], where the oxidation-induced swelling can become uniaxial along the normal direction of the phase boundary. It has been suggested that during Si oxidation, the reactive layer can be as thin as a few atomic layers, and the oxidation

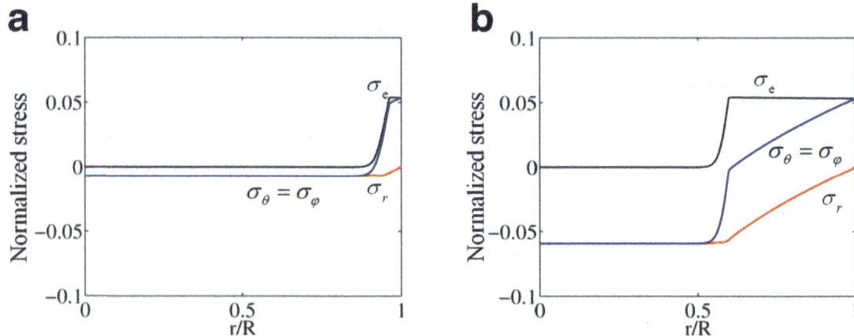

Fig. 16.4 Effects of the unidirectional lithiation strain on stress generation in a two-phase particle with the same radial distributions of Li as Fig. 16.3a (reprinted from [13] with permission). (**a**) Same as Fig. 16.3b except that the lithiation strain is unidirectional. (**b**) Same as Fig. 16.3c except the lithiation strain is unidirectional

occurs through a series of bond reconstructions within the layer. From the standpoint of continuum mechanics modeling, such an interfacial process can be effectively represented by assigning a uniaxial oxidation strain to the material swept by the moving reactive layer. In other words, the oxidation-induced swelling in a particle can be assumed to occur only in the radial direction, normal to the core–shell interface, and further straining of the material behind the moving phase boundary is mainly caused by the push-out effect, as discussed in our previous publications.

Motivated by the above analysis of the oxidation strains, we further study the effect of a radial lithiation strain on stress generation in a two-phase particle, while the exact atomic processes of strain transformation at the phase boundary are not yet clear. In order to compare with the results obtained from a dilational lithiation strain as shown in Fig. 16.3, we take the same volume increase of $\sim$100 %, i.e., the lithiation expansion coefficients $\beta_r = 1$, and $\beta_\theta = \beta_\phi = 0$. Figure 16.4a, b shows the calculated stress distributions that correspond to the two Li profiles in Fig. 16.3a, respectively. In both plots of Fig. 16.4a, b, one can identify the approximate location of the phase boundary where the hoop stress begins to deviate from a constant value in the core.

Comparing Fig. 16.3 with Fig. 16.4, we note the following similarities and differences in stress states between the cases of dilational versus radial lithiation strains in a two-phase particle. As lithiation proceeds, the lithiated shell can readily achieve plastic yielding in both cases, underscoring the importance of plasticity in stress generation in the high-capacity electrode undergoing large lithiation-induced swelling. However, in the case of *radial* lithiation strain, Fig. 16.4 shows that the pristine core is subjected to hydrostatic compression. Within the lithiated shell, the hoop stress is compressive near the phase boundary, but it increases sharply and continuously, such that the tensile plastic yielding is achieved within a small radial distance, and the hoop tension further increases all the way to the particle surface. In contrast, in the case of *dilational* lithiation strain, Fig. 16.3 shows that the core is first subjected to hydrostatic tension, and later changed to hydrostatic compression as lithiation proceeds. Within the lithiated shell, near the phase boundary the hoop stress decreases sharply and almost discontinuously, such that the hoop stress becomes compressive and plastic yielding is achieved. This hoop compression is reversed to tension at the larger radial distance because of the push-out effect.

The above results, regarding both the stress states in the core and near the phase boundary for the early stage of lithiation, agree qualitatively with those from an elastic model of oxidation-induced stresses, for both cases of radial and dilational lithiation strains. Hsueh and Evans [34] have clearly explained why the core is subjected to hydrostatic compression in the case of radial oxidation strain, while the hydrostatic tension develops in the core in the case of dilational oxidation strain. Since a similar stress development is predicted in the early stage of lithiation, it is instructive to briefly review their explanations here. The key to understanding the stress buildup in the core is to first consider the deformation of the oxide (i.e., equivalent to the lithiated shell) in its "unconstrained" state, i.e., the oxide shell is assumed to detach from the core. In the case of dilational oxidation strain, the tangential component of the oxidation strain induces an unconstrained outward radial displacement, and the hydrostatic tension in the core results from the positive traction imposed to achieve displacement continuity at the core–shell interface. In contrast, the radial oxidation strain would induce the mixed tangential tension and compression in the unconstrained state of oxide. The hydrostatic compression in the core results from the negative traction imposed to achieve displacement continuity at the core–shell interface. Here it should be noted that while the analysis by Hsueh and Evans [34] provides a direct physical appreciation of the stress states under different assumptions of oxidation strains, their work is only limited to *elastic* deformation. In contrast, the present model accounts for the *plastic* deformation that can readily develop during the lithiation of candidate high-capacity anode materials (Si, Ge, SnO_2, etc.), as shown in recent experiments, and thus enables a more accurate quantification of stress evolution during lithiation.

Finally, we note that both dilational and radial lithiation strains represent the limiting cases of coarse-graining the effective transformation strains associated with Li insertion at the atomically sharp phase boundary. The main difference in the predicted stress states between the two cases only lies in the early stage response, as discussed above. In fact, the contrast of the stresses in the core predicted by the two cases, i.e., hydrostatic tension versus compression, suggests an experimental route of testing which assumption of lithiation strain is more appropriate, e.g., by measuring the elastic lattice strain in the core through high resolution TEM or electron diffraction experiments. Nevertheless, despite the somewhat different stress distributions between the two cases, the two-phase models capture the physical effect of large swelling at the phase boundary. Namely, the hoop tension develops in the surface layer due to the push-out effect, in a manner similar to inflation of a balloon causing its wall to stretch. The buildup of large hoop tension in the surface layer provides the main driving force of surface cracking, as observed in TEM experiments of Fig. 16.2b.

16.4 Atomistic Modeling of Lithiation in Amorphous Si Thin Film

In this section, we present an atomistic study of the mechanical responses of a-Li_xSi alloys under different loading modes and stress states. Currently, an effective method of measuring the mechanical properties of a-Li_xSi is to test the lithiation stress response of an Si thin film on an inactive substrate [10, 11]. Upon electrochemical lithiation, the substrate constrains the volume expansion in the lithiated Si thin film, thus generating a biaxial compressive stress in the film. The film stress can be determined from the substrate curvature change based on Stoney's equation. The lithiation stresses measured from the thin film experiments have been often taken as the yield strengths of a-Li_xSi alloys in the recent literature. To understand these experiments, we raise the following questions: Can these thin-film stresses represent the plastic properties of a-Li_xSi alloys? In other words, does the loading sequence (concurrent versus sequential chemical and mechanical loading) result in different material properties of the same final product? Furthermore, what is the relationship between the yield strength measured from the thin-film experiment during lithiation and that from other mechanical deformations of a-Li_xSi after lithiation, such as biaxial compression, uniaxial tension and compression? With the limited available experimental data, here we address these questions using MD simulations. The results provide insight into the effects of composition, loading sequence, and stress state on the mechanical properties of a-Li_xSi alloys.

We employ a newly developed reactive force field (ReaxFF) in our MD simulations of a-Li_xSi alloys [14]. This ReaxFF provides accurate predictions of a set of fundamental properties for Li_xSi alloys, such as Li composition-dependent elastic modulus, open-cell voltage, and Li-insertion induced volume expansion. Compared to ab initio calculations, the computational efficiency of ReaxFF enables large-scale MD simulations of a-Li_xSi, which is crucially important for sampling amorphous structures and thus obtaining their statistically meaningful properties. In addition, ReaxFF is capable of describing various bonding environments, essential for improving the chemical accuracy of simulated structures and properties spanning a wide range of Li compositions. Using the ReaxFF, we calculate stress generation in a-Si thin films during lithiation and analyze the plastic properties of a-Li_xSi alloys. We then compare the results with those obtained from pure mechanical loading and discuss whether the loading sequence affects the measurements of the mechanical properties of a-Li_xSi alloys.

We first perform MD simulations to study the stress generation in an a-Si thin film constrained by a substrate during Li insertion. The atomic structure of a-Si is created by melting and quenching of single-crystal Si. To simulate the lithiation response, we take a supercell of a-Si as a representative volume element (RVE) in an a-Si thin film, as schematically shown in Fig. 16.5. This supercell is subjected to periodic boundary conditions. Its size is free to change in the out-of-film direction, but fixed in the plane of the thin film, mimicking an RVE in a thin film constrained by the substrate. We randomly insert 20 Li atoms at a time and then relax the system

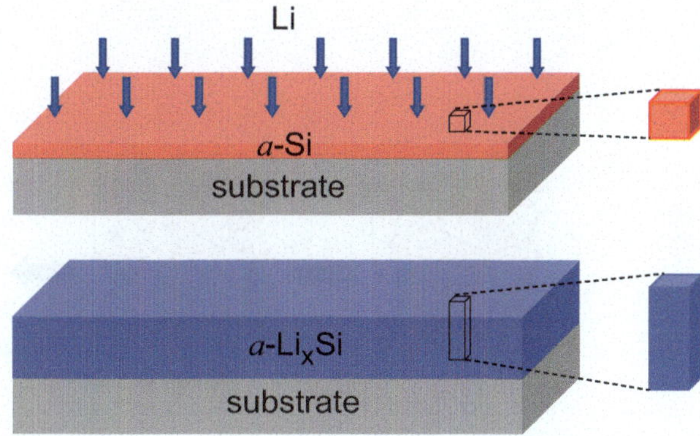

Fig. 16.5 Schematic illustration of lithiation in an *a*-Si thin film constrained by the substrate, showing a representative volume element (RVE) in the film before and after lithiation. As a result of the in-plane geometrical constraint, this volume element undergoes colossal expansion in the out-of-film direction (reprinted from [14] with permission)

at 300 K for 10 ps. The Li concentration is approximately uniform in the system, excluding the concentration gradient-induced stress. This process of insertion and relaxation is repeated until the Li concentration reaches a desired level. The in-plane biaxial stress is calculated after each step of Li insertion and relaxation. The MD simulations show that both the atomic structures and associated stresses are insensitive to the specific scheme of Li insertion, i.e., the number of Li atoms at each insertion and frequency of insertion.

Figure 16.6 shows the atomic structures of pristine *a*-Si, a partially lithiated phase of *a*-LiSi, and fully lithiated *a*-$Li_{3.75}Si$. The starting system of *a*-Si consists of 432 Si atoms, and the final one of *a*-$Li_{3.75}Si$ a total of 2052 Li and Si atoms in the supercell. Figure 16.7 shows the structural characteristics of both *a*-Si and *a*-$Li_{3.75}Si$ in terms of radial distribution function (RDF), angular distribution function (ADF), and ring statistics; here RDF and ADF provide the measure of the short-range order and ring statistics the medium-range order in random-network models of amorphous solids [35]. The dominant features of RDF, ADF, and ring statistics in Fig. 16.7 are consistent with previous quantum calculations [23, 26]. During MD simulations, the volume increase due to Li insertion is mostly accommodated by free expansion in the out-of-film direction. Figure 16.8 shows that the *a*-Li_xSi thin film initially experiences a sharp increase in the in-plane biaxial compressive stress with increasing Li concentration. This stage of stressing is dominated by elastic compression, which arises from the geometrical constraint on the supercell that suppresses the lithiation-induced expansion in the film plane. The biaxial compression reaches a maximum of −6.5 GPa at $x \sim 0.2$, indicative of the onset of plastic yielding. Further lithiation causes a decrease in the magnitude of the film stress, indicating the decrease of the yield stress with increasing Li concentration. Such plastic softening can be attributed to the decreasing fraction of the strong covalent Si–Si bonds that resist against bond switching and ensuing plastic flow, as well as the concomitant increase of the weak metallic Li–Li bonds that facilitate plastic flow. Overall, the trend of change of the film stress from MD is similar to that from experimental measurements [23]. However, the magnitude of the MD-simulated compressive film stress is higher at the corresponding Li

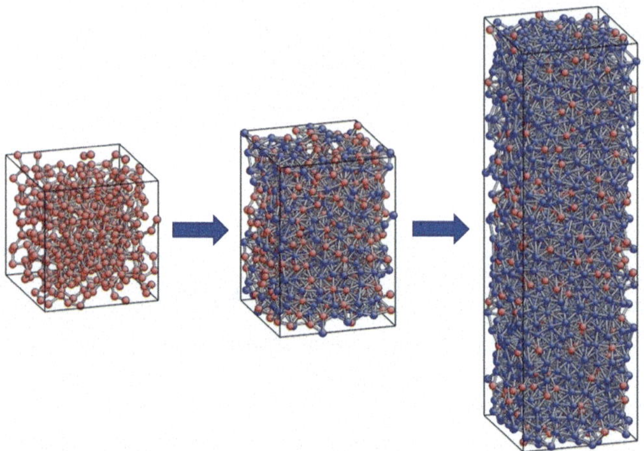

Fig. 16.6 MD-simulated atomic structures of pristine *a*-Si, *a*-LiSi, and *a*-$Li_{3.75}Si$, showing the colossal volume expansion (reprinted from [14] with permission). *Red* and *blue spheres* represent Si and Li atoms, respectively. During lithiation, in-plane dimensions of the system remain the same

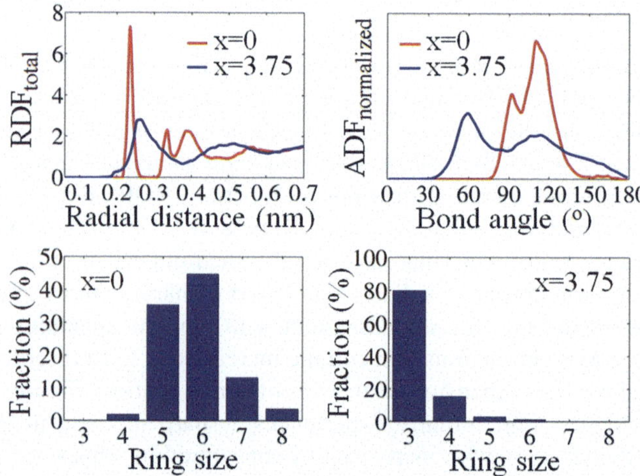

Fig. 16.7 Characterization of atomic structures of *a*-Si and *a*-$Li_{3.75}Si$ in terms of radial distribution function (RDF), angular distribution function (ADF), and ring statistics (reprinted from [14] with permission)

concentration. For example, the experimentally measured peak value was about −1.9 GPa at $x = 0.2$ [29]. This quantitative difference can be attributed to the short MD simulation time scale that limits the extent of stress relaxation. Nevertheless, MD simulations capture the dominant feature of plastic softening during lithiation.

During the constrained thin-film lithiation, stress generates owing to the concurrent chemical reaction and mechanical deformation, both of which involve the

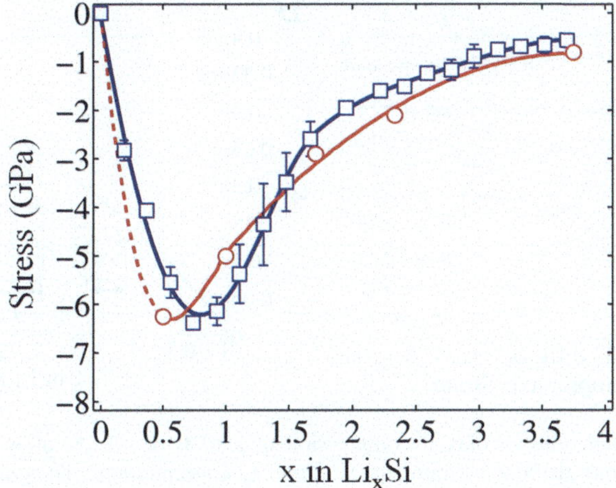

Fig. 16.8 Lithiation-induced biaxial compressive stress in the film as a function of Li to Si ratio, x (reprinted from [14] with permission). *Squares* represent the MD results of film stress generated during lithiation, and the *blue curve* is the numerical fitting. The *error bars* are due to the results of several configurations at a fixed Li concentration. *Circles* represent the yield stresses extracted from an alternative loading mode of MD simulations, i.e., biaxial compression of an already lithiated film

similar atomic processes of bond breaking, switching, and reforming. Alternatively, one can evaluate the mechanical properties of a-Li$_x$Si alloys through chemical lithiation followed by mechanical deformation. That is, a-Si is first lithiated to a specific Li concentration under a stress-free condition, and then is subjected to biaxial compression while keeping the Li concentration fixed. Figure 16.9a shows the simulated stress–strain curves of biaxial compression under a strain rate of 5×10^8/s at 300 K for several representative Li concentrations. A salient feature in these stress–strain responses is the yield-point phenomenon, featuring the peak stress, load drop, and steady flow with limited strain hardening. Notably, the degree of the yield drop varies with Li concentration. In Li-poor phases, the high peak of yield stress is observed with a pronounced load drop. The large peak stress indicates the need of a high load to create enough plastic carriers during the early stage of deformation, a process usually referred to as rejuvenation for creating a deformed amorphous state amenable to plastic flow. In contrast, the yield-point phenomenon is less pronounced in the Li-rich phases, which implies the existence of a high fraction of easily flowing components and thus leads to a support to constitutive modeling of the Li-rich phase as a visco-plastic solid.

Although the detailed atomic processes underlying the yield-point phenomenon in a-Li$_x$Si alloys warrant further study, we focus on extracting the yield strength of σ_Y^m from the above mechanical (m) deformation of biaxial compression. The results enable us to correlate σ_Y^m with the yield strength of σ_Y^{cm} evaluated from thin-film lithiation that involves a concurrent chemomechanical (cm) process. Note

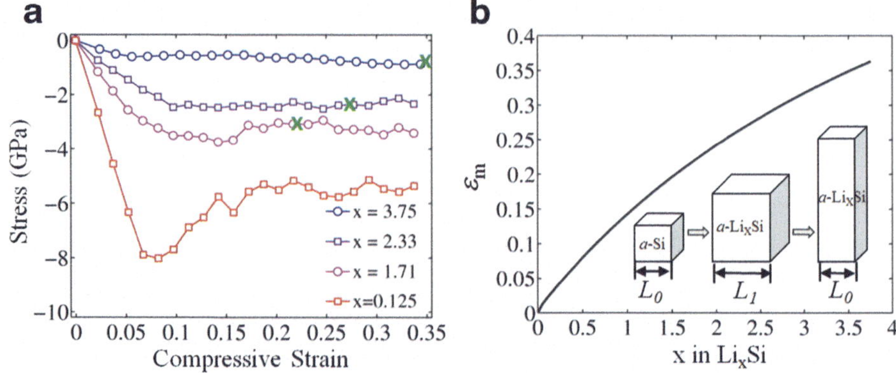

Fig. 16.9 MD simulations of biaxial compression of a RVE of a-Li$_x$Si alloy (reprinted from [14] with permission). (**a**) Stress–strain curves at four Li concentrations. The *green cross* symbol indicates the respective yield strengths of $\sigma_{\mathrm{Y}}^{\mathrm{m}}$, which correspond to the compression strain, ε_{m}, specified in (**b**) for each selected concentration. (**b**) Strain-composition curve for a-Li$_x$Si alloys. The inset describes a sequential loading mode, which consists of a stress-free chemical expansion from a-Si with an initial size of L_0 to a-Li$_x$Si with a size of L_1, and a biaxial compression on a-Li$_x$Si under an external compression strain of $\varepsilon_{\mathrm{m}} = (L_1 - L_0)/L_1$, so that a-Li$_x$Si has the same in-plane dimensions as a-Si

that in order to determine $\sigma_{\mathrm{Y}}^{\mathrm{m}}$ at a specific Li concentration from the stress–strain curve, one needs to determine the corresponding strain at yielding, as shown in Fig. 16.9b. Here we define the strain at yielding as $\varepsilon_{\mathrm{m}} = (L_1 - L_0)/L_0 < 0$, where L_1 and L_0, respectively, denote the supercell size of the stress-free a-Li$_x$Si and the initial a-Si, as schematically shown in the inset of Fig. 16.9b. Application of such a strain at yielding to a-Li$_x$Si cancels the in-plane expansion induced by the stress-free lithiation, consistent with the boundary condition of rigid in-plane constraints during thin-film lithiation. Given ε_{m} at a specific Li concentration, we determine the associated compressive yield strength $\sigma_{\mathrm{Y}}^{\mathrm{m}}$ from the biaxial compressive stress–strain curve in Fig. 16.9a. As indicated by the cross symbols in Fig. 16.9a, $\sigma_{\mathrm{Y}}^{\mathrm{m}}$ corresponds to the characteristic plateau stress of plastic flow. We note that the value of $\sigma_{\mathrm{Y}}^{\mathrm{m}}$ cannot be accurately extracted in the case of low Li concentration (e.g., $x = 0.125$) because of the emergence of a strong yield peak as shown in Fig. 16.9a; we define the yield strength as the steady state flow stress, which should not be sensitive to the value of yield peak. To compare $\sigma_{\mathrm{Y}}^{\mathrm{m}}$ with $\sigma_{\mathrm{Y}}^{\mathrm{cm}}$ we add in Fig. 16.8 $\sigma_{\mathrm{Y}}^{\mathrm{m}}$ as a function of Li concentrations (the red solid line). It is seen that the yield strengths of $\sigma_{\mathrm{Y}}^{\mathrm{m}}$ from biaxial compression are consistent with those of $\sigma_{\mathrm{Y}}^{\mathrm{cm}}$ from thin-film lithiation, by noting that the time scales achievable in MD simulations are similar for the two cases studied. This consistency indicates that the steady state flow stress during the *mechanical* deformation of an a-Li$_x$Si alloy can be effectively taken as its yield strength, even under the *chemomechanical* loading condition.

The magnitudes of above yield strengths are at variance with previous simulation results. The discrepancy of our computed yield strengths with references

[21, 22] arises mainly from the different definitions of yield strength. As shown in Fig. 16.9a, the stress–strain curve from MD simulation usually exhibits the yield-point phenomenon, which involves a peak stress that is followed by stress drop (i.e., softening) and subsequent steady state flow at a plateau stress. The value of the yield peak depends sensitively on various factors such as the applied strain rate and temperature, sample preparation (e.g., annealing temperature and rate), as well as chemical composition of the alloys. The conventional 0.2 % offset yield stress, as defined by Cui et al. [22], is closely correlated with the yield peak, thus being dependent on the aforementioned factors. On the other hand, the stress-controlled energy minimization is performed by Zhao et al. [21], and the yield strength is defined as the critical stress at the load drop or strain jump, which is also closely correlated to the yield peak. In addition, the simulations in reference [21] are performed in a small amorphous system with a highly nonuniform atomic structure and thus their results likely depend sensitively on the system size, an effect that has not been studied systematically by Zhao et al. [21]. In contrast, we define the yield strength as the steady state flow stress, which is not so sensitive to the value of yield peak. More importantly, our definition facilitates a correlation of the yield strength between biaxial deformation and thin-film lithiation. This is because Fig. 16.9b shows that the typical compressive strains during thin-film lithiation are mostly larger than 10 %, and hence the steady state flow stress in biaxial deformation, which prevails at large strains, is more relevant to the yield strength in thin-film lithiation. In addition, the steady state flow stress obtained in our simulations is insensitive to the system size, which ranges from a supercell with side length of 3 nm (~2000 atoms) to side length of 6 nm (~16,000 atoms).

Hertzberg et al. [16] reported the experimental measurements of hardness of a-Li$_x$Si alloys using the indentation method. Due to the well-known strain rate effects associated with MD, it is not possible to quantitatively compare MD with experimental results. However, we note that while these measurements represent the first experimental effort to quantify the yield strength of a-Li$_x$Si alloys, there are several uncontrolled factors in experiments, such as uncertainty on the spatial distribution of Li considering the recently discovered two-phase mechanism in the first lithiation of a-Si, as well as uncertainty on the structural and composition changes when the lithiated samples were transferred to mechanical testing. Hence, the well-controlled experimental studies are critically needed in the future research.

16.5 Summary

In this chapter, we present examples of continuum and atomistic modeling of the deformation and stress generation during electrochemical lithiation of Si electrodes. The main results are summarized as follows.

We have studied the stress generation during lithiation of a spherical particle using a continuum model, which accounts for plastic deformation and a sharp and curved phase boundary between Li-rich and Li-poor phases. We find that a tensile hoop stress is developed in the lithiated shell during Li insertion. This hoop tension originates from the lithiation-induced swelling at the sharp phase boundary with finite curvature, which often arises during lithiation of the crystalline material with curved geometry, such as Si nanoparticles and nanowires. The large hoop tension in the outer layer can consequently trigger the morphological instability and fracture in electrodes. Our work also reveals how the predicted stress states are affected by different assumptions of lithiation and deformation mechanisms, including the dilational versus radial lithiation strain, and the two-phase versus single-phase lithiation mechanism. Our results of lithiation-induced stresses provide a mechanics basis for further studying a wide range of lithiation-related phenomena, e.g., anisotropic swelling in Si nanowires, nanoporosity evolution in Germanium (Ge) nanowires, and size effect on fracture in Si nanoparticles.

We have also conducted an atomistic study of the mechanical properties of a-Li_xSi alloys with a reactive force field. Several chemomechanical deformation processes are simulated by MD, including the constrained thin-film lithiation and biaxial compression. The results reveal the effects of loading sequence and stress state, which are correlated to the atomic bonding characteristics. Since the a-Li_xSi alloys undergo a large range of change in compositions during electrochemical cycling in LIBs, their mechanical behaviors are extremely rich, exhibiting the features of both covalent and metallic amorphous solids. The detailed analyses of the MD results provide atomistic insights for understanding the mechanical properties and designing the future testing of lithiated Si alloys.

We close this chapter by noting that Sect. 16.3 deals with a continuum model of chemomechanics of lithiation in Si particles and Sect. 16.4 with MD simulations of Si thin films. The continuum and atomistic methods have not been tied together at this moment. In the future, it would be interesting to apply both methods to study the same problem such as lithiation in Si particles and Si thin films. The correlation between the simulation results from the two methods could generate new insights into the chemomechanical phenomena of lithiation that would be valuable to the design of reliable Si anodes for high-performance Li-ion batteries.

References

1. J.M. Tarascon, M. Armand, Issues and challenges facing rechargeable lithium batteries. Nature **414**(6861), 359–367 (2001)
2. L.Y. Beaulieu, K.W. Eberman, R.L. Turner, L.J. Krause, J.R. Dahn, Colossal reversible volume changes in lithium alloys. Electrochem. Solid-State Lett. **4**(9), A137–A140 (2001)
3. C.K. Chan, H.L. Peng, G. Liu, K. McIlwrath, X.F. Zhang, R.A. Huggins, Y. Cui, High-performance lithium battery anodes using silicon nanowires. Nat. Nanotechnol. **3**(1), 31–35 (2008)
4. R.A. Huggins, W.D. Nix, Decrepitation model for capacity loss during cycling of alloys in rechargeable electrochemical systems. Ionics **6**, 57–63 (2000)
5. X.H. Liu, H. Zheng, L. Zhong, S. Huang, K. Karki, L.Q. Zhang, Y. Liu, A. Kushima, W.T. Liang, J.W. Wang, J.-H. Cho, E. Epstein, S.A. Dayeh, S.T. Picraux, T. Zhu, J. Li, J.P. Sullivan, J. Cumings, C. Wang, S.X. Mao, Z.Z. Ye, S. Zhang, J.Y. Huang, Anisotropic swelling and fracture of silicon nanowires during lithiation. Nano Lett. **11**(8), 3312–3318 (2011)

6. X.H. Liu, L. Zhong, S. Huang, S.X. Mao, T. Zhu, J.Y. Huang, Size-dependent fracture of silicon nanoparticles during lithiation. ACS Nano **6**(2), 1522–1531 (2012)
7. X.H. Liu, J.W. Wang, S. Huang, F. Fan, X. Huang, Y. Liu, S. Krylyuk, J. Yoo, S.A. Dayeh, A.V. Davydov, S.X. Mao, S.T. Picraux, S. Zhang, J. Li, T. Zhu, J.Y. Huang, In situ atomic-scale imaging of electrochemical lithiation in silicon. Nat. Nanotechnol. **7**(11), 749–756 (2012)
8. M.T. McDowell, S.W. Lee, W.D. Nix, Y. Cui, 25th Anniversary Article: understanding the lithiation of silicon and other alloying anodes for lithium-ion batteries. Adv. Mater. **25**(36), 4966–4984 (2013)
9. J.W. Wang, Y. He, F. Fan, X.H. Liu, S. Xia, Y. Liu, C.T. Harris, H. Li, J.Y. Huang, S.X. Mao, T. Zhu, Two-phase electrochemical lithiation in amorphous silicon. Nano Lett. **13**(2), 709–715 (2013)
10. V.A. Sethuraman, M.J. Chon, M. Shimshak, V. Srinivasan, P.R. Guduru, In situ measurements of stress evolution in silicon thin films during electrochemical lithiation and delithiation. J. Power Sources **195**(15), 5062–5066 (2010)
11. X. Wang, F. Fan, J. Wang, H. Wang, S. Tao, A. Yang, Y. Liu, H.B. Chew, S.X. Mao, T. Zhu, S. Xia, High damage tolerance of electrochemically lithiated silicon. Nat. Commun. **6**, 8417 (2015)
12. Z. Ting, Mechanics of high-capacity electrodes in lithium-ion batteries. Chin. Phys. B **25**(1), 014601 (2016)
13. S. Huang, F. Fan, J. Li, S.L. Zhang, T. Zhu, Stress generation during lithiation of high-capacity electrode particles in lithium ion batteries. Acta Mater. **61**, 4354–4364 (2013)
14. F. Fan, S. Huang, H. Yang, M. Raju, D. Datta, V.B. Shenoy, A.C.T. van Duin, S. Zhang, T. Zhu, Mechanical properties of amorphous Li_xSi alloys: a reactive force field study. Model. Simul. Mater. Sci. Eng. **21**(7), 074002 (2013)
15. X.H. Liu, Y. Liu, A. Kushima, S.L. Zhang, T. Zhu, J. Li, J.Y. Huang, In situ TEM experiments of electrochemical lithiation and delithiation of individual nanostructures. Adv. Energy Mater. **2**(7), 722–741 (2012)
16. B. Hertzberg, J. Benson, G. Yushin, Ex-situ depth-sensing indentation measurements of electrochemically produced Si-Li alloy films. Electrochem. Commun. **13**(8), 818–821 (2011)
17. M.J. Chon, V.A. Sethuraman, A. McCormick, V. Srinivasan, P.R. Guduru, Real-time measurement of stress and damage evolution during initial lithiation of crystalline silicon. Phys. Rev. Lett. **107**(4), 045503 (2011)
18. S.K. Soni, B.W. Sheldon, X.C. Xiao, M.W. Verbrugge, D. Ahn, H. Haftbaradaran, H.J. Gao, Stress mitigation during the lithiation of patterned amorphous Si islands. J. Electrochem. Soc. **159**(1), A38–A43 (2012)
19. A. Kushima, J.Y. Huang, J. Li, Quantitative fracture strength and plasticity measurements of lithiated silicon nanowires by in situ TEM tensile experiments. ACS Nano **6**(11), 9425–9432 (2012)
20. V.B. Shenoy, P. Johari, Y. Qi, Elastic softening of amorphous and crystalline Li-Si phases with increasing Li concentration: a first-principles study. J. Power Sources **195**(19), 6825–6830 (2010)
21. K.J. Zhao, G.A. Tritsaris, M. Pharr, W.L. Wang, O. Okeke, Z.G. Suo, J.J. Vlassak, E. Kaxiras, Reactive flow in silicon electrodes assisted by the insertion of lithium. Nano Lett. **12**(8), 4397–4403 (2012)
22. Z.W. Cui, F. Gao, Z.H. Cui, J.M. Qu, A second nearest-neighbor embedded atom method interatomic potential for Li-Si alloys. J. Power Sources **207**, 150–159 (2012)
23. V.L. Chevrier, J.R. Dahn, First principles model of amorphous silicon lithiation. J. Electrochem. Soc. **156**(6), A454–A458 (2009)
24. V.L. Chevrier, J.R. Dahn, First principles studies of disordered lithiated silicon. J. Electrochem. Soc. **157**(4), A392–A398 (2010)
25. Q.F. Zhang, W.X. Zhang, W.H. Wan, Y. Cui, E.G. Wang, Lithium insertion in silicon nanowires: an ab initio study. Nano Lett. **10**(9), 3243–3249 (2010)
26. S. Huang, T. Zhu, Atomistic mechanisms of lithium insertion in amorphous silicon. J. Power Sources **196**(7), 3664–3668 (2011)

27. H. Kim, C.Y. Chou, J.G. Ekerdt, G.S. Hwang, Structure and properties of Li-Si alloys: a first-principles study. J. Phys. Chem. C **115**(5), 2514–2521 (2011)
28. P. Johari, Y. Qi, V.B. Shenoy, The mixing mechanism during lithiation of Si negative electrode in Li-ion batteries: an ab initio molecular dynamics study. Nano Lett. **11**(12), 5494–5500 (2011)
29. K.J. Zhao, W.L. Wang, J. Gregoire, M. Pharr, Z.G. Suo, J.J. Vlassak, E. Kaxiras, Lithium-assisted plastic deformation of silicon electrodes in lithium-ion batteries: a first-principles theoretical study. Nano Lett. **11**(7), 2962–2967 (2011)
30. M.K.Y. Chan, C. Wolverton, J.P. Greeley, First principles simulations of the electrochemical lithiation and delithiation of faceted crystalline silicon. J. Am. Chem. Soc. **134**(35), 14362–14374 (2012)
31. S.C. Jung, J.W. Choi, Y.K. Han, Anisotropic volume expansion of crystalline silicon during electrochemical lithium insertion: an atomic level rationale. Nano Lett. **12**(10), 5342–5347 (2012)
32. M. Haghi, L. Anand, Analysis of strain hardening viscoplastic thick-walled sphere and cylinder under external pressure. Int. J. Plast. **7**(3), 123–140 (1991)
33. N.F. Mott, S. Rigo, F. Rochet, A.M. Stoneham, Oxidation of silicon. Philos. Mag. B **60**(2), 189–212 (1989)
34. C.H. Hsueh, A.G. Evans, Oxidation-induced stress and some effects on the behavior of oxide-films. J. Appl. Phys. **54**(11), 6672–6686 (1983)
35. D.S. Franzblau, Computation of ring statistics for network models of solids. Phys. Rev. B **44**(10), 4925–4930 (1991)

Chapter 17
Multiscale Modeling of Thin Liquid Films

Han Hu and Ying Sun

17.1 Introduction

Thin liquid films, the intermediate liquid phase between two surfaces with nanoscale separation, are commonly present in lubrication [1, 2], wetting and spreading [3, 4], phase change [5, 6], gas adsorption [7, 8], chemical and food processing [9], and among others. Figure 17.1 shows three common types of thin liquid films: (a) a free-standing film as encountered in nucleate boiling, (b) a thin liquid film sandwiched between two solid surfaces, as in lubrication, and (c) a thin liquid film sandwiched between a solid surface and its own vapor, important for wetting and spreading as well as liquid-vapor phase change. The stability of these thin liquid films is of vital importance in their applications because the breakup of the thin films may cause dramatic changes in their desired properties. For example, the rupture of thin films can lead to several orders of magnitude reduction in heat flux in thin film evaporation. Similarly, the breakup of thin liquid films between bubbles during nucleate boiling causes an insulating vapor layer blanketing the surface, which reduces the heat transfer rate by more than 100 times [10]. In addition, the liquid-infused surfaces designed for anti-fouling and self-cleaning purposes can lose their super slippery properties if the infused thin liquid film drains from the surface [11]. It is therefore very important to understand the properties of thin liquid films so as to better control their stability.

Recently, nanostructures have been introduced to control the stability of thin liquid films. For example, superhydrophilic nanostructures, such as nanoporous alumina and CuO nanostructures, have been used in thin film evaporation to enhance

H. Hu • Y. Sun (✉)
Department of Mechanical Engineering and Mechanics, Drexel University, Philadelphia, PA 19104, USA
e-mail: YSun@coe.drexel.edu

C.R. Weinberger, G.J. Tucker (eds.), *Multiscale Materials Modeling for Nanomechanics*, Springer Series in Materials Science 245,
DOI 10.1007/978-3-319-33480-6_17

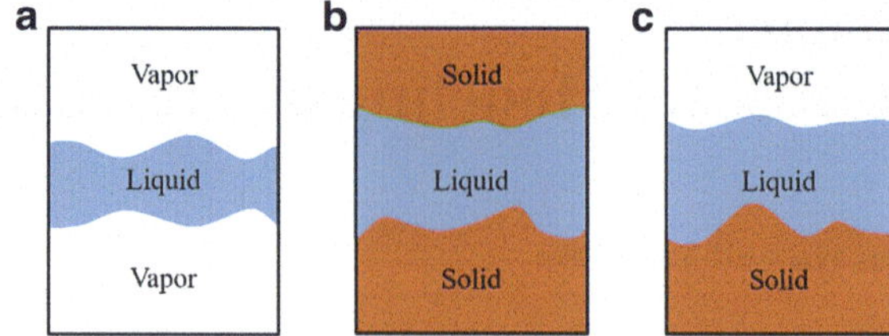

Fig. 17.1 Types of thin liquid films: (**a**) a free-standing liquid film between two vapor phases, (**b**) a thin liquid film between two solid surfaces, and (**c**) a thin liquid film between its own vapor and a solid surface

the liquid delivery to the thin film region where heat flux is ultra-high [12, 13]. Nanofiber networks have been applied to liquid-infused surfaces to better anchor the infused liquid film so as to make the slippery surface more robust [14]. Despite such success, the current design of nanostructures still relies heavily on empirical knowledge as well as costly and time-consuming experiments. There is still a fundamental lack of understanding regarding how nanostructures affect the stability of thin liquid films.

A key physical quantity that affects the stability of a thin liquid film is the disjoining pressure, i.e., the pressure difference between the thin liquid film and the bulk liquid [15]. Disjoining pressure arises from the intermolecular forces between the liquid and its neighboring phases, and is responsible for many nanoscale phenomena. For instance, for a free-standing thin liquid film shown in Fig. 17.1a, the disjoining pressure leads to spontaneous thinning and rupture of the thin film [16]. For evaporation of a thin liquid film on a solid surface, the disjoining pressure suppresses evaporation and finally leads to a non-evaporating thin film on the solid surface [17]. As for a thin liquid film residing on a nanostructured surface, the morphology of the solid surface affects the meniscus shape, which in turn determines the pressure of the liquid film. Accurate predictions of the meniscus shape and disjoining pressure are hence important to better design nanostructures for the enhanced stability of the thin film.

Modeling of thin liquid films using continuum-level equations has been developed for almost 70 years. Hamaker developed the original theory for disjoining pressure of a thin liquid film assuming (1) pair-wise intermolecular interactions and (2) van der Waals interactions dominant [18]. Recall that the long-range van der Waals potential (i.e., the dispersion potential) between two atoms scales with r^{-6} where r is the distance between the two atoms. Integrating the dispersion forces between all interacting atoms gives the disjoining pressure of a thin film as a function of the thin film thickness following:

$$\Pi_{\mathrm{vdW}} = \frac{A}{6\pi\delta_0^3} \tag{17.1}$$

where δ_0 is the liquid film thickness and A is the Hamaker constant. As a result of the disjoining pressure, the equilibrium vapor pressure near the thin liquid film deviates from the saturation pressure. Hamaker's original theory was further developed to include the effects of many-body interactions and intervening medium by Lifshitz,

Ninham, and others [19–21]. Based on the Lifshitz theory, the Hamaker constant can be calculated using the refractive index and dielectric constant of the materials in the system. These continuum-level models do not depend on specific material systems, and can be applied to broad time and length scales. However, these models are often built upon idealistic assumptions, such as an atomically smooth surface, which may not be satisfied in practice.

Experimental studies have also been performed to investigate the disjoining pressure in thin liquid films using surface force apparatus (SFA) [22], atomic force microscopy (AFM) [23], quartz crystal microbalance (QCM) [8], and interferometry [24]. Results show that the classic theory is inadequate in predicting the disjoining pressure of thin liquid films [8, 24, 25]. For example, for a polar liquid such as water, the disjoining pressure obtained from experiments is stronger than that predicted by the classic theory and does not simply scale with ${\delta_0}^{-3}$. Researchers have suggested that this inadequacy in the classical theory is due to various nanoscale effects, e.g., thermal fluctuations of the liquid-vapor interface [7, 8], nanoscale surface roughness [24], non-van der Waals forces [8], structural changes in the liquid medium [25], and ordering of liquid molecules in both position and orientation [22].

On the other hand, molecular-level modeling is not limited to having perfectly smooth surfaces, and has proven to be a powerful tool in investigating the role of nanoscale effects on the disjoining pressure of nanoscale thin liquid films [26–29]. Existing studies using molecular dynamics (MD) simulations have shown to capture the molecular-level details of thin liquid films, e.g., liquid layering, chemical bond breaking and forming, etc. The effects of the out-of-plane and in-plane liquid layering on disjoining pressure are examined by Carey and Wemhoff [26] and Hu and Sun [28], respectively. The reactive adsorption of water thin films on ZnO surfaces has also been studied using MD simulations [27]. Weng et al. investigated the effect of film thickness on the stability of thin liquid films and analyzed the pressure distribution inside the films [29]. Despite their proven success, molecular modeling can be limited in two aspects. Firstly, molecular models depend on specific potentials used and geometries studied, and it is thus difficult to generalize the results. Secondly, the time and length scales of molecular modeling are highly limited. For example, common length scales of MD simulations are smaller than 20 nm, and time scales are less than 10 ns. However, in applications such as thin film evaporation on nanostructured surfaces, the thin liquid film thickness ranges from 2 to 1000 nm and the nanostructure size is on the order of 100–500 nm, computationally very challenging for MD simulations.

As neither continuum nor molecular-level modeling itself is adequate to fully resolve the disjoining pressure of thin liquid films, multiscale models that take advantages of both continuum-level and molecular-level approaches are thus more suitable to reveal the missing physics of thin liquid films, such as the role of nanostructures and electrostatic interactions on the meniscus shape and disjoining pressure. In the multiscale modeling approach, molecular-level modeling is utilized to examine various nanoscale effects (e.g., non-van der Waals forces between the liquid film and the solid surface, nanoscale surface roughness) on disjoining pressure and meniscus shape using a specific material system at small length and time scales. The results from the MD simulations are used to verify the assumptions in the continuum-level modeling. The verified continuum-level model can then be applied to predict the disjoining pressure and meniscus shape for more general materials systems and geometries, e.g., to predict the driving force of liquid delivery onto a nanostructured surface during thin film evaporation.

In this chapter, multiscale modeling is utilized to investigate the effects of nanostructures and electrostatic interactions on meniscus shape and disjoining pressure of a thin liquid film (equilibrated with its own vapor) on a solid surface with nanoscale roughness. The detailed derivation of the continuum-level model is given in Sect. 17.2, and the molecular modeling procedures are described in Sect. 17.3. The results of the continuum-level modeling and MD simulations are compared in Sect. 17.4 for the meniscus shape and Sect. 17.5 for the disjoining pressure, followed by conclusions and outlook for further study in Sect. 17.6. The results presented here can be used to guide nanostructural designs for enhanced thin film stability, as well as to accurately predict the critical heat fluxes in thin film evaporation and boiling heat transfer. The modeling approaches described in this chapter can also be used for other studies associated with thin liquid films.

17.2 Theoretical Modeling of Thin Liquid Films

In this section, a theoretical model is developed for a thin liquid film residing on a two-dimensional periodic nanostructure. The meniscus shape of the thin film is determined by minimization of the system free energy. With the calculated meniscus shape, the mean disjoining pressure in the film is then determined based on the Derjaguin approximation [30]. Figure 17.2 shows the schematic of a thin liquid film of thickness δ_0 on a two-dimensional periodic nanostructured surface of wavelength L and depth D. The basic assumptions of the model are:

(i) The liquid film is in contact with its own vapor, and the vapor near the thin film is assumed to behave as an ideal gas;

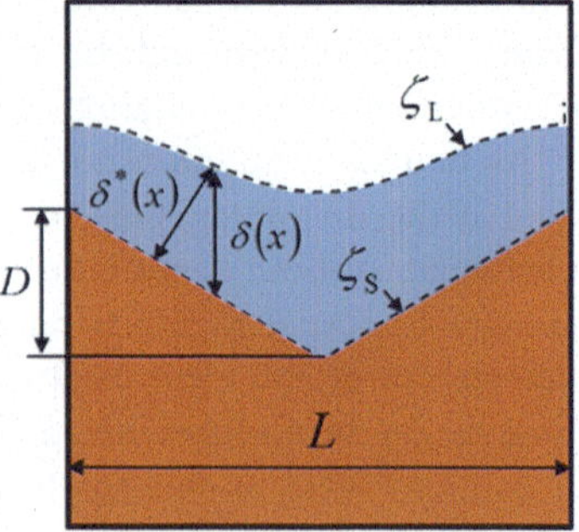

Fig. 17.2 Schematic of a thin liquid film on a triangular nanostructured surface

(ii) The interactions between the liquid and the solid are non-bonded interactions where chemical reactions are not considered;
(iii) The meniscus shape, ζ_L, is periodic with the same wavelength L as that of the substrate surface ζ_S; and
(iv) The liquid film wets the nanostructured surface in the Wenzel state, in which liquid is in full contact with the nanostructured surface.

Before modeling a thin liquid film on a nanostructured surface, we first review the disjoining pressure theory on a flat solid surface. Equation (17.1) gives the disjoining pressure of a thin liquid film assuming the van der Waals forces to be the dominant interactions between the solid and liquid. For circumstances where the electrostatic interactions between the liquid and solid are also important, the electrostatic contribution to the disjoining pressure is derived based on the Poisson–Boltzmann equation as [16]

$$\Pi_{\mathrm{e}} = 2\varepsilon_0\varepsilon\left(\frac{k_{\mathrm{B}}T}{ze}\right)^2 K^2 \tag{17.2}$$

where ε_0 is the vacuum permittivity, ε the relative permittivity, k_{B} the Boltzmann constant, T the temperature, e the electron charge, z the valence, and the factor K is related to the surface charge density σ with electroneutrality following $2k_{\mathrm{B}}TK\tan(K\delta_0)/(ze) = -\sigma/(\varepsilon_0\varepsilon)$. The expression of Π_{e} in Eq. (17.2) reduces to $\Pi_{\mathrm{e}} = -\sigma k_{\mathrm{B}}T/(ze\delta_0)$ at $\delta_0 \to 0$, and $\Pi_{\mathrm{e}} = \pi^2 k_{\mathrm{B}}^2 T^2 \varepsilon\varepsilon_0/\left(2z^2e^2\delta_0^2\right)$ in the limit of $\delta_0 \to \infty$. The disjoining pressure of a thin liquid film on a flat surface is then given as $\Pi_{\mathrm{flat}} = \Pi_{\mathrm{vdW}} + \Pi_{\mathrm{e}}$, where Π_{vdW} and Π_{e} are determined by Eqs. (17.1) and (17.2), respectively.

For a thin liquid film residing on a nanostructured surface, two mechanisms affect the shape of the meniscus. As shown in Eqs. (17.1) and (17.2), both the van der Waals and electrostatic contributions to the disjoining pressure are strong functions of the liquid film thickness such that any gradient in the film thickness leads to a large unbalanced disjoining pressure. The disjoining pressure therefore keeps the film thickness uniform and the film conformal to the substrate. On the other hand, the capillary pressure resulting from the surface tension force flattens the meniscus to minimize the system free energy. In equilibrium, the competition between the capillary pressure and the disjoining pressure determines the meniscus shape, which can be characterized by a scaled healing length, ξ/D, where $\xi^2 = \gamma/(d\Pi_{\mathrm{flat}}/d\delta_0)$ [31] and D is the nanostructure depth. When only the van der Waals interactions are considered between the solid and liquid, the scaled healing length takes the form of [32]

$$\frac{\xi}{D} = \frac{\delta_0^2}{\sqrt{A/(2\pi\gamma)}D} \tag{17.3}$$

where γ is the liquid surface tension. When the electrostatic interactions are taken into consideration, $\Pi_{\mathrm{flat}} = \Pi_{\mathrm{vdW}} + \Pi_{\mathrm{e}}$, the scaled healing length becomes [33]

$$\frac{\xi}{D} = \frac{1}{D}\left[\frac{A}{2\pi\gamma\delta_0^4} + \frac{\varepsilon\varepsilon_0}{\gamma\delta_0^3}\left(\frac{\pi k_B T}{ze}\right)^2\right]^{-1/2} \tag{17.4}$$

Here, the electrostatic contribution Π_e takes its form in the limit of $\delta_0 \to \infty$ for simplicity. As the scaled healing length $\xi/D \to 0$, the solid–liquid intermolecular interactions dominate, and when $\xi/D \to \infty$, the surface tension dominates. As the scaled healing length ξ/D increases, the surface tension force becomes stronger, and a flatter film is expected. While the scaled healing length shows qualitatively how the meniscus shape changes with the liquid film thickness, material system, and nanostructure size, an accurate prediction of the meniscus shape and disjoining pressure using a closed-form model is discussed in Sect. 17.2.1. Simplifications of the closed-form model are then presented in Sect. 17.2.2.

17.2.1 Closed-Form Model

Now we describe the meniscus shape of a thin film, $\zeta_L(x)$, as a periodic function of period L (assumption iii) using the Fourier series

$$\zeta_L(x) = \frac{a_0}{2} + \sum_{n=1}^{\infty} a_n \cos\left(\frac{2\pi n}{L}x\right) + b_n \sin\left(\frac{2\pi n}{L}x\right) \tag{17.5}$$

where $a_0 = \frac{2}{L}\int_{-L/2}^{L/2} \zeta_L(x)\mathrm{d}x = 2\delta_0$, $a_n = \frac{2}{L}\int_{-L/2}^{L/2} \zeta_L(x)\cos\left(\frac{2\pi n}{L}x\right)\mathrm{d}x$, and $b_n = 0$ due to symmetry. Thus, Eq. (17.5) reduces to

$$\zeta_L(x) = \sum_{n=1}^{\infty} a_n \cos\left(\frac{2\pi n}{L}x\right) + \delta_0 \tag{17.6}$$

which follows

$$\zeta_L'(x) = -\sum_{n=1}^{\infty} \frac{2\pi n}{L} a_n \sin\left(\frac{2\pi n}{L}x\right) \tag{17.7}$$

For thin liquid films on a nanostructured surface, as shown in Fig. 17.2, it is convenient to introduce a modified film thickness $\delta^*(x)$ that represents the shortest distance between the meniscus profile and structured solid surface as $\delta^*(x) = \delta(x)/r$, where r is the Wenzel roughness ratio between the actual and flat surface areas [32]. Compared with the traditional definition of film thickness $\delta(x)$ which is the vertical thickness defined as the difference between the solid and liquid surface following $\delta(x) = \zeta_L(x) - \zeta_S(x)$ [31], the modified film thickness $\delta^*(x)$ is more relevant to the present model because the local disjoining pressure is directly related with $\delta^*(x)$.

The total free energy of the system, W_{total}, consists of the surface component, i.e., the surface excess energy W_γ, and the volumetric component, i.e., the van der Waals interaction energy W_{vdW} and the electrostatic interaction energy W_e. Minimizing the system total free energy, W_{total}, by setting $\delta W_{\text{total}} = 0$, the meniscus shape can be determined. Using $\delta W_{\text{total}} = (\partial W_{\text{total}}/\partial \zeta_L)\,\delta\zeta_L + (\partial W_{\text{total}}/\partial \zeta_L')\,\delta\zeta_L'$ gives

$$\delta W_{\text{total}} = \sum_{n=1}^{\infty}\left[\frac{\partial W_{\text{total}}}{\partial \zeta_L}\frac{\partial \zeta_L}{\partial a_n} + \frac{\partial W_{\text{total}}}{\partial \zeta_L'}\frac{\partial \zeta_L'}{\partial a_n}\right]\delta a_n \tag{17.8}$$

where δa_n is the variance of a_n and $\delta a_n = \frac{2}{L}\int_{-L/2}^{L/2} \delta a_m \cos\left(\frac{2\pi m}{L}x\right)\cos\left(\frac{2\pi n}{L}x\right)\mathrm{d}x$. Due to the orthogonality of the Fourier series, δa_n and δa_m are independent for any $n \neq m$, which yields

$$\frac{\partial W_{\text{total}}}{\partial \zeta_L}\frac{\partial \zeta_L}{\partial a_n} + \frac{\partial W_{\text{total}}}{\partial \zeta_L'}\frac{\partial \zeta_L'}{\partial a_n} = 0 \tag{17.9}$$

The surface excess energy is given as $W_\gamma = \gamma A_{L-V}$, where A_{L-V} is the liquid-vapor interfacial area. For a two-dimensional meniscus, the liquid-vapor interfacial area is calculated as $A_{L-V} = \int_{-L/2}^{L/2}\left[1 + \left[\zeta_L'(x)\right]^2\right]^{1/2}\mathrm{d}x$. Thus, the surface excess energy can be expressed as

$$W_\gamma = \gamma\int_{-L/2}^{L/2}\left[1 + \left[\zeta_L'(x)\right]^2\right]^{1/2}\mathrm{d}x \tag{17.10}$$

As for the van der Waals and electrostatic contributions to the system free energy, $\partial W_{\text{vdW}}/\partial \zeta_L = (\partial W_{\text{vdW}}/\partial \delta^*)/r$ and $\partial W_e/\partial \zeta_L = (\partial W_e/\partial \delta^*)/r$, where r is the Wenzel roughness ratio. The disjoining pressure is the gradient of the free energy due to the solid–liquid interactions with respect to film thickness [34], i.e., $\partial\left(W_{\text{vdW}} + W_e\right)/\partial \delta^* = \Pi_{\text{vdW}}(\delta^*) + \Pi_e(\delta^*)$. Thus,

$$\frac{\partial W_{\text{vdW}}}{\partial \zeta_L} + \frac{\partial W_e}{\partial \zeta_L} = \frac{Ar^2}{6\pi}\int_{-L/2}^{L/2}\frac{1}{\delta^3}\mathrm{d}x - \frac{2\varepsilon\varepsilon_0}{r}\left(\frac{\pi k_{\text{B}}T}{ze}\right)^2\int_{-L/2}^{L/2}K^2\mathrm{d}x \tag{17.11}$$

Substituting Eqs. (17.10) and (17.11) into Eq. (17.9), it follows that

$$\int_{-L/2}^{L/2} \frac{\frac{2\pi n}{L}\sin\left(\frac{2\pi n}{L}x\right)\sum_{m=1}^{\infty} a_m \frac{2\pi m}{L}\sin\left(\frac{2\pi m}{L}x\right)\mathrm{d}x}{\sqrt{1+\left[\sum_{m=1}^{\infty} a_m \frac{2\pi m}{L}\sin\left(\frac{2\pi m}{L}x\right)\right]^2}} - \frac{Ar^2}{6\pi\gamma}\int_{-L/2}^{L/2} \frac{\cos\left(\frac{2\pi n}{L}x\right)\mathrm{d}x}{\left[\sum_{m=1}^{\infty} a_m \cos\left(\frac{2\pi m}{L}x\right)+\delta_0-\zeta_S(x)\right]^3} - \frac{2\varepsilon\varepsilon_0}{r\gamma}\left(\frac{\pi k_B T}{ze}\right)^2\int_{-L/2}^{L/2} K^2\cos\left(\frac{2\pi n}{L}x\right)\mathrm{d}x = 0 \tag{17.12}$$

where K is related with the surface charge density σ based on electroneutrality as

$$\frac{2k_B TK}{ze}\tan\left(K\delta^*\right) = -\frac{\sigma}{\varepsilon_0\varepsilon} \tag{17.13}$$

The closed-form relation given in Eq. (17.12) predicts the wave amplitudes, i.e., a_n ($n = 1, 2, \ldots \infty$), and hence the meniscus shape of a thin liquid film of thickness δ_0 on a nanostructured surface with a prescribed profile $\zeta_S(x)$.

Using the Derjaguin approximation, the local disjoining pressure of a curved surface can be approximated by the disjoining pressure of a flat substrate of the same local film thickness, i.e., $\Pi_{\mathrm{Rough}}(\delta^*(x)) \cong \Pi_{\mathrm{Flat}}(\delta^*(x))$ [30]. Averaging the total disjoining pressure along x gives the following mean disjoining pressure of a thin liquid film on a nanostructured surface

$$\overline{\Pi}_{\mathrm{rough}} = \frac{Ar^3}{6\pi L}\int_{-L/2}^{L/2}\left[\sum_{n=1}^{\infty} a_n\cos\left(\frac{2\pi n}{L}x\right)+\delta_0-\zeta_S(x)\right]^{-3}\mathrm{d}x + 2\varepsilon\varepsilon_0\left(\frac{\pi k_B T}{ze}\right)^2\int_{-L/2}^{L/2} K^2\mathrm{d}x \tag{17.14}$$

where the first and second terms of the right-hand side of Eq. (17.14) denote the van der Waals and electrostatic contributions to the disjoining pressure, respectively. For a thin liquid film on a curved surface in equilibrium with its own vapor, the vapor pressure, P_v, near the thin film is affected by both the disjoining pressure and the curvature of the liquid meniscus [31]

$$P_C + \Pi_{\mathrm{rough}} = -\rho k_B T\ln\left(\frac{P_v}{P_{\mathrm{sat}}}\right) \tag{17.15}$$

where P_C the capillary pressure, ρ the density of the liquid, P_v the vapor pressure, and P_{sat} the saturation pressure. Averaging both sides of Eq. (17.15) along x

and substituting Eq. (17.14), the vapor pressure near a thin liquid film on a nanostructured surface can be determined as

$$\frac{P_{\mathrm{v}}}{P_{\mathrm{sat}}} = \exp\left\{-\frac{Ar^3}{6\pi L\rho k_{\mathrm{B}}T}\int_{-L/2}^{L/2}\left[\sum_{n=1}^{\infty} a_n \cos\left(\frac{2\pi n}{L}x\right) + \delta_0 - \zeta_{\mathrm{S}}(x)\right]^{-3}\mathrm{d}x \right. \\ \left. -\frac{2\varepsilon\varepsilon_0}{\rho L k_{\mathrm{B}}T}\left(\frac{k_{\mathrm{B}}T}{ze}\right)^2\int_{-L/2}^{L/2} K^2 \mathrm{d}x\right\} \tag{17.16}$$

where the capillary pressure, P_{C}, is canceled out due to symmetry. Thus, a closed-form relation for the vapor pressure of a thin film of thickness δ_0 on a nanostructured surface with profile $\zeta_{\mathrm{S}}(x)$ is given in Eq. (17.16), where the Fourier coefficients a_n are calculated from Eq. (17.12). According to Eq. (17.16), both van der Waals and electrostatic contributions to disjoining pressure, represented by the first and second terms on the right-hand side of Eq. (17.16) respectively, lead to reduction of vapor pressure. The greater the disjoining pressure effect is, the smaller the vapor pressure becomes.

For a system where the van der Waals interactions dominate and the electrostatic contribution to system free energy is negligible, the closed-form model to calculate the meniscus shape reduces to

$$\int_{-L/2}^{L/2} \frac{\frac{2\pi n}{L}\sin\left(\frac{2\pi n}{L}x\right)\sum_{m=1}^{\infty} a_m \frac{2\pi m}{L}\sin\left(\frac{2\pi m}{L}x\right)\mathrm{d}x}{\sqrt{1+\left[\sum_{m=1}^{\infty} a_m \frac{2\pi m}{L}\sin\left(\frac{2\pi m}{L}x\right)\right]^2}} \\ -\frac{Ar^2}{6\pi\gamma}\int_{-L/2}^{L/2}\frac{\cos\left(\frac{2\pi n}{L}x\right)\mathrm{d}x}{\left[\sum_{m=1}^{\infty} a_m \cos\left(\frac{2\pi m}{L}x\right) - \zeta_{\mathrm{S}}(x) + \delta_0\right]^3} = 0 \tag{17.17}$$

The mean disjoining pressure becomes

$$\overline{\Pi}_{\mathrm{rough}} = \frac{Ar^3}{6\pi L}\int_{-L/2}^{L/2}\left[\sum_{n=1}^{\infty} a_n \cos\left(\frac{2\pi n}{L}x\right) + \delta_0 - \zeta_{\mathrm{S}}(x)\right]^{-3}\mathrm{d}x \tag{17.18}$$

And the vapor pressure is

$$\frac{P_{\mathrm{v}}}{P_{\mathrm{sat}}}=\exp\left\{-\frac{Ar^3}{6\pi L\rho k_{\mathrm{B}}T}\int_{-L/2}^{L/2}\left[\sum_{n=1}^{\infty}a_n\cos\left(\frac{2\pi n}{L}x\right)+\delta_0-\zeta_{\mathrm{S}}(x)\right]^{-3}\mathrm{d}x\right\} \tag{17.19}$$

When van der Waals force dominates the solid–liquid interactions, Eq. (17.19) gives a closed-form relation for the vapor pressure of a thin film of thickness δ_0 on a nanostructured surface with profile $\zeta_S(x)$ using Fourier coefficients a_n, calculated from Eq. (17.17).

17.2.2 Simplifications of Closed-Form Model

To determine the meniscus shape using the closed-form model, the equations of the Fourier coefficients a_n, Eq. (17.12), are solved in conjunction with the transcendental equation for K and δ^*, i.e., the electroneutrality condition represented by Eq. (17.13). Practically, the term $tan(K\delta^*)$ can be approximated with the third-order Taylor series, giving $K=\left[-3+\left[9-(\sigma ze\delta^*)/(\varepsilon\varepsilon_0k_{\mathrm{B}}T)\right]^{1/2}\right]^{1/2}(\delta^*)^{-1}$. The closed-form model can also be simplified under certain conditions. For example, at $\delta_0\rightarrow\infty$, the electrostatic disjoining pressure represented with Eq. (17.2) can be reduced to $\Pi_{\mathrm{e}}=\pi^2k_{\mathrm{B}}^2T^2\varepsilon\varepsilon_0/\left(2z^2e^2\delta_0^2\right)$. Molecular dynamics simulations have shown that the simplified electrostatic disjoining pressure at $\delta_0\rightarrow\infty$ agree well with Eq. (17.2) for water thin liquid films on alumina surface with film thickness greater than 3 nm [33]. Thus, this simplification in Π_{e} is valid for a broad range of liquid film thickness. Using the simplified Π_{e} at $\delta_0\rightarrow\infty$ rather than Eq. (17.2) for the expression of Π_{e} in Eq. (17.9) gives the following equation of Fourier coefficients

$$\int_{-L/2}^{L/2}\frac{\frac{2\pi n}{L}\sin\left(\frac{2\pi n}{L}x\right)\sum_{m=1}^{\infty}a_m\frac{2\pi m}{L}\sin\left(\frac{2\pi m}{L}x\right)dx}{\sqrt{1+\left[\sum_{m=1}^{\infty}a_m\frac{2\pi m}{L}\sin\left(\frac{2\pi m}{L}x\right)\right]^2}}-\frac{Ar^2}{6\pi\gamma}\int_{-L/2}^{L/2}\frac{\cos\left(\frac{2\pi n}{L}x\right)\mathrm{d}x}{\left[\sum_{m=1}^{\infty}a_m\cos\left(\frac{2\pi m}{L}x\right)+\delta_0-\zeta_S(x)\right]^3}-\frac{r\varepsilon\varepsilon_0}{2\gamma}\left(\frac{\pi k_{\mathrm{B}}T}{ze}\right)^2\int_{-L/2}^{L/2}\frac{\cos\left(\frac{2\pi n}{L}x\right)\mathrm{d}x}{\left[\sum_{m=1}^{\infty}a_m\cos\left(\frac{2\pi m}{L}x\right)+\delta_0-\zeta_S(x)\right]^2}=0 \tag{17.20}$$

The mean disjoining pressure becomes

$$\begin{aligned}\overline{\Pi}_{\text{rough}} &= \frac{Ar^3}{6\pi L}\int_{-L/2}^{L/2}\left[\sum_{n=1}^{\infty} a_n\cos\left(\frac{2\pi n}{L}x\right)+\delta_0-\zeta_S(x)\right]^{-3}\mathrm{d}x\\ &+\frac{\varepsilon\varepsilon_0 r^2}{2}\left(\frac{\pi k_B T}{ze}\right)^2\int_{-L/2}^{L/2}\left[\sum_{n=1}^{\infty} a_n\cos\left(\frac{2\pi n}{L}x\right)+\delta_0-\zeta_S(x)\right]^{-2}\mathrm{d}x\end{aligned} \tag{17.21}$$

The vapor pressure is thus

$$\begin{aligned}\frac{P_v}{P_{\text{sat}}} &= \exp\left\{-\frac{Ar^3}{6\pi L\rho k_B T}\int_{-L/2}^{L/2}\left[\sum_{n=1}^{\infty} a_n\cos\left(\frac{2\pi n}{L}x\right)+\delta_0-\zeta_S(x)\right]^{-3}\mathrm{d}x\right.\\ &\left.-\frac{\varepsilon\varepsilon_0 r^2}{2\rho L k_B T}\left(\frac{\pi k_B T}{ze}\right)^2\int_{-L/2}^{L/2}\left[\sum_{n=1}^{\infty} a_n\cos\left(\frac{2\pi n}{L}x\right)+\delta_0-\zeta_S(x)\right]^{-2}\mathrm{d}x\right\}\end{aligned} \tag{17.22}$$

By using the simplified expression of Π_e at $\delta_0 \to \infty$, the Fourier coefficients still need to be determined by solving Eq. (17.20) implicitly. However, Eq. (17.20) can be further simplified to become an explicit expression of the Fourier coefficients. In the limit of $\left|\zeta_L'(x)\right| << 1$ and $|\zeta_L(x)-\zeta_S(x)-\delta_0| << \delta_0$, which corresponds to small wave amplitudes and nanostructured depths, Eq. (17.20) becomes

$$\begin{aligned}&\int_{-L/2}^{L/2}\frac{2\pi n}{L}\sin\left(\frac{2\pi n}{L}x\right)\sum_{m=1}^{\infty}a_m\frac{2\pi m}{L}\sin\left(\frac{2\pi m}{L}x\right)\mathrm{d}x+\left[\frac{Ar^2}{2\pi\gamma\delta_0^4}+\frac{r\varepsilon\varepsilon_0}{\gamma\delta_0^3}\left(\frac{\pi k_B T}{ze}\right)^2\right]\times\\ &\int_{-L/2}^{L/2}\cos\left(\frac{2\pi n}{L}x\right)\left[\sum_{m=1}^{\infty}a_m\cos\left(\frac{2\pi m}{L}x\right)+\delta_0-\zeta_S(x)\right]\mathrm{d}x=0\end{aligned} \tag{17.23}$$

Using the orthogonality of the Fourier series, i.e., $\int_{-L/2}^{L/2}\sin\left(\frac{2\pi n}{L}x\right)\sin\left(\frac{2\pi m}{L}x\right)=0$ and $\int_{-L/2}^{L/2}\cos\left(\frac{2\pi n}{L}x\right)\cos\left(\frac{2\pi m}{L}x\right)=0$ for any $m \neq n$, Eq. (17.23) reduces to

$$\begin{aligned}&\int_{-L/2}^{L/2}a_n\left(\frac{2\pi n}{L}\right)^2\sin^2\left(\frac{2\pi n}{L}x\right)\mathrm{d}x+\left[\frac{Ar^2}{2\pi\gamma\delta_0^4}+\frac{r\varepsilon\varepsilon_0}{\gamma\delta_0^3}\left(\frac{\pi k_B T}{ze}\right)^2\right]\times\\ &\int_{-L/2}^{L/2}\left[a_n\cos^2\left(\frac{2\pi n}{L}x\right)-\zeta_S(x)\cos\left(\frac{2\pi n}{L}x\right)\right]\mathrm{d}x=0\end{aligned} \tag{17.24}$$

Solving Eq. (17.24) gives

$$a_n = \frac{\frac{2}{L}\int_{-\frac{L}{2}}^{\frac{L}{2}} \cos\left(\frac{2n\pi}{L}x\right)\zeta_S(x)\mathrm{d}x}{1 + \left(\frac{2\pi n}{L}\right)^2 \left[\frac{Ar^2}{2\pi\gamma\delta_0^4} + \frac{\varepsilon\varepsilon_0 r}{\gamma\delta_0^3}\left(\frac{\pi k_B T}{ze}\right)^2\right]^{-1}} \tag{17.25}$$

Equation (17.25) can give us some direct hints of how liquid film thickness, δ_0, and nanostructure depth, D, affect the meniscus shape. It is evident from Eq. (17.25) that the Fourier coefficients, a_n, for all $n = 1, 2, \ldots\infty$ decrease with the liquid film thickness increases, indicating a flatter meniscus for larger film thickness. As the liquid film thickness increases, based on Eqs. (17.1) and (17.2), both the van der Waals and electrostatic contributions to disjoining pressure decrease rapidly, and the surface tension force thus becomes dominant and makes the meniscus flatter. The nanostructure depth, D, is not explicitly included in Eq. (17.25), but is adsorbed in the terms Wenzel roughness ratio, r, and solid surface profile, $\zeta_s(x)$. For a given nanostructure shape (e.g., square nanostructure, triangular nanostructure, etc.), a larger nanostructure depth, D, gives a larger $|\zeta_S(x)|$ (magnitude of $\zeta_S(x)$) and a larger roughness ratio, r (because solid surface area increases), which in turn leads to larger a_n based on Eq. (17.25). This indicates that the wave amplitude of the meniscus increases with nanostructure depth.

For systems where van der Waals interactions dominate, using the definition of the scaled healing length, i.e., $\xi^2 = 2\pi\gamma\delta_0^4/A$, Eq. (17.25) becomes

$$a_n = \frac{\frac{2}{L}\int_{-\frac{L}{2}}^{\frac{L}{2}} \cos\left(\frac{2n\pi}{L}x\right)\zeta_S(x)\mathrm{d}x}{1 + \left[2\pi n\xi^2/(rL)\right]^2} \tag{17.26}$$

The validity of $\left|\zeta_L'(x)\right| << 1$ and $|\zeta_L(x) - \zeta_S(x) - \delta_0| << \delta_0$ in deriving Eqs. (17.25) and (17.26) is related with the aspect ratio and shape of the nanostructure. For thin liquid films on triangular nanostructured surfaces, Eq. (17.26) agrees well with the closed-form model represented with Eq. (17.17) for nanostructure aspect ratio, $2D/L$ of 0.25, 0.33, and 0.5, but deviates with the closed-form model for larger aspect ratio of nanostructure ($2D/L = 1.0$) [32]. As for square nanostructured surfaces, the simplified model represented with Eq. (17.26) deviates with the closed-form model for the smallest aspect ratio tested (i.e., $2D/L = 0.25$). This is because the square nanostructures are featured with sharp changes in morphology, violating the assumption of $|\zeta_L(x) - \zeta_S(x) - \delta_0| << \delta_0$.

17.3 Molecular Modeling of Thin Liquid Films

Molecular dynamics (MD) is a powerful tool to explore physical phenomena in the nanoscale, and has been applied in many areas of science and engineering, such as materials science, fluid mechanics, heat transfer, and biomechanics.

MD simulations have two basic assumptions: (1) the motions of the atoms obey the laws of classic mechanics; and (2) the interactions of the atoms can be well described with empirical potential functions, which are usually developed based on experiments and quantum mechanics simulations. Since the quantum effects may not be critical in a wide range of applications, the accuracy of MD simulations depend highly on the potential functions used to describe the intermolecular forces. It is thus very important to validate or verify the potential functions used in MD simulations. For example, a number of studies have been performed to validate or verify the TIP4P-Ew water model to accurately reproduce thermophysical properties of thin liquid films, including temperature-dependent density, surface tension, adsorption structure on alumina surface, and interaction energy with gold surface.

The key advantages of MD simulations are their capability of providing atomic scale details of physical phenomena that are difficult to capture otherwise, such as liquid layering on a solid surface, thermal fluctuations of a liquid-vapor interface, temperature discontinuity at an interface, etc. Furthermore, macroscopic thermodynamic parameters (e.g., temperature, pressure, and density) can be obtained from the atomic trajectories in MD simulations using statistical thermodynamics. As such, the results of the MD simulations can be compared with theoretical modeling of macroscopic properties. The statistical fluctuations of the macroscopic properties are very sensitive to the number of atoms in the system. For example, for a system of N atoms, the relative fluctuations in temperature are of order $1/\sqrt{3N-3}$ [35]. The fluctuations in MD simulated thermodynamic properties can also be reduced by performing time average in the equilibrium state.

As complete descriptions of the concepts and algorithms of MD simulations are available in other publications [35, 36] as well as Chap. 1 of the present book, this section focuses on MD simulations and analysis directly related to obtaining thermophysical properties of thin liquid films. Section 17.3.1 describes the MD simulation setups for calculating meniscus shape and disjoining pressure of a thin liquid film on a solid surface, and the grid method for analyzing the meniscus shape. Section 17.3.2 discusses the method to characterize the solid–liquid interaction strength. Section 17.3.3 describes an energy-based method to calculate liquid surface tension. Section 17.3.4 details the method of characterizing the vibrational properties of a material. The present chapter uses a water–gold system and a water–alumina system as examples for MD simulations. While the CLAYFF potential [37] is used to describe the intermolecular interactions in the water–alumina system, the TIP4P-Ew model [38], a standard 12-6 Lennard-Jones potential [39], and an EAM potential [40] are used to describe the water–water, water–gold, and gold–gold interactions, respectively, in the water–gold system. A number of open source and commercial packages are available for MD simulations. The MD simulations presented in this chapter are using the open source package LAMMPS [41].

17.3.1 Molecular Modeling of Meniscus Shape and Disjoining Pressure

The initial conditions of the MD simulations for determining the meniscus shape and disjoining pressure of a thin liquid film on a solid surface are shown in Fig. 17.3a using the example of a water–alumina system. The setup consists of two water films that are separated 20 nm apart. The water films are adsorbed on alumina surfaces, which are frozen at each end of z direction. Periodic boundary conditions are applied in all directions. The simulation system is equilibrated in an NVT (N the number of atoms, V the volume, and T the temperature) ensemble for 2 ns (or 1 ns after the meniscus shape stops changing with time). The local pressure tensor is calculated with contributions from the kinetic motion of the molecules and contributions from the intermolecular forces [42]. The vapor pressure P_v is calculated by summing up the local pressure tensor in the vapor region. It is noted that although the described setup can be used to determine the meniscus shape and disjoining pressure at any temperature between the melting point and critical point of the liquid, the fluctuations in the vapor pressure is very large for relatively low temperatures due to the small number of molecules in the vapor phase. Here, simulations are conducted at 400 K. Another method to reduce pressure fluctuations is to average the vapor pressure in a relative long simulation time (e.g., 0.8 ns in equilibrium state). Although the water–alumina system is used here as an example (Fig. 17.3a), the described setup and method can be used to study the meniscus shape and disjoining pressure for other solid–liquid systems, e.g., a water–gold system.

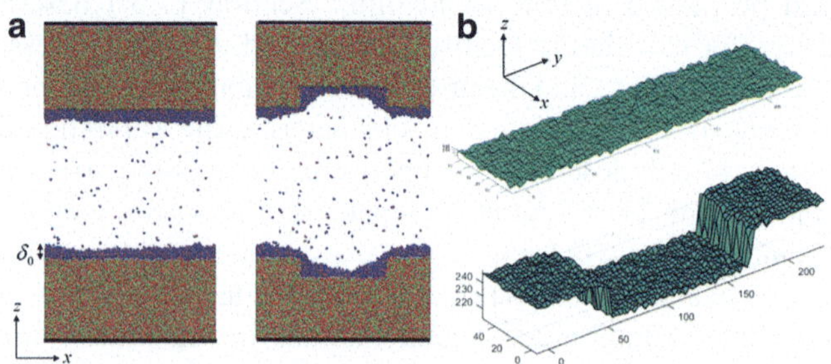

Fig. 17.3 (**a**) Molecular dynamics simulation setups for water films on a flat and a square nanostructured amorphous alumina surface; (**b**) morphology of the flat and square nanostructured amorphous alumina surfaces analyzed with the three-dimensional grid method

Generally, a crystalline solid surface can be created by cleaving a bulk solid at a pre-set orientation, e.g., the α-alumina (0001) surface, gold (100) surface, etc. As for materials of which the amorphous phase is more commonly used (e.g., amorphous alumina), their amorphous phase can be generated through a melt-quenching method [43]. Taking the amorphous alumina as an example, in the melt-quenching method, an equilibrated bulk crystalline α-alumina substrate is heated up to 3500 K in 1 ns with pressure isotropically at 1 atm, held at 3500 K for 1 ns, and then quenched to 300 K in 1 ns. Two free surfaces are cleaved from the quenched bulk alumina and a subsequent NVT equilibration is performed at 300 K for 1 ns. For both crystalline and amorphous solid surfaces, the nanostructures can be created by cleaving the flat substrate, with a subsequent NVT equilibration at 300 K for 1 ns. It is noted that the amorphous solid surface will not be as smooth as the crystalline surface. Here, a three-dimensional grid method is used to characterize the intrinsic roughness of the amorphous solid surface. In the grid method, the simulation domain is divided into grids of 0.3 Å in size and the density of each grid is calculated to generate a three-dimensional density profile, where the grids with density of $\rho_s/2$ (where ρ_s is the density of the bulk solid) is assumed to be the solid surface. Figure 17.3b shows the morphology of a flat amorphous alumina surface (top panel) and a square nanostructured amorphous alumina surface (bottom panel) using the 3D grid method. The peak–peak roughness of the amorphous alumina surface is ~0.4 nm.

Figure 17.4a shows the density profile of a water film on a flat gold surface obtained using the one-dimensional grid method, where a liquid-vapor interfacial region is observed. Following Carey and Wemhoff [26], the thickness of the water film (δ_0) is defined as the distance from the wall ($z = 0$) to the middle of the interfacial region where the density $\rho = (\rho_l + \rho_v)/2$. As for a liquid film on

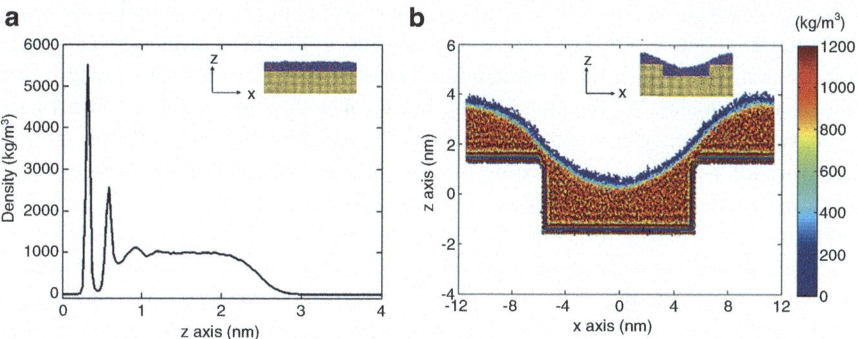

Fig. 17.4 (**a**) One-dimensional density profile of a water film with thickness of 2.48 nm on a flat gold surface ($z = 0$ represents the location of wall) analyzed with the one-dimensional grid method; (**b**) two-dimensional density profile of a water film with the thickness of 2.48 nm on a square nanostructured gold surface with the depth of 2.85 nm analyzed with the two-dimensional grid method

a nanostructured solid surface, a two-dimensional density profile is necessary to analyze the meniscus shape. Figure 17.4b shows the density contours of a water film on a squared nanostructured gold surface, where a liquid-vapor interfacial region is clearly observed. Following the same definition, the interface location is defined at the locations where $\rho = (\rho_l + \rho_v)/2$. Connecting all the interface grids gives the meniscus shape on a nanostructured surface.

The grid method used here can also be applied for other physical problems with moving interfaces, e.g., droplet growth during condensation, bubble growth during nucleate boiling, and droplet morphology change during spreading.

17.3.2 Molecular Modeling of Interaction Strength

The interaction strength between solid and liquid can be characterized with the interaction energy per actual area, W. In order to calculate W, three NVT systems are required, i.e., a system with solid substrate only, a system with a free-standing liquid film, and the merged system where the liquid film is adsorbed on the solid surface. Figure 17.5a shows the three example systems for characterizing the interaction strength between water and amorphous alumina. The total energies of the solid system (E_S), the liquid system (E_L), and the merged system (E_{merged}) are then used to calculate the interaction energy per actual area using

$$W = \frac{\langle E_S \rangle + \langle E_L \rangle - \langle E_{merged} \rangle}{A} \tag{17.27}$$

where A is the actual area of the solid–liquid interface, and the symbol $\langle\rangle$ represents time average.

Figure 17.5b shows the MD-calculated interaction energy per actual solid–liquid interfacial area between water films and triangular gold and alumina nanostructured surfaces as a function of the nanostructure aspect ratio. For the water–gold system where only van der Waals interactions between water and gold are considered, the interaction energy per actual surface area does not obviously change with the nanostructure aspect ratio, indicating that the strength of van der Waals interactions is not affected by nanostructures. On the other hand, for the water–alumina system where both van der Waals and electrostatic interactions are considered, the interaction energy per actual surface area decreases with the nanostructure aspect ratio. Thus, it is evident that the strength of electrostatic attenuates as nanostructure aspect ratio increases. By using the normal area instead of the actual solid–liquid contact area for A, Eq. (17.27) gives the work of adhesion, which is an important physical quantity in characterizing the wettability of a surface [27] and interfacial thermal resistance at a solid–liquid interface [44].

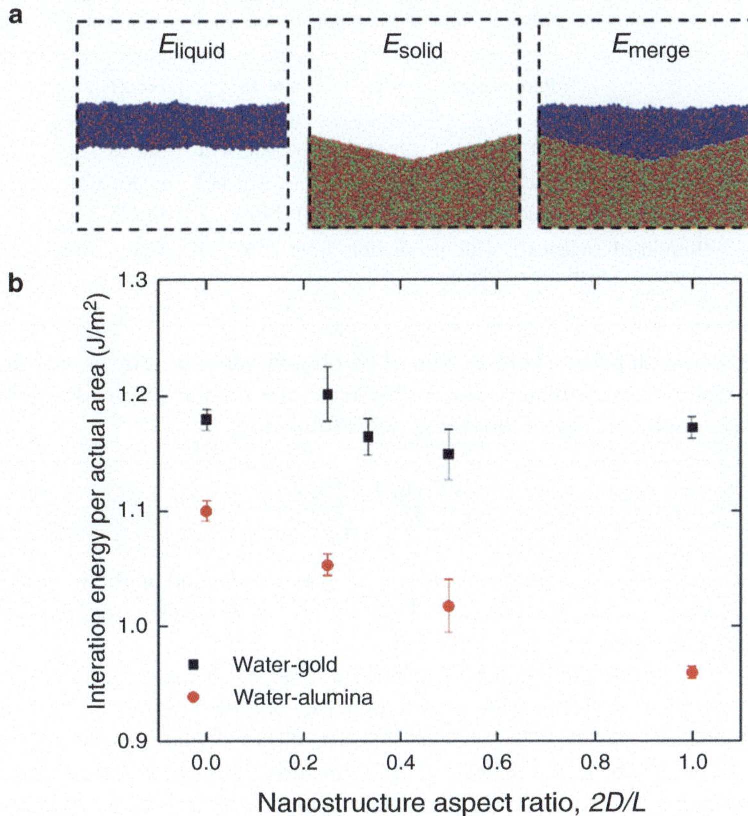

Fig. 17.5 (**a**) Molecular dynamics simulation setups for calculation of interaction energy per actual area for a water–alumina system; (**b**) interaction energy per actual solid–liquid interfacial area as a function of the aspect ratio of the nanostructures for water thin films on triangular nanostructured gold and amorphous alumina surfaces. Reprinted (adapted) with permission from [33]. Copyright (2015) American Chemical Society

17.3.3 Molecular Modeling of Liquid Surface Tension

The surface tension is an important property of liquid that affects the meniscus shape and stability of thin liquid films. The liquid surface tension in a thin film does not deviate obviously from the bulk value even when the film thickness goes to nanoscale. However, other conditions may affect surface tension, e.g., the inclusion of nanoparticles in the liquid. In MD simulations, multiple methods are available to estimate the surface tension, e.g., using Young–Laplace equation [45], and calculating the normal and tangential pressure difference [29]. However, the methods using pressure calculations usually lead to large fluctuations. Here, an energy-based method is used to determine the liquid surface tension. Two NVT

Table 17.1 Liquid-vapor surface tension of water and water–gold nanofluids of different volume fractions based on the free energy method

	Pure water	$n=9$	$n=18$	$n=27$
φ	0	3.43 %	6.77 %	9.81 %
$<E_w>$ (eV)	−1789.081	−2841.828	−3885.474	−4943.323
$<E_{w/o}>$ (eV)	−1796.242	−2853.841	−3910.417	−4969.034
γ_{lv} (N/m)	0.06793	0.1002	0.1109	0.1175

systems, one with a free-standing film of two liquid-vapor interfaces, and the other with bulk liquid are simulated. The surface tension is then determined by taking the difference in total energy of the two systems following:

$$\gamma = \frac{\langle E_w \rangle - \langle E_{w/o} \rangle}{A} \tag{17.28}$$

where E_w and $E_{w/o}$ are the total energy of water with and without liquid-vapor interfaces, respectively, A is the area of the liquid-vapor interface, and the symbol $\langle\rangle$ represents time average.

Table 17.1 shows an example of using Eq. (17.28) to calculate surface tension of water-based nanofluids with gold nanoparticles with volume fractions of 0 % (pure water), 3.43 %, 6.77 %, and 9.87 %. The MD-calculated surface tension of pure water at 300 K is $\gamma = 0.06793$ N/m, with less than 6 % error compared with the experimental value of 0.072 N/m. As the nanoparticle loading increases, due to van der Waals forces between particles at the liquid–gas interface, the surface tension of nanofluids increases, reaching a value of 0.1175 N/m for $\varphi = 9.87$ %.

17.3.4 Molecular Modeling of Vibrational Density of States

The vibrational density of states (VDOS) characterizes the vibrational properties of a given material, and is the spectrum of the velocity auto-correlation function (VACF).

$$F(\omega) = \int_{-\infty}^{+\infty} Z(t) e^{-j\omega t} \mathrm{d}t \tag{17.29}$$

where t is time, ω frequency, and $Z(t)$ is VACF defined as [47]

$$Z(t) = \langle \mathbf{v}(t) \cdot \mathbf{v}(0) \rangle \tag{17.30}$$

where **v** is the velocity vector of an atom. The VDOS is usually taken as the square modulus of $F(\omega)$, VDOS $= |F(\omega)|^2$, or the real part of $F(\omega)$, Re$[F(\omega)]$.

VDOS has a lot of useful applications in molecular dynamic studies. Firstly, VDOS can be used to qualitatively predict the thermal transport properties, such as thermal resistance at an interface, and effective thermal conductivity of nanocomposites. Because the interfacial thermal resistance arises from the mismatch in vibrational properties of two contact materials, a smaller difference in VDOS leads to a smaller interfacial thermal resistance [44]. Also, since phonons (quantized vibrations) at low frequency have larger mean free path and thus contribute more to thermal transport, the material with higher VDOS at low frequency will have higher thermal conductivity [48]. Furthermore, VDOS can be used to calculate the self-diffusion coefficient of a material, D^*. By definition, at zero frequency, the real part of the VDOS becomes [49]

$$\mathrm{Re}\,[F(0)] = 6D^* \tag{17.31}$$

Last, but not the least, VDOS can be used to characterize chemical species. Figure 17.6 shows the example of using VDOS to characterize the adsorption species in water monolayers adsorbed on ZnO surfaces with and without oxygen vacancies. Since the vibrational properties of the oxygen atoms in different states (e.g., H_2O versus OH) will also be different, the peaks in the VDOS of oxygen can reveal different water absorption structures. As shown in Fig. 17.6, the VDOS for water monolayer adsorbed on ZnO surface with vacancy reveals one more peak at 2.92 THz than that on ZnO surface without vacancy, indicating a new form of water absorption structure. With further analysis, the structure is recognized as the coordinate hydroxyl, where the hydroxyls coordinate into the oxygen vacancies and bond with three lattice zinc atoms [27].

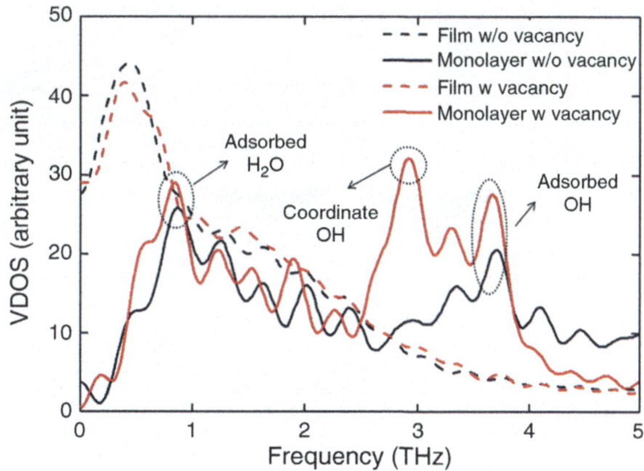

Fig. 17.6 Vibrational density of states (VDOS) of oxygen from adsorbed water monolayer and water film on ZnO surfaces. Reproduced from [27] with permission from the PCCP Owner Societies

17.4 Meniscus Shape of Thin Liquid Films

In this section, the meniscus shape of a liquid film on a solid nanostructured surface is compared between theoretical modeling and molecular modeling. Sections 17.4.1 and 17.4.2 investigate the effect of electrostatic interactions and nanostructures on meniscus shape, respectively.

17.4.1 Effect of Electrostatic Interactions on Meniscus Shape

The effect of electrostatic interactions on meniscus shape is studied by comparing the meniscus shapes of a water thin film of thickness $\delta_0 = 2.48$ nm on a triangular nanostructured amorphous alumina surface of depth $D = 6.08$ nm with and without considering the electrostatic forces between water and alumina in Fig. 17.7. In order to make comparisons between the continuum-level model and the MD simulations, surface tension of TIP4P-Ew water (i.e., 65.2 mJ/m^2 at 300 K [50]) is used in the continuum-level model. As qualitatively shown in the snapshots in Fig. 17.7a, the meniscus for the case with electrostatic interactions is more conformal than the case without electrostatic interactions. The model-predicted and MD-calculated first 20 Fourier coefficients for cases with and without electrostatic interactions are compared in Fig. 17.7a. The model predictions with and without electrostatic interactions are calculated with Eqs. (17.12) and (17.17), respectively. Good agreements are obtained between the theoretical model and MD simulations

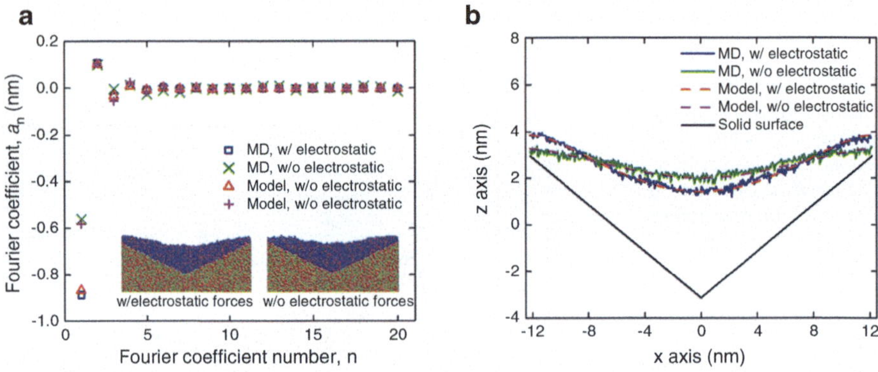

Fig. 17.7 (**a**) Comparison of MD simulation and model predictions for the first 20 Fourier coefficients of the meniscus shape for water film with thickness of 2.48 nm on a triangular nanostructure with depth of 6.08 nm with and without electrostatic interactions between alumina and water. The inset shows the equilibrium menisci from MD simulations with and without electrostatic interactions between alumina and water. (**b**) Comparison of meniscus shape from MD simulations and model predictions. Reprinted (adapted) with permission from [33] Copyright (2015) American Chemical Society

both with and without electrostatic interactions. It is noticed that the leading order coefficient a_1 is at least three times larger than other Fourier coefficients, and hence a_1 plays an important role in determining the wave amplitude of the meniscus. When the electrostatic interactions are turned off, the leading order coefficient a_1 decreases by 33 %, demonstrating a flatter meniscus without electrostatic interactions.

In Fig. 17.7b, the theoretical model (dashed line) is compared with MD simulations (solid lines) for meniscus shape of a water film on a triangular nanostructure with and without electrostatic interactions. The blue and green solid lines are obtained from MD simulations using the grid method. The red and pink dashed lines are obtained using the theoretical model with Eqs. (17.12) and (17.17), respectively. The average relative discrepancy in the meniscus shape is 3 % for with electrostatic interactions and 4 % for the case without electrostatic interactions. As shown in Fig. 17.7b, the electrostatic interactions enhance the disjoining pressure of a nanostructured surface, which in turn makes the meniscus shape more conformal.

17.4.2 Effect of Nanostructures on Meniscus Shape

In order to study the effect of nanostructures on meniscus shape, Fig. 17.8a shows the snapshots of meniscus shapes in equilibrium state for water films on square nanostructured gold surfaces at 300 K. The thickness of the water thin films ranges from 1.25, 2.48, 3.72, to 4.75 nm, and the square nanostructured gold surfaces are featured with a wavelength of 22.84 nm and depths of 2.85 and 5.71 nm. As the nanostructure depth increases or film thickness decreases, the meniscus shape changes from flat to "funnel-like," and finally becomes more and more conformal to the square nanostructures. It is noted that for very thin film ($\delta_0 = 1.25$ nm) on deep square nanostructure ($D = 5.71$ nm), the strong disjoining pressure finally causes the meniscus to break up.

The theoretical model (black dashed lines) is compared with MD results (solid lines) for meniscus shapes for water film of varying thicknesses on square nanostructures with depth of $D = 2.85$ nm and 5.71 nm in Fig. 17.8b, c, respectively. The theoretical model is plotted in Eq. (17.17). Good agreements are reached between model predictions and MD results for all cases, with average discrepancy between the model predictions and MD results $\sim$5 %. It is noted that the theoretical model is capable of capturing the abrupt change in meniscus shape at the edges of the square nanostructured surfaces (i.e., at $x = -5.71$ nm and 5.71 nm).

In Fig. 17.9a, the theoretical model and the MD simulations are compared quantitatively for the first 20 Fourier coefficients for water film of thickness $\delta_0 = 1.25$ nm on a square nanostructure of $D = 2.85$ nm are plotted. Good agreement is reached between the model and MD results, indicating that the model is capable of accurately predicting the meniscus shapes on nanostructures with abrupt profile changes.

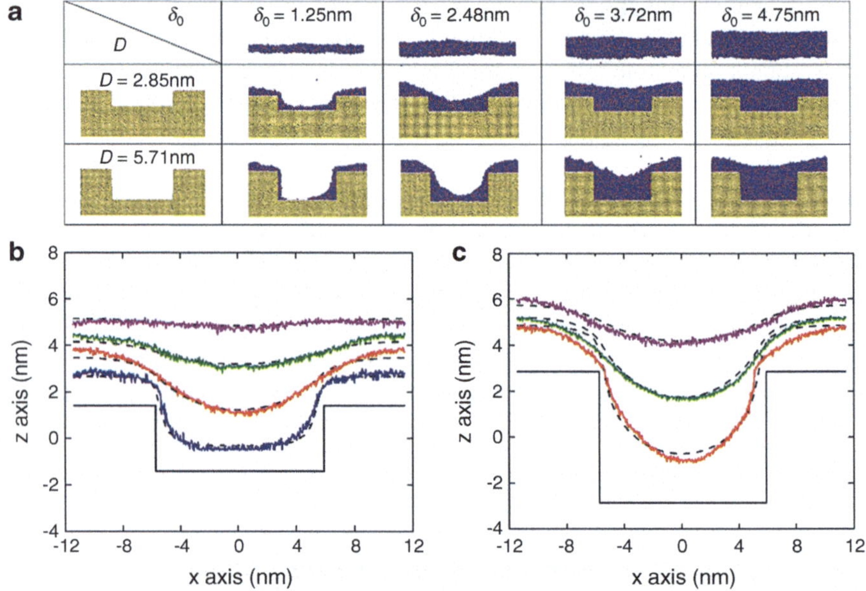

Fig. 17.8 (**a**) Meniscus shape of water films of varying thickness on square gold nanostructures with different depths at 300 K. (**b, c**) Comparison of MD simulation and model prediction for meniscus shape on square nanostructures with (**b**) depth of 2.85 nm and (**c**) depth of 5.71 nm. The blue, red, green, and pink lines represent the menisci of MD simulations for film thickness of 1.25, 2.48, 3.72, and 4.75 nm, respectively. The black solid line represents the solid surface, and the black dashed lines are based on the closed-form model. Reprinted (adapted) with permission from [32]. Copyright (2015) American Chemical Society

The wave amplitude A_m, i.e., the largest distance between the curved and flat menisci, is used to characterize the meniscus shape. The representation of the wave amplitude with Fourier coefficients is

$$A_m = \zeta_L(0) = \sum_{n=1}^{\infty} a_n \tag{17.32}$$

The maximum wave amplitude is obtained when the meniscus is completely conformal to the solid nanostructure, i.e., $A_m|_{\max} = D/2$. In order to include the effect of both the liquid film thickness and nanostructure depth on meniscus shape, the scaled wave amplitude, $A_m/A_m|_{max}$, for a water thin film on square nanostructured gold surfaces is plotted as a function of the scaled healing length, ξ/D in Fig. 17.9b. As defined in Eq. (17.2), the scaled healing length characterizes the competition between capillary pressure and disjoining pressure. In Fig. 17.9b, the solid lines represent the model prediction and symbols represent MD results at 300 K. Good agreement is achieved between the model and MD results. For each nanostructure depth D, as thin film thickness decreases or nanostructure depth

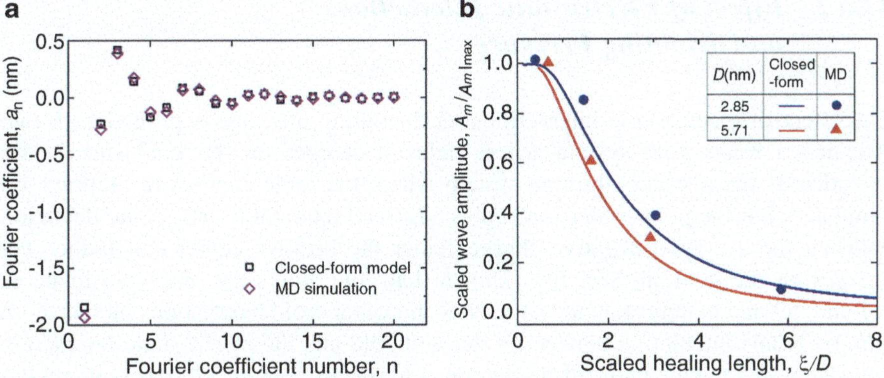

Fig. 17.9 (**a**) Comparison of MD simulation and model prediction for the first 20 Fourier coefficients of the meniscus shape for water film with thickness of 1.25 nm on a square nanostructure with depth of 2.85 nm. (**b**) Comparison of MD simulation and model prediction for scaled wave amplitude as a function of the scaled healing length on square nanostructures with depth of 2.85 and 5.71 nm. Reprinted (adapted) with permission from [32]. Copyright (2015) American Chemical Society

increases, ξ/D decreases, resulting a stronger disjoining pressure effect, which makes the meniscus more conformal to the solid morphology. As a result, the wave amplitude increases monotonically with decreasing scaled healing length, ξ/D. In the limit of $\xi/D \rightarrow 0$, the scaled wave amplitude converges to $2A_m/D = 1.0$, corresponding to a completely conformal film whose wave amplitude equals to $D/2$. In the limit of $\xi/D \rightarrow \infty$, the scaled wave amplitude converges to 0, indicating a flat liquid-vapor interface. It is noted that, due to the numerical instability while calculating the Fourier coefficients, the model predictions stop at $\xi/D = 0.2830$ (corresponding to $\delta_0 = 1.1$ nm) for $D = 2.85$ nm and $\xi/D = 0.4222$ (corresponding to $\delta_0 = 1.9$ nm) for $D = 5.71$ nm, respectively.

17.5 Disjoining Pressure of Thin Liquid Films

In this section, the disjoining pressure of a liquid film on a solid nanostructured surface is compared between theoretical modeling and molecular modeling. Sections 17.5.1 and 17.5.2 investigate the effect of electrostatic interactions and nanostructures on disjoining pressure, respectively.

17.5.1 Effect of Electrostatic Interactions on Disjoining Pressure

The effect of electrostatic interactions on disjoining pressure is probed with two systems: a water–gold system where induced charges on the gold surface are considered, and a water–alumina system where the ionic charges in alumina are simulated. For the polar water molecules adsorbed on a gold surface, the deviation between the averaged negative charge center the positive center can induce net charges on the gold surface. It is known that gold atoms are able coordinate as +1 and +3 and to deprotonate water. Now the water–gold interactions include both van der Waals interaction between water and gold and the electrostatic interaction between water and the induced charges on gold surface. Figure 17.10a compares the scaled vapor pressure of a water thin film on a flat gold surface between with and without surface changes at 400 K. The thickness of the water films ranges from 0.70 to 6.25 nm, and the applied charge density is 0.024 C/m^2. For small film thicknesses ($\delta_0 < 3.15$ nm), it is evident that the scaled vapor pressure decreases when surface charges are present, demonstrating a stronger disjoining pressure with induced charges. As for thicker films ($\delta_0 > 3.15$ nm), the variations in P_v/P_{sat} due to surface charges are within the error bars of the MD simulations, indicating that the enhancement in disjoining pressure due to induced charges weakens as film thickness increases. This enhancement of the disjoining pressure effect due to electrostatic forces shown in Fig. 17.10a is consistent with experimental results of Panella et al. for polar thin liquid films on metal surfaces [8].

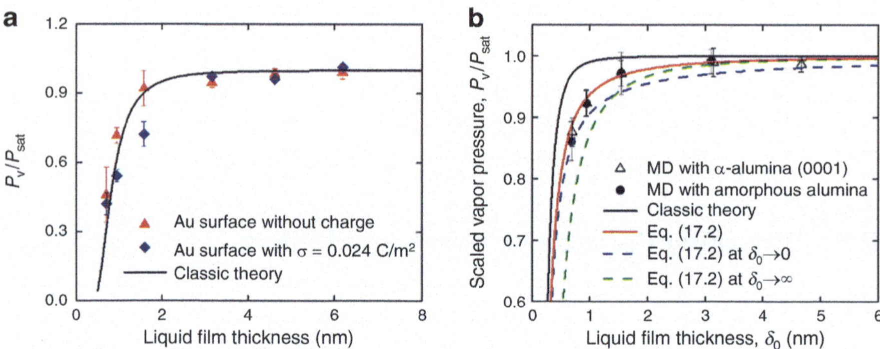

Fig. 17.10 Comparison between MD simulations and model predictions for scaled vapor pressure as a function of the liquid film thickness for water thin films on (**a**) a flat gold surface and (**b**) a flat amorphous alumina surface at 400 K. (**a**) is reprinted (adapted) with permission from [28]. Copyright (2015) American Institute of Physics. (**b**) is reprinted (adapted) with permission from [33]. Copyright (2015) American Chemical Society

As for the water–alumina system, Fig. 17.10b shows the comparison between the MD-calculated (symbols) and model-predicted (lines) scaled vapor pressure as a function of the film thickness for water thin films of varying thickness on flat crystalline and amorphous alumina surfaces at 400 K. The black line represents the prediction of the classical theory and the red line represents the prediction with electrostatic interactions accounted for using Eq. (17.2). The triangles and circles represent MD results for water thin films on crystalline α-alumina (0001) surface and on amorphous alumina, respectively. It is evident that the crystalline structure of the alumina surface does not affect the disjoining pressure significantly because the deviation in scaled vapor pressure between crystalline and amorphous alumina is within the error bars of MD simulations. Moreover, the classic theory overestimates the scaled vapor pressure (or underestimates the disjoining pressure) at all film thicknesses investigated by neglecting the electrostatic contributions. On the other hand, the vapor pressure from MD simulations agrees well with the model with electrostatic interaction accounted for. As is consistent with the water–gold system, the results of the water–alumina cases also support that the electrostatic interactions between the solid and liquid enhance the disjoining pressure.

The blue and green dashed lines represent the prediction of simplified electrostatic disjoining pressure in the limit of $\delta_0 \to 0$ and $\delta_0 \to \infty$, respectively. As shown in Fig. 17.10b, for very thin water films ($\delta_0 < 0.5$ nm), the simplified electrostatic disjoining pressure at $\delta_0 \to 0$ agrees well with the Eq. (17.2) while the simplification at $\delta_0 \to \infty$ overestimates the disjoining pressure. As for thick films ($\delta_0 > 3$ nm), the simplification at $\delta_0 \to \infty$ agrees well with Eq. (17.2) while the simplification at $\delta_0 \to 0$ greatly overestimates the disjoining pressure. For a water film with intermediate thickness (0.5 nm $< \delta_0 <$ 3 nm), both simplifications overestimate the disjoining pressure. Since the thin liquid films are usually larger than 3 nm in most applications, such as thin film evaporation, etc., the simplified disjoining pressure at the limit of $\delta_0 \to \infty$ has a wide range of applications.

17.5.2 Effect of Nanostructures on Disjoining Pressure

Figure 17.11 shows the comparison between the theoretical model (lines) and MD results (symbols) for the scaled vapor pressure as a function of the film thickness for water thin films on flat and square nanostructured (a) gold surfaces and (b) amorphous alumina surfaces. The square nanostructured gold surface has a wavelength of $L = 22.84$ nm and depths of $D = 2.85$ and 5.71 nm, and the square nanostructured amorphous alumina surface has a wavelength of $L = 24.32$ nm and depths of $D = 3.04$ and 6.08 nm. For the water–gold system, only van der Waals forces are considered for liquid–solid interactions, while for the water–alumina system, both van der Waals and electrostatic forces are considered. The solid lines are predictions of the theoretical model using Eqs. (17.16) and (17.19) for water–gold and water–alumina, respectively, and the symbols are MD results. Good agreements between

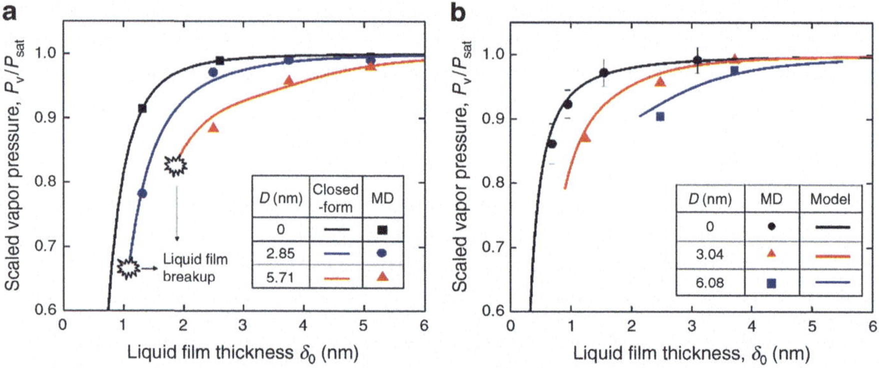

Fig. 17.11 Comparison between MD simulations and model predictions for scaled vapor pressure as a function of the liquid film thickness for water thin films on flat and square nanostructured (**a**) gold surfaces and (**b**) amorphous alumina surfaces at 400 K. (**a**) is reprinted (adapted) with permission from [32]. Copyright (2015) American Chemical Society. (**b**) is reprinted (adapted) with permission from [33]. Copyright (2015) American Chemical Society

the theoretical model and MD simulations are obtained for all cases. For both the water–gold and water–alumina systems, as the nanostructure depth D increases, the scale vapor pressure decreases, indicating a stronger disjoining pressure effect. But this enhancement in disjoining pressure with nanostructure depth attenuates with increasing liquid film thickness.

It is noted that, for thin water films on square nanostructured surface, the model predictions stop at small film thickness due to numerical instability while calculating the Fourier coefficients. For the water–gold system, the model predictions stop at $\delta_0 = 1.1$ nm and 1.9 nm for $D = 2.85$ nm and 5.71 nm, respectively, and for the water–alumina system, the model predictions stop at $\delta_0 = 0.9$ nm and 1.9 nm for $D = 3.04$ nm and 6.08 nm, respectively. Although there may exist a critical film thickness below which the film breaks up, the present model based on the assumption of continuous meniscus and thus cannot capture this critical film thickness. However, the model still gives a useful upper bound of the critical film thickness. For example, it is safe to conclude that the meniscus for water film with thickness greater than 0.9 nm will be stable on a square nanostructured amorphous alumina surface with depth of 3.04 nm.

17.6 Summary

In this chapter, a multiscale modeling approach is introduced to examine the effects of nanostructures and electrostatic interactions on the equilibrium meniscus shape and disjoining pressure of a thin liquid film on nanostructured surfaces. A general continuum-level model is developed based on the minimization of free energy, the Derjaguin approximation, and the disjoining pressure theory for a flat surface. Molecular dynamics simulations are performed for water thin films of varying film thickness on gold and alumina surfaces with both triangular and square nanostructures of varying depth. Good agreements are obtained between the continuum-level modeling and MD simulations. The results show that the wave amplitude of the meniscus increases with decreasing thin film thickness and increasing nanostructure depth. The electrostatic interactions are shown to enhance the disjoining pressure and make the meniscus more conformal to the nanostructured surfaces. Furthermore, both van der Waals and electrostatic contributions to the disjoining pressure increase with the nanostructure depth and decrease with the film thickness. The results demonstrate the capability of the continuum-level model in accurately capturing the effects of electrostatic forces and nanostructures with sharp edges on meniscus shape and disjoining pressure of a thin liquid film.

The model developed here will enable more accurate predictions of the heat flux of thin film evaporation on various nanostructured surfaces. Moreover, the model can guide nanostructural designs for enhanced thin film stability in boiling, evaporation, liquid-infused self-cleaning surfaces, and among others. While this chapter investigates the equilibrium meniscus shape and disjoining pressure of thin liquid films, future work is needed to advance the fundamental understanding of thin liquid films away from equilibrium, e.g., evaporating thin liquid films, wetting kinetics of thin films, etc. In addition, efforts can be made to extend the continuum-level model to three-dimensional nanostructures.

Nomenclature

Definition	Symbol	Unit	Definition	Symbol	Unit
Hamaker constant	A	J/m^3	Surface tension	γ	J/m^2
Fourier coefficient	a_n	m	Local film thickness	δ	m
Nanostructure depth	D	m	Modified film thickness	$\delta*$	m
Electron charge	e	C	Mean film thickness	δ_0	m
Boltzmann constant	k_B	J/K	Relative permittivity	ϵ	1
Nanostructure length	L	m	Vacuum permittivity	ϵ_0	F/m
Capillary pressure	P_C	Pa	Solid surface profile	ζ_S	m
Saturated vapor pressure	P_{sat}	Pa	Meniscus shape	ζ_L	m
Vapor pressure	P_V	Pa	Healing length	ξ	m
Wenzel roughness ratio	r	1	Disjoining pressure	Π	Pa
Temperature	T	K	Liquid density	ρ	kg/m^3
Free energy	W	J			
Valence	Z	1			

References

1. C.M. Mate, Spreading kinetics of lubricant droplets on magnetic recording disks. Tribol. Lett. **51**(3), 385–395 (2013). doi:10.1007/s11249-013-0171-5
2. K. Ono, Diffusion equation for spreading and replenishment in submonolayer lubricant film. Tribol. Lett. **57**(2), 1–14 (2015). doi:10.1007/s11249-014-0455-4
3. Y. Li, F. Wang, H. Liu, H. Wu, Nanoparticle-tuned spreading behavior of nanofluid droplets on the solid substrate. Microfluid. Nanofluid. **18**(1), 111–120 (2015). doi:10.1007/s10404-014-1422-y
4. D. Wasan, A. Nikolov, K. Kondiparty, The wetting and spreading of nanofluids on solids: role of the structural disjoining pressure. Curr. Opin. Colloid Interface Sci. **16**(4), 344–349 (2011). doi:10.1016/j.cocis.2011.02.001
5. J.L. Plawsky, A.G. Fedorov, S.V. Garimella, H.B. Ma, S.C. Maroo, L. Chen, Y. Nam, Nano- and microstructures for thin-film evaporation—a review. Nanoscale Microscale Thermophys. Eng. **18**(3), 251–269 (2014). doi:10.1080/15567265.2013.878419
6. R. Xiao, S.C. Maroo, E.N. Wang, Negative pressures in nanoporous membranes for thin film evaporation. Appl. Phys. Lett. **102**(12) (2013). doi:10.1063/1.4798243
7. K.R. Mecke, J. Krim, Adsorption isotherms and thermal fluctuations. Phys. Rev. B Condens. Matter **53**(4), 2073–2082 (1996). doi:10.1103/PhysRevB.53.2073
8. V. Panella, R. Chiarello, J. Krim, Adequacy of the Lifshitz theory for certain thin adsorbed films. Phys. Rev. Lett. **76**(19), 3606–3609 (1996). doi:10.1103/PhysRevLett.76.3606
9. K.S. Birdi, *Surface and Colloid Chemistry: Principles and Applications* (CRC Press, London, 2009)
10. V.P. Carey, *Liquid Vapor Phase Change Phenomena: An Introduction to the Thermophysics of Vaporization and Condensation Processes in Heat Transfer Equipment*, 2nd edn. (Taylor & Francis, London, 2007)
11. J.S. Wexler, A. Grosskopf, M. Chow, Y. Fan, I. Jacobi, H.A. Stone, Robust liquid-infused surfaces through patterned wettability. Soft Matter **11**(25), 5023–5029 (2015). doi:10.1039/C5SM00611B
12. S. Narayanan, A.G. Fedorov, Y.K. Joshi, Heat and mass transfer during evaporation of thin liquid films confined by nanoporous membranes subjected to air jet impingement. Int. J. Heat Mass Transfer **58**(1–2), 300–311 (2013). doi:10.1016/j.ijheatmasstransfer.2012.11.015
13. Y. Nam, S. Sharratt, G. Cha, Y.S. Ju, Characterization and modeling of the heat transfer performance of nanostructured Cu micropost wicks. J. Heat Transfer **133**(10), 101502 (2011). doi:10.1115/1.4004168
14. T.-S. Wong, S.H. Kang, S.K.Y. Tang, E.J. Smythe, B.D. Hatton, A. Grinthal, J. Aizenberg, Bioinspired self-repairing slippery surfaces with pressure-stable omniphobicity. Nature **477**(7365), 443–447 (2011)
15. B.V. Derjaguin, Y.I. Rabinovich, N.V. Churaev, Direct measurement of molecular forces. Nature **272**(5651), 313–318 (1978)
16. J.N. Israelachvili, *Intermolecular and Surface Forces: Third Revised Edition* (Elsevier Science, San Diego, 2011)
17. P.C. Wayner Jr., Y.K. Kao, L.V. LaCroix, The interline heat-transfer coefficient of an evaporating wetting film. Int. J. Heat Mass Transfer **19**(5), 487–492 (1976). doi:10.1016/0017-9310(76)90161-7
18. H.C. Hamaker, The London—van der Waals attraction between spherical particles. Physica **4**(10), 1058–1072 (1937). doi:10.1016/S0031-8914(37)80203-7
19. I.E. Dzyaloshinskii, E.M. Lifshitz, L.P. Pitaevskii, The general theory of van der Waals forces. Adv. Phys. **10**(38), 165–209 (1961). doi:10.1080/00018736100101281
20. E.M. Lifshitz, The theory of molecular attractive forces between solids. Sov. Phys. - JETP **2**(1), 73–83 (1956)
21. B. Ninham, V.A. Parsegian, G. Weiss, On the macroscopic theory of temperature-dependent van der Waals forces. J. Stat. Phys. **2**(4), 323–328 (1970). doi:10.1007/BF01020441

22. M.M. Kohonen, H.K. Christenson, Adsorption from pure and mixed vapours of n-hexane and n-perfluorohexane. Eur. Phys. J. E **6**(4), 315–323 (2001). doi:10.1007/s10189-001-8046-4
23. A.P. Bowles, Y.-T. Hsia, P.M. Jones, L.R. White, J.W. Schneider, Quasi-equilibrium AFM measurement of disjoining pressure in lubricant nanofilms II: effect of substrate materials. Langmuir **25**(4), 2101–2106 (2009). doi:10.1021/la8024638
24. M. Ojha, A. Chatterjee, G. Dalakos, P.C. Wayner, J.L. Plawsky, Role of solid surface structure on evaporative phase change from a completely wetting corner meniscus. Phys. Fluids. **22**(5) (2010). doi:10.1063/1.3392771
25. B.V. Derjaguin, N.V. Churaev, Structural component of disjoining pressure. J. Colloid Interface Sci. **49**(2), 249–255 (1974). doi:10.1016/0021-9797(74)90358-0
26. V.P. Carey, A.P. Wemhoff, Disjoining pressure effects in ultra-thin liquid films in micropassages—comparison of thermodynamic theory with predictions of molecular dynamics simulations. J. Heat Transfer **128**(12), 1276–1284 (2006). doi:10.1115/1.2349504
27. H. Hu, H.-F. Ji, Y. Sun, The effect of oxygen vacancies on water wettability of a ZnO surface. Phys. Chem. Chem. Phys. **15**(39), 16557–16565 (2013). doi:10.1039/C3CP51848E
28. H. Hu, Y. Sun, Molecular dynamics simulations of disjoining pressure effect in ultra-thin water film on a metal surface. Appl. Phys. Lett. **103**(26), 263110 (2013). doi:10.1063/1.4858469
29. J.G. Weng, S. Park, J.R. Lukes, C.L. Tien, Molecular dynamics investigation of thickness effect on liquid films. J. Chem. Phys. **113**(14), 5917–5923 (2000)
30. B.V. Derjaguin, N.V. Churaev, V.M. Muller, *Surface Forces* (Consultants Bureau, New York, 1987)
31. M.O. Robbins, D. Andelman, J.-F. Joanny, Thin liquid films on rough or heterogeneous solids. Phys. Rev. A **43**(8), 4344–4354 (1991)
32. H. Hu, C.R. Weinberger, Y. Sun, Effect of nanostructures on the meniscus shape and disjoining pressure of ultrathin liquid film. Nano Lett. **14**(12), 7131–7137 (2014). doi:10.1021/nl5037066
33. H. Hu, C.R. Weinberger, Y. Sun, Model of meniscus shape and disjoining pressure of thin liquid films on nanostructured surfaces with electrostatic interactions. J. Phys. Chem. C **119**(21), 11777–11785 (2015). doi:10.1021/acs.jpcc.5b03250
34. H.J. Butt, K. Graf, M. Kappl, *Physics and Chemistry of Interfaces* (Wiley, Weinheim, 2003)
35. D. Frenkel, B. Smit, *Understanding Molecular Simulation: From Algorithms to Applications* (Elsevier Science, San Diego, 2001)
36. M.P. Allen, D.J. Tildesley, *Computer Simulation of Liquids* (Oxford University Press, Oxford, 1991)
37. R.T. Cygan, J.-J. Liang, A.G. Kalinichev, Molecular models of hydroxide, oxyhydroxide, and clay phases and the development of a general force field. J. Phys. Chem. B **108**(4), 1255–1266 (2004). doi:10.1021/jp0363287
38. H.W. Horn, W.C. Swope, J.W. Pitera, J.D. Madura, T.J. Dick, G.L. Hura, T. Head-Gordon, Development of an improved four-site water model for biomolecular simulations: TIP4P-Ew. J. Chem. Phys. **120**(20), 9665–9678 (2004). doi:10.1063/1.1683075
39. K. Tay, F. Bresme, Hydrogen bond structure and vibrational spectrum of water at a passivated metal nanoparticle. J. Mater. Chem. **16**(20), 1956–1962 (2006). doi:10.1039/B600252h
40. M.S. Daw, S.M. Foiles, M.I. Baskes, The embedded-atom method—a review of theory and applications. Mater Sci Rep **9**(7–8), 251–310 (1993)
41. S. Plimpton, Fast parallel algorithms for short-range molecular-dynamics. J. Comput. Phys. **117**(1), 1–19 (1995)
42. A. Harasima, Molecular theory of surface tension, in *Advances in Chemical Physics*, I. Prigogine and P. Debye, (Wiley, New York, 2007), pp. 203–237
43. S.P. Adiga, P. Zapol, L.A. Curtiss, Atomistic simulations of amorphous alumina surfaces. Phys. Rev. B **74**(6), 064204 (2006)
44. H. Hu, Y. Sun, Effect of nanopatterns on Kapitza resistance at a water-gold interface during boiling: a molecular dynamics study. J. Appl. Phys. **112**(5), 053508-1–053508-6 (2012)
45. A.E. Ismail, G.S. Grest, M.J. Stevens, Capillary waves at the liquid-vapor interface and the surface tension of water. J. Chem. Phys. **125**(1), 014702 (2006). doi:10.1063/1.2209240

46. G. Lu, H. Hu, Y. Duan, Y. Sun, Wetting kinetics of water nano-droplet containing non-surfactant nanoparticles: a molecular dynamics study. Appl. Phys. Lett. **103**(25), 253104 (2013). doi:10.1063/1.4837717
47. A. Rahman, Correlations of in the motion of atoms in liquid argon. Phys. Rev. **136** A405–A411 (1964)
48. Z. Tian, H. Hu, Y. Sun, A molecular dynamics study of effective thermal conductivity in nanocomposites. Int. J. Heat Mass Transfer **61**, 577–582 (2013). doi:10.1016/j.ijheatmasstransfer.2013.02.023
49. N.H. March, M.P. Tosi, *Liquid State Physics* (World Scientific, Hackensack, 2002)
50. C. Vega, E. de Miguel, Surface tension of the most popular models of water by using the test-area simulation method. J. Chem. Phys. **126**(15) (2007). doi:10.1063/1.2715577

Appendix A
Available Software and Codes

One important goal of writing this book was to provide some practical perspective on multiscale modeling and its uses in nanomechanics. For the reader who is learning to do this, one big problem that you might face is identifying codes and software to conduct the simulations or analysis that you desire. In some of the chapters, methods have been described and well accepted software or codes may exist. However, in other cases the tools to conduct the work exist in multiple forms. Here, we will attempt to point the reader in the right direction to obtain or access the tools or software. One caveat is that while this list is reasonably accurate at the time of production, the availability of such tools is quite dynamic as new methods are continuously being developed across the modeling community.

For the simulations using classical atomistic methods, LAMMPS [10] is one of the most widely used codes which is freely available and easily extendable, distributed under the GNU Public License. An alternative for simulating condensed matter is IMD [9], The ITAP Molecular Dynamics Program, which as similar capabilities as well as licensing. Many of the other widely used MD codes target primarily soft-matter simulations, especially biological systems and are discussed in Chap. 1.

For continuum approximations, which are discussed in Chap. 3, the most commonly used simulation software used is the finite element method (FEM). Two commonly used commercial FEM codes are ANSYS [3] and Abaqus [1], the latter being much more common in research. COMSOL [8] is another commonly used commercial software package, which is marketed as a multi-physics package. There are a number of freely available multi-physics FEM packages available as well including Alabany [2] which has built in uncertainty quantification and optimization, as well as MOOSE [12] developed at Idaho National Laboratory, and Tahoe [19]. DAMASK [6] is a toolkit that has been developed specifically to simulate material microstructure using crystal plasticity and uses either commercial solvers like ABUQUS or its own spectral solver.

C.R. Weinberger, G.J. Tucker (eds.), *Multiscale Materials Modeling for Nanomechanics*, Springer Series in Materials Science 245,
DOI 10.1007/978-3-319-33480-6

There are a number of freely available codes for simulating the discrete motion of dislocations, i.e. discrete dislocation dynamics as discussed in Chap. 2. This includes PARADIS [14], and microMegas [11] discussed therein.

For density functional theory simulations, there are a wide number of different simulation packages available that range from freely distributed to commercial packages. The Vienna Ab-Initio Simulation Package (VASP) [20] is perhaps the most widely used DFT package, but it is available via purchase from the University of Vienna. Quantum ESPRESSO [17] is an open source freely available DFT code that is also widely used. There is a number of other codes that vary from commercial codes to open source and freely distributed including Soccorro [18] CASTEP [5] and a relatively comprehensive list exists on Wikipedia [23].

For the analysis and visualization of simulation data, there is a wide range of commercial and freely available software. Visual Molecular Dynamics (VMD) [21] is a popular open-source software used to visualize molecular systems. However, other open-source visualization software for condensed matter have emerged including Atomeye [4] and OVITO [13]. These visualization software have also integrated some capabilities to perform post processing analysis on the data that make them particularly powerful tools. Atomistic visualization in general is discussed in Chap. 1 and OVITO in great detail in Chap. 10. Paraview [15] is an open-source data analysis and visualization platform which can visualized processed data from essentially any simulation with particular capabilities to handle FEM data.

In Chap. 6, numerous methods of accelerated timescale molecular methods are described and these methods have been implemented in a number of software including standard freely distributed molecular simulation codes. For example, LAMMPS [10] has temperature accelerated dynamics and parallel replica dynamics as part of its main distribution. Eon [7] is a software package that has the capability to conduct hyperdynamics, kinetic monte carlo, parallel replica dynamics among others and includes interfaces to LAMMPS and VASP. PLUMED [16] is another freely available software package that can interfaces with standard molecular simulation software and is capable of conducting hyperdynamics, metadynamics and umbrella sampling among others. For a better understanding of accelerated timescale molecular dynamics methods, please see Chap. 6.

Finally, for the concurrent multiscale methods described in this book freely available codes exist for the quasi-continuum method(QC) [22] and the atoms-to-continuum (AtC). There is a QC development webpage that hosts a freely distributed version of the QC method [22]. The AtC package is an a user-contributed package that comes with LAMMPS and can be used if compiled with the correct flags.

References

1. Abaqus, http://www.3ds.com/products-services/simulia/products/abaqus/
2. Albany, https://github.com/gahansen/albany
3. ANSYS, http://www.ansys.com

4. Atomeye, http://li.mit.edu/archive/graphics/a3/a3.html
5. CASTEP, http://www.castep.org/
6. DAMASK, http://damask.mpie.de
7. Eon, http://theory.cm.utexas.edu/eon/
8. COMSOL Multiphysics, http://www.comsol.com
9. IMD, http://imd.itap.physik.uni-stuttgart.de/
10. LAMMPS, http://lammps.sandia.gov/
11. microMegas (mM), http://zig.onera.fr/mm_home_page/
12. MOOSE, http://mooseframework.org/
13. OVITO, http://www.ovito.org/
14. PARADIS, http://paradis.stanford.edu/
15. PARAVIEW, http://www.paraview.org/
16. PLUMED, http://www.plumed.org/
17. Quantum ESPRESSO, http://www.quantum-espresso.org/
18. SOCORRO, http://dft.sandia.gov/socorro/about.html
19. Tahoe, http://tahoe.sourceforge.net
20. Vienna Ab-Initio Simulation Package, https://www.vasp.at/
21. VMD, http://www.ks.uiuc.edu/research/vmd/
22. http://qcmethod.org/
23. https://en.wikipedia.org/wiki/list_of_quantum_chemistry_and_solid-state_physics_software/

Index

A
Accelerated dynamics, 206–207
Acceleration factor (boost), 204
Activation energy, 202, 344, 350, 377, 379, 381–389, 391–394, 399, 405, 448–450, 452, 457, 459
Activation volume, 344, 381, 383–385, 387, 391
α–shape method, 329
Amorphous materials, viii, 323, 324, 453, 454
Analysis, viii, 43–45, 90, 110, 133, 140, 141, 150, 153, 160, 163, 173, 183, 212, 213, 227, 232, 249, 281, 297–314, 317–335, 376, 394, 415, 419, 452, 453, 459, 461, 463, 494, 496, 497, 519, 525, 537, 538
Angular distribution function (ADF), 499, 500
Arrhenius equation, 444
Asymptotic expansion, 102–104
Atom-continuum coupling, 234
Atomic control, 239
Atomic neighborhood, 44, 169, 298, 299, 304
Atomistic, vii, 1, 2, 12, 29, 40–43, 46, 47, 54, 58, 79–81, 159–169, 172–176, 179, 181, 182, 185, 187, 188, 195, 223–225, 235–237, 241, 261, 262, 264, 267, 272, 277–278, 283, 284, 286, 288, 290, 292, 293, 297–309, 314, 317, 318, 322, 323, 328–330, 332, 335, 374, 377, 382, 386, 387, 389, 392–396, 399, 402–404, 407, 408, 442, 448, 490, 498–504, 537, 538
Atomistic simulations (AS), vii, 1–47, 66, 68, 74, 79, 161, 174, 188, 209, 261–293, 297–314, 317–335, 376, 377, 382, 384–386, 390, 392–395, 397, 398, 404, 407, 408, 442

B
Barium titanate ($BaTiO_3$), 284, 469–471, 475, 476, 478, 483–485
Becker Doring theory, 378
Boltzmann distribution, 17, 21, 209, 211
Burger(s) circuit, 331–334
Burgers vector, 57–59, 64, 65, 68, 70, 72, 74, 76, 78, 118, 301, 331–335, 343, 348, 354, 360, 375, 384, 393–395, 397, 399–403, 417, 432

C
CAC method. *See* Concurrent atomistic-continuum (CAC) method
Canonical ensemble (NVT), 17, 20, 23, 174
Capillary pressure, 511, 514, 515, 528, 533
Cauchy–Born, 94–95, 100, 169, 173, 235, 239, 245, 252
Cauchy–Born (rule), 28, 29, 235
Cauchy pressure, 36, 37
Caucy stress, 27, 94–96, 98, 99, 110, 111, 225, 229, 230, 416, 431
Centrosymmetry, 44, 45, 172, 178, 180, 298–302, 307, 309, 325
Centrosymmetry parameter, 179, 286–288, 290, 325, 383

C.R. Weinberger, G.J. Tucker (eds.), *Multiscale Materials Modeling for Nanomechanics*, Springer Series in Materials Science 245,
DOI 10.1007/978-3-319-33480-6

Charged optimized many-body potentials (COMB), 41, 42
CNA. *See* Common neighbor analysis (CNA)
Coarse-grained (CG), 154, 161, 164, 166, 168, 169, 171, 173–177, 185–187, 227, 237, 239, 242, 246, 247, 249, 251, 262, 278, 281, 286, 287, 338, 357, 358, 363, 413, 427–429, 433–436, 443, 446, 447, 465
Coarse-graining, 26, 161, 164, 176, 183, 185, 187, 223–255, 261, 283, 338, 420, 443, 446–448, 452, 465, 497
Cohesive energy, 35, 38, 132, 135, 136, 140–144, 148, 153
Collective variables (CVs), 199–201, 203, 210–213
Common neighbor analysis (CNA), 44, 45, 298, 300–302, 307, 309, 311–313, 323, 325–328, 332, 335
Computational homogenization, 104, 112–114, 119
Computational materials science, 318
Concurrent atomistic-continuum (CAC) method, viii, 262–268, 272–275, 278, 281, 283, 284, 288, 292, 293
Conjugate gradient, 6–9
Constraint, 14, 16–18, 21, 59, 60, 135, 136, 161, 164, 196, 203, 207, 236–238, 240–243, 245, 470, 471, 474–480, 482–485, 499, 502
Contact mechanics, 453, 461, 465
Continuum, vii, viii, 2, 23–32, 70, 89–124, 159, 223, 262, 297–314, 321, 339, 374, 442, 489–490, 508
Continuum metrics, 44, 47, 297–314
Couple stress, 96
Coupling principles, 223–255
Cross-slip, 59, 67–68, 80, 82, 115, 343–345, 349–351
Crystal plasticity, viii, 47, 69, 80, 90, 94, 114–119, 187, 339–357, 413–437, 537
Crystal structure, 23, 107, 109, 116, 117, 264, 276, 300, 327, 342, 361, 471, 472, 480, 484
Curie temperature, 469, 473, 479

D

Debye frequency, 68, 378, 392, 444, 448, 452, 457
Deformation, 6, 25, 28–30, 44, 53, 54, 69, 71, 72, 75–77, 79, 81, 82, 90–97, 99, 112, 114–116, 119, 149–152, 154, 167, 169, 172, 173, 177, 178, 181, 183, 185, 211, 223, 232, 239, 248, 249, 255, 261, 264, 266, 269, 271, 281, 282, 285, 291, 292, 298, 300–308, 310–313, 320–324, 334, 335, 338, 339, 341, 343, 347, 348, 351, 353, 356–358, 360, 361, 363, 365, 373–377, 413–419, 421, 423, 427–429, 433, 434, 436, 441, 442, 447, 454–456, 458, 459, 461, 465, 480, 485, 486, 490–494, 497, 498, 500–504
  gradient, 28–30, 92, 93, 115, 116, 172, 173, 232, 249, 303–305, 321–324, 416
  map, 303, 454–456
  mechanisms, 115, 119, 177, 178, 181, 185, 281, 282, 311, 323, 339, 343, 363, 373, 377, 429, 504
Density functional theory (DFT), vii, viii, 2, 32–35, 131–154, 159, 160, 176, 273–275, 279, 293, 395, 408, 492, 538
Derjaguin approximation, 510, 514, 533
Detailed balance, 445, 446, 450
DFT. *See* Density functional theory (DFT)
Diffusion, 9, 27, 28, 30, 59, 67, 119, 205, 207, 211, 212, 215, 223, 232, 365, 457, 489, 491–494
Diffusivity, 254, 492, 494
Discrete dislocation dynamics. *See* Dislocation dynamics
Disjoining pressure, 508–522, 527–533
Dislocation(s)
  climb motion, 59, 60
  core, 47, 53, 58, 79, 81, 82, 164, 312, 323, 330, 332–334, 394
  core energy, 58, 62, 68, 79
  core force, 58, 59, 62
  cross-slip, 59, 67–68, 80, 82, 115, 343–345, 349
  density, 54, 67, 72, 77, 78, 80, 90, 94, 114, 117–119, 334, 337, 340–343, 345–347, 349, 352, 353, 359, 362, 363
  drag coefficient, 59, 60, 79
  emission, 309
  energy barrier, 66, 68, 79, 80, 344, 374, 382–384, 386, 396, 407
  forest, 74, 80, 343, 344
  Frank–Read source, 65, 73–76, 342, 356
  glide, 67, 284, 323, 365
    motion, 67, 68, 284, 323, 350, 361
    plane, 55, 59, 68, 76, 82, 302, 331, 335, 343, 344, 399

image stress field, 58, 67, 76, 77
intersections, 55, 65–66, 70, 74, 80, 344, 413
jog, 65, 66, 76, 80, 82, 334
junction, 55, 61, 64–66, 69, 72, 291–293, 308, 309, 334, 338
junction strength, 66, 79
kink, 65, 82, 334, 350
kink pair, 60
mobility, 47, 56, 59–60, 62, 63, 66, 79, 82, 330, 365
nucleation, viii, 47, 80, 205, 212, 284–292, 344–346, 356, 373–408, 436
Orowan looping, 82
Peach–Koehler force, 57, 63, 71, 79, 346
plastic strain, 71, 72, 74, 75, 78, 117, 361, 362, 416, 417
self force, 58, 63
single-arm source, 73, 75–77, 342, 363
source activation, 73, 75–77, 393
spiral-arm source (*see* Single-arm source)
Taylor hardening, 80, 118
thermally-activated process, 60, 66, 68, 174, 269, 343, 344, 437
viscous drag, 59

Dislocation dynamics
boundary conditions, 66–67, 71–72
codes, 57, 60, 62, 66, 69–70, 81, 82
MDDP, microMegas, MODEL, NUMODIS, ParaDiS, PARANOID, TRIDIS, 62, 70
collisions, 64, 65
elastic anisotropy, 81, 418
free surface, 67, 69, 183, 291, 310, 333, 344, 345, 373, 374, 383, 398, 399, 402
grain boundaries, 82, 164, 261
image solver, 58
kinematics, 81–82
lattice-based discretization, 60–62
mobility law, 59–60, 62, 63, 66, 70
node-based discretization, 56, 60–63
periodic boundary conditions, 5, 67, 72, 76, 77
plastic strain, 71, 72, 74, 75, 78
point defects, 82, 338, 363
precipitates, 58, 69, 82, 337, 339
principle of maximum dissipation, 66
remeshing, 55, 60–63, 70, 73, 185
strain-controlled loading, 71, 72, 75
stress-controlled loading, 71, 72, 74, 75
time integration, 55, 57, 63–64, 81
two-dimensional, 55, 68, 69, 81, 164, 338, 341

Dislocation extraction algorithm (DXA), 331–335
Dispersion effects, 136, 140
Displacement vector, 100, 101, 234, 268, 269, 320–321
Dissipative fluxes, 25
Domain, 5, 47, 67, 80, 81, 90, 92, 101, 115, 160, 161, 166, 168, 172–174, 179, 182, 188, 203, 235, 237, 238, 240–243, 246, 250, 262, 287, 303, 338, 470–485, 521
Domain wall, 470, 476, 478, 479, 483, 484
Domain wall energy, 482, 483
DXA. *See* Dislocation extraction algorithm (DXA)

E

EAM. *See* Embedded atom method (EAM)
Effective properties, 90, 101, 119, 159
Elastic energy, 184, 470, 472–478, 482
Elasticity, 58, 59, 79, 81, 94, 108, 239, 261, 405, 416, 418, 473
Elastic–plastic model, 184, 185, 493
Electrochemical lithiation, 489, 492, 503
Electrostatic interactions, 36, 162, 171, 269, 509–511, 513, 522, 526–527, 529–533
Embedded atom method (EAM), 36–39, 42, 171, 175, 179, 252, 519
Energy minimization, 1, 5–9, 12, 246, 382, 484, 503, 510, 532–533
Engineered domains, 485
Engineered property, 485
Enhanced sampling, 195–216
Ensembles, 2, 17–23, 28, 43, 46, 160–165, 169, 170, 173–175, 185, 188, 197, 200–201, 215, 232, 238, 262, 338, 363, 380, 383, 520
Equilibrium, 2, 4, 19, 21, 23–25, 27–30, 34, 35, 38, 78, 79, 89, 92, 95, 99, 113–115, 121, 139, 140, 144, 149, 151, 153, 154, 168, 173, 175, 196, 210, 232, 272–275, 293, 326, 376, 377, 380, 382, 443–446, 448, 450, 457, 473, 474, 476, 478, 480, 508, 511, 514, 519, 520, 526, 527, 532, 533
Eshelby-inclusion, 419, 447
Ewald (sum), 36, 269
Exchange-correlation functional(s), 131, 132, 134–138

F
FEM. *See* Finite element method (FEM)
Ferroelectrics, viii, 161–162, 172, 173, 176, 177, 469–486
FES. *See* Free energy surface (FES)
Fick's law, 27
Finite element (FE), 43, 67, 70, 90, 95, 98, 108, 110, 112, 115, 118, 160, 161, 164, 165, 167, 172, 174, 223–225, 235, 237–240, 243, 248, 249, 252–254, 261, 262, 269, 271–274, 279, 282, 283, 285, 288, 292, 340, 342, 348, 350–352, 356, 416, 420, 429–431, 446–448, 452, 453, 461, 463, 493, 494, 537, 538
  analysis, 90, 160, 452, 453, 461, 463
  mesh, 108, 110–111, 253, 272, 279, 282, 446–448, 452, 453
Finite element method (FEM), 67, 90, 95, 112, 115, 164, 165, 172, 174, 248, 261, 269, 292, 340, 342, 348, 350–352, 356, 416, 537, 538
First (1st) Piola–Kirchhoff stress, 25, 28, 225, 230, 245, 248
Flow stress, 78, 80, 118, 341, 346, 359, 375, 423, 502, 503
Forest hardening, 342
Fracture, 47, 81, 112, 118, 139, 146, 177, 215, 223, 235, 247, 248, 275, 278, 280–282, 293, 347, 413, 489, 491, 492, 504
Free energy surface (FES), 196, 198–200, 203, 205, 211–216
Free volume, 453, 456–459, 461

G
Generalized gradient approximation (GGA), 132, 135–136, 138, 148, 153
Generalized stacking fault (GSF) energy curve, 394, 395, 397, 400, 404, 408
Geometrically necessary dislocations (GND), 353, 355, 356
GGA. *See* Generalized gradient approximation (GGA)
Ghost forces, 168, 169
Grain boundaries (GBs), 44, 45, 47, 82, 115, 118, 119, 164, 177, 207, 215, 255, 261, 262, 272–276, 278–282, 284, 288–293, 297, 299, 301, 305, 306, 310–313, 323, 326, 331, 349, 355, 357–363, 413–415, 417–419, 426, 437
  deformation, 305, 306, 312–313
  migration, 312, 313
  sliding, 312, 313
Green–Kubo theory, 24, 27, 28, 30–32, 223
Green–Lagrange strain, 305
GSF energy curve. *See* Generalized stacking fault (GSF) energy curve

H
Hall–Petch, 357, 422
Hamiltonian, 12–14, 16, 19, 20, 23, 32, 133, 162, 164, 166, 168, 170, 175, 185
Hardening matrix, 117, 118
Harmonic transition state theory (HTST), 199, 201, 202, 206, 208, 210
Hessian, 4, 7–9, 200, 202, 204
Heterogeneous dislocation nucleation, 378, 379, 398–404
Heterostructure, 469–486
Homogeneous deformation, 28, 239, 266, 324, 441, 454, 456, 458, 459
Homogeneous dislocation nucleation, 212, 378, 379, 394, 396–398, 408
Homogenization, 90–92, 101–114, 119, 160
HotQC, 173, 174, 177
HTST. *See* Harmonic transition state theory (HTST)
Hyperelastic, 93, 94, 114
Hyperstress, 102–104, 111

I
Images, 9–11, 36, 43, 58, 67, 72, 76, 77, 160, 297, 301, 302, 306, 310, 321, 331, 332, 340–342, 357, 382, 490
Indentation, 177–179, 181, 282, 283, 285, 353–357, 461, 462, 503
Inhomogeneous deformation, 441, 454, 456, 459
Interface, 44, 66, 67, 91, 99, 100, 160, 161, 168, 169, 174, 177, 215, 224, 225, 235–237, 240, 262, 272, 275–278, 281, 284, 286, 287, 292, 293, 297, 300, 311, 312, 318, 319, 333, 334, 341, 347, 355, 357–363, 470, 472, 476–478, 480, 484, 491, 494, 496, 497, 509, 519, 522, 524, 525, 529, 538

K
Kinematics, 25, 81–82, 92, 173, 297, 298, 301–308, 310, 312–314, 416
Kinetic Monte Carlo, 207–209, 441–465, 538

L
Landau–Devonshire description, 470, 471, 475
Latent hardening, 118
Lattice constant, 38, 141–145, 153, 273, 389, 392, 404
LDA. *See* Local density approximation (LDA)
Length-scale bridging, viii
Linear elastic fracture mechanics (LEFM), 248
Line search minimization, 6–9
Local density approximation (LDA), 132, 135, 136, 138, 141, 143–146, 148, 153
Long timescales, 196, 216

M
Markovianity, 197–198
Materials modeling, vii, viii, 79, 112, 159, 452, 453, 463
Maximum strength, 132, 136, 148
MEAM. *See* Modified embedded atomic method (MEAM)
Mechanical coupling, 15, 154, 235, 238–241, 252, 483
Mechanochromic materials, 150–152
Meniscus shape, 508–516, 518–522, 526–529, 533
Mesh convergence, 102, 104–107
Mesoscale, 54, 79, 159, 271, 374, 442–446, 448, 455, 458, 460, 463, 465
Metallic glass, 323, 324, 454, 460, 461, 463, 465, 492
Metallic glass matrix (MGM) composites, 453, 454, 463–465
Meyer–Neldel rule, 386, 388, 396, 404, 405, 407
MGM composites. *See* Metallic glass matrix (MGM) composites
Microcanonical ensemble (NVE), 17
Micromorphic, 89, 91, 92, 95–98, 264
Micropolar, 95
Microrotation, 305–307, 309–313
Minimum energy path, 4, 9–11, 42
Modified embedded atomic method (MEAM), 37–39, 492
Molecular dynamics (MD), viii, 1, 2, 9, 12–24, 28, 32, 33, 41–43, 46, 47, 80, 81, 89, 101, 139, 140, 144, 154, 159–161, 174, 179, 181, 188, 195–216, 223–225, 235, 237–240, 243, 245, 250, 254, 261, 262, 269, 272, 275, 276, 278, 282–290, 293, 297, 310, 317, 318, 320, 321, 323, 324, 326, 330, 331, 335, 338, 339, 345, 346, 348, 349, 355, 361, 363, 376, 377, 392, 414, 415, 417, 418, 421, 437, 492, 498–504, 509, 510, 516, 519, 520, 523, 525–533, 537, 538
Monte Carlo (MC), 6, 209, 210, 215, 443–446, 448
Multiscale, 1, 70, 90, 223, 261, 262, 298, 330, 339, 346, 374, 407, 489, 490, 507–533, 537, 538
Multiscale simulation, 339

N
Nanocrystalline (NC), 44, 119, 159, 177, 301, 302, 308, 309, 323, 328, 337, 357, 360, 413–437
Nanoindentation, 69, 81, 119, 161, 177–182, 284, 292, 337, 340, 355–357, 453, 454, 461, 462, 465, 491, 492
Nanomechanics, vii, viii, 1, 89, 203, 211–213, 298, 308, 318–320, 374, 441–465, 469–486, 537
Nanopillars, viii, 211, 310, 311, 345, 376, 392, 397, 400
Nanoscale, vii, 2, 46, 47, 89, 91, 94, 99–101, 119, 159, 167, 177, 181, 223, 254, 297, 301, 310, 313, 317, 373, 374, 407, 415–418, 461, 462, 485, 507–510, 519, 523
NEB. *See* Nudged elastic band (NEB)
Newton–Raphson, 6, 8, 9
Nonlocal, 89, 91, 92, 100–101, 113, 118–120, 122, 135, 136, 138, 141, 144, 153, 154, 165, 166, 168–170, 173, 177, 178, 181, 187, 269, 271
Nosé–Hoover, 18–20
Nucleation sites, 345, 346, 378, 379, 381, 392, 393, 397–400, 436, 460
Nucleation strength/nucleation stress, 373, 374, 376, 377, 379, 381, 383, 388–390, 392–394, 397, 398, 400, 408
Nudged elastic band (NEB), 10, 11, 206, 208, 374, 382, 448

O

Open visualization tool (OVITO), 43, 44, 301, 318–321, 329, 331, 538
Orthorhombic phase, 470, 471, 478, 479, 484
OVITO. *See* Open visualization tool (OVITO)

P

Partial dislocation, 44, 285, 301, 310, 328, 373–376, 378, 395–397, 399, 414
Partition function, 18, 20, 24, 175, 201, 443
Periodic boundary conditions, 5, 12, 36, 67, 72, 76, 77, 104, 181, 246, 269, 321, 359, 414, 483, 498, 520
Perturbation theory, 102
Phase diagram, 476, 484, 495
Phase field method, 481–482
Phase transition, 173, 469, 471, 472, 476, 479, 481, 485, 486
Piola–Kirchhoff stress, 25, 28, 93, 225, 230, 245, 248, 416
Plastic deformation, 53, 54, 69, 72, 76, 77, 114–116, 269, 291, 292, 308, 323, 339, 377, 413, 423, 433, 436, 461, 490, 493, 497, 503
Plastic slip, 94, 114, 115, 117
Polarization, 24, 104, 152, 177, 233, 234, 250–252, 469–476, 478, 479, 481–485
Polarization stress, 112, 113
Polycrystalline strontium titanate, 262
Potential energy, 2–7, 9–12, 21, 27, 32, 33, 37, 39, 41, 44, 46, 93, 94, 96, 98, 100, 152, 162, 165, 166, 168, 169, 171, 174, 198–200, 202–204, 214, 230, 263, 269, 298, 299, 301, 307, 309
Potential energy surface, 203–211
Probability density, 18, 418, 430
Probability distribution, 22, 175, 212, 213, 418, 430, 443
Probe sphere, 329, 330

Q

QC codes, 176, 178
Quantized slip, 119, 414, 415, 418–419, 421, 422, 425, 426, 436
Quasicontinuum (QC), viii, 47, 159–188, 205, 235, 538
Quasi-harmonic (approximation), 163, 173
Quasi-harmonic (model, approximation), 30

R

Radial distribution function (RDF), 23, 459, 499, 500
Rate independent model, 118
RDF. *See* Radial distribution function (RDF)
Reaction pathways, 3, 4, 9–11
Reactive force field (ReaxFF), 41, 498, 504
Representative atom, 161–165
Representative volume element (RVE), 112, 498, 499, 502
Residual force, 167, 168
Rhombohedral phase, 470
RVE. *See* Representative volume element (RVE)

S

Sampling atom, 165–167, 169, 172, 178–180, 187
Sampling rule, 162, 165, 167, 185
Schmid factor, 373, 375, 376, 384, 396, 398, 420, 424, 429, 430, 432, 436
Schmid tensor, 115
SHAKE, 14, 16–17
Shear band, 323, 324, 441, 442, 454, 456, 458–461, 465
  nucleation, 461
  propagation, 460
Shear localization, 459
Shear transformation zone (STZ), 324, 441, 442, 446–465
Single-arm source, 73, 75–77, 342, 363
Single crystalline, 178, 181, 182, 352
Size effects, 27, 30, 118, 159, 167, 177, 179, 181, 337, 339–343, 346, 348, 351–355, 358, 361–363, 392, 393, 421–422, 504
Slip systems, 38, 70, 72, 80, 90, 114–118, 284, 349, 353–355, 359, 417, 418, 420, 421, 426, 429, 430, 436, 437
Slip vector, 298–302, 306, 309–313
Solubility gap, 491
Spurious forces, 165–172
Stacking fault, 41, 44, 215, 286–288, 292, 301, 302, 326, 328, 335, 350, 376, 383, 394–396, 402, 403, 408, 414
Stillinger–Weber, 39, 42
Strain(s), 5, 70, 91, 144, 182, 211, 239, 277, 305–306, 321–324, 341, 374, 413, 441, 470, 491
Strain gradient, 91, 92, 95, 101, 103, 104, 108, 109, 112, 356, 363, 486
Strain hardening. *See* Work hardening

Strain tensor, 93, 94, 96, 99, 103, 305, 309, 321–324
Stresses, 5, 56, 89, 144, 177, 225, 263, 307, 320, 340, 373, 414, 441, 473, 489
Stress tensor, 93, 95, 96, 98, 101, 111, 115, 145, 225, 239, 263, 267, 307
String method, 374, 382, 383
Strong convergence, 104–107
STZ. *See* Shear transformation zone (STZ)
Substrate, 205, 253, 272, 341, 355, 470–485, 498, 499, 511, 514, 521, 522
Summation rule, 161, 165–172, 178, 181, 182, 185, 187
Surface area, 329, 330, 335, 379, 518, 522
Surface effects, 89–92, 95, 99, 100, 109, 111, 119, 122–124, 181–183
Surface energy, 23, 99, 249
Surface roughness, 346, 509
Surface stress, 89, 91, 92, 99–100, 119, 373, 376, 377, 389–391, 393, 397, 398, 404
Surface tension, 99, 511, 512, 518, 519, 523–524, 526, 533, 535
Symplectic integrators, 14, 16, 23

T

Tensile strength, 140, 141, 145, 148, 149, 153, 281, 282, 492
Tersoff, 39–42, 253
Tetragonal phase, 471, 475, 476, 483
Thermal coupling, 241–243
Thermostat, 17–21, 23, 30, 173, 174, 197, 236, 238, 243
Thin film, viii, 67, 70, 182, 207, 339–341, 351, 355, 359, 362, 363, 469–486, 491, 492, 498–504, 507–510, 512, 514–516, 523, 526–528, 530–533
Timescale separation, 197–198
Torsion, 337, 339, 348–351
Transition state (TS), 3, 4, 9–11, 198, 200–205, 207–209, 212, 213, 444–446, 448, 455
Transition state theory (TST), 201–203, 206, 208, 210, 215, 378, 444–446
TrussQC, 176, 178, 183, 187, 188
TST. *See* Transition state theory (TST)
Twin boundaries (TBs), 82, 179, 302, 310–312, 326, 337, 357, 360–362
  migration, 310–311

U

Uncertainty quantification (UQ), 225, 235, 537
Unit cell, 91, 92, 95, 96, 98, 102–105, 107–114, 123, 140, 145, 154, 173, 246, 262, 264, 268, 269, 272, 273, 276–279, 282, 285

V

Van der Waals, 36, 136, 138, 149, 508, 509, 511, 513, 515, 516, 518, 522, 524, 530, 531, 533
Verlet, 9, 14–17
Vibrational density of states (VDOS), 524–525
Virial (stress), 27
Viscoplastic, 117, 501
Vorticity, 307, 309, 310, 313

W

Weak convergence, 102, 105, 106
Weakest link, 342
Weighted histogram analysis method, 212
Work conjugate, 93, 96, 416
Work hardening, 65, 69, 73, 77, 78, 343
Work hardening rate, 341, 342

Y

Yield-point/yield strength/yield stress, 78, 308, 352, 357–359, 364, 428, 460–464, 492–494, 498, 499, 501–503

The manufacturer's authorised representative in the EU is Springer Nature Customer Service Centre GmbH, Europaplatz 3, 69115 Heidelberg, Germany. If you have any concerns regarding our products, please contact ProductSafety@springernature.com

Printed and bound by CPI Group (UK) Ltd, Croydon, CR0 4YY
15/07/2026
02167620-0005